FOURTH EDITION

Biological Science

AN INQUIRY INTO LIFE

Prepared by the
Biological Sciences Curriculum Study

THE AUTHORS OF FOURTH EDITION

Faith M. Hickman, *Supervisor*
Manert H. Kennedy, *Revision Coordinator*
Eunice A. Combs, *BSCS Editor*
Catherine M. Marble, *Project Assistant*

Bruce Criley
Illinois Wesleyan University
Bloomington, Illinois

H. Edward Drexler
Pius XI High School
Milwaukee, Wisconsin

Bentley Glass
State University of New York
Stony Brook, New York

Edwin Phillips
Pomona College
Claremont, California

S. Arthur Reed
University of Hawaii
Honolulu, Hawaii

Gordon E. Uno
University of California
Berkeley, California

CONTENT REVIEWERS OF FOURTH EDITION

Ann Egan
Coast Community College District
Costa Mesa, California

Education Division
The Population Institute
Washington, D.C.

HARCOURT BRACE JOVANOVICH

New York Chicago San Francisco
Atlanta Dallas *and* London

FOURTH EDITION

Biological Science

AN INQUIRY INTO LIFE

AUTHORS OF PREVIOUS EDITIONS

John A. Moore, *Supervisor* (1st, 2nd, 3rd)
Donald H. Bucklin (1st)
J. Maxwell Davis (1st)
Edward F. Degenhardt (2nd)
H. Edward Drexler (3rd)
Bentley Glass (1st, 2nd)
Luke Hallenbeck (2nd)
Manert H. Kennedy (2nd)
William V. Mayer (1st, 2nd, 3rd)
Tamsen G. Meyer (2nd)
Ingrith Deyrup Olsen (2nd, 3rd)
George Schwartz (1st)
Wilson N. Stewart (1st, 2nd)

THE BSCS STAFF

Jay Barton II, *Chairman, Board of Directors*
William V. Mayer, *Director*
Manert H. Kennedy, *Associate Director*
George M. Clark, *Secretary-Treasurer*
Don E. Meyer, *Managing Editor*
Linda Fetterman, *Controller*
Jane Larson, *Art Director*

Staff Associates:

Elliott Asp
Janet Bishop
William Callahan
Faith Hickman
Ronald Ladd
Mary McConnell
Joseph McInerney
James Robinson
Norris Ross
John Tarp
Richard Tolman

Illustration Staff:

Bill Border
Eileen Burke
Debra Jones
Lawrence Strand
John B. Thornton
Roy M. Udo
Robert F. Wilson

Photo Researcher:

Carlye S. Calvin

Photographs: Front cover, Coral polyps at night, Doug Wallin/Taurus Photos; Back cover, Safflower, W. H. Hodge/Peter Arnold; Title page, Water hole, Africa, Clem Haagner/Bruce Coleman Inc.

Printed in the United States of America

ISBN 0-15-360755-6

Acknowledgments: Copyrighted Materials: Page 34, © 1962 Parke, Davis & Co.; page 110. Rattner, J. B. and S. G. Phillips, 1973, "Independence of Centrioles and DNA Synthesis," *J. Cell Biology* 57:359-372; page 127 top, © Dr. Giuseppe Mazza; page 186, adapted from *Genetics* by Robert P. Levine, © 1962, reprinted by permission of Holt, Rinehart & Winston, Inc.; page 208 top right, from J. D. Watson, *The Double Helix*, Atheneum, New York, p. 206, © 1968 by J. D. Watson; page 212, © Biology Media; page 219, Miller, O. L., Jr. and B. R. Beatty, 1969, "Portrait of a Gene," *J. Cell Physiol.*, 74, Suppl. 1: 225-232; page 223 right, © Biology Media; page 228 both, © Dr. Giuseppe Mazza; page 321 left, © Bob Campbell; page 343 bottom, © Dr. Giuseppe Mazza; pages 360-61, R. H. Whittaker, *Science*, 163: 150-60, Fig. 3, Jan 10, 1969. Copyright 1969 by American Association for the Advancement of Science; page 364 top, © Parco Scientific Co., M. I. Walker, Phillip Harris Biological, Ltd., London; page 365 bottom left, © D. P. Wilson; page 366 top right, © Dr. Giuseppe Mazza; page 368 top left and top right, © Biology Media; page 374 top left and bottom left, © Biology Media; page 376 left and right, © D. P. Wilson; page 377 top left, top right, bottom left, © D. P. Wilson; page 378 bottom left, © Parco Scientific Co., Dr. Andrew White; page 379 top right and bottom left, © D. P. Wilson; page 380 top left and top right, © D. P. Wilson; page 383 left 2nd from top, © D. P. Wilson; page 386 bottom center, © Dr. Giuseppe Mazza; page 386 bottom right, © Biology Media; page 388 left center, © Dr. Giuseppe Mazza; page 410 right, © Biology Media; page 466, © 1959 Parke, Davis & Co.; page 493 right, © Parco Scientific Co.; page 520, © Biology Media; page 539, Reproduced with permission from *Birth Atlas*, published and copyright by Maternity Center Association. New York; page 573 top left, © Allison Wilson, bottom left, © D. P. Wilson; page 601 bottom, © Dr. Giuseppe Mazza; page 622, Adapted from *The Dancing Bees* by Karl von Frisch. Copyright 1955. Reproduced by permission of Harcourt Brace Jovanovich, Inc., New York, and Metheun & Co., Ltd. Publishers, London, page 624, *Animals, Animals* © Charles Palek; page 640 bottom left and bottom right, © D. P. Wilson; page 699, Sturla Fridriksson, © 1975, *Surtsey: Evolution of Life on a Volcanic Island*, Butterworths, London.

Key: (t) top; (c) center; (l) left; (r) right; (b) bottom

Part one: Page 1(t), Werner Schulz; (b), Hans Pfletschinger/Peter Arnold; 2(l), USDA-APHIS PHOTO; 5(b), John A. Moore; 6(1st,3rd), Commentaries de M. Pierre Andre Mathiole, Lyon, 1598; (2nd,4th), Johann Von Cube, Hortus Sanitasus, 1485; 7(l), A. Laveran, *Paludism*; (r), J. B. Jensen, Rockefeller University; 10(tl), HBJ Photos; 11(tl), HBJ Photos; 12(tl), Marg and Bill Staley; (tr), Dekalb AgResearch; (bl), Grumman; (br), USDA photo by Murray Lemmon; 13(tl), HEW Center for Disease Control, Atlanta, Georgia; (br), Courtesy American Veterinary Medical Association; 14(tl), Calvin Davis; (tr), Jack Goodman/Science and Mankind; 15(t), Nebraska Game and Parks Commission; (bl), Courtesy William K. Redican; (br), Edward S. Ross; 16(tl), Courtesy Jewel Industries Inc., Chicago, Ill.; (tr), U.S. Department of Commerce NBS; (b), Nebraska Game and Parks Commission; 25, Courtesy of J.G. van Clittert-Eymers, Utrecht University Museum; 26, Courtesy American Optical Corporation; 27, HBJ Photos; 28, HBJ Photos; 40(r), Courtesy Argonne National Laboratory; 41, The Bettmann Archive; 45(br), Bloom Agency; 46, John A. Moore; 59(l), Warren N. Marble; (r), Edward S. Ross; 64, Medical Research Council; 74(both), Photomicrographs provided by Ripon Microslides, Inc.; 75(tl, r), *Micrographia*, The New York Public Library; (bl), S.E.M. by Dr. Delbert E. Philpott; 77, Michelle Marks, Pitzer College; 79(all), HBJ Photos; 81, Sargent-Welch Scientific Company, D. A. Skrupky and R. E. Arnold; 82(tl), John A. Moore; (tr), Bruce Criley; (br), John A. Moore; 83, Keith Porter; 84, Keith Porter; 85, Keith Porter; 86, Courtesy of Dr. K. G. Murti, Labs of Virology, St. Jude

(continued on page 742)

Foreword

We live in an age of science and technology. Because of advances in these fields, contagious diseases are being eliminated and food is grown in former deserts. Our air and water are becoming cleaner, our cities better planned, and our recreational facilities enlarged. The continuation of life on planet Earth depends on our scientific successes. Of all of the sciences, biology is the most relevant to our daily lives. It touches on almost every human action. As the science of life, biology examines all aspects of people. It looks at our genetic and evolutionary history, our behavior, our relationship to our environment, and our future potential. However, we cannot gain a complete understanding of humans from the study of a single discipline. Biology must be considered in terms of our social systems and our cultural heritage. We must understand the social and humanitarian aspects of science in order to make the best possible applications of our biological knowledge.

No one quarrels with the knowledge of orders of insects, parts of flowers, or the process of digestion as the facts of biology. However, there are differences of opinion regarding the application of biological facts. We live in a world of controversy and difference of opinion. Biology alone cannot settle controversies. But biological knowledge can provide a background of information that will help us to make intelligent choices among the alternatives. As citizens of a democracy, the decisions you make may influence our society for generations. The responsibility for such decision making is great. The more information and background we have, the more likely we are to make correct decisions. The more uncritical and uninformed a citizen may be, the more likelihood there is of being swayed by erroneous data or unwarranted conclusions.

One of the purposes of this text is to provide background information. Another is to induce a critical and inquiring attitude until adequate data are obtained. Facts are important. But more important will be the development of the questioning attitude necessary to ensure the proper application of the facts concerned. As you read the text, keep an open and inquiring mind. Ask questions; make sure you understand. Consider alternative courses of action. Then select the one you feel is most soundly based and the one that will result in the greatest benefit to society.

We most earnestly solicit your opinion of these materials. Send us your comments and suggestions. They aid in the realization of our goal to improve biological education in all levels. Such comments may be sent to the Director at the address below.

Jay Barton II, *Chairman*
Board of Directors
Biological Sciences
Curriculum Study
West Virginia University
Morgantown, West Virginia
26505

William V. Mayer, *Director*
Biological Sciences
Curriculum Study
P.O. Box 930
Boulder, Colorado 80306

Preface

> The way a book is read—which is to say, the qualities a reader brings to a book—can have as much to do with its worth as anything the author puts into it.
>
> Norman Cousins

A book is many things to many people. Perhaps the best it can be is an invitation to discovery. The authors of this book hope it leads you to other books on particular topics that you feel are interesting and important.

But, as Norman Cousins suggests, the true worth of a book depends only partially on what the authors chose to write. The most important measure of the worth of this book is what you as a beginning biology student bring to your study of this text. What you read is probably not as important as how you read. Try to be selective—spending more time on topics that you find interesting or challenging. Develop a certain degree of healthy skepticism. Ask questions like "How can I know that is true?" or "Is there a better interpretation than the one given here?" Ask yourself also how knowledge is generated. Perhaps more important is finding out how you can apply the information you read to your real life situations.

This book has been carefully designed to provide you with some of the basic ideas in biology. But biology, like any other area of human endeavor, is constantly changing. What is assumed to be true today may be proved false tomorrow. And the real science of biology is not contained in textbooks. It is in the minds of biologists around the world who work each day to extract some meaning from their observations of the natural world.

The authors of *Biological Science: An Inquiry into Life,* offer this invitation to discovery as only one avenue for exploring the fascinating world of the life sciences. We hope you will use this book well. Read it creatively, and begin forming your own concept of the value of the life sciences in our complex and changing world.

Faith Hickman
Supervisor, Yellow Version,
Fourth edition

Contents

PART ONE UNITY

3 Cell Structure and Function, 74

4 Cell Reproduction, 108

PART TWO CONTINUITY

5 Patterns of Heredity, 128

6 The Chromosome Theory of Heredity, 165

PART THREE DIVERSITY

PART FOUR INTERACTION

PART ONE
UNITY

1 Biology–Investigating the Living World

The twentieth century has been a time of great change. In 1900, the use of nuclear energy, either for wartime or peaceful purposes, was unimagined. Radio was used only for wireless telegrams. Television was undreamed of. No vitamin had been identified. No hormone had been found. Basic laws of heredity had been discovered, but lay forgotten. The new kinds of grain that have so greatly increased our food supply were not even thought possible. Antibiotics were unknown. Because nothing was known about the chemistry of life, little could be determined about the causes and treatments of diseases and birth defects. The average life expectancy in the United States was 47 years – it is now 72 years.

All of these changes have taken place since 1900 – about 80 years ago. This may seem like plenty of time for so many changes. But if we consider all of human history, these changes have taken place in a very short time. It took over 2,000 years to find the cause of malaria. The debate over the origin of living things lasted more than 300 years. Yet, in our own century, advances of equal importance have come very quickly. Some of these came within a decade or two. What can explain this enormous growth in our knowledge and in our ability to change the natural environment?

One part of the answer is the activity we call science. Science is a way of seeking answers to questions. Like detectives, scientists search for clues. They try to find information that will lead them to the answers. Science is not just a disorganized pile of facts. It is a way of investigating the natural world.

Scientific knowledge, when used for some practical purpose, is called *technology* or "applied science." For example, knowledge of radio waves is a part of science. We enjoy the practical application of this knowledge when we listen to music on the radio. The discovery of an antibiotic is science. We benefit from this discovery when a doctor treats pneumonia with penicillin.

Often, we see only the results of science and technology. But let's start at the beginning. What really happens when a scientist tackles a problem? In 1962, James Watson and two other scientists received the Nobel Prize for Physiology or Medicine. They earned this prestigious award for their work on the basic molecule of heredity. A few years later, Watson published *The Double Helix: A Personal Account of the Discovery of the Structure of DNA*. In the preface of the book, Watson wrote, "Science seldom proceeds in the straightforward, logical manner imagined by outsiders. Instead, its steps forward (and sometimes backward) are often very human events in which personalities and cultural traditions play major roles." Watson wanted to show what goes on "behind the scenes." Instead of emphasizing the end result of the work, he stressed the process. The book showed that scientists are affected by their personalities, their friends, and their co-workers. Science is something people *do,* not just a mass of facts.

Biology is one kind of science. Biologists try to answer questions about living things. While one biologist studies the behavior of wolves, another investigates the chemistry of the cell. Still another is interested in the interaction of plants and animals with their environments. The subjects of biology range from plant breeding for agriculture to drug research for medicine. Biologists study the smallest virus on earth. They also wonder about the chance of life on distant planets.

During this course, you will conduct many biological investigations. You also will study some of the topics biologists investigate and some of the questions they try to answer. Let us begin with a short introduction to what biology is and how biologists look for answers.

BIOLOGY AS A FIELD OF INQUIRY

All science is **inquiry.** The "way of knowing" that we call science has grown and changed over many centuries. Perhaps science first began when, thousands of years ago, one of our ancestors asked a question about the living world. Today's science is a well-structured way to inquire—to ask questions and look for answers.

What does it mean to inquire? Part of the answer can be found by determining the **field of inquiry**—the subject you ask questions about. Suppose you ask a friend if it will rain tomorrow. Your field of inquiry is the weather. The field of inquiry explored by biologists is life. Such a broad field of inquiry leads to many kinds of questions. Biology often overlaps with other subjects. For example, dealing with problems such as food production requires some knowledge of chemistry, geology, meteorology, and mathematics. Many topics are included in a biologist's field of inquiry.

The usual definition of biology is the study of living things—**organisms.** Biologists, then, ask questions about organisms and anything that affects organisms. But does the subject matter of biology stop there?

Often, the knowledge built by biologists turns out to be important in our everyday life. A biologist may begin working on a problem out of curiosity. But such work often has practical results that are important to our health and well-being. Even research that seems to have little to do with people may be helpful to them. For example, scientists found that some female insects produce chemicals that attract males of the same **species,** or kind. This fact alone might seem unimportant. But further research on such chemicals has led to new ways of controlling pests that destroy crops. The payoff for us may be an increase in food production. And this method of pest control does not use dangerous chemicals that may pollute the environment.

So biology is the search for answers about living things. In many cases, these answers have some practical use for everyone. Let's consider one example of a biological problem and the history of its solution.

A Field of Inquiry—Malaria

Malaria is an example of one of the topics biologists investigate. Malaria has killed more people than has any other disease. It occurs most often in the warmer parts of the world. But a century ago it was found as far north as Ohio and New York and in some parts of Europe.

1-1. The irregular line of the graph shows changes in body temperature that occur in one common type of malaria. *How do the symptoms of the disease relate to the temperature cycle?*

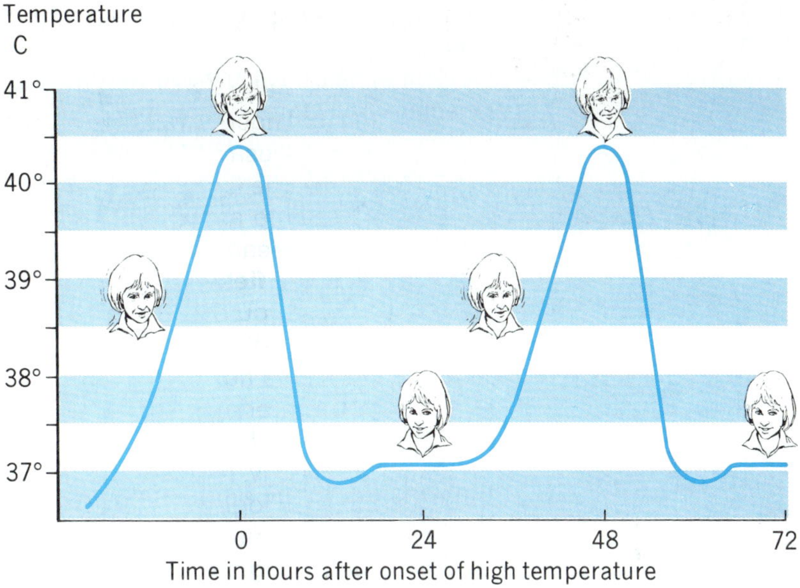

Malaria is a dramatic disease. In a typical attack, the victims first have severe chills. They feel very cold, but their temperature is above the normal 37° Celsius (C). Several hours later, they may feel very hot and may have a fever as high as 41° C (Figure 1-1). Nausea and a headache usually go along with the attack. In a few hours, victims may feel much better—although weak and tired. Without treatment, they may have attacks every two or three days for several weeks. Then the disease seems to go away. But without medical care, the victims probably will become sick again. Clearly, malaria upsets the way the human body works. But how? And why?

Physicians in Egypt, Iraq, India, China, Greece, and Italy described malaria over 2,000 years ago. They noted that the disease was most frequent among people living on low, marshy lands (Figure 1-2). They thought that the dirty water of the marshes somehow poisoned

1-2. Early Greeks and Romans noted that malaria occurred most often among people living near swampy areas. *What was the value of this observation?*

the air. This belief was an early step toward solving the problem of malaria.

Some historians believe that malaria was partly to blame for the decline of Rome in the third century A.D.—and perhaps for the decline of ancient Greece even earlier. After the fall of ancient Rome, Western Europe entered a period of more than a thousand years called the Dark Ages. We have few records from that period. During this time, little progress was made in solving biological problems.

Treatment Without Knowledge of Cause

A "rebirth" of scientific investigation followed the Dark Ages. Many substances were tried as medicines, not only for the treatment of malaria but for all kinds of illnesses (Figure 1-3). Some of the substances were not effective. Others were even harmful. After the discovery of the New World, many plants, previously unknown, were sent back to Europe to be tried as medicines. One of these, the bark of a certain kind of tree, turned out to be useful in treating malaria. The bark contains quinine (KWY·nine), a chemical that works against malaria. Quinine was the only known remedy for malaria from the seventeenth to the twentieth century. But physicians in the seventeenth century treated malaria without knowing its cause. The only clue was the observation that malaria was in the "bad air" of marshes. But some volunteers drank marsh water from places where malaria was common. They did not develop malaria. What, then, could be the cause of the disease?

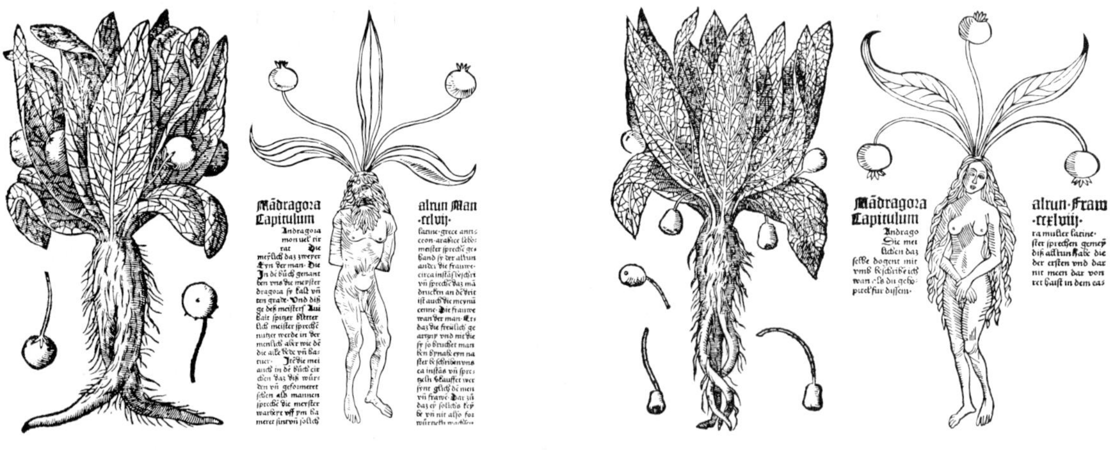

1-3. Before the age of scientific medicine, plants that looked like human beings were thought to have special healing powers.

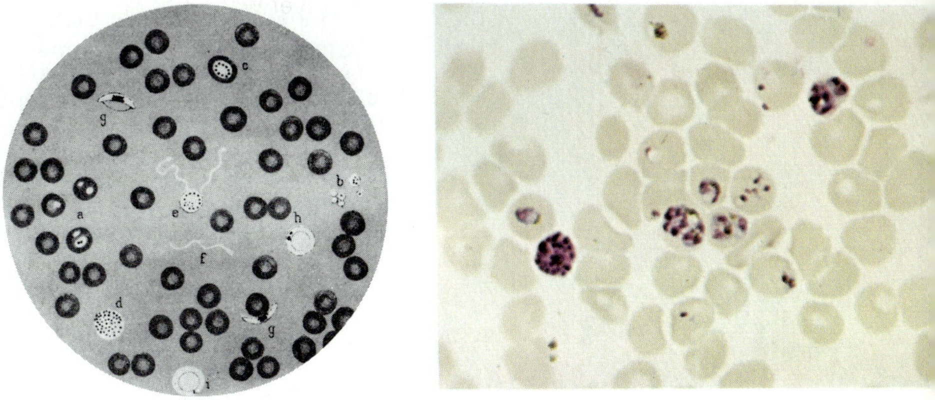

1-4. Laveran's drawing of *Plasmodium* in human blood **(left)** still compares favorably with today's photograph **(right).**

The Search for a Cause

In 1878, a French physician named Laveran (luh·VRAHN) began to study malaria. He used a microscope to study samples of blood taken from people with malaria. In the samples, he noticed some tiny organisms. They were long and narrow like little hairs and so small they could hardly be seen with the microscope (Figure 1-4). Could these tiny organisms be the cause of malaria?

At this point, investigators trying to find the cause of malaria had only four facts to work with:

1. Malaria is associated with marshes.
2. Quinine is an effective remedy.
3. Tiny organisms are present in the blood of people with malaria.
4. Drinking marsh water does not cause malaria.

Using the first fact, a simple **deduction** was made and tested: *If* malaria comes from marshes, *then* getting rid of marshes should remove malaria. This is called a deduction. In a deduction, if the first part is true, then the second part is also true. The deduction was put to the test. Some marshes were drained. As a result, in some cases, fewer people caught malaria. In others, no one did. The deduction was supported. But the question of cause was far from answered. It was known that drinking the marsh water did not cause malaria. Perhaps something else about marshes was the cause.

As early as 1717, someone had suggested that malaria might be carried by mosquitoes. This idea came from **observations.** If marshes were drained, the number of cases of malaria declined. So did the size of the mosquito population. Later in the nineteenth century, three other observations seemed to support the idea:

1. People who lived near a marsh and slept outdoors were more likely to get malaria than people who did not.
2. Others who slept under fine nets were less likely to get malaria than those who slept in the open.
3. Those who slept near a smoky fire usually did not get the disease.

These and other observations did not prove that mosquitoes carry malaria. But they were important because they suggested a **hypothesis** that could be tested.

Testing the Hypothesis

A hypothesis is a simple, but powerful tool. It is a possible explanation—one that looks promising given all available information. Ronald Ross was a British Army doctor working in India in the 1880s. Ross formed a hypothesis based on two ideas: Mosquitoes might carry malaria; and the tiny organisms Laveran had observed might cause the disease. Ross made a deduction from this hypothesis: *If* the organisms cause malaria, and *if* mosquitoes carry the disease, *then* the organisms should be found in a mosquito that has bitten a person who has malaria.

To test the deduction, Ross allowed a mosquito to bite a person with malaria. Some days later, he killed the mosquito and found the organisms in its stomach. But, more important, the tiny organisms were increasing in number. To find the organisms in the mosquito was one thing. To find them multiplying was much more significant. Why?

Ross' hypothesis was supported. (What do you think Ross would have done if the hypothesis had not been supported?) Another deduction was now possible. *If* mosquitoes carry malaria, *then* a healthy person bitten by an infected mosquito should get the disease. But Ross could not test this deduction. Scientists are reluctant to use people for experiments. And Ross felt that he should not experiment on people with a disease as serious as malaria.

Many different animals can get malaria. The tiny organisms Laveran had seen—called plasmodia (plaz·MOH·dee·uh; singular, plasmodium)—are found in birds as well as people. Ross obtained some sparrows with plasmodia in their blood. He allowed mosquitoes to feed on them. Some of these mosquitoes were killed and studied at various times. In each mosquito, Ross found plasmodia multiplying in the stomach.

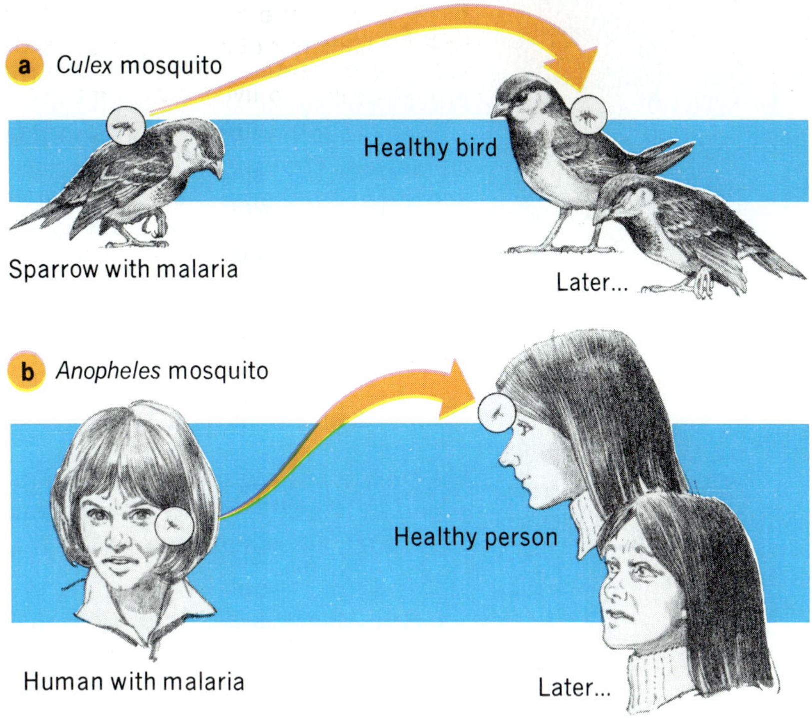

1-5. Different kinds of mosquitoes carry malaria to people and sparrows.

Ross did not kill all the infected mosquitoes. Some were kept and allowed to bite healthy sparrows. Some days later, Ross examined the blood of the healthy birds. Plasmodia were in their blood. The birds had the disease.

Before this experiment, plasmodia had been observed in the blood of people with malaria. But Ross' experiment provided evidence that the disease was caused by plasmodia carried by mosquitoes. Later trials with human volunteers showed that the disease is similar in people and birds. However, different kinds of mosquitoes are involved (Figure 1-5).

After the Cause Was Found

After nearly 2,000 years, both the cause of malaria and the way it is carried were finally known. Once the cause was determined, people found ways to prevent the disease. They drained puddles, marshes, and pools where mosquitoes laid eggs. Sometimes they put a layer of kerosene on top of the water to kill the young mosquitoes. They covered doors and windows with screens and used insect sprays.

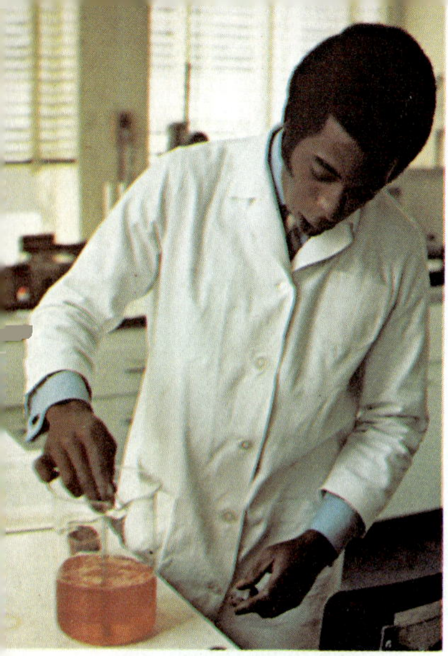

1-6. Quinine does not always cure malaria. Here, scientists continue the search for drugs to fight this disease.

People infected with malaria were kept in screened buildings so the disease would not spread to others. People who already had the disease were treated with quinine. And scientists developed better drugs.

Even though we now know how to control malaria, the problem is not solved. True, public health measures in the United States, Italy, and some other countries have nearly wiped out the disease. But in other parts of the world, malaria continues to be a major problem. In 1976, over one million people died from malaria in Africa alone. Thus, even today, the search continues for new drugs to fight malaria and for new ways to prevent it (Figure 1-6).

Malaria as a Field of Inquiry

The story of malaria is an example of biology as a field of inquiry. We began by considering what biologists ask questions *about.* What has the malaria example shown? Five points deserve special attention:

1. The objects of study were living organisms. The search for answers about malaria involved people, plasmodia, and sparrows.
2. Biologists, like all scientists, look for causes. They look for objects, events, or processes that cause a specific result. There is seldom a single, simple cause for a biological effect. In the case of malaria, the primary cause is the organism plasmodium. But the disease is spread only by certain mosquitoes that live in marshes. And the mosquitoes must first bite individuals who already have the disease. When biologists learned the causes, they could develop a treatment and suggest some ways to prevent the disease.
3. Biology is a search for relationships. Biologists try to find out how one thing affects something else. They try to put two or more ideas together. Ross was searching for relationships between plasmodia, mosquitoes, and malaria.
4. Biologists try to find relationships that are always true. Just because plasmodia were present in the blood of *some* malaria victims did not mean that plasmodia would be found in *all.* But after much time and many observations, it became clear that plasmodia must be present in the blood of *all* malaria victims.
5. Biologists start with what is already known or believed to be true. The field of inquiry has a starting point. In the case of malaria, the observations of the early Greeks and Romans set the stage for centuries of inquiry.

SCIENCE AT WORK: Research on Malaria

The fight against malaria is far from over. Today, scientists all over the world continue the battle. Some scientists are searching for new and better drugs to treat the disease. Others are trying to find safe, yet effective, ways to control mosquitoes. Still other scientists hope to develop a vaccine that will protect people against malaria.

Since 1912, researchers have been trying to grow plasmodia, the malaria parasites, *in vitro*. *In vitro* means "in glass"—in other words, in some laboratory apparatus. Growing an organism *in vitro* is necessary for the development of a vaccine. Some scientists had managed to grow plasmodia *in vitro* for a few days. But in 1976 scientists at Rockefeller University in New York kept plasmodia alive and multiplying outside the body of an organism for several months.

The technique took 34 years to perfect. The malaria parasites are grown in a shallow layer of human red blood cells. Nutrient materials flow slowly and continuously over the cells.

The plasmodia are then killed or weakened and injected into monkeys. If the monkeys show some resistance to the disease, we may be on our way to a vaccine for humans. If they do not, other approaches must be tried.

Scientists at Rockefeller University are quick to point out that there is much yet to be learned. Their technique still depends on cells from a living organism—red blood cells that must be supplied fresh every day or two.

Scientists hope to grow plasmodia in an artificial medium where they can control the amount and types of nutrients supplied. If they do, they will be able to determine what the organisms actually use as they grow and multiply. Until plasmodia can be grown in such a controlled medium, researchers will not be sure what nutrients these organisms use in the human blood stream.

FIELDS OF INQUIRY IN MODERN BIOLOGY

Biologists study many subjects besides disease. The following pages show only a few of the many subjects biologists study. As you continue your work in biology, you will encounter many more.

1-7. The health and well being of the human population is of concern to biologists. One important consideration is food. Even today, there is not enough food to feed all the world's people. If population continues to rise, the situation may become even worse. Biologists attack the problem in a number of ways. The wheatfield shown in **(a)** represents the cereal plants such as oats, rice, and corn that provide more than half our food. Biologists and plant breeders try to develop new kinds of cereal plants that are disease and pest resistant, better yielding, higher in nutritional value, or stronger **(b).** Biologists also seek ways of controlling the insect pests that consume acres of crops each year. Some kinds of pesticides **(c)** may be harmful to animals, including humans. Researchers are looking for natural controls, such as the predator species being observed **(d).**

a

b

c

d

1-8. Much of biology is related, either directly or indirectly, to health. As with malaria, the search is often for causes. Biologists may also investigate the treatment of disease. Shown in **(a)** are some biologists at the Center for Disease Control in Atlanta, Georgia who are investigating the rubella virus. Still other scientists **(b)** are involved in cancer research at the AMC Cancer Research Center in Denver, Colorado. As shown in **(c),** we now have available a wide variety of drugs for treatment of disease. Biologists sometimes test medicines for safety and effectiveness **(d).**

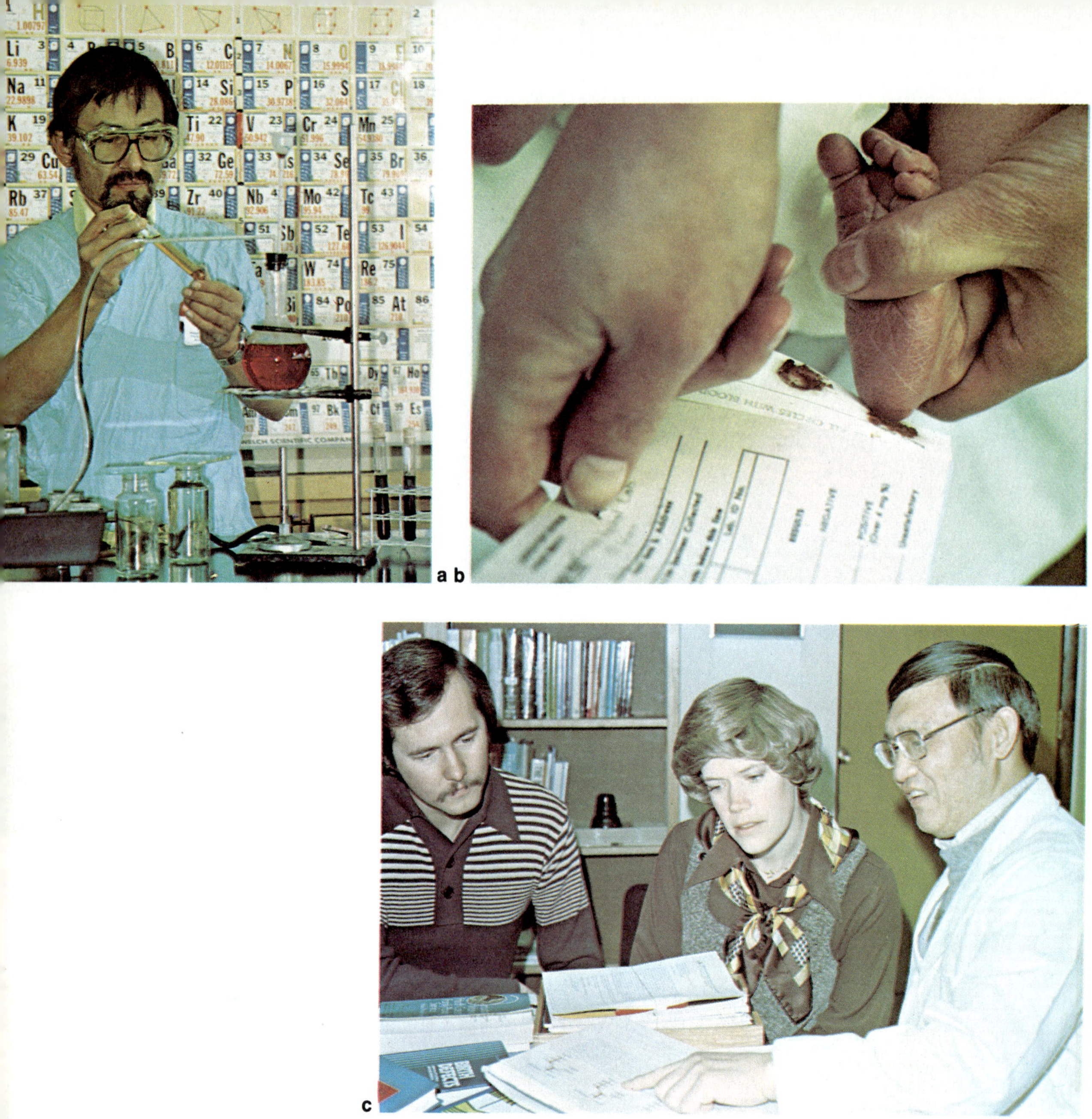

a b

c

1-9. One class of human diseases is of particular interest to biologists. These are the inherited, or genetic, diseases. Since the cracking of the genetic code (see Chapter 7), it has become possible in many instances to identify a genetic disease. The search sometimes looks more like chemistry than biology **(a).** But the results have practical application, such as the testing of infants for PKU **(b),** a genetic disease that can be controlled through diet. Couples with a history of genetic disease may now consult a genetic counselor for information they may need to make decisions about having children **(c).**

1-10. One of the newest and most rapidly growing fields in biology is behavior. In the case of animals, including humans, most behavior is made possible by the nervous system. Complex behaviors may be caused by the effects of the environment on systems of the body. In learning about such things as migration of birds **(a),** parent-offspring relationships in monkeys **(b),** and the social orders of bees **(c),** we may come to understand our own behavior better.

a

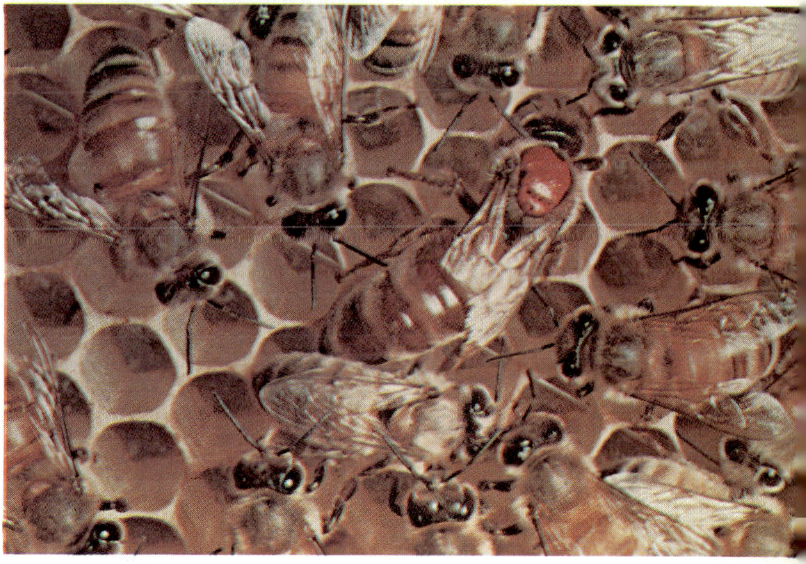

b c

a

b

c

1-11. In the last two decades, we have become more aware of our dependence on our environment. The human population places large demands on the earth's natural resources. Human activities can pollute water and air and kill plants and animals. Today, we are trying to learn more about our environment and our effect on it. The effects of air pollutants on crop plants can be studied in the laboratory **(a).** Ways of maintaining a pure water supply are being explored **(b).** Biologists are studying how and where plants and animals live **(c).**

PRINCIPLES OF INQUIRY

It is not really possible to separate the *subject* of a biological question from the *methods* used to answer it. You read about the methods of observation and hypothesis in the malaria story. In the next section, we'll use another example to examine more closely the methods that biologists used in another field of inquiry. Such methods are called **principles of inquiry.** But first, use principles of inquiry yourself, by doing **Inquiry 1-1.**

INQUIRY 1-1 Asking Questions—Seeking Answers

How can we use clues to seek answers to biological questions? In this inquiry you will have a chance to ask a question and search for an answer using an indicator. An **indicator** is a substance in which you can see a change happen.

MATERIALS (per team)

Days 1 and 2

2 test tubes with phenol red indicator
9 empty test tubes, 18 × 150 mm
9 brass screws
4 corks
straw
phenol red indicator
bunsen burner or hot plate
large beaker
small amount of dry ice
medicine dropper
desk lamp or other light source
2 sprigs of a water plant, such as elodea
2 live grasshoppers or other large insect

Day 3

FILM: *Mimicry*

Procedure (Day 1)

Your teacher will give you two test tubes. Both were set up before class. The test tubes were treated the same with one exception. It is your job to determine that single difference. First, examine the two

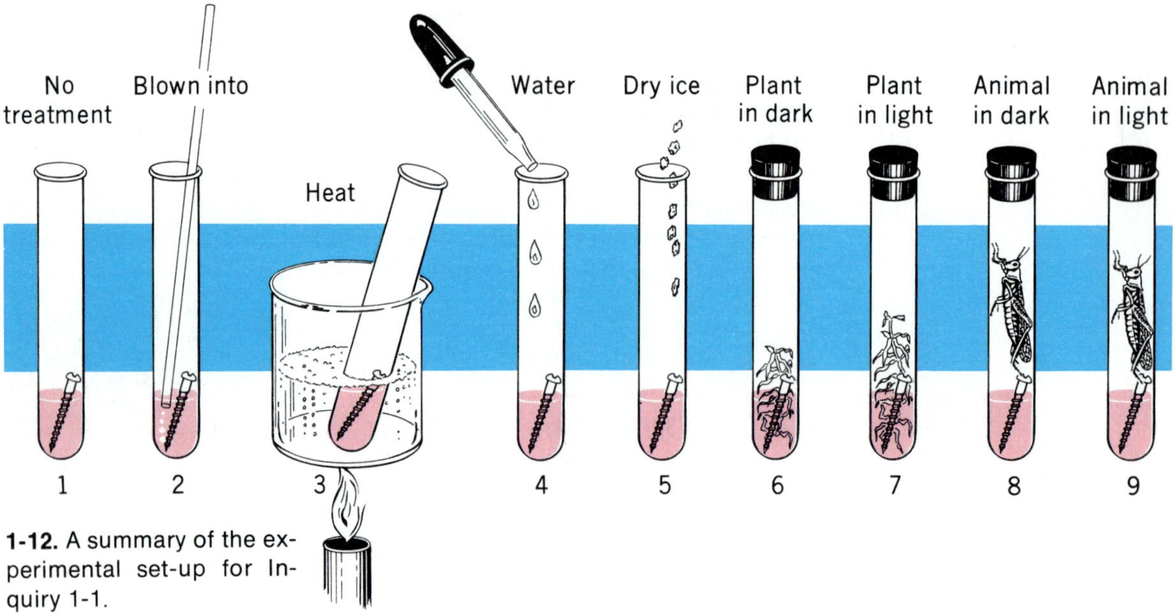

1-12. A summary of the experimental set-up for Inquiry 1-1.

tubes. **1.** What difference can you observe? **2.** List as many hypotheses as you can to account for the difference.

Next, prepare and number nine tubes of indicator solution as instructed by your teacher. Add a brass screw to each one. Figure 1-12 is a summary of the treatment all tubes will receive. **3.** What similarities and differences can you see between the tubes you prepared and the other tubes you observed?

Leave the first tube as it is. Blow gently through a straw into the liquid in Tube 2 for several minutes. (Be careful not to pull the liquid into your mouth. Also, blowing too hard will force the liquid out of the tube.) **4.** What happens? **5.** What is it about your breath that might produce the change? List as many possible explanations for the change as you can.

Perhaps you listed heat as one possible cause. Your hypothesis is that heat from your breath caused the change. Form a deduction from your hypothesis. You can test this deduction by an experiment. **6.** State your deduction about heat causing the change by completing the following statement: *If* heat causes a change in the indicator, *then . . . .*

Put a small amount of water in a large beaker. Set Tube 3 in the beaker. Heat the water in the beaker over a flame or hot plate. **7.** What happens to the solution in the tube? **8.** What can you say about your hypothesis? **9.** You may know your breath contains water vapor. Form a new hypothesis. Make an "If . . . then" deduction that relates water to the observed change.

Using the medicine dropper, add a few drops of water to Tube 4. **10.** Do the results support your hypothesis about water? Why or why not?

Dry ice is frozen carbon dioxide. Add a *tiny* amount of dry ice to Tube 5. **11.** What happens? **12.** How does this observation relate to the changes you observed when you breathed into the tube? **13.** Devise a statement that explains both changes. **14.** Why is the hypothesis that the cold of the dry ice causes the change not acceptable?

Before leaving the laboratory, set up the remaining tubes. You will observe these at your next class meeting. In Tubes 6 and 7 place sprigs of elodea or other water plant. Put one grasshopper in Tube 8 and put the other one in Tube 9. (The brass screws will serve as supports for the animals.) Cork all four tubes. Place Tubes 6 and 8 in a dark place such as a closed cabinet. Place Tubes 7 and 9 in continuous light.

Procedure (Day 2)

Examine the four tubes you prepared earlier. **15.** What has happened to the tubes in the dark (6 and 8)? **16.** What has happened to Tubes 7 and 9? **17.** What hypothesis can you suggest to account for any observed differences? **18.** How do these new observations relate to the results you got before, especially when using your breath and the dry ice? **19.** What was the purpose of Tube 1? **20.** Why were brass screws placed in all nine tubes, even though they were only needed in Tubes 8 and 9? **21.** What questions might you investigate now? Design and carry out new investigations to look for answers to your questions.

Procedure (Day 3)

This inquiry may have given you some understanding of the principles of inquiry used in scientific investigations. You can gain more experience with such principles if your teacher can arrange for you to view and discuss the film, *Mimicry*. If you use the film, you will formulate hypotheses, design experiments, and interpret results. Your teacher can provide more information.

Principles of Inquiry—Spontaneous Generation

Until the last part of the nineteenth century, scientists had three hypotheses about the origin of living things. One of these said that or-

a Reproduction of offspring from parents of the same kind

b Transformation of one kind of living thing into another

c Spontaneous generation from mud or earth or remains of dead plants and animals

1-13. Until the end of the 19th century, there were three hypotheses used to account for the origin of living things. Observations and experiments, however, suggest that only **(a)** is accurate.

ganisms could reproduce others of their own kind (Figure 1-13a). Another suggested that organisms of one kind could produce organisms of a very different kind (Figure 1-13b). The third hypothesis suggested that organisms could be formed spontaneously (without a known cause), from materials that were not alive (Figure

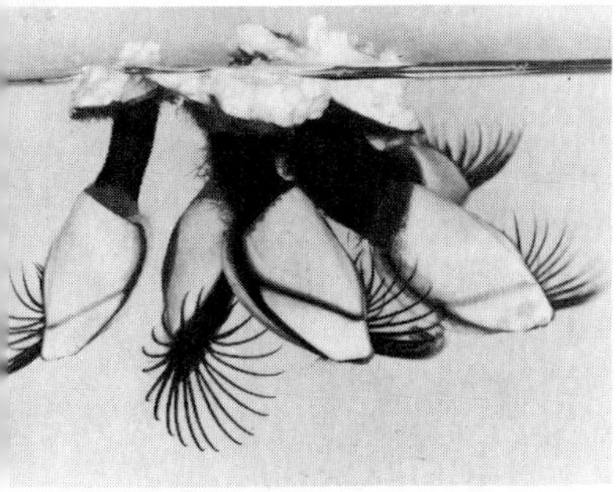

Carolina Biological Supply House

1-14. A cluster of goose barnacles. *Can you see the similarity to birds, as in Figure 1-13b?*

1-13c). These three hypotheses relate to a basic biological problem: the origin of living things.

The idea that offspring are produced by parents of their own kind, or species, is a familiar one. "Like produces like."

The second idea, "like produces *un*like," was popular for many centuries. The origin of the barnacle goose is an example. Goose barnacles often grow on logs floating in the sea. The goose barnacle looks a little like a bird's head (Figure 1-14). This probably accounted for the hypothesis about the origin of the barnacle goose (Figure 1-13b). But observation and study of the barnacle proved that it could not develop into a goose. The barnacle goose breeds in the far north. It lays eggs, and young geese hatch from them, as is the case with any other goose. Barnacles do not change into geese; they are distinctly different animals.

The hypothesis of **spontaneous generation** (Figure 1-13c) stated that living things can form from nonliving materials. Some biologists, most of whom lived before the end of the nineteenth century, believed in spontaneous generation. Aristotle, the Greek philosopher of the fourth century B.C., was one. Here is a simplified version of one of Aristotle's statements:

> The great majority of fish develop from eggs. There are some fish, however, that develop from mud and sand. A pond near Cnidos dried up—even the mud at the bottom. Later the pond was filled by rain. It was then observed that the pond contained many tiny fishes. From this fact it is clear that certain fishes come spontaneously into existence, not being derived from eggs.

Nearly 2,000 years after Aristotle wrote of spontaneous generation, biologists still believed in it. In 1652, Jean-Baptiste van Helmont (zhon bah·TEEST van HEL·mahnt) wrote that if wheat grains and a dirty shirt were put in a pot, mice would be formed.

Probably few biologists of van Helmont's day followed this recipe for making mice. Yet nearly all of them would have explained some other observations as spontaneous generation. The most familiar examples had to do with decaying meat. (There were no refrigerators in those days.) In warm weather, meat rapidly became covered with tiny, white worms called maggots. Where did the maggots come from? Most biologists of the seventeenth century would have answered, "from the meat, through spontaneous generation."

Testing the Spontaneous Generation Hypothesis

Francesco Redi (fran·CHESS·ko RAY·dee) was a seventeenth-century Italian physician and biologist. He decided to put spontaneous generation to the test by controlled experiment and observation. He began by placing three dead snakes in a box to decay. In three days the snakes were covered with maggots. Eventually only the snakes' bones were left. On the nineteenth day, some of the maggots stopped moving. They seemed to shorten and become oval shaped, like an egg. Later they became hard little balls.

Redi put some of these little balls in a glass jar and covered the jar with paper. After eight days the little balls broke open and out of each came a gray fly. At first each fly moved very slowly, and its wings were closed. But, after a few minutes, the wings began to unfold and the flies appeared normal. Redi wondered if the maggots could be produced by flies. Perhaps they had not come from the decayed meat. He had seen flies around the meat before the maggots appeared. And these flies were of the same kind as those that later emerged from the little balls.

Redi's hypothesis was that the maggots were the offspring of flies. How might the hypothesis be tested? Redi made a deduction: *If* maggots are the offspring of flies, *then* meat protected from flies would *not* develop maggots. Here is what Redi did:

PROCEDURE: He put a dead snake into a wide-mouthed flask. Into another flask he put some fish, into still another some eels, and into the last a piece of veal. He left these flasks open. Another four flasks were filled in the same way. These flasks, however, were closed and sealed.

Why did Redi use sets of four flasks, with different materials in each flask? Why did he cover the flasks in one set and not in the other?

OBSERVATIONS: It was not long before the meat and fish in the open flasks were covered with maggots. Flies entered and left at will. In the flasks that had been closed and sealed, no maggots appeared, even after many days.

Write down as many conclusions as you can that could be drawn from these observations. Then read how Redi proceeded with a new hypothesis and a new procedure.

HYPOTHESIS: From these experiments, Redi thought he had proved that the flesh of dead animals could not generate maggots spontaneously. The maggots could only arise if flies laid eggs on the meat and these hatched into maggots.

Can you think of any sources of error in Redi's experiment? Redi could.

> There is a possible objection to the experiment. The flasks in which no maggots appeared had been closed and the air could not penetrate or circulate. Possibly it was for this reason, and not for the absence of flies, that maggots were not spontaneously generated. Therefore, in order to remove all doubt, I carried out another experiment.

PROCEDURE: Redi placed some meat and fish in a large jar and covered it with a fine net. This allowed air but not flies to enter the jar. As a further protection against flies, he placed the jar in a frame covered with the fine net.

OBSERVATIONS: Maggots never appeared in the meat. The flies did land on the net over the frame. They deposited eggs there.

On the basis of this and similar experiments, Redi concluded that when meats and fish are protected from flies, they do not generate maggots.

Did Redi's results show that maggots could not be formed by any sort of material (other than meats, fish, and so forth)? Did they prove that organisms other than flies could not arise by spontaneous generation? Redi thought not. He still believed that spontaneous generation could and did occur in some cases (Figure 1-15).

Redi's work put arguments about spontaneous generation to rest for a while. But the question was reopened by the discoveries of Anton van Leeuwenhoek (LAY·ven·hook). Leeuwenhoek made many observations with microscopes that he built (Figure 1-16). He discovered tiny organisms that had never been seen before.

In water from rain, marshes, and the river flowing near the Dutch town where he lived, he saw numerous tiny living things. Here was a

1-15. Even after his experiments, Redi believed spontaneous generation could occur.

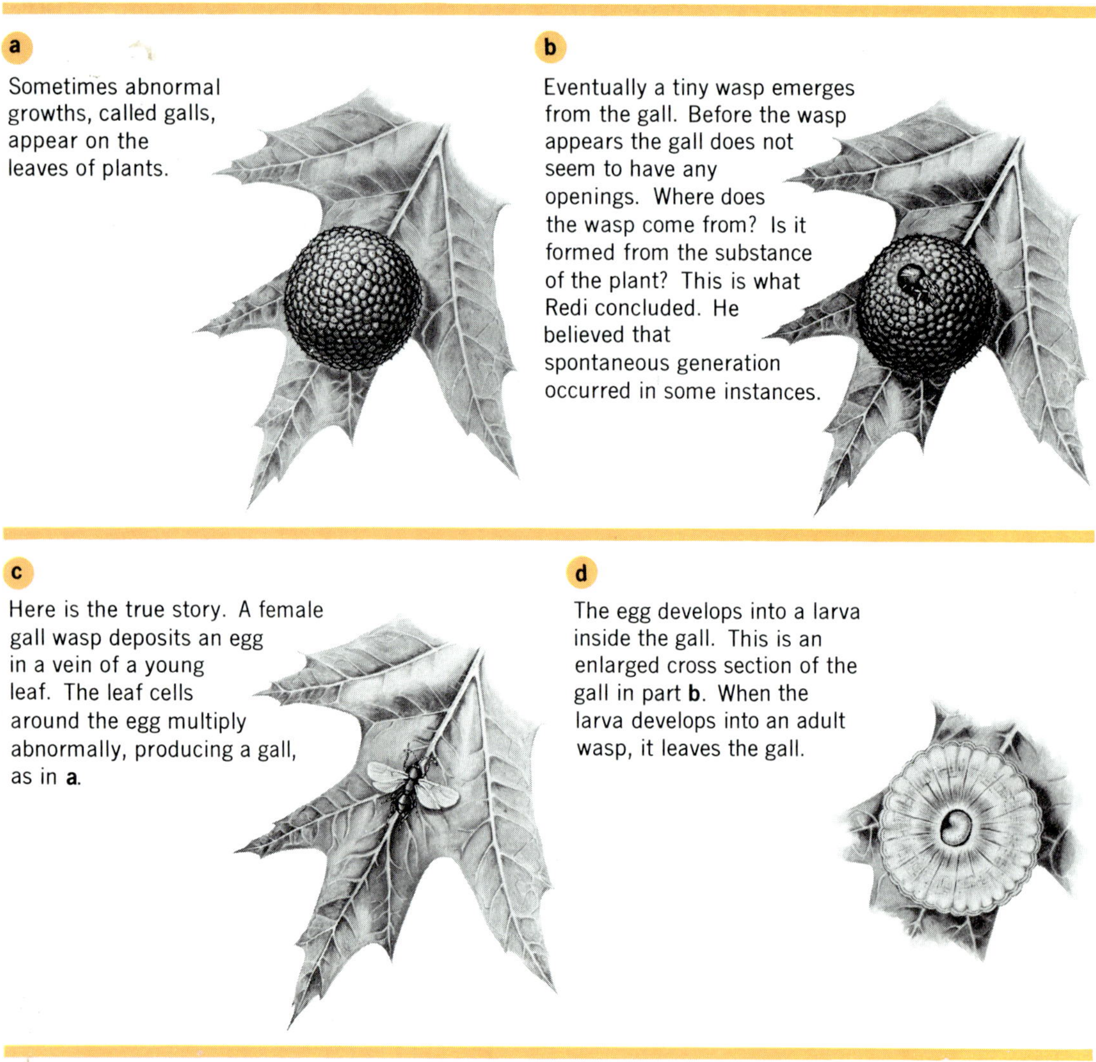

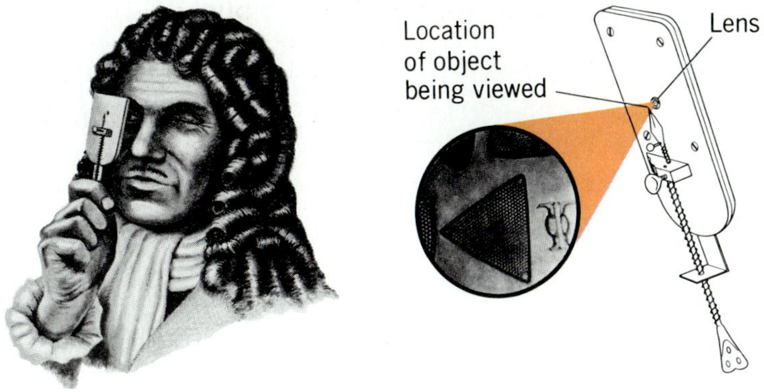

1-16. Leeuwenhoek and one of his microscopes. Notice how different the microscope is from those of today. The photograph shows diatoms (microscopic plants) in pond water. It was taken recently through one of Leeuwenhoek's microscopes.

whole world of organisms, tiny animals and plants that no one had ever seen before. No one *could* have seen them until the necessary tool – a microscope – was built. You will be able to appreciate Leeuwenhoek's discoveries better if you now do **Inquiry 1-2.**

INQUIRY 1-2 The Compound Microscope – A Scientific Tool

To answer some of the questions posed in this course, you will need to use a microscope. A microscope is a tool that allows you to see things too small to be seen with the unaided eye. It magnifies them – makes them look bigger.

Biologists use different kinds of microscopes. Some can magnify an object 1,000 times (1000 ×). Most school microscopes magnify objects from 100 × to 430 ×. School microscopes are sometimes called light microscopes because light passes through the object being observed. They also are called compound microscopes because they use more than one set of lenses.

MATERIALS (per team)

1 or more microscopes
lens paper
several microscope slides
cover glasses
dissecting needle
soft, clean cloth
medicine droppers
piece of newspaper
silver polish
pond water
yeast in a weak molasses solution
small amount of soil
test tube with stopper

1-17. TYPICAL COMPOUND MICROSCOPE

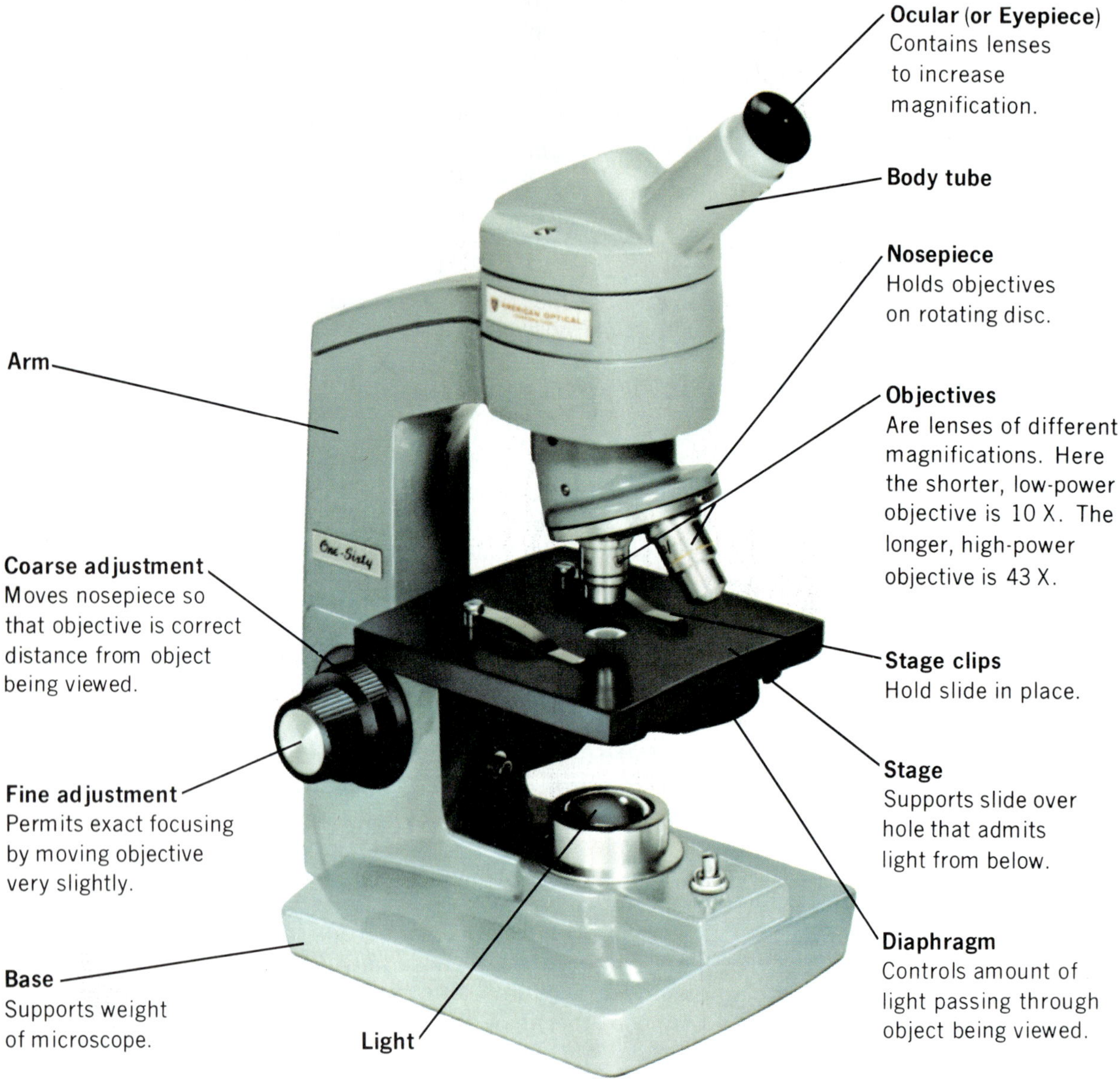

Procedure (Day 1)

Setting up the Microscope. Carefully remove the microscope from its case. Hold the microscope *arm* with one hand, and place your other hand under the *base* (Figure 1-17). Set the instrument down gently with the base several centimeters in from the edge of the table.

Locate on your microscope the parts shown in Figure 1-17. There are several kinds of microscopes, but the basic parts are similar.

Rotate the *nosepiece* so that the *low-power objective* (the shortest one) is in line with the *body tube.* The objective should click or snap into place.

Open the *diaphragm* to its largest opening. If your microscope has a *mirror* instead of a light, you must look through the *ocular* (eyepiece) and move the mirror until it reflects light upward through the opening in the stage. Use the flat side of the mirror, not the curved side. Do not let direct sunlight strike the mirror. While looking through the ocular of the microscope, adjust the mirror (if any) and the diaphragm. The round *field of view* should be lighted evenly and without glare.

If the ocular or objective is cloudy or dusty, wipe the lenses gently in one direction with a piece of *lens paper.* Do not use any other kind of paper or cloth. Optical glass is soft and scratches easily.

How to Prepare Materials. Place the materials you will study on a glass *microscope slide.* In most cases, you will cover the material with a thin piece of glass or plastic. This is a *cover glass.* Both the slide and cover glass should be as clean as possible.

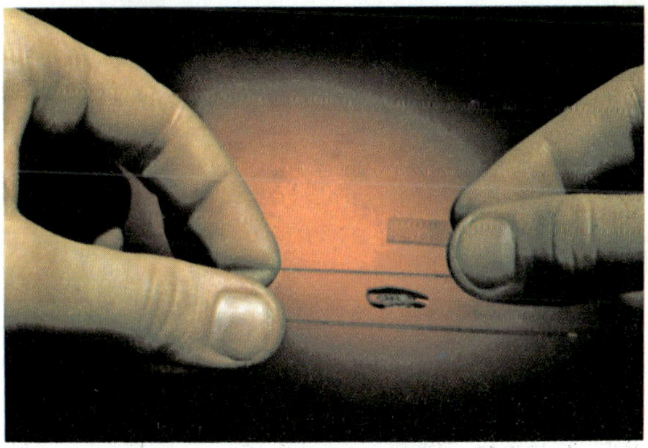

1-18. How to hold a slide and cover glass.

Always hold slides and cover glasses by their edges (Figure 1-18). Hold the slide this way as you clean it with soap and water. Rinse completely. Dry it thoroughly with a soft cloth that has no loose fibers or lint. A cover glass is fragile. Handle it with care. Clean both sides of the cover glass at the same time using the soft cloth. Be careful not to touch the surfaces of the cover glass with your fingers.

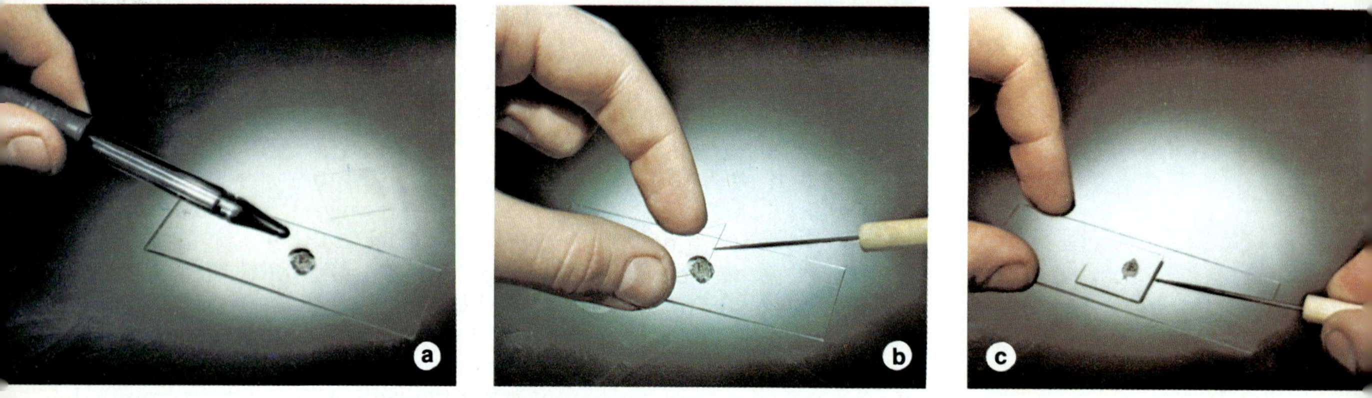

1-19. Preparing a wet mount. **(a)** Place a drop of water on the specimen. **(b)** Hold the cover glass at a 45° angle. **(c)** Lower the cover glass with a dissecting needle.

Lean the slides against the edge of a book or other object on your laboratory table. Lean the cover glass against the slide.

In a newspaper find some fine print that includes the letter "e." Cut this letter from the newspaper in a piece about 1 centimeter (cm) square. Place the piece of newspaper on the center of the slide, with the letter "e" right side up.

With a medicine dropper put 1 drop of water on the piece of paper. Let the water soak into the paper. Place the cover glass over the paper. The cover glass must be placed so that no air bubbles are trapped in the drop of water. To do this, hold the cover glass at approximately a 45° angle to the slide. Gently lower it with a dissecting needle until it covers the water, as shown in Figure 1-19. You can remove bubbles by tapping gently on the cover glass with the eraser of a pencil. You have now prepared a *wet mount* of the piece of paper.

How to Focus the Microscope. Place your slide on the stage of the microscope. Use the low-power objective. Center the letter "e" on the slide over the opening in the stage.

Different kinds of microscopes are focused in different ways. But the procedure is always the same. Change the distance between the objective and the stage until you can see the letter clearly. On a microscope like the one in Figure 1-17, you do this by moving the adjustment knobs that control the height of the nosepiece. In some other microscopes, the coarse and fine adjustment knobs move the stage. If you need help with this adjustment, ask your teacher.

Now look at the slide on the stage of the microscope *from one side.* Use the *coarse adjustment* to lower the nosepiece (or raise the stage). Turn it until the low-power objective is about 2 cm above the cover glass *or* until you reach the stop that prevents the stage or objective from being moved any farther. (CAUTION: *Do not lower the nosepiece while looking through the ocular. If you accidentally ram the objective into the slide, you may break the slide and damage the lens of the objective.*)

While looking through the ocular, slowly turn the coarse adjustment to raise the nosepiece (or lower the stage) until you can see the printed letter. If necessary, move the slide until the whole letter is in the field of view. Then turn the *fine adjustment* to make the focus as sharp as possible. **1.** How does the image that you see through the microscope differ from the actual position of the letter on the slide?

Now slowly move the slide on the stage from right to left. **2.** Which way does the image that you see through the microscope move?

Move the slide away from you. **3.** Which way does the image seem to move?

As you look through the ocular, keep both eyes open. You may have trouble at first, but you will soon learn to ignore what you see with the eye that is not looking through the microscope. Closing one eye may cause eyestrain.

Now observe the letter with high power. Move the letter on the slide back to the center of the field. Ask your instructor if your microscope is *parfocal.* (In parfocal microscopes both the low- and high-power objectives are adjusted to the same focus. You can switch carefully from low to high power without refocusing.) If your microscope is not parfocal, raise the nosepiece very *slightly* (or lower the stage). Then change from the low-power to the high-power objective. Always use this procedure for microscopes that are not parfocal. Be sure that the high-power objective does not touch the slide. Next, focus the image using the fine adjustment only. NEVER turn the coarse adjustment when using the high-power objective. Always use the coarse adjustment cautiously. **4.** Does the field of view show a larger or smaller segment of the letter? **5.** Is the brightness of the field greater or less than with low power? You can adjust the diaphragm to provide even lighting without glare.

Care of the Microscope. Microscopes are expensive instruments. They must be given proper care. Always carry a microscope by the arm and in an upright position with one hand under the base. Place it on the table several centimeters from the edge to prevent an accidental fall.

Some microscopes have an *inclination joint.* They can be tilted toward you. Tilt the instrument only if you must because of your height. The microscope is off-balance when tilted. Do not leave it in this position untended.

At the end of a laboratory period, turn the low-power objective into position and adjust it to approximately 1 cm above the stage. Be sure the stage clips do not extend beyond the stage. Return the microscope to its storage place. Clean all used slides and cover glasses.

Procedure (Day 2)

Using the Microscope. The following steps will help you use the microscope properly as you work through the next part.

A. Make certain the low-power objective is in place.

B. Adjust the diaphragm to its largest opening.

C. Look through the ocular and adjust the diaphragm for an even, well-lighted field of vision.

D. Lower the nosepiece (or raise the stage) until the objective is about 2 cm above the stage.

E. Place the slide on the stage and center the object you are going to study over the stage opening.

F. While looking through the ocular, slowly raise the nosepiece (or lower the stage) with the coarse adjustment until the object comes into focus.

G. Sharpen the focus with a slight turn of the fine adjustment.

H. If it is necessary to change to a high-power objective after Step G, switch to the high-power objective and sharpen the focus with the fine adjustment only. If your microscope is not parfocal, remember to raise the nosepiece (or lower the stage) slightly before switching to high power.

Observations of Living (or Once-Living) Things. If the pond water looks clear, stir up materials from the bottom. Place a drop of the pond water on a microscope slide. Cover it with a cover glass. Look at it through your microscope. What do you see? Pond water usually contains many kinds of organisms. Perhaps the water contains organisms like those shown in Figure 1-20. You probably cannot name the organisms, but you may be able to describe some of them. **1.** Describe any organisms you see. Include such information as relative size, color, shape, movement, and so on.

Make a wet mount of the smallest possible amount of silver polish mixed with a drop of water. Use the microscope to observe the once-living organisms—diatoms. **2.** Make simple sketches of

1-20. Pond water often contains many different microorganisms. Only a few are shown.

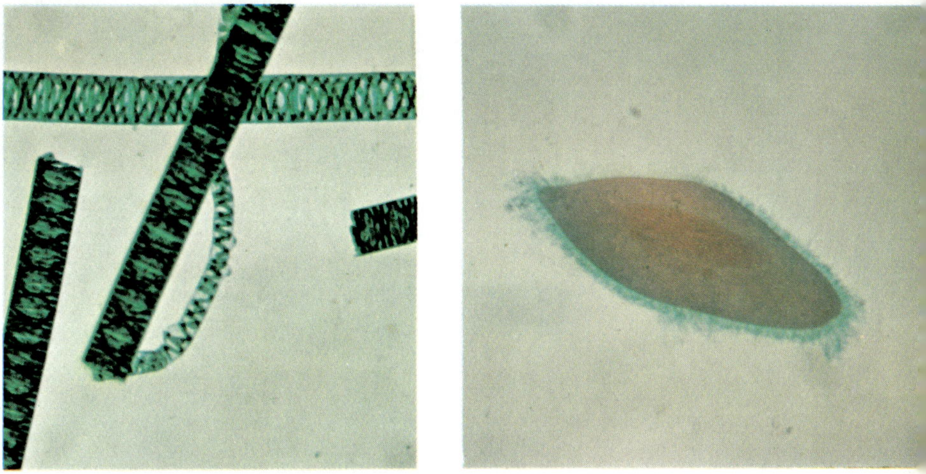

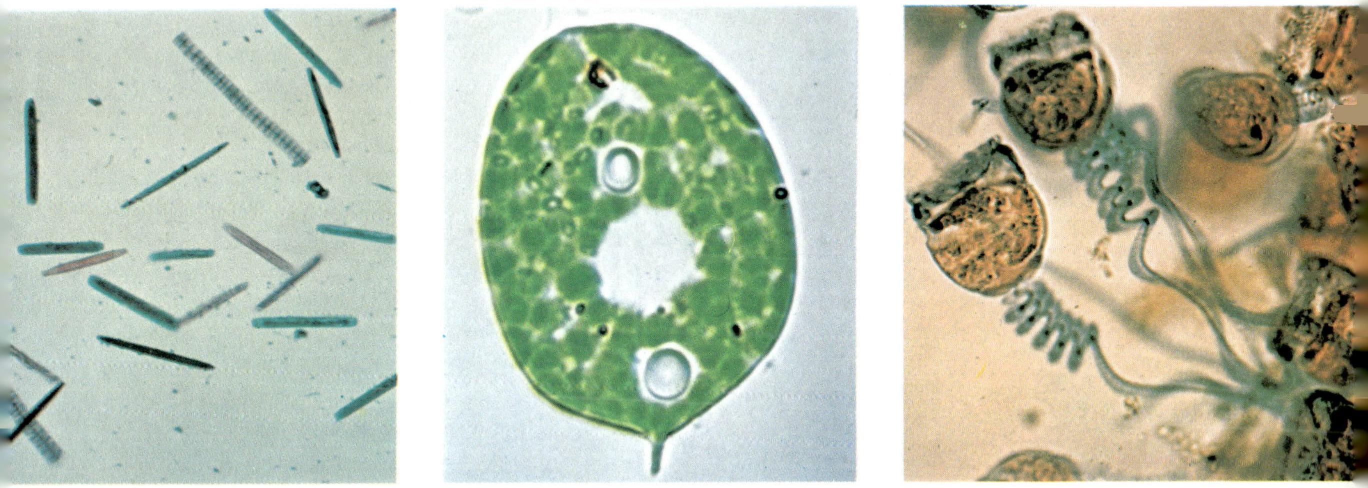

Carolina Biological Supply House *Carolina Biological Supply House*

what you see. (For more information on diatoms see Chapters 12 and 23.)

Place a drop of yeast and molasses solution on a slide and cover with a cover glass. **3.** What do you see through the microscope? Some of the yeast will probably have small bulges on them. These are called buds. Budding is one type of reproduction (see Chapter 19). Then the bud grows and breaks off. It is then a new yeast cell.

Place a tiny amount of soil in a test tube. Add 10 to 15 drops of water. Stopper the tube. Shake it vigorously until the soil is well broken up. Put a drop of the liquid on a slide, cover with a cover glass, and observe under the microscope. **4.** Describe any organisms you see. Can you find any nematodes (roundworms) like those

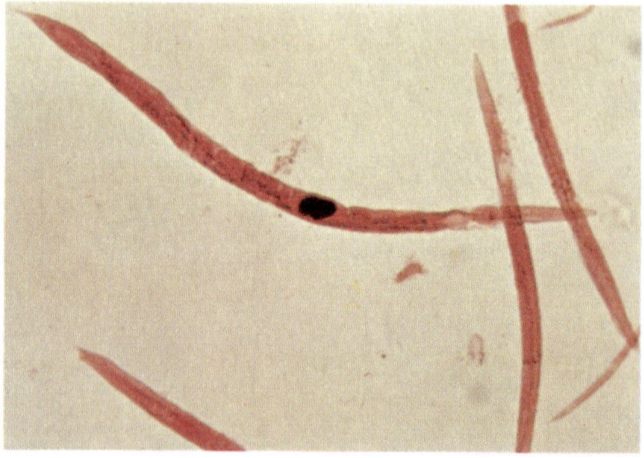

1-21. Nematodes are round worms. The stained forms shown here are commonly found in soil.

shown in Figure 1-21? (For more information on nematodes, see Chapter 12.)

Review the procedure for caring for the microscope. Clean and store the microscope.

The 300-Year Debate

What was the origin of the tiny animals and plants that Leeuwenhoek saw under the microscope? Leeuwenhoek had an opinion about these **microorganisms** (MY·kro·or·gan·izms). He felt that the discovery "must surely convince all of the absurdity of these old opinions that living creatures can be produced from corruption or putrefaction."

Other biologists were not convinced. Most had given up the idea that larger animals—mice, worms, maggots, geese, and flies—could be produced spontaneously. But what about microscopic forms? Biologists knew that if small amounts of chopped hay or a few seeds were put in pure rainwater a tremendous number of microscopic animals and plants would soon appear. Were these organisms generated from materials put into the water?

For the next 300 years biologists debated the question of the origin of microorganisms. One group thought that they were produced by spontaneous generation. Others thought that microorganisms could be produced only by other microorganisms.

Those who rejected the idea of spontaneous generation thought that natural air always contained living microorganisms. The French scientist, Pouchet (poo·SHAY), made artificial "air" in one experiment. He prepared a hay infusion, which is simply hay boiled in water. He then added nitrogen and oxygen to form "air" above the infusion. In a few days the infusion was rich with microorganisms. For another experiment, the scientist made water by burning hydrogen gas in air to form water. The water was then boiled with hay to make the infusion. In spite of all this care, the infusion was soon filled with organisms. Pouchet thought that spontaneous generation had occurred.

Louis Pasteur (LEW·ee pas·TOUR) of France believed that these conclusions were wrong. Pasteur made infusions of yeast and sugar boiled in water. When exposed to air, the infusions soon had a rich growth of organisms. When he was careful to keep air from reaching the infusions, however, no organisms appeared.

Both scientists were doing the same, or nearly the same, experiments. Yet spontaneous generation seemed to occur in one but not in the other.

On June 22, 1864, Pouchet and Pasteur appeared before members of the French Academy of Science. Each was to do a series of experiments. Pasteur began by showing the evidence of past experiments. Four years before, he had thought that the air of the high mountains would be relatively free of microorganisms. To test this hypothesis, he boiled infusions in a number of glass flasks. While they were still hot, he sealed the necks of the flasks. Every infusion was *sterile,* that is, free of organisms.

Pasteur packed some of the sealed, sterile flasks on a mule and set off for the French Alps. At a point high in the Alps he cut the necks of the flasks open. The infusions were exposed briefly to the air and then resealed. The flasks were returned to the laboratory.

Pasteur showed three of these flasks to the members of the Academy. They found no organisms in them. But the members of the Academy wanted to know more. Was Pasteur's report really accurate? And was there really oxygen, which they knew to be necessary for life, in the flasks? They opened Flask 1 and analyzed the air. It contained the normal amount of oxygen. Flask 2 was opened to the air. In three days it contained a huge number of organisms. They did nothing to Flask 3. This flask remained sterile—as it had been for the four years since it was opened in the Alps.

Pasteur then began a new demonstration. A total of 60 flasks were prepared with an infusion of yeast and sugar. The infusions in all the flasks were boiled for two minutes. Then the necks of 56 of the flasks were sealed. The 56 flasks were used in three groups.

Group A: 19 flasks were opened briefly in the amphitheater where the Academy was holding its hearings. The flasks were then resealed and set aside.

Group B: 19 flasks were opened briefly high up in the dome of the amphitheater. These were then resealed and set aside.

Group C: 18 flasks were opened outside the building, then resealed and set aside.

1-22. Pasteur is shown examining a flask that is cloudy with microorganisms. The fluid in the gooseneck flask in his left hand is clear and sterile.

Pasteur thought that the air inside the building would contain fewer microorganisms than air outside the building. Growths of organisms developed as follows:

Group A: 5 flasks (26 percent)
Group B: 6 flasks (32 percent)
Group C: 16 flasks (89 percent)

In every group, some infusions remained sterile. *Most* of the infusions exposed indoors remained sterile, but only a *few* of those exposed outdoors did.

It was the four flasks not yet accounted for that proved most interesting of all. Pasteur had heated the glass neck of each of these four flasks. He drew the necks out into a long curve (Figure 1-22), and left the necks open. Pasteur then boiled the infusions in the flasks and let steam flow out of the neck of each one. The steam would kill any organisms in the neck of the flask.

Then the flasks were allowed to cool. As air entered the long, curved necks of the flasks, particles would settle on the curved walls of the glass. In some cases, at least, no particles should be expected to pass through the curved passage and reach the infusions. As a matter of fact, all four flasks remained sterile. Yet none of the four was sealed.

This concluded Pasteur's demonstrations before the Academy members. Although somewhat critical of Pasteur's results, Pouchet refused to appear before the Academy. He did not want to expose his work to the judgment of the Academy.

The Modern View

After Pasteur's demonstrations, few biologists believed in spontaneous generation. Did the long line of scientists from Redi to Pasteur actually prove that spontaneous generation cannot happen? Not at all! Perhaps this comes as a shock to you. All that was demonstrated *scientifically* was:

1. All observations and experiments thought to be examples of spontaneous generation were shown to be false.

2. No experiment showed convincingly that spontaneous generation could occur.

Most biologists accept the view that, on the earth today, life comes only from life. But if life comes only from life, does this mean that there was *always* life on the earth? Evidence suggests that the world was once without life—that life appeared later. How? We think it was by a spontaneous, natural process. But we believe this process was very different from the hypothesis of spontaneous generation. We must wait until Chapter 12 to continue this part of the story.

The Spontaneous Generation Story Reconsidered

Today the idea that living things might arise from rotting meat or dirty shirts seems almost funny. But was the debate over spontaneous generation foolish? Were Aristotle, Redi, Pasteur, and others simply unable to discover an obvious fact? No. These scientists are among the greatest of all time. They understood principles of inquiry and applied them well.

Consider first the observation of Aristotle that fish could be found when dried-up ponds filled up again. The conclusion that the fish arose spontaneously was reasonable given what he had seen. Aristotle was simply doing what all scientists do—drawing a conclusion from limited information. All conclusions are tentative. Any conclusion is considered accurate until different or more complete evidence is found. Aristotle's "error" set up the conditions that led to much more inquiry.

Consider next the experiments that Redi performed with maggots. For the first time, the question of spontaneous generation was attacked by direct experimentation. Previous ideas had been accepted or not accepted only on the basis of observations. But Redi did more. In each experiment, all conditions were kept the same *except one.* The one condition that was different in each experimental setup was the **variable.** For example, in Redi's first experiment two sets of jars were prepared. One set was left open to the air. The other was closed. Thus, one set could be used as a basis for comparison with the other. Redi knew what would happen in the open jars. This set was the **control.** Redi could compare the closed jars with the open ones and see if blocking out air made any difference. Exposure to air was the only variable in the experiment. Several variables in the experiment would have made it impossible to tell which one was responsible for any change that occurred.

Leeuwenhoek's work showed the importance of observation. What Leeuwenhoek saw reopened the debate over spontaneous generation. But there is another important point here. Leeuwenhoek's observations were made with the new microscopes he had designed and built. For the first time, observations of tiny living things were possible. With the development of the microscope, an entirely new field of inquiry opened up. Tools such as the microscope are perhaps as important to the work of science as are hypotheses and experiments.

The work on spontaneous generation deserves special attention. Pasteur and Pouchet both formulated hypotheses. Both designed and carried out controlled experiments. Both were respected by the scientific academy of their day. But how could two capable scientists come to such different conclusions?

Chance also plays a part in scientific studies. Pasteur's infusions of yeast and sugar were boiled in water to kill microorganisms. These experiments gave strong evidence against the theory of spontaneous generation. Likewise, the experiments with hay infusions were boiled. Why were the results so different? Today we know that some kinds of microorganisms can survive boiling for many minutes or even hours. Dried hay, as used in the hay infusion, might contain spores, or inactive forms, of such microorganisms. Pasteur's yeast and sugar solution would be much less likely to contain spores. Thus, although both scientists did their experiments carefully, they reached opposite conclusions.

It is also important to consider two other aspects of the controversy over spontaneous generation. First is personality. Both scientists were invited to demonstrate their work before the French Academy. But after Pasteur's demonstration, Pouchet "lost his nerve." If Pouchet had presented his findings, the outcome of the hearings might have been very different. In fact, the arguments *for* spontaneous generation probably would have been very convincing in 1864.

Second, consider the kind of society in which these biologists worked. The nineteenth century was a time of rapid advance in science. There was much interest in scientific inquiry. The fact that the French Academy reviewed a specific question and judged the evidence shows this interest in inquiry. The age of science was surely upon us.

In Brief

In this chapter, you have considered two examples of what biologists investigate and how they work. What can you now say about the importance of questions, observations, hypotheses, evidence, and deductions in biology?

Recall the quotation from the beginning of this chapter: "Science seldom proceeds in the straightforward, logical manner imagined by outsiders. Instead, its steps forward (and sometimes backward) are often very human events in which personalities and cultural traditions play major roles." Do you think Watson is right? Why?

Guide Questions and Problems

1. Consider how many years passed between the discovery of plasmodia by a French army physician (Laveran) and acceptance of the hypothesis "Plasmodia cause malaria."
   (a) Why were scientists slow to accept this hypothesis?
   (b) How is a hypothesis tested?
   (c) What can be done to find out if a particular organism causes a particular disease?

2. Suppose one of the early volunteers who drank swamp water had developed malaria. How would this have affected the search for the cause of malaria?

3. Many people and many birds with malaria parasites in their blood do not show symptoms of the disease. Suggest a hypothesis to explain this and a way to test deductions from it.

4. How does each of the following contribute to solving a scientific problem: hypothesis, deduction, experiment, observation, equipment?

5. Nineteen of the 20 sterile flasks Pasteur exposed briefly to the "pure" air of the Alps remained sterile. Microorganisms later began to grow in only one. Could Pasteur just as reasonably have said that spontaneous generation occurs, but only rarely—about 5 percent of the time? Explain.

6. Can spontaneous generation be disproved entirely? Why are biologists today no longer doing experiments like Redi's and Pasteur's?

7. One reason spontaneous generation was "observed" by Aristotle and "confirmed" by so many others for 2,000 years is that people believe what they can see for themselves. Is "seeing believing"? Give as many examples as you can to support your answer.

8. The experiments on spontaneous generation caused many arguments. Similar debates occur today. They are important to the development of biological knowledge. Why are such disagreements so important?

9. The photographs on pages 12–16 suggest a few of the fields of inquiry and principles of inquiry of biology. What more can you learn about what is happening in these pictures? What other activities might be a part of biology?

Ideas for Independent Study

Listed at the end of each chapter in this book are some topics that you may want to study more deeply. You also may develop ideas of your own for independent study. Perhaps you will not wish to start this sort of work at once. But, in any case, we suggest that you read over the topics to see if they interest you. Later you can come back to any that seem worthwhile.

1. Plants such as the mandrake (Figure 1-3) were long believed to be useful for many purposes. In some cases, they were actually quite useless. In other cases, there was a sound basis for their use. If you would like to look up some information on these substances, here are a few names to start with: digitalis, belladonna, rauwolfia, mandrake, salicylic acid.

2. You may find it interesting to find out about the opinions that people have today about spontaneous generation. You could begin by carrying out a survey among people of different ages and experiences. Include some young children in your survey. Describe one or two observations that you have made or read about. It will take some skill on your part to get people to consider the questions seriously. But it may be worth the effort to see how they think organisms arise.

3. If you live near a college, university, research laboratory, national park, or large industry, you may be able to interview some working biologists. Find out what he or she does (field of inquiry). Ask about the questions biologists study and how they look for answers (principles of inquiry). Report the results of your interview to your class or some other interested group.

Related Readings

BOOKS

Asimov, I. *A Short History of Biology.* Garden City, N.Y.: Natural History Press, 1965.

Bedini, S. A. *Thinkers and Tinkers: Early American Men of Science.* New York: Scribner's, 1975.

Bova, B. *Through Eyes of Wonder: Science Fiction and Science.* Reading, Mass.: Addison-Wesley, 1975.

Burns, J. M. *Biograffiti: A Natural Selection.* New York: Quadrangle/The New York Times Book Co., 1975.

Cottrell, A. *Portrait of Nature: The World as Seen by Modern Science.* New York: Scribner's, 1975.

De Kruif, P. *Microbe Hunters.* New York: Harcourt Brace Jovanovich, 1966.

Lanham, U. *Origins of Modern Biology.* New York: Columbia University Press, 1968.

Newton, D. E. *Science and Society.* Boston: Holbrook Press, Inc., 1973.

Reid, R. *Microbes and Men.* New York: E. P. Dutton, 1975.

Ternes, A. (ed.) *Ants, Indians, and Little Dinosaurs: Selections from* Natural History. New York: Scribner's, 1975.

Werner, V. *Scientist Versus Society.* New York: Hawthorn, 1975.

White, T. H. *The Bestiary: A Book of Beasts.* New York: Putnam, 1960.

ARTICLES

Alvarado, C. A. and L. J. Bruce-Chwatt. "Malaria." *Scientific American,* May 1962, page 86.

Gould, S. J. "On Heroes and Fools in Science." *Natural History,* August-September 1974, page 30.

Hawking, F. "The Clock of the Malaria Parasite." *Scientific American,* June 1970, page 123.

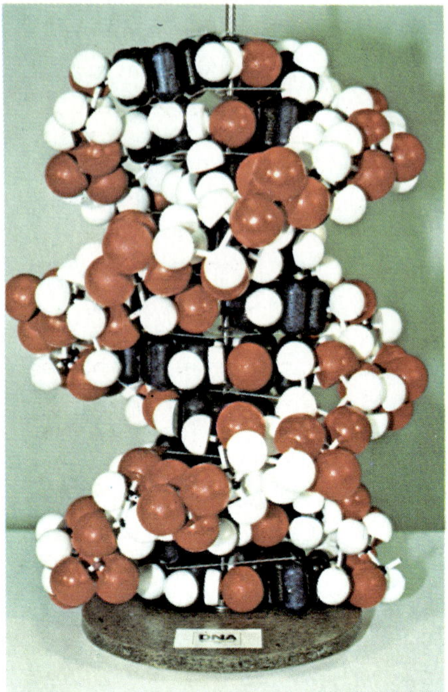

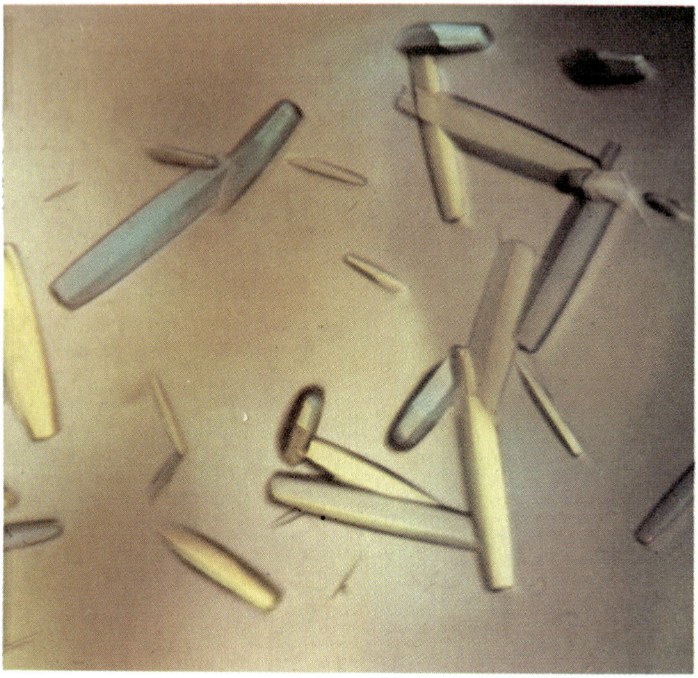

2 The Molecules of Life

Life and death have puzzled human beings for many thousands of years. Until recently, people had little to help them answer these questions. They knew that when an organism died, it decayed. What was left seemed to be the same as other substances in the soil. At one moment, the substance was alive; at another, it was not.

EXPLANATIONS OF LIFE

Because the difference between the living and nonliving seemed so great, people believed that there must be some special mysterious force in living things. The belief in a special "life force" is called *vitalism*.

Vitalism

The early Greeks and Romans thought that all matter was made of different amounts of earth, water, air, and fire (Figure 2-1). They thought that these substances could not be changed, destroyed, or divided. These four substances were called **elements.**

In the eighteenth century, the results of some experiments suggested that earth and air are not elements. Rather, they seemed to be composed of many elements. Scientists began to learn about the elements that we know today. These are oxygen, carbon, hydrogen, nitrogen, and others. An English chemist suggested that all matter was composed of **atoms** – tiny particles too small to be seen. Oxygen, for example, might be combined with hydrogen to form water. Or oxygen might be combined with carbon, forming carbon monoxide. But the oxygen, hydrogen, and carbon atoms would not change. If separated from each other, they would still be oxygen, hydrogen, and carbon.

Biologists learned that living materials are made up of combinations of the same elements as nonliving materials are made of. What is passed on from one living organism to another? Those who believed in vitalism thought that a "vital force" was the answer.

2-1. An 18th century picture of Aristotle's four elements – fire, air, water, and earth.

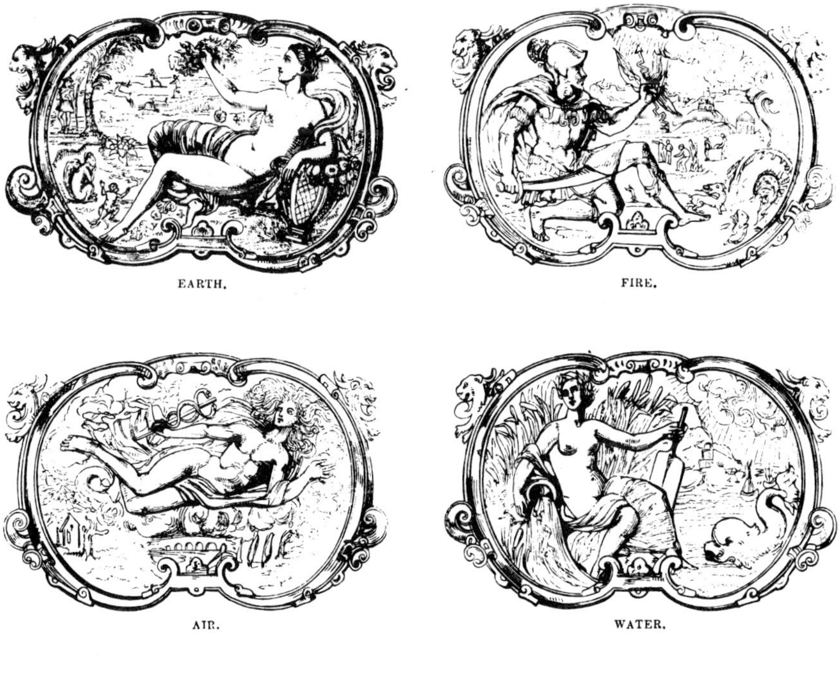

Mechanism

In the early 1600s, scientists were studying the movements of planets, the swing of the pendulum, and the rise and fall of tides. And they could describe these natural events in mechanical and mathematical terms. René Descartes (reh·NAY day·KART) tried to explain living organisms in the same mechanical or mathematical ways. The movement of fluids in the body, the action of muscles, and the pumping of the heart seemed to resemble the action of machines. Descartes thought that life could be explained with the same laws that applied to nonliving things. This hypothesis—that life can be explained in mechanical and mathematical ways—is called *mechanism.*

Vitalism and mechanism are very different hypotheses. The first assumes a vital force, the second no special force. The hypothesis of mechanism has one advantage—*it can be tested.* That is, scientists can carry out experiments to see if living things behave in some of the same ways as nonliving things. There is no direct way to test vitalism.

Since Descartes' time, mechanism has become a basic principle in the study of science. The reason for this is simple. Modern scientists can test their hypotheses about physical laws that operate in organisms.

COMBINATIONS OF ATOMS

We now know much about the **molecules** that make up living things. Molecules are combinations of two or more atoms. The atoms may be the same element, or they may be different elements. When scientists through the centuries investigated molecules and atoms, they found many of the same kinds of molecules in living and nonliving things. For example, living substances contain hydrogen and oxygen. Water also is made up of hydrogen and oxygen. As Figure 2-2 suggests, atoms of the elements that are found in organisms also are found in nonliving things.

Let's look at how the atoms in three different substances—air, water, and diamonds—are put together.

Air

We cannot see *pure* air. And we cannot smell or taste it. Air is very light in weight. Yet, if we pump it into tires, it will hold up an automobile. Air is a gas. At first, chemists believed it was an element. But when Antoine Lavoisier (an·TWAN luh·VWAH·see·ay) removed

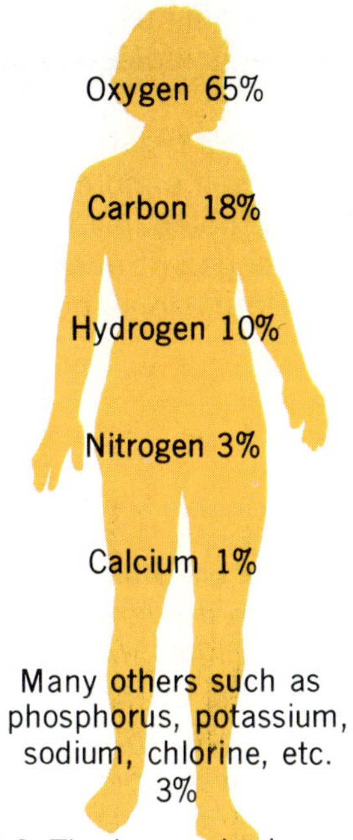

2-2. The human body contains the same elements that are found in air, water, rocks, and other nonliving substances. *What, then, accounts for the unique properties of life?*

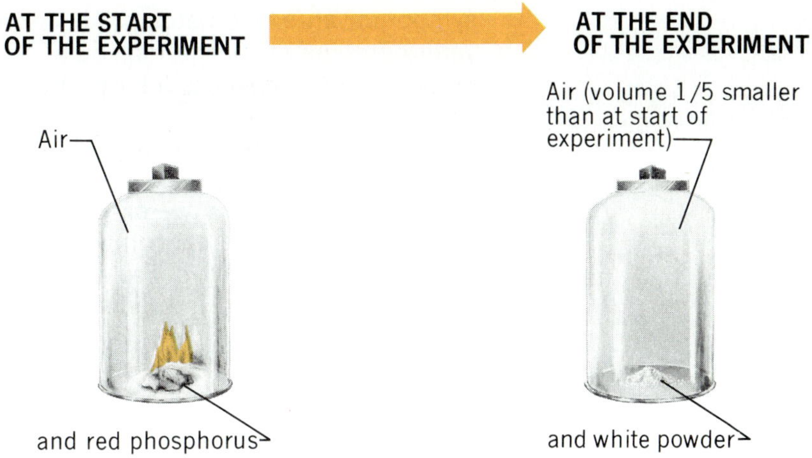

2-3.

oxygen from the air, there was still some gas left. So Lavoisier knew that air must be made up of more than one element.

One of Lavoisier's experiments is shown in Figure 2-3. He burned red phosphorus in a closed container of air. As the phosphorus burned, it changed into a white powder. Lavoisier found that when the burning was over, one fifth of the volume of the air had disappeared. By using different amounts of phosphorus, he learned that only a certain amount of the phosphorus would burn.

Lavoisier was a careful experimenter. Among other things, he weighed the substances before and after an experiment. This was something few others of that time did. Lavoisier found that the gain in weight of burned phosphorus was always exactly equal to the loss in weight of the air.

These results suggested that something in the air combines with the phosphorus. But why did only one fifth of the air disappear when the phosphorus burned? Why didn't the addition of more phosphorus cause all of the air to be used up? Lavoisier reached a conclusion: Air consists of two parts—one that will combine with phosphorus, and another that will not. Today we know the part of the air that combines with the phosphorus is oxygen. The part that does not combine is mainly nitrogen.

Analysis later showed that there are small amounts of other gases in air. Carbon dioxide is present in traces—very small amounts. Traces of other gases, such as helium, neon, and argon, also are present. And air contains water vapor.

Is air, then, a mixture of nitrogen atoms, oxygen atoms, and a few others? The answer is partly yes, partly no. Some rare gases, such as helium, neon, and argon can exist in air as individual atoms. Nitrogen and oxygen, however, are *not* present as individual atoms. The oxygen atoms in air always occur in pairs—as O_2. The "O" is the symbol for an oxygen atom. The "2" means that two atoms of oxygen combine to form a molecule. The nitrogen atoms of air also occur in pairs. Chemists refer to the nitrogen in air as molecules of N_2.

Air, then, is made up of molecules of nitrogen and oxygen, along with atoms of helium, argon, and neon. There are some molecules of carbon dioxide and water, too. The atoms and molecules in air are *separate*. They do not combine, nor do they change one another. Thus, air is a **mixture** of gases, because the gases found in air do not react (affect each other) chemically.

Water

Water was thought to be an element until Lavoisier burned hydrogen in the presence of oxygen and got water. In this and other experiments, he showed that water is a compound made up of oxygen and hydrogen. A **compound** is formed when two or more elements are combined chemically. These elements become something very different. For example, hydrogen and oxygen—two gases—combine to form water—a liquid. In water, two atoms of hydrogen are combined with each atom of oxygen. We can represent *any* water molecule as H_2O. As in all compounds, the number of each kind of atom is the same in every molecule.

A molecule of H_2O is different from a molecule of O_2. H_2O is a compound, O_2 is not. What is the difference? A molecule is formed by a chemical reaction between two or more atoms. These atoms may be of the same element, as in O_2, or of different elements, as in H_2O. A compound is *always* made up of atoms of *different* elements. A water molecule is a compound in which hydrogen and oxygen atoms are chemically **bonded,** or attached to each other.

The same elements that form one compound may react in a different way, forming another compound. Oxygen and hydrogen, for example, may form water, or they may form hydrogen peroxide (H_2O_2). A molecule of hydrogen peroxide is made up of two atoms of hydrogen and two atoms of oxygen.

Diamonds

With a diamond, we have a new situation. We cannot separate a diamond into other substances. A diamond is made up of a single element—carbon. Scientists believe that the atoms of carbon are bonded together in a repeating pattern—a crystal. Both ice and diamonds are crystals. Figure 2-4 shows a comparison of the crystal

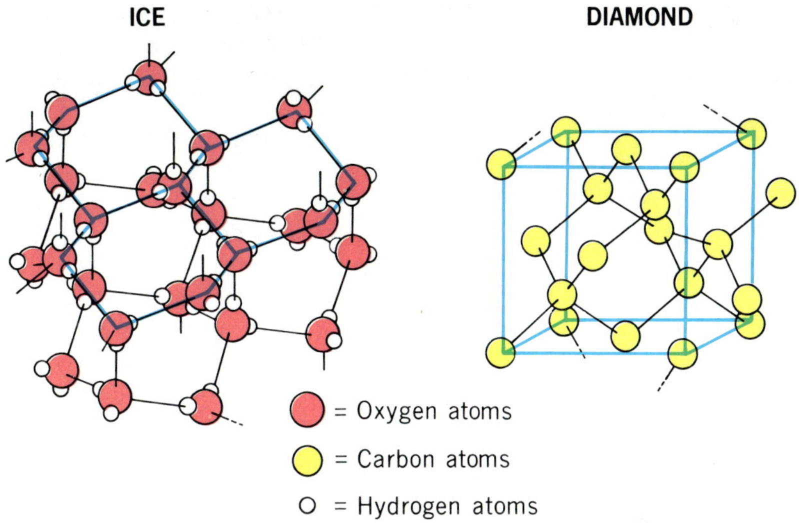

2-4. Ice is a crystal of the *compound* water. A diamond is a crystal of the pure element *carbon.* The blue lines show the basic six-sided shape of the ice crystal. They also show the cube shape of the diamond crystal.

patterns in ice and in a diamond. These drawings are models of how we think atoms are arranged in these crystals.

DALTON'S THEORY

John Dalton made observations of various types of **matter.** In 1808, he proposed a theory about the composition of matter. A **theory** is a broad set of principles used to explain observations. A general hypothesis that stands repeated testing over a long period of time may become accepted as a theory. Dalton's *Atomic Theory* has stood the test of time. Here is what Dalton suggested:

1. Matter is composed of particles that cannot be destroyed and cannot be seen—atoms.
2. All the atoms of one element are the same. For example, all oxygen atoms are alike, especially in having the same mass. (**Mass** is a measure of the amount of matter. The mass of a substance causes it to have weight in a field of gravity. An atom of oxygen would have the same mass on the earth as it would on the moon. But it would weigh less on the moon because of a weaker force of gravity.)
3. Atoms of different elements are different from one another.

That is, oxygen atoms, hydrogen atoms, and carbon atoms differ, especially in mass.

4. Molecules are combinations of atoms.
5. Compounds are combinations of different kinds of atoms.

BASIC CHEMISTRY

Studies of air, water, and a diamond give different results. Air is made up of oxygen, nitrogen, and other gases. It is a mixture. Water is composed of hydrogen and oxygen and is a compound. The diamond is pure carbon. Carbon is an element. No chemical method can divide it into simpler substances. Nor can oxygen, nitrogen, and hydrogen be divided into simpler substances by chemical means. So far, scientists have found over a hundred elements. These occur in different patterns in both living and nonliving things (Figure 2-5).

2-5. The earth is made up of 92 natural elements. **(a)** Most of the land is made up of oxygen, silicon, aluminum, iron, calcium, sodium, potassium, magnesium, titanium, phosphorus, hydrogen, manganese, and sulfur. All other elements are found only in traces. **(b)** Water contains mostly oxygen and hydrogen. The sea also has large amounts of dissolved chlorine, sodium, magnesium, sulfur, and calcium. **(c)** In the air, there are nitrogen, oxygen, argon, carbon, and hydrogen. Dust in the air contains traces of many more.

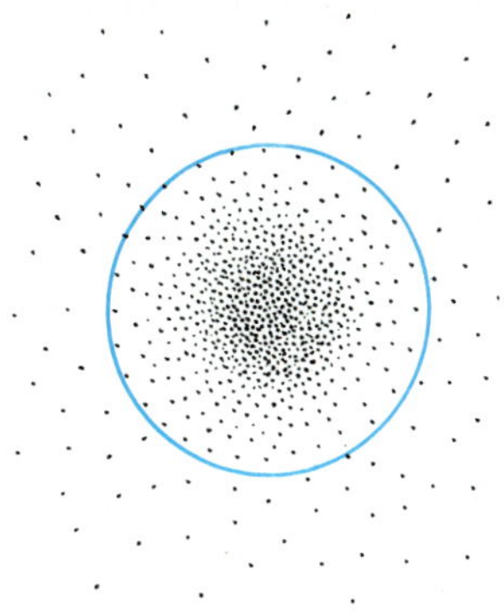

2-6. The "cloud model" of a hydrogen atom. The dots represent *possible* positions for a single electron. The circle represents the area in which the electron will probably be located 90 percent of the time.

Some elements occur in nature. Others are produced only in the laboratory.

The Anatomy of an Atom

Atoms are so small they cannot be seen even under the most powerful microscope. In fact, they are so small we do not know what an atom really looks like. In a case like this, we must build a model to explain something we cannot see.

Atoms appear to be made up of tiny particles. Hydrogen is the simplest atom known. It consists of a single proton and a single electron. Each of these particles carries a tiny electrical charge. A proton carries a positive (+) charge. An electron carries a negative (−) charge.

The center, or **nucleus,** of the hydrogen atom in our model contains the proton. The electron is whirling about the nucleus at great speed. Figure 2-6 shows the many positions where the electron might be found at any given time. This part of the model is called the electron cloud.

Other atoms are more complex than hydrogen. Atoms of all other elements have a third type of particle, the neutron. The neutron is found in the nucleus. The neutron has no electrical charge. Atoms of all other elements also have more than one electron outside the nucleus. The number of electrons outside the nucleus is equal to the number of protons within the nucleus.

Interactions of Atoms

Reactions between atoms depend on the number of electrons the atoms have. Reactions may involve an electron moving from one atom to another. Let's consider what happens when atoms of sodium and chlorine react. When these two kinds of atoms react, an explosion takes place. The product of the reaction is sodium chloride—ordinary table salt.

An atom of sodium has 11 electrons; an atom of chlorine has 17 electrons. During the reaction between sodium and chlorine, each sodium atom gives up an electron to a chlorine atom. The number of electrons in a sodium atom no longer equals the number of protons. There is one more proton than electrons. Therefore, the sodium atom has a positive charge. This charged sodium atom is called a sodium **ion.** An ion is a charged particle.

The electron from the sodium has been captured by the chlorine atom. The chlorine atom now has 18 electrons, one more electron

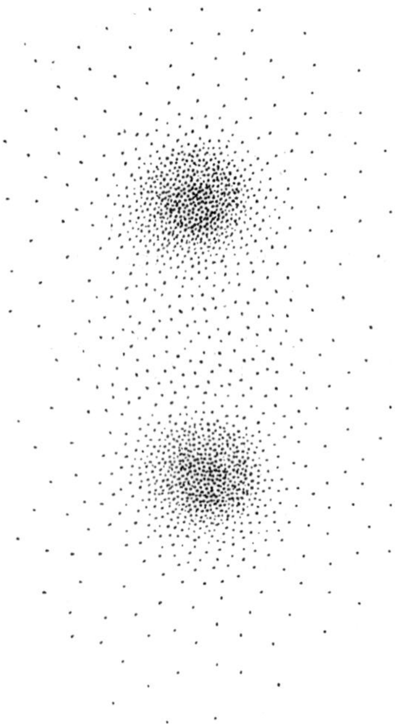

2-7. Covalent bond between two hydrogen atoms in a hydrogen molecule.

than it originally had. The number of protons and electrons in the chlorine atom is no longer equal; there is one more electron than proton. As the chlorine atom now has a negative charge, it, too becomes an ion—the chloride ion.

The positively charged sodium ions and the negatively charged chloride ions are attracted to each other. These particles of unlike charge come together in an organized pattern—a crystal of sodium chloride. The forces, or bonds, that hold the crystal together are called **ionic** (eye·ON·ik) **bonds.** Ionic bonds are formed when positive ions and negative ions attract each other. If a crystal of sodium chloride is dropped into water, the crystal dissolves. The ions are separated in the water. The ions still attract and repel one another to a certain extent. But the ions are in a looser organization—a solution—rather than a crystal.

This explanation contains ideas important to biology. We know an explosion occurs when sodium and chlorine atoms react. What is an explosion? It is a process in which a lot of energy is released. When atoms of sodium and chlorine combine, energy is released. An explosive reaction between sodium and chlorine atoms does not occur in life processes. The energy released in such a reaction would destroy the delicate organization of living things. But living things do obtain energy from some chemical reactions.

Often, when atoms react, they do not gain or lose electrons. Instead, they *share* electrons. A new molecule is formed, because the atoms are bonded together by the shared electrons. This kind of bond is called a **covalent** (co·VAY·lent) **bond.** One example of a covalent bond is the bond between the hydrogen atoms in a hydrogen molecule (Figure 2-7). Covalent bonds also are formed between two hydrogen atoms and an oxygen atom in a molecule of water.

In biology, we have to "think big" when using chemical ideas such as these. Molecules in organisms often are made up of hundreds or even thousands of atoms. These are combined by ionic, covalent, and other kinds of bonds. Energy is used, or released, as molecules are built up or broken down in life processes.

ENERGY

Over the years, scientists began to see a relationship between heat, light, movement, and electricity. *All* are forms of energy. And each form can be changed to any other form. In addition, these changes often take place during chemical reactions. For instance, under certain conditions, two chemicals, which we will call A and B, can combine. They become AB. Under other conditions, AB can break down into A and B. We can express these relations as follows:

$$A + B \longrightarrow AB$$
$$AB \longrightarrow A + B$$

Generally, if energy is *used* when A and B combine, then energy is *released* when AB breaks down. Some chemical processes give off light. Some move objects, as in an explosion. Still others can generate electricity. Thus, we can see that chemical energy, too, can be changed to every other form of energy. To describe the reaction between A and B more correctly, we should include energy. Let's say that e is the energy change involved in combining A and B. Then e also is the energy change that results when A and B separate. We can write this as

$$A + B + e \longrightarrow AB$$
$$AB \longrightarrow A + B + e$$

What is energy? No one can define energy in a simple word or two. But energy is so basic to our lives that we all know something about it. We buy stored chemical energy at the grocery store (food) and at the service station (gasoline). That chemical energy is changed to heat and movement in our bodies and in our cars. Energy is stored in ammunition and explosives. Electric energy lights buildings and runs machines.

Although it is hard to define, energy is easy to measure. For instance, energy can be measured as a change in heat. A change in heat will cause a change in temperature. **Inquiry 2-1** will give you a chance to measure energy while you study a process in which chemical energy changes to heat energy.

INQUIRY 2-1 Food Energy

All foods contain energy. How much energy? Do equal amounts of different foods contain the same amount of energy? In this inquiry you will use a tin-can calorimeter (Figure 2-8) to measure food energy in calories.

A **calorie** is the amount of heat required to raise the temperature of 1 gram (g) of water 1°C. Calorie values of food in diet charts are listed in **kilocalories** (1,000 calories). The symbol for kilocalories is "C."

Heat energy that is released when a known amount of a substance is burned can be transferred to a measured amount of water. The temperature of the water rises and can be measured.

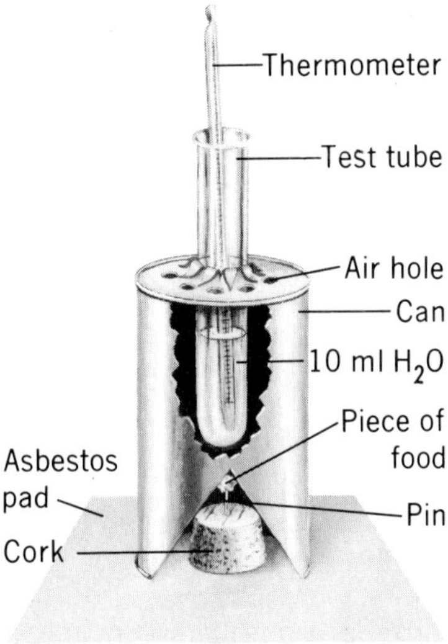

2-8. A calorimeter made with a tin can.

MATERIALS

(for making the calorimeter)

soft drink cans or frozen fruit juice concentrate cans
tin snips
hammer and nail
heat-resistant test tubes to hold about 25 ml
cork
metal foil
straight pin or needle
thermometer
asbestos pad

(for testing foods)

peanuts, walnuts, or other foods
graduated cylinder, 10-ml
matches

Procedure (Day 1)

Construct your calorimeter, using Figure 2-8 as a guide. (CAUTION: *Watch out for the sharp edges of the can.*)

Each piece of food has a mass of 0.2 g. Place a piece on the pin held up by the cork. Put 10 milliliters (ml) of water in the test tube. Record the temperature of the water. Remove the thermometer.

Start the food burning, and place it under the test tube inside the calorimeter. As soon as the food is completely burned, measure and record the temperature of the water. Subtract the temperature before burning from the temperature at the end of burning. Record this increase.

If the food does not burn completely, remove it from the calorimeter. Record the increase in temperature at this point. Relight the food, and record the new starting temperature. Repeat, if necessary, until all the food is burned. Then, add the temperature increases for each relighting. Do this test three times with each type of food. Use fresh water in the tube each time.

Calculate the average difference in temperature for the three peanut burnings and the three walnut burnings. Use the following equation to calculate the number of kilocalories per gram produced:

$$\begin{matrix}\text{average difference} \\ \text{in temperature}\end{matrix} \times \begin{matrix}\text{ml of} \\ \text{water used}\end{matrix} \times 5 = \begin{matrix}\text{calories} \\ \text{per gram}\end{matrix}$$

$$\frac{\text{calories per gram}}{1000} = \text{kilocalories per gram}$$

$$\text{kilocalories per gram} \times 100 = \text{kilocalories per 100 grams}$$

1. Why was it necessary to multiply by 5 in the first equation? **2.** How do your data compare with the measurements for 100 grams of the same or similar foods listed by your teacher? **3.** How can you account for any differences? **4.** Which of the foods you tested seems to be the best energy source? **5.** Do you think energy is released in the human body in a similar way?

Basic Laws of Energy

Experiments in the eighteenth century suggested that matter cannot be destroyed. But it can be changed. The total quantity of matter (mass), however, remains the same. This idea is called the *Law of Conservation of Mass: Matter can neither be created nor destroyed.* The total mass of substances is the same before and after a reaction. Consider, for example, burning wood. The total mass of the ashes, water, and smoke produced by the burning is equal to the mass of the original wood.

After they learned how to measure energy, scientists made another important discovery. The total amount of energy always stays the same, no matter how often it is changed from one form to another. This idea became known as the *Law of Conservation of Energy: Energy can neither be created nor destroyed, but it may be changed from one form to another.* For example, when gasoline is burned in a car, the stored chemical energy is changed to the energy of movement and heat. The total *amount* of energy is the same before and after the gasoline is burned, but the *form* of the energy is different.

There is one additional fact to consider. Whenever an energy change takes place in a chemical reaction, some of the energy is always turned into a form that cannot be used again. For instance, part of the energy of the car's gasoline is changed to heat. The heat is released. This heat cannot be trapped and turned into any other form. We often speak of this change to a useless form as "loss of energy." But this does not mean a real loss. "Energy loss" only means that some energy is changed to a form that cannot be used.

Although we do not know why such energy loss occurs, we accept it in our daily lives. We know that a car's engine changes part of the chemical energy of gasoline to mechanical energy that turns the wheels. But we also know that part of the energy is changed to heat, because we must pay attention to the car's cooling system.

Energy and Life

Living things can use only one form of energy. This is chemical energy. Chemical energy is stored in food. Plants make their own

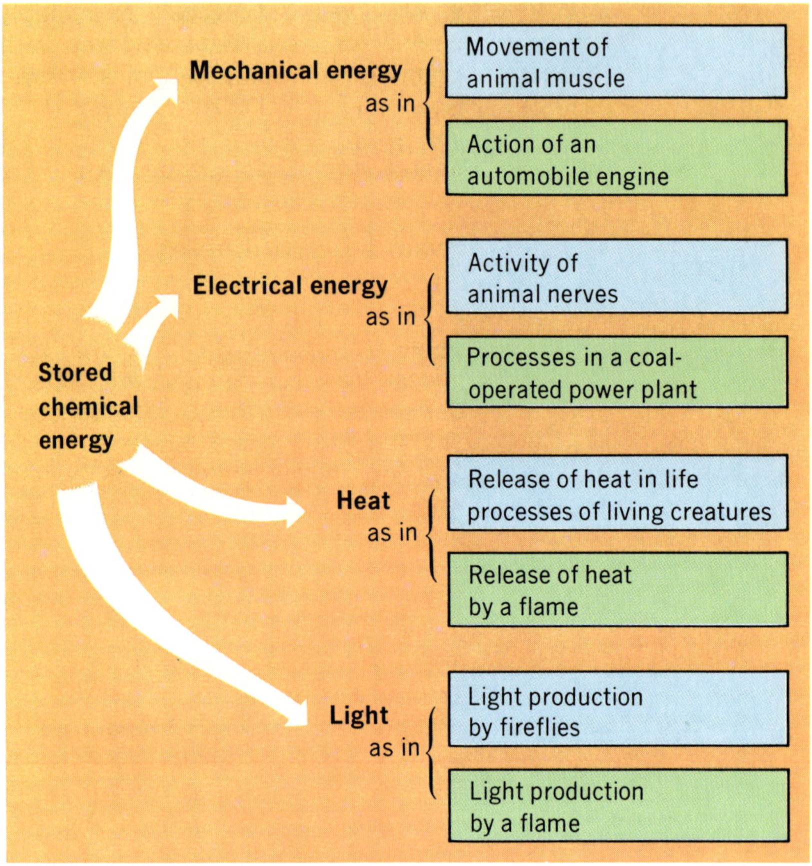

2-9. Chemical energy can be changed to many forms. Changes that occur in organisms are shown in blue. Changes that occur in nonliving systems are shown in green.

food using the radiant energy of the sun. The sun's energy is trapped by plants and stored as food (chemical) energy. Living things use the chemical energy they get from their food in many different ways. Chemical energy is used to build their bodies. Animals change chemical energy into mechanical energy as they move about. Heat energy keeps the bodies of some animals warm. Growth in plants, animals, and all other organisms uses energy. The change of one form of energy to another occurs in both living and nonliving systems. Figure 2-9 compares some energy changes.

Sources of Energy

Where does the energy used by living things come from? The sun is the direct source of most of the energy used by organisms. But green plants (and a few microorganisms) are the only organisms that can capture the energy of sunlight and turn it into chemical energy. A plant's chemical energy is stored in compounds made from carbon dioxide and water. Thus, plants change light energy to chemical energy as they build foods (Figure 2-10). And plants become food (energy) for other organisms. The food-making activity of green plants is called **photosynthesis.** As green plants make food, they give off oxygen to the air. You will read more about photosynthesis in later chapters.

THE CHEMISTRY OF LIFE

What molecules are found in living organisms? We cannot answer this question fully. Only a very small fraction of all living things has been studied. But we do have some useful ideas about what chemicals occur in organisms and what they do. Fortunately, scientists

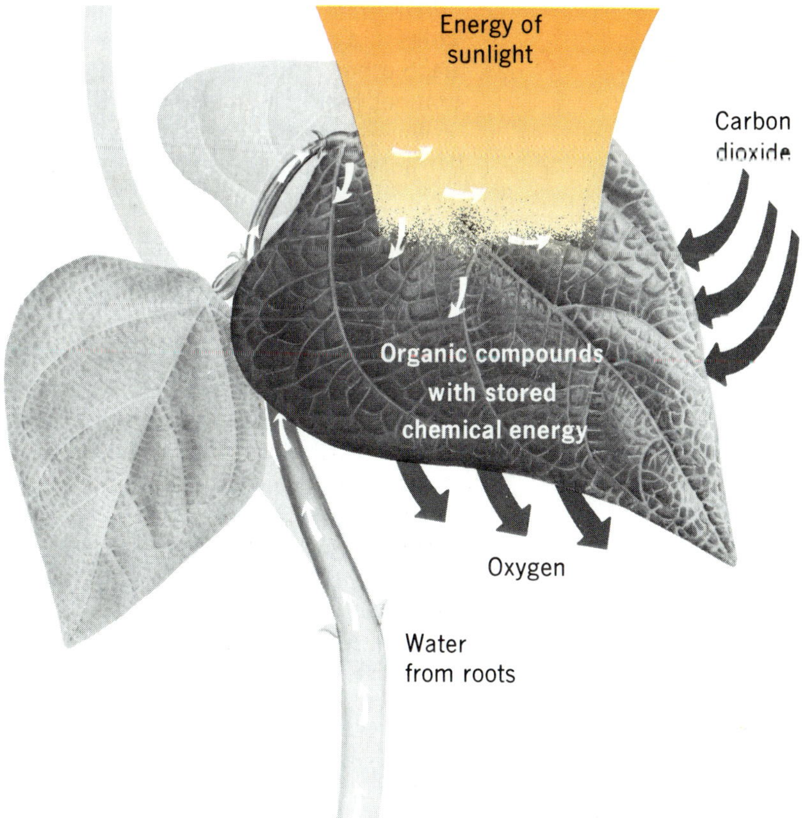

2-10. The energy in sunlight becomes chemical energy during photosynthesis.

working for many years have found ways to separate, test for, describe, and name different chemical substances. The scientists found that living organisms are made up mostly of water, along with other compounds.

Water in Cells

All organisms are made up of microscopic units called cells. Most cells are at least 65 percent water. Thus, water is by far the most abundant compound in organisms. It plays a major part in the processes of life.

Water is the best **solvent** known. More chemical substances dissolve in it than in any other liquid. Why is this important? In most cases, substances react best if they are mixed thoroughly and are in very small particles, such as molecules or ions. Both separation and mixing occur when substances are dissolved. Water, therefore, makes possible many chemical reactions.

Mixing is partly explained by the motion of molecules. The molecules in a water solution are always moving about and bumping into each other. Thus, they have many chances to react. Also, many substances become charged particles, or ions, when they are dissolved in water. Ions can react more readily. Their ions are charged and are attracted to particles of the opposite charge.

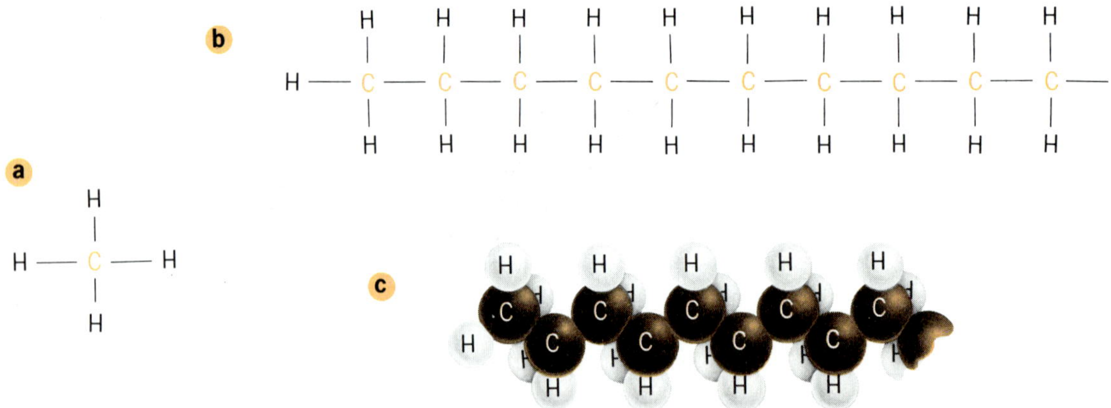

2-11. We can describe ways in which carbon atoms link with other atoms in structural formulas such as these. The letters stand for atoms: C is carbon, H is hydrogen, and O is oxygen. In each formula, a short line means that the atoms share electrons, forming a covalent bond. **a** shows a complete molecule. **b** is a part of a molecule. **c** is another way to show part of the same molecule that is shown in **b.** **c** should remind you that all chemistry takes place in three-dimensional space.

Organic Compounds

The molecules that make up most of the substances of living cells, other than water, are called **organic** molecules. All organic molecules contain carbon atoms, in addition to other elements.

Carbon plays a central part in the chemistry of life. Carbon atoms bond readily with one another. As a result, molecules with chains and rings of these atoms are built up easily (Figure 2-11). Organisms contain thousands of molecules. These are formed on a framework of carbon atoms linked to one another and to other atoms. Some of the other atoms found in organic molecules are hydrogen, oxygen, nitrogen, and sulfur. Some organic molecules you may be familiar with are the sugars and fats in food, the organic compounds of the clothes you wear, and the special molecules that make up your body.

There is another way to look at the key roles that carbon atoms play in life processes. Some organic molecules, like those in your bones, skin, and hair, are hard to break down. Some, such as sugars and fats, break down more easily. When these molecules are broken down, energy is released. This energy is then available to life processes.

The following sections describe a few kinds of organic molecules.

Urea

Early in the nineteenth century, chemists could not make any of the organic compounds known to occur naturally in organisms. At that time, scientists thought it was impossible to synthesize (or make) these substances in the laboratory. They thought this synthesis could only occur in living things.

In 1828, Friedrich Wohler (VER·luh) discovered one organic compound that could be produced in the laboratory. By heating a solution of ammonium cyanate, he produced the compound **urea** (yoo·REE·ah). Urea, which is found in urine, is produced by many animals. Until Wohler did his work, urea had been produced only by living cells.

But Wohler had not produced an organic compound using only *in*organic materials. Although ammonium cyanate is classed as an inorganic compound, Wohler had made it from substances obtained from animals. It remained for one of Wohler's students, Hermann Kolbe (KOHL·beh), to finish the work. Adding the elements carbon, sulfur, and chlorine to water, he produced acetic (ah·SEE·tik) acid. For the first time, an organic compound was made by a chemist from inorganic materials.

Although the syntheses (SIN·the·seas; plural of synthesis) of urea and acetic acid were humble beginnings, the work showed that at least *some* organic compounds could be made in the laboratory. Gradually, chemists and biologists began to suggest that all organic compounds could be synthesized.

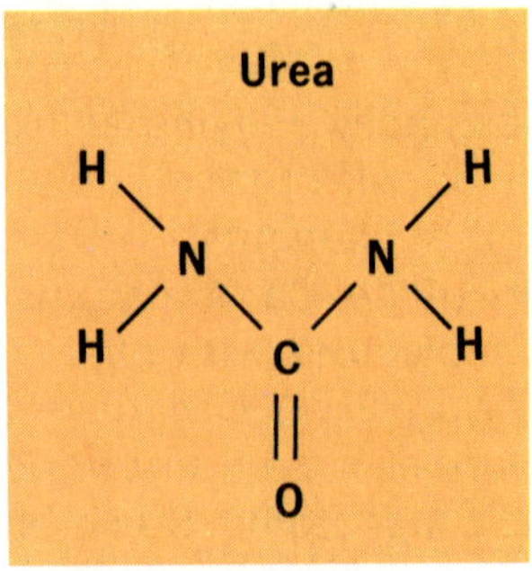

2-12. A structural formula for urea. It shows the positions of the eight atoms. The two lines between carbon and oxygen represent a strong, double bond.

A Note on Chemical Formulas

Urea is made of one carbon (C) atom, one oxygen (O) atom, two nitrogen (N) atoms, and four hydrogen (H) atoms. This is represented by the formula, CON_2H_4. But an organic molecule has other characteristics besides the numbers and kinds of atoms. Where the atoms are located—their positions—in a molecule also is very important. Here, we can compare words and molecules. The same letters, *a, e,* and *t,* can be combined one way to form "ate" and another way to form "tea." Just as the positions of letters determine the word, so the positions of atoms determine the molecule.

The formula CON_2H_4, for urea does not indicate the way the atoms fit together. A better way of showing the urea molecule is to use a **structural formula** (Figure 2-12). One NH_2 group is shown at each end. These are bonded to a central carbon atom. A single oxygen atom is also bonded to the carbon. There are many compounds that have the same number and proportion of atoms. They are different compounds because their atoms are arranged differently.

Structural formulas are useful, but they have one serious limitation. Molecules have three dimensions, while the paper on which structural formulas are shown is only two-dimensional. Even if shown in three dimensions, a structural formula would not be a true picture. Think of a urea molecule dissolved in water. Imagine each of its atoms, with fast-moving electrons circling in an orbit around a nucleus. Where two atoms are shown by letters connected with lines, you must imagine instead the quick back-and-forth movements of shared electrons. Each electron and atom vibrates and rotates. All of the atoms are in constant motion in water.

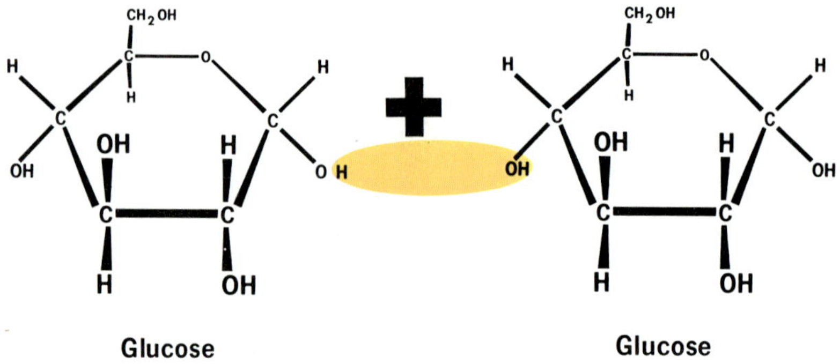

2-14. Simple sugars such as glucose link together chemically to form the double sugar, maltose. A molecule of water also is produced. Many steps are involved, but only the overall reaction is shown here.

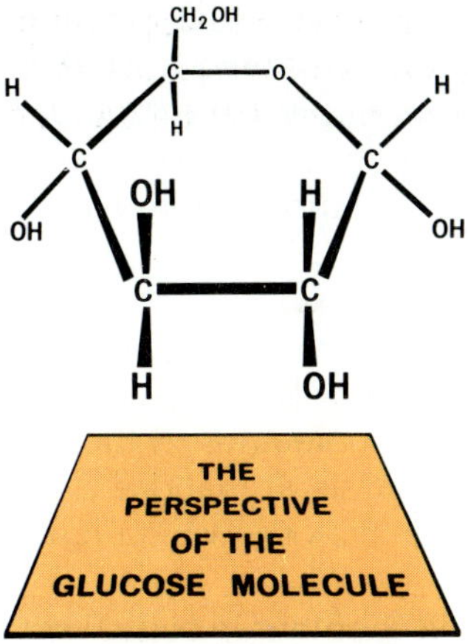

2-13. The simple sugar glucose. One can imagine the molecule as a six-sided plate to which the H, OH, and CH_2OH groups are attached.

Later, as we describe other molecules, we will not be able to give this same sort of description. We will have to continue showing atoms and molecules in two dimensions, using letters and lines. Can your imagination supply the action? If so, you will have a much better understanding of chemistry and some of the processes of living things.

Carbohydrates

Carbohydrates are a class of organic compounds found in all organisms. In addition to carbon, they contain the same ratio of hydrogen to oxygen atoms as water contains. Carbohydrates in their simplest form are called **simple sugars.** These sugars play an important role in the energy relations of all organisms. The simple sugars can be linked together to form complex carbohydrates. These carbohydrates are used in building cells or as storage molecules for chemical energy.

Glucose (GLUE·kohs) is a common carbohydrate that is very important as an energy source. It is the sugar that causes the sweet taste of honey and fruits, such as grapes. It can be found in almost all living organisms. Glucose occurs in your blood. The amount of glucose in human blood is critical. If it falls to half the normal amount, the person rapidly loses consciousness and dies unless treated at once. Glucose has the formula $C_6H_{12}O_6$. The 24 atoms of each of its molecules are arranged as shown in Figure 2-13. When simple sugars combine, more complex carbohydrates are formed (Figure 2-14). For example, when two molecules of glucose combine, a molecule of maltose and a molecule of water are produced. One glucose molecule gives up an atom of hydrogen. The other gives up one hydrogen atom and one oxygen atom. The two hydrogen atoms and one oxygen atom form a molecule of water. The formation of complex organic molecules by the removal of water is a **synthesis reaction.**

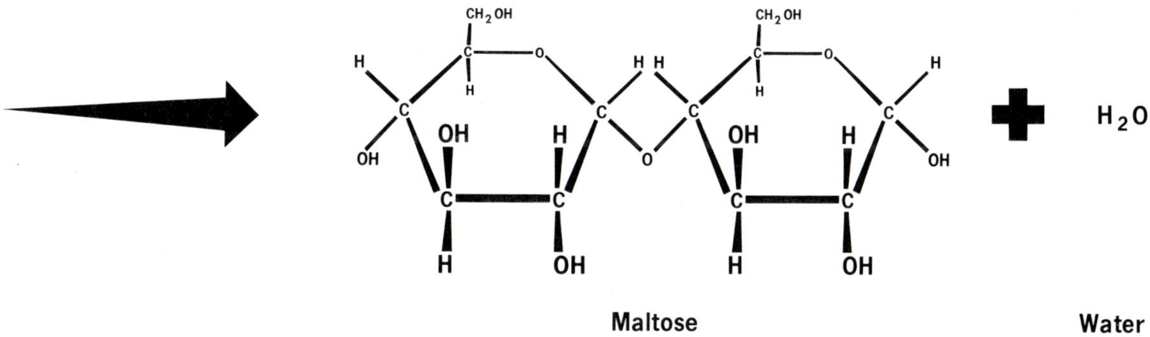

Since a molecule of maltose is formed by a synthesis reaction from two molecules of simple sugar, maltose is called a **double sugar.** Sucrose (table sugar) is obtained from sugar cane or sugar beets. A molecule of sucrose is formed when one molecule of glucose and one of fructose (FROOK·tohs), another simple sugar, combine. A molecule of water is also formed.

Both in the laboratory and in living cells, it is possible to break down a double sugar into two molecules of simple sugars. In order to do this, you must put water back into the double-sugar molecules. This breaking down of complex organic molecules by the addition of water is called **hydrolysis.** The hydrolysis of maltose is almost the reverse of the synthesis reaction that was shown in Figure 2-14.

Glucose and similar simple sugars can link together in cells. They form large molecules that may contain from 100 to 10,000 glucose units. These molecules are very difficult to break down. Two such molecules are **plant starch** and **glycogen** (GLY·ko·jen). Glycogen is animal starch. Plant starch and glycogen break down into smaller carbohydrate molecules when they undergo hydrolysis.

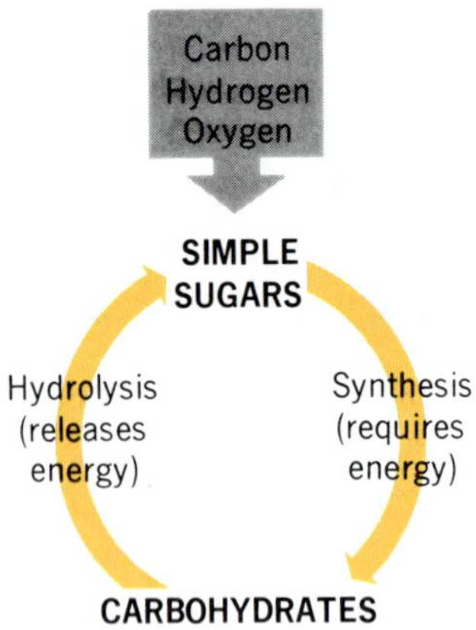

2-15. The synthesis of carbohydrates takes energy. But energy is released when carbohydrates break down (hydrolyze) into simple sugars.

Carbohydrate molecules can act as storage molecules. For example, if too many glucose molecules are present in plant cells, they are synthesized into starch. At some later time, the starch is hydrolyzed back to glucose. A summary of these reactions is shown in Figure 2-15.

Some carbohydrate molecules are important energy sources. Others are the building materials in the solid protective and supporting parts of organisms. One example is **cellulose** (CELL·yoo·los). Cellulose is made of many glucose units. It is the most abundant carbohydrate in all of nature. Cellulose is found in the cells of plants. Cotton fibers are pure cellulose. Wood is mostly cellulose, as is the paper of this page.

Lipids

Lipids are another major group of organic compounds found in living organisms. Like carbohydrates, they are made up of carbon, hydrogen, and oxygen. In lipids, however, the ratio of hydrogen atoms to oxygen atoms is greater than two to one. There are many different kinds of lipids with a variety of functions. Fats are one kind of lipid. Butter and the fat we see in meat are this kind of lipid. Energy is stored in fats. Another group of lipids, the steroids, is important in sexual development. Steroids also affect other body functions. Other lipids are waxes and oils. Hydrolysis breaks down lipids into smaller organic compounds and water. Energy is released in these hydrolysis reactions.

SCIENCE AT WORK: Of Proteins and Moths

Amino acids and moths may seem to have little in common. The fact is that amino acids are the building blocks of the proteins that make up all living things. Although some animals can make at least a few of their own amino acids, most must obtain them from the food they eat.

Where do ideas for research about the chemistry of organisms come from? Many ideas are based on research already done by others. Other ideas can spring from seemingly unrelated observations.

While sitting in a bathtub reading a magazine, one scientist came upon a picture of a moth drinking at a pool of blood. Blood contains many amino acids. Moths also drink nectar. Could these two facts be related? It was well known that nectar contains sugars, but could nectar also contain amino acids? This question prompted a team of scientists at the University of California, Berkeley, to carry out a series of studies both in the field and in the laboratory.

Sitting in a field and watching insects is one way to find out about their food habits. What insects visit one species of flower? How frequent are the visits? Do the insects take nectar from the flower? All these questions were answered in the field. But nectar also was collected and taken to the laboratory to be analyzed. The scientists wanted to know what chemicals it contained.

In the laboratory, it was found that nectars of different flowers contain amino acids. They also contain sugars, proteins, and other chemical compounds. This discovery has led to many new questions. Research is now directed toward answering questions about how the amino acids are actually used in the insects' bodies.

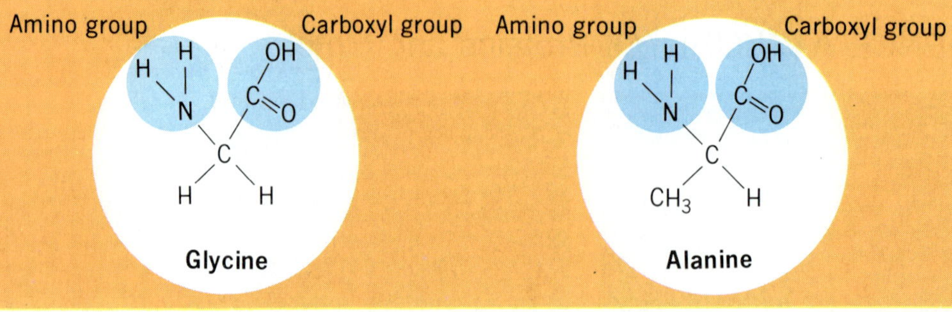

2-16. Glycine and alanine are two amino acids. Each amino acid molecule has an *amino group* and a *carboxyl group.* These are bonded to the same carbon atom. But the *other* atoms bonded to the carbon atom are different in glycine and alanine.

Proteins

Most of the organic compounds in plant and animal cells are **proteins.** Some are building blocks for the bodies of organisms. Others are important to the control of chemical reactions in cells.

Hemoglobin (HEE·moh·glow·bin) is a protein. This compound carries oxygen in red blood cells. Every cell in our bodies depends on hemoglobin for oxygen. Other proteins play a key role in muscle movement, food digestion, and blood clotting. Indeed, there seems no limit to the number of things that proteins do.

2-18. Two amino acids link together. The blue circle shows the source of the water. An amino group is intact at one end and a carboxyl group at the other end. At these points, other amino acids can be joined to the chain. The building up of protein molecules requires much energy.

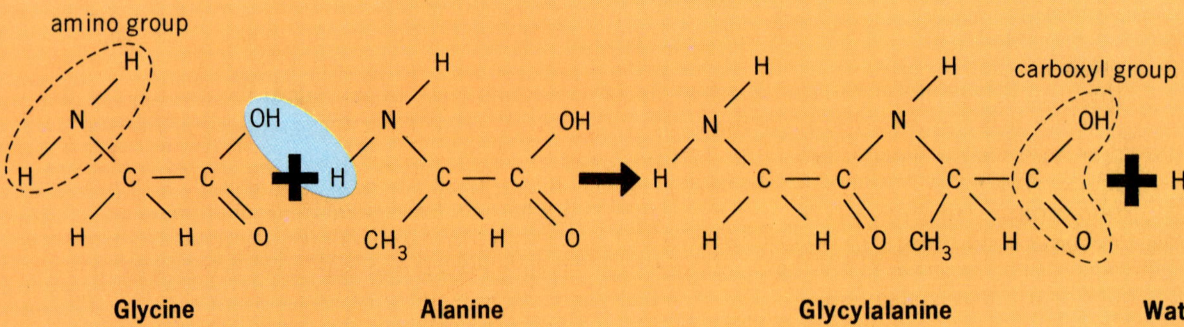

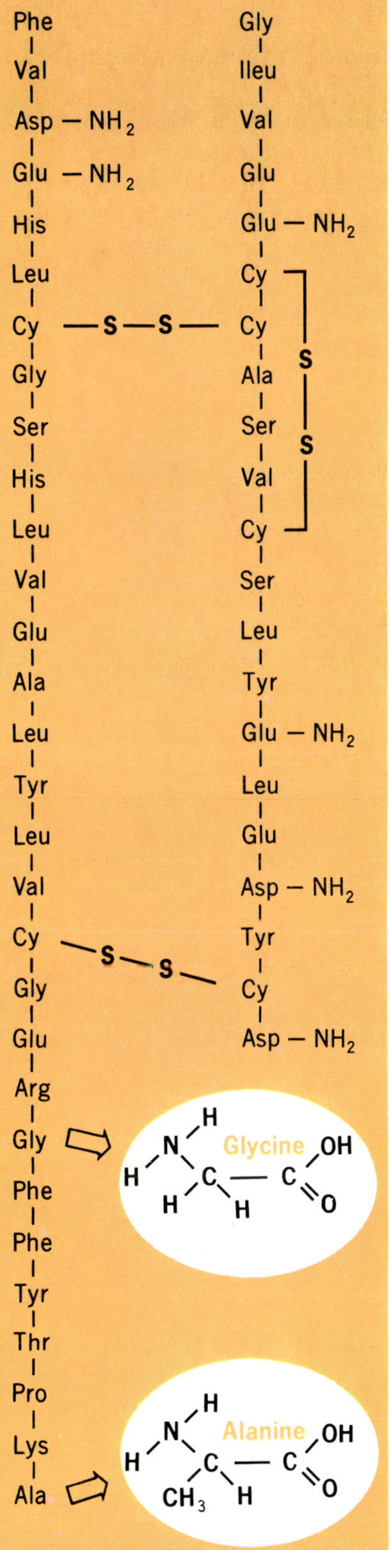

2-17. The structure of the insulin molecule. The two chains of amino acids are coiled in the molecule. They are presented here as being flat in order to show all the amino acids. Two of the amino acids, glycine and alanine, also are shown separately. Insulin is a rather small protein molecule with only 51 amino acids. It is easier to study than larger proteins. "--S--S--" shows a sulfur bond.

Proteins are very large and complex. For example, the formula for human hemoglobin is $C_{3032}H_{4816}O_{872}N_{780}S_8Fe_4$. The total of all the atoms in this single molecule is 9,512.

Proteins are built of combinations of more than 20 kinds of organic compounds. These compounds are called **amino** (ah·MEE·noh) **acids** (Figure 2-16). Each kind of amino acid consists of two groups of atoms chemically bonded to the same carbon atom. The *amino group* contains a nitrogen atom and two hydrogen atoms bonded together ($-NH_2$). The *carboxyl group* is made up of carbon, two atoms of oxygen, and hydrogen ($-COOH$). The simplest amino acid is **glycine** (gly·SEEN), one of the two shown in Figure 2-16. Note the amino and carboxyl groups.

Insulin is a protein that regulates the use of sugar in our bodies. When the body does not produce enough insulin, the disease diabetes (die·uh·BEE·teez) results. Insulin is one of the "simple" proteins. But it contains 16 kinds of amino acid units. Figure 2-17 gives you some idea of the complexity of an insulin molecule. Each of the abbreviations (*Cy, Val, Ileu*) represents a different kind of amino acid. There are 51 amino acids in the two long chains. These chains are held together by covalent bonds. The bonds occur between sulfur atoms.

Figure 2-17 looks very complex but it is not nearly as complex as this molecule really is. For example, suppose that in place of the abbreviations of the amino acids, we showed the structural formula for each one. We have done this in the figure for just two amino acids. There would not be room on a page to "map" the whole insulin molecule. And insulin is a relatively small protein molecule.

Amino acids are chemically linked in a certain manner, shown in Figure 2-18. In the reaction, one molecule of water is removed. The carboxyl group ($-COOH$) contributes one atom of oxygen and one atom of hydrogen (OH). The amino group ($-NH_2$) contributes one atom of hydrogen (H). When these are removed, a chemical bond, called a **peptide bond,** is formed between the carbon of the carboxyl group and the nitrogen of the amino group.

Proteins are made of many amino acids linked by peptide bonds into chains. This method of "chain-building" always leaves a free amino group at one end and a free carboxyl group at the other. More peptide bonds may be formed at these points to build an even larger molecule. Proteins can be broken down into amino acids by hydrolysis. Hydrolysis is almost the reverse of the process shown in Figure 2-18.

About 20 kinds of amino acids are present in the human body, almost as many as there are letters in the English alphabet. It is possible to build millions of words from 26 letters. But the average number of letters in a word is fewer than ten. Many proteins contain 300 to 3,000 amino acid units. Thus, millions of combinations are possible.

We can learn something more from this comparison between words and protein molecules. A word has a particular meaning because of the way its letters are put together. And a protein molecule works in a particular way because of the way its amino acids are put together. For example, biologists have discovered an abnormal type of hemoglobin. In this hemoglobin, just one of the 574 amino acids in the molecule is different from normal. This alone may upset the ability of the hemoglobin to carry oxygen to the cells of the body.

You have been introduced to three major classes of organic compounds—proteins, carbohydrates, and lipids. Before you go on to a more detailed study of proteins and of another major class of compounds, do **Inquiry 2-2.** In this inquiry, you can test for the presence of organic compounds in their pure forms and in ordinary foods.

INQUIRY 2-2 Compounds of Living Organisms

MATERIALS (per team)

beaker
test tubes, 13 × 100 mm
medicine dropper (pipette)
graduated cylinder, 100-ml
bunsen burner or other heat source

Foods

egg white
liver
apple
orange
onion
potato
other foods of your choice

Compounds

glucose
sucrose
cellulose
vitamin C
sodium chloride
butter

Solutions

5 ml 1% gelatin
5 ml 1% glucose
1 ml 1% starch suspension
5 ml 5% vitamin C
5 ml 2% sodium chloride
butter

Test reagents

1% silver nitrate
Lugol's iodine
0.1% indophenol
Biuret reagent
Benedict's solution
isopropyl alcohol
brown wrapping paper or paper bag

Procedure (Day 1)

Find out whether the test compounds can dissolve in water. Place a small amount (0.5 to 1.0 g) of one of the compounds in a test tube. Fill the tube halfway with water. Place your thumb over the opening of the tube and shake vigorously. Observe the mixture. Repeat this procedure for all samples. **1.** How can you tell whether the material has dissolved in the water? **2.** Why is it important in the life of a cell that these substances can dissolve in water?

Scientists use **reagents** to detect the presence of certain organic compounds. Conduct the following tests with reagents, using the solutions and the butter your teacher has provided. **3.** Follow the directions for each reagent test and record your results on a separate sheet of paper using a table similar to the one below.

Substance	*Reagent Test*	*Results*
Gelatin solution (protein)	Biuret reagent	
Glucose solution	Benedict's solution	
Starch suspension	Lugol's iodine	
Vitamin C solution	Indophenol	
Sodium chloride solution (chloride ions)	Silver nitrate	
Butter	Paper	

Biuret test. Place the gelatin solution (protein) in a test tube. Add five to ten drops of Biuret reagent. Biuret reagent changes color in the presence of protein. Record the change. (CAUTION: *If you get the reagent on your skin or your clothes, rinse immediately with water. Biuret reagent contains sodium hydroxide, which is very harmful.*)

Iodine test. Add two or three drops of Lugol's iodine to the starch suspension. Iodine changes color in the presence of starch and glycogen. Record the color change.

Benedict's solution test. Add 3 ml of Benedict's solution to 5 ml of glucose solution. Benedict's solution indicates the presence of glucose. Place the test tube in a beaker of boiling water and heat for 5 minutes. Record the color change.

Indophenol test. Add four to eight drops of indophenol to the vitamin C solution. Record the color change. This change indicates the presence of vitamin C.

Silver nitrate test. Add three to five drops of silver nitrate to the sodium chloride solution. The silver nitrate indicates whether chloride ions are present. Record what occurs.

Paper test for fat. To test for fat, rub the butter on a piece of brown wrapping paper. If the butter has been dissolved in water, pour it onto the paper and allow the paper to dry. Hold the paper up to the light to detect the oil spot.

Procedure (Day 2)

Try the reagent tests on the foods. You will need six small samples of each food. Place each sample in a test tube with some water.

When food contains a very small amount of fat, it may not be removed by the method just described. If no fat has been detected, place the food in 10 ml of a fat solvent such as isopropyl alcohol. Allow the fat to dissolve in the solvent for about five minutes. Then pour the solvent on brown paper. The spot should dry in about ten minutes. Check the paper for a "grease" spot. This indicates the presence of fat. **4.** For each food that you will test, write a prediction of what substances you think you will find.

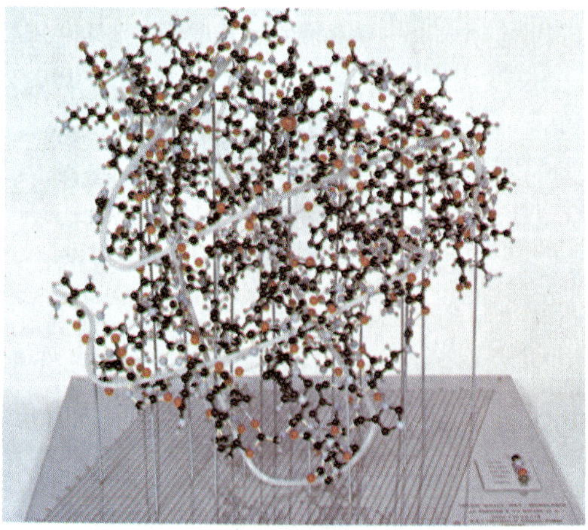

2-19. A model of a myoglobin molecule. This protein is small and simple compared with many other proteins in your body. The light-colored, rodlike structure bending back and forth through the model represents the chain of amino acid units linked together.

Test Results (+ or −)

Substance	protein	glucose	starch	vitamin C	chloride	fat
Egg Potato etc.						

Use a table similar to the one above to record your observations. Test each food for the six different chemical substances. Indicate the presence (+) or absence (−) of each substance. **5.** Which of your predictions were totally correct? **6.** Which food contains all the compounds for which you tested? **7.** How might the original colors of the tested materials affect the results? **8.** On the basis of your tests, which food could be used as a source of protein? starch? glucose? vitamin C? fat? sodium chloride?

A CLOSER LOOK AT PROTEINS

Of the many proteins known, only a few have been studied carefully. But proteins are among the most important of all organic molecules. Much of what cells do, how they look, and what they contain depends on proteins. A full understanding of the processes in organisms must be based on a knowledge of proteins. But this important branch of biology is still very young.

The *size* of a protein molecule depends on what kinds and how many amino acid molecules are bonded together. The *shape* of the molecule, however, depends on how chains of amino acids are connected. It also depends on which amino acids are in the chains. Some proteins are long, thin fibers, such as those that form hair and silk. Others are round, such as hemoglobin. The amino acid chains are curled into a ball. Still other molecules resemble sausages. Often, protein molecules are folded back and forth many times. These are held together by bonds between the folded chains. Figure 2-19 shows the protein, myoglobin (MY·oh·glow·bin), which is found in muscle cells.

Enzymes

Enzymes (EN·zimes) are proteins that speed up chemical reactions in living cells. This speeding-up is necessary for the cell to live. Chemists could not make some reactions happen in the laboratory until they could separate enzymes from living organisms. You will work with an enzyme in **Inquiry 2-3.**

INQUIRY 2-3 A Chemical Reaction of Living Systems

MATERIALS (per team)

raw liver
enzyme solution
medicine dropper
graduated cylinder
glass rod
6 test tubes, 13 × 100 mm
2 test tubes, 18 × 150 mm
3% hydrogen peroxide
boiling water bath
warm water bath (35°C)
ice bath
0.1 M sodium hydroxide
0.1 M hydrochloric acid
pH paper
test-tube rack
straw

Procedure (Day 1)

Put 20 drops (1 ml) hydrogen peroxide in a small, clean test tube. Put the tube in the rack. (CAUTION: *If you spill any peroxide on you or your clothes, wash it off immediately.*)

Add five drops of stock enzyme solution to the peroxide. Mix with the glass rod. **1.** Record your observations.

Feel the bottom of the test tube. **2.** What temperature change do you feel? What might be the source of the change? **3.** What are some possible explanations for the bubbles?

The stock enzyme solution was made by crushing a piece of liver in a small amount of water. This released many molecules of the enzyme from the liver into the water. The reaction mixture in your test tube contained hydrogen peroxide and the enzyme. The enzyme in this reaction is called *catalase* (CAT·uh·lace). To explain the reaction, three hypotheses are possible: (a) Catalase and hydrogen peroxide react together to form one or more new products; (b) the hydrogen peroxide is changed in the presence of the enzyme, but the enzyme does not change; or (c) the enzyme is changed in the presence of hydrogen peroxide, which does not change. **4.** Using your knowledge of chemical reactions and enzymes, choose one of the three hypotheses. Design a simple experiment to test your hypothesis.

If time and equipment allow, conduct the experiment you designed. Or, carry out the experiment outlined below. But think through your experiment and try to determine what the results might have been.

To test these hypotheses, use raw liver instead of the stock enzyme solution. Use the two large test tubes and a piece of liver about the size of a rice grain. With a glass rod, push the liver to the bottom of one of the tubes. Put the tube in the rack and add 2 ml hydrogen peroxide. (You may have to keep pushing the liver down with the glass rod.)

Allow the reaction to continue until all bubbling stops. Stir gently with the glass rod until the bubbles disappear. Pour this reaction liquid into the other test tube. Put another piece of liver (of the same size) into this liquid. Observe. Pour 2 ml hydrogen peroxide onto the liver in the first tube. Observe. **5.** Record your observations about both tubes. **6.** Does hydrogen peroxide, the enzyme, or both change after the reaction? Explain the reason for your answer. **7.** What does this tell you about what an enzyme does in a chemical reaction? **8.** Which of the three hypotheses is supported by these results?

Hydrogen peroxide is a poisonous chemical that is continually formed by reactions in living cells. It could kill a cell. The chemical formula for hydrogen peroxide is H_2O_2. When hydrogen peroxide is placed with catalase, water (H_2O) and oxygen gas (O_2) are formed. **9.** What role does catalase play in the cell? **10.** Is this enzyme found only in the liver? Hydrogen peroxide will bubble if placed on a cut finger. What does this reaction tell you about the enzyme?

Carbon dioxide also is produced in reactions in living cells. Does catalase cause the same reaction with carbon dioxide or carbonic acid as it does with hydrogen peroxide? (Carbonic acid is formed when carbon dioxide dissolves in water.) Put 3 ml of distilled water in a test tube. Use the straw and blow into the water (carefully) for about 1 minute. This should form a lot of carbonic acid, just as you saw in **Inquiry 1-1.** Place a small piece of liver in the test tube. Watch what happens. **11.** From your observations, compare the action of catalase on carbon dioxide with the effect of catalase on hydrogen peroxide.

If left open in the sunlight, a bottle of hydrogen peroxide will slowly change into water and oxygen. Under these conditions it takes a very long time for the reaction to happen. On the other hand, when catalase is placed in the hydrogen peroxide, the reaction takes place almost immediately. **12.** What does this tell you about the function of an enzyme?

Procedure (Day 2)

Water ionizes only very slightly into hydrogen (H^+) and hydroxide (OH^-) ions. But these few ions account for the importance of water solutions. In pure water, the concentrations of H^+ and OH^- ions are equal. Pure water is considered to be *neutral*—neither *acid* nor *base.* Solutions that have more H^+ ions than OH^- ions are acidic. Basic (or alkaline) solutions have more OH^- than H^+ ions. The **pH**

scale (Figure 2-20) is used to measure the H^+ and OH^- ions in a solution. The pH scale ranges from 0 (acidic) to 14 (basic). Each unit on the scale represents ten times as many hydrogen ions as the higher unit. A solution of pH 1 has ten times as many hydrogen ions as a solution at pH 2. The same relationship holds all along the scale. The halfway point, 7, is neutral. Any value less than 7 is acidic, and any value more than 7 is basic. **13.** Write a hypothesis about the effects that a very low and a very high pH will have on the activity of an enzyme.

You can test your hypothesis. You will need six test tubes. Label three of them *A, B,* and *C* and leave three unmarked. To Tube A add 2 ml distilled water. To Tube B add 2 ml hydrochloric acid. To Tube C add 2 ml of sodium hydroxide (basic) solution. (CAUTION: *If you spill the acid or the base on you or your clothes, wash it off immediately.*) To Tubes A, B, and C add 1 ml stock enzyme solution.

Test the pH of all three solutions with the pH paper. (Review Figure 2-20, if necessary.) Record the pH.

To each of the three unmarked tubes add 1 ml hydrogen peroxide. Pour the contents of these tubes, *all at the same time,* into Tubes A, B, and C. **14.** Record your observations. Note especially the *rate* of activity. **15.** What effect does pH have on the activity of this enzyme?

Does temperature affect the rate of the reaction? You will need six test tubes. Mark three of them *A, B,* and *C.* Leave the other three unmarked. To each of Tubes A, B, and C add 1 ml stock enzyme solution. All at the same time, place Tube A in the boiling-water bath, Tube B in the warm-water bath, and Tube C in the ice bath. Wait 4 minutes. While you are waiting, put 1 ml hydrogen peroxide into each of the three unmarked tubes. At the end of the 4 minutes, remove all three tubes from their water baths at the same time. Place them in the test-tube rack. Quickly pour the hydrogen peroxide from the unmarked tubes into Tubes A, B, and C. **16.** Observe and record the reactions. Note especially the rate of reactions. **17.** Based on your observations, what effect does temperature have on the activity of this enzyme? **18.** Describe an enzyme, based on your observations in this inquiry.

2-20. The pH scale.

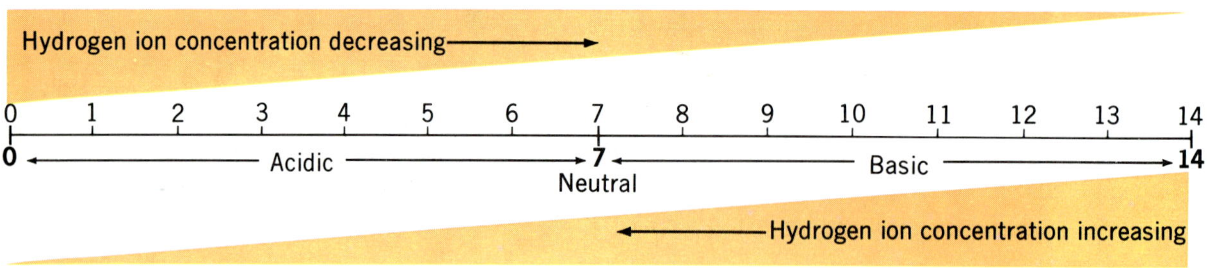

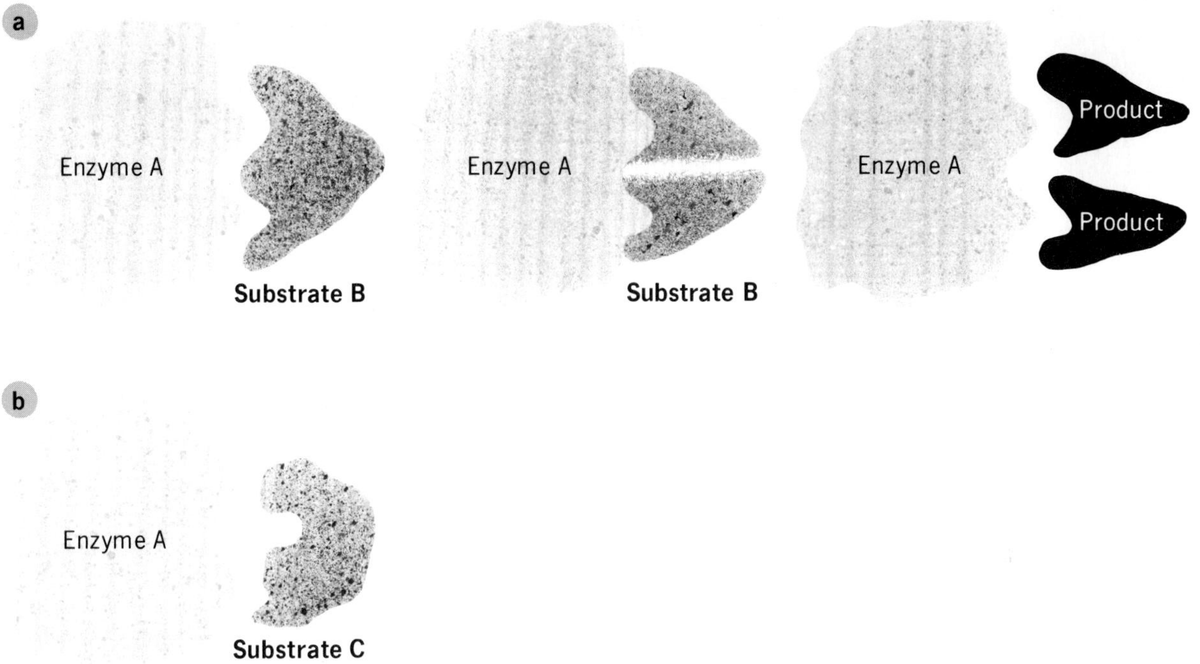

2-21. The "lock and key" model of enzyme action. In **a,** molecules of Enzyme A and Substrate B fit together. When combined with Enzyme A, Substrate B breaks down into two molecules. In other reactions, these two molecules combine to become Substrate B. In **b,** Enzyme A is shown with a molecule of Substrate C. The two surfaces do not fit. *What will happen?*

Catalysts

Enzymes act as biological catalysts. A **catalyst** is a substance that speeds up a chemical reaction by lowering the energy required for a chemical reaction to get started. The catalyst, however, is not changed or used up during the reaction. Without enzymes, many chemical reactions could not take place except at very high temperatures. But with enzymes, animals and plants carry out millions of chemical reactions at much lower temperatures.

As you saw in **Inquiry 2-3,** enzymes are very specific—each enzyme catalyzes only one kind of reaction. And the same enzyme molecule can catalyze the same reaction over again, often thousands of times per minute. Enzymes are active even when present only in tiny amounts. All of the synthesis and hydrolysis reactions mentioned earlier need enzymes.

An enzyme molecule combines with a reacting molecule, which is called the **substrate.** One model of this combination is shown in Figure 2-21. This is a very simple model for enzyme action. Other things, such as the electrical charges of the enzyme and substrate molecules, play important roles. Scientists now believe that the shape of the enzyme molecule is not as rigid as the "lock-and-key" model suggests. In some cases, the substrate molecule seems to change the shape of the enzyme slightly. This model of enzyme action is called the "induced-fit" model.

OTHER COMPOUNDS

There is another group of compounds that is of special importance in living cells. Because we will study them again, our discussion here will be brief.

Only recently have biologists come to understand the importance of **nucleic** (new·KLAY·ik) **acids** and **nucleotides** (NEW·kle·o·tides) in the life of cells and organisms. Like proteins, nucleic acids are long chains of simpler units. But the units are not amino acids, as in proteins. They are nucleotides. Each nucleotide molecule is built from rings of carbon and nitrogen. Each molecule also contains atoms of hydrogen, oxygen, nitrogen, and phosphorus. Each nucleic acid is built from only four kinds of nucleotides. As in proteins, the order of the units in the chain is very important.

There are two major types of nucleic acids. **Deoxyribonucleic** (dee·OKS·ee·ry·boh·new·KLAY·ik) **acid,** or **DNA,** is found in the cell nucleus. **Ribonucleic** (ry·boh·new·KLAY·ik) **acid,** or **RNA,** is found in many places in the cell. DNA is the substance that is passed from one generation to the next and determines what a new organism will be like. Every organism has its own particular kind of DNA.

Scientists believe that DNA molecules build RNA molecules. RNA molecules, in turn, build proteins. Do you remember the hemoglobin molecule with one different amino acid? The change in the protein was caused by a change in DNA. In this instance, the change was a fatal one for the organism.

Nucleotides play other roles in the cell. In the next chapter, we will see how some of them play key roles in providing the energy used in cells.

In Brief

Four classes of organic compounds—proteins, carbohydrates, lipids, and nucleic acids—play two major roles in cells. They are important in building the cell, and they supply energy the cell uses.

The chemistry of cells is a very lively business. Molecules are constantly being built, torn apart, rebuilt, and moved from one place to another. Enzymes control these processes so that hundreds of thousands of reactions go on in a single cell, all at the same time.

We know a great deal about living chemistry. Biological processes cannot be explained without some knowledge of chemistry. The question "What is a living thing made of?" can be answered in terms of atoms and the characteristic molecules they form. Another important question is "What does a living thing do?" This can be answered by looking at how atoms share or exchange electrons, bond together, and attract or repel one another. How atoms store energy and change stored energy from one form to another is another area of investigation.

Although we know quite a bit about the chemistry of living things, there is still much to be learned. We do not even know all of the chemicals in a toenail, a hair, or a blood cell. As a matter of fact, we cannot describe a single organism completely in terms of its chemistry.

We need still other kinds of information. How do chemical substances form living cells? How do cells work? Then we must ask how cells are put together in different organisms. We will start to look for answers to these questions in Chapter 3.

Guide Questions and Problems

1. Compare the ideas of vitalism and mechanism. Which had more influence on the development of biology as a science?

2. Living and nonliving things are made up of atoms. How can they differ in such an important way—one with life and one without? Use what you have learned about chemistry to try to answer this question.

3. Which of these forms of energy could you use to replace your stored energy after a long day's work? (1) light energy in sunshine; (2) heat energy from an open fire; (3) chemical energy in a loaf of bread. You learned that energy in any form can be changed into any other form. Why can't your body use all forms equally?

4. The growing need for energy (such as electricity, chemical energy in gasoline, and natural gas) is a major problem of our society. Could we solve this problem simply by recycling all energy? Could we change it from one form to another and reuse it without waste? Why?

5. What common activities of your own show a relationship between chemical and mechanical energy?

6. Describe the work of Wohler and Kolbe.

7. The sugars glucose, galactose (guh·LAK·tohs), and fructose all occur in living organisms. All have the same formula—$C_6H_{12}O_6$. They are different in their chemical reactions and behave differently. How can you account for this?

8. Body fat is built up of stored fatty materials in cells. A person who eats more carbohydrates (bread, rice, spaghetti, candy, or cake) than the body needs for energy is very likely to become fat. What

does this show about the chemical relationship between carbohydrates and fats?

9. Catecholase (CAT·uh·kul·ace) is the enzyme that causes the cut surfaces of such foods as potatoes and peaches to turn black when exposed to the air. The enzyme catalyzes a reaction in which catechol (an organic molecule) combines with oxygen. If a tiny amount of the enzyme is placed in a test tube and some catechol is added, the solution will turn dark rapidly. If the enzyme is repeatedly separated from the solution and used again, the same reaction happens. What does this suggest about enzymes?

Ideas for Independent Study

1. The word energy was not used until early in the nineteenth century. And it was a long time before scientists had clear ideas about how energy fits into the natural world. Many of our modern ideas have come from technical developments, such as the steam engine and electrical generators. But people have always needed energy for work. Find out about the kinds of energy used in the past for transportation, lighting, heating, communication, and warfare. How many of these energy sources were biological, or of biological origin? How would you answer that question in terms of our life today?

2. Plants change the energy of light into chemical energy by building organic compounds. Solar batteries change light energy into electrical energy. Find out how solar batteries work, and how they are being used today. Do you think that this sort of energy may become one answer to our energy needs?

3. Find out as much as you can about the chemical makeup of materials with biological origins. Some examples are wool, leather, cotton, linen, glue, paper, wax, tallow, paraffin, natural gas, gasoline, and coal. What is in the molecules in these materials that makes them useful?

4. You can discover many things about living chemistry by a careful study of foods. Begin by asking questions. Is there any relationship between the *cost* of foods and their chemical makeup (for instance, their protein or carbohydrate content)? What are "dietetic foods"? What are "organic foods"? Could a food be "nonorganic"? Could any person live on nonorganic materials? What must be taken into account in preparing the foods that astronauts take into space?

Related Readings

BOOKS

Albersheim, P., J. Dowling, and J. Hopkins III. *The Molecular Basis of Metabolism.* A BSCS Laboratory Block. Lexington, Mass.: D. C. Heath, 1968.

Barry, J. M., and E. M. Barry. *Introduction to the Structure of Biological Molecules.* Englewood Cliffs, N. J.: Prentice-Hall, 1969.

Biological Sciences Curriculum Study (BSCS). *Biology in Three Dimensions.* Northbrook, Ill.: Hubbard Scientific, 1975.

Breed, A., T. Rodella, and R. Basmajian. *Through the Molecular Maze: A Helpful Guide to Chemistry for Beginning Life Science Students.* Exp. ed. Los Altos, Calif.: William Kaufmann, 1975.

Dickerson, R. E., and I. Geis. *The Structure and Action of Proteins.* Menlo Park, Calif.: W. A. Benjamin, 1969.

White, E. H. *Chemical Background for the Biological Sciences.* Englewood Cliffs, N. J.: Prentice-Hall, 1970.

Wilson, M. *Energy.* New York, N.Y.: Time-Life, 1967.

Bio-organic Chemistry: Readings from Scientific American. San Francisco: Freeman, 1968.

ARTICLES

Dickerson, R. E. "The Structure and History of an Ancient Protein." *Scientific American,* April 1972, page 58.

Koshland, D. "Protein Shape and Biological Control." *Scientific American,* October 1973, page 52.

Kretchmer, N. "Lactose and Lactase." *Scientific American,* October 1972, page 70.

Lambert, J. B. "The Shapes of Organic Molecules." *Scientific American,* January 1970, page 58.

Pastan, I. "Cyclic AMP." *Scientific American,* August 1972, page 97.

Stroud, R. "A Family of Protein-Cutting Proteins." *Scientific American,* July 1974, page 74

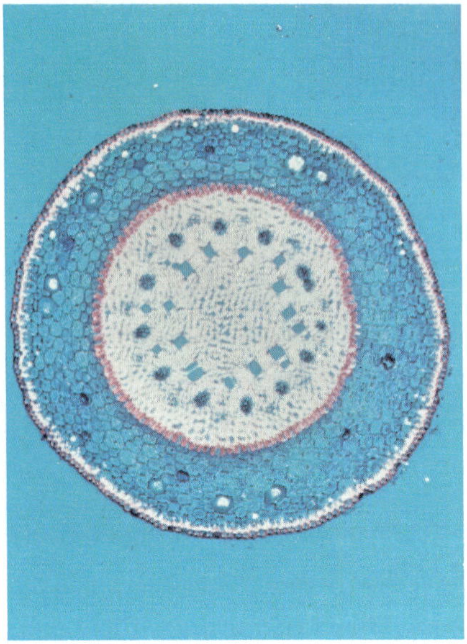 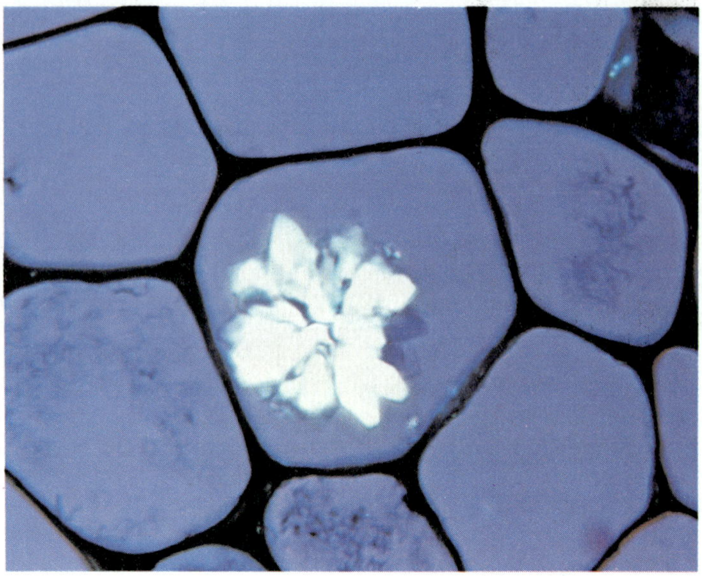

3 Cell Structure and Function

The basic question in the spontaneous generation story was, "Where do organisms come from?" Once this question was settled, biologists began to search for the natural laws that operate in living systems. One question biologists then tried to answer was, "What is the basic unit of structure and function in living things?"

Structure and **function** are important words in this book. You will find *structure* used to refer to the way the parts of something are put together. It also is used to refer to one of those parts. The *function* of a structure is simply what it does.

Today we know that cells are the basic units of life. This knowledge grew out of a long series of investigations.

THE CELL THEORY

Simple microscopes were first made in Holland, shortly before 1600. By about 1610, Galileo (gal·ih·LAY·oh), an Italian scientist, had made a microscope. This microscope was powerful enough to be used for observing living things that were too small to be seen with the naked eye.

Later in the seventeenth century, Robert Hooke, an English scientist and inventor, built a microscope. This new tool gave much better results than did previous microscopes. Hooke examined many things—minerals, cloth, and small plants and animals. He published a report in 1665 after examining cork. In his report, he described the cork cells as a honeycomb. He thought the cells were passageways through which materials move (Figure 3-1).

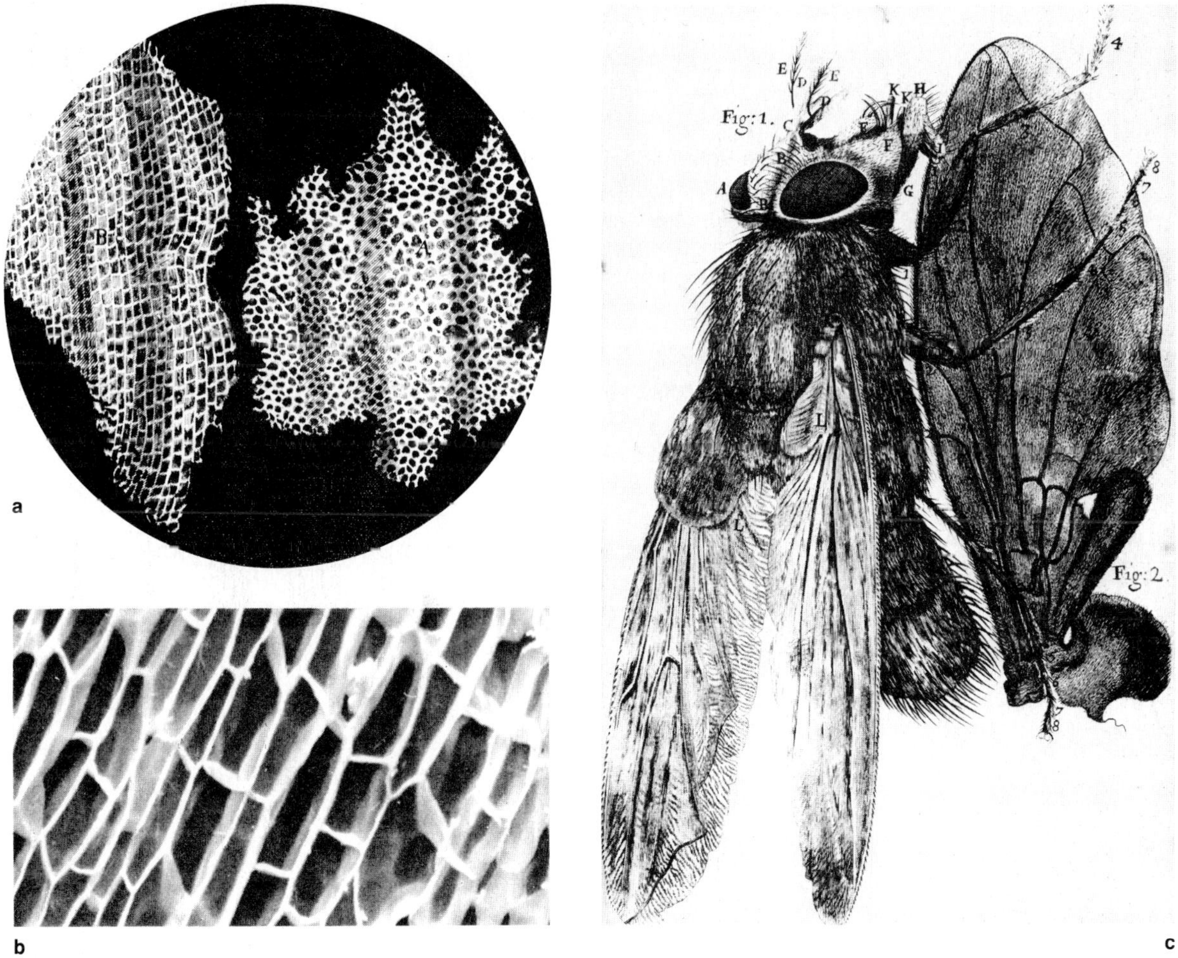

3-1. In the 17th century, Robert Hooke published the drawing of cork cells shown in **a.** Compare it with the recent photomicrograph shown in **b.** Hooke also studied many common objects, such as the fly shown in **c.**

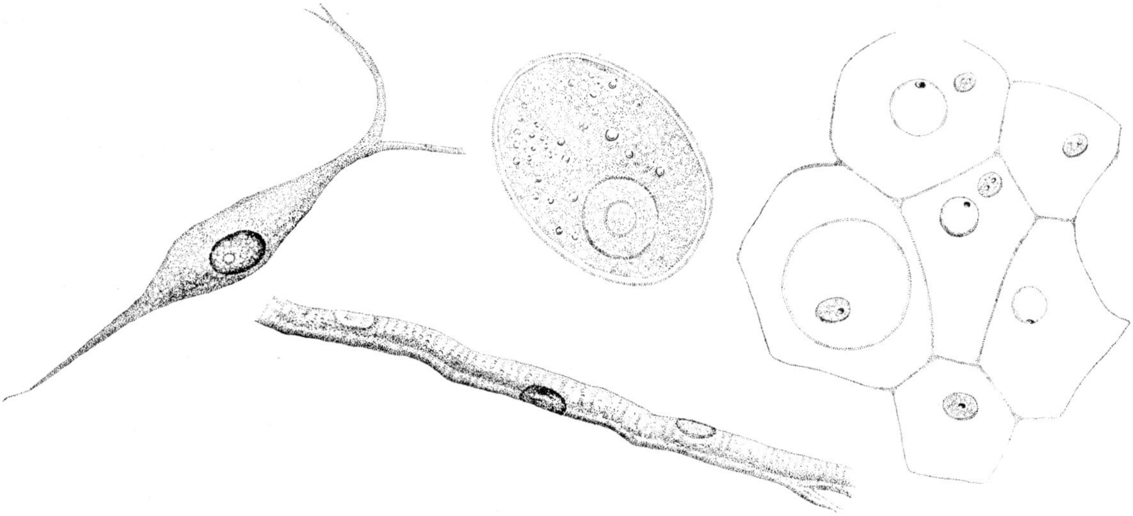

3-2. The microscopic structure of small parts of animals as seen by Schwann. The group of six cells at the right is from a fish. The oval cell is from the nervous system of a frog. The long cell at the bottom is from the muscle of an unborn pig. The spindle-shaped cell at the left is also from an unborn pig. *Do you think you would have seen or suspected that things that look so different could be very similar?*

More scientists began to use microscopes. They examined many different animals and plants. Often they saw structures that reminded them of the cork cells that Hooke described. About 150 years after Hooke's observations, scientists began to ask, "Could it be that *all* living things are made up of cells?"

As time went on, many plants were found to be made of heavy-walled, boxlike cells. But what about animals? Are they also made of cells? When biologists examined bits of animal matter, they could not find cells like those they saw in plants. Indeed, it was difficult to see anything at all in the animal matter.

In 1839, a German zoologist, Theodor Schwann (SHVON or SHWON), interpreted his observations of cells in a new way. Previous investigators had emphasized the walls of the boxlike cells. Schwann emphasized what was inside the "box."

Earlier, Robert Brown, a botanist, had observed a little sphere in orchid cells. This spherical structure was later named the **nucleus** (NEW·klee·us; plural, nuclei—NEW·klee·eye). When Schwann looked at little animal cells under the microscope, he saw nuclei but not cell walls (Figure 3-2). Each nucleus was surrounded by a nearly transparent substance. The animal cells had a very thin outer layer, or **membrane.** These cells did not look very much like plant cells with their thick cell walls.

But Schwann thought that what was inside a plant cell might be similar to what he saw in animal cells. He found nuclei, for example, in both animal and plant material. A membrane surrounded each kind of cell. And there was a clear substance around the nucleus in both. In fact, Schwann saw more similarities than differences between animal and plant cells (Figure 3-2). Schwann described the cell as a structure with a nucleus, not a structure with a thick wall. He then wrote a hypothesis:

> The bodies of animals and plants are composed of cells and the products of cells.

A German botanist, M. J. Schleiden (SHLY·den) was working with similar cells at about the same time. Schleiden agreed with Schwann's hypothesis. From the observations of Schwann and Schleiden, and those of other scientists, the cell theory was born.

Schwann had proposed a hypothesis—the cell hypothesis. It led to a theory—the cell theory. What is the difference? A hypothesis is a statement to be tested. It is a possible explanation for what is observed. Schwann and others tested the cell hypothesis by observing the microscopic structure of many animals and plants. All the animals and plants they examined seemed to be made up of cells.

When a hypothesis is found to be true after years of testing, it is called a theory. The cell hypothesis became the cell theory only after cork, frogs, carrots, people, wheat, fish, and many other living things that had been studied were found to be made up of cells. A theory is not a fact, however, and it continues to be tested.

The Origin of Cells

A new question arose. Where do cells come from? Using the methods available in the early 1800s, biologists could not be sure. Schwann believed that new cells formed spontaneously from living substances. Other biologists noticed that sometimes one cell seemed to divide into two (Figure 3-3). Was this the way new cells

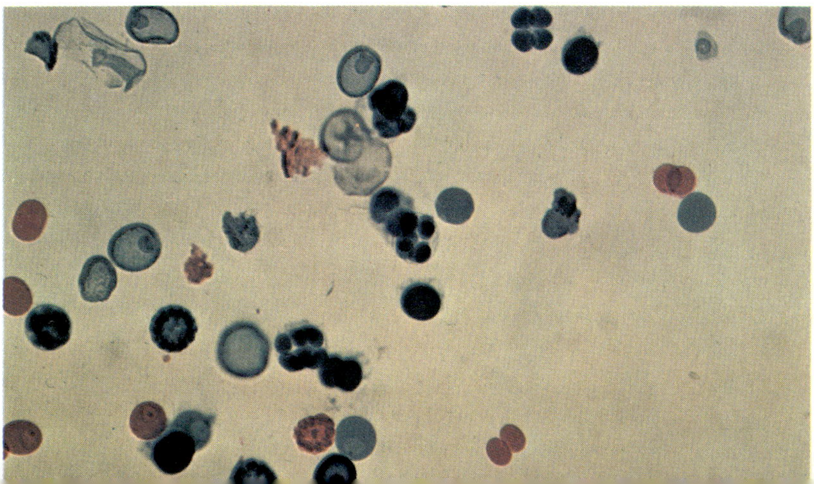

3-3. *Can you find a cell that has divided into two in the picture taken through a microscope?*

were formed—by the parent cell dividing into two parts? Many biologists thought so. One of these was the German biologist Rudolf Virchow (VEER·kow). He made many observations and studied reports. He decided that cells always arise by division of parent cells. In 1855, Virchow expressed this thought as "all cells from cells." Some observers thought that cells could form in different ways. Today we know that cells are the basic structural unit of living things. And we know that all cells arise from other cells.

Before you go on to study the details of cell structure, you may want to observe cells for yourself by doing **Inquiry 3-1.**

INQUIRY 3-1 Living Cells

MATERIALS (per team)

onions
elodea
iodine or methylene blue stain
finger bowls or other containers, 10 cm in diameter
slide and cover glass
medicine dropper
forceps
light microscope
dissecting needles
new, sharp razor blade
paper towels
toothpick

Procedure

Cut an onion bulb into quarters (Figure 3-4). Each quarter will separate neatly into layers. Hold one of the layers with both hands, so that it curves toward you. Push your thumbs away from you and pull the two edges of the onion layer together. Starting at one of the torn edges, pull the transparent, paper-thin layer of **epidermis** (ep·ih·DER·mis), or outer skin, from the surface. It comes off in a sheet. Make a wet mount of the epidermis as follows:

Use the razor blade to cut a piece of epidermis about 1 cm square. Place this piece in a drop of water on a clean slide. Smooth out wrinkles in the epidermis with a dissecting needle. Put on the cover glass. Be sure the water comes to its edges, but not beyond. Add water at the edge with a medicine dropper, or remove excess water with a paper towel, as necessary.

Look at the cells of the epidermis under low power. **1.** What is their general shape? **2.** If you did not know their origin, how could you tell they are plant cells?

Observe the cells under high power. **3.** What evidence is there that these are living cells?

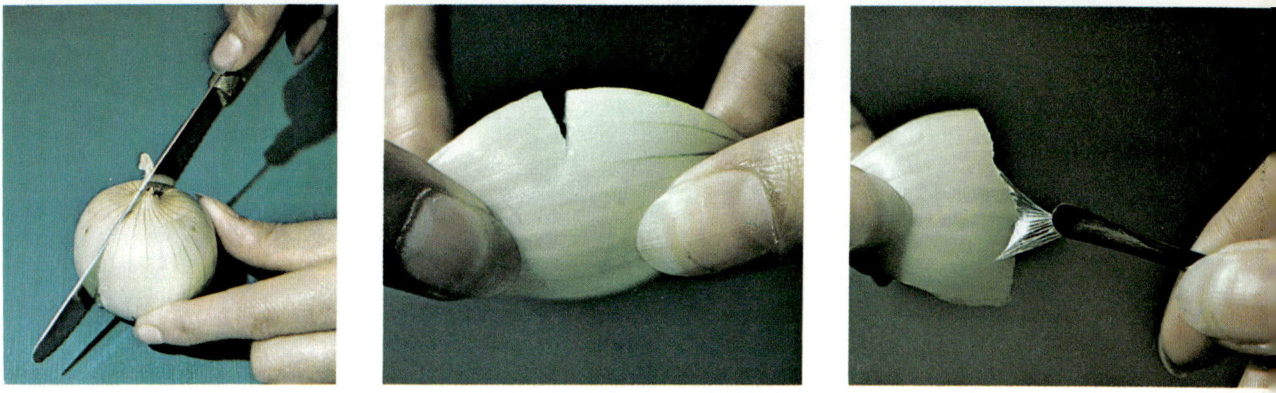

3-4. Preparing onion epidermis for study.

Remove the slide from the microscope stage. Place a drop of the stain (coloring) at one edge of the cover glass. Draw it under the cover glass by touching a piece of paper towel to the *opposite* edge (Figure 3-5). (CAUTION: *Stains can cause permanent damage to clothes and laboratory surfaces. Handle carefully.*) Under low power look for a nucleus that is stained brown or blue. **4.** What is its shape?

Switch to high power and look at the nucleus again. **5.** Where is the nucleus located in the cell? **6.** How can you tell that the cell has depth?

Look at the area between the nucleus and the edges of the cell. You will find a fine granular (grainlike) substance that has been stained lightly. **7.** Does the substance fill the entire cell? **8.** Describe what it looks like. **9.** How might this substance be different in cells that have not been stained?

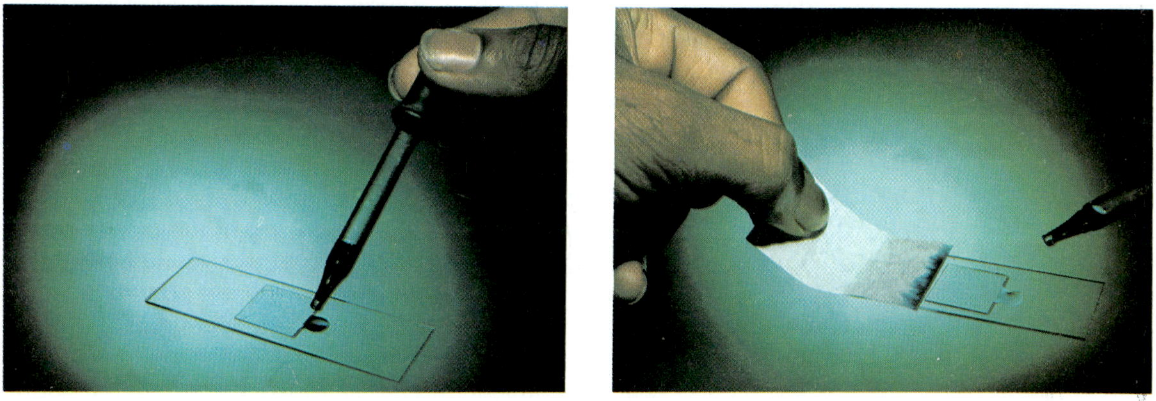

3-5. How to run a liquid under a cover glass.

Elodea (sometimes called *Anacharis*) is a common flowering plant that lives in fresh water. The green cells of elodea are similar in general structure and function to those in most other plants. Take one plant from the bowl. Break off one of the younger leaves near the tip. Place the leaf with its underside up, in a drop of water on a clean slide. Put on a cover glass.

Look at this leaf under low power. Some cells seem to be packed with small green bodies. These bodies are called **chloroplasts** (KLOR·o·plasts). **10.** Do you think that these elodea cells have the same function as those of the onion you looked at? Why or why not?

The part of the onion from which you took cells is usually found below the ground. The elodea plant is found where sunlight strikes the plant. **11.** What does this suggest about the function of chloroplasts? **12.** Where are the chloroplasts located in the cell? **13.** What is the shape of a chloroplast? **14.** Describe the movement of chloroplasts.

Chloroplasts have no means of moving about. **15.** How might their movement be explained?

Other structures are present in cells of elodea. But all of these, except the cell wall, are hidden by chloroplasts. With patience and careful observation of many cells, however, you may be able to find a nucleus and see other parts of the cell. **16.** What methods might you use to see these structures better? **17.** Compare and contrast the two kinds of living cells you have seen. How are they alike? How do they differ?

Place a very small drop of tap water on a clean slide. Rinse your mouth out with water. With the broad end of a toothpick, *gently* scrape the inside of your cheek. Deposit a little of the scraping in the drop of water by rolling the toothpick in it.

Stir the scraping into the drop of water with the toothpick until all of the drop is cloudy. Add a *small* drop of stain to the scraping. Place a cover glass on the slide. Observe with low power. You may find that some of the cells are folded over or piled on top of one another. Some may be broken. Find one or two that you can see clearly. Center these for viewing. **18.** How does the edge of this kind of cell compare with the plant cells you observed?

Turn to the high-power objective. The nucleus of each cell should be clearly visible. You may have to adjust (usually reduce) the light. Using the fine adjustment, focus up and down on one cell to determine its shape. **19.** How does the shape of this animal cell compare with the shape of the plant cells you observed? **20.** How would you be able to tell that this cell was from an animal and not a plant? **21.** How are the plant and animal cells you have seen alike? **22.** Do your observations support the ideas of both Schwann *and* Virchow? Why or why not? **23.** Would you expect

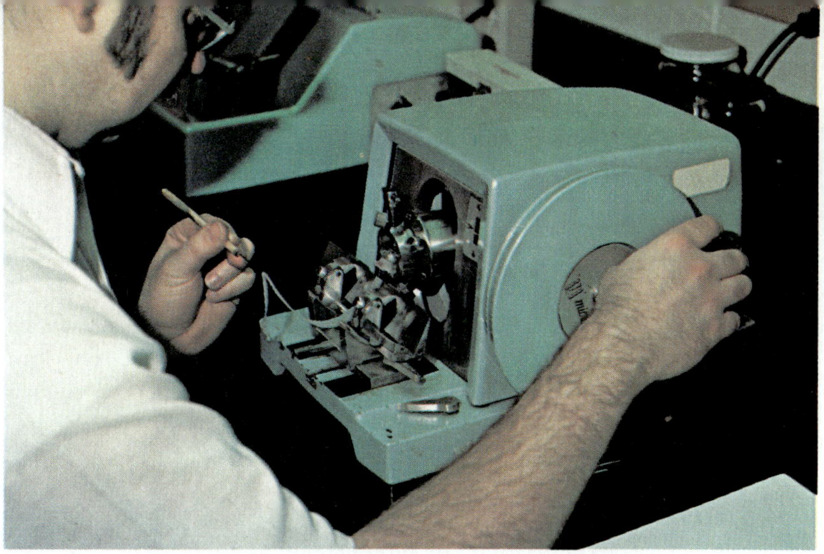

3-6. A microtome is used to prepare thin slices of material for observation with a microscope.

cells from other parts of your body to be exactly like the cells from inside your cheek? Why or why not? **24.** An organism may have many small cells rather than a few large cells. Suggest one or more possible explanations for this.

STUDYING CELLS

We have been able to learn most of what we know about cells because of developments in technology. New ways of studying cells have been worked out. New hypotheses and experiments resulted. Let us look at some of the ways that modern biologists study cells.

Microscopy

Microscopes have changed a great deal since Hooke looked at cork cells. Most microscopes provide magnifications of between 10 and 1,500 times (10 × and 1500 ×). Much of our early knowledge of cells and cell parts was based on observations with such microscopes. Most material studied with a light microscope must be prepared carefully. It must be thin and transparent. The basic procedures involve first **fixing** (killing and preserving) the cells. Fixing hardens soft cell structures and locks them in place. First, the cell material is soaked in alcohol to remove water. Then it is embedded in heated paraffin wax which replaces the alcohol. The wax cools and hardens. Then the material is sliced into very thin sections (often ten micrometers or less). This is done with a mechanical slicing instrument called a microtome (Figure 3-6). The thin sections are placed on slides, and the wax is dissolved away. The remaining cell materials are stained with special dyes to make them easier to see.

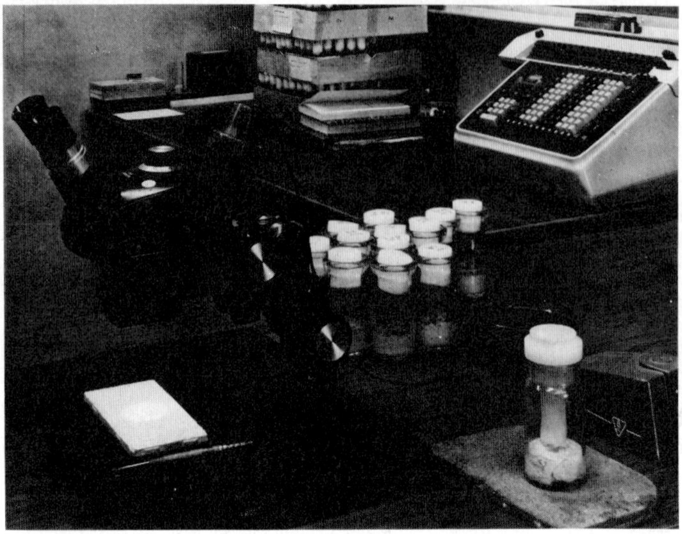

3-7. A stereoscopic microscope is useful for studying the structure of small plants or animals.

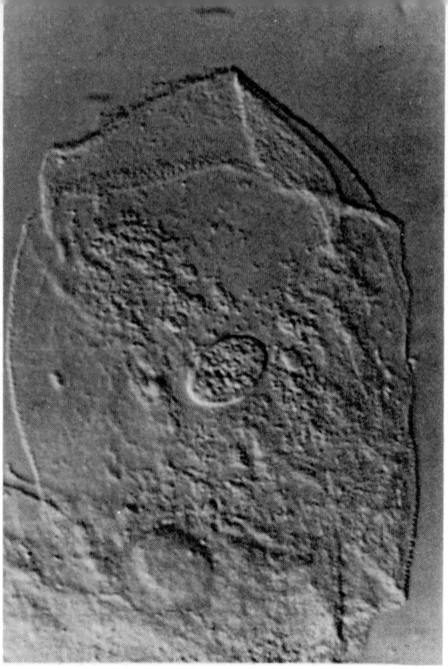

3-8. A phase contrast view of a human cheek cell.

3-9. An electron microscope can magnify as much as 1,000,000 X. The range used for biological materials is usually 1000 X to 100,000 X. This valuable instrument is now widely used in medical and biological research.

3-10. This giant high voltage electron microscope is used to examine specimens too thick to be observed with ordinary light microscopes.

The stereoscopic microscope (Figure 3-7) is used to study small animals and plants, or their parts. This kind of microscope gives depth to the image. The range of magnification usually varies from 6 to 50 times. The microscope in the figure is being used in a study of the fruit fly, *Drosophila.* We can see a fruit fly with the unaided eye. However, without magnification, we cannot see all the details of its structure.

A phase-contrast microscope is often used to study living cells. This instrument greatly increases the differences in light and dark areas. Cell structures become easier to see. Most stains are poisonous to living cells. But the phase-contrast microscope allows observation of living cells without the use of stains. Figure 3-8 shows a phase-contrast view of a human cheek cell.

An electron microscope (Figure 3-9) is used for much higher magnifications. This instrument is used widely in biological research. It has increased our knowledge of the structure of cells and organisms by magnifying them up to 100,000 times. In this case, cells are fixed in plastic, so the cell material can be sliced very thin. The electron microscope shown in Figure 3-10 is three stories high. Such

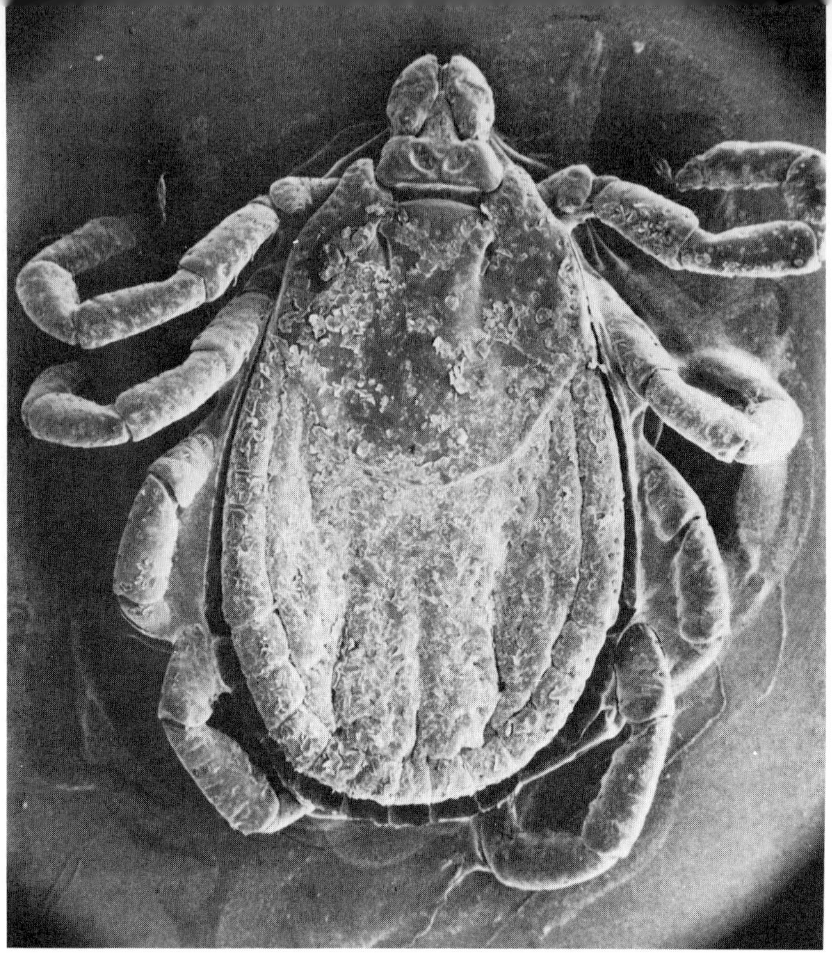

3-11. Scanning electron microscopes allow detailed observations of the surface of biological specimens. Shown here is a tick of the kind that carries Rocky Mountain spotted fever.

microscopes are used to observe the internal structure of thicker materials. In recent years, a new type of electron microscope, called the scanning electron microscope, has come into use. It allows the study of the surface of cells and tissues that are too thick for study with electron microscopes (Figure 3-11).

Figure 3-12 shows the magnification ranges of some microscopes in relation to some common organisms and cells. Note that single molecules cannot be seen, even with the most powerful microscopes.

Cell Culture

Since the beginning of this century, biologists have been able to grow animal and plant cells in glass dishes. This method is called

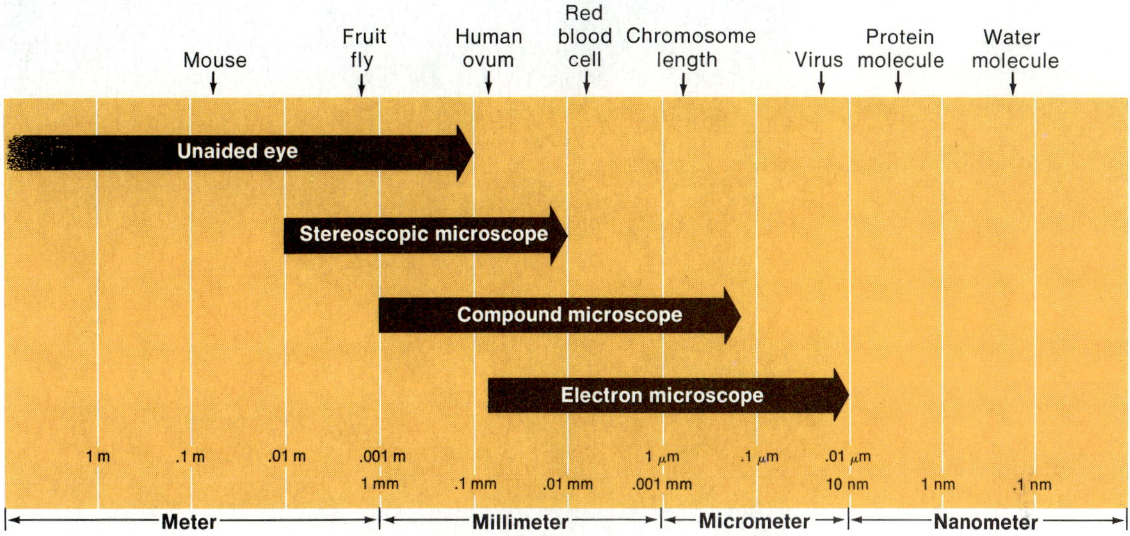

3-12. Microscope capabilities in comparison to the sizes of some common organisms and cells. Note that larger protein molecules are too small to be seen even with the electron microscope.

cell culture. Biologists can study cells under much simpler conditions than those found in an organism. These cells are grown in a fluid called a culture medium (plural, media). Culture media may contain nutrients such as glucose or other sugars, amino acids, lipids, vitamins, salts, and others. As these materials are used up, the cells are placed in fresh media. In this way, cells can be kept alive and studied over long periods of time (Figure 3-13).

3-13. A scanning electron microscope view of nerve tumor cells grown in cell culture.

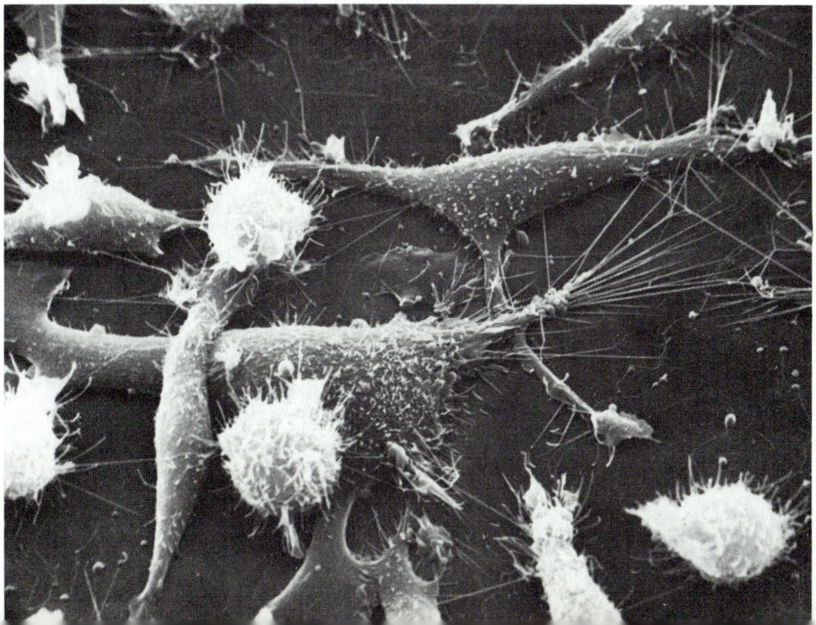

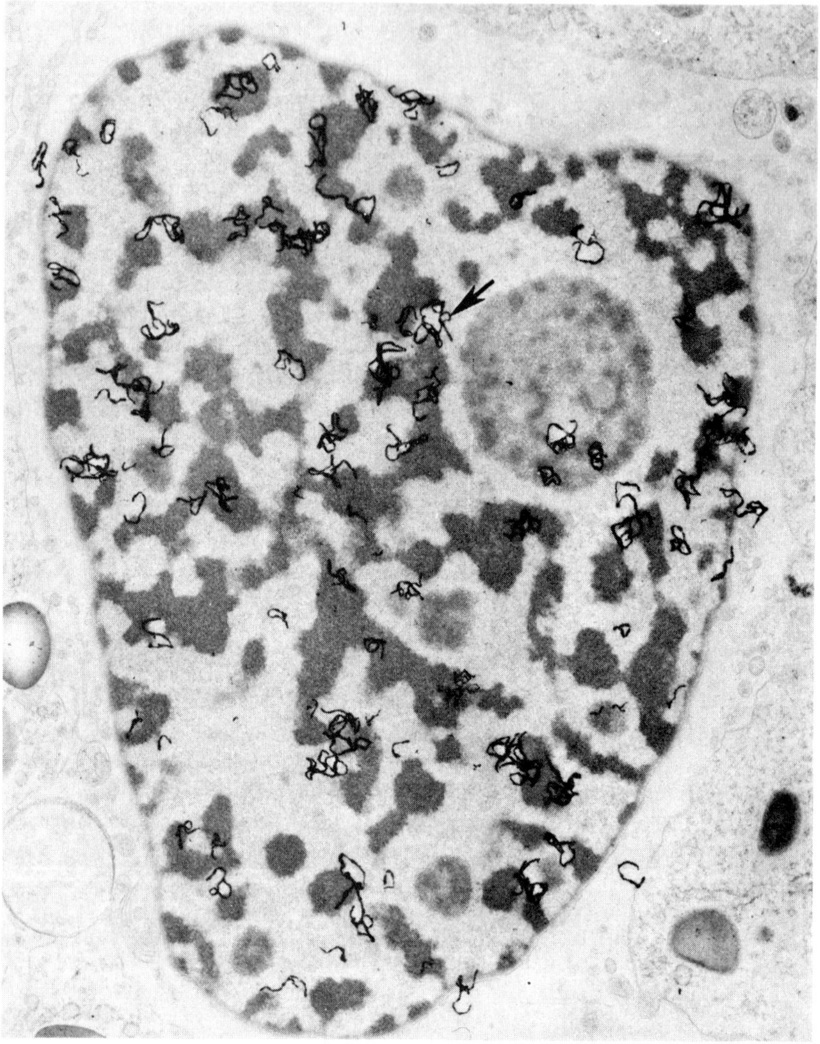

3-14. Labeled DNA in the nucleus of a single-celled organism. The arrow shows one of the DNA strands that was made when an isotope was included in the medium.

Autoradiography

The number of neutrons is not always the same in the nuclei of different atoms of an element. For example, most carbon atoms have six neutrons, six protons, and six electrons. But some carbon atoms have five, seven, or eight neutrons. When the atoms of an element occur in different forms, these forms are called **isotopes.** The atoms have the same numbers of electrons and protons; only the number of neutrons is different.

Some isotopes do not break down easily. They are stable. Others are unstable. Unstable isotopes break down and release particles.

Unstable isotopes are said to be **radioactive.** They are used in autoradiography, an important method for studying cell processes.

In cells, large molecules are made up of smaller molecules. Radioactive isotopes—called **tracers**—are added to a cell culture. The cells in the culture take in the tracer molecules. Some of the cells are removed and fixed at different times. In this way, it is possible to *trace* radioactive molecules as they are used by the cell parts. With this method, scientists have learned much about the structure and function of cell parts and how they work together.

The method used to find the radioactive substance in the cell is quite simple. The cells are placed on microscope slides that are covered with a coating like that on photographic film. The isotope gives off radiation. This radiation affects the coating just as light exposes photographic film. When the film is developed, black spots appear over the parts that contain the tracer isotope. By labeling molecules used to make nucleic acids, for example, we can find what part of the cell is making DNA or RNA (Figure 3-14).

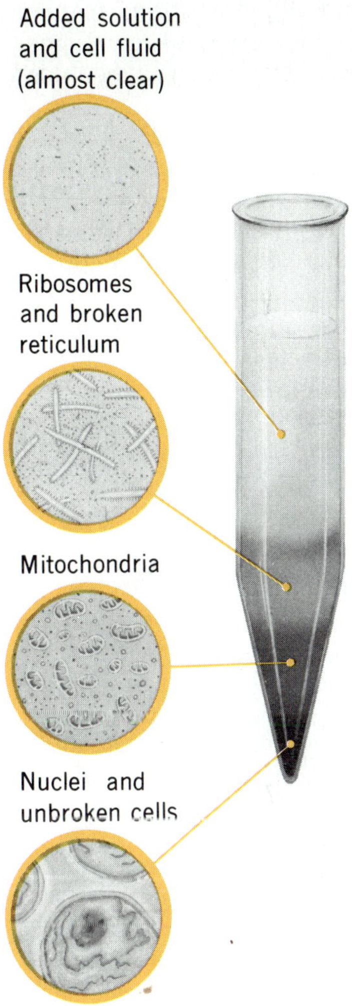

3-15. The separation of cell structures by a centrifuge. Cell fractions are in different layers.

Cell Fractionation

To learn more about parts of cells and their functions, scientists grind up the cells in a liquid. The parts are separated by a centrifuge (sen·tri·fewjuh). A centrifuge is an instrument that separates substances of different **densities.** (Density is a measurement of how tightly the parts of something are packed together in a certain amount of space.) A tube with a sample is placed in the centrifuge. As the centrifuge spins, groups of particles are separated out according to their densities. The heaviest particles fall to the bottom of the tube and the lighter particles form layers in the tube. This process is called fractionation.

For example, we can use fractionation to study the liver cells of a mouse (Figure 3-15). We put the liver in a solution of salts or sugars of known density. We then grind up the liver, so that most of the cells are broken. We put this solution in a tube and spin it in the centrifuge. The parts of the liver cells separate into several layers within the tube. We can examine the material in the different layers with a light microscope or an electron microscope. Or, we can add chemicals that show whether the parts contain proteins, fats, or other kinds of organic compounds.

A LOOK INTO CELLS

Today we know a great deal about the structures inside a cell. Biologists now realize that cells are more than just structural units.

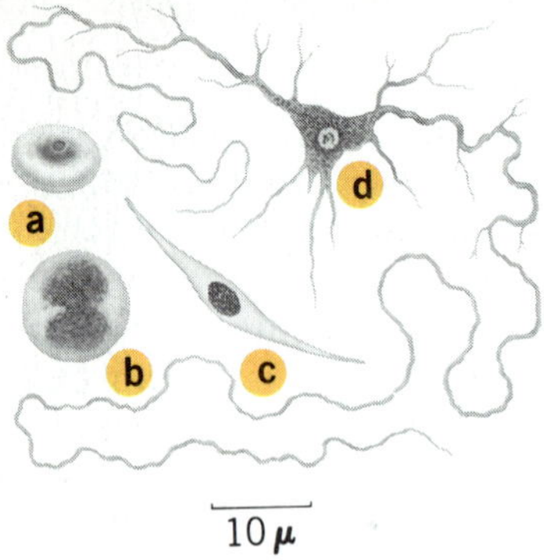

3-16. Cells range in size from 8 to 40 micrometers in diameter. **a,** red blood cell; **b,** white blood cell; **c,** smooth muscle cell; **d,** nerve cell.

They are also units of function. Each kind of cell performs specific tasks. For example, muscle cells contract and nerve cells carry messages from one part of the body to another. Gland cells in the intestine help in the digestion of foods.

Most cells are very small (Figure 3-16). Most human cells are about 8 to 40 micrometers in diameter. But there are exceptions. The width of a nerve cell at its widest point is only a few micrometers. But it may be more than a meter in length! Some single nerve cells, for example, extend from the lower end of your spinal cord to the muscles that move your toes.

Why are most cells so very small? Why don't cells continue to grow? You can begin looking for answers to these questions by doing **Inquiry 3-2.** You also will gain some exprience with an important scientific method—making models.

INQUIRY 3-2 Cell Size—An Exercise in Making Models

When cells reach a certain size, their rate of growth slows down. Then they stop growing. Each of these cells divides into two smaller cells, which begin to grow. What causes this? An easy way to investigate such questions is to build a model. A model is often thought of as a small copy of something large. Here we'll reverse the process and build a large model of something small.

MATERIALS (per team)

block of phenolphthalein agar
100 ml 4% sodium hydroxide
beaker, 250-ml, or culture dish
millimeter ruler
razor blade
plastic spoon
plastic table knife
paper towel

Procedure

With the razor blade, cut the block of phenolphthalein agar into cubes. The first cube should be 3 cm on each side; the second, 2 cm on each side; and the third, 1 cm on each side. Measure carefully and trim away waste. Examine the cubes. Think of them as giant models of tiny cells. **1.** Which of your three different-sized "model cells" do you think would be most likely to survive? Why do you think so?

Materials used during cell activity and growth enter the cell from the outside. Waste products go through the cell surface to the outside. **2.** Do you think the cell with the greatest *total surface* area will do the best job of moving materials in and out of the cell? Why or why not? **3.** Calculate the total surface area of each of your three models by using the following formula: Surface area of cubes = length × width × number of surfaces. **4.** Which cell model has the greatest surface area—the 3-cm cube, the 2-cm cube, or the 1-cm cube? **5.** The "cells" have different surface areas. And they have different volumes—amounts of material inside. Calculate the volume of each cube. Volume of cube = length × width × height. **6.** Think about the prediction you made in question 1. Do these new calculations change any of your earlier ideas? Why or why not?

Place the cubes in the beaker. Pour in enough sodium hydroxide solution to cover them. Record your starting time. Use the plastic spoon to turn the cubes often for the next 10 minutes. Be careful not to cut or scratch the surface of the cubes. Blot them dry. Allow them to dry completely. Slice each cube in half with the plastic knife. **7.** What differences do you observe? **8.** Measure the outer colored zones with the millimeter ruler. What similarities do you notice? **9.** Which "cell" was apparently most efficient in receiving materials from the outside? Which was least efficient? **10.** Recall your calculations of surface area and volume. Does either calculation explain what you have observed in your cell models? Why or why not?

Calculate a *ratio* of surface area to volume for each cube. To do this, for each cube size, divide its surface area by its volume. The number obtained can be expressed as a ratio. For example, a surface area of 24 cm^2 divided by a volume of 3 $cm^3 = 24/3 = 8/1 = 8:1$ (a ratio of "8 to 1"). **11.** Record the ratio of surface area to volume for each of the cubes. **12.** Relate your calculated ratios to what you observed with the cell models. **13.** How do your observations and your calculations relate to the question of why cells are usually very small? **14.** Calculate the ratio of surface area to volume in a cell model that is 0.1 cm^3. **15.** What predictions can you make about the cell model in question 14? **16.** In what other ways could models be used in scientific research?

Within Cells

A living cell, when examined with a light microscope, seems almost transparent. Very little of what is inside the cell can be seen. But other methods can be used to find out more about the structure of cells. Fixed cells can be stained with different colored dyes. The dyes mark parts of the cell that we usually cannot see. In this way, biologists have learned that the many kinds of cells have a great

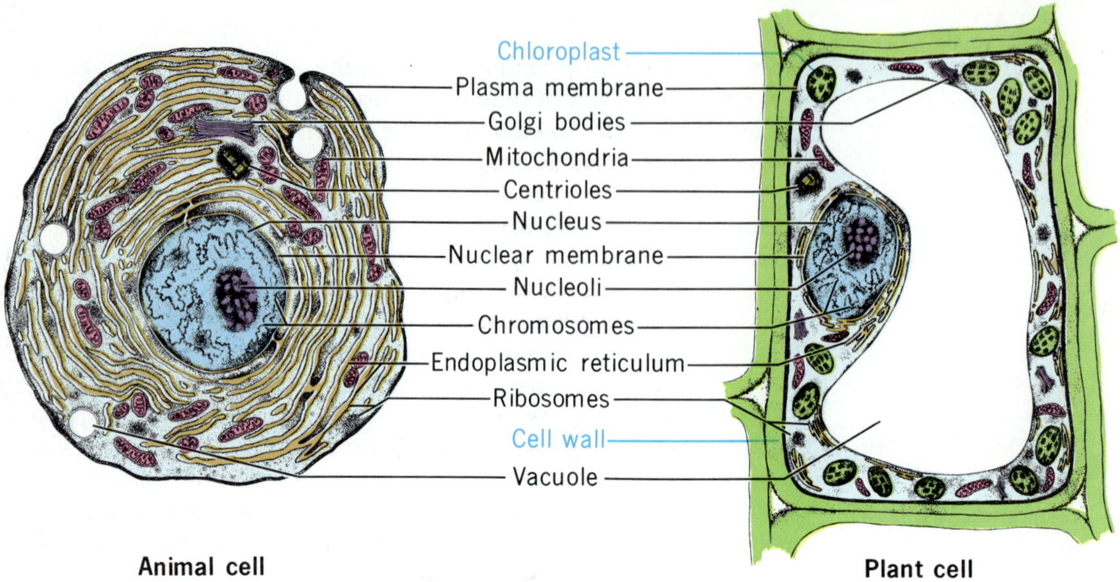

3-17. All cells have similar structures. But differences exist, as shown in this comparison of a plant and an animal cell. Structures found in both are labeled in black. Those found only in plant cells are labeled in blue. No real cells look exactly like these diagrams. *Compare them with what you see using your light microscope.*

deal in common. What are some of these common features? Let us begin with the cells shown in Figure 3-17.

When we examine plant and animal cells, we see certain similarities. But there are also some major differences. Generally, plant cells have thick walls and large **vacuoles** (VAK·yoo·ohlz). These are large liquid-filled structures enclosed in a membrane. Animal cells do not have thick walls, but they frequently have many small vacuoles. Plant cell walls were first observed in cork. Cell walls form a rather hard outer case for the rest of the cell. The vacuoles often are in the middle of the cell. They may be so large that most of the other cell material is pushed next to the cell wall. The microscopic chloroplasts in green plant cells contain the green pigment, **chlorophyll** (KLOR·o·fill). Chloroplasts absorb the light energy that plants use as they make their food. This food is used not only by plants, but by animals that eat plants.

The Cell Membrane

The **cell membrane** is the outer surface of animal cells. In most plant cells, the cell membrane is just inside the cell wall. It was once thought that the membrane did little more than hold the cell together. We now know that the membrane of a living cell allows materials to move into and out of the cell. It also helps control the environment inside the cell.

Cell membranes are so thin it has been difficult to find out much about them. The electron microscope reveals that they have two layers. Chemical tests show that these layers are made up of proteins and lipids. But no one is certain how the lipids and proteins are arranged in the membranes of living cells.

Scientists have used models to understand the structure and function of the membrane. One model proposes that the membrane is a double layer of lipid and protein molecules (Figure 3-18). The structure and composition of the membrane control what enters and leaves the cell. In general, very large molecules cannot pass in or out. But molecules that can be bonded to lipids—such as alcohol—pass through easily even if they are fairly large.

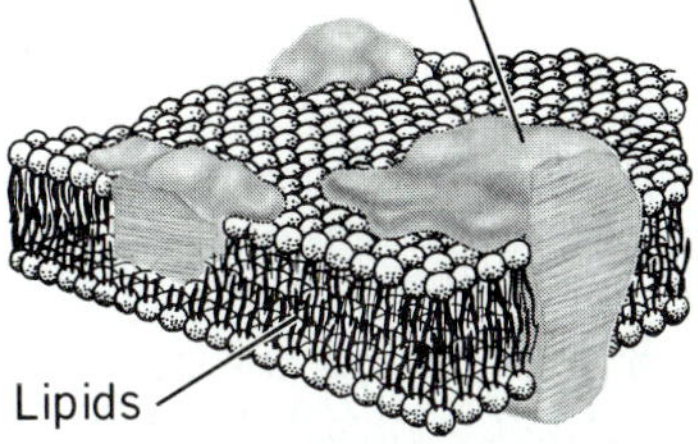

3-18. Fluid-mosaic model of the cell membrane.

Because cell membranes allow only certain molecules to pass through them, we say that they are **differentially permeable** (PUR·me·uh·b'l). Consider, for example, the movement of oxygen and carbon dioxide molecules. Cell membranes are permeable to both of these substances. That is, both oxygen and carbon dioxide can pass into and out of the cell. As the cell uses oxygen, the amount of oxygen inside the cell decreases. Then, a higher concentration of oxygen molecules exists *outside* the cell. Because of this difference in concentration, oxygen molecules tend to move into the cell. Carbon dioxide, on the other hand, is produced within the cell. It is at a higher concentration *inside* the cell than outside it. Therefore, more carbon dioxide leaves the cell than enters it.

This movement of molecules from a region of higher concentration to one of lower concentration is called **diffusion.** Diffusion results in a more even distribution of each kind of molecule. You saw diffusion at work in **Inquiry 3-2.** In **Inquiry 3-3,** you will observe diffusion across a membrane and will study the functions of cells in changing environments.

INQUIRY 3-3 Movement of Molecules

MATERIALS (per team)

Day 1

2 carrots
50 ml 50% sucrose
glycerine
water
2 flasks, 1000-ml, or containers, 20 cm deep
2 graduated pipettes, 1-ml
2 #1 one-hole rubber stoppers
petroleum jelly
marking pencil
cork borer
paring knife or potato peeler

Day 2 (A)

5 ml suspension of yeast cells
Congo red solution
microscope slide
cover glass
2 test tubes
beaker, 400- to 600-ml
2 medicine droppers
test-tube rack
test-tube holder
light microscope
bunsen burner
marking pencils

Day 2 (B)

sprig of elodea
5% sodium chloride
light microscope
microscope slide
cover glass
beaker of water
beaker for sodium chloride
medicine dropper
paper towels or filter paper

Procedure (Day 1)

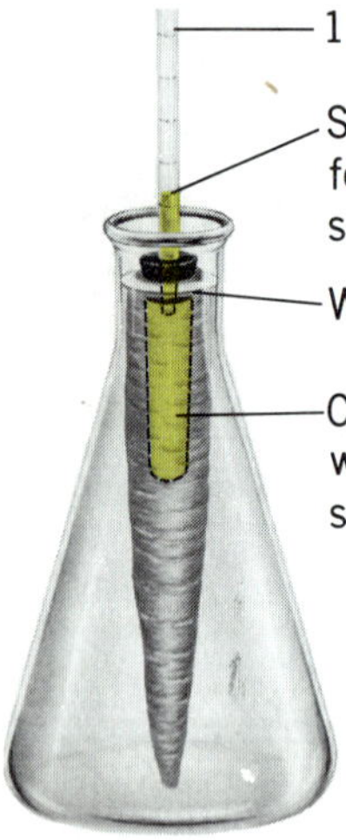

3-19. Experimental set-up for Inquiry 3-3.

The part of the carrot that you eat is the root of the carrot plant. What functions might you expect the carrot root to carry out? **1.**

Arrange two setups as pictured in Figure 3-19. Using the cork borer, cut a hole 5 to 7 cm down the center of each carrot. Do not break through the side of the carrot with the cork borer. If necessary, enlarge the hole with a paring knife or potato peeler. Use the same tool to scrape off a thin layer of the epidermis from the carrots.

Fit a one-hole rubber stopper snugly into the cavity in one carrot. Rub glycerine along the tip of a 1-ml graduated pipette. Carefully slide the pipette into the hole of the rubber stopper until about 3 cm of the tip sticks out from the bottom of the stopper. (CAUTION: *Be very careful not to break the pipette.*) Do the same with the other carrot.

Use the marking pencil to label one carrot *A* and the other *B*. Into Carrot A pour the sugar solution to a point 1 cm from the top. Fit the stopper and pipette into the carrot so that the pipette tip dips into the fluid. (If the stopper slips out, wrap a thin strip of paper towel around it.) When the stopper is in the carrot, you should be able to see the sugar solution in the lower part of the pipette. If the liquid is not high enough to be measured, remove the stopper and add a little more solution. Measure the position of the solution as your starting point for Carrot A. Seal the seam between the stopper and the carrot with petroleum jelly even if you have used paper toweling.

Fill Carrot B with water to a point 1 cm from the top. Fit the second stopper into this carrot. When the stopper is in position you should be able to see the water in the pipette. If not, add more water. Measure and record this starting level. Seal the seam between the carrot and stopper.

Rinse the carrots with water. Fill both flasks to the 1000-ml mark with water. Place one carrot in each flask. Add enough water to bring the level *almost* up to the top of the carrot. Write down the starting time for both carrots and observe the level of water in each pipette every ten minutes during the rest of the period. **2.** What was the purpose of Carrot B in this experiment? **3.** Write a hypothesis to account for the change in the level of sugar solution in the pipette of Carrot A. **4.** What does the change in the level tell you about the movement of water molecules? of sugar molecules?

You scraped off the epidermis of the root to speed up the movement of molecules. A root hair sometimes grows out from an epidermal cell in a root. The root hair increases the surface area through which substances may pass into and out of the cell. **5.** Write a hypothesis about the movement of water into the root through root hairs and epidermal cells. **6.** To kill weeds, some people put salt on the ground near the base of the plant. Suggest a hypothesis to explain the cause of death of the weeds.

Procedure (Day 2)

A. Effects of Temperature on Living Cells. Place 1 ml of yeast suspension into each of two test tubes. Add three drops of Congo red solution to each tube. Label one tube *A* and the other *B*. Place Tube A in a beaker of water. Allow the water to begin to boil, then turn off the burner. Do not heat Tube B.

Use different droppers to prepare wet mounts of both yeast suspensions. Label the slides *A* and *B* according to the solution that is used. Observe under low and high power. **7.** Describe the differences you observe in the two suspensions. **8.** What effect does heating have on the yeast cells?

Like all cells, yeast cells have a cell membrane which allows only certain molecules to pass through. **9.** What do the results of this experiment suggest about the effect of heat on the condition of a membrane and its permeability?

B. Effects of Salt Concentration. The environment inside an elodea cell is closely regulated. It contains a solution of salts and sugars similar to a 0.9 percent sodium chloride (table salt) solution. **10.** What do you expect would happen to the movement of water molecules if an elodea cell were placed in a solution with the same concentration of salts as that inside the cell?

Place a leaf from a growing tip of elodea in a drop of tap water on a clean slide. Add a cover glass. Study the slide under low and high power. Draw what you see. **11.** Design a hypothesis to predict what would happen if you placed a leaf of elodea in a salt solution of a concentration *higher* than 0.9 percent sodium chloride.

You will now have an opportunity to test this hypothesis. Use the wet mount that you prepared. Leave it under the microscope. Place a small bit of paper towel or filter paper at one edge of the cover glass to draw the water off the leaf. Add a drop of salt solution on the opposite side of the cover glass. It will be drawn under the cover glass as the water is drawn off by the paper. What happens in the cells as the salt water moves over the leaf? **12.** Was your hypothesis supported, or do you have to make some changes? Explain.

What will happen to the cells if the leaf is washed and placed in pure water? Remove the leaf and place it in a glass of water. After 5 minutes, study it again with the microscope. **13.** Record any changes that you observe. **14.** What can you conclude about the ability of water molecules to move in and out of the cells? Compare this to the movement of Congo red solution. **15.** Suppose a single-celled, ocean plant was transplanted to a freshwater pond. Predict whether the cell would shrink, swell, or remain the same size. Explain why.

How Substances Get into Cells

You have seen evidence that molecules move across the cell membrane. But there are no apparent openings between the cell and its environment. *How* do water and other substances get into and out of cells?

Osmosis. The diffusion of water across a membrane between solutions of different concentration is called **osmosis.** Your observation of elodea cells in **Inquiry 3-3** showed that osmosis was taking place. The movement of water molecules from areas of higher concentration to areas of lower concentration results in some important biological problems. Consider one-celled animals that live in fresh water.

Fresh water contains dissolved materials, but in very small amounts. We can say, then, that in fresh water the concentration of water molecules is almost 100 percent. The amount of dissolved material is greater in one-celled animals than in the water surrounding them. The concentration of water molecules in the one-celled animals is about 98 percent. Thus, more water molecules will move from fresh water into the animal cell than will leave it. If this went on for very long, the cell would swell and probably burst. But one-celled organisms can get rid of the extra water. As a result, they remain about the same size.

Not all cells have a means of getting rid of water. For example, blood cells, placed in pure water, will swell and burst. In contrast, if cells are surrounded by water containing large amounts of dissolved materials, water may begin to leave the cells. Without water, the activities of cells slow down and the cells may die.

Active Transport. Diffusion is not the only way in which materials can cross the cell membrane. Many cells use energy to move substances. The process in which energy is used to move materials into

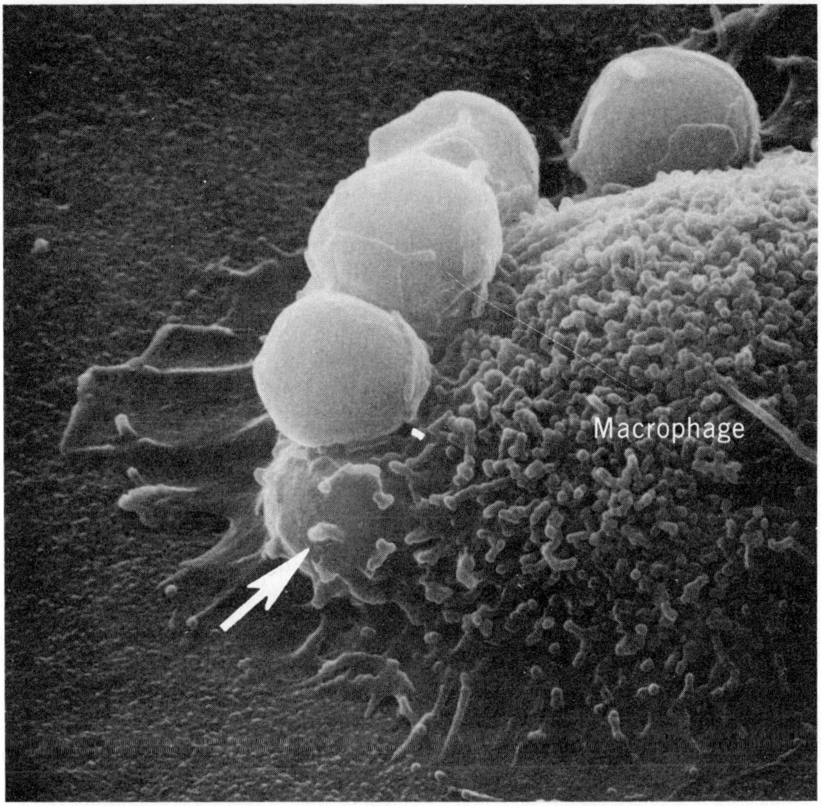

3-20. Activity of the membrane can be very dramatic in some cells. This macrophage (magnified more than 5,000 times) was taken from the blood of a mouse. When red blood cells from a sheep were introduced, the macrophage began to engulf the foreign cells. One red blood cell is already surrounded (arrow) and four others are about to be engulfed. *What advantages might such a system have for the organism?*

or out of a cell is called **active transport.** We do not know exactly how substances are transported in this way. But it seems likely that proteins or enzymes in the membrane are involved. (Diffusion is not "active"—it uses no cellular energy.) Active transport plays a major part in the function of cells and organisms.

The fluid-mosaic model of membrane activity (Figure 3-18) helps explain the transport of large molecules into the cell. Such materials attach to specific places on the outside of the cell membrane and concentrate there. In some cases, they are moved into the cell by an infolding of the membrane (Figure 3-20). But not all active transport of materials into the cell involves infolding.

The places on the outside of the membrane that selectively bond materials are known as **receptor sites** and have a very important biological role. The proteins of the membrane and some carbohydrate molecules that attach to these proteins are thought to make up receptor sites. All cell membranes are similar. But the special kinds of proteins, carbohydrates, and lipids at a surface of a cell make a difference. They determine what that cell will contain and how it will function. These differences are at least part of the reason there are different kinds of cells.

The Nucleus

The nucleus is the most noticeable structure in either a plant or an animal cell. In some cells the nucleus is round. In others, it is irregular in shape. The nucleus of a *living* cell, when examined with a light microscope, may appear to have nothing inside it.

3-21. An electron micrograph of the nuclear membrane. Notice the pores just to the left of center.

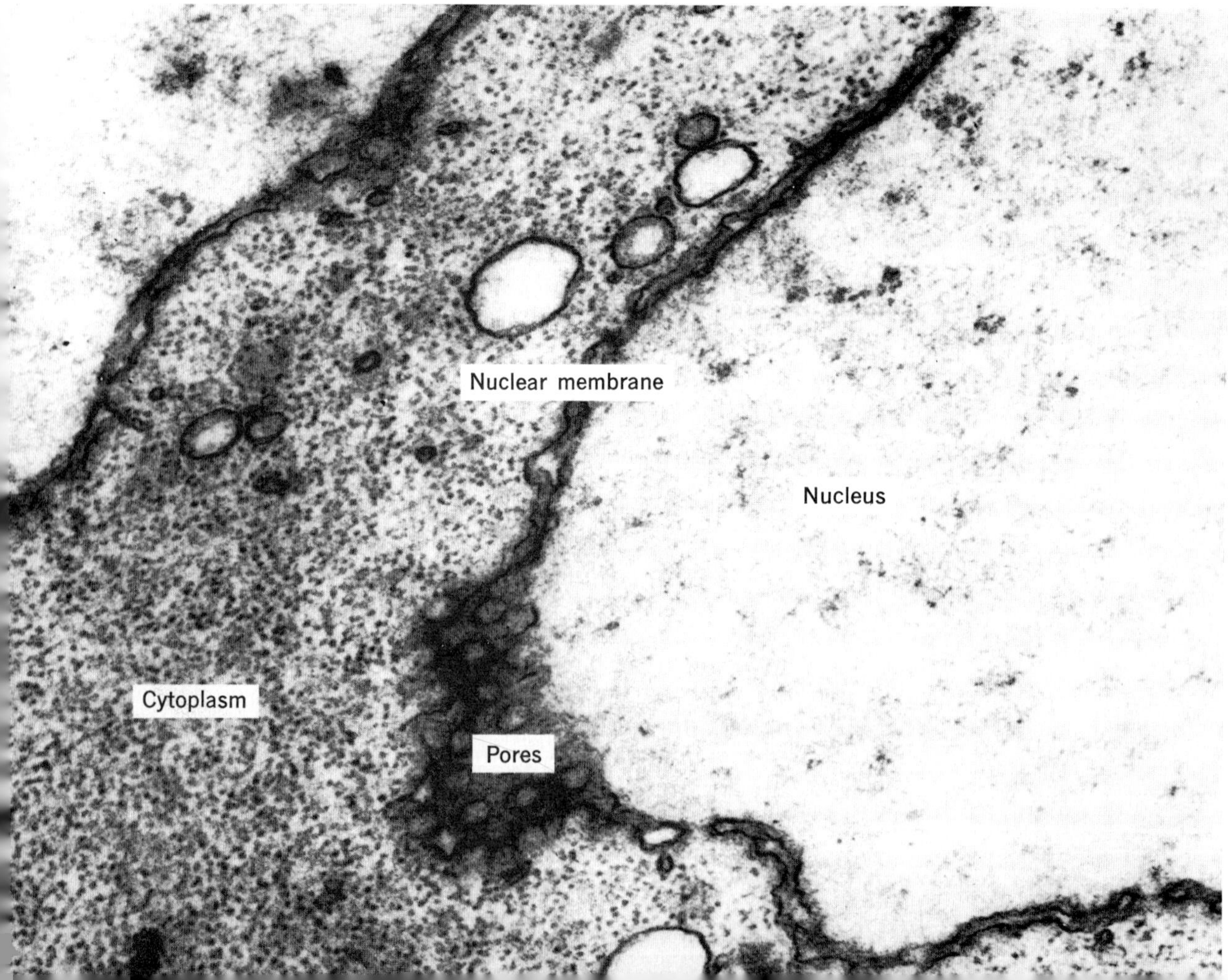

The nucleus is surrounded by a double membrane called the **nuclear membrane.** There are many small openings, called pores, throughout the nuclear membrane (Figure 3-21). These pores allow materials to enter and leave the nucleus. What kinds of materials do you think might pass through these pores?

The nucleus controls the life and activities of the cell. Most of the controls are in the **chromosomes** (Figure 3-17). Chromosomes are made up largely of protein and DNA. Biologists believe that DNA is one of the most important substances in our cells because of the part it plays in inheritance. Inheritance is the transfer of characteristics from one generation to a following generation. Human DNA carries only human characteristics. The DNA in frog cells transfers characteristics that are quite different. But how can DNA and its protein carry out such important functions? They do so by controlling the synthesis of certain types of RNA. RNA molecules leave the nucleus and work with other cell parts. Together, DNA and RNA control the synthesis of all types of proteins made by cells, including enzymes.

The nuclei of most cells in an animal or a plant have the same number of chromosomes. For example, most of the cells of your body contain 46 chromosomes. The cells of a frog have 26. Those of a fruit fly have 8.

Other Cell Structures

All the cell substances outside the nucleus and within the cell membrane make up the **cytoplasm** (SY·toe·plaz'm). The cytoplasm is mostly fluid. It contains a variety of structures. These structures can be identified on the basis of their shape, chemistry, and appearance when stained.

Ribosomes. These small, granular structures are found in all cells. There are two types of **ribosomes.** Some are attached to membranes and some float freely in the cytoplasm. Both types are involved with protein synthesis, and both are necessary for the normal functioning of the cell. Free-floating ribosomes are thought to manufacture proteins that are used within the cell. The ribosomes that are attached to membranes probably synthesize proteins that leave the cell.

Proteins are made in the cell from amino acids. But the process is more complicated than just hooking amino acids together. Every cell contains many kinds of proteins. Every protein has amino acids chemically bonded to one another in definite ways. The arrangement of amino acids takes place on ribosomes. No other cell part is known to make proteins. Proteins for growth and enzymes are made continuously in cells.

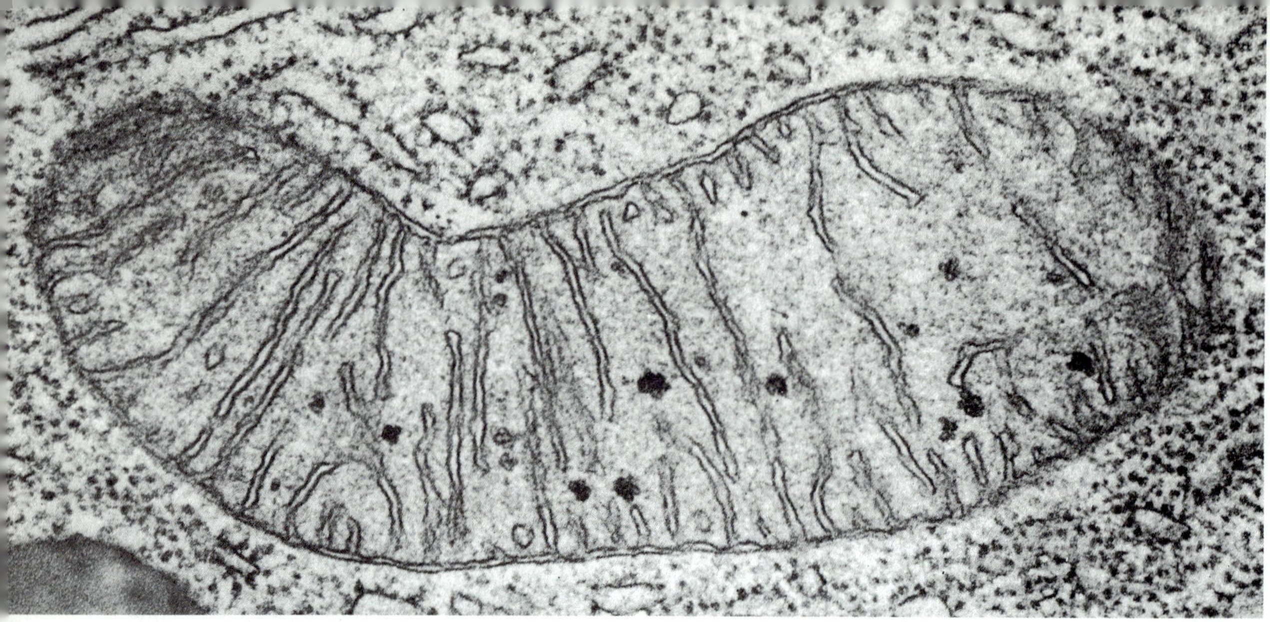

3-22. An electron micrograph of a mitochondrion in an animal cell (77,500 X). The mitochondrion has a double membrane. The inner layer of the membrane turns inward and stretches almost completely across the mitochondrion. The membranes contain enzymes involved in energy-transferring reactions.

Nucleoli. When a cell has been fixed and stained, one or two small round bodies may be seen within the nuclei. These are the **nucleoli** (new·KLEE·o·lye; singular, nucleolus—new·KLEE·o·lus). The nucleoli make ribosome parts from RNA and protein. The nucleoli can be thought of as "ribosome factories."

Mitochondria. These are common structures in the cytoplasm. **Mitochondria** (my·toe·KON·dree·uh) have different shapes, depending on the cells in which they are found. The function of mitochondria has been discovered only recently. They are important in providing energy for the cell.

An electron microscope shows the complex folding of the membranes inside the mitochondria (Figure 3-22). The membranes are mostly protein and lipid molecules and contain many enzymes. Biologists thought the reactions that produce energy occurred in the mitochondria even before they had electron microscope pictures of these structures.

Mitochondria are not the only places where these reactions occur. Enzymes in other cell structures also may carry out reactions that provide energy. But mitochondria are the places where most of the energy becomes available to the cell.

Many of the reactions that take place in mitochondria release the energy that is used in the cell's activities. Biologists refer to the main kind of energy-releasing process as **cellular respiration.** One such reaction is the breakdown of glucose. We can summarize this breakdown as follows:

$$\underset{\text{glucose}}{C_6H_{12}O_6} + \underset{\text{oxygen}}{6\ O_2} + \text{enzymes} \rightarrow \underset{\text{carbon dioxide}}{6\ CO_2} + \underset{\text{water}}{6\ H_2O} + \text{energy} + \text{enzymes}$$

(This equation represents only the raw materials and products of the reaction. It does not account for the many chemical reactions that occur. At least 22 compounds are involved in this process.)

You have read about the molecules that take part in this reaction. They are glucose, oxygen, carbon dioxide, and water. But exactly how does energy fit into the picture? Much of the stored chemical energy of glucose is used in building compounds that break apart easily.

ATP. The buildup of unstable, energy-transfer compounds is one of the most important things that happens in mitochondria. **Energy transfer** is the movement of energy from one kind of molecule to another. This occurs as some compounds break down and others are formed. The best known and perhaps the most important of these compounds is **adenosine** (ah·DEN·oh·seen) **triphosphate, ATP.** ATP is a nucleotide. It occurs in every cell of every living organism. Energy is stored in ATP. When glucose breaks down, energy is released. This energy is captured in the formation of ATP.

How does this happen? Figure 3-23 shows the structural formula of three compounds. These consist of a molecule of adenosine plus one (mono-) phosphate **(AMP),** or two (di-) phosphates **(ADP),** or three (tri-) phosphates **(ATP).** The phosphates are held to the molecule of adenosine by unstable bonds. Each bond is a potential source of energy. AMP is the simplest compound of the three, but ADP and ATP occur more frequently in cells. When ATP breaks

3-23. The structural relation of ATP, ADP, and AMP. The basic nucleotide is the same in all three. The difference is one or more phosphate groups. (A single phosphate group is shown in yellow.) Each of the last two phosphate groups shown at the left forms an unstable bond (red line) with the rest of the molecule. Breaking the bond transfers energy.

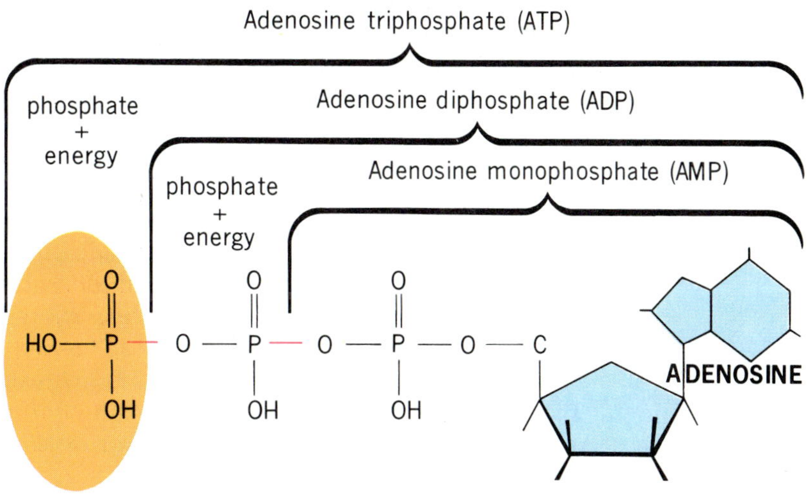

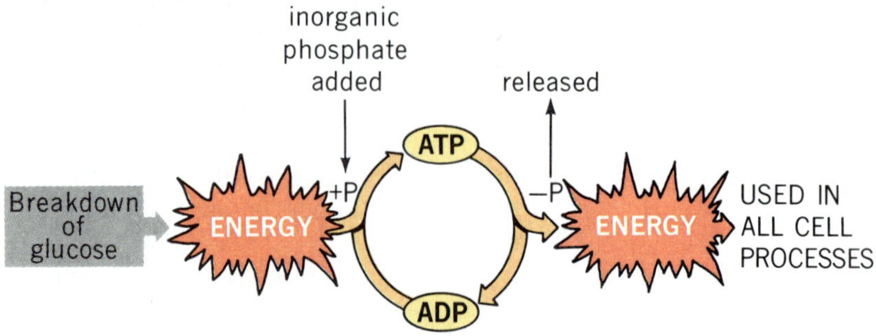

3-24. Energy is stored or released in the formation of ATP and ADP.

down to ADP, it loses a phosphate and energy is released. This energy can be used by the cell. When ADP gains a phosphate, it gains energy. Energy is stored in the ATP that is formed. ATP may then break down later, in other parts of the cell, forming ADP and again releasing energy. The ADP may reenter the cycle and again become ATP (Figure 3-24).

Glucose is not the only source of energy for living cells. Proteins, fats, and many other organic compounds also are energy sources. Like glucose, these compounds react in the presence of enzymes. Many of these enzymes are located in mitochondria. The breakdown of different compounds proceeds to a certain point. At this point, the molecules that are formed are the same as (or similar to) those produced in the breakdown of glucose and other carbohydrates. The breakdown of these molecules is similar to that of glucose. Thus, different kinds of organic compounds are used in similar ways in cells. The sum total of a very large number of continuing chemical reactions in cells is called **metabolism.**

Supplying energy to cells is not the only function of mitochondria. They can *build* organic compounds as well as break them down. For instance, enzymes in mitochondria can make fats. Building up fat compounds takes energy. The energy is supplied when ATP molecules break down.

When chemical energy has been stored in ATP, how does it get from mitochondria to the other places where it is used in the cell? Biologists are trying to answer this question. They already know that mitochondria can move from one place to another within a cell. But they do not know *how* the mitochondria move to areas of the cell that need energy. Nor do they know how the ATP molecules get from mitochondria to other cell structures.

Chemical reactions in your body take place at different rates. How is the *rate* of the activity of mitochondria regulated? When you are running or working hard, you need more energy. At such times, energy is released in your cells at a faster rate than when you are resting. But what causes these chemical reactions to speed up? We do not know the answer, but we have some clues. For instance, it seems that the products that are formed when ATP is broken down to ADP actually control the function of enzymes necessary for ATP formation. As more ATP is broken down to ADP, releasing energy, more of these products are formed. As a result, even more ATP is formed and is broken down, releasing even more energy. Problems of the control of processes in living organisms are an exciting area in biological research. There are many important questions to be answered.

Endoplasmic reticulum. The **endoplasmic reticulum** (en·doh·PLAZ·mik rih·TIK·yoo·lum) is a network of membranes within the cell. Biologists do not know the details of its function. However, they think it serves as a transport system that moves materials around in the cell. Some enzyme activities also occur in the endoplasmic reticulum.

Many tubelike structures that extend throughout the cytoplasm are part of the endoplasmic reticulum. The membranes of the endoplasmic reticulum are similar to the cell membrane. Some of these membranes have ribosomes on their outer surfaces. Such membranes are called "rough" endoplasmic reticulum, because they look bumpy in electron micrographs (Figure 3-25). The ribo-

3-25. An electron micrograph of a small, central portion of an animal cell. The existence of the endoplasmic reticulum and the detailed structure of mitochondria were unknown until the electron micrograph was developed.

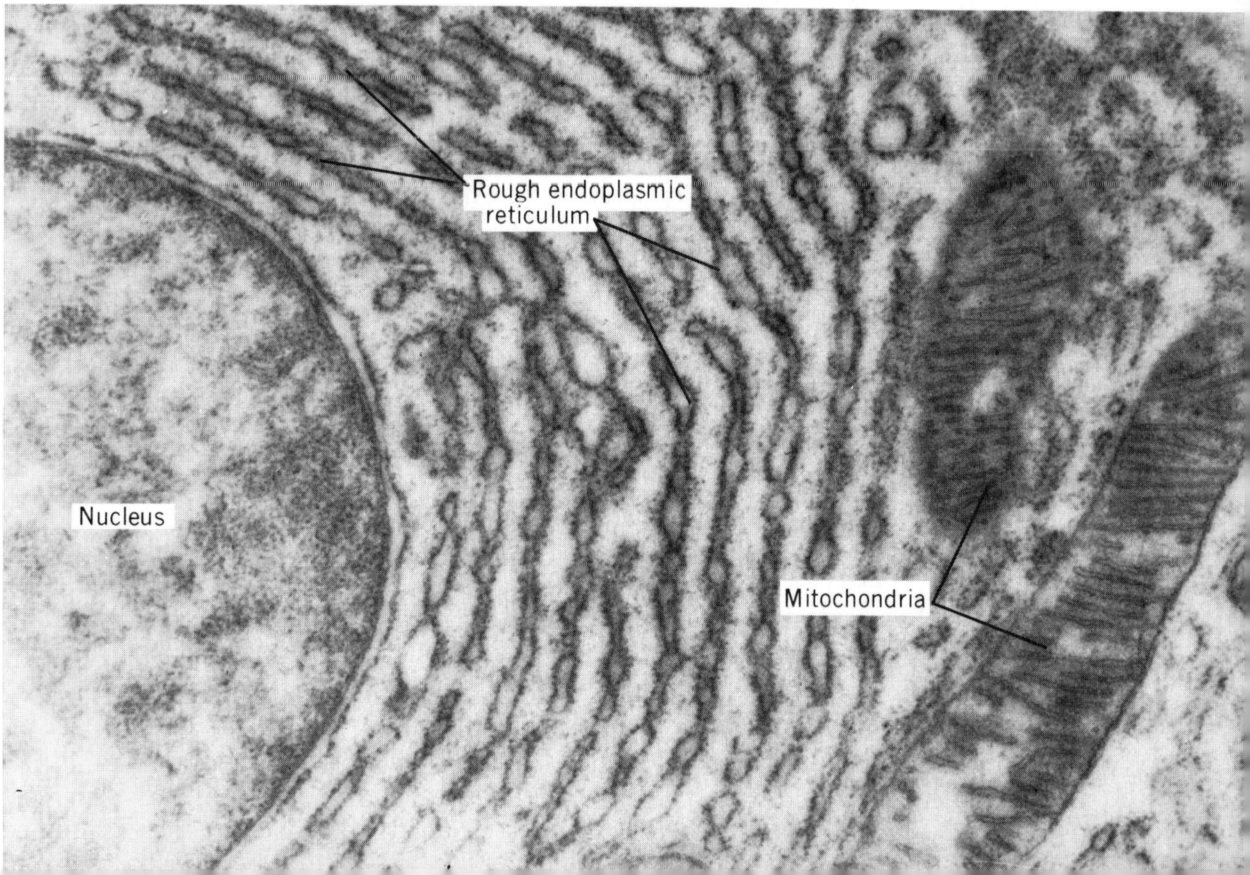

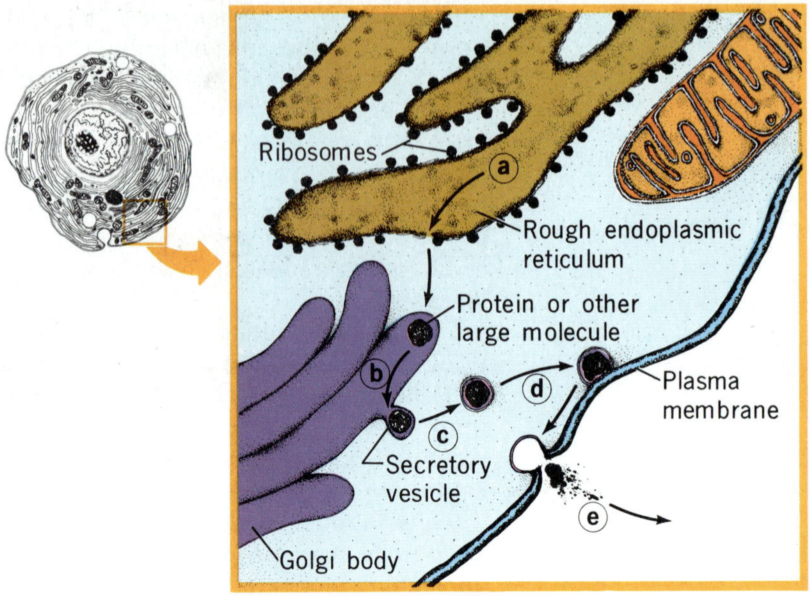

3-26. Golgi bodies are found in secretory cells. They seem to be involved in packaging the materials that are secreted. The steps are: **(a)** Molecules move through the endoplasmic reticulum to the Golgi body; **(b)** they are packaged in a secretory vesicle; **(c)** the secretory vesicle pinches off from the Golgi body and moves out into the cell; **(d)** the vesicle membrane fuses with the plasma membrane; **(e)** the molecules are released to the outside of the cell.

somes of the rough endoplasmic reticulum make proteins. These proteins are carried to various parts of the cell through channels in the membranes.

Membranes that do not have ribosomes are part of the "smooth" endoplasmic reticulum (not shown). The smooth endoplasmic reticulum makes and sends out lipids.

Golgi bodies. Both types of endoplasmic reticulum move their products to the **Golgi** (GOL·jee) **bodies** of the cell. The Golgi bodies are flattened sacs in the cytoplasm. These sacs form cell **secretions** (sih·KREE·shunz). Secretions are products formed within the cell and then passed out of it. Golgi bodies are thought to "package" materials that are moved along by the endoplasmic reticulum. Golgi bodies also make some complex sugars. Proteins and lipids from the two types of endoplasmic reticulum, along with sugars, are collected in tiny "bags" that form at the ends of Golgi bodies. These are called **secretory vesicles.** The vesicle moves through the cytoplasm and becomes attached to the cell membrane. The contents of the package are released to the outside of the cell (Figure 3-26). The molecules released in this way are carried to other parts of the body.

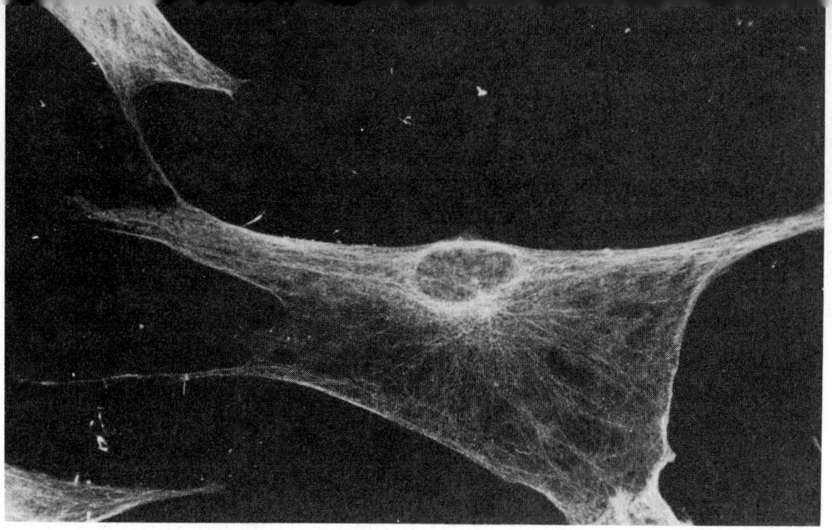

3-27. Microtubules seem to serve as the "skeleton" of the cell. The filaments shown here are responsible for the shape of the cell and perhaps for the movement of some materials through the cell.

The contents of one kind of secretory vesicle remains inside the cell. These are called **lysosomes** (LIE·suh·sohmz). Lysosomes contain digestive enzymes. These enzymes are produced by the rough endoplasmic reticulum and packaged by the Golgi bodies. The enzymes in lysosomes break down large molecules brought into the cell by the infolding of the membrane. Lysosomes also break down worn-out cell parts. The products from this process are used in making new cell parts. No one knows why the membrane of the lysosome is not digested by the enzymes it contains.

Microtubules. These long thin cylinders are found in the cytoplasm of many cells. They are so tiny they can be seen only under an electron microscope. The walls of the tubules are made up of proteins (Figure 3-27). **Microtubules** break down and reassemble rapidly in response to changes in temperature, pressure, and chemical environment. They seem to play an important role in maintaining or changing cell shape and in transporting cell materials. They are parts of special structures that are active in cell division and movement.

SPECIALIZED CELL FUNCTION: SECRETION

Individual cell structures do not work independently; rather cell processes are coordinated in a highly organized manner. The cells of complex animals do more than carry out functions to keep individual cells alive. They perform vital functions in the organism of which they are a part. For example, bone cells build up the hard substance of bone. Red blood cells carry oxygen from one part of the body to another. Some cells of the digestive system secrete enzymes that digest food.

One of the main functions of some cells is to make and secrete products that are used elsewhere in the organism. We will consider secretion as one example of how the parts of a living cell work together.

Assume that the secretion of the cells we are taking as an example is a protein. From chemical studies, biologists know that proteins are made of amino acids. And these amino acids can be labeled with a radioactive tracer such as tritium (an isotope of hydrogen). These labeled molecules will be picked up by cells in culture. The cells are then removed from the culture at different times.

Using an electron microscope, biologists can find the radioactive material in the cell. They usually find the radioactive material attached to amino acids at the ribosomes of the rough endoplasmic reticulum. This suggests that the amino acids are being used to make proteins. The newly formed protein is passed from the ribosomes to the inside of the endoplasmic reticulum.

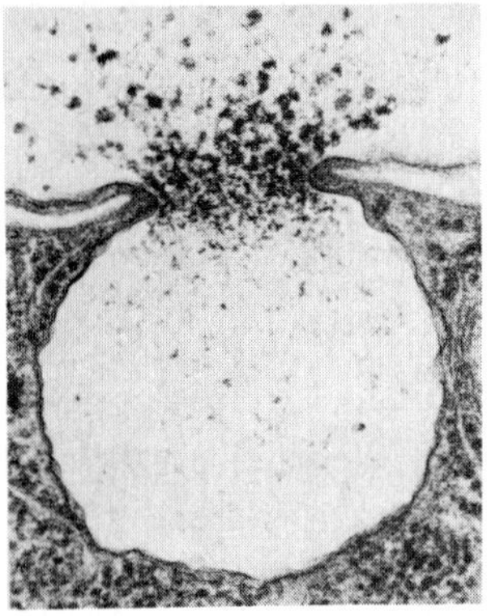

3-28. Discharge of materials to the outside of the cell is evident in this micrograph. Note that the membranes of the cell and the vesicle have fused.

From the endoplasmic reticulum, the protein moves into secretory vesicles on the Golgi bodies. The membrane of the vesicle fuses with the cell membrane and the product is discharged to the outside of the cell (Figure 3-28).

Now, let us see how the cell that will use the proteins handles them. Large protein molecules cannot diffuse into the cell as the amino acids did. They must be brought in by active transport or by infolding of the membrane. In the infolding, the protein molecules concentrate at receptor sites on the cell membrane. Part of the cell membrane then sinks inward and forms a vesicle around the protein molecules. A lysosome attaches to the vesicle and breaks down the protein molecules.

In Brief

The idea that the cell might be the basic unit of structure in living organisms was first advanced over 300 years ago. We now know a great deal about what living cells look like and what they do.

Our tools for studying cells have improved from simple microscopes to electron microscopes. And methods have changed from simple observation with a light microscope to such advanced methods as cell fractionation and autoradiography.

We have gained much knowledge of cellular structure and function from over 500 years of scientific work. But we still have many questions to answer. We now know, for example, quite a bit about what happens in mitochondria. But we do not know what controls the rate of energy-release in mitochondria. Nor do we know how energy, released from the ATP molecule, is transferred from mitochondria to other cell parts. Research continues into these and many other questions about cell structure and function. In fact, cell research is one of the most active areas of biological inquiry today.

SCIENCE AT WORK: The Action of the Cell Membrane

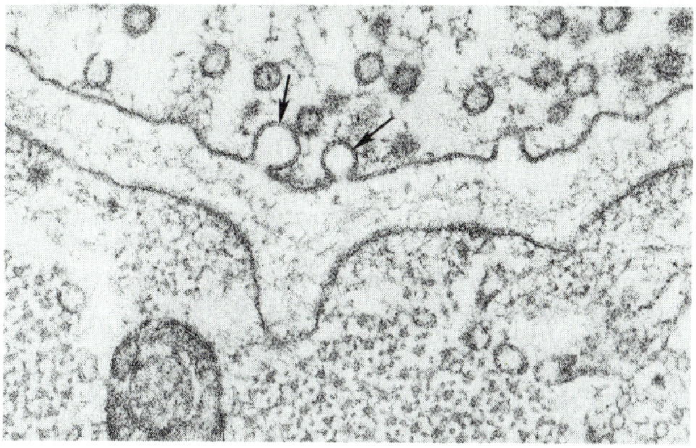

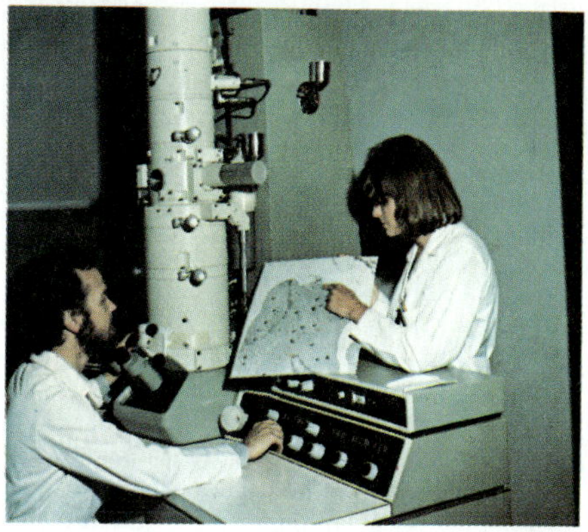

When you touch your finger to a hot stove, your hand immediately jerks back, even before you feel the pain of the burn. Why? The answer is not simple. This reaction involves many interactions among your nerves, muscles, spinal cord, and your brain. It also involves the nerve cell membranes.

Since the 1930s, we have known nerve cells function by releasing tiny amounts of chemicals into spaces between the nerve endings. Not until 1976, however, could scientists see *how* the chemicals are released.

The process seems to be similar to what we think happens in a secretory cell. Research in the 1950s suggested that the chemicals are in tiny "packages" surrounded by a membrane inside the cell. At that time, scientists thought these membranes fused with the cell membrane in response to an electrical signal. This fusion would release the chemicals. The idea seemed like a good explanation, because the "packages" could be seen inside a cell. But no one had ever seen the membrane of the "package" fusing with the cell membrane and releasing chemicals.

Scientists at the University of California Medical Center in San Francisco tackled this problem. First, they had to find a way to stop the action of a cell at the exact time the chemicals were released. Standard chemical methods of preserving specimens take much too long—many seconds or minutes. The fusion of the membranes and the release of chemicals occur much faster—they take less than one thousandth of a second! How could the action be stopped?

The scientists built a rapid-freeze machine. "We mount the tissue at the end of a plunger and let the plunger drop through the air," one researcher explained. "We stimulate it just before it hits a cold metal plate. Smacking tissue against cold metal is the fastest way to draw out heat. It takes us less than one millisecond (one thousandth of a second) to freeze even the deepest parts of the tissue."

Once the material was fast frozen, the researchers could take electron microscope pictures of the activities of the cell membrane at the moment of stimulation. The photograph shows the membranes of the "packages" fusing with the cell membrane (see arrows). The contents are then released into the space between nerve cells. The hypothesis of nerve cell action was confirmed.

The methods these scientists developed can be applied to other kinds of research, such as how muscles or sense organs like the eyes actually work.

Guide Questions and Problems

1. What is the difference between a hypothesis and a theory? The cell *hypothesis* of Schwann and Schleiden is now considered the cell *theory.* What happened since their time that caused the change?
2. In what ways does a theory differ from a fact?
3. What are the advantages of fixing and staining cells for microscopic examination? What are the disadvantages?
4. In what ways are plant and animal cells alike? How are they different? What is the significance of the differences?
5. Normally your red blood cells are in the shape of a flat disc, almost hollowed out in the center. In a 2 percent sodium chloride solution, these cells shrink and change shape. Why? The same cells placed in only a 0.1 percent sodium chloride solution will swell, become round, and finally burst. Why?
6. What is the importance of ATP?
7. Structure often helps us understand function, and vice versa. How does the microscopic structure of the endoplasmic reticulum help us understand its function?
8. What is thought to be the relationship between ribosomes, endoplasmic reticulum, Golgi bodies, and the cell membrane in a secretory cell?

Ideas for Independent Study

Scientists sometimes make a model of a particular cell part. By doing so, they hope to understand cell function. They study the behavior of the model, which is always much simpler than that of the real cell part. **Inquiry 3-2** was based on an agar-block model of the cell. Here is an idea for a model of the cell membrane.

The basic design is a glass funnel set in a flask. Fit two pieces of filter paper in the funnel. Pour a measured amount of water through the filter paper. The time needed for the water to run through is a measure of the permeability of the "membrane." Test the effect of a thin layer of organic material between the filter-paper sheets on the permeability of the "membrane." Because real cell membranes are made up of fats and proteins, test both of these in your model. You can place a thin, continuous layer of margarine between the two layers of filter paper. Then compare this setup with the same experiment using a concentrated gelatin solution. Part of your analysis should be a careful comparison between the model and known features of real cell membranes.

The model will show some aspects of differential permeability. You can test whether your various membrane models are permeable to dyes. (You can use your imagination here. Food dyes might be tested. Dyes of various sorts are found in the laboratory. For instance, we had success in our experiments with methylene blue and sudan IV.) How is the membrane model affected by fat solvents?

Related Readings

BOOKS

Chappell, B. and S. Rees. *Mitochondria.* Oxford Biology Reader. London: Oxford University Press, 1972.

Dyson, R. D. *Essentials of Cell Biology.* Boston: Allyn and Bacon, 1975.

Jordon, E. G. *The Nucleolus.* Oxford Biology Reader. London: Oxford University Press, 1971.

McElroy, W. D. and C. P. Swanson. *Modern Cell Biology,* 2nd ed. Englewood Cliffs, N.J.: Prentice-Hall, 1975.

Pfeiffer, J. *The Cell.* New York: Time-Life, 1964.

Swanson, C. P. *The Cell.* Englewood Cliffs, N.J.: Prentice-Hall, 1969.

Thomas, L. *The Lives of a Cell: Notes of a Biology Watcher.* New York: Viking Press, 1974.

The Living Cell: Readings from Scientific American. San Francisco: Freeman, 1965.

ARTICLES

Fox, C. F. "The Structure of Cell Membranes." *Scientific American,* February 1972, page 30.

Galston, A. W. "Bios: The Membrane Barrier." *Natural History,* August-September 1974, page 86.

Hopson, J. L. "Bacterial Blitz: Storming the Achilles Pore." *Science News,* May 1, 1976, page 283.

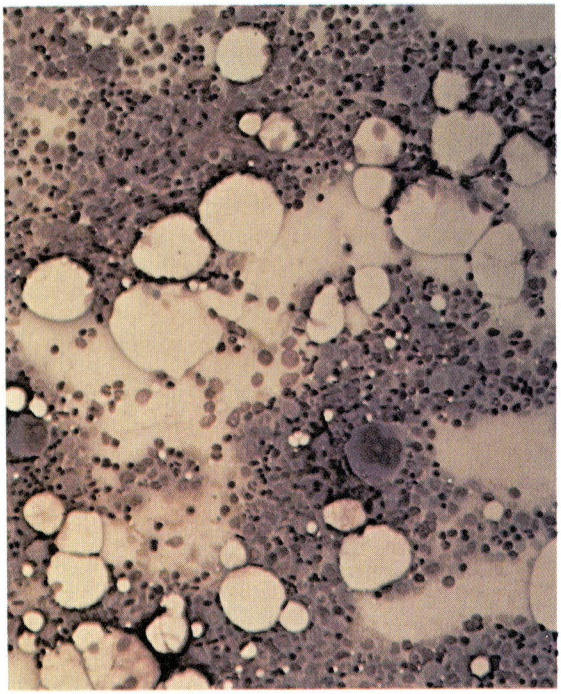

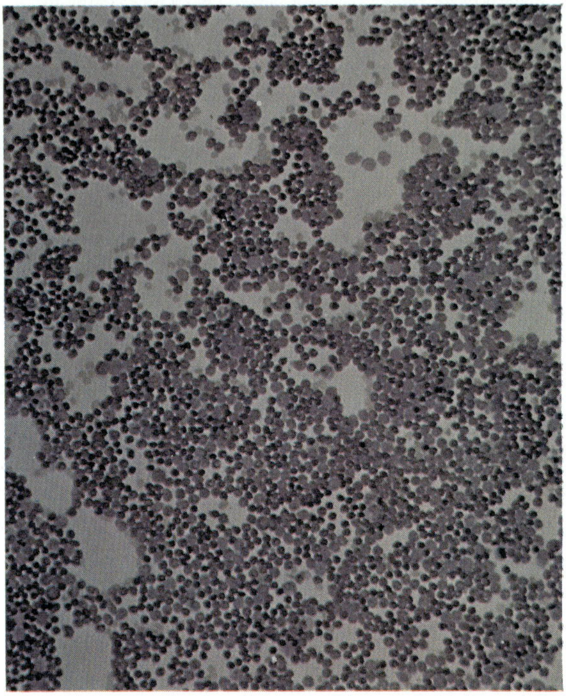

4 Cell Reproduction

You began life as a single cell – a fertilized egg. All the cells of your body are descended from it. That fertilized egg came from the union of two cells: an egg cell from your mother and a sperm cell from your father. The fertilized egg divided into two cells, the two into four, and the four into eight. Eventually, this process of **cell division** produced the billions of cells that make up your body. It may seem a simple process to have a single cell divide. But, cells don't just split into two. Each cell must receive all the parts it needs to function, such as a nucleus, mitochondria, and ribosomes. In some parts of your body, cell division is still occurring. In other parts, cells stopped dividing when you were born.

Many biologists have studied, and are studying, the processes of cell division. We now know a great deal about the events that occur when cells divide.

MITOSIS

When a cell divides, one of the most obvious events is the division of its nucleus. The division of the nucleus is called **mitosis** (my·TOE·sis). The first person to give an accurate and detailed account of mitosis was a German biologist, Walther Flemming. In the 1880s, Flemming could use improved tools and methods to observe the process of cell division. Light microscopes were well developed, and methods for fixing and staining cells were good.

Flemming examined many kinds of animal and plant cells. He selected those that showed the details of cell division most clearly. Cell division is easier to observe in some cells than it is in others. Flemming was interested in events in *living* cells. However, the structures in fixed and stained cells were easier to see. Flemming made it a rule never to believe what he saw in fixed and stained cells until he had observed the same thing in living cells.

Flemming noted that a dividing nucleus passes through an orderly series of changes (Figure 4-1). In all the cells he studied, both

4-1. Walther Flemming described mitosis and cell division almost a century ago. His drawings of mitosis in stained cells of a lily are reproduced.

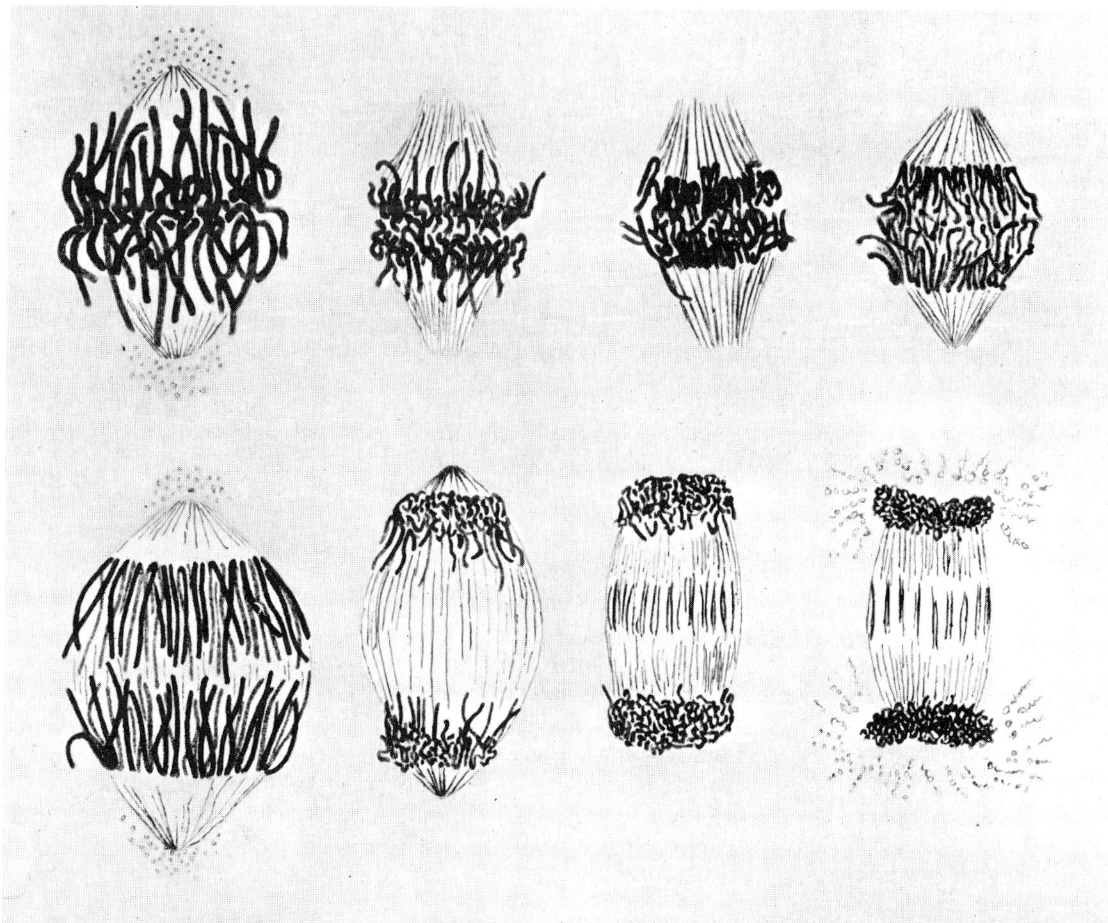

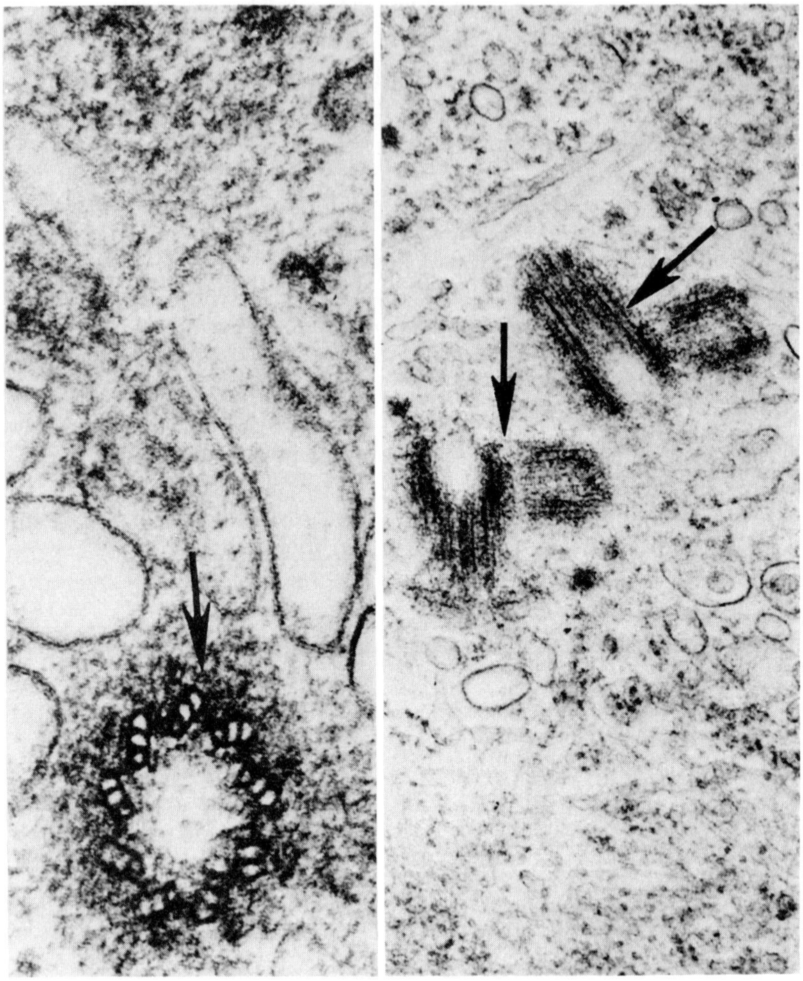

4-2. This photomicrograph shows centrioles of mouse cells grown in culture. **(Left)** A cross-section of a centriole before cell division has begun. **(Right)** Two pairs of centrioles.

animal and plant, the events of mitosis were similar. In mitosis, as in other activities of living organisms, there seems to be a unity of life.

How Cells Undergo Mitosis

Just before an animal cell divides, a nucleus with one or more nucleoli can be seen. Usually no chromosomes can be seen within the nucleus. Just outside the nucleus is the **centriole.** The centriole is found in a darkened area called the **centrosome** (SEN·tro·zome).

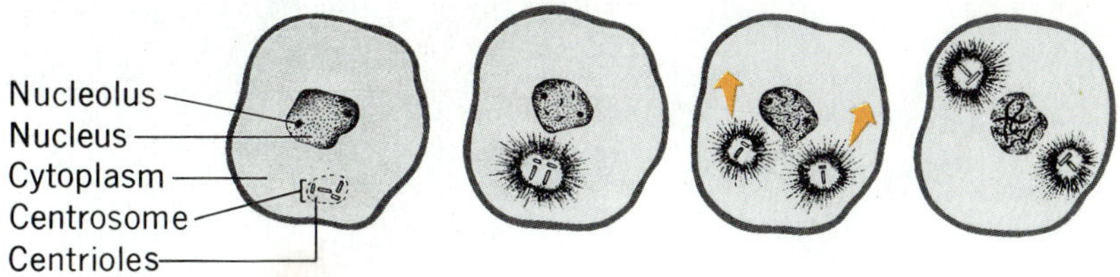

4-3. As cell division begins, the centrioles move to opposite sides of the cell. The nucleolus and nuclear membrane disappear.

To biologists using a light microscope, the centriole appears to be a tiny granule.

As seen with the electron microscope, the structure of the centriole is actually quite complex (Figure 4-2). The first indication that a cell is about to divide is the division of the centriole. When cell division begins, the two centrioles, each with its surrounding centrosome, move to opposite sides of the nucleus (Figure 4-3). Any nucleoli within the nucleus gradually disappear.

Figure 4-4a shows a cell at the beginning of mitosis. The nucleus appears as a ball containing darkly stained threads. The two darker spots, to the right and left of the nucleus, are the centrosomes. (The centrioles in the centrosomes do not show in the photographs.)

The most noticeable events in mitosis involve the chromosomes. Chromosomes become visible as mitosis begins. In Figure 4-4b, the chromosomes look like thin, dark rods. (They are dark because they have been stained.) The membrane around the nucleus has disappeared. We can see fibers that reach from each centrosome, forming what is known as the **spindle.** The chromosomes are now grouped in the center of the football-shaped spindle.

In Figure 4-4c, the chromosomes seem to be separating into two equal groups. One group is moving along the spindle toward one centriole; the other group is moving toward the other centriole.

Looking closely at the part of the cell around each centriole, we can see other fibers. These are easiest to see in Figure 4-4c. This

4-4. This set of photographs shows mitosis in cells of a whitefish embryo. All of these photographs were taken through a light microscope. (Continued on page 112.)

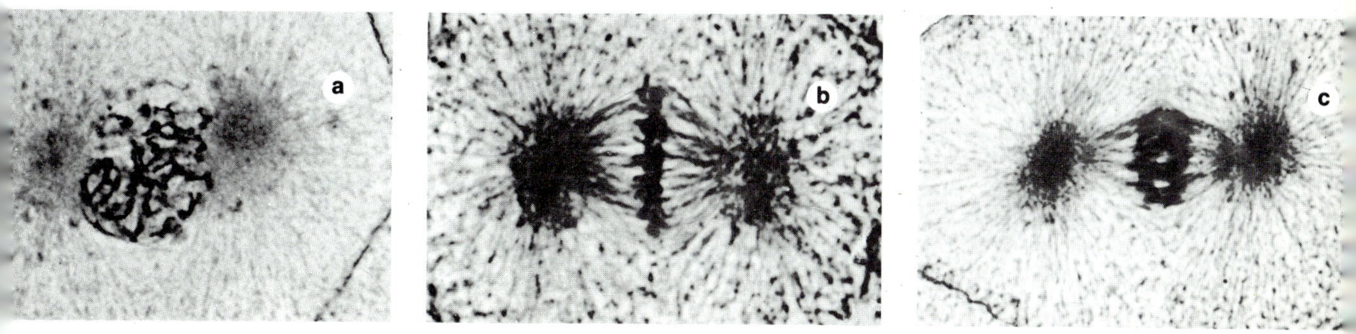

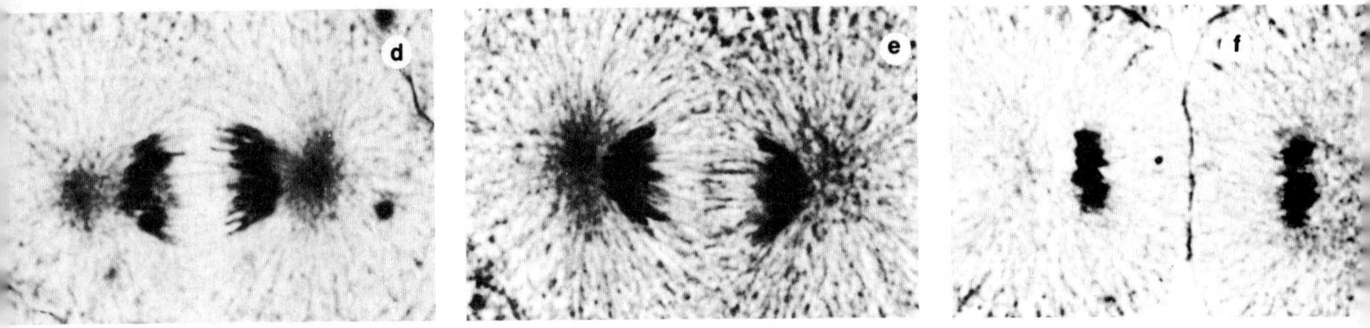

4-4. (continued)

group of fibers around each centriole forms the **aster.** The asters actually can be seen in a stage earlier than the one shown in this photograph. Look for them in Figures 4-4a and b. The centrioles, the spindle, and the asters, together, make up the **mitotic** (my·TOT·ik) **apparatus.**

The mitotic apparatus is a separate structure in the cell. If the cytoplasm around it is dissolved away, the mitotic apparatus remains intact. Little is known about how the mitotic apparatus functions. But it certainly is involved with the movement of chromosomes.

In Figures 4-4d and e, the two groups of chromosomes have moved farther apart. Figure 4-4f shows what happens near the end of mitosis. The first cell has divided into two. Each new cell has a full set of chromosomes. The asters and the spindle are no longer easy to see.

At a stage somewhat later than that shown in Figure 4-4f, a nuclear membrane forms around the chromosomes in each new cell. Once again, the chromosomes become hard to see. Nucleoli form in each new nucleus. The asters and spindle disappear. The centrioles become smaller. Eventually, the cell pinches in and becomes two separate cells. Each new cell has a full set of chromosomes. In some cells, the processes of mitosis and cell division are completed in about 15 minutes. In other cells, they may take hours.

We do not know what causes cells to divide. Or what causes them to stop. Cancer is a form of uncontrolled cell division. If we had a better understanding of what controls cell division, we might learn to control cancer.

Cell division in many plants (Figure 4-6) is much the same as in animal cells. However, there are no asters in plant cells and most do not have centrioles. And the cell does not pinch in two. Instead, a wall, or partition, forms across the middle of the spindle. As the wall begins to form, tiny swellings appear on the fibers at the middle of the spindle. These swellings join, forming a cell wall between the two new cells. The cytoplasm of each cell forms a new cell membrane next to the cell wall.

4-5. A cancer researcher at the American Medical Center in Denver, Colorado.

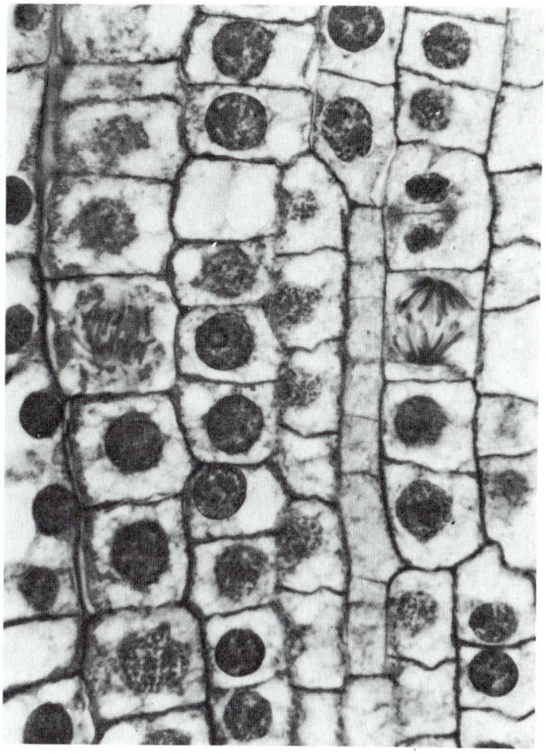

4-6. Mitosis in cells of the growing tip of an onion root. When the cells were fixed and stained, some were dividing. Note the prominent nucleoli in the nuclei of nondividing cells. *Why do some cells appear to be without nuclei?*

Chromosome Replication

The more Flemming and others studied mitosis, the more convinced they became that the nucleus does not just divide. At least this seemed to be true of the chromosomes, since the nucleus of each new cell had a complete set of chromosomes. If the nucleus merely divided in half, each new cell would contain only half a nucleus, *with half the number of chromosomes.* Yet, as far as the early observers could tell, the number and size of chromosomes seemed to be the same in each new cell.

Flemming noticed, in some very thin microscopic preparations, that each chromosome seemed to be double. He thought this was an important clue. If the number of chromosomes in each cell is the same before *and* after each division, the chromosomes must double at some time before mitosis. As far as Flemming could tell, each

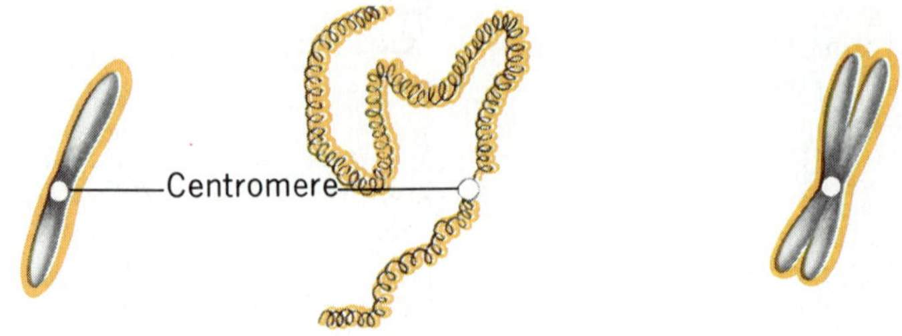

a

As mitotic division ends, each chromosome is condensed in length and easily visible under the microscope.

b

The chromosome uncoils in the new cell and disappears from view. Replication begins at this stage.

c

The chromosome becomes visible when another mitotic division begins. It is now double, with replication completed except for the centromere.

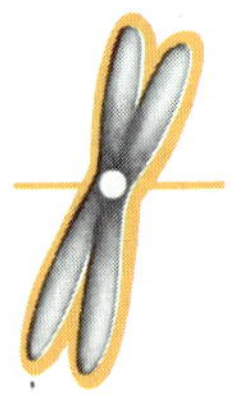

d

The double chromosome becomes attached to the spindle at the centromere. Finally the centromere divides;

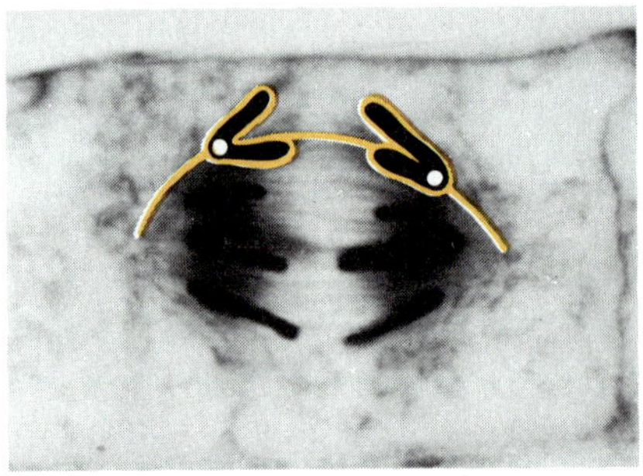

and each half-chromosome becomes an independent chromosome, one going to each pole of the spindle.

4-7. One chromosome becomes two. Five stages of the process are illustrated here. Each chromosome is replicated before the cell divides. Drawing **a** shows a chromosome at the end of one mitotic cycle; **b, c,** and **d** show stages ending in replication of the chromosome early in the next mitotic cycle; the final photograph shows the separation of the replicated chromosome and others in the succeeding cell division.

chromosome was single when it moved to one end, or **pole,** of the spindles. The chromosomes seemed to disappear in the newly formed cells. Yet, at the beginning of the next mitosis, when the chromosomes again became visible, they were double.

Since he could not see the chromosomes before mitosis, all Flemming could do was to form a hypothesis to explain the double

SCIENCE AT WORK: Cancer Drugs

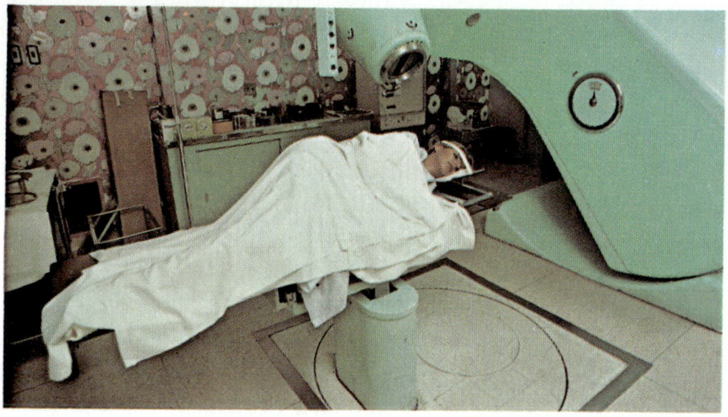

Cancer kills almost as many people as heart attacks do. Cancer is an abnormal type of cell division. For some unknown reason, these abnormal cells get the available food for growth before normal cells do. As a result, the normal cells may starve to death. Or the body may become so weak that infection or hemorrhaging and blood clots occur. Dead or dying cells may cause an organ (liver, stomach) to fail. Any of these conditions can cause death of the organism.

Some cancers are tumors or masses of abnormal cells. Other cancers are not confined to a single location. Leukemia, for example, is a form of cancer in which the number of white cells in the blood increases dramatically. Bone marrow is progressively destroyed. The photographs at the beginning of this chapter show normal bone marrow (left) compared with bone marrow from a leukemia patient (right).

Surgery and radiation are the preferred methods of treating tumors. Recently, progress has been made in treating other forms of cancer. In the early 1950s, cancer researchers began looking into the possible use of chemicals to fight this disease. Early results were disappointing. People had begun to expect "miracle" cures. Antibiotics were new then, and these were effective against many bacterial diseases. But these early disappointments in cancer research led to continued research. Today we have a promising new weapon against cancer.

Over the last 20 years, 300,000 chemicals were tested as possible treatments for cancer. Today, about 40 drugs are being used to treat cancer. Treatment with drugs has been particularly successful with childhood leukemia. About 80 percent of those treated seemed to have been cured.

Testing thousands of chemicals to find only a few that are effective is an example of an important part of scientific research. The combined efforts of many people working in different parts of the country were responsible for the success of this research. You can learn more about cancer research by looking for magazine and newspaper articles about the research centers that have played important roles in finding drugs that fight cancer. Some of these are Children's Cancer Research Foundation, Boston, Massachusetts; Anderson Hospital and Tumor Institute, Houston, Texas; Mt. Sinai School of Medicine, New York City; Memorial Sloan-Kettering Cancer Center, New York City; Southern Research Institute, Birmingham, Alabama; and The National Cancer Institute, Bethesda, Maryland.

chromosomes. He believed that each chromosome doubled—or as we say, replicated—between mitotic events. For example, the 46 chromosomes in human cells replicate, making 46 double chromosomes. During mitosis these double chromosomes would separate, one going to each of the newly formed cells. As a result, each of the two new cells would have 46 chromosomes.

Observations and experiments since Flemming's time have shown that such a hypothesis is a good explanation of the observations. Figure 4-7 shows how we think the chromosomes replicate and then separate. At the beginning of mitosis, each chromosome replicates. The two new chromosome strands, called **chromatids** (CROW·muh·tids), do not separate. Rather, the chromatids remain attached at one point. This point is the **centromere** (SEN·tro·meer).

During the later stages of mitosis, the centromere divides. Then the two chromatids separate, each going to one end of the spindle. Thus, two complete sets of chromosomes are formed—one in each new cell. As a result of mitosis, each new cell has the same number of chromosomes that the parent cell had.

In the laboratory you can observe dividing cells. By doing **Inquiry 4-1,** you will see for yourself some of the events we have described.

INQUIRY 4-1 Observing Mitosis

MATERIALS

Day 1

FILM: *Mitosis*

Day 2 (per team)

root from fresh onion bulb or cuttings of Tradescantia *or* Zebrina
prepared slide of onion root tip
prepared slide of Ascaris *or whitefish eggs*
0.5% toluidine blue 0 stain
1 N hydrochloric acid
light microscope
slide and cover glass
bunsen burner or alcohol lamp
forceps
razor blades
paper towels

Procedure (Day 1)

There is a great deal of movement and constant change going on during mitosis. A film of this process will help you to see the activity of the cell-division process. If such a film is available, your teacher will suggest what you should observe most closely.

Procedure (Day 2)

Select a root that is 2 to 3 cm long. Place it on a clean slide. Cut off 1 to 2 cm of the root tip. Discard any remaining upper portion.

Place three drops of 1 N hydrochloric acid on the root tip. Wait about 1 minute. While you are waiting, heat the slide by passing it back and forth over a flame. (CAUTION: *Hold the slide with forceps so you won't burn your fingers. Do not allow the liquid on the slide to boil.*) Heat the slide in this way two or three times. Carefully blot off the acid with a strip of paper towel.

Cover the root tip with two or three drops of toluidine blue 0 stain. Warm the slide for about one minute as you did before. Carefully blot off the excess stain.

Add a drop of fresh stain. Apply a cover glass. Place the slide in a folded paper towel on a hard surface. Press the cover glass with your thumb, using a steady, firm pressure. *Be careful not to break the cover glass.*

Clean the stain from the slide and cover glass. Examine under low power of your microscope. When you have located the cells, switch to high power. The cells may be stained so darkly that you cannot see the individual parts. If this happens, dilute the stain by placing 1 or 2 drops of water at one edge of the cover glass. Place a piece of paper towel at the opposite edge to pull the water through. Repeat this until you can see the stained portions of the cell.

Find these stages of mitosis:

(a) The nuclear material forms long, slender threads that are stained. At this stage, the separate chromosomes cannot be identified. They form a loose ball of tangled and twisted threads.

(b) The individual chromosomes are much shorter and thicker. At this stage, each chromosome has two strands, or chromatids. The strands are held together at the centromere.

(c) The chromatids have separated and are now two separate groups of chromosomes.

(d) The two nuclei are present in the cell, with strands of chromosomes still visible.

1. Make simple sketches of what you find. Describe the structures you see. **2.** What problems did you have in making this study?

Study a prepared slide of dividing cells in the onion root tip. (The stain used for the slide is different from the one you used.) Scan the entire length of the section of root tip, using low power. **3.** Where do you observe cells dividing? **4.** Where are the cells not dividing?

on slide

Locate some cells in division. Switch to high power and look for the following stages:

(a) Early appearance of chromosomes in the nucleus. Find round structures that stain at this early stage. These are the nucleoli.
(b) The chromosomes are easy to see and are grouped in the middle of the cell.
(c) Chromosomes have separated and have moved toward the poles.
(d) A new cell wall is formed between the new cells.

5. What has happened to the nucleoli? **6.** What is the condition of the nuclear membrane?

Find the spindle. The ends of the spindle fibers that are near the center of the cell are attached to the centromeres of the chromosomes. **7.** How many chromosomes can you see? **8.** What part do you think the spindle fibers might play in moving the chromosomes?

Look for the beginning of the new cell wall. **9.** Where does it appear?

Cells divide rapidly in **embryos** (young organisms, at very early stages of development). These cells are good material in which to study mitosis. Carefully examine a prepared slide of either *Ascaris* (AS·kar·is) or whitefish embryos. (*Ascaris* is a worm.) Look for the following stages:

(a) The chromosomes are long and threadlike.
(b) The double chromosomes are attached to spindle fibers at the center of the cell. Look at the poles of the spindle and compare them with those of the plant cells you studied.
(c) The chromosomes are separating and the cell is pinching in two.

10. How many individual chromosomes are in one cell? **11.** How do the poles of the spindle of an animal cell differ from those of a plant cell? **12.** Compare the separating of chromosomes and dividing of these cells with what you saw in plant cells. **13.** What structures do you see in the dividing animal cell that were present in the dividing plant cell? no aster

MEIOSIS

Biologists began to wonder whether the number of chromosomes *always* remains the same when cells divide. They knew that one kind of reproduction results from the combining of sperm and egg cells. If these cells were formed by mitosis, the combined cell would have

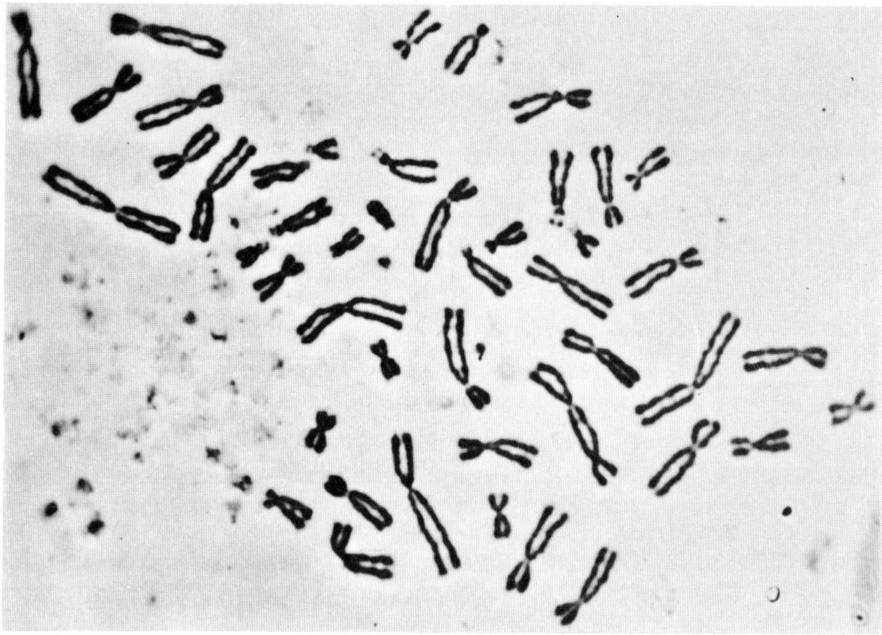

4-8. The chromosomes of a normal human male.

twice as many chromosomes as it should have. It is possible to make an "If . . . then" statement to test the hypothesis:

> *If* the number of chromosomes is the same in all cell divisions (including those that produce sperm and egg cells) . . . *then* individuals of the next generation will have twice the number of chromosomes their parents had.

To test this hypothesis, biologists over the years carefully examined many kinds of plants and animals. In each species, they found that individuals of each new generation always have the same number of chromosomes (Figure 4-8). The number does not double. These results failed to support the hypothesis. Some important factors must have been overlooked.

Biologists had studied developing embryos and the growing roots of plants. But they had studied only *some* cells during a *limited time* in each kind of organism. Some of these biologists assumed that the number of chromosomes was the same in *all* parts of the organism at *all* times.

August Weismann (WISE·mun) was one of the biologists who did not agree with this assumption. Because his eyesight was poor, Weismann could not use a microscope to study cells. He could not collect more facts. But he made an important contribution by analyzing and interpreting the data collected by others. Weismann considered the facts:

1. In each new generation, individuals of a species have the same number of chromosomes.

2. In cell division, the number of chromosomes remains the same.

He realized that both statements could not *always* be true. Perhaps not *all* cell divisions are the same. Weismann suggested this hypothesis:

> There must be a kind of cell division in which each new cell gets only *half* as many chromosomes as the parent cell had.

Let's see what this hypothesis means in terms of human cells. As a result of mitosis, each new cell receives 46 chromosomes. But what *if* there were a different means of nuclear division that cut the number of chromosomes in half. *Then* each new cell would have only 23 chromosomes.

If this type of division occurs during the formation of sperm or egg cells, all the facts could be explained. In humans, each of these reproductive cells would have 23 chromosomes. When they combined, the resulting cell, a fertilized egg, would have 46 chromosomes. Then this fertilized egg would divide by mitosis, and each new cell would have 46 chromosomes.

Such a process has indeed been observed. It is called **meiosis** (my·OH·sis). Meiosis takes place in many animals and plants. In animals it happens before egg and sperm cells are formed. In plants it usually happens when reproductive cells called spores are formed.

The reproductive cells have half the number of chromosomes found in other cells of the organism. This is the **monoploid** (MON·o·ploid) number, often abbreviated *n*. When these monoploid cells combine, a cell with the **diploid** (DIP·loid) number (2*n*) of chromosomes results. Egg cells and sperm cells are monoploid. When they unite, they form a fertilized egg cell, with the diploid number of chromosomes.

The Events of Meiosis

Meiosis begins with the diploid cells. Each cell undergoes *two* cell divisions. (There is only *one* cell division in mitosis.) The result of meiosis is four cells, but each cell has only the monoploid number of chromosomes. The first division reduces the number of chromosomes to half. The second division is similar to mitosis. Meiosis is shown in the drawings in Figure 4-9.

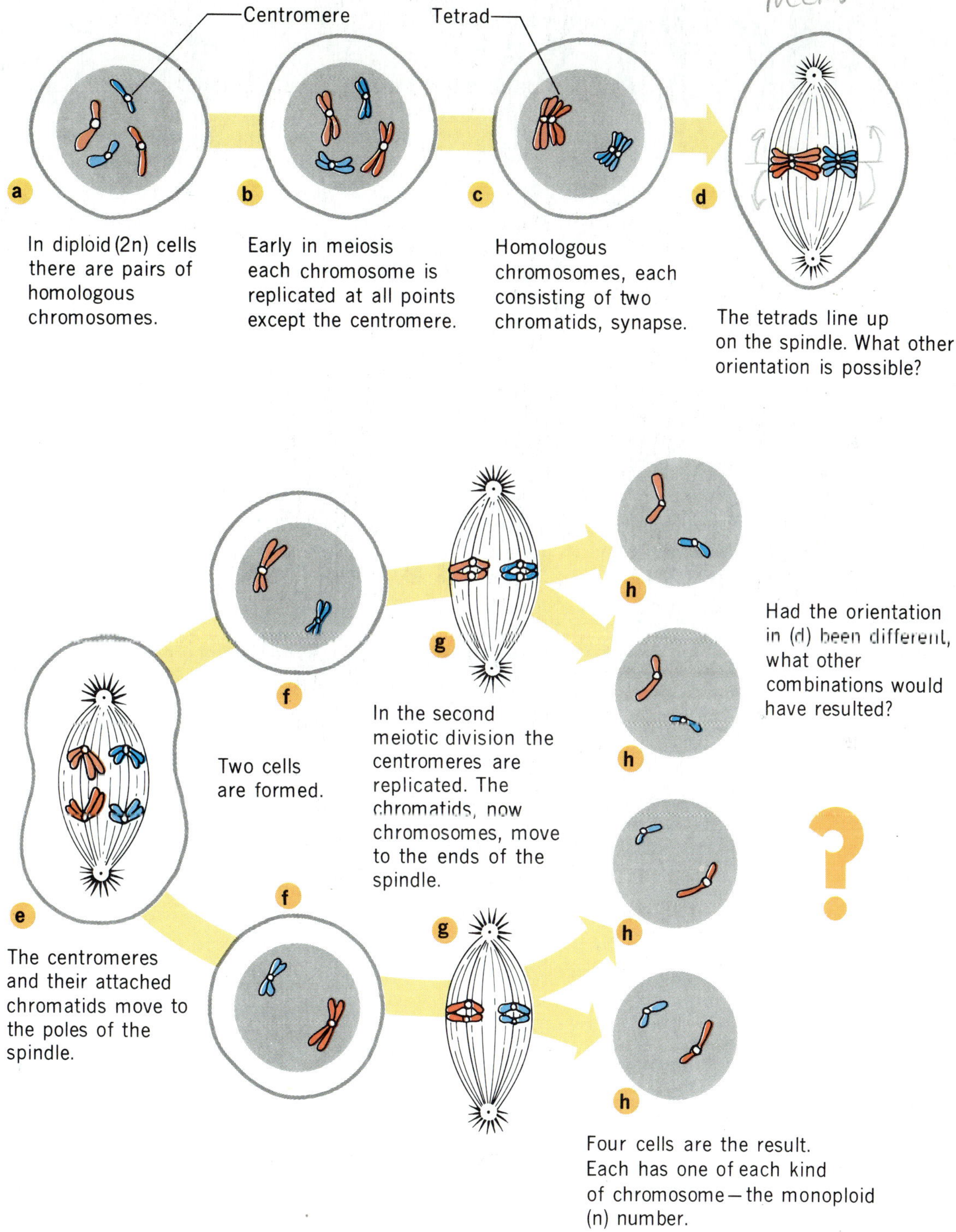

4-9. Monoploid reproductive cells are formed as a result of meiosis.

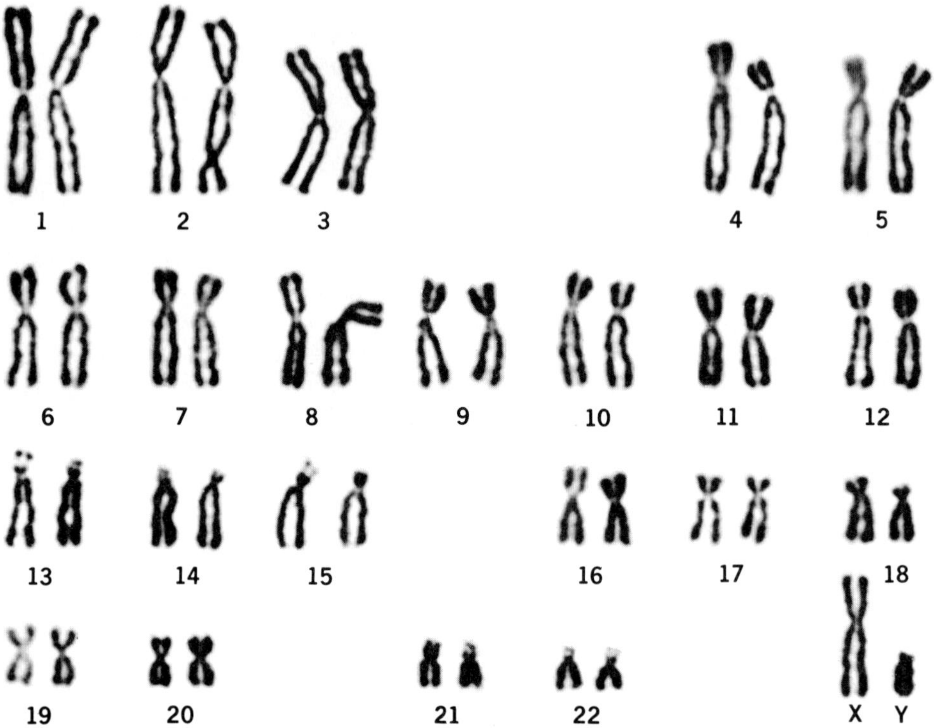

4-10. Karyotype of the chromosomes of a normal human male.

In a male human being, meiosis occurs during the formation of sperm cells. The cell that will form sperm cells contains 46 chromosomes—the diploid number—as in other cells. A close examination of the chromosomes at the beginning of cell division shows that they seem to vary greatly. However, if we arrange the chromosomes according to their structural features, we find they can be paired. That is, there are not 46 *different* kinds of chromosomes but only 23. (This is not strictly true in the human male, but we'll return to that point in Chapter 6.) The pairs of chromosomes are shown in Figure 4-10. Here the chromosomes have been arranged according to size and structural features such as positions of the centromeres. This arrangement is called a karyotype (CARE·ee·oh·type).

To simplify, we will concentrate on only two pairs of chromosomes in a nucleus undergoing meiosis. If we find out what happens to these pairs, we can understand what happens to other pairs as well. In Figure 4-9, one chromosome pair is represented in orange and the other pair in blue. The two orange chromosomes are said to be **homologous** (ho·MOL·oh·gus) chromosomes. Similarly, the blue chromosomes are homologous. The two members of a homologous pair are similar, though not identical.

Now follow the events of meiosis shown in Figure 4-9. As in mitosis, each chromosome replicates (or doubles). That is, each chromosome consists of two chromatids joined at the centromere.

The next major event that occurs is **synapsis** (sin·NAP·sis). The homologous chromosomes, which are now replicated, come together in pairs (Figure 4-9c). In our example, the orange ones come together and the blue ones come together. Each pair of homologous chromosomes consists of four chromatids and two centromeres. Such a structure is called a **tetrad** (TEH·trad).

Next, the tetrads line up in the center of the spindle (Figure 4-9d). One of the centromeres with its two attached chromatids moves to one pole of the spindle. The other centromere, with its attached chromatids, moves to the other pole (Figure 4-9e). As a result, each new cell has one member of the orange pair and one member of the blue pair. The first stage of meiotic division is now complete (Figure 4-9f).

In the second stage of meiosis, only the centromere divides. The chromosomes *do not* undergo replication again during this stage. Thus, in each of the two cells, the orange chromatids separate, becoming two independent orange chromosomes. The blue chromatids also separate to become two independent chromosomes (Figure 4-9g).

In each of the two cells, one orange and one blue chromosome go to one pole; the other orange and blue chromosomes go to the opposite pole. The result is four cells, each with *one* member of each homologous pair of chromosomes (Figure 4-9h). No matter how many chromosomes a cell has, each cell that results from meiosis has one of each pair of chromosomes. Thus, at the end of meiosis, the chromosome number has been halved.

CHROMOSOMES AND GENETIC INFORMATION

When one of your bone cells divides, it produces two bone cells. A dividing skin cell produces two skin cells. Why does cell division produce individuals that are exactly like the parent cell or organism? The information about what a cell or an organism will be must be carried from the parent or parents to the next generation. This information is called the *genetic* (je·NET·ik) *information.* It determines what a cell or organism will be and what it will do.

Scientists have learned that chromosomes carry the genetic information. In the years since Flemming's time, but especially since 1953, we have learned a great deal about the structure of chromosomes. They are composed mainly of protein and DNA. DNA contains the genetic information that controls the activities of the cells of an organism. Human cells contain human DNA. A frog's cells contain another kind of DNA.

Meiosis occurs as reproductive cells are formed. These cells contain one of each type of chromosome that was present in the parent cell. As a result, they contain one half of the genetic information. When two reproductive cells combine, the genetic information from both parents is combined. This produces new *combinations* of genetic information. For example, in a herd of deer, some parents may have longer legs or stronger muscles than others. As a result of new combinations of genetic information, some young deer may have both long legs and strong muscles. The chances for survival of long-legged, strong-muscled deer probably would be higher.

In Brief

Biologists today have some understanding of the structure and function of living cells. They are beginning to understand the mechanisms of mitosis and cell division. Somehow a living, organized cell divides into two cells. And, in nearly every case, the new cells have all the structures and functions of the parent cell.

In mitosis, each chromosome is doubled. The two chromosomes that result separate during mitosis. One goes to each new cell. When a chromosome replicates, its DNA also replicates. The rest of the cell is more or less equally divided. Each of the two new cells contains half of the original cell material. But each gets a *full set* of genetic information in the DNA. With this information, each new cell grows and lives as the parent cell did.

In organisms that reproduce by the combining of two reproductive cells, the chromosome number remains the same in each generation. This is made possible by another kind of division—meiosis. Meiosis reduces the number of chromosomes from diploid to monoploid. The combining of two monoploid cells results in one diploid cell. This cell has a combination of the genetic information from the two parent organisms.

Guide Questions and Problems

1. Suppose you are talking with a friend who has not studied biology. You say, "I am a human being because my cells have DNA of the human type." Your friend replies, "Nonsense. You are a human being because your father and mother are human beings." How can this argument be settled?

2. What is mitosis? Describe the process.

3. We still do not understand the cause (or causes) of cancer. But we do know that cancer involves uncontrolled cell division. How might an understanding of mitosis help in the research on cancer?

4. Before scientists discovered and observed meiosis, they had predicted this type of division. How was it possible to predict that meiosis occurred before there was evidence for it?

5. What is the difference between the monoploid and diploid numbers of chromosomes?

Ideas for Independent Study

1. Invent a way of showing what happens to chromosomes during mitosis or meiosis. You might want to create models of some kind. Or, perhaps, you could design a game that would help the players learn the basics of mitotic and meiotic processes.

2. The history of cancer research is important because of the effects of the disease. And it shows much about the processes scientists use to investigate a biological problem. Consult a current *Readers' Guide to Periodical Literature* and begin a library search for articles on cancer research. Present your findings to others, perhaps in the form of a display or slide presentation.

3. Are you genetically unique? First, consider the number of genetic types of sperm or eggs one parent can produce. Two pairs of homologous chromosomes can yield four genetically different types of sperm or eggs. Three pairs can yield eight genetic types. How many can 23 pairs of homologous chromosomes produce? Express this as a fraction: $1/x$ (the figure you calculate). You also must calculate the unlikely probability that *one particular type of sperm* and *one particular type of egg* will come together at fertilization. To do this, multiply the fraction by itself.

There is a process known as crossing-over. This process results in a single pair of homologous chromosomes producing possibly hundreds of different gene combinations. In how many of the pairs of your parents' chromosomes could crossing-over have occurred?

Consider all the parents in the world who possibly could produce sperm or eggs genetically like those that led to you—perhaps involving crossing-over. The probability that such parents exist should be multiplied by the probability that they might meet and marry. The result can then be multiplied by your probability figure. Finally, this product can be added to the probability figure obtained for your parents.

Now, return to the original question: Are you genetically unique?

Related Readings

BOOKS

Beadle, G. and M. Beadle. *The Language of Life.* Garden City, N.Y.: Doubleday, 1966.

Bonner, D. M. and S. Mills. *Heredity.* Englewood Cliffs, N.J.: Prentice-Hall, 1964.

John, B. and K. R. Lewis. *Somatic Cell Division (Mitosis).* Oxford Biology Reader. London: Oxford University Press, 1972.

John, B. and K. R. Lewis. *The Meiotic Mechanism.* Oxford Biology Reader #65. Boulder, Colo.: EPIC, P.O. Box 3406, 1972.

Mazia, D. *Cell Division.* A BSCS Pamphlet. Boulder, Colo.: BSCS, Dept. BEM, P.O. Box 930, 1963.

Moore, J. A. *Heredity and Development.* Fair Lawn, N.J.: Oxford University Press, 1972.

Patt, D. I. and G. R. Patt. *An Introduction to Modern Genetics.* Reading, Mass.: Addison-Wesley, 1975.

Pilgrim, I. *The Topic of Cancer.* New York: Crowell, 1974.

Prescott, D. M. *Cancer: The Misguided Cell.* Indianapolis: Pegasus, 1973.

Stahl, F. W. *The Mechanics of Inheritance.* Englewood Cliffs, N.J.: Prentice-Hall, 1969.

ARTICLES

Mazia, D. "The Cell Cycle." *Scientific American,* January 1974, page 54.

PART TWO

CONTINUITY

5 Patterns of Heredity

The production of offspring in each generation that are exactly, or very much, like the parent or parents is called **genetic continuity.** We can put a frog's egg and a toad's egg in the same culture dish and give them the same kind of food. But in every case the frog's egg becomes a frog, and the toad's egg becomes a toad.

Scientists who study how offspring inherit characteristics from a parent are **geneticists.** The study of the processes and effects of inheritance (or heredity) is called **genetics.** The information about what a cell or an organism will be must be carried from the parent or parents to the next generation. This is the genetic information.

Geneticists have learned that chromosomes carry the genetic information. They also have learned a great deal about the structure of chromosomes. Chromosomes are composed mainly of DNA and protein. DNA is the material in the chromosomes that contains the genetic information that controls the activities of the cell. Individual bits of the information contained in DNA are referred to as **genes.** What a cell is and what it does depends on the kind of genes it contains. The genes in your cells make you human. A frog's cells contain other kinds of genes.

Studies through the years have shown certain logical patterns that govern the inheritance of characteristics. You can discover some of these patterns for yourself by beginning **Inquiries 5-1** and **5-2.**

INQUIRY 5-1 Drosophila Technique

The common fruit fly *Drosophila melanogaster* (droh·SOF·i·luh meh·LAN·oh·gas·tur) is found on spoiled fruits. You will see them near markets, food warehouses, and garbage pails during the summer. *Drosophila* is an excellent organism for genetic studies. It is easy to raise, completes its life cycle in a short time, and is easy to observe.

As these flies develop, they go through several stages. The first is the egg. The form that hatches from the egg is called a larva (plural, larvae). The next stage is a period during which the larva changes form. During this stage, it is called a pupa (PEW·puh; plural, pupae —PEW·pee). In the final stage, the pupa changes into an adult fly. These stages are shown in Figure 5-1. You also can observe these stages in the bottle that contains flies in these different forms. (The eggs may be hard to see.)

MATERIALS (per team)

bottle of Drosophila melanogaster *in various stages*
vial of 30 to 40 adult Drosophila
stereoscopic microscope, or good hand lens
ether
etherizer
re-etherizer (optional)
medicine dropper
small, tapered brush or bent-tipped dissecting needle
large index card
morgue
water pan, if necessary

Procedure

Place several drops of ether on the cotton lining of the etherizer. Use only enough to dampen the cotton. Contact with liquid ether

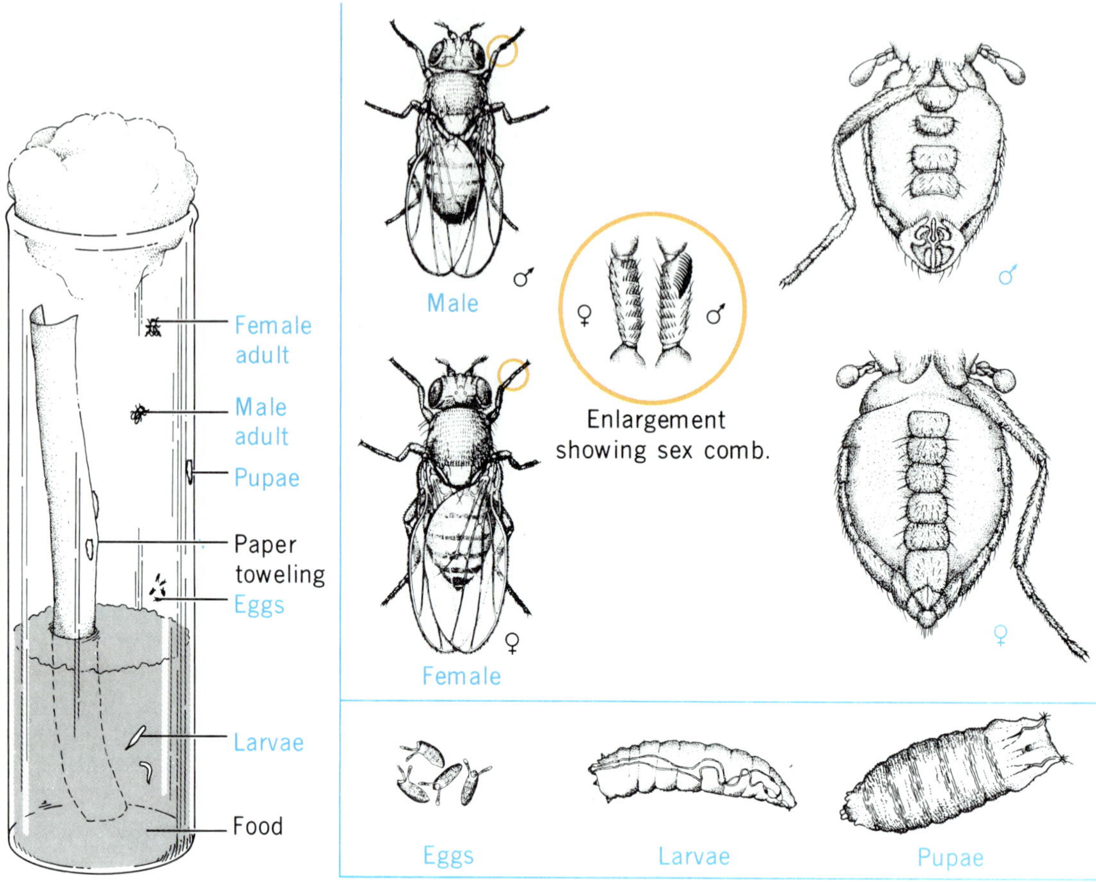

5-1. The stages of development in the life cycle of *Drosophila melanogaster.* Male and female *Drosophila* and their identifiable characteristics are also shown.

will kill the flies. (CAUTION: *No flames should be allowed in the room. Use ether only in a well-ventilated room and breathe it as little as possible.*)

Gently tap the bottom of the glass vial of flies on the table or on your thigh. The flies should drop to the bottom of the vial (Figure 5-2b). Quickly remove the plug from the vial. Invert the vial and place it firmly on the etherizer (Figure 5-2c).

Hold the vial and the etherizer together firmly. Bump the bottom of the etherizer on the table or your thigh once or twice to loosen the remaining flies.

The flies will be overcome within a few seconds and will fall to the bottom of the etherizer. Do not leave them in the etherizer for more than a minute!

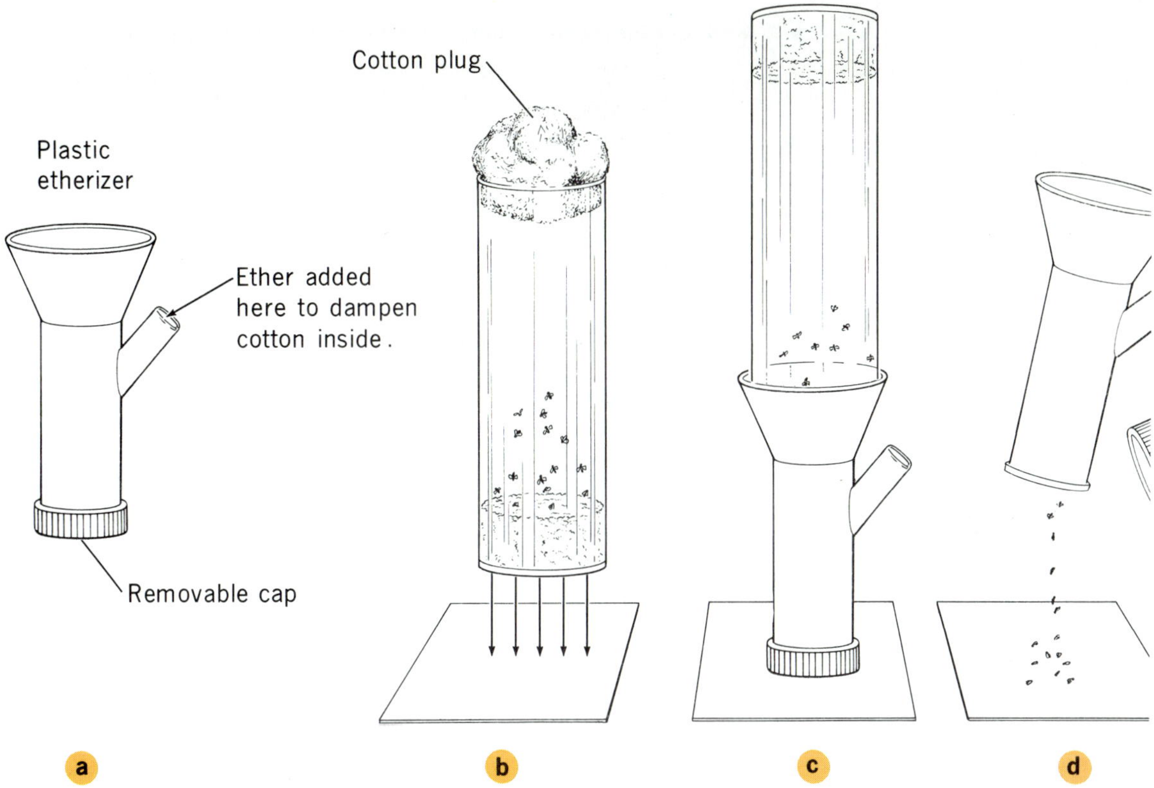

5-2. ANESTHETIZING *DROSOPHILA*

Wait until the last fly stops moving in the etherizer. Then remove the vial and replace its plug. Remove the bottom of the etherizer. Pour the flies onto the index card for examination (Figure 5-2d). Place any dead flies in the morgue.

The flies can be injured easily. Move them around with the brush or dissecting needle. Examine the flies using a stereoscopic microscope or hand lens. The flies will begin to move in about 5 minutes. They may begin to move before you have completed your examination. If necessary, use the re-etherizer.

Using Figure 5-1 to help you, try to separate the male and female flies. Make a copy of Figure 5-3, in which some of the important characteristics for male fruit flies have been listed. As you study the flies, fill in those characteristics for females.

If time permits, test each other's ability to identify male and female flies. You and another student may exchange flies whose sex you have determined. Do you agree on the sex of both groups of flies?

After you examine each fly, place it back in the vial from which it came. Lay the vial on its side until the flies recover. This will prevent the flies from sticking to the food.

5-3. Model chart for recording sex differences in *Drosophila*

Distinguishing Features	*Comparisons*	
	♂ *(Male)*	♀ *(Female)*
1. Relative size	small	
2. Kind of banding on abdomen	tip of abdomen is solid black	
3. Relative size of abdomen	small—five segments	
4. Shape of tip of abdomen	rounded	
5. Sex combs (present or absent)	present	
6. External genitalia (describe)*	black bristles around genitalia—genitalia obvious	

*Turn flies onto their backs.

INQUIRY 5-2 Mendelian Genetics

In this inquiry you will perform a **cross** (mating) between a normal (wild) type of fruit fly and a **mutant** (changed) type. The wild type has normal wings and red eyes. The mutant you will use has dumpy (short) wings and sepia (dark reddish-brown) eyes.

MATERIALS (per team)

basic equipment from **Inquiry 5-1**
stock cultures, wild type and mutant, parental strains
F_1 *(first generation) offspring from a cross between the pure wild and pure mutant flies*
culture vials (or bottles) containing fresh medium

Day 21

vials marked "sepia eyes," "red eyes," "dumpy wings," and "normal wings" containing adult flies (sorted on days 11 to 20)

Procedure (Day 1)

Etherize and examine a few flies from stock cultures of the two parent types as you did in **Inquiry 5-1.** The male parents are of the pure wild variety. The females are of the pure mutant variety. Describe the differences you can see between the two types. Refer to Figure 5-4.

The vial marked F_1 is the first generation of offspring from the parental types you observed. Etherize the vial of F_1 flies as you did in **Inquiry 5-1.** Classify each of these flies as male or female and as mutant or wild type. Copy the tally sheet in Figure 5-5. Enter your data on the tally sheet.

Combine your data on the chalkboard with the data from the other teams to obtain a class total. **1.** How many flies with mutant characteristics did your team find? How many did the class find?

Label a culture vial of medium with your team name, the date, and F_2. Take four or five of the F_1 males and four or five F_1 females and place them in the "F_2" vial. Lay the vial on its side until the flies awaken. Place your culture vial in a warm place, but away from windows, radiators, and direct sunlight. The temperature should not change greatly from night to day.

Check the vial each day to see if the flies are in good condition. If the vial is kept at about 25°C, you should be able to see the F_2 larvae within a week.

Procedure (Day 8 or 9)

Etherize the F_1 flies in your vial. Put them in the morgue. The F_2 generation of your cross should begin to appear as adult flies on the tenth day.

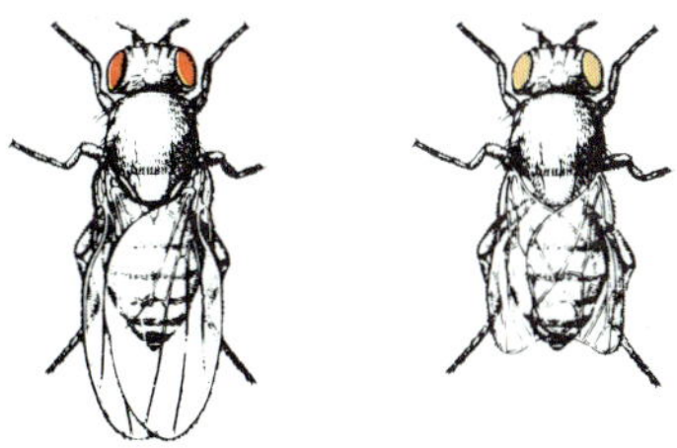

Normal-winged, red-eyed female

Dumpy-winged, sepia-eyed female

5-4. Wild and mutant type female *Drosophila.*

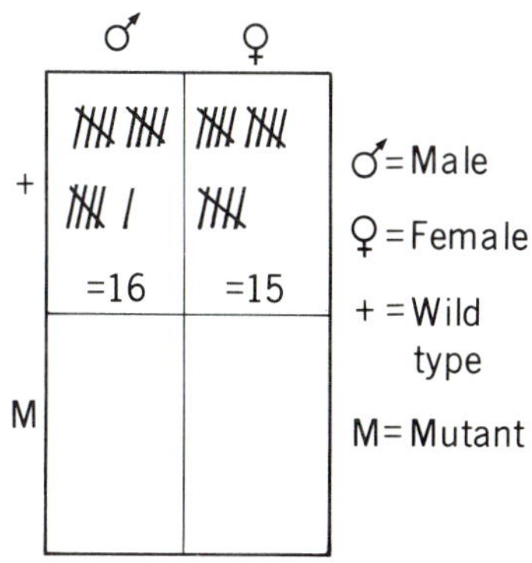

5-5. Sample of F_1 offspring of wild type x mutant.

Procedure (Days 11 to 20)

All flies in the vials are now F_2 flies. Your teacher will divide the teams into two groups, I and II, for this part of the inquiry.

Group I. Each team should label two vials "dumpy wings" and "normal wings." As the F_2 flies appear sort them into these vials according to type of wings. Record the total number of each type. Combine the data from all Group I teams.

Group II. Label two vials "sepia eyes" and "red eyes." As the F_2 flies appear, put them into these vials according to eye color. Record the number of each type. Combine the data from Group II teams.

Place these vials in a warm place. Leave them on their sides if the flies are not awake.

Procedure (Day 20)

Place the combined data from Groups I and II on the chalkboard. **2.** How many different phenotypes were there in the F_1 generation? **3.** How many different phenotypes did each group find in the F_2 generation? **4.** What evidence do you have that the genotypes of the F_2 flies were different from those of the F_1 flies? **5.** What is the ratio of normal- to dumpy-winged F_2 flies? What is the ratio of red- to sepia-eyed F_2 flies? **6.** How do these ratios compare with the 3:1 ratio that you would expect if each abnormal trait was caused by a single mutant gene? How do you account for any differences? **7.** Of the four traits represented in the F_2 generation, which are dominant and which are recessive? How do you know?

Procedure (Day 21)

Crosses involving two pairs of traits are called **dihybrid crosses.** In the original dihybrid cross made for you, *Drosophila* with dumpy wings and sepia eyes were crossed with *Drosophila* with normal wings and red eyes.

For this part of the investigation, you will need the vials into which you sorted the mutant and wild F_2 flies. Group I should take the vials marked "sepia eyes" and "red eyes." Group II should take the vials marked "dumpy wings" and "normal wings." Each group should label four vials "sepia, dumpy," "sepia, normal," "red, dumpy," and "red, normal."

Group I: Separate the sepia- and red-eyed flies into two groups according to normal or dumpy wings. Place them in the proper vials. Count and record the flies of each phenotype.

Group II. Separate the dumpy- and normal-winged flies according to eye color. Place them in the proper vials. Count and record the number of flies of each phenotype.

After sorting and recording your data, place all flies in the morgue. Combine the data of the two groups and answer the following questions. **8.** What fraction of the total number of flies were "sepia, dumpy"? What fractions were the other phenotypes? **9.** Recall the Product Rule of Probability. What fraction of the total number of flies would be expected to have both normal wings and red eyes? What fraction would be expected to have sepia eyes and dumpy wings? **10.** Use the checkerboard method to predict the expected genotypes and phenotypes. Use *D* (normal) and *d* (dumpy) to stand for the genes for wings. Use *S* (red) and *s* (sepia) for the genes for eyes. You may want to refer to Figures 5-15 and 5-16. **11.** How many different genotypes are expected in the F_2 generation? **12.** How many different phenotypes are expected in the F_2 generation? What are they? In what ratio do you expect to find them? **13.** How close are your experimental results to the phenotypic ratio that you predicted?

THE WORK OF GREGOR MENDEL

The modern story of genetics began with Gregor Mendel, an Austrian monk. Mendel kept a small garden plot at the monastery (Figure 5-6), and he experimented with pea plants. These experiments threw the first clear light on the explanation of heredity. When Mendel's results were published in 1865, no one paid much attention to the work or realized its importance. Around 1900, other researchers began working on similar studies. When they searched the scientific literature for information on genetics studies, they found Mendel's report. Their research confirmed Mendel's results.

5-6. Growers now produce high-yielding varieties of peas.

Mendel was curious about certain traits in pea plants, such as wrinkled or round seeds, yellow or green seeds, and long or short stems. He wondered how these traits were passed from one generation of pea plants to the next. He also wanted to know why one form of a trait (such as wrinkled seeds) sometimes appeared in the next generation, while another form (round seeds) appeared at other times.

Mendel spent several years raising plants that always produced offspring that showed the same form of a trait. He thus obtained **pure varieties** of pea plants. For example, one variety always produced wrinkled seeds. Another variety always produced round seeds.

Pea plants are easy to cross, or mate, and they grow quickly. Mendel could grow many generations of plants in a short time. And the

5-7. Small French Marigolds are crossed with the larger-flowered American Marigolds to produce Red Nugget Hybrid Marigolds. The plants with the yellow flowers are the female parents. Bees transfer pollen from the small, red marigolds to the larger, yellow female flowers. The picture at the right shows a close-up of Red Nugget Marigolds in the foreground.

offspring of crosses between the different varieties of peas are fertile. That is, they can reproduce. It also was important to Mendel's research that the pea flower contains both male and female reproductive parts. Although the pea plant usually self-pollinates under natural conditions, it is possible to interfere with the process. One can collect pollen grains (which give rise to male reproductive cells) from one plant and transfer them to another plant. Thus, the seed that will become a new plant is the offspring of two parent plants, not just one. This is what Mendel did, and plant breeders continue to use this technique today (Figure 5-7).

Mendel next began experiments involving crosses of the peas. These experiments were different from those carried out by other scientists of that time. First, Mendel repeated the crosses of different pure varieties many times to be sure of the results. Second,

Mendel used mathematics to analyze that data and to arrive at a hypothesis. And he limited the study of each cross to one trait at a time. An example is a cross between plants with round and wrinkled seeds *or* between tall and short plants. When Mendel began experimenting, biologists knew very little about chromosomes or cell division. His conclusions were based only on the results of experiments.

Mendel made hundreds of crosses to study certain traits. One of these was a cross between pea plants whose seeds were always round and plants whose seeds were always wrinkled. When a cross is made between two varieties, the parent generation is called the P_1 generation. The offspring of the P_1 cross are the *first filial* (FILL·ee·ul), or F_1, *generation.*

In every case, Mendel found that all the F_1 plants had round seeds. When one form of a trait appeared in the F_1 of a cross of pure varieties, Mendel called this the **dominant** trait. The trait that did not show was called the **recessive** trait. The offspring in all of Mendel's crosses were exactly like one of the parents. There were never any in-between types, nor any that showed a mixture of the parents' traits.

Mendel let the F_1 plants self-pollinate. In the next generation, the F_2, the dominant trait (round seeds) appeared in 75 percent of the offspring. In 25 percent, the recessive trait (wrinkled seeds) *reappeared.* This gives a ratio of three dominant traits to one recessive in the F_2 generation. Mendel observed the same results in crosses involving other traits.

Figure 5-8 gives some of Mendel's results. The ratios of plants showing a dominant form of a trait to those with the recessive form was always about the same. Only the dominant form of the trait appeared in the F_1 generation. In the F_2 generation, the ratio was always three dominant to one recessive.

Mendel looked for a hypothesis to explain these results. He assumed that some part of the reproductive cell was responsible for carrying traits to offspring. Geneticists now call these "parts" genes. And we now know they are carried on chromosomes. Mendel assigned letters to stand for the genes that cause each trait. For example, he used a capital *R* to stand for the gene that caused the dominant trait—round seeds. A small *r* stood for the recessive gene that caused wrinkled seeds.

In this example, we are assuming that the appearance of round seeds is caused by a certain gene, *R.* We use the word **phenotype** (FEE·no·type) when we refer to the appearance of the trait. The phenotype in the cross we have been discussing would be *round* or *wrinkled* seeds.

We use the word **genotype** (GEE·no·type) to refer to the combination of genes that causes the trait. The male and female reproduc-

5-8. MENDEL'S RESULTS WITH TWO GENERATIONS OF GARDEN PEAS

Traits Selected for P_1 Cross	*F_1 Plants*	*F_1 Self-pollination*	*F_2 Plants*	*Actual F_2 Ratio*
1. Round × wrinkled seeds	all round seeds	round × round	5,474 round seeds 1,850 wrinkled seeds 7,324 Total	2.96:1
2. Yellow × green seeds	all yellow seeds	yellow × yellow	6,022 yellow seeds 2,001 green seeds 8,023 Total	3.01:1
3. Colored × white seed coats	all colored seed coats	colored × colored	705 colored seed coats 224 white seed coats 929 Total	3.15:1
4. Inflated × wrinkled pods	all inflated pods	inflated × inflated	882 inflated pods 299 wrinkled pods 1,181 Total	2.95:1
5. Green × yellow pods	all green pods	green × green	428 green pods 152 yellow pods 580 Total	2.82:1
6. Axial × terminal flowers	all axial flowers	axial × axial	651 axial flowers 207 terminal flowers 858 Total	3.14:1
7. Long × short stems	all long stems	long × long	787 long stems 277 short stems 1,064 Total	2.84:1

tive cells that combine to produce the offspring may be different genetically. Each male or female reproductive cell contains only one of each kind of gene. When the two cells combine, the new cell contains two genes—one from the male and one from the female.

Mendel hypothesized that the reproductive cells of a *pure,* round-seed plant would carry only *R* genes. Thus, its genotype would be *RR.* And those of a pure, wrinkled-seed plant would carry only the *r* genes (genotype *rr*). Why?

The result of a cross between plants with pure round seeds and plants with pure wrinkled seeds would be an F_1 generation of plants that had *both* the *R* and *r* genes. As the *R* gene is dominant, the phenotype of the F_1 generation would be round seeds. But we know that the *r* gene was present in the F_1 plants because the wrinkled-seed trait reappeared in the F_2 generation.

A MODEL OF INHERITANCE

We can predict the genotypes that will occur in the F_1 generation from a cross of pure round- and wrinkled-seed plants (genotypes *RR* and *rr*). To do this, we use a checkerboard as shown in Figure 5-9. The percentage in each box is obtained by multiplying the percentage directly above each box by the percentages to the left of it. The genotypes of the F_1 generation are all *Rr*. All of the F_1 plants show the round-seed phenotype. Why?

		Eggs	
		50% R	50% R
Sperm	50% r	25% Rr	25% Rr
	50% r	25% Rr	25% Rr

5-9. Checkerboard for a cross of RR x rr.

To predict the genotypes in the F_2 generation, we must assume that half (50 percent) of the male cells produced by the F_1 plants will carry the *R* gene and half will carry *r*. The same ratio will occur in the female cells. In this case, our checkerboard looks like Figure 5-10. Note that there are three genotypes but only two phenotypes. Why? Does this model explain the ratio of phenotypes that Mendel observed in his F_2 generations?

		Eggs	
		50% R	50% r
Sperm	50% R	25% RR	25% Rr
	50% r	25% Rr	25% rr

5-10. Checkerboard for a cross of Rr x Rr.

Figure 5-11 shows a cross of another pair of traits—terminal (end) and axial (side) flowers. Construct a checkerboard to show the results of the F_1 self-pollination. Calculate the percentages of F_2 genotypes.

5-11. INHERITANCE OF TERMINAL AND AXIAL FLOWERS IN PEA PLANTS

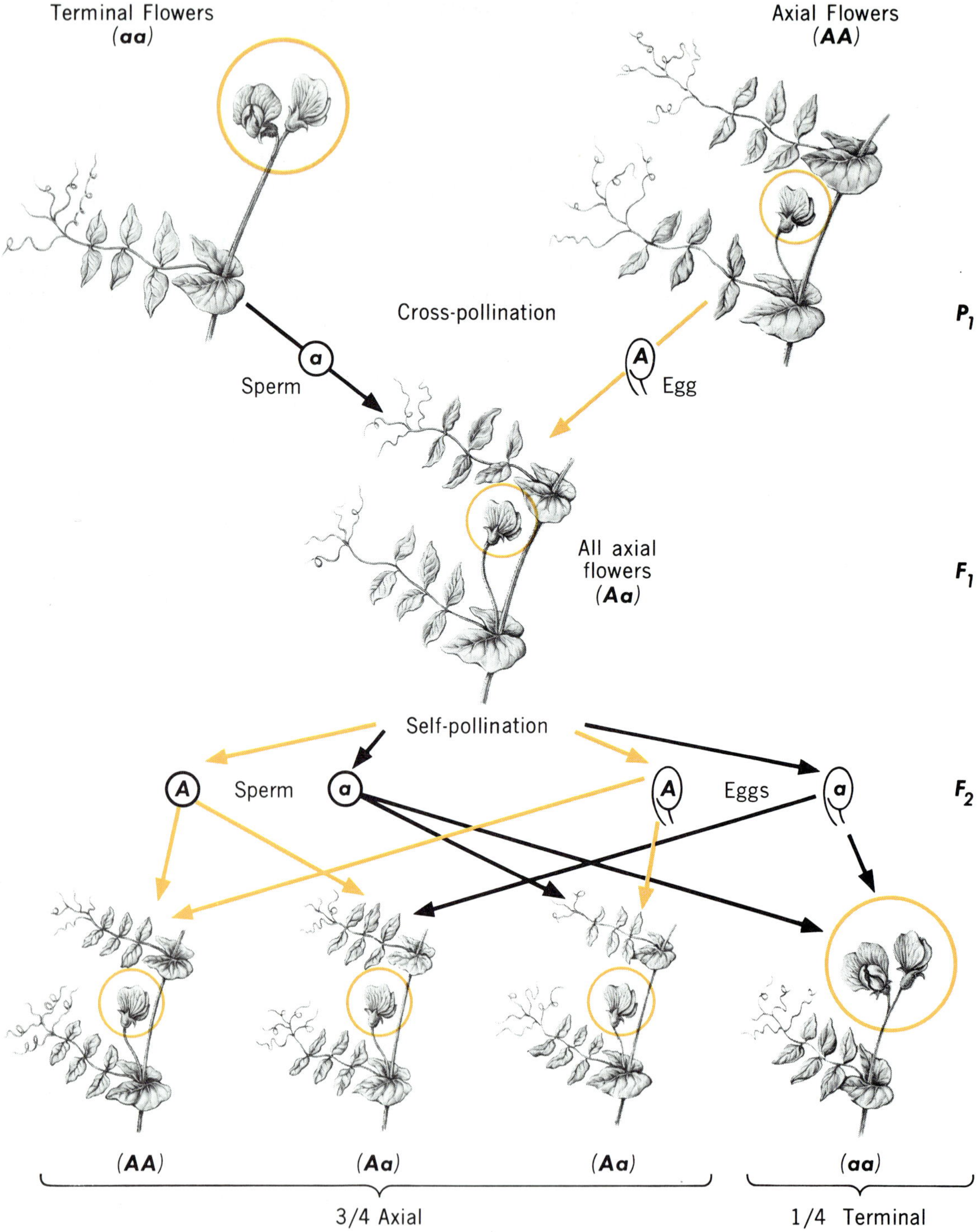

INQUIRY 5-3 A Study of Genetic Probability

Much of what happens in genetics studies is a matter of chance. "Chance" refers to any situation in which there are many factors affecting the outcome. With so many factors, we can never determine one cause. The expression "choosing at random" means choosing by chance.

When scientists must deal with chance, they figure the **probabilities** of the different things that might happen. In this way, they determine what is most likely to happen. There are some basic rules that are used in figuring probabilities. One of these states that *the results of one trial of a chance event do not affect the results of later trials of the same event.* For example, no matter how many times in a row a coin comes up heads, the next tossing of the coin will not be affected by the results of earlier tosses. Each toss of a coin is *independent* of the other tosses. Any two or more tosses are said to be **independent events.** If we toss a coin nine times and obtain nine heads, we may feel that a tail is "due" on the next toss. But the probability of a tail on the tenth toss is still one out of two, or 1/2.

The basic question in probability is "How often should we expect a particular event to occur in a given number of events?" You will learn how to determine this as you consider the probabilities of genetic events in this inquiry.

MATERIALS (per team)

dice (of different colors)

Procedure (Part A)

1. How many sides does each die (singular of dice) have? **2.** What is the chance, on any throw of the die, that a particular number will come up?

$$\text{probability} = \frac{\text{number of chances for a particular event}}{\text{total number of equally likely events}}$$

3. What is the probability that an even number—any even number—will come up? **4.** What is the probability that an odd number will come up?

Use only one die, and work in pairs. One person will roll the die, while the other records the results. Assume a heterozygous parent (*Aa*). Let all even numbers stand for the probability that the gene for the dominant trait (*A*) will go into one reproductive cell. Let all odd numbers stand for the probability that the gene for the recessive trait (*a*) will go into another reproductive cell.

Prepare a score sheet with two columns, "EVEN (A)" and "ODD (a)." Throw the die 100 times. After every tenth throw, draw a line across the score sheet and total the two columns. When you have finished, find the grand totals. **5.** How do the results of each group of ten throws compare with the expected ratio of 1/2 even:1/2 odd? **6.** How does the total for 100 throws compare with the expected ratio?

The difference between what you obtain and what you expect is known as the **deviation.** For example, in each lot of ten throws, you expect by chance to get five odd and five even. If you get six odd and four even, the deviation was one (more) odd and one (less) even. This is a total deviation of two, or 20 percent of the ten throws. **7.** Calculate the deviations for each of the ten groups and for the grand totals. Write them as percentages. **8.** Is the deviation for a group of ten throws greater or less, as a rule, than the deviation for 100 throws? **9.** What happens to deviation when the number of throws is large? How does this show the importance of the number of trials used in studies involving probabilities?

Assume both a male and female parent, each with the genotype *Aa.* Use two dice, and let one represent the female cell and the other represent the male cell. Use dice of different colors to stand for the two kinds of cells. The combination that turns up with a roll of the dice will stand for the random probability of the *A* (even) and *a* (odd) female cells combining with *A* or *a* male cells.

There are four possible combinations of odd and even numbers when you roll the two dice. What genotypes do they stand for? Prepare a score sheet to record the genotypes that turn up when you roll the dice. Use the following headings: "BOTH EVEN (AA)"; "FEMALE EVEN/MALE ODD (Aa)"; "FEMALE ODD/MALE EVEN (aA)"; "BOTH ODD (aa)." Make a total of 40 throws. **10.** How many times does each genotype show up? **11.** Why do all combinations show up with nearly equal frequencies? **12.** What common fraction best expresses the probability of any one of the four particular combinations? **13.** What is the relation of the combined probability of *A* and *A* (*AA*) to the individual probabilities of *A* in one cell and *A* in the other? **14.** In what way might you figure the combined probability from the individual ones?

You have demonstrated another of the principles of probability: *The probability that two independent events will occur together is the product of their individual chances of occurring separately.* (This is sometimes called the Product Rule of Probability.) Let's see how the rule works. When rolling dice, the chance of rolling an even number on one die is 50 percent, or 1/2. The chance of rolling an even number on the other die is the same—50 percent, or 1/2. Using the Product Rule, we can calculate the probability that *both* dice will come up even. The probability is the two chances multiplied together; that is, $1/2 \times 1/2 = 1/4$ (25 percent).

Using the Product Rule, you should be able to calculate the probabilities of the four classes in the preceding example by simple multiplication. Use *A* as a symbol for even and *a* as a symbol for odd. You can then express the probabilities as shown in Figure 5-12. **15.** Copy and complete the checkerboard. What are the probabilities of each combination? Remember *A* is dominant over *a*. Then the phenotype of the combination *Aa* (or *aA*) could not be distinguished from *AA*. **16.** What would be the ratio of dominant to recessive phenotypes? **17.** To what genetic ratio does this correspond?

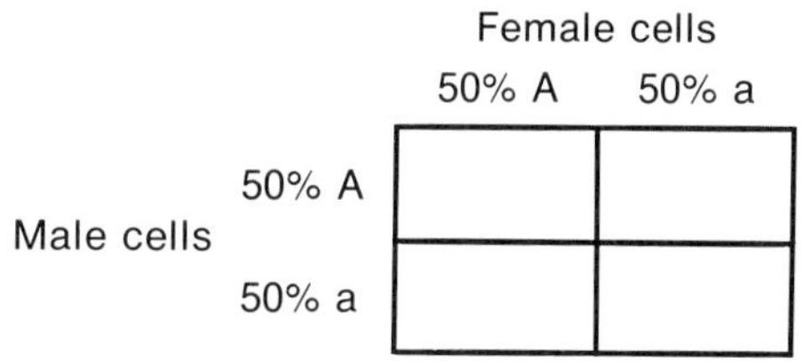

5-12. Probability checkerboard.

Procedure (Part B)

Is the sex of an individual determined by chance? You now can apply what you know about probability to the study of the inheritance of sex. **18.** Recall the Product Rule of Probability. What two probabilities multiplied together will give a probability of 1/2? **19.** What genotype would one parent have to be to produce this ratio? **20.** What genotype would the other parent have to be?

How many brothers and sisters do you have? You may have two brothers and one sister. Including yourself, this is a total of four children. Combine the totals of the families of all your classmates. Add up the numbers of males and females as well as the total number of children. **21.** Compare the totals with an expectation of 1/2 males, 1/2 females. How well do your data agree with the prediction? **22.** Among two-child families in your class, what is the ratio of families that have one boy and one girl to those that have two children of the same sex? **23.** What might cause differences between class data and the mathematical prediction?

You can now apply what you have learned about probabilities of combinations of independent events. Is the sex of an individual really independent from one birth to another in the same family?

Use the dice to work out the probabilities for families of different sizes. **24.** Suppose a family has three girls. What is the probability that the fourth child will be a girl? What is the probability that the fourth child will be a boy? Remember that

$$\text{probability} = \frac{\text{number of chances for a particular event}}{\text{total number of equally likely events}}$$

INHERITANCE IN ANIMALS

5-13. *What are the genotypes of these two guinea pigs?*

Mendel worked out his ideas using one kind of organism, the garden pea. Since Mendel's time, geneticists have learned a great deal more about inheritance. But they have found that Mendel's basic ideas are true in all species of plants and animals that reproduce by the combining of two reproductive cells.

For example, we can cross a pure, black variety of guinea pig (*BB*) with a pure, white variety (*bb*). All of the F_1 offspring from this cross are black. The black trait is dominant to white. But the F_1 offspring have two genes for coat color—one for black and one for white (*Bb*). When the F_1 male and female guinea pigs form reproductive cells, 50 percent of the cells carry *B* and 50 percent carry *b*. The expected ratio of genotypes in the F_2 generation is 1/4 (25 percent) *BB*, 1/2 (50 percent) *Bb*, and 1/4 (25 percent) *bb*. What is the ratio of phenotypes in this generation?

In genetic studies, the word **allele** (uh·LEEL) refers to one of the forms of a gene. In the gene pair *Rr*, *R* is one allele, *r* is the other. An individual organism in which both genes are the same allele (*RR* or *rr*) is described as **homozygous** (hoh·moh·ZY·gus). If the alleles are different (*Rr*) the individual is **heterozygous** (het·er·oh·ZY·gus).

The Test Cross

Homozygous and heterozygous individuals may have the same phenotype, even though their genotypes are different. How, then, can their genotypes be determined? One way is to make a test cross. Suppose you had a black guinea pig whose parents were not known to you. How could you find out whether the guinea pig was homozygous or heterozygous for black? Figure 5-14 shows how a test cross can answer the question. The black guinea pig is mated to a white one (homozygous, recessive, *bb*). Several litters from this pair are observed. If all the offspring in these litters (F_1) are black, the black parent guinea pig is, in all probability, homozygous for the dominant trait (*BB*). If some of the F_1 offspring are white, the black parent must be heterozygous (*Bb*). Can you see why? Use a check-

5-14. Making a test cross.

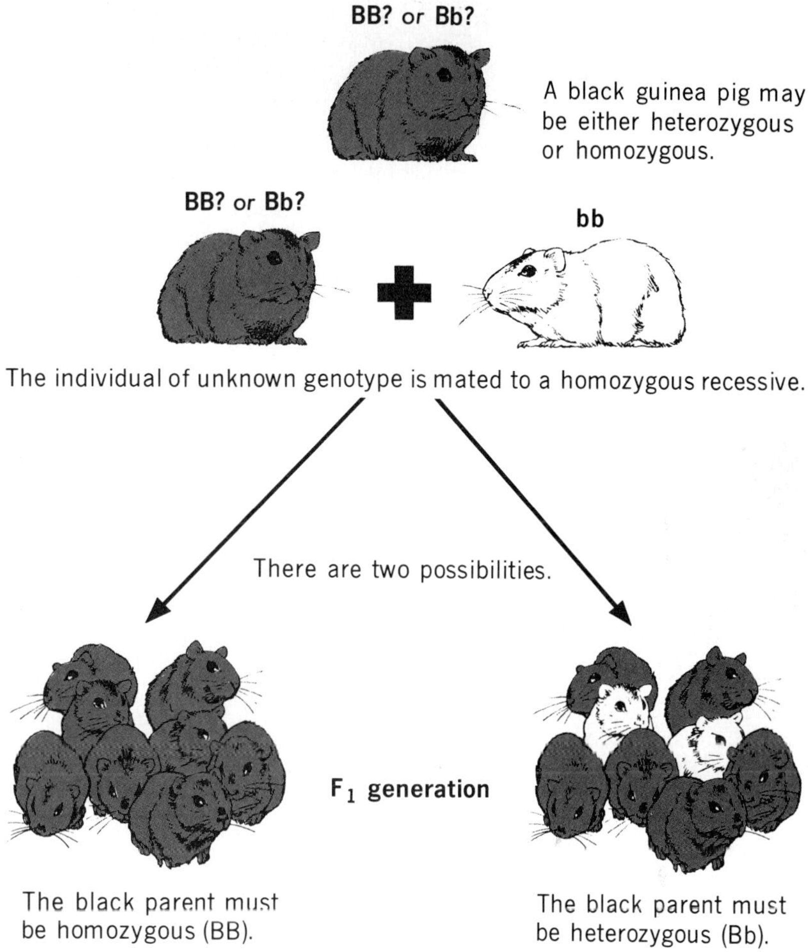

erboard to calculate the two possible crosses (*BB* × *bb* and *Bb* × *bb*).

PROBLEM 1: How would you carry out a test cross to determine the genotype of a round-seed pea plant?

PROBLEM 2: In sheep, white coat is dominant to black. Occasionally, a black sheep appears in the flock. The farmer cannot sell black wool. How can the gene for black coat be eliminated from the flock?

Inheritance of Two Traits

Mendel also studied the inheritance of two different traits at the same time. For example, he crossed a pure variety of plants that had round, yellow seeds with a pure variety that had wrinkled, green seeds. All the F_1 plants had round and yellow seeds. The F_1 was allowed to self-pollinate. The F_2 generation consisted of 315 plants with round, yellow seeds; 108 plants with round, green seeds; 101 with wrinkled, yellow seeds; and 32 with wrinkled, green seeds.

In this F_2 generation, then, there are *four* phenotypes. Two are the same as those of the original parents. But two are not—round, green seeds and wrinkled, yellow seeds. The two pairs of traits seem to be combining in all four possible ways. Apparently, these traits are inherited independently. The inheritance of *R* and *r* (for round and wrinkled) has no effect on the inheritance of *Y* and *y* (for yellow and green).

We know that the ratio of round- to wrinkled-seed phenotypes in the F_2 generation will be 3/4 to 1/4 (3:1). The ratio of yellow to green phenotypes also will be 3/4 to 1/4.

Of the 3/4 of the F_2 generation that are round, 3/4 should also be yellow, and 1/4 should also be green. Therefore,
3/4 of 3/4, or 9/16, should be round and yellow
1/4 of 3/4, or 3/16, should be round and green

Of the 1/4 of the F_2 generation that are wrinkled, 3/4 should be yellow and 1/4 should also be green. Therefore,
1/4 of 3/4, or 3/16, should be wrinkled and yellow
1/4 of 1/4, or 1/16, should be wrinkled and green

Eggs

Sperm	**RY**	**Ry**	**rY**	**ry**
RY	**RRYY**	**RRYy**	**RrYY**	**RrYy**
Ry	**RRYy**	**RRyy**	**RrYy**	**Rryy**
rY	**RrYY**	**RrYy**	**rrYY**	**rrYy**
ry	**RrYy**	**Rryy**	**rrYy**	**rryy**

5-15. A genetic checkerboard of Mendel's cross involving two traits. Genotypes of sperm cells and eggs of F_1 individuals appear at the left side and above the checkerboard. Genotypes of F_2 individuals appear in the squares. *What is the ratio of phenotypes in F_2 individuals?*

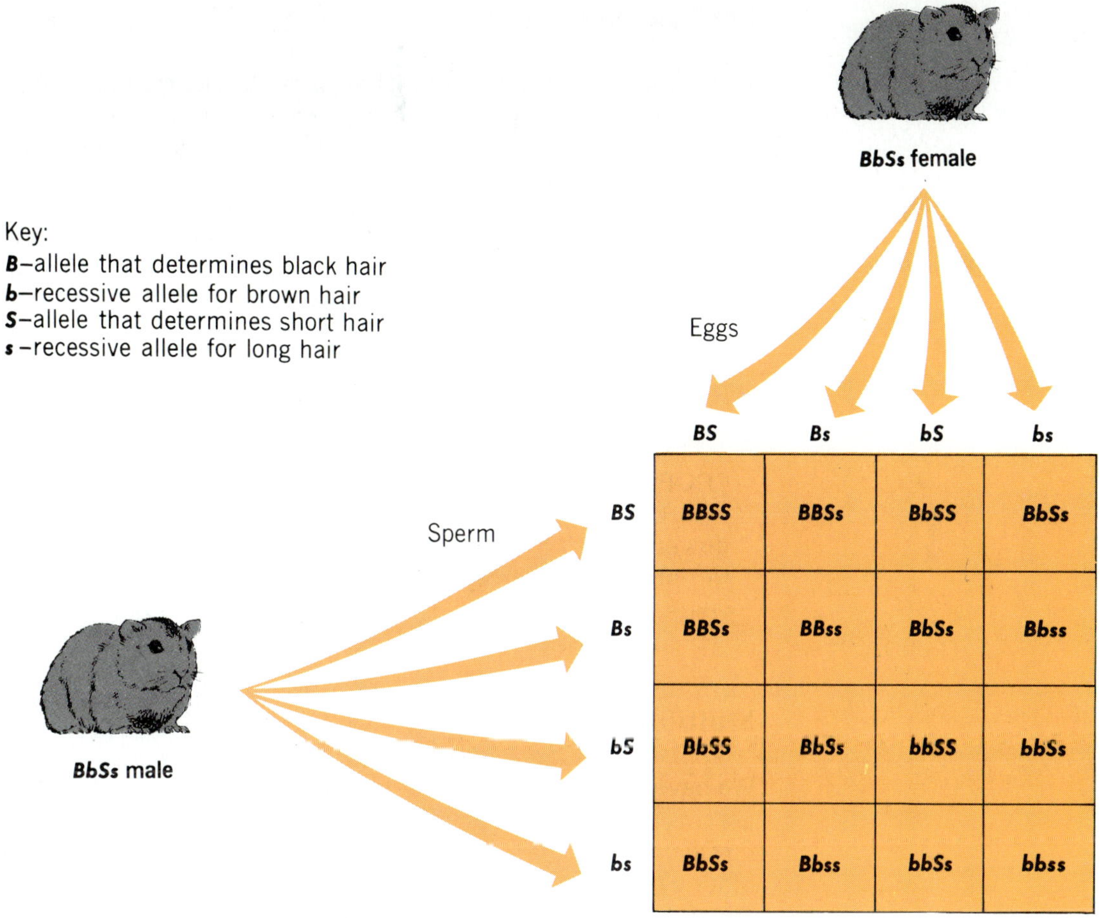

5-16. Expected genotypes in offspring. Both parents are heterozygous for two traits. *How many different genotypes and phenotypes can be found in the offspring?*

This is a ratio of 9/16 to 3/16 to 3/16 to 1/16, or 9:3:3:1. Check to see how closely Mendel's results came to these expected results. The cross is diagrammed in Figure 5-15. Mendel found that this was true for all combinations of the seven pairs of traits in his pea crosses.

PROBLEM: In guinea pigs with colored hair, the hair may be black (*B*) or brown (*b*). Black is dominant to brown hair. Also, the hair may be short (*S*) or long (*s*). Short hair is dominant to long hair. In each cross described so far, we have begun with homozygous individuals. This time we will cross animals of heterozygous genotypes: *BbSs* × *BbSs*. What is their phenotype? What will be the genotypes and phenotypes of the F_1 offspring? Try this problem on your own. (Check your answer with Figure 5-16.)

Lack of Dominance

So far, we have spoken of traits as dominant or recessive. But there are other possibilities. In shorthorn cattle, for instance, when a red bull (*RR*) is crossed with a white cow (*rr*), the heterozygous offspring (*Rr*) are neither red nor white. Instead, they are *roan,* which means they have mixed red and white hairs. In a cross between a roan bull (*Rr*) and a roan cow (*Rr*), what are the chances that the calf will be white (*rr*)? What are the chances it will be roan (*Rr*), or red (*RR*)? Use the checkerboard method to calculate these.

PROBLEM 1: Work out a cross of roan × red.

PROBLEM 2: A farmer wants to have a pure strain of roan cattle that breeds true. Is this possible? Why, or why not?

PROBLEM 3: If a four-o'clock plant with red flowers is crossed with a white-flowered four-o'clock, the F_1 plants all have *pink* flowers. From this evidence alone, what ratios would you expect to find in these crosses: red × red? red × pink? pink × white? white × white?

Multiple Alleles

We have learned how to deal with genetic problems in which there are two alleles of a gene pair—*A* and *a,* for instance. For many traits, however, more than two alleles of the gene may affect the trait. In other words, there may be **multiple alleles.** Any combination of any *two* of these may occur in the same organism. As always, one allele comes from each parent.

The inheritance of blood type in humans is an example. A person has type A, B, AB, or O blood. This depends on whether a certain type of protein occurs on the red blood cells. These blood proteins are called **antigens** (AN·tih·jenz). There are two antigens, A and B. A person with antigen A on the red blood cells has type A blood. A person with antigen B has type B blood. The blood of someone who has neither antigen is type O. A type AB person has red blood cells that have both antigens.

Three alleles are involved in determining blood type. These are indicated by I^A, I^B, and *i.* (*I* and *i* are used to indicate the dominant and recessive traits.) Any two of the three traits may be combined. Both I^A and I^B are dominant. If they occur together, both A and B antigens are produced. The individual has type AB blood. When neither dominant trait is present, the genotype is *ii.* This person has type O blood. Type A and type B persons may have either of two genotypes, however. In some instances, the genotype may be determined from

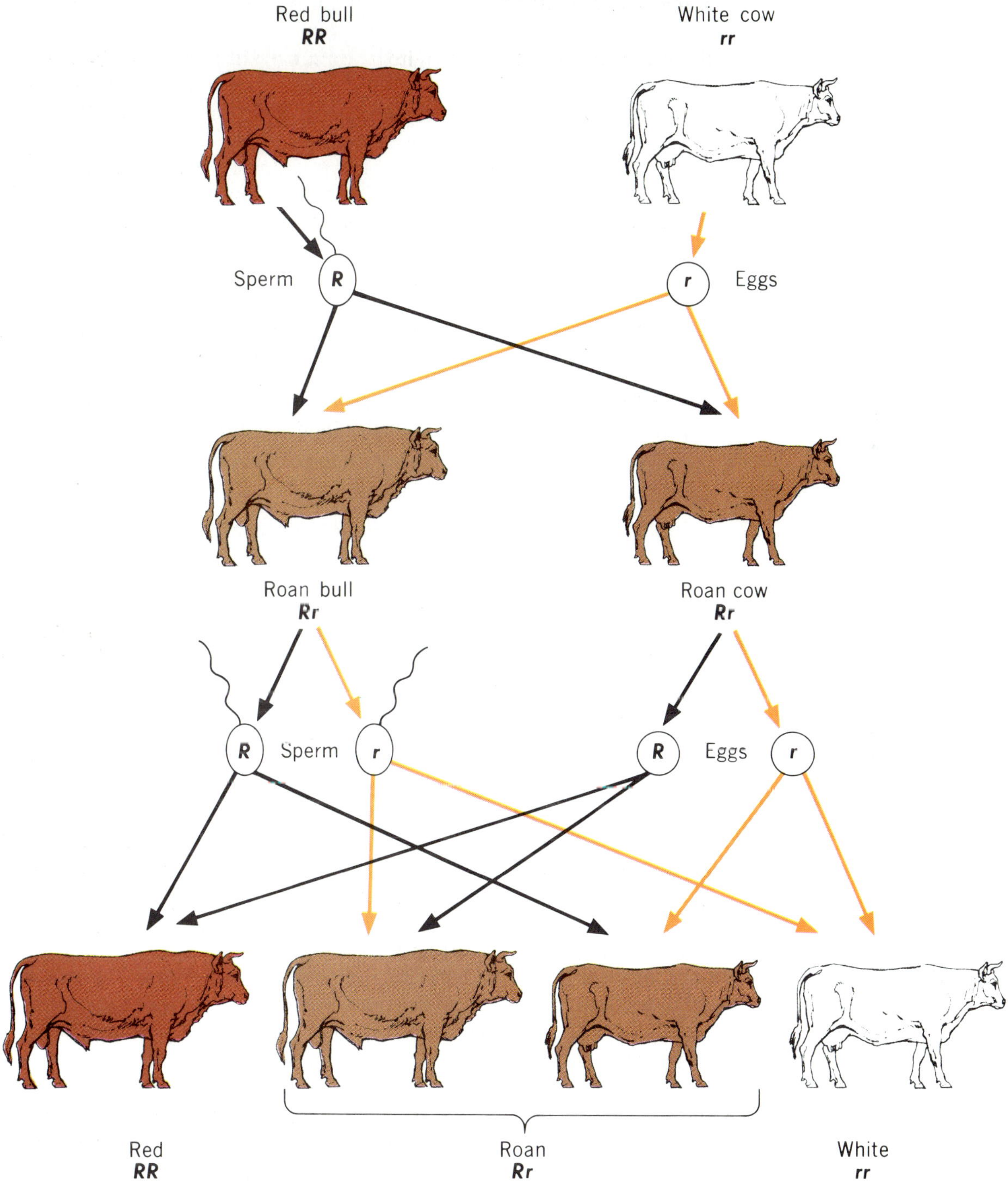

5-17. Inheritance of coat color in shorthorn cattle. Neither white nor red is dominant.

the blood types of the parents or offspring. These relationships are summarized below:

Phenotype	*Genotype*	*Antigen on red blood cells*
Type A	I^AI^A or I^Ai	A
Type B	I^BI^B or I^Bi	B
Type AB	I^AI^B	A and B
Type O	ii	none

Try working out the following problems:

PROBLEM 1: Suppose a woman with type A blood marries a man who has type AB blood. What blood types would you expect to find among their children? In this problem we have no way of knowing whether the woman is homozygous or heterozygous for the I^A trait. What would tell you which of the genotypes she has?

PROBLEM 2: Suppose two newborn babies were mixed up in the hospital. From the following blood types, determine which baby belongs to which parents:

Baby 1	Type O	Mrs. Brown	Type B	Mrs. Smith	Type B
Baby 2	Type A	Mr. Brown	Type AB	Mr. Smith	Type B

What is the genotype of each of the six persons?

Continuously Varying Traits

When Mendel conducted experiments with pea plants, he worked with traits that contrasted distinctly. For example, he studied wrinkled and round seeds, yellow and green seeds, axial flowers and terminal flowers. But what about such traits as the length of a giraffe's neck, or the color of human skin? In such cases as these, there may be thousands of variations. These variations can hardly be accounted for by a single pair of genes. Even when neither trait is dominant, one pair of alleles can yield only three phenotypes. Many such cases of varying traits are due to multiple alleles. But other characteristics seem to have a continuous variability.

Human height, skin color, and intelligence are examples of continuously varying traits. Geneticists have developed the hypothesis that many different pairs of genes may affect the same trait. For example, in a certain organism, gene X produces a certain degree of red coloring and *XX* is redder than *Xx*. Genes *Y* and *Z* also produce red coloring. The recessive alleles of the three genes, *x, y,* and *z,* produce no color. In such a case, an organism with the genotype *xxyyzz* is pure white. An organism with the genotype *XXYYZZ* has six color-forming genes and is very dark red. An organism with *Xxyyzz* has only one kind of red coloring and is a very light shade of pink. Those with *XXyyzz, xxYYzz,* and *xxyyZZ* each have two genes for red and are a darker shade. In this way, as more genes for color occur in the genotype, the coloring becomes darker. For this organism, there are six shades of red that may occur in the phenotype.

Based on such interpretations, new strains of hybrid corn were developed by American geneticists E. M. East, G. H. Shull, and D. F. Jones. Hybrids often are more vigorous than the parent types. That is, the hybrids produce a much higher yield but require no more land, fertilizer, or labor than ordinary strains. To understand this, let's assume that the genes affecting yield in corn are *W, w, X, x, Y, y, Z,* and *z*. Geneticists reasoned that, according to the hypothesis, a cross between pure strains of *WWXXyyzz* and *wwxxYYZZ,* should produce the genotype *WwXxYyZz* in the hybrid. If any of these genes are dominant for the desired trait, the hybrid should be better, in terms of that trait, than either of the parents.

These geneticists selected pure strains of corn and crossed them. The F_1 hybrid plants were much better in yield than either of the pure parent strains. The hypothesis was indeed correct. The F_2 generation, however, was greatly variable. The crops ranged from the very best-yielding to quite worthless types.

Jones then worked out what he called a "double cross." He combined four pure lines to get a productive hybrid. Line A was crossed with line B, and line C was crossed with line D. Then the A-B hybrid was crossed with the C-D hybrid. In this way, recessive, poor-yielding genes became ineffective. Only high-yielding, dominant genes remained effective (Figure 5-18). But such hybrids do not breed true. Thus, the farmer must buy new seeds the next year.

The methods of these geneticists have been used in breeding other important food plants, such as wheat and rice. Rice is the main food of many peoples living in the tropics and subtropics of the Old World. The rice as grown in India, Southeast Asia, or the Philippines has a much higher productivity than wild rice. That improvement occurred over thousands of years of cultivation. Farmers could select the seeds of the better strains for future planting without any knowledge of genetics.

But most cultivated rice has serious defects. Because the stems are long and slender, rice plants often fall in stormy weather and are lost in the mud. Rice also is subject to severe diseases caused by various fungi. By applying the principles of heredity, plant geneticists have produced a "miracle rice." This strain has strong, short

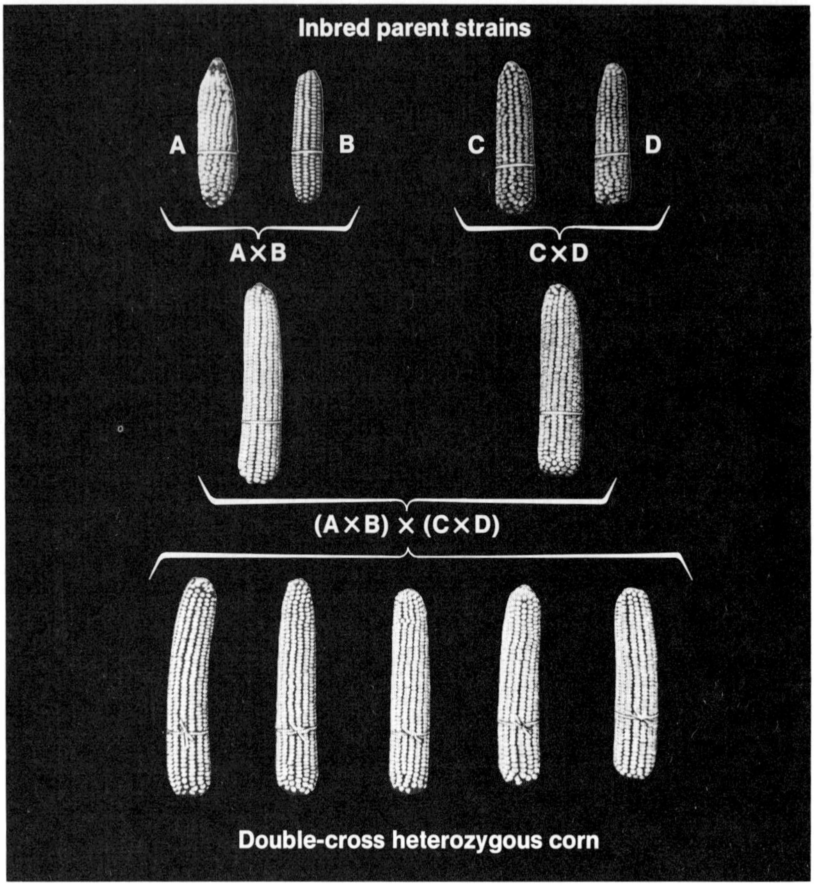

5-18. How hybrid corn is produced. A, B, C, and D are four inbred types. They are crossed to produce F_1 hybrids. The two F_1 hybrids are then crossed to produce the F_2 hybrid corn.

stalks; improved resistance to disease; and high productivity (Figure 5-19). There are still some drawbacks, even to miracle rice. It requires expensive fertilizers, most of which come from scarce petroleum. Also, some people won't eat it because of its different taste and appearance.

Miracle wheats also have been produced for various parts of the world. And a special hybrid corn is changing the diets in many tropical lands. The new corn is high in lysine, one of the necessary amino acids that is nearly absent in ordinary corn.

HUMAN HEREDITY

Soon after 1900, biologists discovered that Mendel's rules apply to humans. Figure 5-20 shows a few of the nearly 3,000 human traits that we know have a simple Mendelian basis.

5-19. A field of high-yielding rice plants, "miracle rice."

Dominant	*Recessive*
A and B antigens	no A or B antigens
Rh antigen	no Rh antigen
brown teeth	normal
cleft chin	no cleft
achondroplastic dwarfism (failure of growth)	normal size
wet, sticky, ear wax	dry, crumbly, ear wax
Huntington's disease	normal
incisor teeth missing	incisors present
mid-digital hair (on fingers)	no mid-digital hair
extreme myopia (short-sightedness)	normal
polydactyly (extra fingers)	normal
normal	albinism (no pigment in skin, hair, eyes)
white forelock	normal
normal	Tay-Sachs disease
normal	cystic fibrosis (a lung disease)
some forms of muscular dystrophy	normal

5-20. Some genes in humans that are inherited in a simple Mendelian fashion.

In spite of the great interest in the subject, the study of human heredity is difficult. The reasons are obvious if you think about what geneticists try to do in the laboratory. They prefer to work with pure lines, which they obtain by breeding animals for many generations that are closely related. In addition, they repeat crosses many times and try to obtain many offspring. And, of course, they try to keep the experimental organisms in a constant environment. Although it is not possible to carry out such experiments on people, geneticists have learned a great deal about human genetics by tracing the inheritance of traits.

Current research in genetics is uncovering evidence that a wide variety of human traits is inherited. We will list a few human traits and indicate briefly what we know of their heredity.

Blood Types

Many different genes affect the blood. For example, a dominant gene (*R*) produces Rh antigen. A person who is *RR* or *Rr* is Rh-positive. These people make up about 85 percent of the Caucasian population. The remaining 15 percent are *rr*, or Rh-negative. The data are not the same for all populations. For instance, in Oriental, American Indian, and Eskimo populations, the Rh-negative allele is almost completely absent.

The Rh factor may be very important to the health of a developing infant, which is called a **fetus.** If the mother is *rr* and the father is *RR,* the offspring will be *Rr* (Rh-positive). If the father is *Rr,* the offspring will be either *Rr* or *rr* (Rh-negative). The red blood cells of the Rh-positive fetus (carrying Rh antigen on their surfaces) may pass accidentally into the mother's blood. If she is *rr,* her body may produce antibodies against the Rh antigen.

When antibodies are produced by the mother's blood, she becomes *sensitized* against Rh-positive blood. This sensitization occurs most often at birth, when blood vessels rupture in the **placenta** (pluh·SEN·tuh), a special structure that develops in the uterus. (Through the placenta, oxygen, carbon dioxide, and nutrients are exchanged between the mother and the fetus.) Some of the Rh-positive blood of the baby then mixes with the Rh-negative blood of the mother. The mother's blood produces antibodies, which have the ability to destroy Rh-positive blood cells.

The blood of the first Rh-positive child of an Rh-negative mother rarely is affected by the antibodies. During a second pregnancy, however, antibodies in the mother's blood can cross the placenta and destroy the blood cells of the Rh-positive fetus (Figure 5-21). If the damage is widespread, the fetus will die. If there is less damage, the baby may live, but it will have Rh disease. Symptoms of Rh dis-

SCIENCE AT WORK: Putting Human Genes on the Map

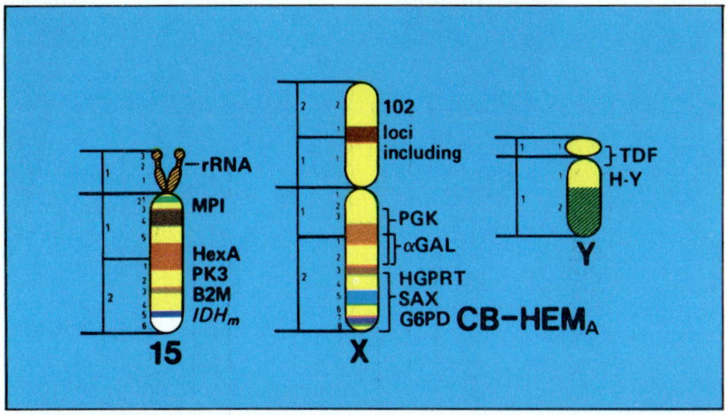

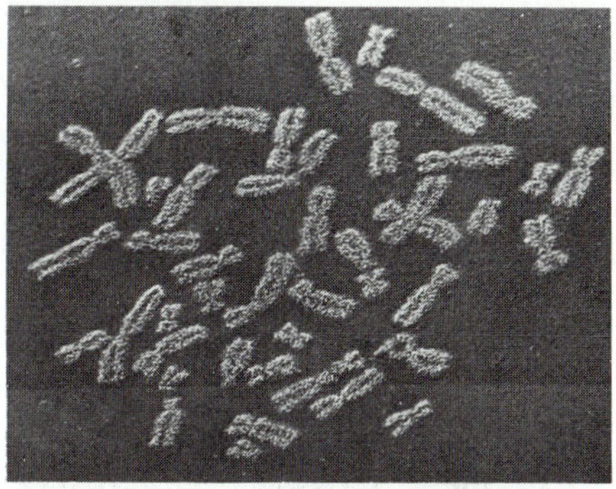

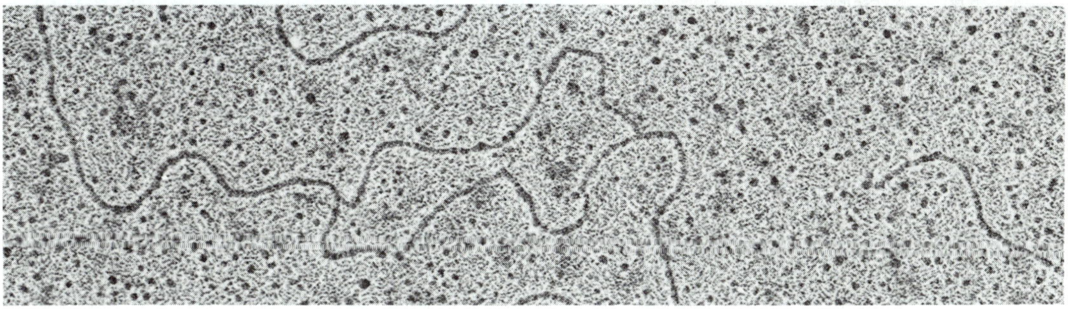

At Yale University, a team of geneticists, biochemists, and cell biologists are in the process of mapping all of the gene locations on the chromosomes in a human cell. Their work is a good illustration of how new discoveries or developments in science usually are built on the knowledge provided by the work of many earlier scientists.

Mapping human genes has become possible only recently. In 1967, scientists at New York University developed a technique to identify the genetic coding that specifies the synthesis of a particular protein. By 1969, scientists were able to match particular genes to particular chromosomes. Finally, in 1973, researchers learned how to locate genes at specific points on a chromosome. In some cases, they could determine the order in which several genes are lined up on a chromosome.

Using the new techniques, the Yale team, and other researchers in the field, have mapped some one hundred genes, so far. But this is only a beginning, since a human cell is believed to contain about 100,000 genes.

One practical application of the techniques used in mapping genes will be identifying and locating genes that cause human diseases. This information may open up many possibilities for the detection and prevention of human genetic disorders.

Other geneticists have isolated and photographed a specific gene on a DNA strand. The gene for 18s rRNA is contained on the loop at the center of the lower photograph. These scientists are now trying to determine the function of other DNA segments contained in the loop. These segments of "spacer DNA" are thought to regulate the expression of the 18s rRNA gene.

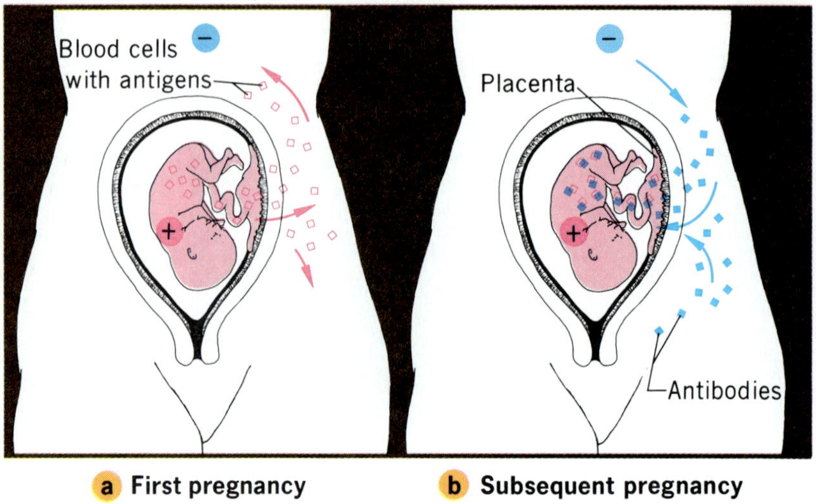

5-21. (a) Accidental mixing of mother's and fetus' blood. Rh antigens (□) on fetus' blood cause production of antibodies (■) in the mother's blood. **(b)** Antibodies in mother's blood destroy fetus' red cells, which carry the antigens.

ease are a shortage of red blood cells and jaundice. (Jaundice is a condition in which the skin is greenish-yellow.) Sometimes, the infant's life can be saved by gradually replacing all of its blood. With modern medical techniques, this can even be done before the child is born.

Another method has proved successful in preventing Rh disease. After the birth of each child or after a miscarriage, the Rh-negative mother is given an injection of a serum called RhoGam, which contains anti-Rh antibodies. This inhibits the production of antibodies in the mother's blood by destroying Rh-positive cells before the mother's body becomes sensitized. Later, the injected antibodies break down. No anti-Rh antibodies are present when the next pregnancy begins. If the next child is Rh-positive, the mother's blood will not carry antibodies that would harm the child.

Eye Color

Eye color is a complex trait that we do not understand completely. It seems that pure blue eye color is due to a single gene that is recessive to a gene for pigmented iris. We must lump together under "pigmented" any eye with an iris that has even the least spot of brown pigment. This includes hazel, green, and gray eyes. Other genes probably also affect eye color.

Skin Color

The amount of dark pigment in human skin seems to be controlled by three to six pairs of genes. The genes assort independently. In each pair, neither allele is dominant. Many shades of skin color exist. Exposure to sunlight also changes skin color.

Height and Weight

There is a definite genetic influence on height and weight. These characteristics probably are influenced by several genes. Environmental differences, such as diet, also play an important role. Occasionally, extreme variations in height occur, such as dwarfism and gigantism. These conditions usually are under genetic control.

Mental Retardation

There are a number of causes for mental retardation. A defect in the function of just one gene may result in an impairment of mental activity. Some conditions of mental retardation are strongly influenced by heredity. Others may be due to injuries before or at birth. One type of mental retardation, *Down's syndrome* (SIN·drome), is usually caused by a single extra chromosome (Figure 6-19). Another well-known type, *phenylketonuria* (fen·il·key·toh·NYOO·rih·a), called PKU, is a simple recessive trait. When this trait occurs, the body fails to produce an enzyme that changes an amino acid in the liver. Tests made soon after birth can detect the disease. It can be controlled by a carefully regulated diet.

Intelligence

Intelligence is not easy to define. It is even more difficult to measure. The most common type of intelligence test is one that gives an intelligence quotient (IQ). In this type of test, intelligence is measured in ways that require memorizing, reasoning, and the ability to visualize objects in space. Ability to use language and vocabulary are important in these tests. Thus, the person's background and education play a large part in determining the results. Judging from IQ tests of identical twins, there is some hereditary basis of intelligence. Environment also plays a major role. It is the interaction of heredity and environment that finally determines the intelligence of an individual. In all probability, many genes contribute to the determination of intelligence.

Human Genetic Diseases

In 1976, 2,500 human genetic diseases had been described, and the list has grown every year since then. Scientists are finding that a "disease" is really the product of interactions between an organism's genetic makeup and its environment. For example, some persons seem to have an inherited resistance to tuberculosis. Under the same conditions of exposure to the disease, one person will develop it and another will not. Diabetes, the disease in which insulin does not properly control the body's use of sugar, is another disease that seems to be inherited—apparently caused by a recessive gene. However, not all forms of diabetes are alike. The disease may be mild or severe, or it may appear very early in life or late in life. The expression of the gene seems to depend on a variety of environmental factors—conditions that may exist either inside or outside the organism.

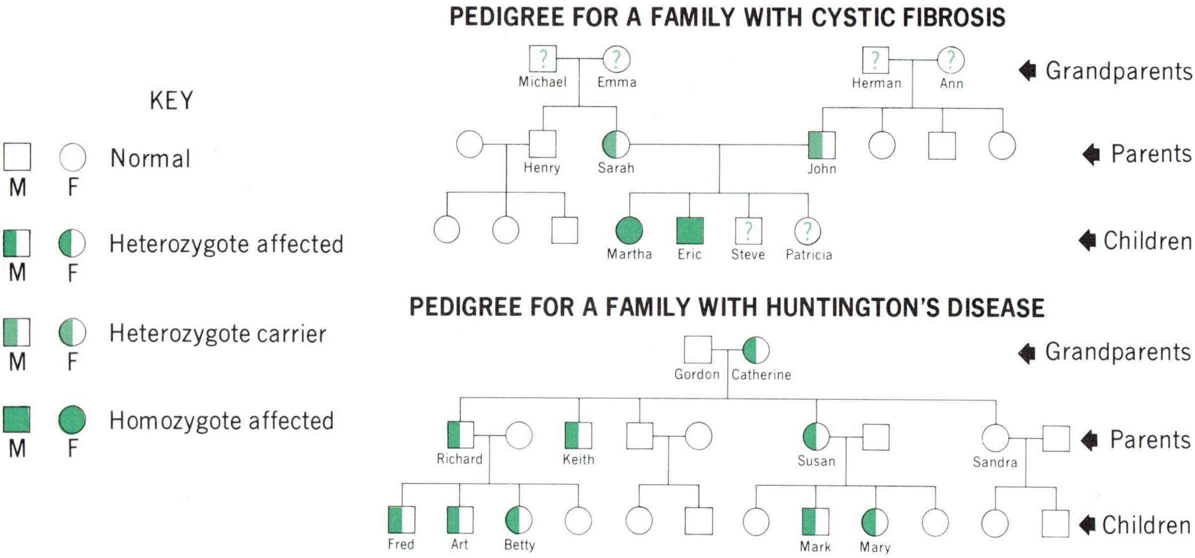

5-22. Pedigrees of families with genetic diseases.

Some human genetic diseases, however, are known to be inherited in a simple Mendelian fashion. As examples, Figure 5-22 shows *pedigree charts* of family histories for cystic fibrosis and Huntington's disease. In these pedigrees, squares represent males and circles represent females. Persons belonging to the same generation are placed on the same horizontal row. Husbands and wives are connected by a horizontal line, and through vertical and horizontal lines to their children in the next generation.

Cystic fibrosis affects children. It is often fatal by the teen or young adult years. In children with cystic fibrosis, a thick, sticky mucus clogs the lungs and breathing passages, making breathing difficult. Cystic fibrosis occurs most often in families of European descent. The disease is inherited as a recessive gene. Thus, a person may carry the gene, but not actually have the disease. The trouble arises when two people who carry the gene marry. They have a 25 percent chance of having an affected child. Study the pedigree for cystic fibrosis in Figure 5-22. Note that the grandparents were not affected but that Sarah and John, the parents in the next generation were both carriers. There is no test to detect a carrier of cystic fibrosis. The fact that the parents were carriers did not emerge until they had their first affected child, Martha. The second child, Eric, also had the disease. Whether or not Steve and Patricia are carriers cannot be determined. The disease will only occur in their children if they also marry a carrier. We cannot know exactly which of the grandparents were carriers. But we can reason that either Michael or Emma must have been a carrier. Either Herman or Ann also must have been a carrier. How do we know this?

The second pattern of inheritance shown in Figure 5-22 is for Huntington's disease. This disease causes a degeneration of the central nervous system. It usually does not appear until a person is in his or her thirties or forties. The disease is characterized by uncontrolled body movements, loss of speech, and mental problems such as recurrent periods of depression and anger. Unlike cystic fibrosis, where two recessive genes must be present before the disease appears, Huntington's disease is caused by a single dominant gene. Thus, the offspring of a parent with the disease have a 50:50 chance of developing the disease. In the figure, the grandmother Catherine was affected by the disease. Three of her five children were also affected. One of the children, Richard, passed the gene to three of his four children. Two of Susan's four children also developed the disease. Special mention should be made of Sandra. At the time this pedigree was made, she was 26 years old. It was too early to tell whether she will develop the disease, and, if she does, whether her children also will develop it.

Recent Advances

Our knowledge of human inheritance has advanced in recent years. Geneticists have identified hundreds of new hereditary disorders. Fortunately, most of them are quite rare. Still, about one third of the admissions to teaching hospitals are children who have some disorder that has a genetic basis.

Much headway has been made in understanding specific genetic disorders. Tay-Sachs disease, for example, causes an apparently healthy baby to undergo a gradual nervous degeneration. The disease is fatal. It is a recessive hereditary trait. The parents of a child with the disease must be heterozygous for the lethal gene. They both carry the recessive form of the gene.

Drs. Okado and O'Brien, at the City of Hope Hospital near Los Angeles, discovered that the problem in Tay-Sachs disease is an inability to break down a waste product formed in the cells of the brain and nervous system. A particular enzyme is needed to dispose of this substance. Like most enzymes, its production is controlled by genes. When the enzyme is not produced, the waste substance accumulates and the child dies. Scientists are now searching for a treatment for this condition.

Through the medical technique of *amniocentesis* (am·nee·oh·sen·TEE·sus), it is possible to detect some genetic diseases

5-23. This couple is planning to have a child. The counselor explains that laboratory tests show that their child will not be at risk for Tay-Sachs disease.

before a baby is born. Cells drawn from the fluid that surrounds the developing baby are grown and analyzed. Some disorders show up in a chromosome analysis. For example, a child that will be affected by one type of Down's syndrome has an extra chromosome. Other disorders, such as Tay-Sachs, show up as abnormal enzyme activity.

It is also possible to detect genetic disorders carried by the parents. Blood tests are used to identify carriers of sickle-cell anemia and Tay-Sachs.

Medical genetics is a new branch of medicine that deals with genetic testing, counseling, and prenatal diagnosis. As our knowledge of medical genetics increases, an increasing number of genetic counselors (Figure 5-23) will be advising parents who are planning their families.

In Brief

The study of modern genetics began with Gregor Mendel. The knowledge gained from Mendel's observations of inheritance in peas holds true for inheritance in other plants and animals. Mendel learned that when two alleles cause different forms of a trait, one trait is often dominant over the other. In a cross of a pure, dominant variety and a pure, recessive variety, only the dominant phenotype appears in the F_1 generation. But the F_1 generation consists of individuals carrying both traits. They are said to be heterozygous for that trait. When the F_1 individuals are allowed to mate, the recessive trait reappears. The ratio of the dominant trait to the recessive trait in the F_2 generation is 3:1. The recessive phenotype in the F_2 generation is caused by the homozygous, recessive genotype.

In some cases, however, neither trait is dominant. The F_1 generation then shows a phenotype that is different from that of either parent. For example, in the flower called four-o'clock, a cross of a plant with red flowers and one with white flowers produces pink flowers in the F_1 generation. When the blood-group alleles I^A and I^B appear together, neither allele is dominant over the other. That is, both A and B antigens are produced.

If two pairs of traits are studied in such a cross, four phenotypes occur in the F_2 generation. The ratio of phenotypes in the F_2 generation is 9:3:3:1. Such studies show that the inheritance of one pair of traits is independent of the inheritance of the other pair of traits. Simple Mendelian rules do not hold for the inheritance of all traits. Some traits, such as skin color, depend on many genes.

We are increasing our knowledge about patterns of human inheritance. New techniques in the rapidly growing field of medical genetics allow diagnosis of some genetic diseases before birth.

5-24. GENEALOGY OF CAPTIVE WHITE TIGERS

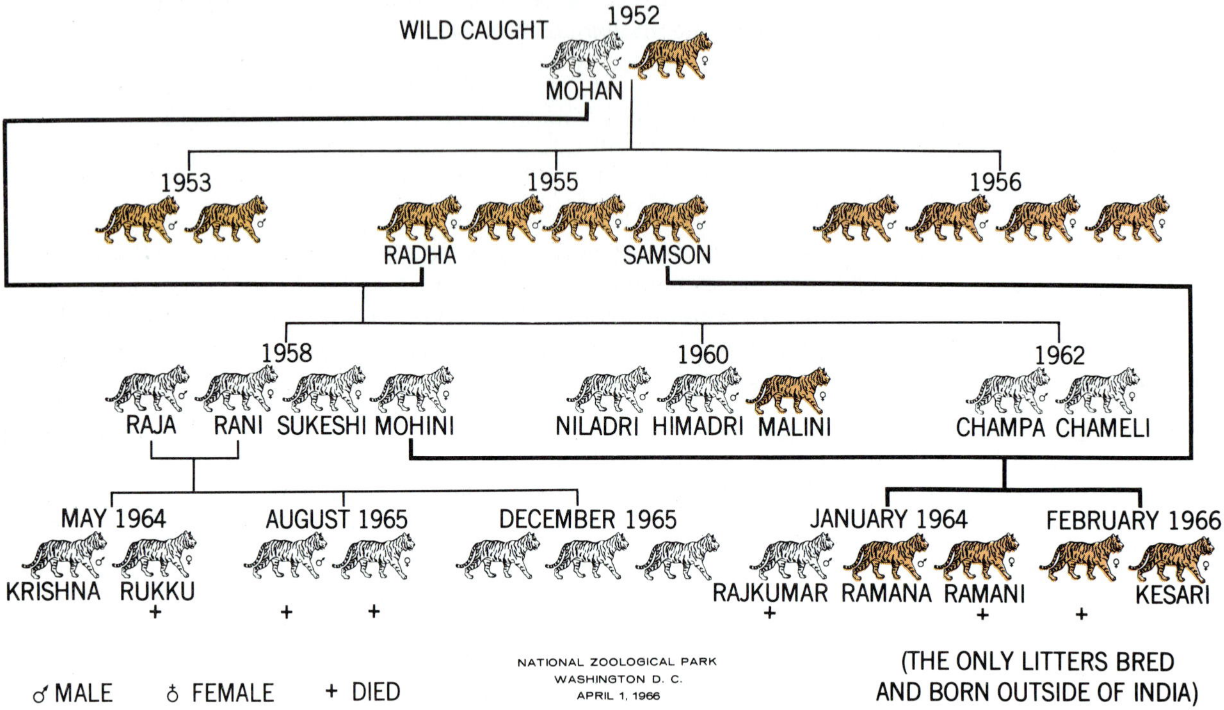

Guide Questions and Problems

1. What is the difference between genotype and phenotype? Why is the distinction between genotype and phenotype important in the study of heredity?

2. In 1952 a male white tiger, Mohan, was captured in India. A family tree of its descendents is shown in Figure 5-24. The heavy black line at the left shows that Mohan was mated with one of his daughters, Radha. The heavy line at the right shows that Samson, one of Mohan's sons, was mated to one of the daughters of Radha (Samson's sister). What is the evidence that the gene for white coat color is recessive in this family of tigers? In what ways do the relatively few offspring *not* reflect expected Mendelian ratios?

3. Suppose you thought you were raising a purebred line of solid black cats. Then you found that some of the cats had spots of white hair at the tips of their tails. How could you determine whether this trait is recessive or dominant? Assume that only one pair of genes is affecting this trait.

4. Two white sheep produce a black offspring. What must the parents' genotype for color be? What is the chance that the next offspring of the two white sheep will be black?

5. Red fruit (*R*) is dominant to yellow fruit (*r*) in tomatoes. Tallness (*T*) is dominant to shortness (*t*) in these plants. Predict the ratios of phenotypes and genotypes for offspring of these parent plants: one is red homozygous and tall homozygous, the other is red heterozygous and tall heterozygous.

6. Among the genotypes listed at the end of this question, which are heterozygous and which are homozygous? Which of the genotypes have the same phenotype? Pick out from the list one or more genes for which only two alleles are given. Also pick out a series of multiple alleles. AA, I^AI^B, ss, Bb, I^Bi, Aa, BB, rr, ii.

7. An extra finger on the human hand occurs rarely. This condition is due to a dominant gene. Suppose one parent is normal and the other parent has an extra finger, but is heterozygous for the trait. What is the probability that their first child will not have an extra finger? their second child? their third?

Ideas for Independent Study

1. Use the checkerboard method to work out the results of a three-trait cross. Let *A* represent the dominant gene for shortness and *a* the recessive gene for tallness. Let *B* represent the dominant gene for black hair and *b* the recessive gene for white. Let *C* represent the dominant gene for curly hair and *c* the recessive gene for straight. What would be the results of a cross of $AaBbCc \times AaBbCc$? What genotypes would occur? What phenotypes would occur? What would be the ratio of phenotypes?

2. All the species of animals and plants around us are products of the processes of inheritance. Yet the mechanisms of inheritance are far from obvious. In the past, people have accounted for the inheritance of special traits in all sorts of ways. For instance, human heredity has been linked to a rather vague idea of transmission of blood from one generation to another. Perhaps you have heard phrases such as "blue blood," "it's in the blood," and so forth. However, you have learned that the reproductive cells carry genetic information from one generation to the next.

Trace some distinctive traits through several (at least three) generations in your family. What can you conclude about the inheritance of those traits?

Related Readings

BOOKS

BSCS. *Food for Humanity.* Environmental Resource Papers. Menlo Park, Calif.: Addison-Wesley, 1975.

Glass, B. *Genetic Continuity.* A BSCS Laboratory Block, Lexington, Mass.: D. C. Heath, 1968.

Hendin, D., and J. Marks. *The Genetic Connection.* Caldwell, N. J.: Morrow & Co., 1978.

McKusick, V. A. *Human Genetics.* Englewood Cliffs, N. J.: Prentice-Hall, 1969.

Milunsky, A. *Know Your Genes.* Boston: Houghton-Mifflin Co., 1977.

Moody, P. A. *Genetics of Man,* 2nd ed. New York: Norton, 1975.

Volpe, E. P. *Human Heredity and Birth Defects.* New York: Pegasus, 1970.

ARTICLES

Cerami, A. and C. M. Peterson. "Cyanate and Sickle-Cell Disease." *Scientific American,* April 1975, page 44.

Friedmann, T. "Prenatal Diagnosis of Genetic Disease." *Scientific American,* November 1971, page 34.

Witkop Jr., C. J. "Albinism." *Natural History,* October 1975, page 48.

6 The Chromosome Theory of Heredity

Mendel's work on inheritance in peas helped scientists to understand patterns of inheritance in other organisms. But many questions remained unanswered. For example, where are genes located? What is their chemical composition? How do they function? It took nearly a century of work after Mendel's experiments to answer these questions.

Research in this area was difficult because no one knew what a gene looked like. One of the earliest clues came from experiments like the one outlined in Figure 6-1. Certain single-celled organisms were cut up and observed for many days. Scientists found that only the piece of each organism that contained the nucleus could survive and reproduce. Biologists concluded that the nucleus must be important in cell function and reproduction.

Other observations also suggested that the nucleus was involved in inheritance. Although egg cells are much larger than sperm cells,

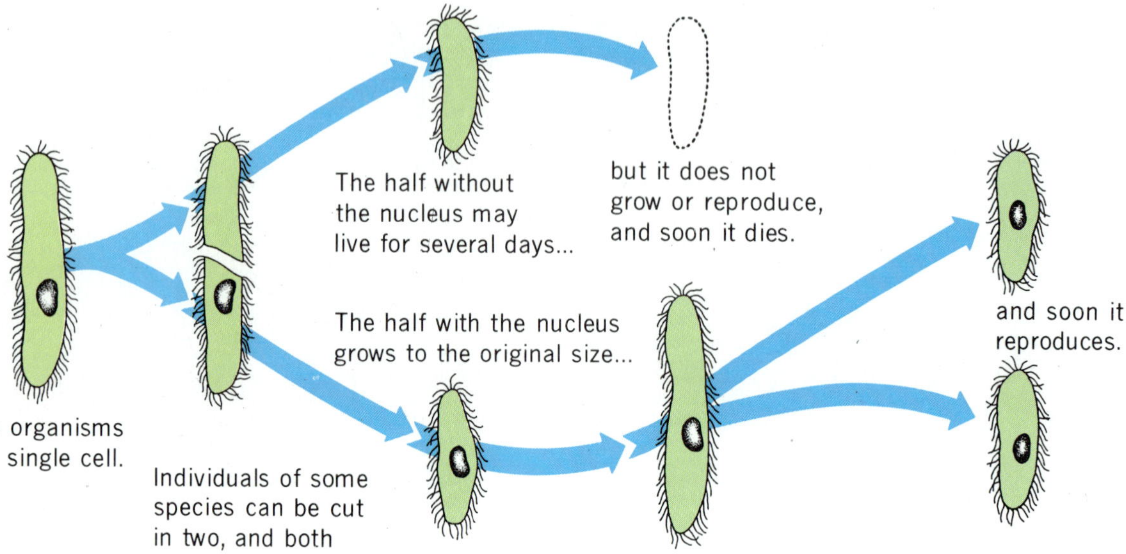

6-1. *What seems to be the role of the nucleus in cell growth and reproduction?*

biologists believed that the two parent cells contributed equally to the heredity of their offspring. The discovery that the nuclei of the sperm and egg are the same size suggested that the nucleus was the seat of inheritance.

If the nucleus is important, what in the nucleus is involved? When this question was first asked, biologists could detect only a fluid and the solid, threadlike chromosomes in the nucleus. Could the chromosomes be involved? Biologists began to learn more about chromosomes and their behavior before and during fertilization.

The upper part of Figure 6-2 is a photograph of the chromosomes of a human female. There are 46 of them. In this illustration, each chromosome is double, since the nucleus is about to undergo mitosis. Biologists learned that, with some exceptions, all cells in the same individual had the same number of chromosomes. And they found that the form of the various chromosomes was the same in every cell. This chromosome constancy held true for all members of the same species. Such observations led to an important conclusion: In any particular species, the chromosomes are the same in all cells (with a few exceptions). For example, the normal number for humans is 46; for potatoes, 48; for hydras and cherry trees, 32; for garden peas, 14; and for the fruit fly, 8.

Slowly, as microscopes and means of studying cells improved, other things were learned about chromosomes. In studies with *Drosophila,* which has only eight chromosomes, observers noticed that the female fruit fly had only four *kinds,* or pairs, of chromosomes. Much later, when it became possible to study human chro-

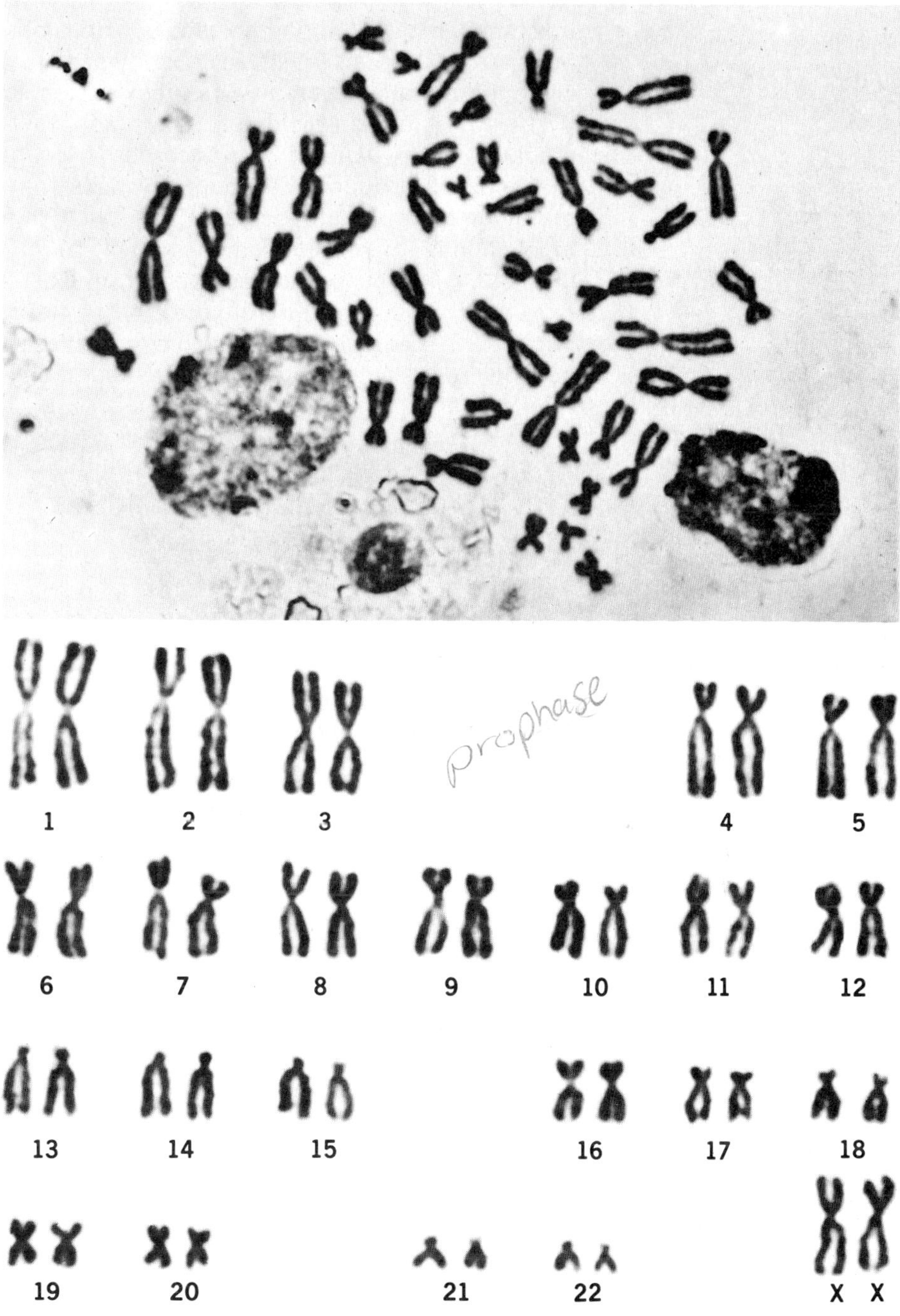

6-2. Two views of the chromosomes of a human female. **(Top)** A photo of the chromosomes of a cell about to divide — note that each chromosome is double. **(Bottom)** A second print of the photograph was cut up to isolate the chromosomes. These were then arranged according to size. *Try to match some of the chromosomes in the two illustrations.*

mosomes, biologists learned that our chromosomes also occur in pairs. The lower part of Figure 6-2 shows the female's chromosomes arranged in pairs. Such an arrangement is a karyotype.

The next question biologists asked was, "What could be the significance of the *pairs* of chromosomes?" In Chapter 4, you learned that meiosis produces reproductive cells that contain the monoploid number of chromosomes. Each sperm or egg has one member of each pair of chromosomes. When sperm and egg unite, the fertilized egg that results has two chromosomes of each kind. One chromosome is inherited from the mother, the other from the father. When the nucleus of the fertilized egg divides during mitosis, all the new cells in the organism have identical pairs of chromosomes in their nuclei.

THE RELATIONSHIP OF GENES AND CHROMOSOMES

A hundred years ago, biologists were aware that male and female parents made an equal contribution to a new individual. They also knew that the chromosome number was the same in all individuals of a species. Some biologists thought the chromosomes carried hereditary characteristics from one generation to another.

In 1902, a young American, Walter S. Sutton, was observing chromosomes in meiosis in grasshoppers. Sutton observed that each reproductive cell formed during meiosis has one chromosome of each kind—one from each pair of homologous chromosomes. Similarly, Mendel's work suggested that only one allele for a particular characteristic was present in each reproductive cell. The union of an egg with a sperm restores the diploid number of chromosomes in the **zygote**—the fertilized egg. Here the chromosomes are in pairs. Mendel's work showed that the genes occur in pairs in the offspring. Each pair of chromosomes assorts, or separates, independently from all other pairs. The possibilities are illustrated in Figure 6-3. Four kinds of reproductive cells are produced by the assortment of two pairs of chromosomes. Similarly, Mendel found that separate pairs of genes assort independently. Two pairs of genes will assort independently to give four kinds of phenotypes, as you observed in the *Drosophila* inquiry in Chapter 5.

Sutton calculated the number of combinations possible for several pairs of *chromosomes*. It was the same as the number of *gene* combinations that explain independent assortment of hereditary characteristics in pea plants. Sutton then proposed the hypothesis that genes are located on the chromosomes.

The grasshoppers Sutton was studying had only 12 pairs of chromosomes. But a grasshopper must have more than 12 pairs of

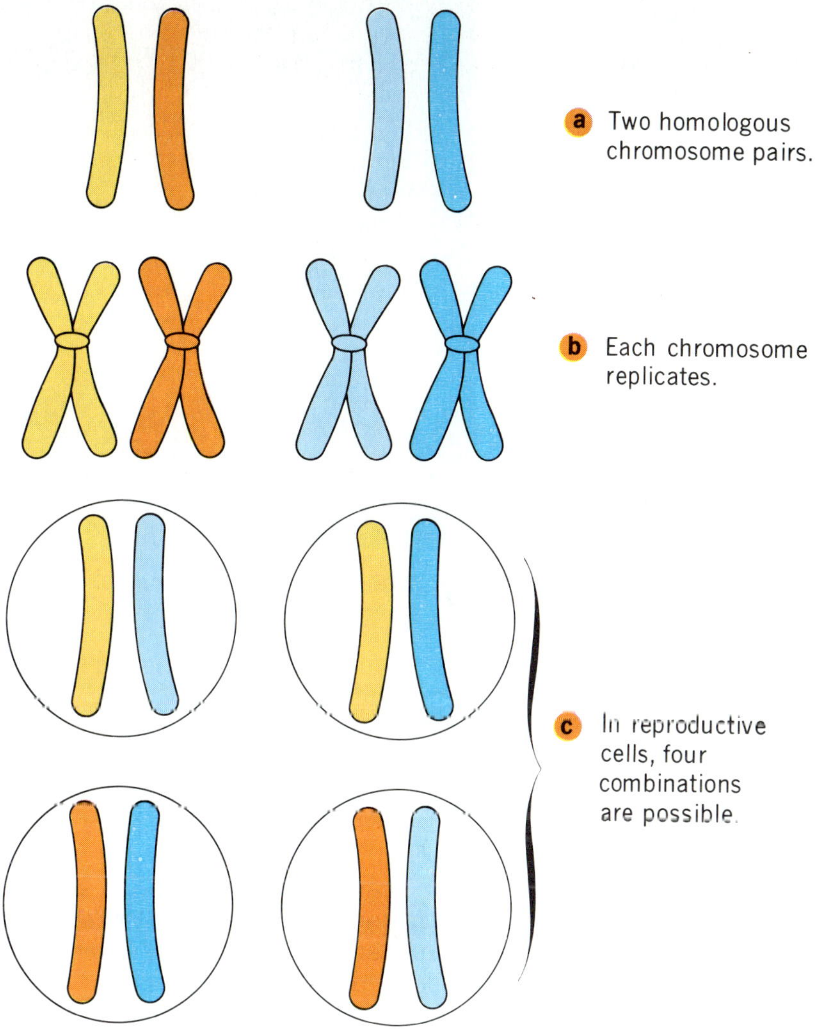

6-3. Four kinds of reproductive cells are produced during meiosis.

genes to control all its hereditary characteristics, just as a human must have more than 23 pairs of genes. Sutton realized that more than one gene pair must be located on each pair of chromosomes.

Sutton also realized that pairs of genes located on the same pair of chromosomes could not assort independently. He therefore made a prediction on the basis of these deductions:

> Pairs of genes carried on the same pair of chromosomes will tend to be inherited together, in a common **linkage group.** There should be, in any species, just as many linkage groups as there are pairs of chromosomes.

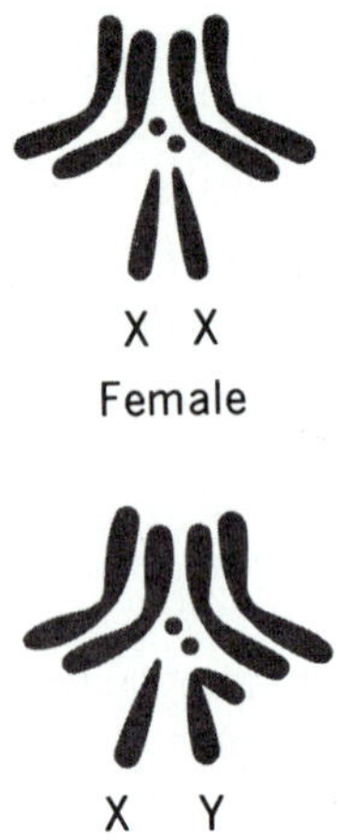

6-4. The four pairs of chromosomes of the diploid cells in female and male *Drosophila*.

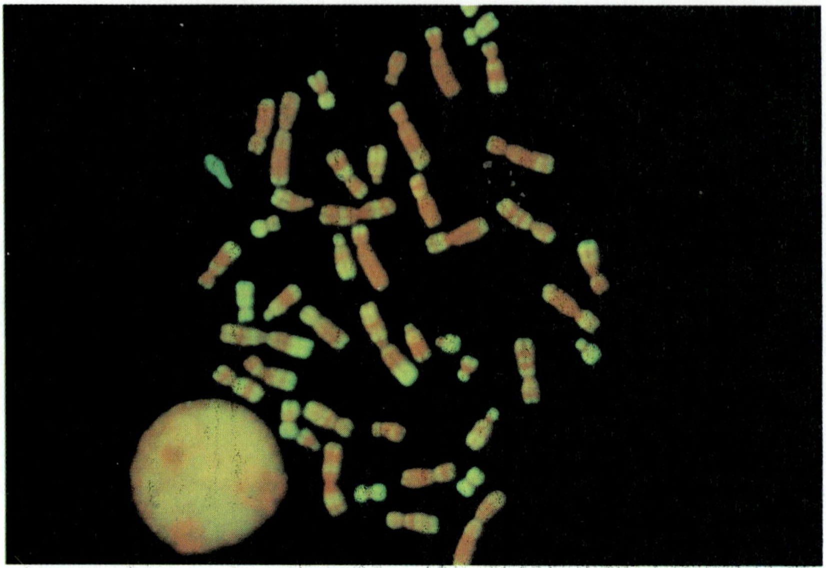

6-5. Banding patterns in human chromosomes.

More than a decade passed after Sutton made this prediction before genetic studies with the fruit fly *Drosophila melanogaster* confirmed the prediction. A fruit fly has four pairs of chromosomes. Its genes are inherited in four linkage groups: two large groups, one of medium size, and one of very few genes. And, as Figure 6-4 shows, *Drosophila* has two pairs of large chromosomes, one pair of medium size, and one pair of small chromosomes.

The same relationship soon was proved true for other species. Maize (Indian corn) has ten linkage groups of genes and ten pairs of chromosomes. The pea has seven of each. Even in humans, the relationship between the number of chromosomes and linkage groups of genes has been confirmed. It is now possible to assign particular genes to a particular pair of chromosomes (Figure 6-5). All the human genes fall into 23 linkage groups, just as would be expected on the basis of Sutton's prediction.

THE NATURE OF SCIENTIFIC PROOF

What is meant by "convincing evidence"? We have seen how evidence for or against a hypothesis accumulates in biological studies. But a hypothesis can never be "proved"—at least not in the same way that a theorem in geometry can be proved. Mathematical proof depends on stating and assuming the truth of a generalization. Then a specific case that is clearly included within that generalization is taken to be proved. Consider this example: "All Olympic athletes are amateurs." If we assume that this generalization is true, and we then say, "Jill Stone is an Olympic athlete," it follows that Jill is an amateur.

Evidence in science is of a different sort. Scientists gradually gather data. Then they try to form a hypothesis to explain those data. Scientists make predictions about new data that may be acquired based on the hypothesis. Thus the hypothesis is tested and discarded if it does not hold up.

When Sutton suggested a relationship between genes and chromosomes, he stated a hypothesis of far-reaching importance. Because much testing of the hypothesis has confirmed it, we now speak of the "chromosome theory of heredity."

This theory, like all theories in science, continues to be tested. A theory is a useful explanation and serves as a framework for further research. But a scientific theory can never be proved completely. The best that can be done is to test it continually. If the results of tests support the theory, it is said to be confirmed. Yet even one case that does not support the theory may discredit it, unless the theory can be changed to include the exception. Scientists move forward by discarding or modifying theories, as well as by forming them.

The chromosome theory of heredity is one of the cornerstones of modern genetics. It applies to all species that have been investigated. Yet it could never be said to be *proved,* unless every species of plant and animal and microorganism in the world were analyzed.

DISCOVERY OF THE SEX CHROMOSOMES

At first, Sutton's chromosome theory of heredity was supported strongly by the results of research at Columbia University, where Sutton had been a student. But a young embryologist, Thomas Hunt Morgan, had no faith in Sutton's theory. He thought it would be quite easy to refute. Morgan also picked the fruit fly *Drosophila melanogaster* to breed for this purpose.

Morgan raised thousands of red-eyed flies. When he examined these flies, he found one fly that had *white* eyes instead of red eyes. What was the basis of this variation? Could it be due to genetic differences, such as Mendel had found in peas?

The white-eyed fly was a male. Morgan mated it with a red-eyed female. The F_1 generation consisted entirely of red-eyed flies. According to Mendel's work, this meant that the white-eye trait is recessive and the red-eye trait is dominant. The F_1 red-eyed flies were mated to produce an F_2 generation. Among these offspring, a ratio of 3/4 red-eyed flies to 1/4 white-eyed flies was obtained. This, too, agreed with Mendel's results for single-trait inheritance in the F_2 generation.

But then Morgan noticed that all the white-eyed flies were males. About half the males had white eyes and half had red eyes. All the F_2 females had red eyes. The numbers of males and females in the F_2 generation were about equal, but 3/4 of the F_2 generation had red eyes and 1/4 had white eyes. If this had been a true Mendelian ratio, both the males and females should have been 3/4 red-eyed and 1/4 white-eyed. But they were not.

How could these results be explained? Just before Morgan's discovery of the white-eyed, male fly, some careful work had been done on the cells of *Drosophila.* This work showed a difference between the chromosomes of males and females. A *Drosophila* cell has four pairs of chromosomes. Three of these are the same in males and females. But one pair is different (Figure 6-4). In the female, the fourth pair is made up of identical, rod-shaped chromosomes called X chromosomes. In the male there is only one X chromosome. The other member of the pair is a hook-shaped chromosome called the Y chromosome. The X and Y chromosomes are called **sex chromosomes.** All other chromosomes are called **autosomes** (AW·toe·sohmz). For example, the male *Drosophila* has three pairs of autosomes. Its sex-chromosome pair consists of one X and one Y. The female has three pairs of autosomes and two X sex chromosomes.

After meiosis, each reproductive cell has one member of each pair of chromosomes. Therefore, each egg cell formed by a female *Drosophila* contains one of each pair of autosomes and one X chromosome. The sperm cells formed by the male also receive one member of each pair of autosomes. But since there are two kinds of sex chromosomes in the male, half of the sperm receive an X chromosome and the other half receive a Y chromosome.

The role of sex chromosomes in determining the sex of *Drosophila* is shown in Figure 6-6. Notice that the inheritance of the X and Y chromosomes determines the sex of the offspring. All egg cells contain one X. If a sperm with an X fertilizes an egg, the result is XX—a female. If a sperm with a Y fertilizes an egg, the result is XY—a male. This illustration shows something else about the inheritance of the sex chromosomes. Female offspring receive one X from the male parent and one X from the female. A male offspring, however, inherits an X only from the female. Its Y chromosome must come from the male parent.

Morgan considered this information about sex chromosomes in the light of the results of the cross between a white-eyed male and a red-eyed female. He then hypothesized that the alleles for white and red eyes are located on X chromosomes and that the Y chromosome carries no alleles for eye color. A cross of a homozygous red-eyed

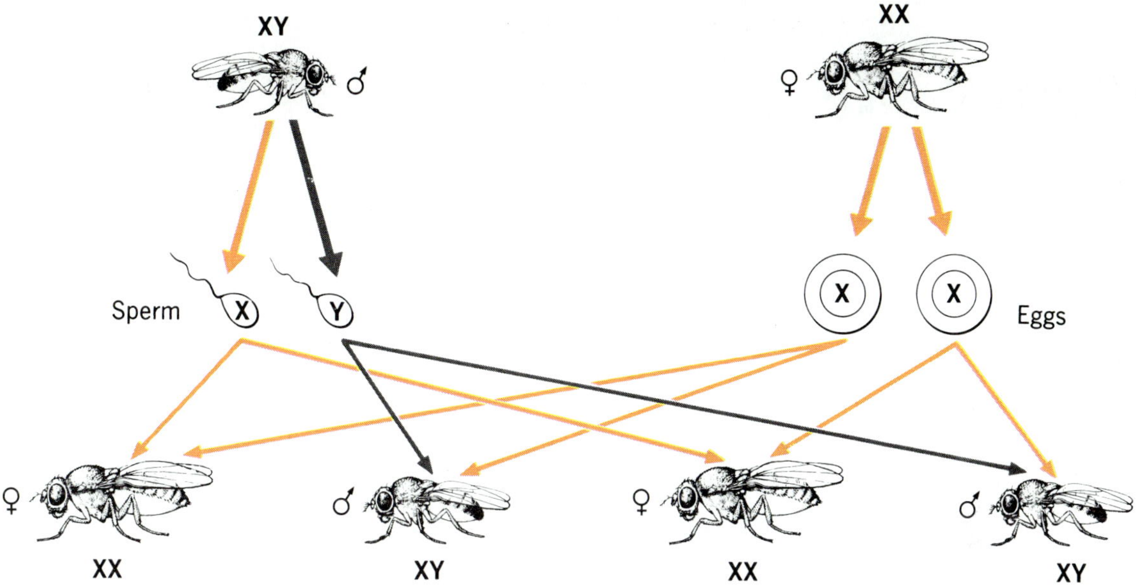

6-6. The relation of chromosomes to sex determination in fruit flies. *Which parent contributes the sex-determining chromosome to the offspring?*

female (*RR*) and white-eyed male (*r*, because it has only one allele for eye color, which is on its X chromosome) should turn out as shown in Figure 6-7. This hypothesis explains the results of Morgan's original experiment.

Morgan predicted that a cross between a red-eyed male and white-eyed female would produce an F_1 generation in which all males would have white eyes and all females would have red eyes. In both males and females of the F_2 generation, half would have red eyes and half would have white eyes. When Morgan made this cross, he obtained the expected results. Make a drawing similar to Figure 6-7 to show how this happens.

Morgan was able to make other deductions. All of them supported the original hypothesis—even though Morgan had set out to discredit Sutton's hypothesis. As far as Morgan could determine, all the data could be explained by the hypothesis that the alleles for white eyes and for red eyes are on the two X chromosomes.

PROBLEM 1: If you had a red-eyed female fly, how would you plan a test cross to find out if she carried a gene for white eye color?

PROBLEM 2: What would be the result of the cross between a white-eyed male and a white-eyed female?

Geneticists have learned that many other genes behave in a way similar to that of the eye-color alleles in *Drosophila*. These alleles

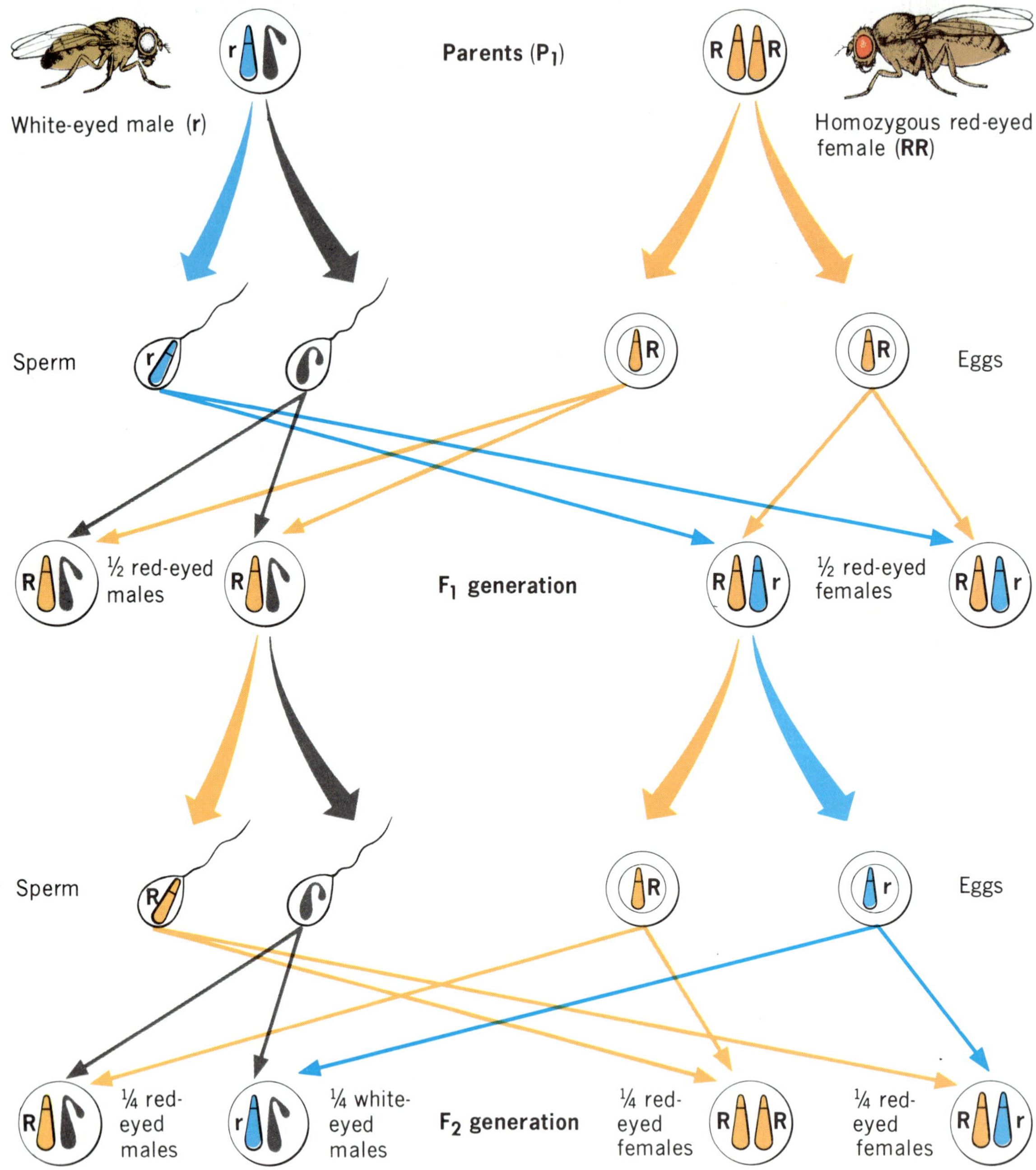

6-7. A diagram of two generations of *Drosophila*, showing inheritance of a trait controlled by a gene on the X chromosome. The gene in this case controls eye color.

are not carried on the Y chromosome. They occur only on X chromosomes. The Y chromosome in *Drosophila* is inert, or inactive, in inheritance. As a result, a gene on the single X chromosome in the male fly determines its phenotype. In studies of other organisms, the Y chromosomes are found to have very few genes or none at all.

You will have an opportunity to examine the inheritance of some traits in humans as you do **Inquiry 6-1.**

INQUIRY 6-1 Patterns of Human Inheritance

Red-Green Colorblindness. People with this trait have difficulty in telling red from green. Doctors detect this type of colorblindness by using special charts made up of a number of colored dots (Figure 6-8).

Hemophilia. Genetic hemophilia (he·mo·FEEL·ya) is a much more serious condition than colorblindness. Fortunately, it is very rare. Hemophilia is a condition in which the blood fails to clot after an injury, or it clots very slowly. People with hemophilia experience frequent and painful bleeding around the joints.

MATERIALS

none

Procedure

The inheritance of red-green colorblindness in humans is well understood. In the pedigrees in Figure 6-9, squares represent males and circles represent females. Members of the same generation are placed on the same horizontal row. Husbands and wives are connected by a horizontal line. They are connected by vertical and horizontal lines to their children in the next generation. If a circle or square is filled in, ● or ■, that individual has red-green colorblindness. **1.** Is the allele for colorblindness dominant or recessive? What evidence supports your answer? **2.** Cite evidence to support the statement that red-green colorblindness is sex-linked.

Predict the expected results of a mating between a colorblind man and a woman who is homozygous for normal color vision. **3.** Among their male offspring, what fraction would have

6-8. Charts for testing color blindness. Persons with any defect of red-green vision have trouble distinguishing the numbers quickly. An accurate test cannot be made with only two charts. However, anyone who fails to read both charts correctly should consult an ophthalmologist.

Reproduced from Dvorine Pseudo-Isochromatic Plates by permission.

6-9. TWO PEDIGREES OF COLOR BLINDNESS

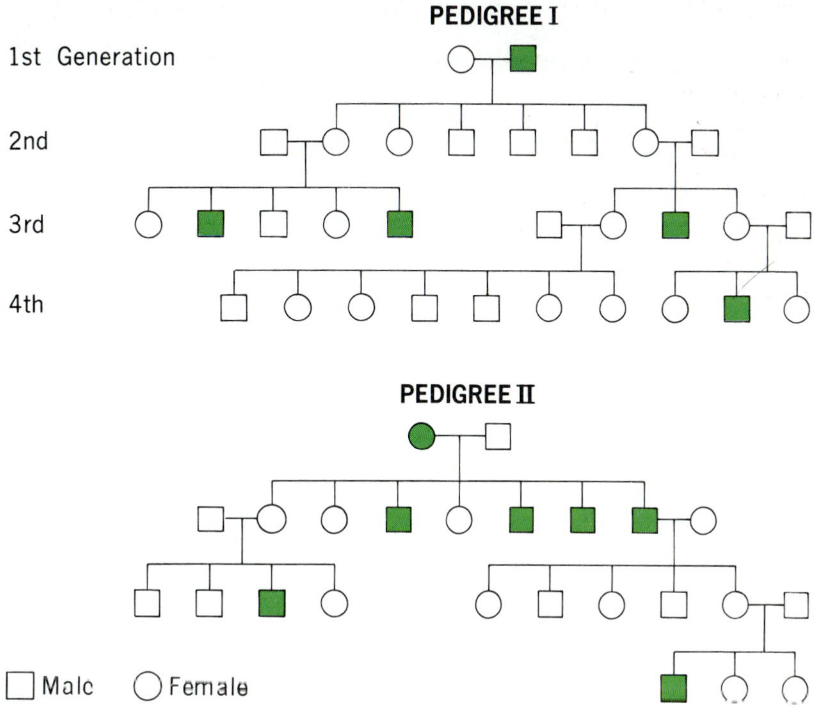

normal color vision? What fraction would be colorblind? **4.** Among their female offspring, what fraction would have normal color vision? What fraction of the female offspring would be carriers (heterozygous)? What fraction of the female offspring would be colorblind?

Repeat the calculations, this time for a mating between a carrier woman and a man with normal color vision. **5.** Among their male offspring, what fraction would have normal color vision? What fraction would be colorblind? **6.** Among their female offspring, what fraction would have normal color vision? What fraction would be carriers? What fraction would be colorblind?

The gene for hemophilia is carried on the X chromosome. As in colorblindness, it takes but one recessive allele for a male to be hemophilic. It takes two recessive alleles to produce the condition in a female. Why?

Figure 6-10 shows, in part, the distribution of hemophilia among certain royal families of Europe. **7.** Study the chart and then list those mothers who *must* have been carriers. **8.** The present British royal family descended from Edward VII. What must be true of the hemophiliac trait in that line of descent?

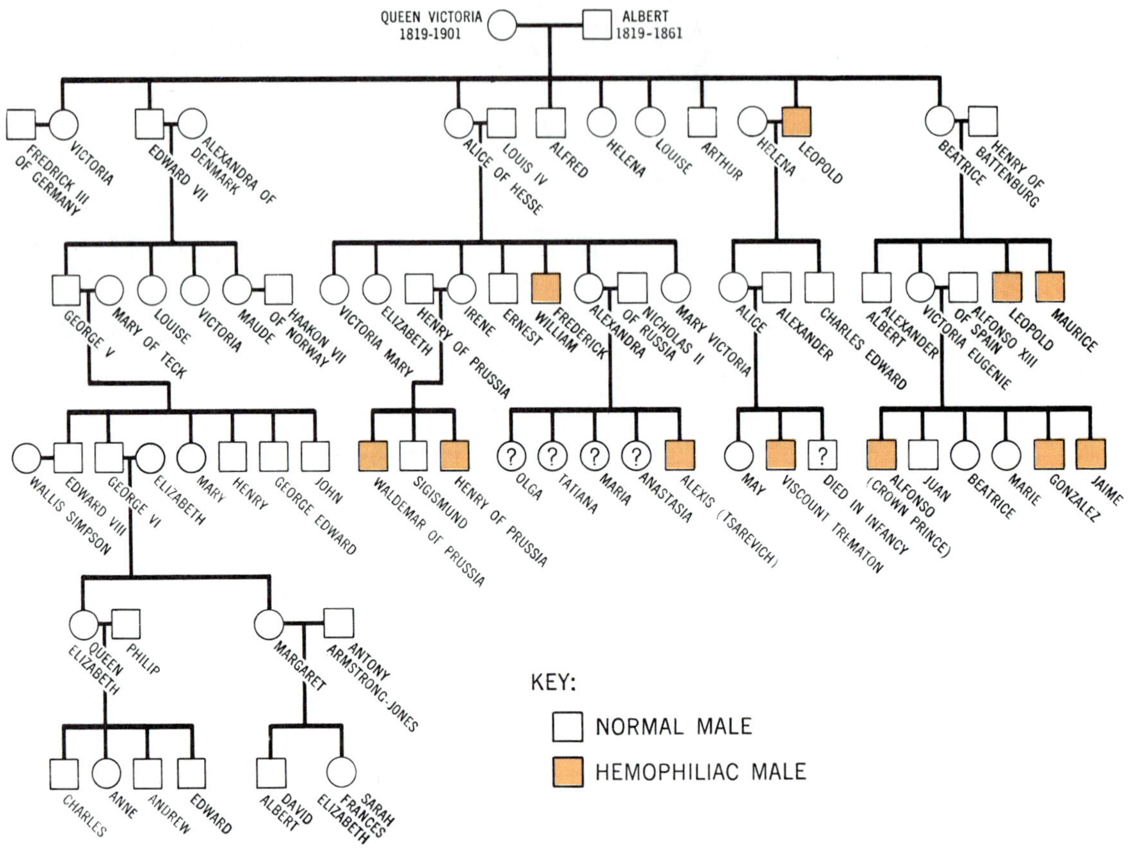

6-10. A pedigree of some of Queen Victoria's descendants, showing the hereditary distribution of hemophilia.

Sex Determination

With improved methods of investigation, geneticists were able to show that sex chromosomes occur in many organisms. In most plant species, the same plant produces both male and female sex cells. Sex chromosomes can be found, however, in plant species that do have separate male and female individuals. Such plants include holly, asparagus, willows, and cottonwoods.

Human cells have 23 pairs of chromosomes—22 pairs of autosomes, and one pair of sex chromosomes. Figure 6-2 shows the chromosomes of a human female. In human males, the Y chromosome is very small, compared with the X chromosome (Figure 6-11). As in *Drosophila,* human males produce two kinds of sperm. Half of them carry an X chromosome; the other half carry a Y chromosome. All of the eggs produced by the female carry an X chromosome. If an egg is fertilized by a Y-bearing sperm, the offspring is a male (XY). If an egg is fertilized by an X-bearing sperm, the offspring is a female (XX).

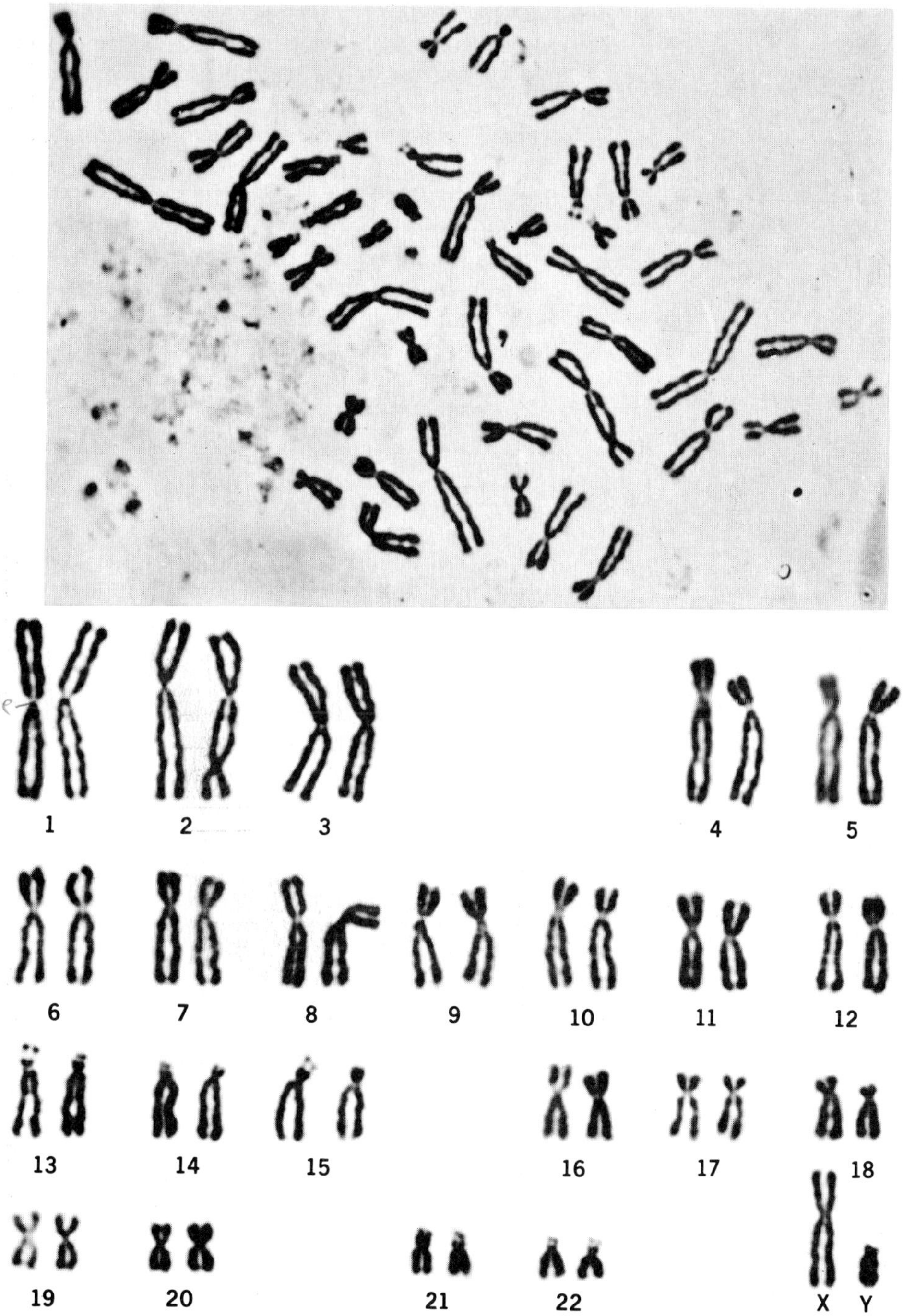

6-11. The chromosomes of a normal human male. A squashed cell is shown above. The karyotype is shown below. *How do the male chromosome pairs differ from those of the female* (Figure 6-2)?

Not all kinds of animals have sex-determining sperm. In some species, such as birds, butterflies, and moths, the female has two kinds of sex chromosomes and produces two kinds of egg cells. In these organisms, all sperm cells carry the same kind of sex chromosomes. The chromosome of the egg cell determines the sex of the offspring.

In still other species, one of the sex chromosomes may be missing entirely. In some grasshoppers, for instance, males have 22 autosomes (11 pairs) and one X chromosome. Females also have 22 autosomes, but they have two X chromosomes. Half the sperm cells have 11 autosomes. The other half have 11 autosomes and one X. The sex of the young grasshopper depends on the kind of sperm that fertilizes the egg. In this case, one X chromosome results in a male grasshopper.

GENETIC RECOMBINATION

Mendel's work indicated that different pairs of genes assort independently during meiosis. As a result, in a study involving two pairs of genes, an F_2 generation will exhibit a 9:3:3:1 ratio of the four possible phenotypes. On the other hand, the chromosome theory suggests that pairs of genes located on the same pair of chromosomes will not assort freely. The genes on the same chromosome will be expected to go together. If that were always so, the kinds of offspring two parents could produce would be restricted by the number of pairs of chromosomes. A species with very few chromosomes, like *Drosophila,* would produce very few kinds of offspring in any particular mating.

This expectation was soon put to the test. Not long after the discovery of the white-eyed *Drosophila,* Morgan found a new mutant type. It produced a yellow body, instead of the normal gray body. Like the gene for white eyes, this gene was located on the X chromosome. A surprise came when both the new genes were involved in the same cross.

Crossing-Over

What do you think would be the result of a cross between a female with white eyes and a male with a yellow body? What would happen if the F_1 males and females mated?

When Morgan did the experiment, he obtained the expected ratios, with one startling exception (Figure 6-12). A few males had both white eyes and yellow bodies. According to the theory of sex-linked inheritance, all these males must have only one X chromo-

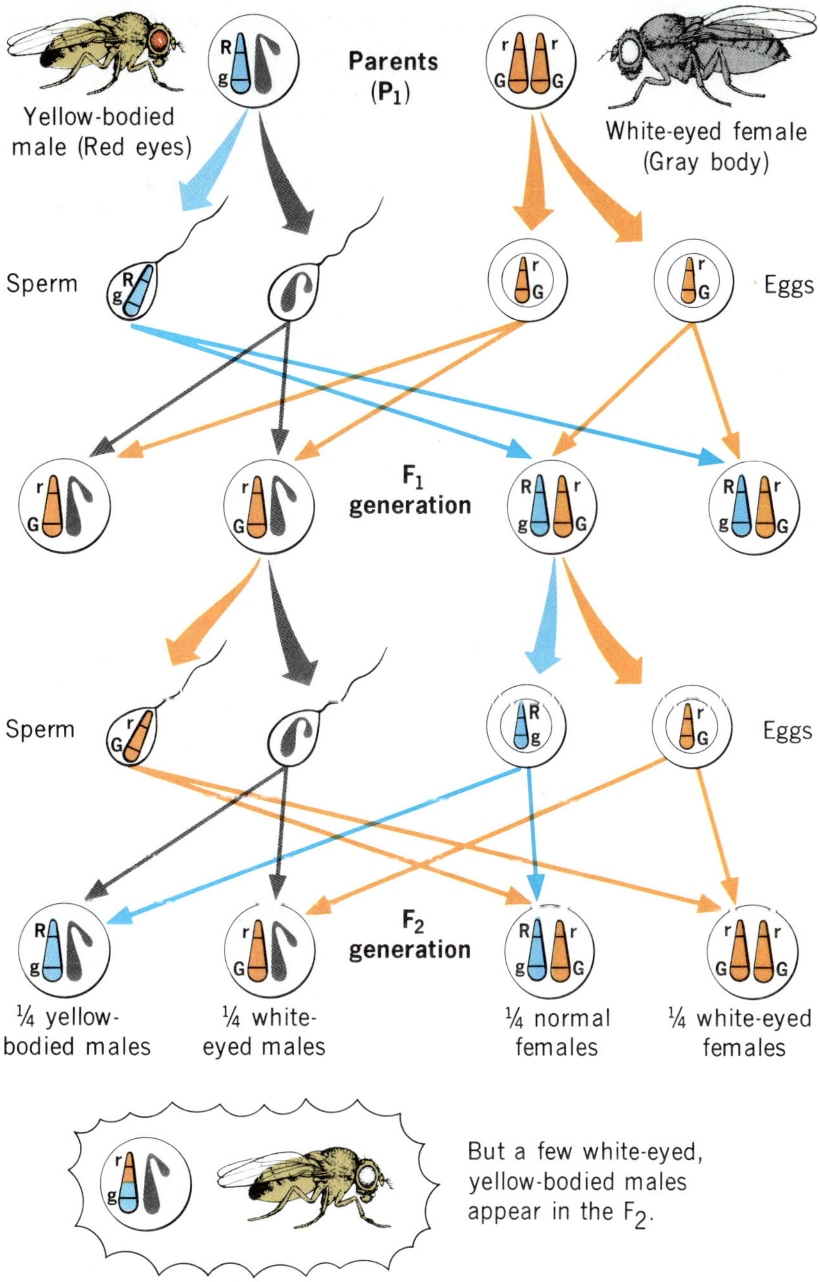

6-12. A cross between a yellow-bodied male and a white-eyed female. Alleles are coded. R = red eyes (dominant); r = white eyes (recessive); G = gray body (dominant); g = yellow body (recessive). Note the unexpected combination of characteristics in a few of the F_2 males. *How could this new combination come about?*

some. The genes for yellow body and white eyes were originally on different X chromosomes. How did these genes get into the same X chromosome?

These males clearly represent a **recombination** of the sex-linked mutant genes. How could such a thing have come about? Some observers had reported cross-shaped figures between homologous chromosomes early in meiosis (Figure 6-13). Maybe, thought Morgan, the homologous chromosomes actually break and exchange segments. This might happen between the locations of the gene for yellow body and the gene for white eyes on the X chromosomes of the female during meiosis. Then a male offspring might inherit an X chromosome that carried both the recessive genes. Morgan named this hypothetical process **crossing-over.**

Consider the X chromosomes in the female after crossing-over has taken place. Morgan predicted that a female would then have one X chromosome with both mutants and one X chromosome with neither mutant. Thus, the male offspring should either be normal or have yellow bodies and white eyes. But if crossing-over were still taking place during meiosis, a few males would have white eyes but not yellow bodies. And a few males would have yellow bodies but not white eyes. Moreover, the percentage of either kind of male in this generation should be the same as that of the males in Morgan's *first* cross that showed the two mutant traits together. Why? That is what actually was observed. In this cross, 1.5 percent of the males had white eyes and 1.5 percent had yellow bodies.

Soon Morgan and his students, especially A. H. Sturtevant, were using crossing-over to determine the locations of genes on chromosomes. They constructed maps of these locations by looking at recombination frequencies in *Drosophila.* When recombination oc-

6-13. Crossing–over occurring in meiosis during formation of grasshopper sperm cell.

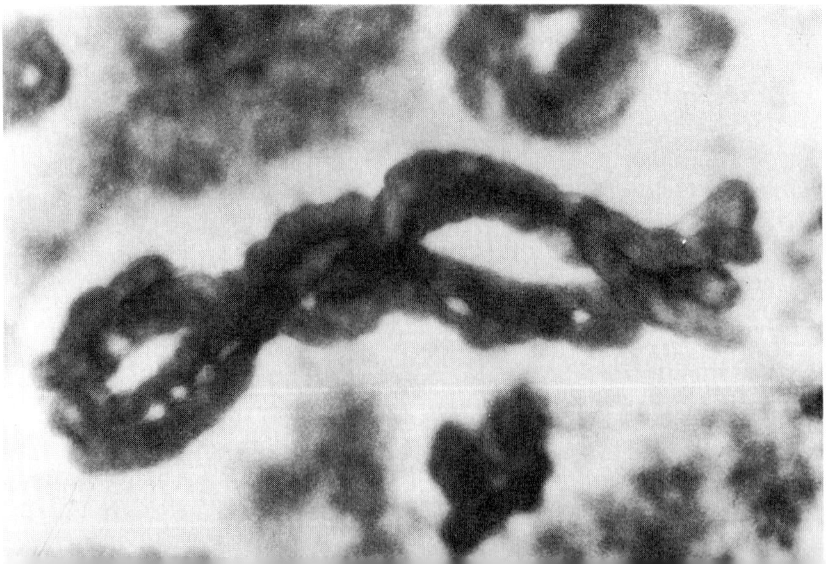

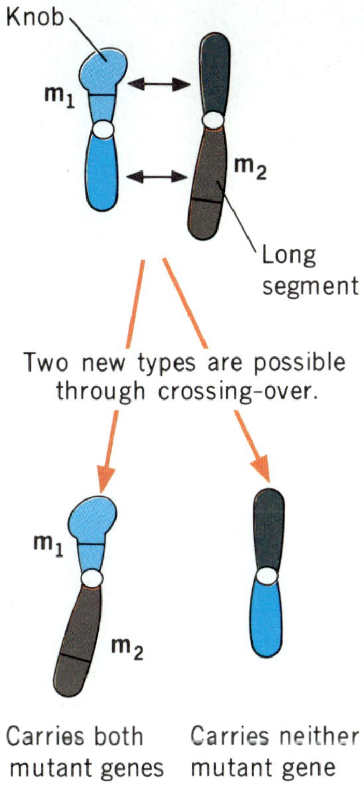

6-14. McClintock and Creighton found that crossing-over can be observed on chromosomes. The symbols m_1 and m_2 mark two mutant genes. Because of crossing-over, some chromosomes carry both mutants. Others carry neither mutant.

curs rarely, the genes probably are very close together on the chromosome. When recombination occurs frequently, the genes are probably far apart. Can you explain the basis for such reasoning? Within a few years, maps of dozens of *Drosophila* genes were prepared. All of them fell into one of the four linkage groups that corresponded to the four pairs of *Drosophila* chromosomes.

In 1930, other investigators showed that when genes are exchanged during meiosis, an exchange of chromosome segments also occurs. Two American scientists, Barbara McClintock and Harriet Creighton, produced a strain of maize in which one chromosome sometimes has a knob on one end and also has an abnormally long segment. The homologous chromosome is normal.

Among other genes, this chromosome pair carries two gene pairs that affect the food storage tissue in corn grains. One of these pairs determines whether a protein will be colored (*C*) or not colored (*c*). The other affects the outer covering; it may be starchy (*Wx*) or waxy (*wx*). The plant with which these scientists were working was heterozygous for both genes. The normal chromosome in this plant contained *C wx*. The unusual chromosome contained *c Wx*.

Crossing-over occurred in some of the eggs formed by this plant (Figure 6-14). As a result of the crossing-over during meiosis, alleles were exchanged *and* the unusual-looking segments of the chromosomes also were exchanged. These researchers could correlate the occurrence of the two traits in the phenotype with an unusual form of chromosome. As a result, they could predict that the unusual chromosomes would occur in certain phenotypes.

This research provided distinct evidence of the exchange of chromosome segments and its relation to the exchange of genes. Further studies showed that recombination occurs in a great number of organisms that reproduce sexually, plants as well as animals. Geneticists mapped the mutant genes of maize, mice, chickens, and many other species. In flies, for some reason still unknown, crossing-over occurs only in the female. But in most species, including humans, it occurs in both sexes.

Genetic Recombination in Microorganisms

Scientists began to wonder if the chromosome theory of heredity would apply to microorganisms, such as bacteria. Bacteria are single-celled microorganisms that reproduce by simple cell division. One bacterium divides and becomes two complete bacteria.

Escherichia coli is a bacterium that lives in the human intestine, where it usually causes no harm. Joshua Lederberg demonstrated, in 1945, that in *E. coli* a sort of sexual union between individuals sometimes occurs. Some years later, electron microscope pictures

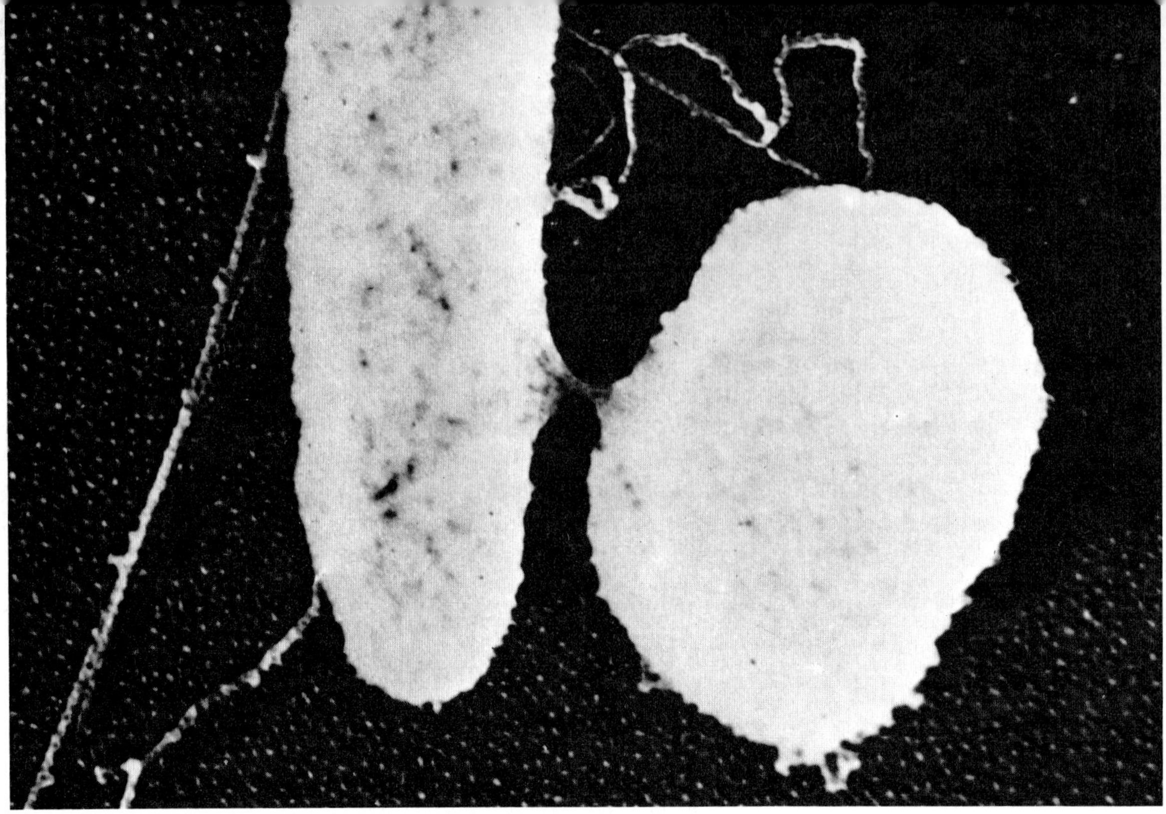

6-15. Visible evidence for the exchange of cell materials between bacteria. This electron micrograph reveals a little tube between bacteria of two different strains. Biologists think that genetic material moves through the tube.

showed a fine tube connecting two such bacteria (Figure 6-15). Each bacterium has only a single chromosome, which breaks up before the tube is formed. Pieces of the chromosomes of both individuals are exchanged through the tube. In this way, genes that occur in different strains of bacteria or that arise through mutation recombine, just as genes in plants and animals do.

A similar process occurs in the reproduction of viruses. The process has been studied most thoroughly in the *bacteriophages* (singular, back·TEER·ee·oh·fahj) – viruses that live and reproduce inside bacteria. The phages, as we call them for short, attach themselves to host bacteria and inject their single chromosome into the cell (Figure 6-16). Here the genes of the virus replicate. The end result is as many as a hundred identical chromosomes of the phage type. The phage genes cause production of a protein coat and tailpiece – the two other parts of a phage. After these parts are assembled, the new phages burst out of the dying host bacterium.

Sometimes two different strains of the same species of phage infect one host bacterium. Then, parts of both chromosomes may combine. In this way, a recombination of genes occurs in phages.

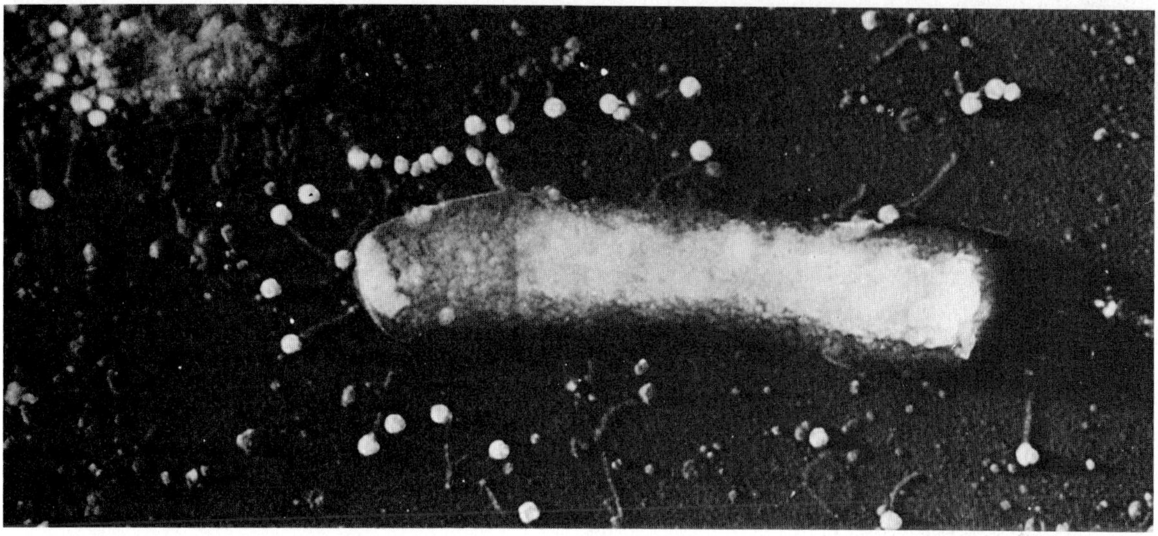

6-16. Two electron micrographs of stages in virus replication. **(Top)** Phage viruses attached by their tailpieces to a bacterium. Note the remains of another bacterium destroyed by the viruses (30,000 X). **(Bottom)** A very thin section through a bacterium containing reproducing phage viruses. Heads of new viruses appear black (55,000 X).

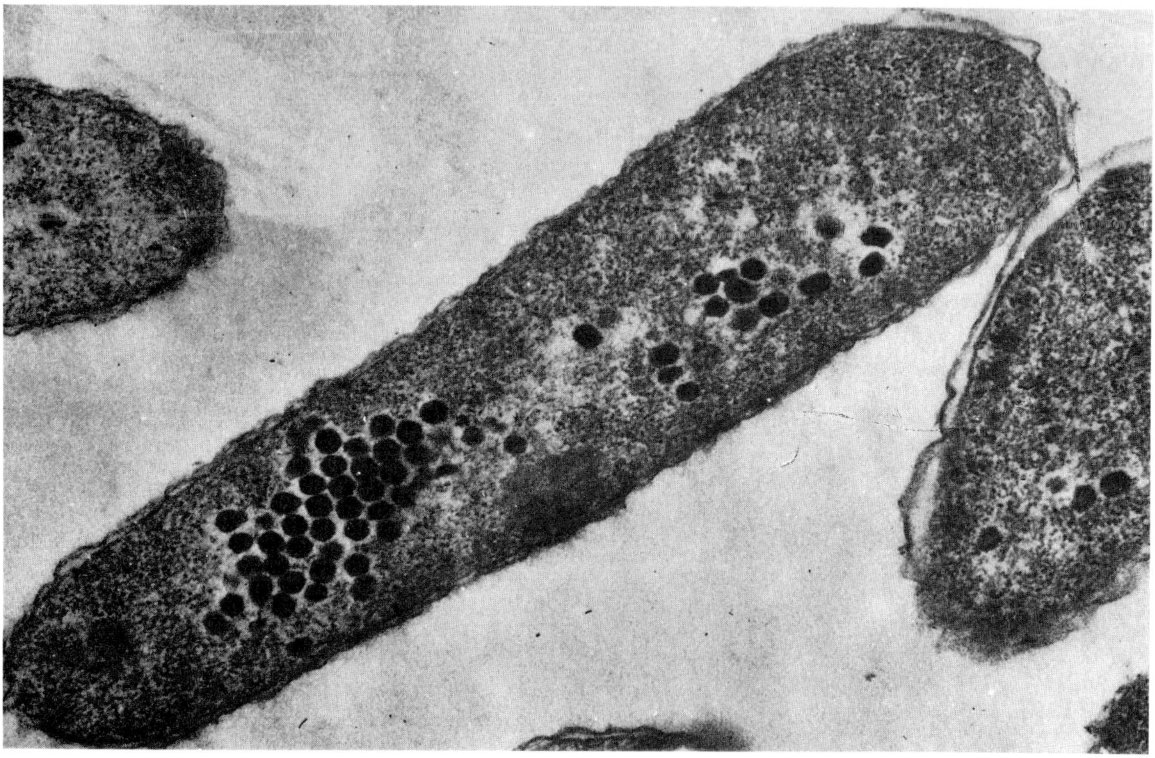

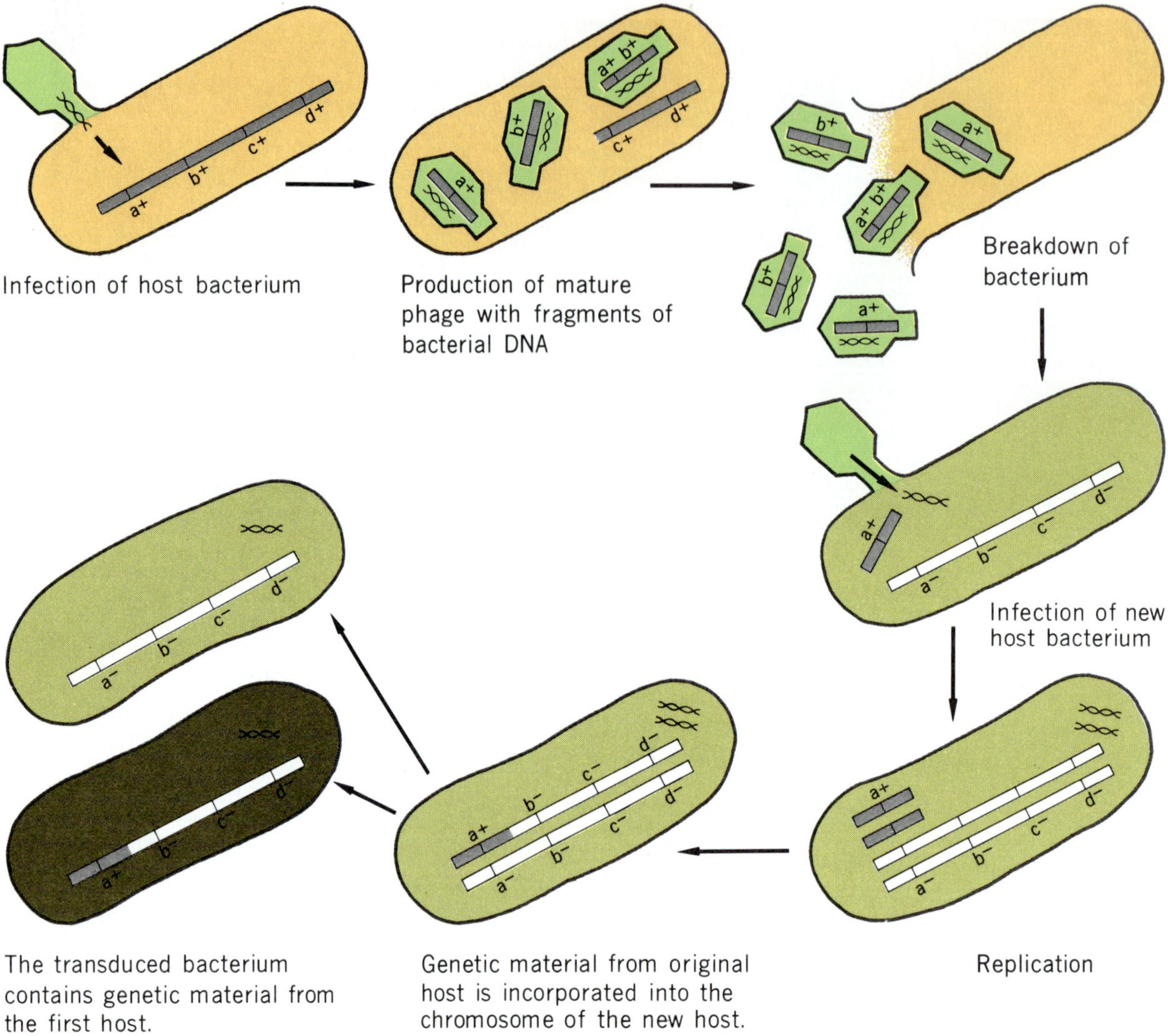

6-17. Transduction in bacteria. Genetic material is carried from one host to another by a phage.

Transduction

There is one other way in which genes can be transferred from one microorganism to another. It has been named **transduction.** The process, discovered by Lederberg and Norton Zinder, is shown in Figure 6-17. When phages multiply within a host cell, a piece of a chromosome of the host cell may become attached to the chromosome of a phage. The genes of the bacterium are carried by the new generation of phages. Later, these phages may infect a new host cell. If the phages do not kill this host, the genes from the first bacterium may be transferred to the new bacterium. In this way, genetic material can be carried from one bacterium to another by a phage.

Knowledge of transduction has led to one of the most promising, yet controversial, developments of modern biology. Suppose we have a virus that can infect both human cells and bacteria. Can we use the virus to transfer some human genes to a bacterium? It does seem to be possible, and many scientists are investigating the process. The benefits of such transfers could be great. For example, genes for the production of insulin, which diabetics cannot manufacture, might be transferred to a harmless bacterium. These bacteria could then be used as an inexpensive way of manufacturing insulin.

Or, genes from a green plant might enable bacteria to carry on photosynthesis—a possibility that could help solve world food shortages.

But, at the present time, we cannot control exactly which genes are transferred. It is possible that we might accidentally develop a bacterium that would cause a disease against which we would have no resistance. An epidemic could result. Scientists and government officials have developed strict guidelines to regulate the safety of this research.

Recombinant DNA

Another very new technique can cause genetic recombination artificially. This is the technique of **recombinant DNA.** The bacterium *E. coli* often contains one or more small loops of DNA called plasmids. The plasmids can be isolated from bacteria and broken open with enzymes. The same class of enzymes also can break apart a DNA sequence from another organism. Then these two DNA strands are mixed. In many cases, the strands will stick together in certain places. The result is a combination of DNA from two different organisms. The plasmid from *E. coli* can be returned to the bacterium, where it will replicate normally. This recombinant DNA may even produce proteins like those produced by the donor (nonbacterial) organism.

Recombinant DNA techniques are powerful tools for research. They provide relatively simple ways of working out the genetic makeup of organisms. Practical applications include the possibility of producing vast quantities of proteins, drugs, and food products. On the other hand, such research might produce a new disease bacteria that would be harmful to humans. Should the new form escape from the laboratory, epidemics with potentially disastrous results could occur. The federal government has put together a series of strict rules governing recombinant DNA research. However, scientists and others continue the debate over the benefits and risks of recombinant DNA.

ABNORMAL CHROMOSOME NUMBERS

Vermilion (scarlet) eyes occur in *Drosophila* as a result of a sex-linked recessive gene. This trait is inherited in exactly the same way that white eyes and yellow body are inherited. One of the early challenges to the chromosome theory of heredity came from some crosses of these mutants with normal, red-eyed flies. C. B. Bridges crossed vermilion-eyed females with normal red-eyed males. In addition to the expected red-eyed females and vermilion-eyed males, about one fly in every 2,000 was a *vermilion-eyed female* or a *red-eyed male*. How could this be?

According to the theory, a vermilion female would have to carry two recessive vermilion genes. She could not carry a normal dominant allele that produces red eyes. But the only way a female could get two vermilion genes would be to get both of the X chromosomes from the mother. Was it possible that the two X chromosomes might sometimes fail to separate? Could both of the chromosomes remain in one egg?

Bridges looked at the chromosomes of these two unusual classes of flies. He found that the vermilion-eyed females had two X chromosomes and one Y chromosome (Figure 6-18). The red-eyed males had an X chromosome but no Y chromosome. (All of these males were sterile.) The unexpected characteristics in the offspring were the result of abnormal meiosis when the egg cells were produced. Bridges called this failure of the chromosome to separate during meiosis **nondisjunction.**

For many years, nondisjunction remained only a scientific curiosity. It was found in every species of plant or animal that was studied with sufficient care. But no practical value was attached to nondisjunction until 1957. At that time, new techniques allowed closer ex-

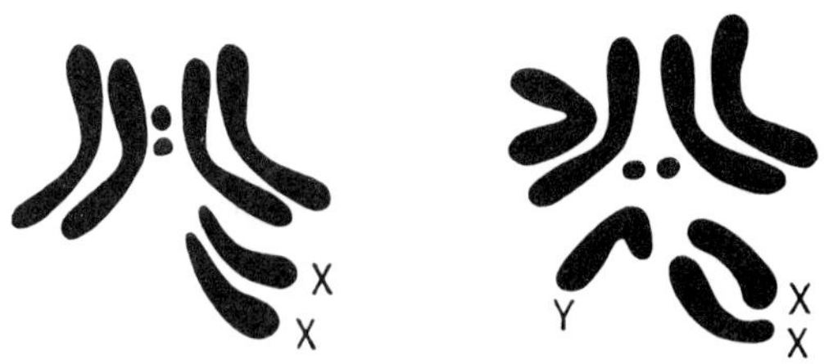

6-18. Chromosomes of a normal female *Drosophila* **(left),** and an exceptional vermilion-eyed female with two X chromosomes and one Y chromosome.

SCIENCE AT WORK: Too Many Chromosomes

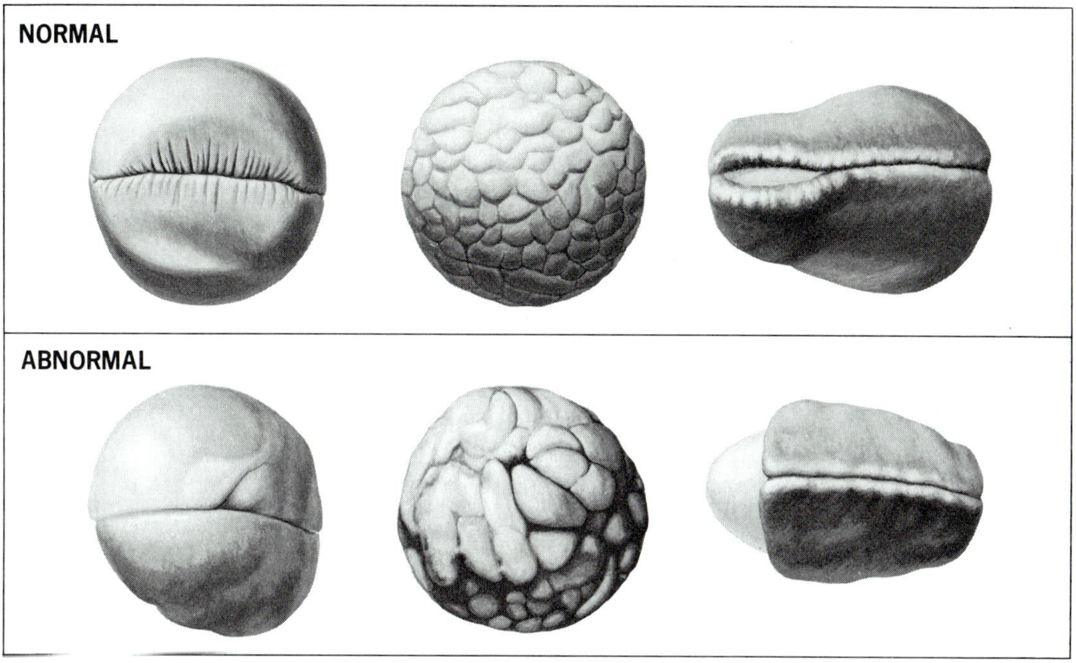

In the early 1950s, a University of Iowa biologist sent a paper to a medical journal describing work done with "old" frog eggs. These eggs had remained four or five days in the body cavity of the female before being released. The aged eggs could be fertilized, but they grew into malformed embryos. The chromosomes in the body cells of these embryos had 13 to 52 chromosomes, instead of the normal 26.

In 1959, three French scientists at the University of Paris published a similar discovery. They had learned that a human malformation also was associated with an abnormal number of chromosomes. The body cells of most children with Down's syndrome have 47 chromosomes rather than 46. This change in chromosome number disrupts development causing mental retardation and abnormal facial features.

A female frog produces a new supply of eggs for each breeding season. But all of the eggs of a human female are present from the moment of her birth. As a result, an egg that is fertilized when a woman is 35 years old has been in her body for 35 years.

The scientists in Paris learned that, as the age of the mother increases, the incidence of Down's syndrome rises markedly. About 7,000 Down's syndrome children are born each year in the United States. Most are born to mothers over 35 years of age. It may be that the "aging" of the eggs is the cause of the increased probability of older mothers having a child with Down's syndrome.

At first glance, the research with "old" frog eggs might have seemed to be of little significance. But the development of the idea and further research showed how important the original research was. Through publication of scientific papers, investigators all over the world can communicate, cooperate, and look for new ways of understanding biological problems.

amination of human cells. Geneticists learned that nondisjunction also occurs in humans, where it often produces serious abnormalities.

Abnormal Chromosome Numbers in Humans

The first type of nondisjunction of chromosomes to be found in humans was Down's syndrome. Down's syndrome occurs about 14 times in every 10,000 births. The child is physically abnormal and mentally retarded. Many of these children die. Persons with the most common form of Down's syndrome have 47 chromosomes, instead of the normal 46. This abnormal number is the result of three Number 21 autosomes as shown in Figure 6-19. Down's syndrome also can occur as a result of a piece of chromosome 21 being hooked to another chromosome. Geneticists refer to this as a translocation.

Turner's syndrome is a rare disorder of human females. It occurs in about one of every 10,000 births. The affected individuals may

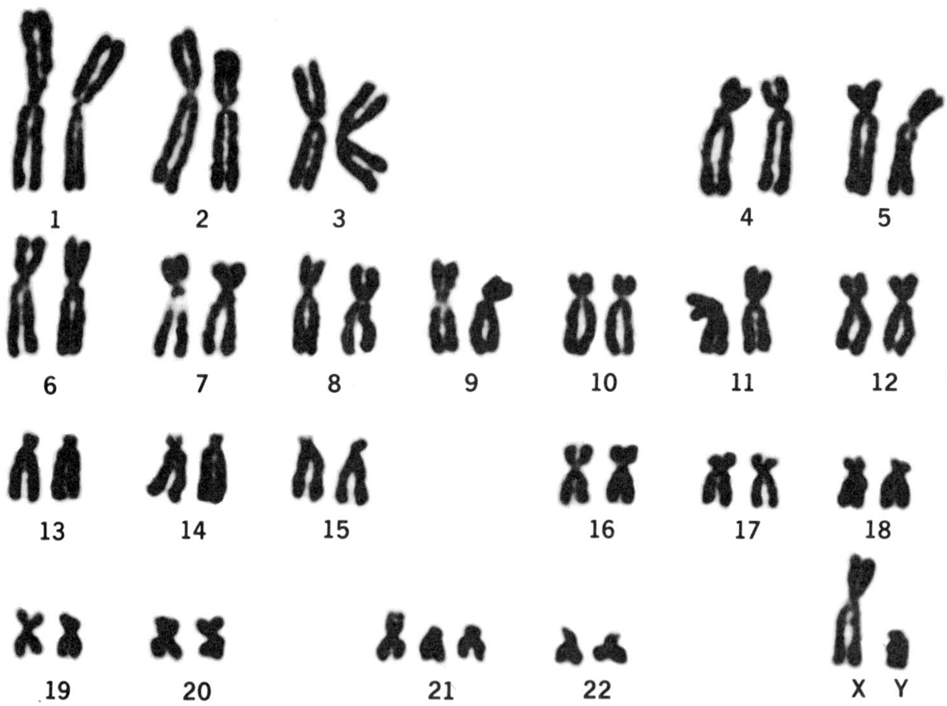

6-19. The chromosomes of a boy with Down's syndrome. Notice the extra autosome Number 21.

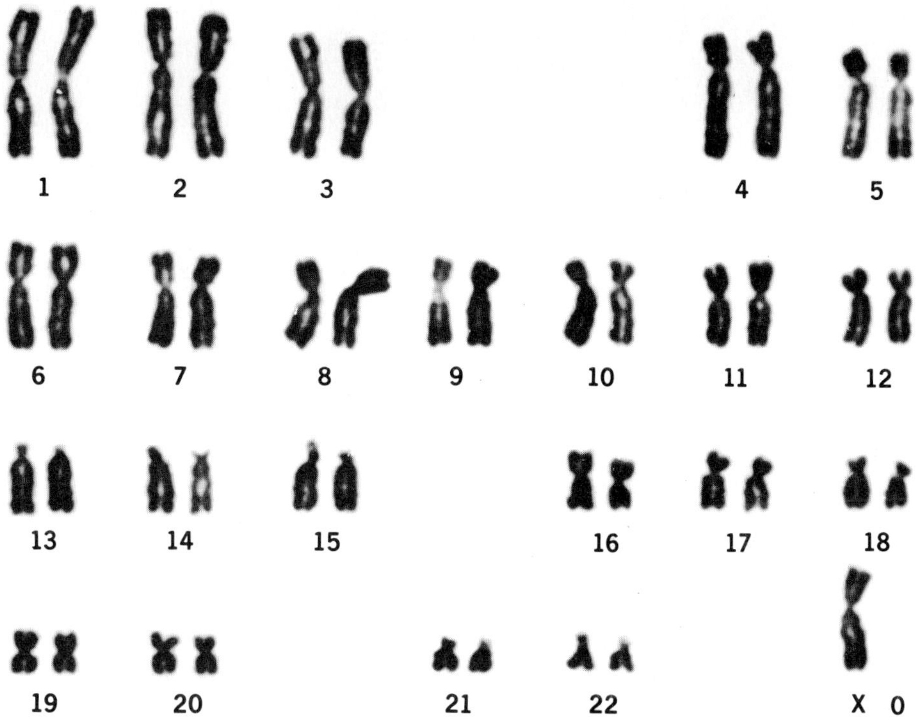

6-20. The chromosomes of a human female with Turner's syndrome, which prevents normal sexual development. Note that one of the two X chromosomes is missing.

become adults. However, these individuals never become sexually mature. Females with Turner's syndrome have only 45 chromosomes. Careful study shows that one of the X chromosomes is missing (Figure 6-20).

Klinefelter's syndrome occurs in about 12 of every 10,000 births. These are males that grow to be adults but fail to mature sexually. Their chromosome count is 47. This is the same number as in Down's syndrome. However, the male has an extra X chromosome. In this case, the sex chromosomes are XXY (Figure 6-21).

Females who have three X chromosomes occur rarely. Although they are usually fertile and physically normal, they often are mentally retarded. Males with two Y chromosomes (XYY) also occur.

The study of human chromosomes is a very active field of biomedical research today. Many abnormal conditions that were difficult to explain have been related to abnormal chromosomes. But little is known about what causes abnormal chromosomes. It is assumed that an accident occurs during meiosis. This causes eggs or sperm with abnormal chromosome numbers to be produced.

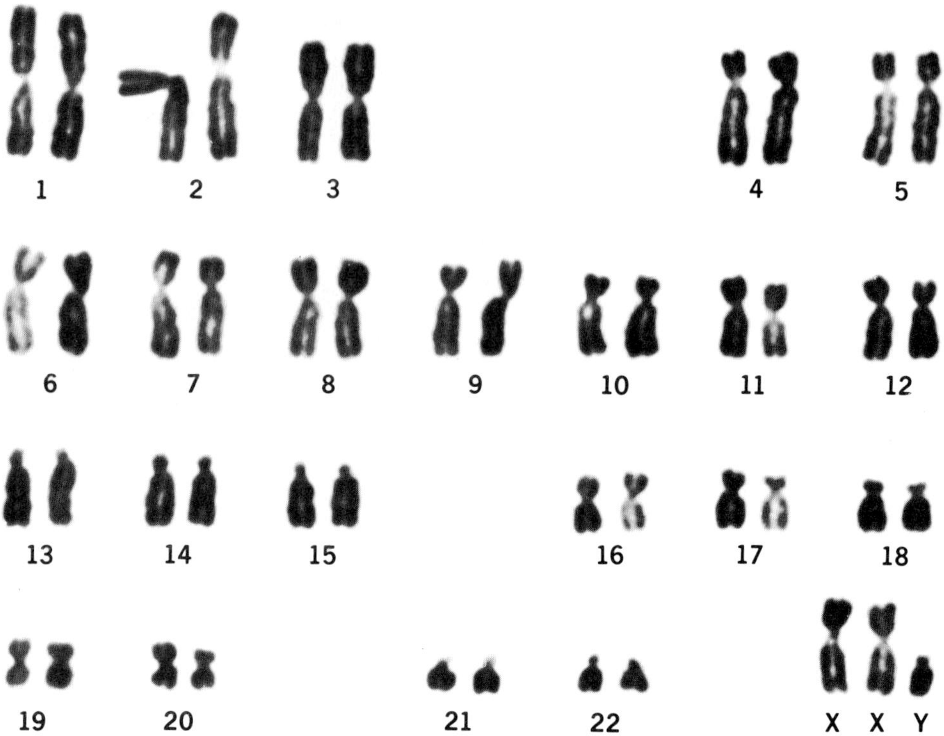

6-21. The chromosomes of a human male with Klinefelter's syndrome. The extra X chromosome prevents normal development.

In Brief

Mendel's principles of inheritance apply to a wide variety of organisms. Mendel observed that traits are inherited separately. As further research was done, however, it became apparent that this is not always true. Some traits are inherited in linkage groups—genes carried on the same chromosome.

As with all research, each answer generates many new questions. Linked genes did not always appear together in the offspring. How could this be? The work of McClintock and Creighton established a relationship between the inheritance of unexpected combinations of characteristics and the crossing-over process that sometimes occurs during meiosis. When parts are exchanged between homologous chromosomes, new combinations of linked traits are possible.

A great deal has been learned through the study of sex-linked characteristics. Crossing-over has been observed between X chromosomes in the female. This can lead to new linkage groups and new patterns of heredity. The study of sex-linked traits in humans shows how certain defects, such as hemophilia, may be passed from one generation to another through the female line. The woman appears normal, but is the carrier of the gene for hemophilia on one of her X chromosomes.

Through transduction, genes may be transferred from one bacterial host to another. Knowledge of this process has led to an important and controversial area of research. Scientists are investigating the possibility of implanting genes that would introduce new capabilities into a microorganism. The possibility of producing potentially harmful bacteria, which might cause human disease, is an issue of considerable importance to our society.

Some human genetic defects are the result of abnormal chromosome numbers. Such defects can be detected in the fetus through the technique of amniocentesis. It also is possible to diagnose in the fetus some inherited enzyme defects. Thus, genetic counseling allows not only the prediction of probabilities, but, for many defects, diagnosis before birth.

Guide Questions and Problems

1. An investigator of heredity early in the present century might have ended a report on an experimental study with the statement, "I therefore reach the conclusion that the nucleus is the seat of heredity." Write a definition for the word *conclusion* used in this sense. Then check with a dictionary. How does the dictionary definition compare with yours?

2. What caused biologists to suspect that chromosomes are involved in the transmission of hereditary material?

3. Suppose that a recessive mutation occurs on an X chromosome in a human egg cell. What kind of sperm must fertilize the egg if the mutation is to be expressed in the F_1 generation?

4. Why is it not possible for a male to be homozygous for colorblindness?

5. If a woman who is not a carrier of genes for hemophilia is married to a man who is a hemophiliac, what percentage of their male offspring could be expected to be hemophiliacs? Why?

6. If a colorblind man marries a woman whose father was colorblind, what is the probability of their first child being a colorblind boy?

7. If a woman carried the recessive gene for colorblindness in one of her X chromosomes and the recessive gene for hemophilia in the other, what would be her phenotype with respect to these

traits? Suppose she marries a man who is neither colorblind nor hemophilic. What would be the expectation among their sons and daughters? Now suppose that they produce a son who is neither hemophilic nor colorblind. How could this be explained? What if they produced a son who was both hemophilic and colorblind?

Ideas for Independent Study

1. The drawing in Figure 6-22 shows the chromosomes from a human cell. Trace the chromosomes, then cut them out and arrange them in groups as in Figures 6-2, 6-11, and 6-19. Are the chromosomes those of a màle or a female? Is the number of autosomes normal? Is the number of sex chromosomes normal? If there is an ab-

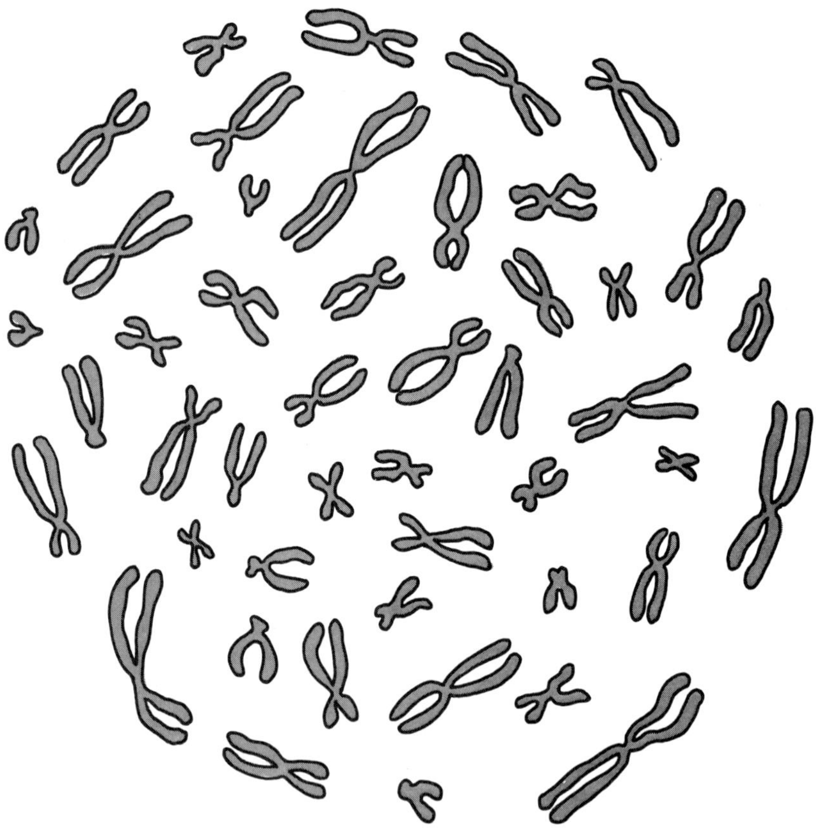

6-22. Human chromosomes.

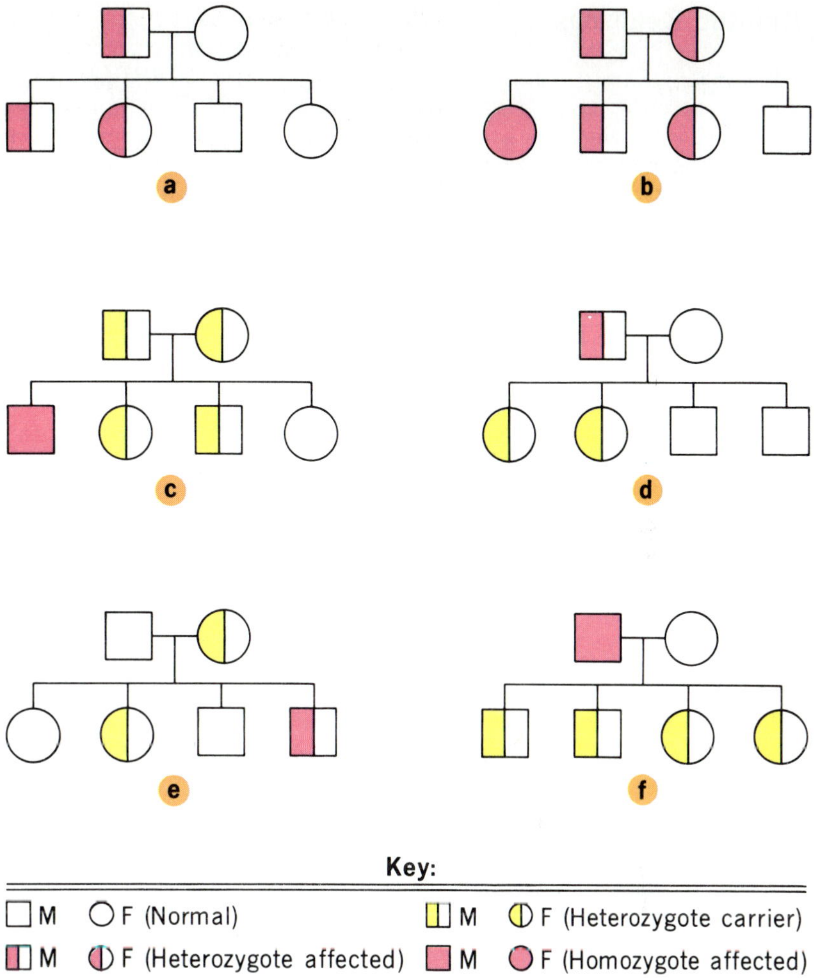

6-23. Pedigrees.

normal number of chromosomes, can you find out what the condition is called?

2. Different kinds of inherited diseases show different patterns of heredity. Consider three types:

(a) a *dominant* condition carried on an autosome (autosomal dominant)

(b) a *recessive* condition carried on an autosome (autosomal recessive)

(c) a *recessive* condition carried on the X chromosome (sex-linked recessive)

Study the pedigrees in Figure 6-23 and indicate the type of inheritance shown by each. In each case, explain your reasoning. State the Mendelian ratio in each case.

Related Readings

BOOKS

Carson, H. L. *Heredity and Human Life.* New York: Columbia University Press, 1968.

Stern, C. *Principles of Human Genetics,* 3rd ed. San Francisco: Freeman, 1973.

Sullivan, N. *The Message of the Genes.* New York: Basic Books, 1975.

Facets of Genetics: Readings from the Scientific American. San Francisco: Freeman, 1970.

ARTICLES

Bodmer, W. F. and L. L. Cavalli-Sforza. "Intelligence and Race." *Scientific American,* October 1970, page 19.

Cerami, A. and C. M. Peterson. "Cyanate and Sickle-Cell Disease." *Scientific American,* April 1975, page 44.

Grobstein, C. "The Recombinant-DNA Debate." *Scientific American,* July 1977, page 22.

7 The Hereditary Materials

In 1869, just four years after Mendel reported his findings on heredity, a young German-Swiss chemist, Friedrich Miescher (MY·shur) decided to look for the hereditary materials. Miescher was not familiar with Mendel's work. But he was aware of the work of cell biologists that showed what happens to chromosomes when cells divide. Miescher analyzed the materials in the nuclei of white blood cells. Later, he used sperm cells from salmon.

Miescher discovered an entirely new kind of chemical material in these cells. It was not a carbohydrate or a fat. It was always associated with protein, but unlike ordinary proteins it contained a great deal of phosphorus. This material was a nucleoprotein. The part that contained phosphorus was a nucleic acid. Miescher seemed quite sure that he had identified the chemical material responsible for heredity. Many other scientists at that time, both chemists and biologists, agreed. But by the time that Mendel's work was being studied, Miescher's work was almost completely forgotten. It was not until the 1940s that his insight was proved correct.

THE SEARCH FOR THE HEREDITARY MATERIAL

Scientists had thought that nucleic acids were too simple in structure to transmit hereditary information. Proteins seemed more likely because of their variety and complexity. Proteins are huge molecules made up of any of about 20 different kinds of amino acids. Each nucleic acid molecule, on the other hand, is made up of only six different kinds of molecules—sugars, phosphates, and four special kinds of **bases.**

There are two kinds of nucleic acids. The first, DNA, contains the sugar deoxyribose. The other nucleic acid, RNA, contains the sugar ribose. Ribose is different from deoxyribose by just one atom. Besides the sugar, nucleic acids contain a simple phosphate and four kinds of bases. In DNA the bases are adenine (ADD·eh·neen), cytosine (SY·toh·seen), guanine (GWAH·neen), and thymine (THY·meen). In RNA, the bases are the same, except that uracil (YOU·reh·sil) is substituted for thymine.

The nucleic acid Miescher found in sperm cells was DNA. Further chemical study of DNA showed that it is composed of four different kinds of molecules, which are nucleotides (Figure 7-1). Each nucleotide consists of one of the bases—adenine, cytosine, guanine, or thymine—plus deoxyribose and phosphate.

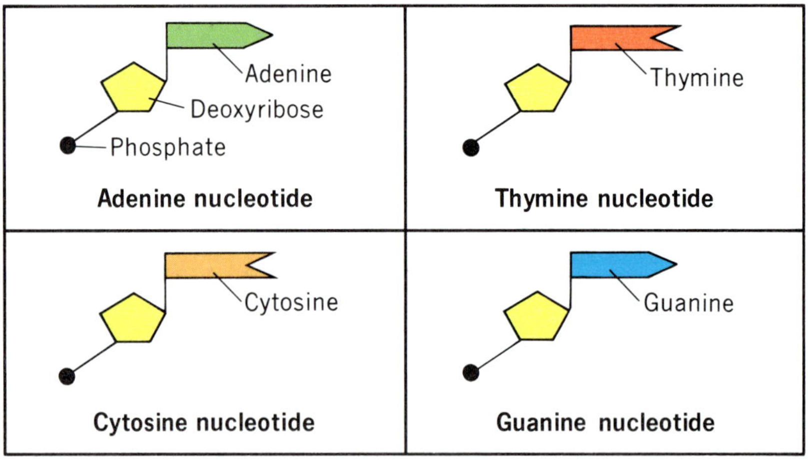

7-1. The DNA molecule is composed of four different kinds of nucleotides.

Chromosomes are made of DNA and protein. If the chromosomes carry the hereditary instructions, it was reasonable to ask whether only the protein, only the DNA, or both were responsible for inheritance. Given this problem, nearly every biologist before 1940 reasoned that, since the life of a cell is very complex, the instructions must be very complex. It seemed to them that more kinds of molecules could be made from the 20 amino acid units in protein than from the four nucleic acid units in DNA. Therefore, they thought that proteins, rather than DNA, were the hereditary materials. At that time, it seemed that nucleic acids were far too simple to account for the millions of different genes that must exist in all the animals, plants, and microorganisms of the world. This conclusion seemed to be very reasonable, even though it later turned out to be quite incorrect.

Genetic Transformation

Almost 50 years ago, an English scientist, Frederick Griffith, was working with two strains of bacteria that cause pneumonia. One strain was the smooth (S) type. On a solid growth medium, this type forms shiny, smooth colonies because each cell is surrounded by a slimy capsule. The S-type pneumonia bacteria are virulent (VEER·uh·lent)—they will cause pneumonia when injected into experimental organisms.

The other pneumonia bacteria used by Griffith was the rough (R) type. Cells of R-type bacteria lack capsules. The R-type colonies appear dry and have a rough surface. When injected into experimental animals, R-type bacteria do not cause pneumonia—they are nonvirulent. When S-type cells divide, they almost always produce new cells that have capsules. Similarly, R-type cells, without capsules, produce offspring without capsules. The presence or absence of capsules, therefore, is inherited.

Griffith injected a small quantity of R-type bacteria into a mouse (Figure 7-2). At the same time he injected a larger quantity of heat-killed, S-type bacteria into the same mouse. The R-type was nonvirulent, and the virulent S-type had been killed with heat. But the mouse developed pneumonia and died.

When Griffith examined the mouse, he made a discovery that was hard to explain. In addition to R-type bacteria, Griffith found living, virulent, S-type bacteria in the dead mouse. Where did they come from? Perhaps the heat-killed, S-type bacteria produced some kind of poison that killed the mouse. To test this hypothesis, Griffith injected only heat-killed, S-type bacteria into a mouse. The mouse did not develop pneumonia, nor were any living S-type bacteria found in its body (Figure 7-2).

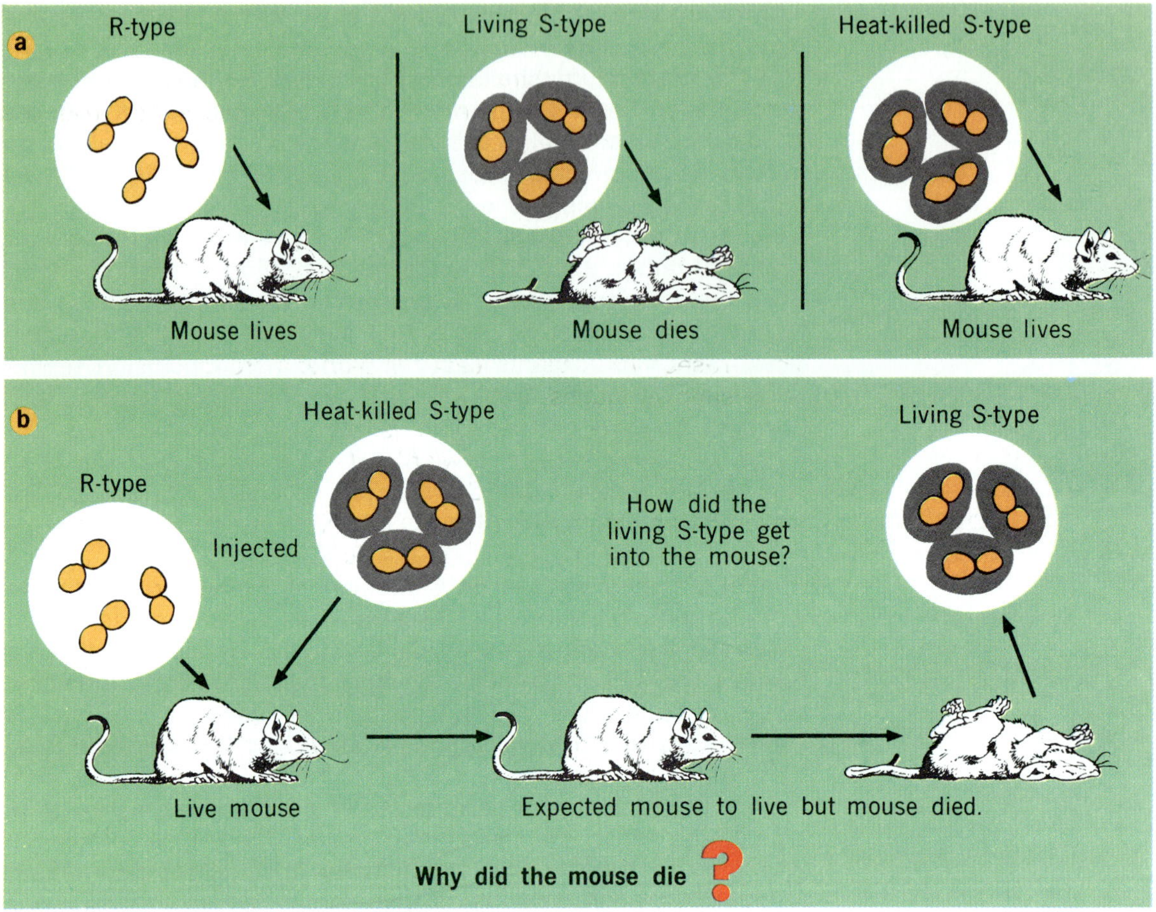

7-2. Genetic transformation of pneumonia bacteria. *What do these experimental results suggest about heredity?*

Griffith concluded that the dead S-type bacteria must contain some substance that transforms some of the harmless R-types into S-types. Griffith removed the newly formed S-types from the dead mouse and grew them in culture. Only S-type cells were formed. This seemed to be proof that the hereditary materials of the R-type had indeed been changed.

What was the cause of this unusual change? Many years of work passed before the chemical nature of the transforming substance was identified. Oswald T. Avery and other scientists at the Rockefeller University in New York City worked on the problem. Finally, they succeeded in separating the different kinds of substances that might be responsible. After separation, the substances could be tested. Fats and carbohydrates were quickly excluded. Protein and DNA were tested for the transforming properties. It was found that DNA, and not protein, produced the change.

But in 1944, biologists were not certain that inheritance in bacteria was the same as inheritance in many-celled plants and animals. At that time, not all biologists agreed that bacteria had chromosomes. The experimental results that convinced almost everyone came some years later.

The Genes of Viruses

Today we hear a great deal about the tiny organisms called viruses (Figure 7-3). Some viruses cause diseases in animals and plants. The viruses that cause diseases in bacteria are bacteriophages, or more simply, phages.

A phage, like a chromosome, is composed of only protein and DNA. Alfred Hershey and Martha Chase set out to determine whether the protein, the DNA, or both were responsible for the transmission of hereditary instructions in phages. They investigated this problem using two different radioactive isotopes.

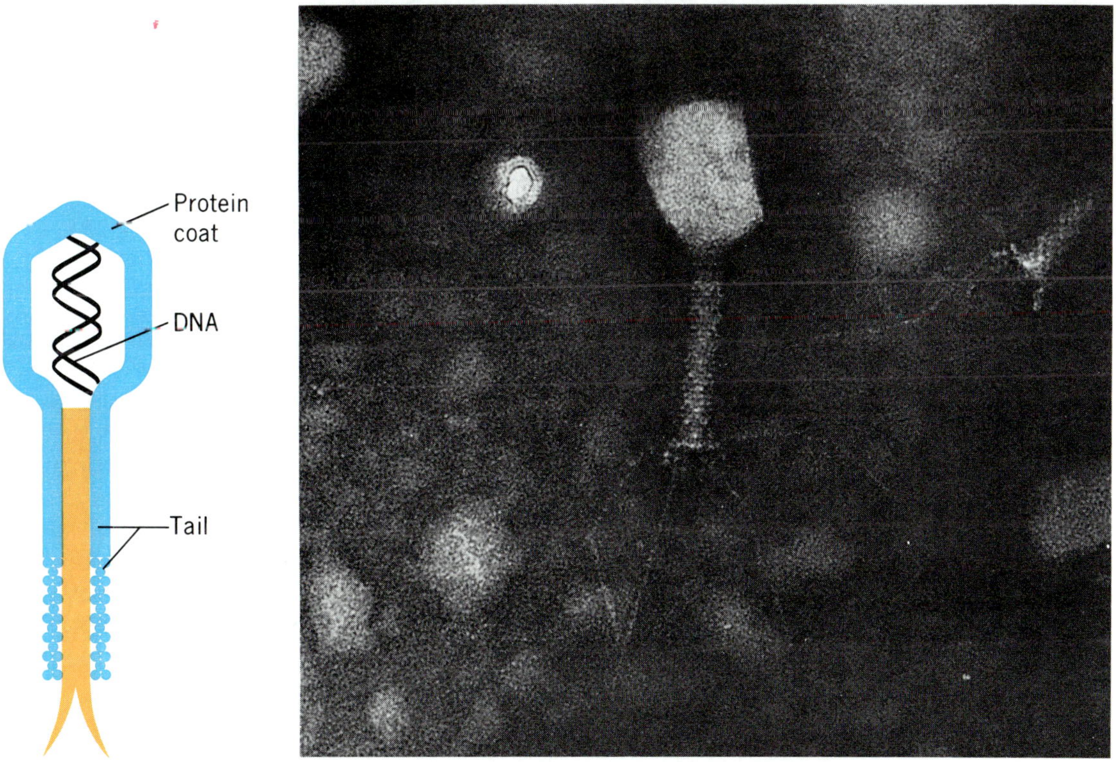

7-3. Compare the electron micrograph of a phage virus with the diagram. *Can you find the protein coat and tail in the picture?*

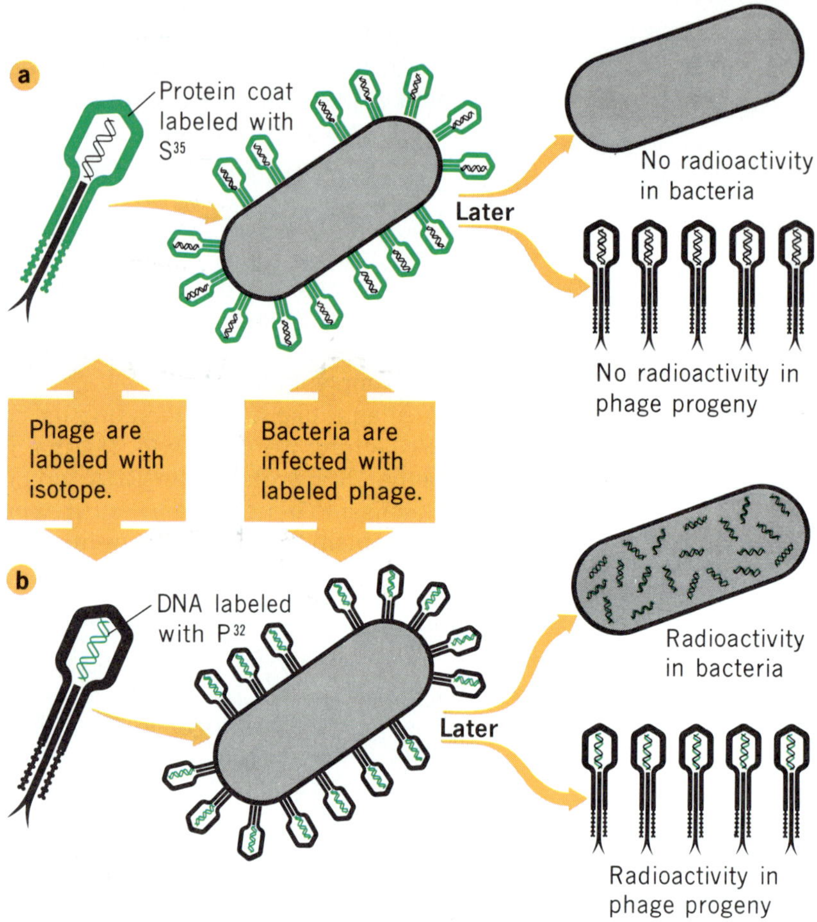

7-4. The experiments of Hershey and Chase showed that the hereditary instructions of viruses are carried by DNA.

Two experiments were performed (Figure 7-4). First, phages with protein coats labeled with an isotope of sulfur, S^{35}, were allowed to infect bacterial cells. Sulfur is found in proteins but not in nucleic acids. In another experiment, phages with their DNA cores labeled with P^{32} (an isotope of phosphorus) were allowed to infect bacterial cells. Phosphorus occurs abundantly in DNA but it is not found in the proteins Hershey and Chase were studying. Later, the bacteria and the phage particles that grew in them were tested for radioactivity.

The bacteria that had been infected with the S^{35}-labeled phage were *not* radioactive. On the other hand, the bacteria that had been infected with the P^{32}-labeled phage were. The presence of radioactivity inside the bacteria that were infected with P^{32}-labeled phage showed clearly that *only* the DNA of the phage entered the bacterial cell.

The same difference was evident in the phage offspring, or **progeny.** Only those that had grown in the bacteria infected by phages labeled with P^{32} were radioactive. The radioactivity of the progeny showed that DNA alone controls the synthesis of new DNA, as well as new protein, for the next generation of phages. Thus, DNA must be the hereditary material.

DNA—THE ANSWER AND THE QUESTION

The results of the Hershey and Chase experiments provided good evidence that DNA was the material of inheritance. The next question was, "How does it work?"

The researchers knew there was some mechanism for reproducing exactly the same instructions at each cell division. All of the instructions are present in a fertilized egg. The fertilized egg divides repeatedly, producing all the cells of the adult. Since some of the adult's cells eventually become sperm or egg cells, the instructions must be in *all* cells at *all* times. And all cells would have to be able to follow all the instructions. Of course, if the instructions programmed a cell to synthesize a molecule of, say, hemoglobin, the cell must contain the materials to do so.

Although the instructions would be reproduced exactly most of the time, there must be some basis for change. We know that mutations are changes, and that they do sometimes occur. Thus, research was directed toward learning how the hereditary material is reproduced, how it acts in the cell, and how it mutates. The first essential step was to learn the structure of DNA.

DNA Structure

Some important advances were made by two scientists, James Watson and Francis Crick. When Watson and Crick began their work, biologists had much data suggesting that DNA was the hereditary material. But no one knew the structure of the DNA molecule. They knew, however, that DNA is made up of nucleotides that contain adenine, guanine, cytosine, or thymine.

Available chemical data suggested that the nucleotides formed long strands that usually were double. But this double strand of DNA was far too small to be seen with a microscope. X-ray studies, however, suggested that the DNA had a uniform diameter. Data also indicated that the nucleotides of each long strand were joined together by bonds between the phosphate of one nucleotide and the sugar of the next. Thus, DNA could be pictured as two long chains of nucleotides. Each chain would be linked together by the sugars and phosphates. The bases—adenine, guanine, cytosine, and thymine—would be sitcking out on the side of each chain.

The problem for Watson and Crick was to show how the double strand of DNA was constructed. How could all the nucleotides fit together to form a double strand of *uniform diameter?* The phosphates in the nucleotides are the same size. So are the sugars. But this is not true of the four bases. The nucleotides containing cytosine and thymine are smaller than nucleotides containing adenine and guanine.

Given this basic information, you can work through the logic Watson and Crick used by doing **Inquiry 7-1.**

INQUIRY 7-1 The Structure of DNA

MATERIALS

Days 1 and 2

tracing paper
pencil
scissors
tape that can be written on

Day 3

FILM: *Regeneration of Acetabularia*

Procedure (Days 1 and 2)

Trace each of the four nucleotide models shown on page 205 onto a separate sheet of paper. Trace each model twice so you will have two sets. Copy the names of the bases on each traced model. After tracing and labeling, cut out your models on the dotted lines. DO NOT CUT THIS BOOK.

Determine how a single strand of DNA might be constructed. Remember, a bond forms between the phosphate (o) of one nucleotide and the sugar (⬠) of the next nucleotide. Place four of your nucleotides (one set) in a vertical column and determine where the bonds must go. Tape the four nucleotides together and draw lines between the phosphate of one nucleotide and the sugar of the next to indicate a bond. You now have a model of a chain of four different nucleotides held together by bonds between the sugars and the

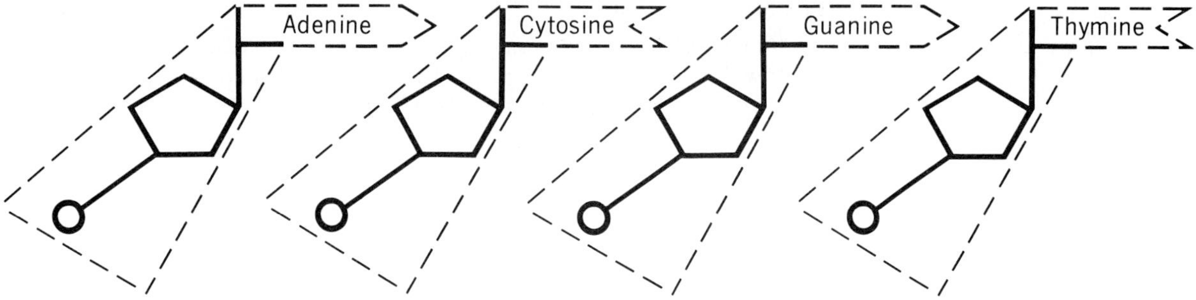

phosphates. **1.** On which side of the model are the bases? **2.** All of the sugars in the chain should be turned in the same direction. Why?

The data indicated that DNA is a *double* strand of *uniform diameter.* Arrange your remaining four nucleotides in a column parallel to the first chain you made. **3.** How can the models be arranged so that the width of the total structure is always the same? **4.** What nucleotides will "fit" next to the thymine nucleotide in your first chain? next to the guanine nucleotide? **5.** What "fits" are possible for the adenine nucleotide? for the cytosine nucleotide? **6.** What characteristics of the models determined your answer to questions 4 and 5? **7.** Use these same characteristics to explain why thymine and cytosine nucleotides *cannot* pair with each other.

So far, you have found that some bases fit together and others do not. You know what some possibilities for fit are, but you have not determined what is the case in a real DNA molecule. From observations of shape and size alone, you know that adenine *could* pair with either thymine or cytosine. But does it pair with one, or the other, or both in a real DNA molecule? **8.** If adenine always pairs with cytosine, how should the relative amounts of adenine and cytosine compare in a DNA molecule? **9.** If adenine always pairs with thymine, what relative amounts of these bases would you expect to find? **10.** If adenine sometimes pairs with cytosine and sometimes with thymine, what relative amounts of adenine, compared to the amounts of the other two, would you expect to find?

Studies by an American biochemist, Erwin Chargaff, showed that DNA in human liver cells is 30.3 percent adenine, 19.5 percent guanine, 30.3 percent thymine, and 19.9 percent cytosine. **11.** Of the possible pairings described in questions 8, 9, and 10, which one seems most likely in light of Chargaff's data? **12.** What do the data suggest about which nucleotide pairs with guanine? Explain your answer.

Now consider more of Chargaff's data. The following table shows the percentages of the four bases found in DNA from different kinds

Tissue and Organism	Adenine	Guanine	Thymine	Cytosine
Thymus cells				
Human	30.9	19.9	29.4	19.8
Sheep	29.3	21.4	28.3	21.0
Pig	30.9	19.9	29.4	19.8
Spleen cells				
Human	29.2	21.0	29.4	20.4
Sheep	28.0	22.3	28.6	21.1
Pig	29.6	20.4	29.2	20.8
Liver cells				
Human	30.3	19.5	30.3	19.9
Sheep	29.3	20.7	29.2	20.8
Pig	29.4	20.5	29.7	20.5
Sperm cells				
Human	30.7	19.3	31.2	18.8
Sheep	28.8	22.0	27.2	21.0
Yeast	31.3	18.7	32.9	17.1

of cells. **13.** Copy the data for the four different kinds of *human* cells. What do you notice about the percentages of any single base in the different kinds of cells? **14.** Which bases pair together in human DNA? Use examples from Chargaff's data to explain your answer. **15.** Does the same pattern of pairing hold true for all organisms? Explain your response. **16.** Position your models for your second DNA chain and tape the entire model together. Summarize briefly your reasons for constructing it as you have.

Procedure (Day 3)

Your teacher may be able to arrange for you to view and discuss the inquiry film loop *Regeneration in Acetabularia. Acetabularia* (Ass·uh·tab·you·LAIR·ih·ya) is a single-celled organism that lives in water. It is a plant that can regrow missing parts. You can learn much about how DNA controls cell activities by viewing this film. As you view the film, try to formulate hypotheses to explain the experimental results that are presented.

The Watson-Crick Model of DNA

As you saw in **Inquiry 7-1,** Chargaff's data supported the idea that adenine always pairs with thymine and that cytosine always pairs with guanine. These molecules form weak chemical bonds, called hydrogen bonds. Adenine and thymine always form two hydrogen

7-5. Notice how adenine always pairs with thymine and cytosine always pairs with guanine. *How does Chargaff's data presented in Inquiry 7-1 support this conclusion?*

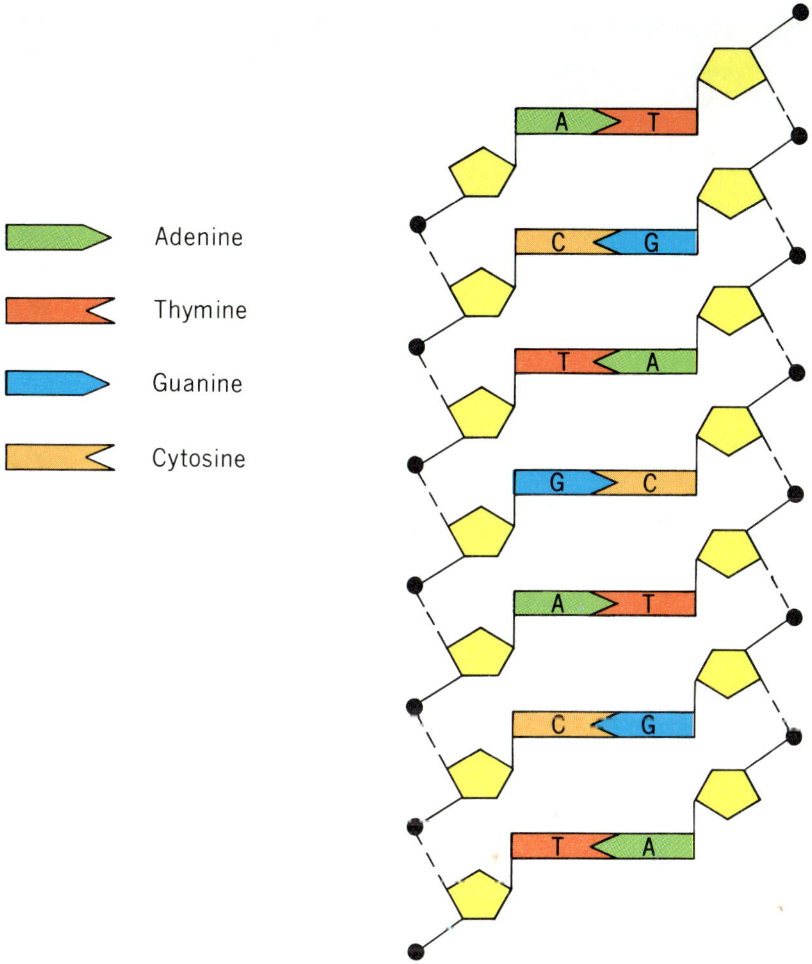

bonds. Guanine and cytosine each form three hydrogen bonds. The atoms that are responsible for forming these bonds are not in the proper positions to make bonds between adenine and guanine or between adenine and cytosine. But adenine and thymine will pair, because their atoms line up and form two hydrogen bonds. Similarly, guanine and cytosine pair together with three hydrogen bonds.

When the order of nucleotides is established for one side of the chain, the order on the other side is determined. An adenine nucleotide must always pair with a thymine nucleotide, and a guanine nucleotide must always pair with a cytosine nucleotide. And the sides of the double chain are the same along the entire length—alternating molecules of sugar and phosphate. Figure 7-5 shows the basic DNA structure of the base pairs, sugars, and phosphates. The

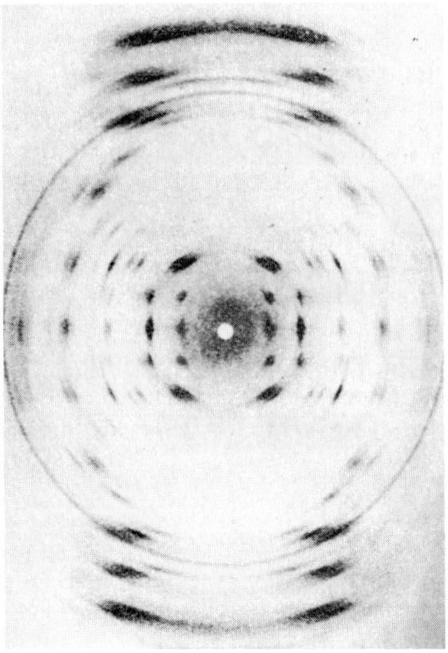

7-6. X-ray diffraction photograph of DNA.

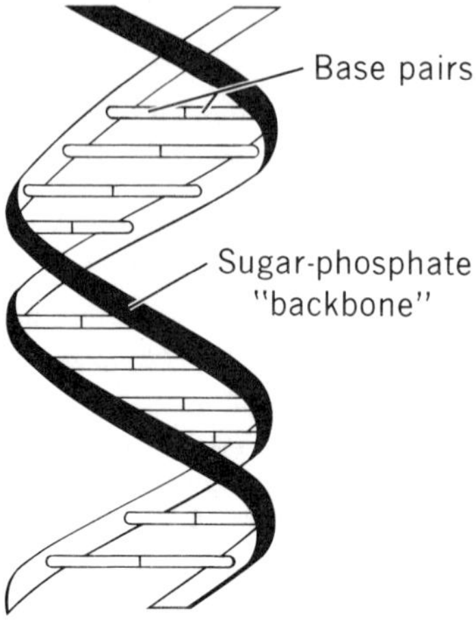

7-7. The double helix of DNA appears as two twisted spirals. Base pairs lie between the sugar and phosphate "backbone."

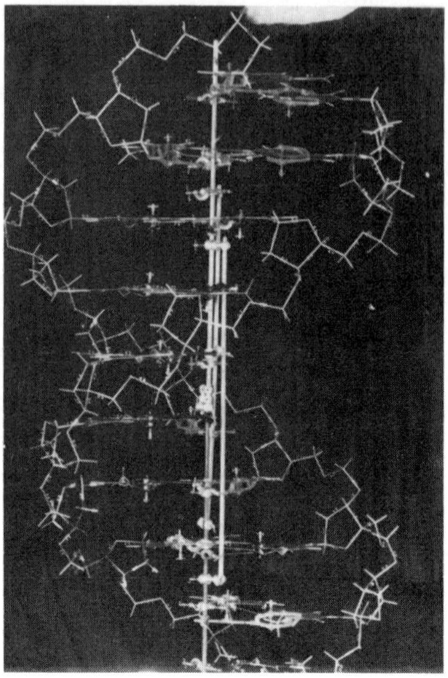

7-8. Watson-Crick model.

distinctive part of each DNA molecule is the particular sequence of base pairs.

But Watson and Crick had additional evidence to consider. Rosalind Franklin, who also was working at Cambridge, had taken X-ray "pictures" of DNA (Figure 7-6). In these pictures, the DNA molecule looked like two strands twisted in the form of a double helix (HEE·lix). A double helix is two spirals twisted together (Figure 7-7). Chemical bonds hold the entire DNA molecule together. The full model worked out by Watson and Crick is shown in Figure 7-8.

Direct Replication of DNA

How can we get two identical DNA molecules—each one a double helix—starting with only one DNA molecule? The Watson-Crick model suggests a mechanism by which DNA may be replicated. First, the two coils of the DNA molecule come apart. The bonds that hold together the double helix are broken, starting from one end, like a zipper. Each base separates from its partner on the other DNA strand. Each separation leaves two unmatched bases, one on each strand.

The cell contains a variety of nucleotides. Each of these has three phosphate groups. From this assortment of nucleotides, a *new* nucleotide of the proper kind is formed to fit each vacant position on both chains. The extra phosphate groups on the nucleotides used in this process are split off. This releases energy. The energy is used to attach the deoxyribose of one nucleotide to the phosphate of the next nucleotide on the same DNA strand. All of this occurs as a result of the activity of specific enzymes that catalyze the various reactions. This process of "unzipping" the DNA molecule and forming new chains is shown in Figure 7-9.

Look carefully at the new molecules (Figure 7-10) that are formed. They are exactly like each other and like the original molecule. Thus,

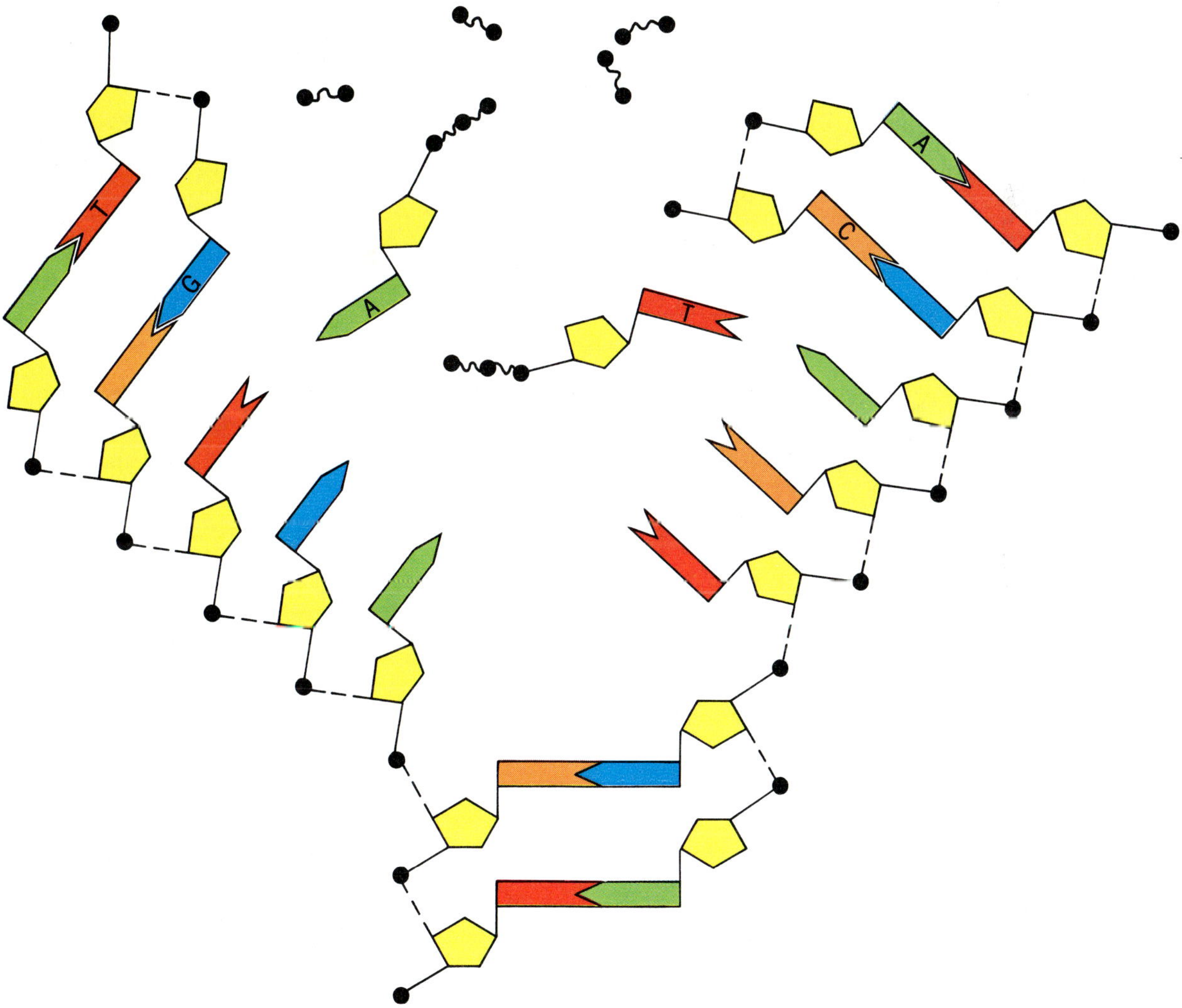

7-9. A diagram of the replication of DNA. A is adenine, T is thymine, G is guanine, and C is cytosine. The two strands of the DNA molecule begin to "unzip." New nucleotides of the proper kind move into place. *What happens when the extra phosphate groups are split off?*

the Watson-Crick model allows for precise duplication, or replication, of the hereditary information.

The replication process has been confirmed experimentally. The mass number of ordinary nitrogen is 14 (N^{14}). In 1958, scientists at the California Institute of Technology used a heavier isotope of nitrogen (N^{15}) to label nucleotides made by bacteria. (All four bases contain nitrogen.) The isotope was provided to the bacteria as they were reproducing and making new DNA molecules. After many generations, the DNA molecules contained only N^{15}. Then, the bacteria were supplied with ordinary nitrogen, N^{14}. The cytoplasm of these cells contained only N^{14}. After one cell division, all the bacteria contained DNA in which the nitrogen was exactly halfway in mass be-

7-10. The replication process is now complete. *How do these two double strands of DNA compare with the molecule shown in Figure 7-9?*

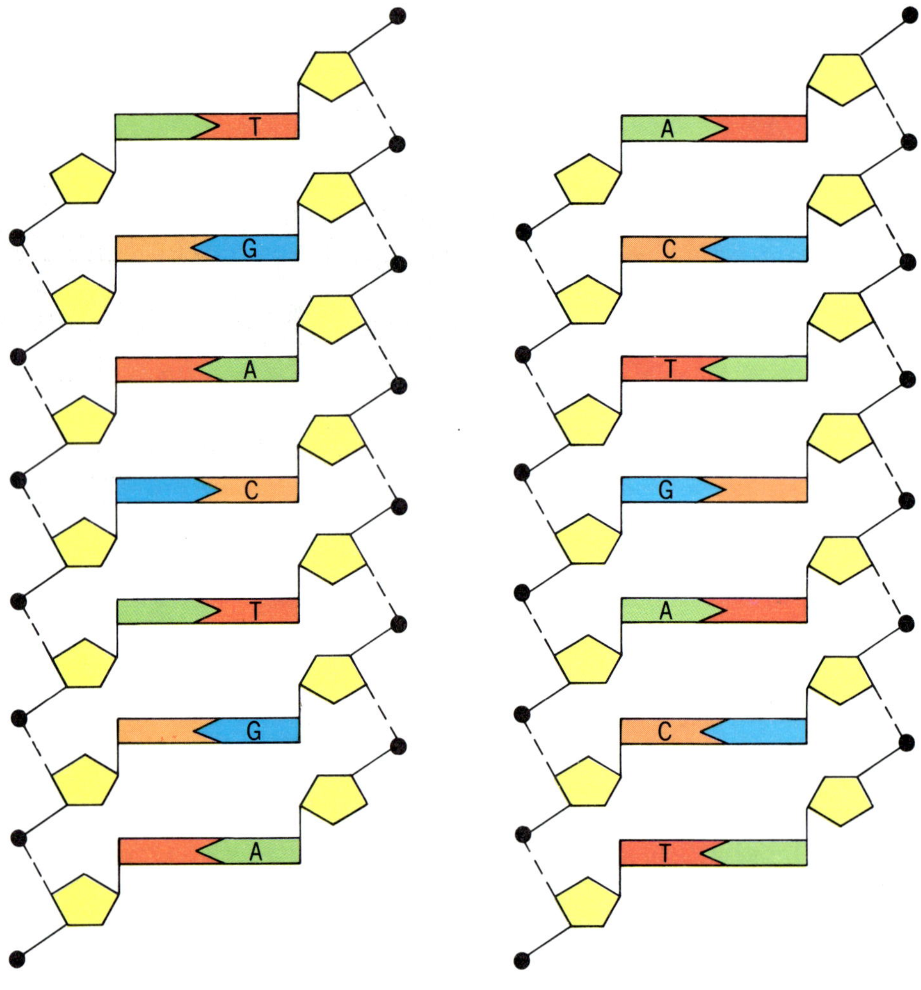

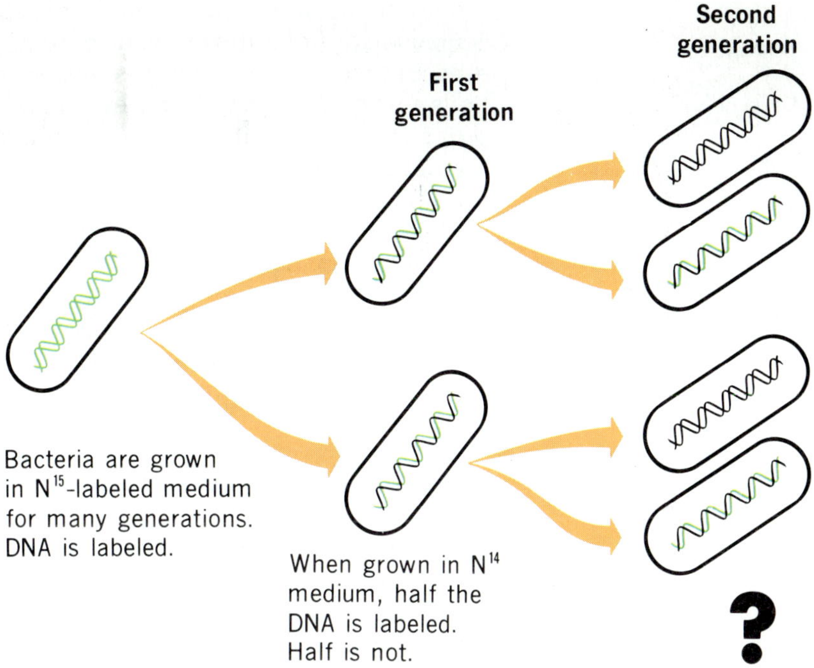

7-11. Confirmation of the replication process. *What proportion of the DNA in the second generation is labeled with N^{15}? How does the Watson-Crick model help explain these results?*

tween the labeled (N^{15}) DNA and ordinary (N^{14}) DNA. Each new molecule of DNA consisted of a double helix made up of one labeled strand and one ordinary strand (Figure 7-11). The scientists concluded that each *new* strand of DNA made in the replication process was made from the material in the cytoplasm that contained ordinary nitrogen.

When this generation of cells divided, half contained the strand of N^{15} DNA plus a strand of N^{14} DNA replicated from nucleotides in the cell material. The other half contained the strand of N^{14} DNA plus an N^{14} strand replicated from the cell material. This is what the model of the replicating DNA molecule predicts.

FUNCTION AND CHANGE IN THE HEREDITARY INSTRUCTIONS

Let's now consider two other aspects of inheritance. First, how do the hereditary instructions act in cells? Second, how do the hereditary instructions change or mutate?

Inheritance of eye color can serve as an example of instructions. Suppose you inherit from your parents the instructions for brown eyes. But how are brown eyes actually made? Somehow the instructions in certain cells of the eye must cause the production and organization of chemicals that form dark pigment. The pigment is then deposited in a particular concentration. The result is brown eyes.

We have already learned in Chapters 5 and 6 that certain parts of chromosomes, called genes, control specific activities of the cell. The portions of the hereditary material that control the formation of brown eyes are known as genes for brown eyes. But how do these segments of DNA actually produce brown eyes? It took studies of organisms other than humans to give us the answer to the question.

Gene Action in Neurospora

Heredity in microorganisms, such as bacteria or microscopic fungi (fun·JEE), is easier to study than heredity in large, complex organisms, such as humans. The salmon-pink fungus *Neurospora* (new·ROS·pur·uh) shown in Figure 7-12 has been widely used for such studies. *Neurospora* grows well on a medium of a dilute mixture of salts, table sugar, and one vitamin—biotin (BY·oh·tin). It reproduces by means of spores. Each tiny spore can grow into a new colony.

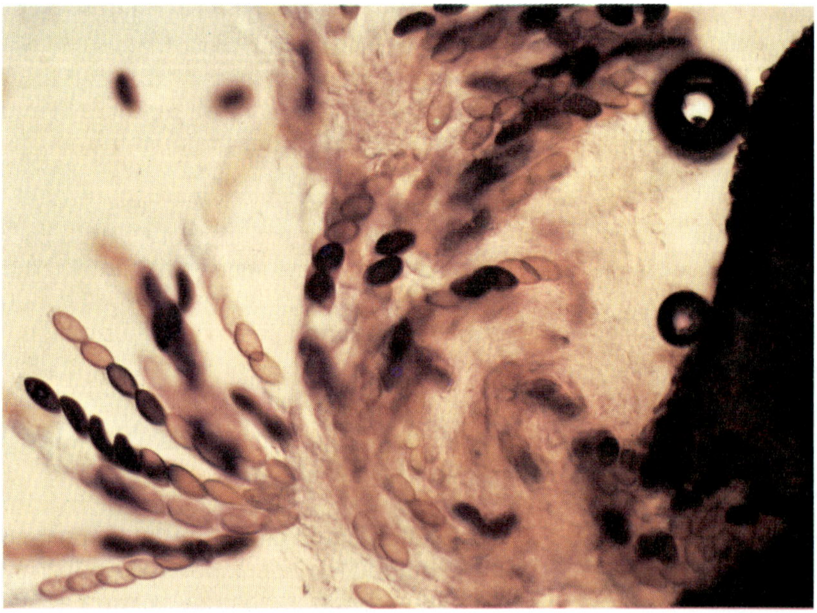

7-12. A light microscope photograph of *Neurospora,* a fungus that has been widely used for studies of gene action.

Neurospora is made up of many complex proteins, carbohydrates, fats, vitamins, nucleic acids, pigments, as well as other materials. The organism synthesizes all of these complex compounds from the simple raw materials in the medium. Furthermore, each individual *Neurospora* closely resembles, in its chemical composition, the parent that originally supplied the spores. This again is an example of genetic continuity.

In the 1940s, biologists learned that microorganisms may lose some of their synthesizing abilities. The hereditary instructions (genes) become changed by **mutation.** A mutation is a change in the structure of a gene. The mutated genes no longer are able to direct the normal syntheses. George W. Beadle and Edward L. Tatum, then at Stanford University in California, treated spores of *Neurospora* with X rays or ultraviolet rays. Then they tested the treated spores to determine if they had been changed in any way. Beadle and Tatum were particularly interested to see if any of *Neurospora's* ability to direct normal synthesis had been changed.

Over a period of time, they found that virtually every substance normally synthesized by the fungus is subject to genetic control. The gene or genes responsible for any particular biochemical synthesis can be damaged or eliminated by radiation. Beadle and Tatum found mutant spores that could not grow at all on the usual medium. These mutant spores could survive, however, if nutritional supplements, such as amino acids or vitamins, were added to the medium.

Beadle and Tatum found some mutations that seemed to affect a series of related chemical syntheses. We might designate the compounds synthesized in such a series as O, C, and A. Careful experiments revealed that Mutant 1 would grow *only* if O, or C, or A was added to the medium. Mutant 2 would grow *only* if C or A was added to the medium. Mutant 3 would grow *only* if A was added to the medium.

The three compounds must have some relationship. Beadle and Tatum suggested that each mutation alters a single gene that controls *one* step in the synthesis of a particular molecule. We might diagram such a chain of reactions as follows:

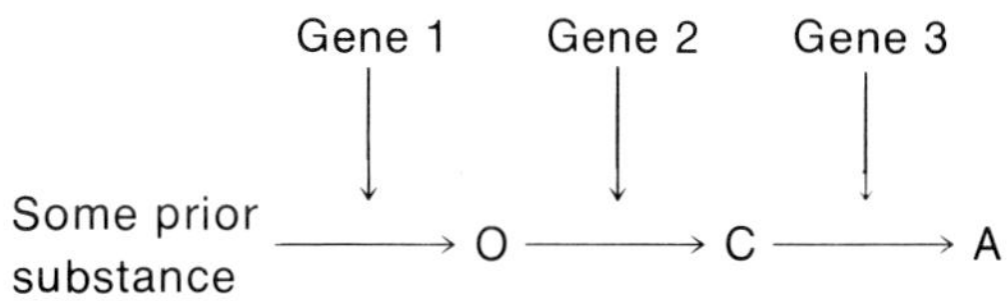

This diagram suggests that Gene 1 controls the conversion of some substance into O, Gene 2 controls the conversion of O into C, and Gene 3 controls the conversion of C into A. If Gene 1 were damaged, the organism could not grow unless the deficiency was corrected. This could be done by adding compound O to the medium thus bypassing the need for Gene 1. But adding C also would solve the problem, because we would be supplying the compound in a more advanced stage of synthesis. We would thus bypass the normal functions of Genes 1 and 2.

If Gene 3 mutates, the deficiency cannot be met by adding substances O or C. Gene 3 converts C to A. There is no alternative except to add A to the medium, thus bypassing the function of all three genes.

Beadle and Tatum identified a series of steps in synthesis. They also showed that a different mutation affected each step. Most importantly, they showed that the functions of genes can be investigated at the chemical level.

Genes and Enzymes

What kinds of chemical reactions do genes normally control? Of special importance is the formation of proteins and particularly one kind of protein—enzymes.

One hypothesis about the relationship between genes and enzymes is called the "one gene—one enzyme" hypothesis. That is, a single gene may control the production of only one kind of enzyme. Although it is thought today that the relationship may not be that simple, the hypothesis provided a framework for suggesting additional experiments on gene action.

We can now refine our earlier diagram to include the enzymes controlled by the genes:

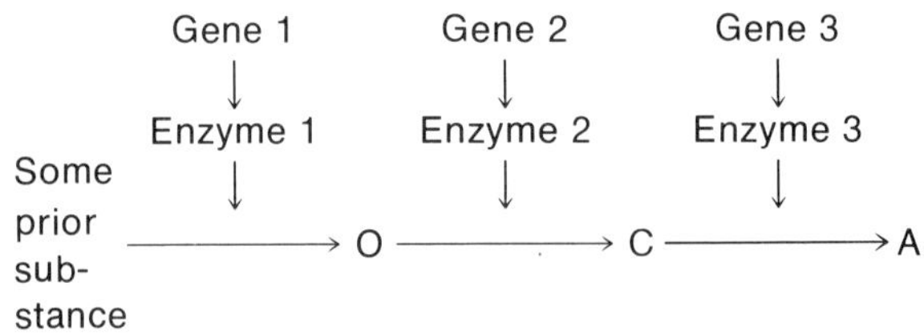

In 1958, Beadle and Tatum received a Nobel Prize for their work on *Neurospora*. When Beadle went to the library to prepare a lecture (to be given upon receiving the Nobel Prize), he found something quite shocking. Much of what he and Tatum had discovered had been worked out by a British physician over 50 years before.

At that time, Archibald Garrod had become greatly interested in some abnormal human hereditary traits. One of these was alkap-

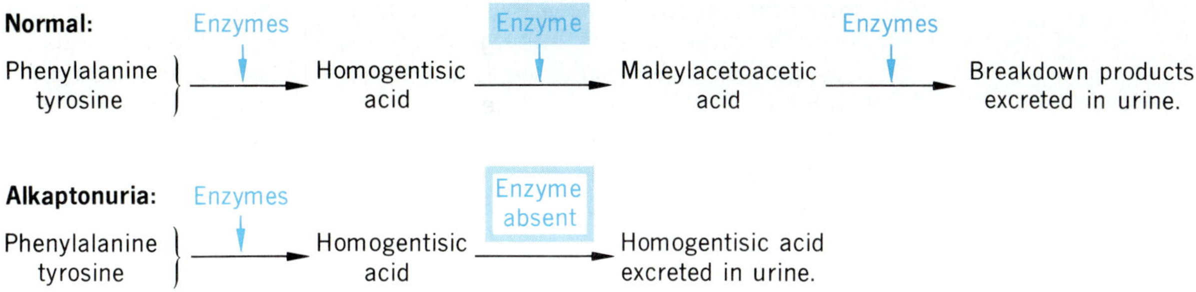

7-13. In 1908, Garrod hypothesized that alkaptonuria ("black urine") was caused by the absence of an enzyme involved in the breakdown of the amino acid tyrosine. Today, we know the entire series of steps that occur. Notice the missing enzyme that causes the "black urine" abnormality.

tonuria (al·KAP·tuh·nur·ih·uh), which means "black urine." Some mothers were terrified when their babies' wet diapers turned black. This happened when the urine contacted the air. The condition actually is quite harmless. It occurs because two amino acids, phenylalanine and tyrosine, are not broken down completely into waste products. A particular chemical change seemed to be blocked (Figure 7-13). Garrod suggested that the enzyme responsible for that chemical step might be absent or inactive.

Garrod gave some lectures about alkaptonuria and similar conditions in 1908. The lectures were published, and then forgotten. The similarity of this story to the story of Mendel's work, and that of Miescher's, is significant. What do these bits of history suggest about the nature of scientific inquiry?

Gene Action in Humans

Genes control the synthesis of all proteins, including hemoglobin. In people affected by the inherited condition called sickle-cell anemia, red blood cells become distorted when the oxygen concentration is low (Figure 7-14). Biologists compared the chemical structures of hemoglobin from normal individuals and from those with sickle-cell disease. They found a slight chemical difference.

Vernon Ingram, working at Cambridge University in England, tried to identify the difference. It was laborious work. Hemoglobin is made up of four chains of amino acids. There are two different kinds of chains. Each chain, containing a large number of amino acids, is called a **polypeptide.** The production of each of the two kinds of polypeptides is controlled by a different gene. At first, every piece of the abnormal hemoglobin seemed to be exactly the same as the cor-

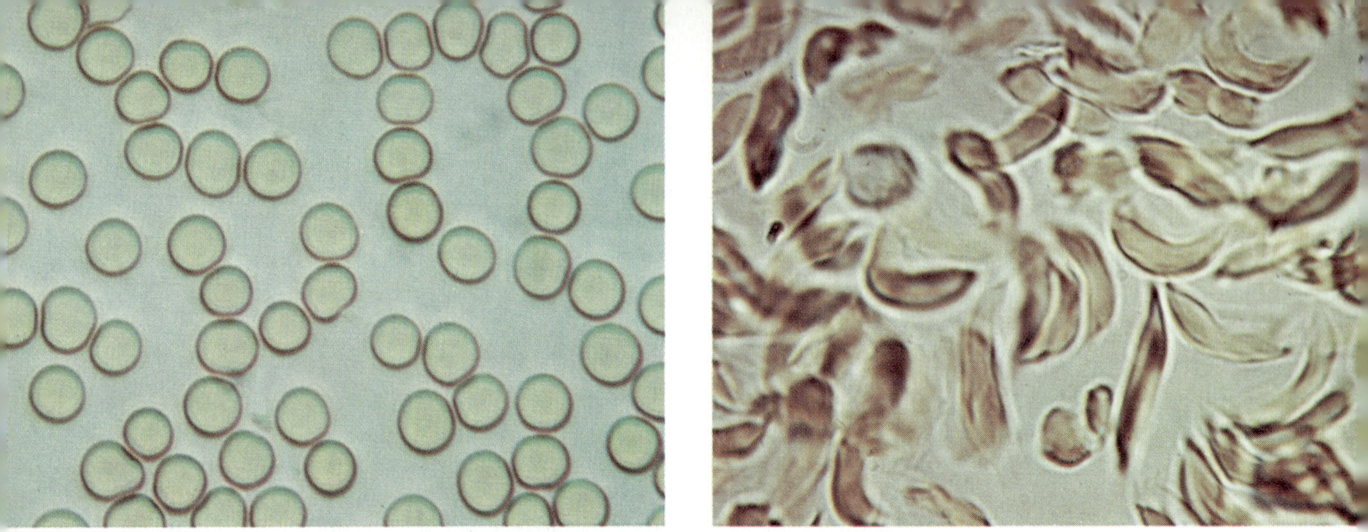

7-14. Normal red blood cells are shown on the left. Cells from an individual with sickle-cell anemia are shown on the right. Notice that the abnormal cells are sickle- or crescent-shaped.

responding piece of normal hemoglobin. But finally Ingram found a difference. At one place in the chain, glutamic (gloo·TAM·ic) acid was replaced by another amino acid, valine (VAY·leen). This single substitution is enough to reduce the oxygen-carrying capacity of the hemoglobin. Sickle-cell anemia is the result.

Part of polypeptide of *normal* hemoglobin	Corresponding polypeptide of *sickle-cell* hemoglobin
|	|
valine	valine
histidine	histidine
leucine	leucine
threonine	threonine
proline	proline
glutamic acid	*valine*
glutamic acid	glutamic acid
lysine	lysine
|	|

As a result of Ingram's work, the "one gene—one enzyme" relationship was modified. When a protein contains more than one kind of polypeptide, the synthesis of each kind is controlled by a different gene—thus, we have "one gene—one polypeptide." Ingram showed that at least one kind of mutation may change only *one* amino acid in a protein molecule.

THE GENETIC CODE

A gene is a segment of a DNA molecule. A gene controls the sequence of amino acids in a polypeptide chain. But how? DNA is found only in the nucleus. Yet synthesis of polypeptides occurs outside the nucleus, in the cytoplasm. The instructions in the DNA must be carried by some kind of "messenger" from the nucleus to the cytoplasm.

Messenger RNA

For several reasons, biologists thought that RNA was the messenger. Unlike DNA, RNA occurs both in the nucleus and in the cytoplasm. Large amounts of RNA are found in the nucleus and in the nucleoli. When egg cells mature and enlarge just before fertilization, they contain enormous amounts of RNA in both the cytoplasm and the nucleus. And when a cell is very active and is synthesizing large amounts of protein, the movement of RNA from the nucleus to the cytoplasm increases.

RNA closely resembles DNA in chemical structure. Three of its four kinds of nucleotides are the same as in DNA. The fourth nucleotide in RNA contains uracil, instead of thymine. And the only difference between ribose and deoxyribose is that ribose has one more oxygen atom.

Two French geneticists, Francois Jacob (Fran·SWAH juh·KOB) and Jacques Monod (Jahk Moh·NAHD), suggested that RNA might be formed from DNA. They predicted that a "messenger RNA" would soon be discovered. And so it was, in 1961.

The making of **messenger RNA (mRNA)** is similar to the replication of DNA. The two strands of DNA first unzip. Messenger RNA is formed with nucleotides that pair with those of one of the DNA strands. Thymine, guanine, and cytosine in the DNA pair with adenine, cytosine, and guanine in the newly forming RNA. Uracil pairs with adenine in the DNA. Imagine a strand of DNA with a sequence of bases like that shown in Figure 7-15. The figure shows the mRNA strand that would be produced by this DNA strand.

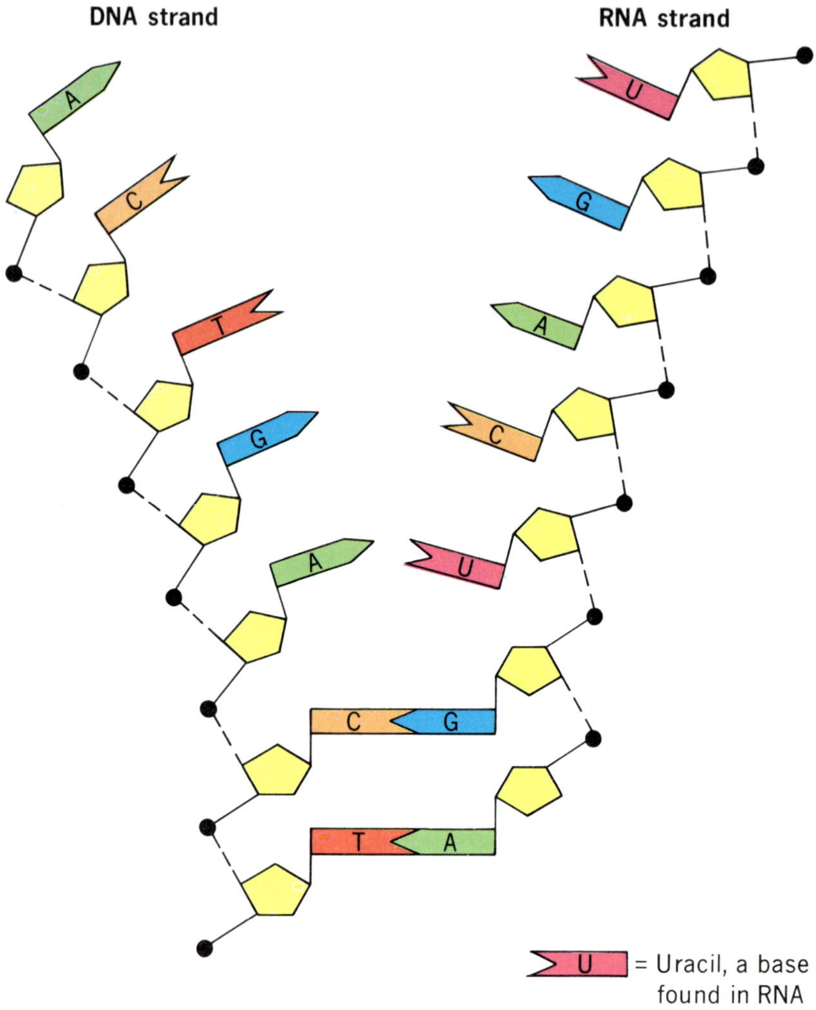

7-15. Messenger RNA is formed on a DNA strand. The process is similar to DNA replication.

A special enzyme starts the production of mRNA. The enzyme is called RNA polymerase (puh·LIM·uh·race). When it becomes attached to a DNA strand, it moves steadily along it. At certain points, it begins production of an mRNA molecule. The mRNA becomes longer until the entire length of the gene has been transcribed. Figure 7-16 shows some genes in an amphibian egg cell in which this is happening.

Cracking the Code

How many bases in the DNA and RNA are needed to specify one message? Each message specifies which amino acid will be put into

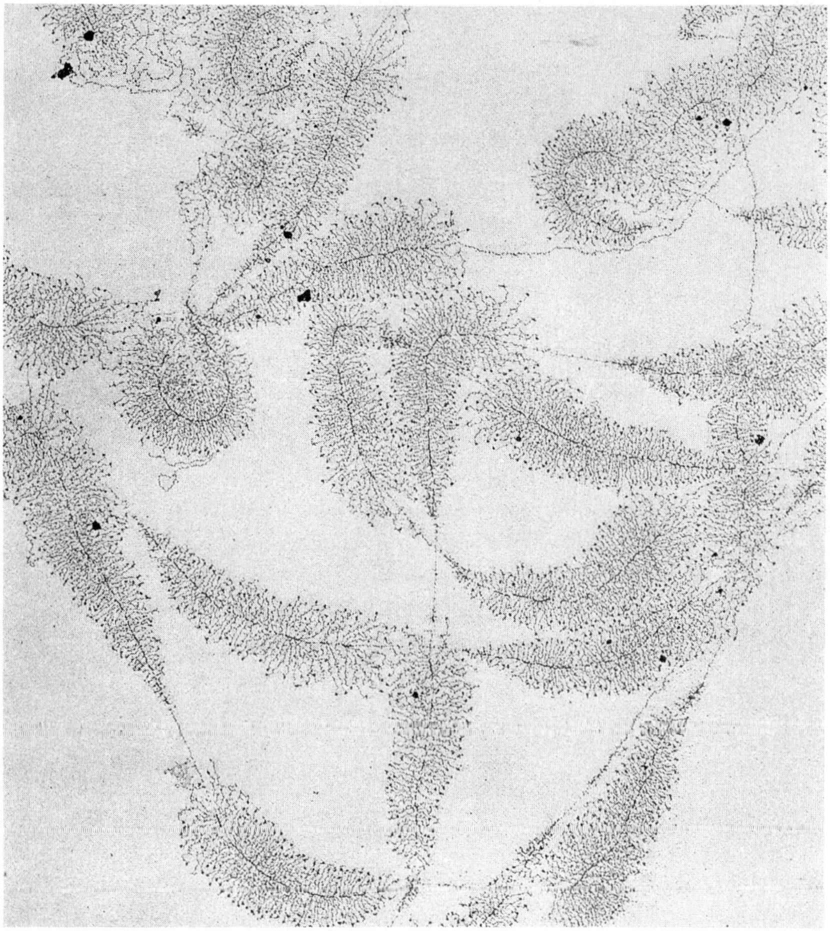

7-16. The long thin strands are DNA. The "branches" on the sides of the DNA are messenger RNA strands lengthening as the RNA polymerase enzyme moves along.

a polypeptide chain. There are about 20 different amino acids, so there must be at least 20 different "code words." Two-letter messages cannot work, because there are only 16 ways any two symbols of the four bases A, C, G, and T, can be put together. If three-letter code words are used, however, 64 different code words are possible – more than enough for the 20 or so different amino acids. Research has shown that the code in DNA is indeed in sequences of three bases. Three-letter code words in DNA are actually the mechanism of gene function. Three-letter sequences in mRNA, called **codons,** specify what particular amino acids will go into a growing polypeptide chain. How is this accomplished?

Transfer RNA

Biologists soon discovered another kind of RNA, which they called **transfer RNA (tRNA).** The mRNA molecules are long, single strands of RNA. The tRNA molecules are smaller, and they are curled into the shape of a cloverleaf. Each tRNA molecule has a particular kind of amino acid attached to its free end. This kind of RNA delivers the amino acid to the ribosomes where polypeptide chains are being made. Each tRNA molecule has, at the side opposite the amino acid, a particular sequence of bases, called the **anticodon.** The sequence is coordinated with the mRNA's codon for a particular amino acid (Figure 7-17). Thus, if the mRNA has a codon -UUU-, the complementary tRNA molecule has -AAA-. If the messenger has a -CAU- codon, the transfer RNA anticodon is -GUA-. It is a perfect mechanism for matching the messages with the material.

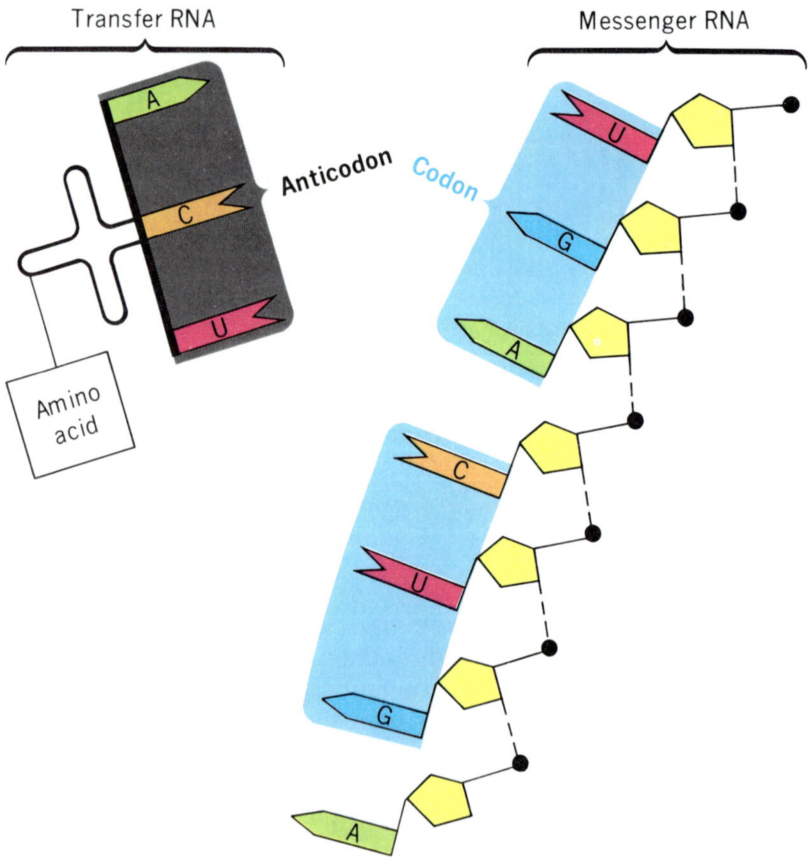

7-17. A diagram of a transfer RNA. Note the match between the anticodon and the codon of the messenger RNA.

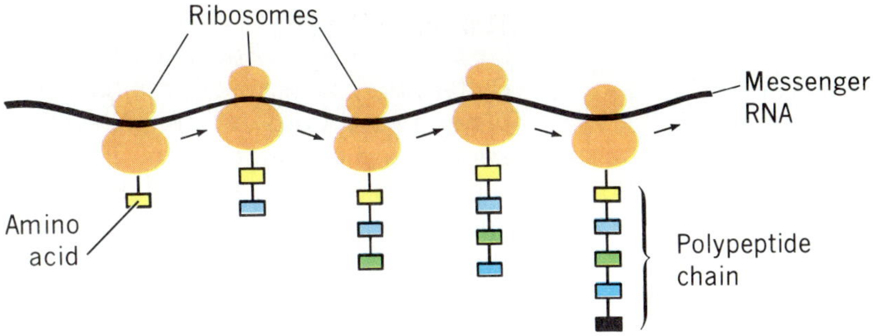

7-18. Ribosomes move along the messenger RNA strand. The ribosomes synthesize a polypeptide chain, adding one amino acid at a time. The amino acids are carried to the ribosomes by transfer RNA.

Putting Amino Acids Together

Now the mechanism for making a protein or polypeptide is complete. As Figure 7-18 shows, the ribosomes move along the mRNA strand. When the first codon of an mRNA molecule comes into place at the first ribosome, it is matched with the anticodon of a particular kind of tRNA. Thus, a particular amino acid is brought into position. The ribosome moves to the next codon. A new tRNA molecule, bearing a second amino acid, moves into place. Enzymes from the cytoplasm join the amino acids together. Enzymes also free the amino acids from the tRNA. Finally, a codon in the mRNA is reached that says simply "stop." This is the end of a particular polypeptide chain.

After the polypeptide is made, the mRNA may be reused or it may be destroyed by an enzyme. Messenger RNA has a short life. The whole process of synthesis would stop if new mRNA were not being supplied constantly by the nucleus. Meanwhile, each tRNA molecule picks up a new amino acid that will become part of some other polypeptide chain.

The complete understanding of the genetic code was worked out a step at a time. In 1961, Marshall Nirenberg and J. H. Matthaei worked out the first actual match between a codon and an amino acid. They began by making an artificial mRNA molecule composed entirely of uracil bases, -UUU-UUU-UUU-UUU-. They put this artificial mRNA molecule into a test tube with all the materials needed to make a polypeptide: ribosomes, tRNA, many different amino acids, the enzyme for synthesis, and ATP to supply energy. The artificial mRNA molecule caused synthesis of a polypeptide. The polypeptide consisted entirely of phenylalanine amino acid units. The biologists concluded that the codon, UUU, specifies the amino acid phenylalanine.

This exciting discovery led to an intense period of work by Nirenberg's group and others. They made many artificial messages and learned the particular codons in the mRNA for each of the amino acids. The complete code was established by 1967. One amino acid may have as many as six different codons. (Remember the 64 possible combinations.) Some codons do not specify an amino acid at all. Rather, they cause synthesis to stop when the polypeptide is complete. The codons for amino acids are the same in every plant, animal, and microorganism that has been studied so far.

Biologists then began making artificial DNA. But they did not know how to put in the gene segments that start and stop protein synthesis. In 1976, Har Gobind Khorana (kor·AH·nah) and his colleagues at the Massachusetts Institute of Technology succeeded in producing a completely functional gene. These biologists constructed, molecule by molecule, a gene of a common bacterium. When the artificial gene was implanted in a living bacterium, it functioned quite normally, producing the appropriate protein.

Using New Genetic Knowledge

Where do all these discoveries lead? Suppose we want to make a large supply of insulin, the hormone that regulates the body's use of sugar. People who lack sufficient insulin have diabetes mellitus. Sugar appears as an unused waste in their urine. At the present time, we must extract insulin from the pancreases of animals. This process is expensive. But through gene manipulation, it might be possible to make insulin cheaply and efficiently. Theoretically, this is possible because we know the code and because we have already succeeded in producing a hormone from the human brain this way. From the sequence of amino acids in insulin, we can determine exactly what messenger RNA is needed to put the ribosomes to work making insulin. From the sequence of codons in the mRNA, we can identify exactly what the DNA sequence of bases must be in the gene that directs the synthesis of insulin. We could then proceed to make such a gene.

Once the gene was made, we might try to introduce it into a bacterium. The bacteria could be grown in giant vats, producing large quantities of insulin. Or we might try to synthesize insulin in an artificial system without any living organisms. Both lines of research are being pursued. Both have advantages and disadvantages. And there are dangers. In introducing one gene we might unwittingly introduce others. The first approach, therefore, might transform a harmless bacterium into a disease-causing organism. The other approach would probably be far more expensive. Much thought and careful testing will be needed before we proceed.

How could we get an artificial gene, or for that matter one that had been extracted from some living cells, into the cells of another organism? Not so long ago that seemed impossible. Now there is a

SCIENCE AT WORK: Making an Artificial Gene

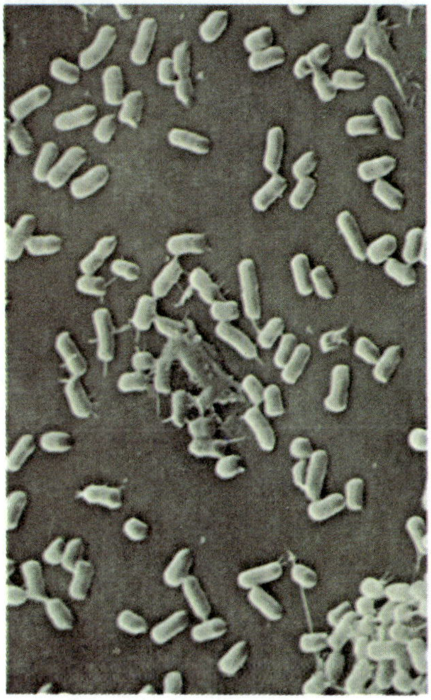

After nine years of work, a team of scientists led by Dr. Har Gobind Khorana, of the Massachusetts Institute of Technology, successfully synthesized an artificial gene. Genes had been synthesized in the past but they lacked the critical regulatory mechanisms that told the gene when to start and stop working.

Dr. Khorana and his colleagues were able to synthesize a specific gene from the common intestinal bacterium *Escherichia coli.* There are 199 pairs of nucleotides in this gene. The first 52 nucleotide pairs are the promoter, which in addition to starting the natural formation of the gene's product, may also determine how often or when the process repeats itself. The next 126 pairs specify the gene's product. The last 21 are the terminator, which may also play a more complex role than simply ordering a halt to the formation of a new molecule.

Synthesizing the gene was one thing, but the big question was, "Will it work?" To answer this question, the synthetic gene was implanted in individual cells of *E. coli.* Later tests indicated that the synthetic gene was functioning. It was doing its job, that is, making a molecule known as tyrosine transfer RNA. Tyrosine transfer RNA then finds a free-floating tyrosine, an amino acid, and transfers it to a ribosome where the tyrosine combines with other amino acids to make a protein molecule.

This successful synthesis of a working gene opened the door for scientists to synthesize genes that produce medically useful substances such as hormones or antibiotics. The first substance synthesized in this way was somatostatin, a hormone produced in the human brain. We may expect to see many more such substances produced in this way in the coming years.

way. It depends on the viruses that pick up genes from one host and carry them to another. We know this transfer of genes works in bacteria. Would it also work in humans? An extensive search is now going on for a harmless virus that might be used to do this. It might be possible to use a virus to introduce genes that control insulin production into the pancreas cells of a person with diabetes.

Yet gene therapy, as the substitution of healthy genes for disabled ones might be called, could involve dangers. For example, it was once thought that the best virus that could be used for the purpose might be a virus that nearly all of us carry in our mouths. It is called the herpes simplex virus. Most of the time this virus produces no symptoms, although it may cause occasional fever blisters. But scientists discovered that if a certain strain of this "harmless" virus gets into the female reproductive system, it may cause cancer. For this and other reasons, great caution and extensive studies will have to be carried out before a "harmless" virus can be used as the carrier of a desirable gene.

Reversing the Sequence

Biologists used to think of the sequence of protein synthesis as going in only one direction: from DNA to RNA to protein. Then, in 1970, Howard Temin and David Baltimore announced independently that they had discovered in certain viruses an enzyme that reverses the process. The enzyme permits a particular piece of RNA to cause a change in the DNA. By means of this enzyme, a virus that has only RNA gets its genetic instructions to dominate the activities of a host cell.

The virus RNA makes a complementary DNA. This DNA then takes over the role of the normal DNA in the cell. It programs the cell's ribosomes to make virus proteins instead of the cell's proteins. The cell becomes transformed into a virus factory. And, if the cell divides, the new cells are abnormal and malignant. Many plant and some animal viruses contain this enzyme. Some cancer viruses of animals and humans also are of this kind.

In Brief

Almost the entire development of our knowledge of hereditary processes and the nature of genetic material has taken place in the present century. It began in 1900, when Mendel's experiments were rediscovered. The concept of the gene as the unit of heredity developed rapidly. And scientists showed that the genes are located in the chromosomes.

In the 1940s, it became clear that a gene is a segment of a DNA molecule. Later evidence showed that genes control characteristics by determining the arrangement of amino acids in both structural proteins and enzymes. Knowledge of the structure of the DNA molecule led to an understanding of the "DNA→RNA→protein" sequence. Scientists now talk about how to substitute genes in living organisms. This ability could lead to genetic therapy for hereditary diseases and disorders. It also could lead to the creation of new forms of life by a radical alteration of the genes. Such possibilities raise questions of safety and ethics that must be considered carefully before decisions dealing with the "new genetics" are made.

Guide Questions and Problems

1. Suppose that the nucleus is removed from a cell. How would this affect the cell's ability to function? Explain your response.
2. How does the Watson-Crick model of DNA fulfill the requirements that the hereditary instructions must be both exact and permanent?
3. Many of the drawings in this chapter, for example Figure 7-15 and Figure 7-17, are models. They are symbol systems we use to explain data gained from experiments. They are not the "real thing." What are some of the values of using such models?
4. Assume a short segment of DNA with bases in the following sequence: C-T-G-A-A-G. Trace the steps through protein synthesis that would be controlled by this segment.
5. Assume the DNA segment C-T-G-A-A-G. What would happen if the T were somehow omitted from the chain during replication? What would be the result if two T's were somehow put in the place of the one T?
6. Explain mutation in terms of the Watson-Crick model of DNA.
7. The genetic code is the same in all organisms that biologists have studied. What is the significance of this fact?
8. Cancer is uncontrolled cell division. Cancerous cells do not stop dividing when they should. How might new research of DNA help us solve the problem of such uncontrolled cell division?
9. Review the information on mechanism presented in Chapter 2. What events reported in this chapter are examples of mechanism? Would such research have been possible if mechanism were not the accepted framework for inquiry in biology? Why or why not?

Ideas for Independent Study

1. Genetics, particularly human genetics, is one of the fastest growing areas in modern biology. In many respects, it is also among the most controversial. Check the *Readers' Guide to Periodical Literature* in your library to find articles about the newest research in the field. Prepare a summary of your findings for your class or some other audience.

2. Invent a protein synthesis game or working model. Try to include in your game or model such elements of the process of DNA, the triplet code, messenger RNA (codons), ribosomes, transfer RNA (anticodons), polypeptide chains, and mutation. Try the game or model with friends, family, or other class members. Use the summary of messenger RNA codons in Figure 7-19 to construct your game or model.

7-19. MESSENGER RNA CODONS

FIRST LETTER	SECOND LETTER				THIRD LETTER
	U	C	A	G	
U	PHENYLALANINE	SERINE	TYROSINE	CYSTEINE	U
	PHENYLALANINE	SERINE	TYROSINE	CYSTEINE	C
	LEUCINE	SERINE	STOP	STOP	A
	LEUCINE	SERINE	STOP	TRYPTOPHAN	G
C	LEUCINE	PROLINE	HISTIDINE	ARGININE	U
	LEUCINE	PROLINE	HISTIDINE	ARGININE	C
	LEUCINE	PROLINE	GLUTAMINE	ARGININE	A
	LEUCINE	PROLINE	GLUTAMINE	ARGININE	G
A	ISOLEUCINE	THREONINE	ASPARAGINE	SERINE	U
	ISOLEUCINE	THREONINE	ASPARAGINE	SERINE	C
	ISOLEUCINE	THREONINE	LYSINE	ARGININE	A
	START-METHIONINE	THREONINE	LYSINE	ARGININE	G
G	VALINE	ALANINE	ASPARTIC ACID	GLYCINE	U
	VALINE	ALANINE	ASPARTIC ACID	GLYCINE	C
	VALINE	ALANINE	GLUTAMIC ACID	GLYCINE	A
	VALINE	ALANINE	GLUTAMIC ACID	GLYCINE	G

Related Readings

BOOKS

Asimov, I. *The Genetic Code.* New York: New American Library, 1962.

Bouk, E. *The Code of Life.* New York: Columbia University Press, 1965.

Patt, D. I. and G. R. Patt. *An Introduction to Modern Genetics.* Reading, Mass.: Addison-Wesley, 1975.

Sayre, A. *Rosalind Franklin and DNA.* New York: Norton, 1975.

Watson, J. *The Double Helix.* New York: Atheneum, 1969.

ARTICLES

Cohen, S. N. "Manipulation of Genes." *Scientific American,* July 1975, page 24.

Temin, H. M. "RNA-Directed DNA Synthesis." *Scientific American,* January 1972, page 24.

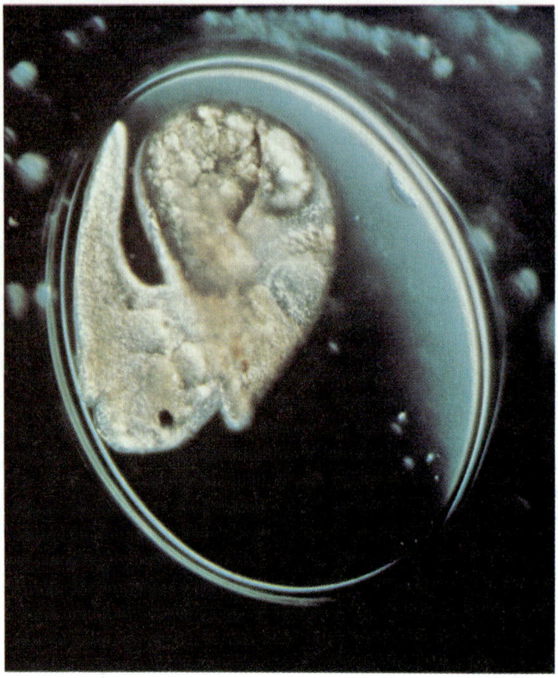

8 An Analysis of Development

How does a single, fertilized egg cell (the zygote) give rise to a multicellular adult organism? This is the question that is studied in a branch of biology called **embryology** (em·bree·OL·oh·jee). Embryologists study the development of an organism from zygote to adult. In this chapter, we will explore some of the basic processes of development.

PROCESSES OF DEVELOPMENT

Many things occur during development. All are interrelated. Development begins with an increase in the number of cells. The zygote is a single cell (Figure 8-1). Its nucleus contains the chromosomes of the sperm plus those of the egg. This diploid nucleus divides into two nuclei by mitosis, and the cell divides into two cells (Chapter 4).

Mitosis, cell division, and other processes continue. Slowly the adult body forms. This may take a few days, weeks, months, or years. The time it takes depends on the species of animal or plant.

Cell division continues in some parts of the body throughout the life of the organism. In our own bodies, certain kinds of cells continue to divide. These include the cells of the lower layers of the skin. Cells of the lining of the intestine and those that produce blood cells also continue to divide. Cells in many other parts of our bodies (for example, the brain) divide rarely or not at all.

Development also involves growth. A zygote is a tiny cell. This first cell divides, and then the two new cells divide. For a while, cell division continues, although the developing organism becomes no larger. Soon, however, the cells begin to take in raw materials from the outside, and the embryo increases in size. (The term *embryo* is used for the early stages of development. For example, look at the snail embryos in the photographs at the beginning of this chapter.)

Many different kinds of cells are produced in the course of development. We call this process **differentiation.** The many cells that eventually arise from the zygote are not mere fragments or duplicates of it. Instead, they change in many ways. Some become muscle cells. Others become skin cells, and still others become nerve or bone cells.

During development, the many kinds of different cells undergo organization into various structures. That is, in the course of development, the cells become grouped according to their function.

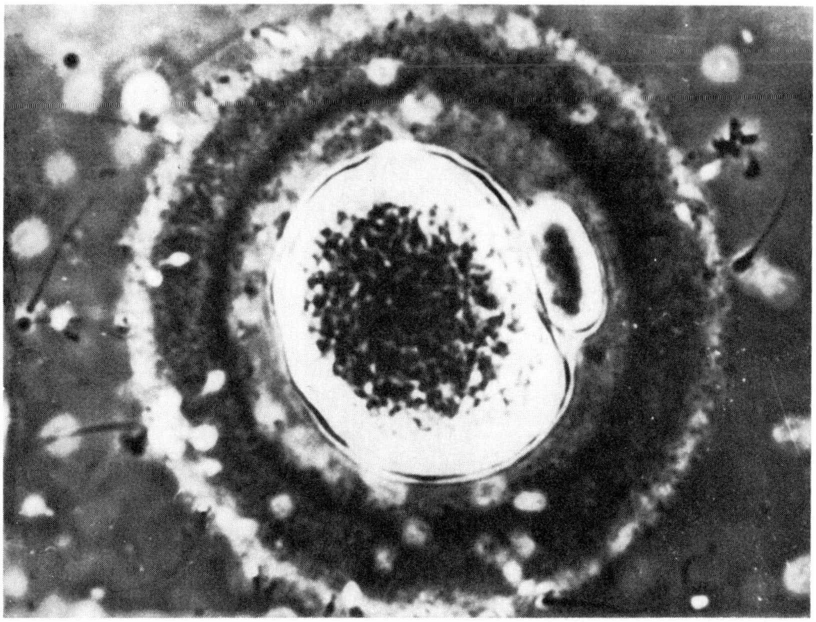

8-1. This fertilized human egg is highly magnified.

In the developing organism, all these processes may be occurring at the same time. Later, you will have a chance to consider descriptions of development in amphibians (like frogs and salamanders) and in humans. But first, you should observe development in a chick embryo by doing **Inquiry 8-1.**

INQUIRY 8-1 Development of the Chick

In this inquiry, you will examine several stages in the development of a chick. Like many other organisms, a chick develops from a fertilized egg cell, or zygote. This zygote is really only one tiny part of the much larger structure we call the hen's egg.

The zygote is, in fact, too small to be seen with the unaided eye. The development of this zygote involves an increase in the number of cells, the differentiation of these cells, the arrangement of the differentiating cells into organs and tissues, and a gradual increase in size. In the course of development, structures such as eyes, brain, legs, and wings are formed.

MATERIALS (per team)

Days 1 and 2

fresh fertilized hen's egg
fertile eggs incubated 48, 72, and 120 hours, and 10 days
physiological saline solution, 7% sodium chloride, warmed to about 38°C
stereoscopic, dissecting microscope or hand lens
finger bowl
fine, sharp-pointed scissors
fine, sharp-pointed forceps
bright light
ink marker
paper towels

Day 3

FILM: *A Study of Frog Development*

Procedure (Day 1)

Pour the warm saline solution into the finger bowl. Select a freshly fertilized egg that is, at most, only a few hours into its development.

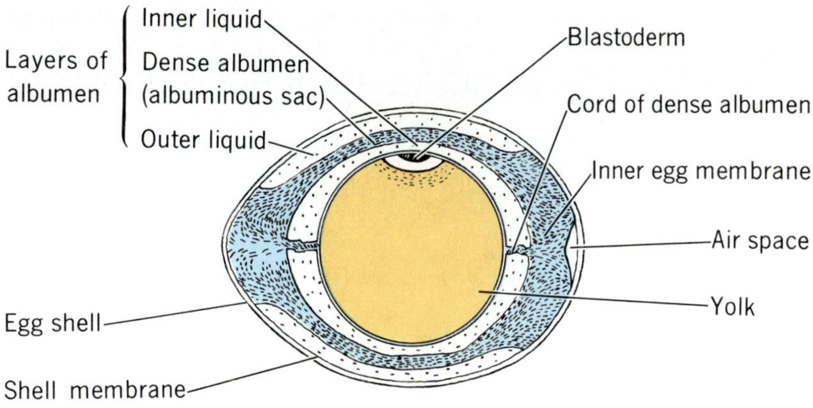

8-2. Internal structure of a hen's egg.

To crack the egg without breaking the yolk, rap it against the edge of the finger bowl. Position the egg close to the bottom of the finger bowl. Carefully open the egg and let the contents flow gently into the solution.

Observe the structure of the egg carefully. Compare it with the internal structure shown in Figure 8-2. See how many parts you can identify. At this stage the embryo is very small. It can be seen as a whitish spot, called the **blastoderm** (BLAS·toh·derm), on the upper surface of the yolk. **1.** Where might a developing embryo obtain its food? **2.** What evidence should be found in later stages if this is true?

Next, examine a chick embryo that is about 48 hours old. Remove an egg from the incubator. Keep the egg in the same position it was in the incubator. This will assure that the embryo will stay floating on top and the heavier yolk will be underneath.

Open the incubated egg as before. Place the dish on the stage of the dissecting microscope or use a hand lens. Locate the embryo. The embryo is still very small, but you will be able to see its tiny head. Most of the head is occupied by the brain. You may be able to see two bumps on the sides of the brain. These are the eyes. Behind the head are two long lines of blocklike structures, called **somites.** These will form the muscles and the backbone. **3.** Locate the embryo's heart. Describe any motion you see. **4.** If you see signs of blood, describe its movement.

Open a fertile egg that has been incubated 72 hours. Observe as before. As you examine the embryo, notice carefully its stage of development. Notice the blood vessels around it. **5.** To what area of the embryo are they connected?

Locate the heart. Then move the embryo until you can see part of the heart and some of the blood vessels. **6.** What is the relationship between the pulsations of the heart and the surges (quick movements) of the blood? **7.** Are blood cells present? **8.** Observe the same structures you observed in the 48-hour embryo. Describe the changes that have taken place in these structures.

The heart is formed by the fusion of two straight tubes. Later it divides into four chambers. This type of heart is characteristic of birds and mammals.

Procedure (Day 2)

Observe an embryo that has been developing for about 120 hours. Crack the egg into a dish of warm saline solution. The embryo should be on the upper surface of the yolk. Partially covering the embryo is a thin, delicate membrane that has blood vessels at its surface. As the embryo develops, this membrane, the **allantois** (ah·LAN·toh·is), will expand until it completely lines the shell. The allantois is important to the exchange of materials. Through the many blood vessels of the allantois pass food and oxygen vital to the life of the chick. Carbon dioxide and other wastes are removed through the blood vessels of the allantois.

Next, look for the small chamber surrounding the embryo. The membrane surrounding and forming this chamber is called the **amnion** (AM·nih·on). Gently probe the surface of the amnion. **9.** With what does the amnion seem to be filled? **10.** What function can you suggest for the amnion? **11.** Which of the world's great environments is suggested by the immediate environment of the embryo?

Now open the amnion with forceps and scissors, being careful not to damage the embryo. Observe the embryo carefully. **12.** What separates the head from the trunk? **13.** Look at the limb buds. Would you describe them, at this stage of development, as wings or legs?

Now you will study an embryo that is 10 to 12 days old. Because this embryo has developed further, many of its internal structures would be damaged if you simply cracked open the egg. To avoid damage, remove the shell as shown in Figure 8-3.

In this stage, the **chorioallantoic** (KOH·rih·oh·a·LAN·toh·ik) **membrane** lines the shell. This membrane is formed from the allantois and another embryonic membrane, the **chorion** (KOH·rih·on).

Carefully pick the shell away to expose the air space. You will see the outer shell membrane sticking to the shell and the inner shell membrane sticking to the egg's contents.

Carefully remove the rest of the shell as shown in the figure. If you are skillful, you can expose the whole chorioallantoic membrane, which lines the shell. The chorioallantoic membrane encloses the embryo, amnion, and yolk sac. It excludes only the albumen of the

a

Crack the large end of the egg with your scissors, and use forceps to pick away the shell. Avoid breaking the membrane.

b

After part of the shell has been removed, place the egg in a finger bowl of saline and pick off the remainder of the shell.

c

The egg with the shell removed.

8-3. Opening an egg containing a 10-day to 12-day chicken embryo.

egg. Observe the blood vessels in this membrane. Note their connections with the embryo through veins and arteries in the membrane. Figure 8-4 will help you identify the various structures.

Carefully remove the chorioallantoic membrane. **14.** What is the relationship of the embryo to the fluid-filled amnion? **15.** What do you notice about the sac surrounding the yolk? **16.** What connects the yolk to the embryo? **17.** What structures do you see that travel up through the yolk

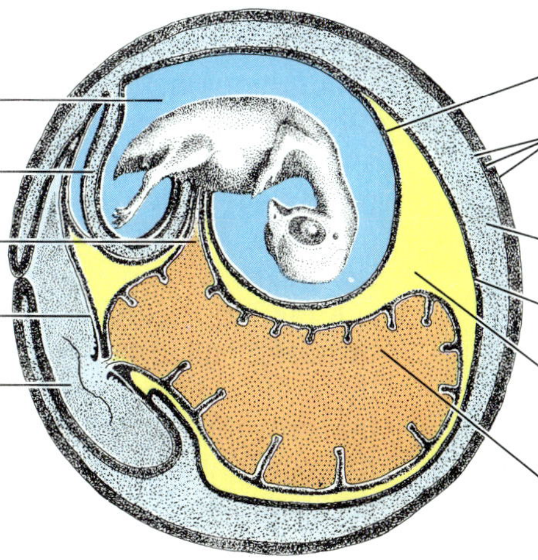

8-4. Diagram of a late-stage chicken embryo.

stalk? **18.** What features make the embryo look like a bird? **19.** How does it differ from other birds with which you are familiar? **20.** Does it look specifically like a chicken or just like a bird in general? **21.** Describe the condition of the eyelids, feathers, and claws. **22.** What feature associated with reptiles (such as snakes and lizards) do you see on the legs and feet? **23.** Note the position of the embryo's beak. Where does the chick get air?

Locate the egg tooth, a hard, calcium mass on the tip of the upper beak. Also find the hatching muscle at the back of the neck. **24.** What is the apparent function of these structures?

Procedure (Day 3)

Now that you have observed the development of a chick embryo, you may want to extend your studies to include the development of other animals. Your teacher may be able to arrange for you to view and discuss the film *A Study of Frog Development*. As you watch the film, compare what you see to what you have already learned. Your teacher can provide information about how to use the film.

DEVELOPMENT OF THE AMPHIBIAN EMBRYO

Much has been learned about development from observations and experiments on chick embryos. Even more has been learned from studies of amphibians (am·FIB·ee·uns) such as frogs and salamanders. Amphibian embryos can be kept easily in the laboratory. A dish of pond water and a relatively cool room (15°-20°C) are all that is necessary. No food is needed. The egg, when it leaves the body of the female, has a large supply of yolk. The yolk of an amphibian egg is much like that of a hen's egg. It consists largely of protein.

About one half of the outer layer of a frog's egg is pale brown; the other half is nearly white (Figure 8-5). The darker half is called the animal hemisphere. The cell nucleus is located in this part. The paler half, which is called the vegetal hemisphere, contains yolk granules. The vegetal hemisphere is heavier than the animal hemisphere.

A few hours after fertilization, the zygote begins a series of cell divisions called **cleavage.** At first, a thin dark line appears across the animal hemisphere. This line quickly becomes a deep groove that cuts the single cell into two cells (Figure 8-6). Inside the cell, mitosis has taken place. Thus each of the two new cells has a diploid nucleus.

The embryo divides again. In each of the two cells, the nucleus undergoes mitosis, and each cell divides. Thus the two-celled

8-5. Fertilized frog eggs.

8-6. Frog embryos, two-cell stage.

Carolina Biological Supply House

embryo becomes a four-celled embryo. The four cells are equal in size. Each contains a portion of the animal hemisphere and a portion of the vegetal hemisphere.

After another interval, the third cleavage occurs. The first two cleavages produced cells of equal size. The third cleavage is different. It divides the embryo into four small and four large cells (Figure 8-7). The two quartets of cells are of unequal size. Each new cell contains a diploid nucleus, but contains different materials in the animal and vegetal hemispheres.

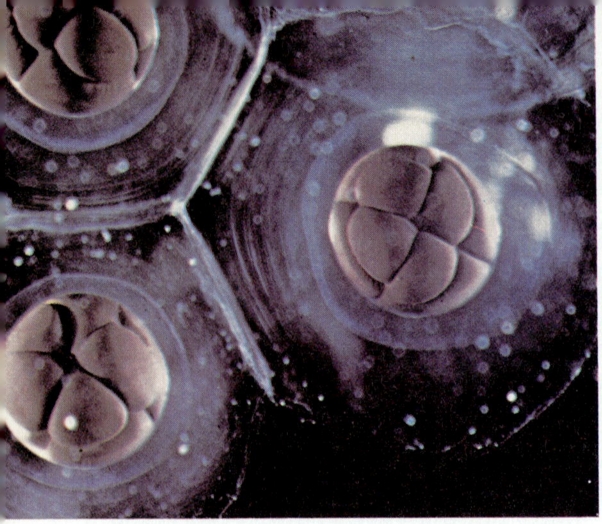

Carolina Biological Supply Company

8-7. Frog embryos, eight-cell stage.

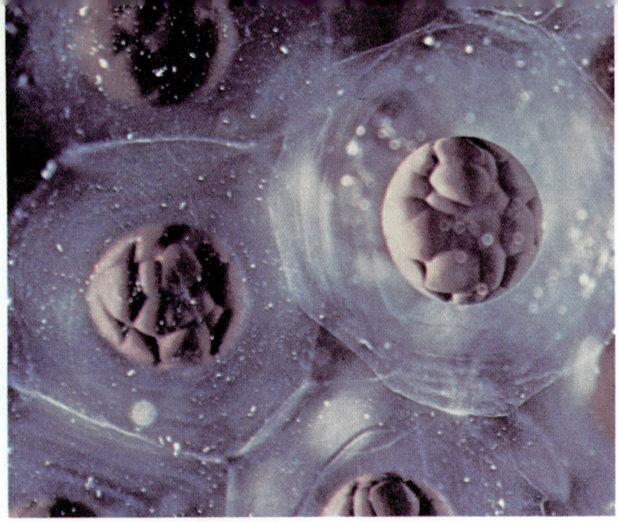

Carolina Biological Supply Company

8-8. Frog embryos, 32 or more cells.

The process of cell division continues. The original material in the one-celled zygote becomes distributed among many cells. More than 32 cells are visible in Figure 8-8. Cell division is slightly more rapid in the animal hemisphere than in the vegetal hemisphere. In Figure 8-9, the cells of the animal hemisphere are quite small, and there are hundreds of them. A cavity occupies much of the interior of the animal hemisphere (Figure 8-14, page 239). It is filled with a fluid, largely water but with some dissolved salts and proteins. At this stage, the embryo is called a **blastula** (BLAS·tyoo·la). The cells are nearly alike in appearance but differ in size and color. Internally, they may be quite different.

A curved groove, known as the **blastopore,** then forms. Figure 8-10a shows how this occurs in a salamander embryo. Some of the cells of the embryo move slowly toward the blastopore and enter it. The movement of these cells into the interior of the embryo is known as **gastrulation** (gas·troo·LAY·shun). The blastopore becomes deeper and larger. Surface cells continue to move inside the embryo. These cells will form the organs of the circulatory, digestive, excretory, reproductive, and respiratory systems.

Carolina Biological Supply Company

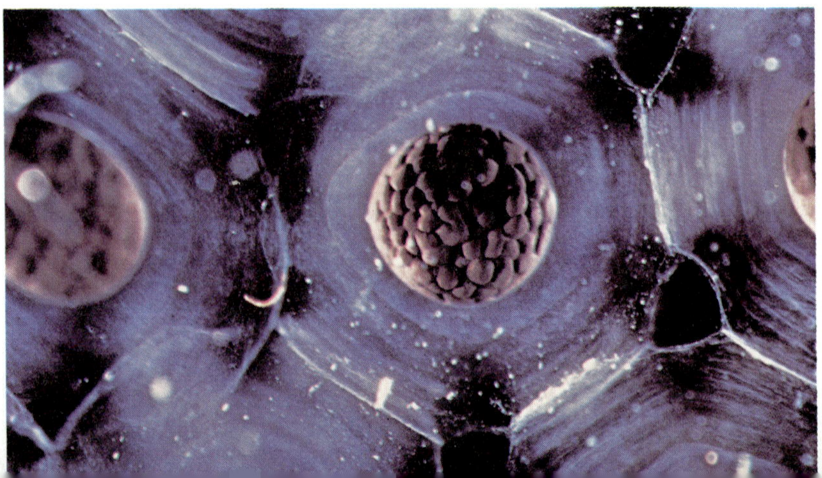

8-9. Frog embryos, early blastula.

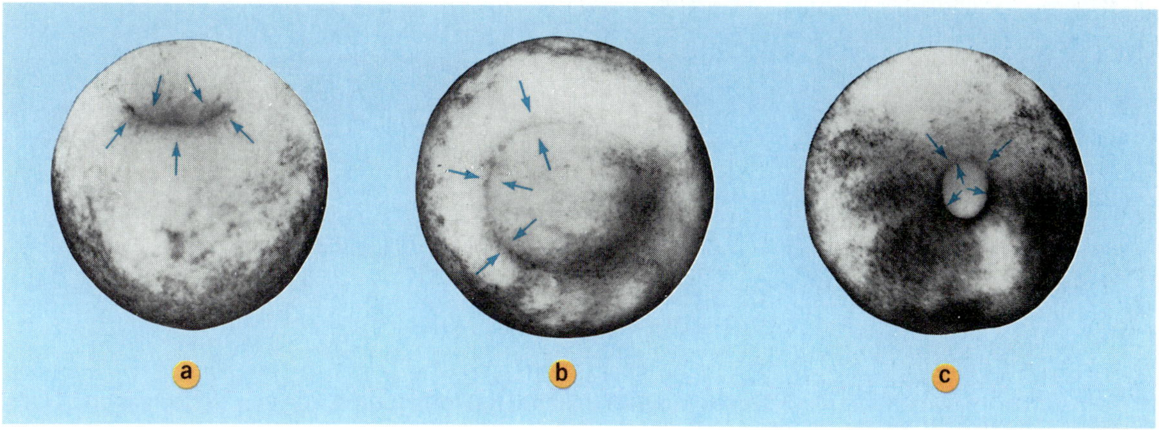

8-10. Formation of the blastopore in a salamander embryo.

In the embryo in Figure 8-10b the blastopore has formed a complete circle. Cells continue moving through the blastopore to the interior of the embryo. The cells that remain on the outside will form the skin and nervous system. Gradually, the blastopore becomes smaller (Figure 8-10c). Eventually, it will disappear entirely. But not too long after this happens, the beginning of the digestive tube forms in nearly the same place.

In Figure 8-11 the nervous system of the frog is starting to form. The beginnings of this process are shown. At this stage, the embryo is called a **gastrula.** The closed blastopore is on the underside of the embryo. Extending up the sides and over the top of the embryo are two lines. These are the neural folds. The neural folds will join at one end, forming an inverted "U." The neural folds become larger and move together. Where the folds meet, they fuse. An internal tube is formed immediately beneath the fused edges. (This process is diagrammed in Figure 8-15, page 239.)

At the stage shown in Figure 8-12, the embryo still relies on the yolk in its cells for food. As yet there are no blood vessels, but the

8-11. Frog embryo, early neural folds.

Carolina Biological Supply Company

8-12. Frog embryo, gastrula.

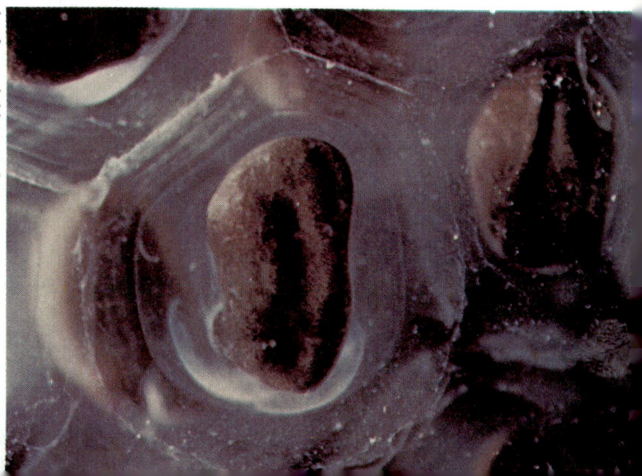

Carolina Biological Supply Company

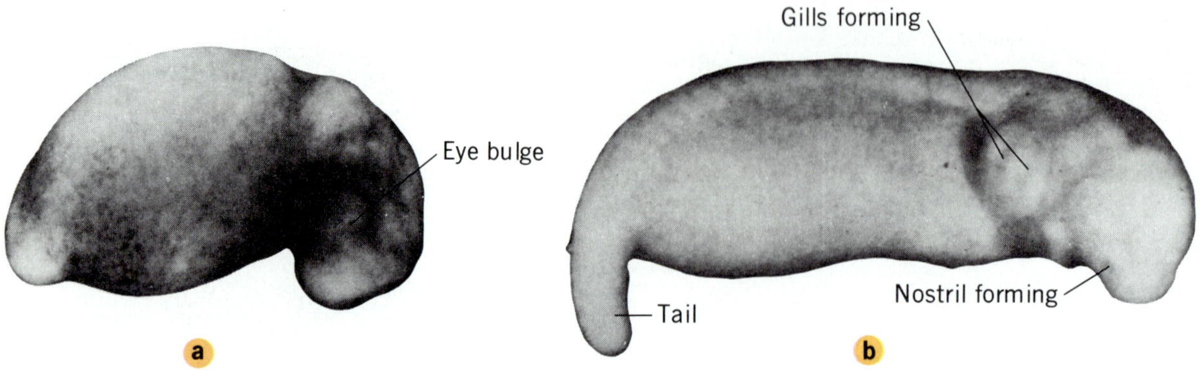

8-13. Developing salamander, **(a)** formation of the eye, **(b)** formation of gills and tail.

heart is forming just behind where the head bends down. In Figure 8-13a, the head of a developing salamander has a bump on the near side of it (and also on the far side). Part of the brain is pushing outward and forming the eye. The dark spot near the front end is a nostril. The big bump, about one quarter from the front end (Figure 8-13b), is where gills are forming. The tail is now fairly well developed.

Internal Development

Now we have seen the external changes in the embryo as it develops from a zygote into a young larva (tadpole). Let us consider some of the events that have been taking place inside the embryo. The processes of early development of the frog include these major events: formation of the zygote (fertilization), formation of a blastula, and formation of a gastrula. After fertilization, cell division begins in the zygote. After many divisions, a hollow ball of cells has formed. This ball of cells is the blastula. Figure 8-14 shows the inside of the blastula.

At gastrulation, cells move inside the blastula through the blastopore. Different cell layers begin to appear. These layers are the **ectoderm** (EK·toh·derm; outer skin), the **endoderm** (EN·doh·derm; inner skin), and between them, a layer called **mesoderm** (MES·oh·derm; middle skin). At this stage, the embryo is known as the gastrula.

The entire outer layer of cells is the ectoderm. This layer of cells will form the nervous system and the outer part of the skin. The mesoderm, which is now inside, will form the muscles, skeleton, kidneys, circulatory system, and reproductive organs. The small pocket of endoderm cells that first forms at the blastopore gradually

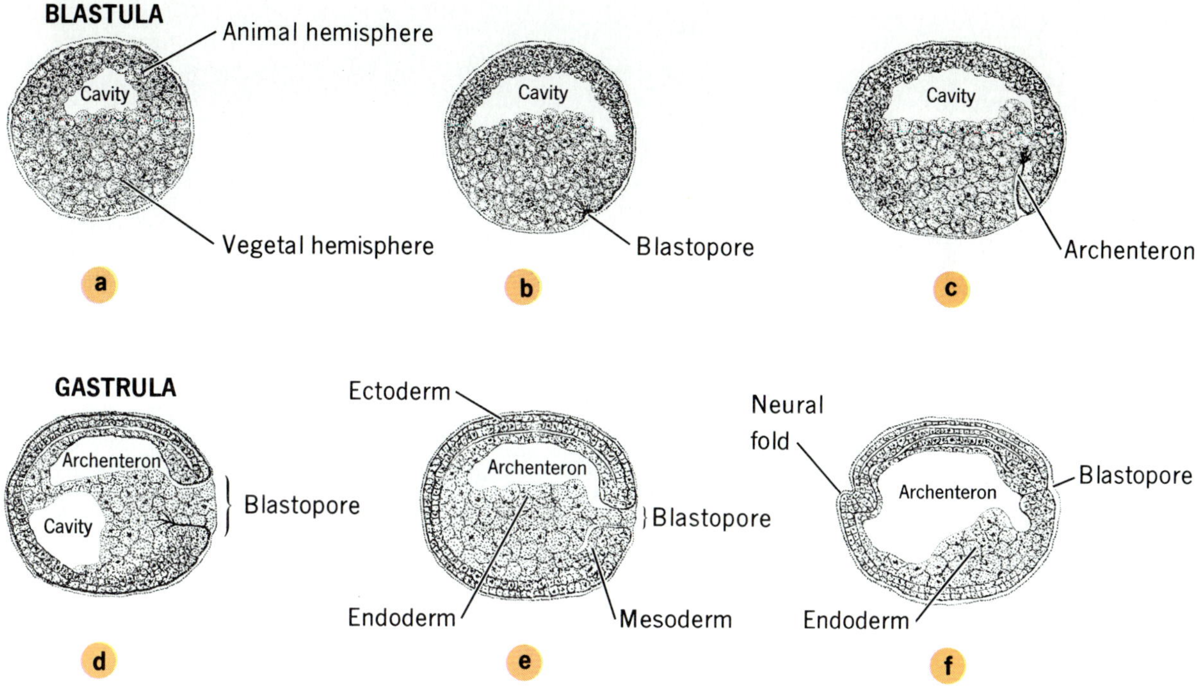

8-14. Early steps in the formation of a salamander embryo.

expands into a large cavity known as the *archenteron* (ar·KEN·ter·on). The archenteron later forms the digestive system. The lungs, liver, and pancreas grow out from its walls.

The drawings in Figure 8-15 show the neural folds as they move together to form the neural tube. The forward part of the neural tube becomes the brain. The rest becomes the spinal cord. All the nerves develop from the neural tube and the ectoderm next to it. The eyes, which looked like mere bumps on the side of the head of the early embryo, develop from the brain region of the neural tube. Some cells of the ectoderm over the eye bulge will become the lens. The eye, then, is formed in part as an outgrowth of the brain and in part from the ectoderm.

8-15. The formation of the neural tube in a salamander embryo.

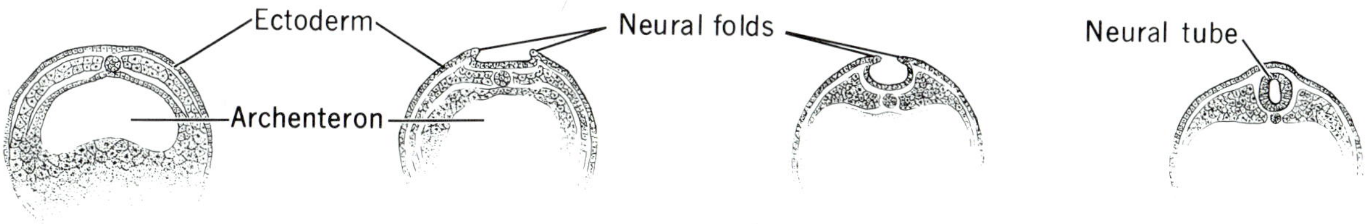

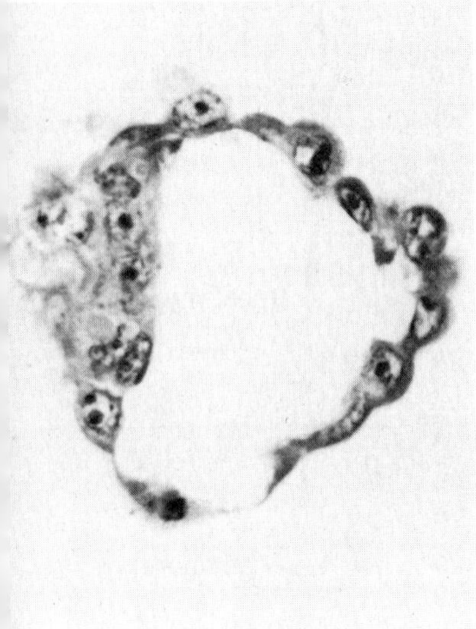

8-16. Photomicrograph of a human blastocyst.

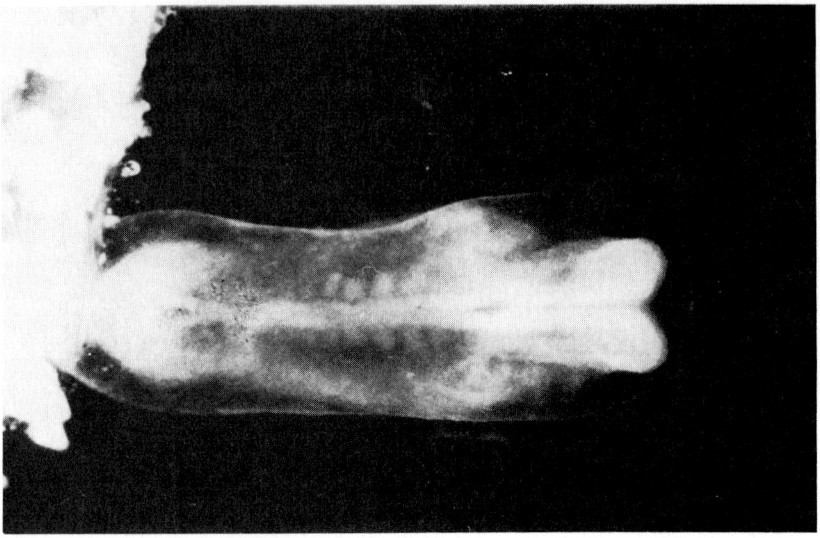

8-17. Three-week-old human embryo showing neural folds.

HUMAN EMBRYOLOGY

Biologists have learned much about development in human embryos. Their studies have led to a rather complete understanding of these events. As in other organisms, the human zygote undergoes cleavage. As a result of cleavage, one cell becomes two, then four, then eight, and so on, until hundreds of cells are formed. The solid cluster of cells that results is called the *morula.* After a few days, the morula changes to a hollow ball of cells. This is the blastula, or, in human embryos, the **blastocyst.** Figure 8-16 shows a section of a human blastocyst. The fluid-filled cavity is called the **blastocoel** (BLAS·toh·seel).

Some of the cells accumulate on one side. This mass of cells is the blastoderm from which the embryo will develop. The remaining cells form the amnion, yolk sac, and allantois. The embryo changes during the next few days. The three layers of cells—the ectoderm, mesoderm, and endoderm—form. These become different systems and organs in humans just as they did in frogs and salamanders. Over the next few weeks, the flattened, three-layered embryo will become tubular.

Figure 8-17 shows a three-week-old human embryo. Extending from the back of the embryo are the neural folds. The tops of the folds fuse along their entire length, beginning at the head end. After four weeks (Figure 8-18a), a number of external changes have taken place. The neural tube is closed over most of its length. The eyes are forming and are visible as bumps on the sides of the head. The arm and leg buds are recognizable as small swellings on the sides of the body wall. By the sixth week (Figure 8-18b), the head is prominent and the limbs and eyes are recognizable.

The first 12 weeks of development are referred to as the embryonic period. These first weeks are the most critical period of development. During this time, the three cell layers differentiate into the beginnings of all the main organ systems.

After the twelfth week, a human embryo is called a **fetus.** This name change does not reflect some abrupt transformation in the developmental process. It indicates a greater degree of differentiation. Development from this point on involves mainly growth and formation of the tissues and organs. Fingernails and toenails form and begin to grow. Much growth occurs during this period. The average, 12-week-old fetus has a mass of about 14 grams. This will increase to almost 3,000 grams at the time of birth.

EXPERIMENTAL ANALYSIS OF DEVELOPMENT

How is it that the many cells of an individual come to perform different functions? Some become stomach cells, while others become skin, liver, or muscle cells. And how do these cells become organized into the different structures that form the body of a human, or a cat, or a grasshopper?

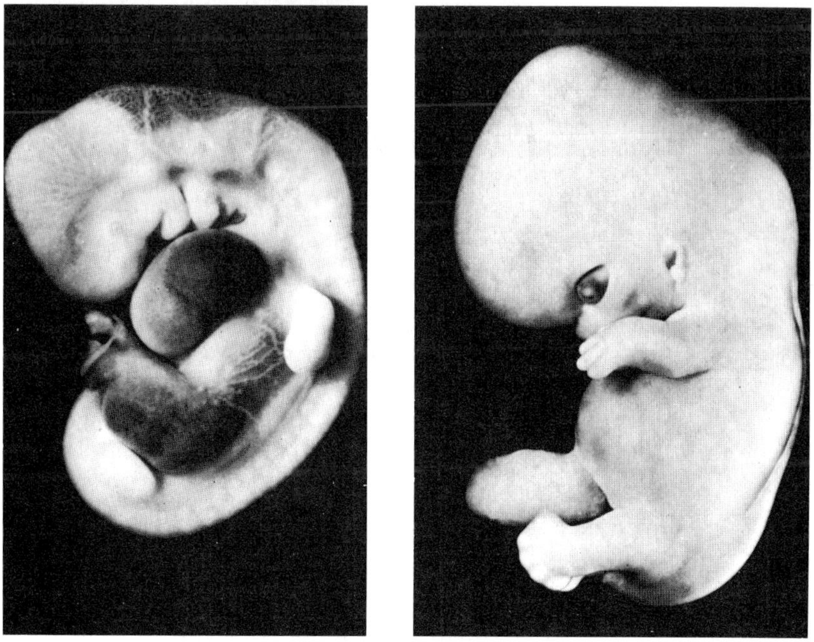

8-18. (Left) Four-week human embryo; eye bumps, arm and leg buds are visible. **(Right)** Six-week human embryo.

At the heart of the question is the idea of levels of organization (Figure 8-19). We can think of a single cell as the first level of organization. At the next level are **tissues.** A tissue is a group of similar cells that perform the same function. Muscle tissue is an example. Two or more tissues that function together may form an **organ.** The stomach, liver, and small intestine are organs. Organs that function together are **organ systems,** such as the digestive, respiratory, and nervous sytems. All the organ systems, functioning in a coordinated way, make up the organism.

What is responsible for the differentiation of cells? How does their organization into the parts of the adult come about? Our answers to

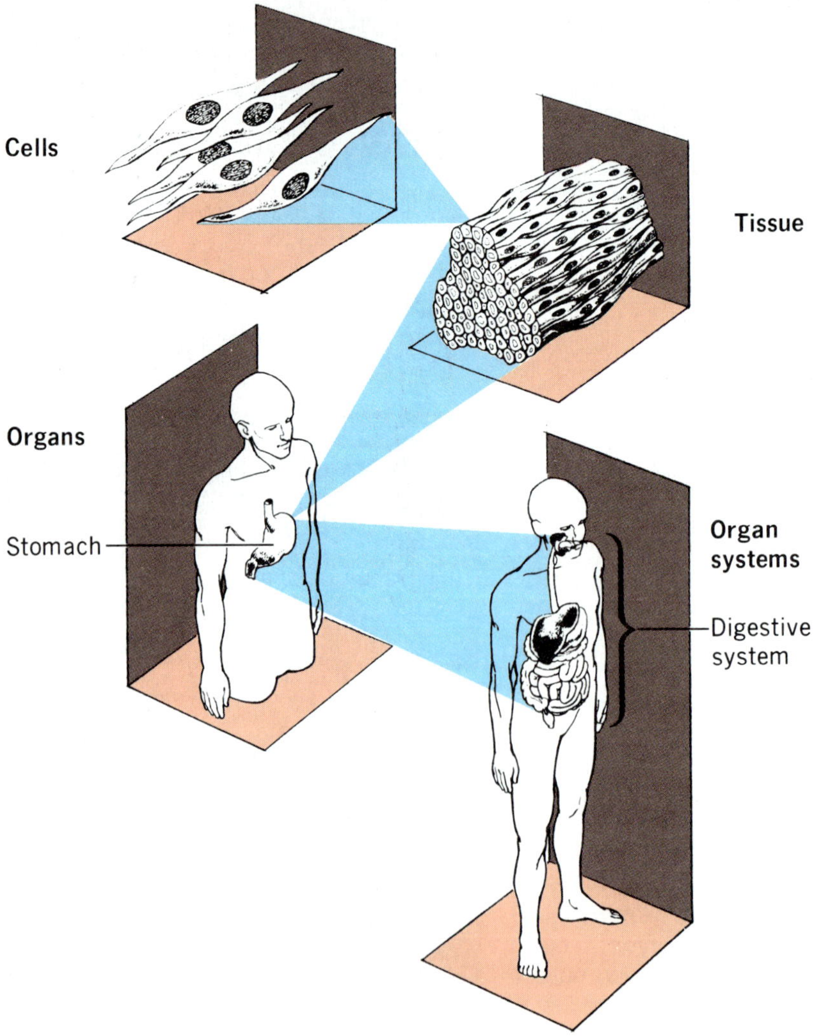

8-19. Basic questions in embryology relate to levels of organization. In the mature, healthy organism, cells, tissues, organs, and organ systems function in a coordinated manner.

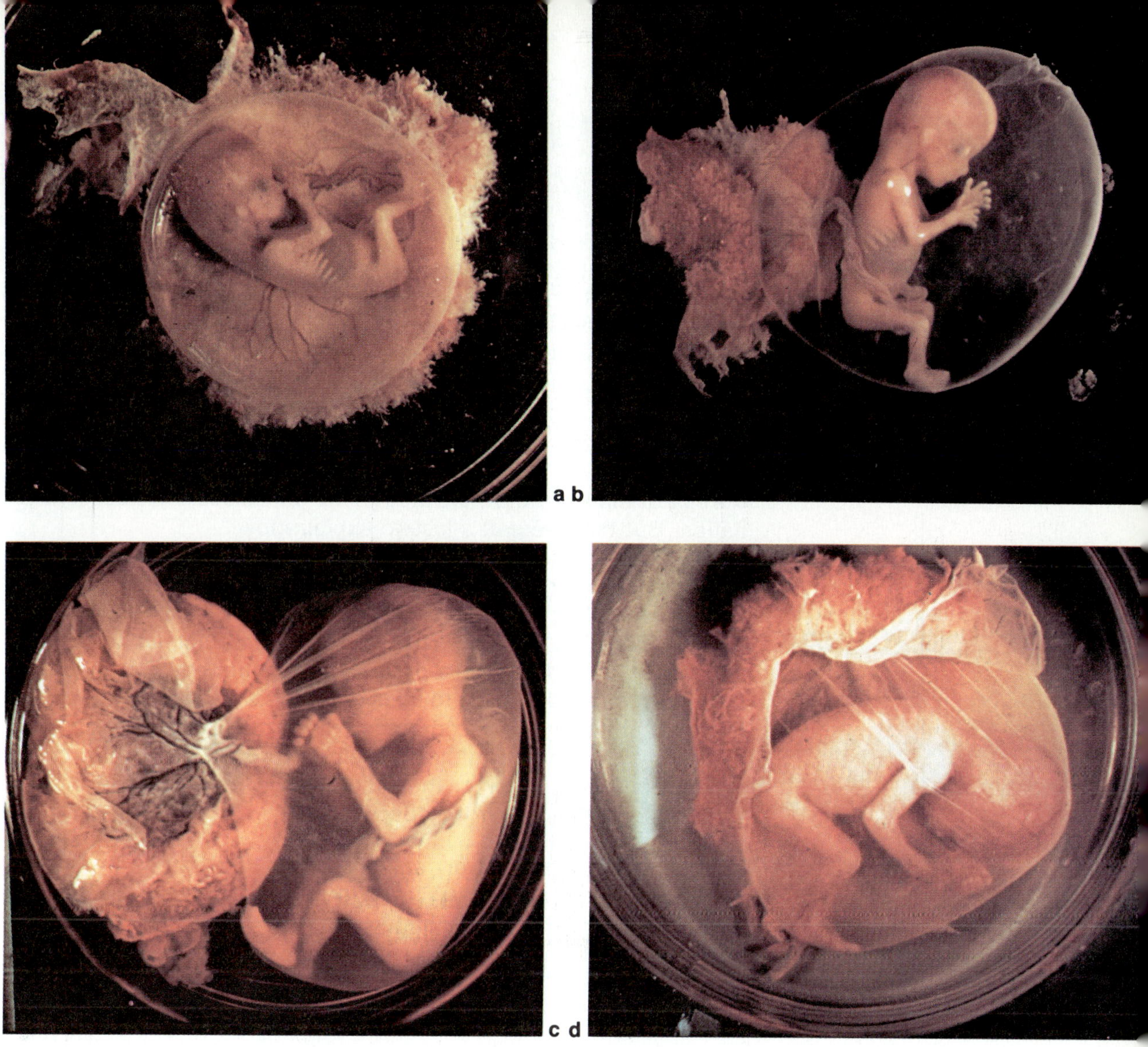

8-20. Development of a human fetus, **(a)** two months, **(b)** three months, **(c)** three and one half months, and **(d)** four months.

these questions are still incomplete. Nevertheless, some observations and experimental results suggest a few answers.

Study the gradual development of a human fetus, as shown in Figure 8-20. Notice the increasing differentiation between the stages. For example, arms and legs are absent in very young embryos. Later, tiny limb buds develop. Still later, fingers and toes form. All this from a single-celled zygote! For thousands of years, people have wondered how an individual that can move, see, hear, digest food, and excrete waste can arise from just a single cell.

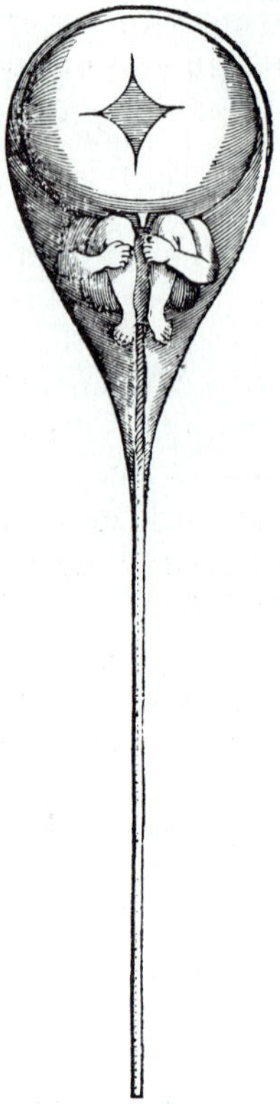

8-21. Some early embryologists imagined that a human sperm contained a tiny individual, shown here with a large head and folded arms and legs.

Development—By Preformation or Epigenesis?

Centuries ago, Aristotle proposed that there are two ways to explain how an embryo arises from an egg. Aristotle's first idea has been called *preformation*. Preformation implies that an egg or sperm contains a new individual already formed (Figure 8-21). Aristotle's other proposal is called *epigenesis* (ep·ih·JEN·e·sis). The idea is that eggs and sperm contain no preformed structures. Instead, these structures somehow develop in their proper positions later, constructed from material in the egg.

For 2,000 years, people argued whether preformation or epigenesis was the correct explanation for development. Scholars who favored preformation thought that the animal was present in either the egg or the sperm. Others thought epigenesis was the best explanation. But they were at a loss to explain how.

During the past 50 to 100 years, these arguments have been resolved, at least in part, by careful experimentation. One of the most important experiments was done by the German scientist Hans Driesch (DREESCH). He separated sea urchin embryos at the two-cell stage. Each of the cells produced an entire embryo (Figure 8-22). This result led Driesch to think epigenesis was a better explanation than preformation. According to preformationist theory, half an egg would produce only half an embryo. But if each half could produce a whole embryo, then the embryo could not have been preformed in the egg or the sperm.

8-22. DRIESCH'S EXPERIMENT WITH SEA URCHIN EGGS

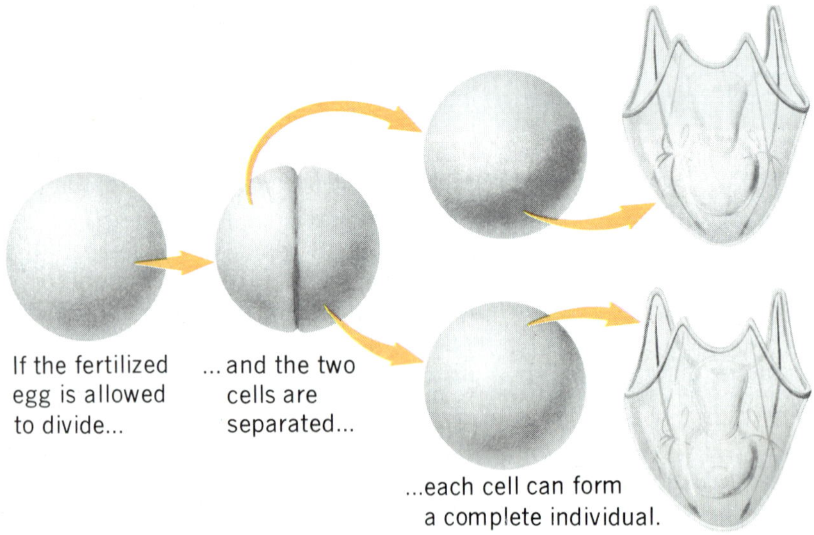

Many observations and experiments have supported Driesch's conclusions. There is no preformed tadpole in a frog egg or sperm. There is no tiny person in a human egg or sperm. The many different structures form during development.

Differentiation

It would be easy to explain how development occurs if preformation were correct. Development would be just the enlargement of an organism already present in the egg or sperm. Since preformation is not the explanation, we have a harder task. We must account for the fact that new structures appear where none was present before.

In previous chapters, you learned how genes control the making of all sorts of proteins. All of these proteins are made according to directions contained in DNA in our genes. We inherited our genes from our parents. Is it reasonable to think that genes might control development?

Obviously, cells that make insulin must be different from cells that make hemoglobin. Yet, as a result of mitosis, most of the new cells have a duplicate of the set of chromosomes that was in the zygote. Most cells have the same genetic information, the same DNA. Then how do they become different?

The answer involves both genes and environment. Studies of *Escherichia coli* showed that even this simple organism does not use all of its genes all the time. When *E. coli* grows in a medium that contains glucose, it does not produce the enzyme that breaks down milk sugar, lactose, to glucose. The genes that produce that enzyme are "turned off." But when *E. coli* is transferred to a medium that contains lactose but not glucose, manufacture of the enzyme begins within a few minutes. If transferred back to the glucose medium, the bacteria quickly stop making the enzyme that digests lactose. They again use the glucose that is already available. The genes that produce the lactose-digesting enzyme appear to be "turned off."

In 1961, Francois Jacob and Jacques Monod developed a model that shows how genes might be turned off or on. Figure 8-23 shows the model in a simplified way. According to the model, the genes that direct the formation of a particular protein form a cluster known as an **operon.** The operon is controlled by a gene that turns it on.

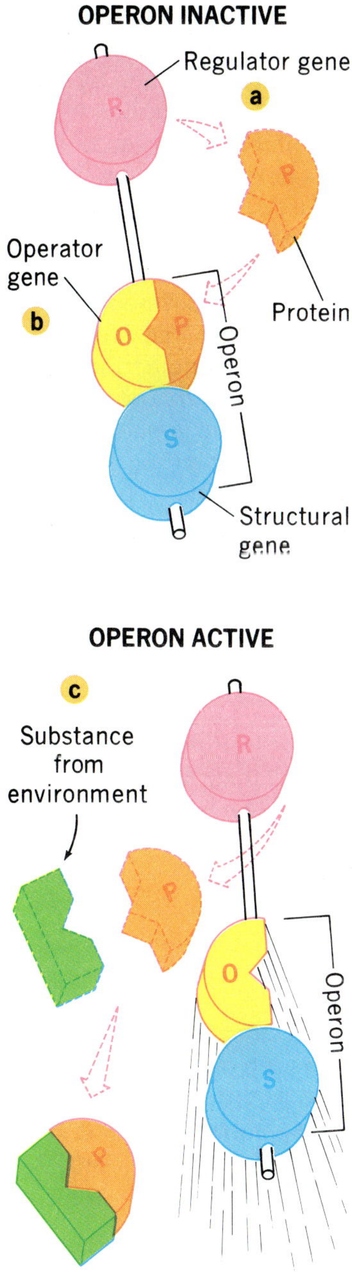

8-23. One possible model of gene action. The regulator gene **(a)** directs the synthesis of a protein. If the protein combines with the operator gene **(b)**, the operon (structural gene) becomes inactive. But if some substance from the environment **(c)** ties up the regulator protein, the operator is free and the operon can act. In some cases, the regulator protein may turn the gene on, not off.

That gene is called an **operator gene.** Each operator gene is controlled by the combined action of a **regulator gene** and substances in the cytoplasm of the cell.

Regulator genes make **repressor proteins.** Some repressor proteins are active, and some are inactive. Active repressors combine with the operator gene and keep it turned off until a substance in the environment combines with the repressor protein. This combination frees the operator gene. The operator gene then activates the rest of the genes in its operon. Sometimes the protein made by the regulator gene is inactive unless it combines with a substance from the environment. When this occurs, the operator gene shuts down the operon. Either way, the presence or absence of some key substance in the environment controls the activity of the genes.

The Jacob-Monod model has provided a good explanation for gene action in bacteria. But what of frogs, chicks, and humans? To date, no one knows for sure that the model explains differentiation in animals. But it has been very useful in two ways. First, it stimulated research into possible relationships between gene action and differentiation. Second, it has led to new hypotheses that go a long way toward explaining differentiation. Although the mechanisms probably are much more complex than those of bacteria, differentiation in animals seems to involve some "turn off—turn on" process. There may be stopping mechanisms at any of the steps that occur between DNA transcription and eventual protein synthesis (Chapter 7). It may even be true that enzyme action is blocked in some way *after* the enzyme is made.

Further Evidence

Evidence that some genes are "turned off," or inactivated, during development has come from experiments in which a nulceus is transferred from one cell to another. It is possible, for example, to remove nuclei from frog eggs and replace them with nuclei from frog embryos that have developed to a later stage. Eggs that receive nuclei from early embryos (up to the blastula stage) usually develop normally. Nuclei from cells at a later stage, such as the gastrula, often cause the embryo to develop abnormally.

These results suggest that, as development progresses, more of the genes in the nuclei of differentiating cells are inactivated. Only those genes that form and maintain the cell in its differentiated state remain active. This means that a cell becomes a nerve cell, for example, when all genes are turned off *except* those that cause formation of nerve cells.

Why would we hypothesize that the genes are turned off, or inactivated? Why not suppose, instead, that the genes are simply miss-

SCIENCE AT WORK: On Again, Off Again

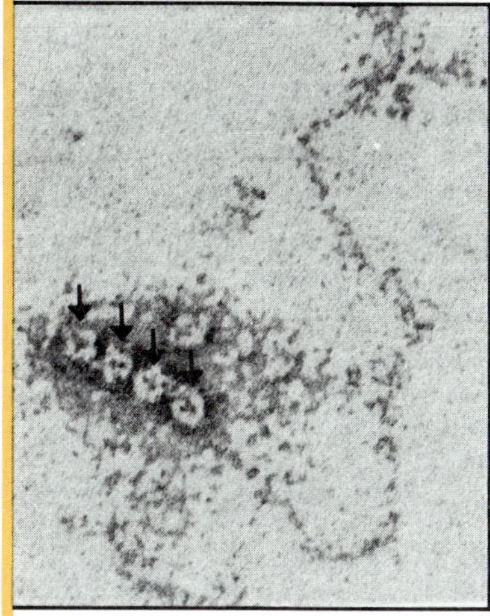

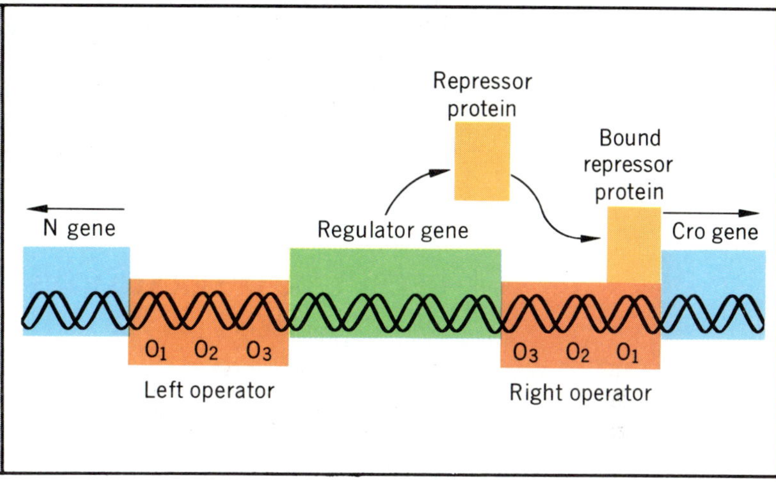

The Jacob-Monod hypothesis has stimulated much research since it was originally proposed in 1961. Since that time, many scientists throughout the world have tried to test the hypothesis.

A great deal of attention has been given to a virus called *lambda* (LAM·duh). Lambda infects the common intestinal bacterium *E. coli.* When lambda infects an *E. coli,* one of two things can happen. Sometimes, the virus immediately takes over protein production in the *E. coli* cell. After only 45 minutes, the bacterial cell will burst, releasing hundreds of new lambda viruses. At other times, the DNA of the virus merely attaches to the DNA of the bacterium. When the bacterial cell divides, the lambda DNA is replicated and passed on to the two new cells. The lambda DNA can remain inactive, being passed on to generation after generation of *E. coli.* Days, weeks, or months later, the lambda DNA may once again become active. The virus DNA takes over the protein production mechanism of the bacterium and produces hundreds of new lambda viruses. These break out of the host cell, destroying it.

The question that has intrigued researchers is how the lambda DNA can be active at some times but not others. The answer, they have found, is in repressor proteins, just as the Jacob-Monod model suggests.

In the photograph, the arrows point to proteins that are bound to the strand of lambda DNA. These proteins are repressors of gene action. The color illustration above shows two genes, which produce different polypeptides, with a regulator gene between them. On each side of the regulator gene are the operator genes. The regulator gene makes repressor proteins that can bind to one of the operator genes. In this case, the operator gene is O_1 on the right side. A bound repressor at O_1 will block the expression of the *cro* gene. This blocking occurs because the enzyme responsible for the production of messenger RNA is prevented from attaching to the DNA. No messenger RNA is produced, and the gene is not expressed.

Is the process the same in higher plants and animals as it is in bacteria and viruses? No one knows for sure, but the research continues.

ing or lost? Consider a recent experiment with frogs. First, the nucleus of the frog's egg was removed. Then the nucleus from an intestinal cell was transplanted into the egg cell. The egg cell developed normally. In this way, a normal adult frog was produced from the genetic information in the nucleus of a single intestinal cell. It appears that a cell retains all of its genes, even though some of them do not function after the cell has differentiated.

The process of producing a complete organism from a single, nonreproductive cell is called **cloning.** Cloning has not been accomplished with any organism more complex than a frog.

CELLULAR INTERACTIONS DURING DEVELOPMENT

Cells become organized into tissues, and tissues are organized into organs. It is obvious that there is some interaction between the cells. Embryologists have learned that, during development, one tissue can influence the way another tissue develops. This process is known as **embryonic induction.** It means that cells in one kind of tissue stimulate the development of cells in other kinds.

Some of the classic experiments on differentiation were performed by Hans Spemann (SHPAY·mahn) in the 1920s. Because of results from earlier experiments with salamanders, Spemann thought that mesoderm might stimulate the ectoderm to develop into the nervous system. To test this hypothesis, Spemann performed the experiment shown in Figure 8-24.

8-24. A classic experiment to test the hypothesis of embryonic induction.

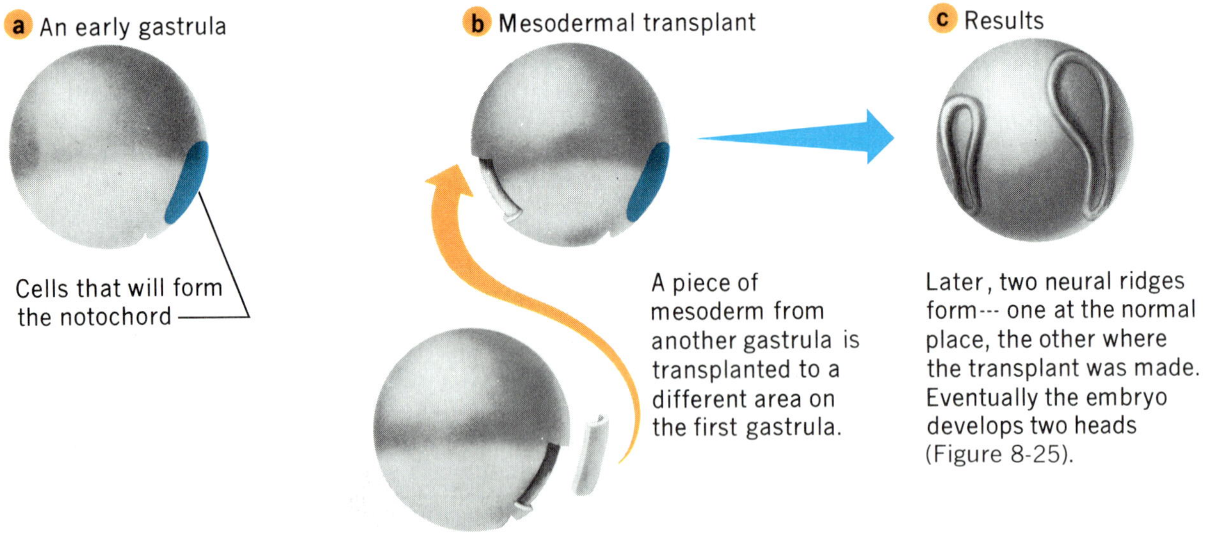

8-25. A salamander following an experiment similar to the one shown in Figure 8-24. Notice the induced second head protruding from the belly.

Embryologists knew that certain cells of the ectoderm normally form a notochord, part of the nervous system. Perhaps the mesoderm caused, or induced, the ectoderm to differentiate in this way. Spemann took some mesoderm from one side of a gastrula (Figure 8-24a) and transplanted it to the opposite side of another gastrula (Figure 8-24b). The embryo formed two notochords—one in the usual place, the other where the transplant was made (Figure 8-24c). The hypothesis was supported. It appeared that the mesoderm did induce the ectoderm to develop into parts of the nervous system. Eventually, embryos treated in this way will develop two heads (Figure 8-25).

It is not known exactly what causes induction. In some instances, chemicals seem to be the cause. Some experiments show that only one kind of tissue can induce the formation of another kind. In other cases, a wide variety of tissues seem to act as inducers.

GENES AND ENVIRONMENT

A few genes seem to produce the same effect no matter what the environment. Most genes, however, are influenced by the environment. The environmental factors that influence gene action may be within an organism, or they may be outside it. In other words, genes determine what an organism *may* become, not what it *will* become. Always, what the developing organism becomes depends on both its heredity and its environment.

An example can be cited. Yellow fat and white fat are genetically determined characteristics in rabbits (Figure 8-26). Rabbits that were homozygous for yellow fat were fed carrots and leafy vegetables, which are rich in yellow pigments. These rabbits developed yellow fat. When they were fed a diet that did not include carrots and leafy vegetables, they developed white fat. Other rabbits that have an allele for white fat have an enzyme that breaks down the yellow pigment. Their fat is always white, on any diet.

The enzyme that breaks down yellow pigment is absent in the yellow-fatted rabbits. On a diet that does not contain yellow pigment, these rabbits develop white fat. Does this change the genetic makeup of these rabbits? Breeding tests show that the alleles are not changed. Furthermore, when these rabbits are given a diet of carrots and leafy vegetables again, their fat turns yellow.

Another example of the effect of environment on the expression of genetic variability is the Himalayan rabbit. This rabbit normally has a black nose, black feet, and black ear tips (Figure 8-27). By altering the temperature of the environment, we can change the appearance of the Himalayan rabbit to resemble that of a white rabbit. However, the change cannot be passed on to the Himalayan rabbit's offspring.

Thus, the external environment can influence gene expression. But what does this have to do with embryological development? In recent years, embryologists have been focusing their attention on

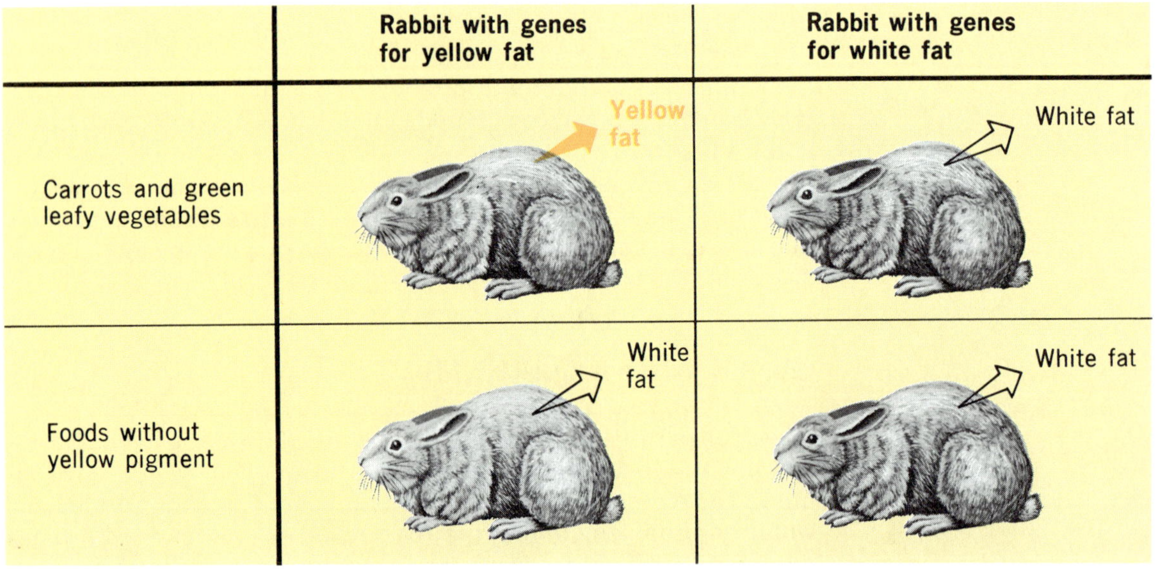

8-26. The expression of genes sometimes depends on environment. Here, food is an environmental influence upon genetic expression in rabbits.

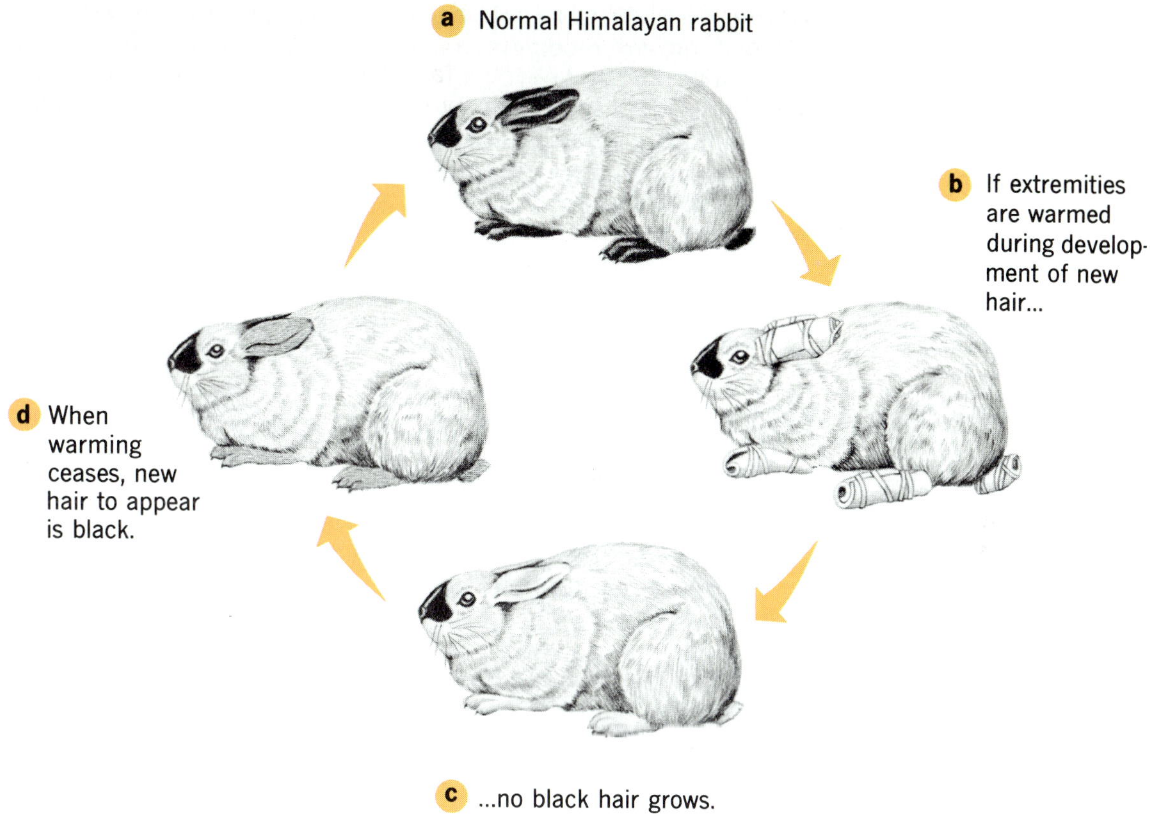

8-27. Environmental factors such as temperature can influence gene expression.

the environment directly surrounding the cell—the microenvironment. Probably no two cells have the same microenvironment even if they are in the same tissue. Within a cell's environment are varying concentrations of oxygen, carbon dioxide, cell secretions, foods, waste products, and the like. Since individual cells can hold and collect materials on their membranes (Chapter 3), considerable differences could exist even in the environment of neighboring cells.

The microenvironment of embryonic cells seems to have a great effect on differentiation. Embryonic cells of the same type tend to cluster together even before they are visibly different. For example, cells that will form brain tissue group with other developing brain cells, and they move away from muscle cells. This grouping seems to be caused by differences in the cell membranes and in the substances attached to them. Once grouped, these cells may form an environment that somehow favors further differentiation. The environment also may influence the cells to remain in their differentiated state. If cells from a tissue are separated, they tend to lose their differentiated state and look more like cells at an early embryonic stage. If they are recombined later, they often regain their differentiated state.

Scientists can now grow cells, tissues, and embryos in culture. The main goal of such studies is to understand cell-to-cell interaction in differentiation and the role of the microenvironment. Cell culture research may lead to many important outcomes. Through such research, we are learning a great deal about how drugs affect cancerous cells. We are learning about the nature of genetic diseases. Such research also aids in the production of vaccines against viruses.

CONTINUING DEVELOPMENT

Development does not stop when a baby chick hatches from an egg or when a human baby is born. It goes on continuously in all organisms. Once an animal has reached adulthood, however, changes in its body form and function occur much more slowly. Yet such changes *do* occur.

Cell Replacement

In many tissues of the adult, cells continue to divide and differentiate, replacing older cells. Thus, the tissue remains constant in form and function. The tissue is maintained by this replacement of cells, and all individual cells remain fairly young. For instance, the red cells in human blood are continually destroyed and replaced—at the rate of seven to ten million cells every second. As a result, there is always a nearly constant number in the circulatory system. Such renewal of cells occurs in many tissues of adult animals. In other tissues, such as the brain, the majority of cells are never replaced.

Aging

Cell replacement does not go on forever. Gradually the adult animal changes. The changes represent a complex process that we call aging. Aging is most easily recognized, in all species where it occurs, by the increased probability of death. Frogs tend to live no longer than 5 to 14 years. The human's potential life span is about 70 years. The actual length of life varies greatly, of course. It depends on the individual's genetic background, diet, disease, accidents, and other factors. Other changes also are a part of aging. Some of these changes are structural, such as loss of weight and altered skin form and hair color in humans. Other changes are functional, such as the loss of the ability to bear children.

Many biologists are as interested in the fundamental mechanisms of aging as they are in embryonic development. Aging is much more difficult to study, for it occurs far more slowly and is less predictable and uniform. Today, much research is being carried on in this field. It is of interest to scientists and it is of practical concern to all humans.

Animals vary greatly in their patterns of cell replacement. In the tiny freshwater animal *Hydra,* there seems to be a continuous replacement of cells. In a sense, *Hydra* never ages and never dies. Why should aging occur so predictably in some species and not at all in others? We do not yet have the answer to this question. Some biologists suggest that aging and death are based on genetic mechanisms, especially mutations. They think that mutations actually "program" the events leading to death—or at least do not protect the organism against them. Other biologists believe that the highly differentiated cells that do not divide—muscle and nerve cells, for instance—represent "points of weakness." In these cells, the effects of accidental damage and disease may build up gradually, destroying the cells and disrupting the normal function of the animal. Viruses, radiation, and diet have been suggested as possible agents that cause such damage.

Abnormal Development

Although the processes of development are repeated with remarkable accuracy over many generations, sometimes mistakes occur. These mistakes may be due to gene mutations or other changes in the hereditary material. If the genes that control the processes do not function properly, abnormal development may result. Examples are some kinds of dwarfism and giantism. Usually these patterns can be traced to a malfunctioning pituitary gland. Abnormal numbers of chromosomes also may result in defects of development. Down's syndrome is an example.

Ordinarily, the rate of cell division drops off gradually after an organism reaches its adult form. New cells are produced only as replacements for those that wear out in normal life activities. If, however, the rate of cell division in an organ suddenly and abnormally increases, more cells are produced than are needed by the body. These rapidly developing cells take raw materials for division and growth from the other cells of the organ.

Any such uncontrolled cell division is termed cancer. Leukemia (loo·KEEM·ee·uh) is one example. In normal blood there are relatively few white blood cells. But in a person suffering from leukemia

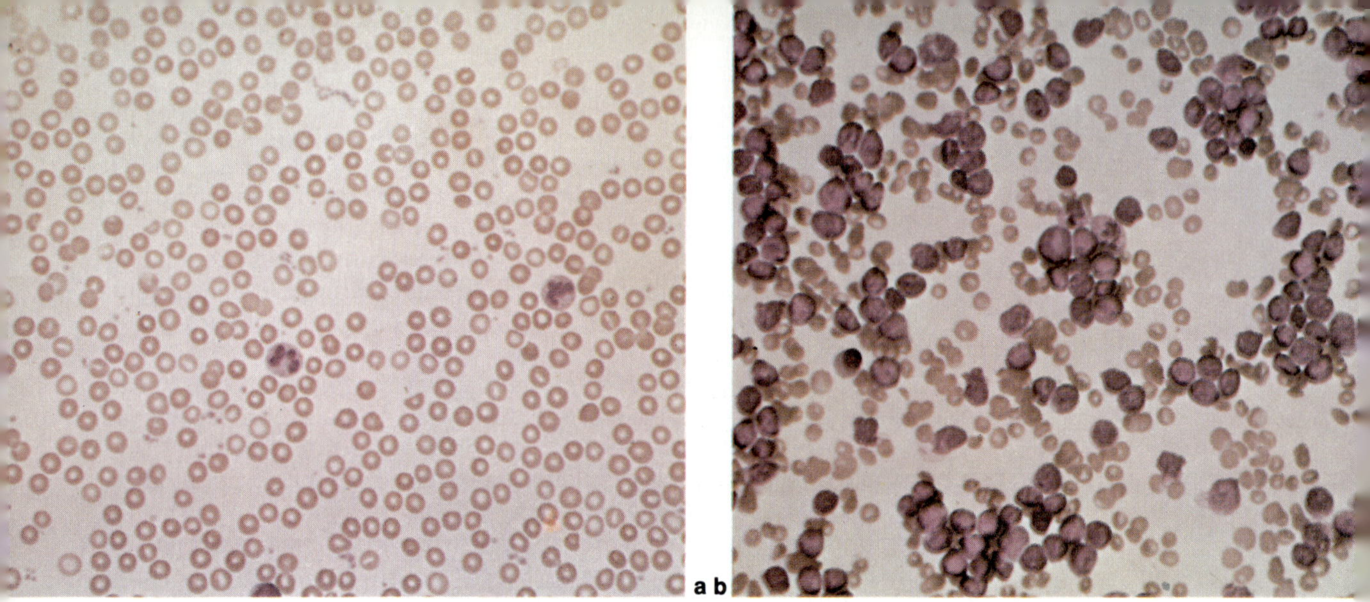

8-28. The white blood cells in these photomicrographs have been stained. They are the larger, purple cells. **(a)** Normal human blood. **(b)** Blood of a person with leukemia.

extremely large numbers of white blood cells are produced and released into the blood (Figure 8-28).

A great deal of progress has been made in cancer research, although there is still much to be learned. Nevertheless, techniques for detecting and treating cancer are being improved each year. Our knowledge will continue to increase. Eventually we may be able to control such abnormal development.

In Brief

Patterns of development from zygote to adult are much the same in all organisms. The zygote is a single cell. The adult may consist of billions of cells. This increase in cell number is brought about by cell division.

Not only does the number of cells increase, but ultimately the cells increase in size. The organism grows larger, obtaining food from the yolk (as in the chick) or from its mother (as in the human).

As the cells increase in number, they also differentiate. That is, cells become different in their structure and function. For example, some cells of the ectoderm differentiate into the cells of the neural tube or into the outer layer of the skin. Others form the cell types found in the eyes. Still others form cells of the ears and nose. The various cell types that differentiate from the three embryonic layers become organized into the specialized tissues and organs of the older embryo and the adult. The cells are organized to form stomach, kidneys, bones, muscles, glands, and other structures.

What is responsible for the differentiation of cells? How does their organization into the different parts of the body come about? These questions have not been completely answered. The question of whether the embryo is a preformed body within the zygote has been settled. It is not preformed. Instead, epigenesis occurs. Now structures appear where none were present before.

At least in part, differentiation is determined by the influence of environmental factors on genes. High temperatures can prevent gene expression of black coat color in Himalayan rabbits. The microenvironment of cells may account for some genes being activated while others are inactivated. Embryonic induction may be thought of as a special case of environmental effects.

At least some of the processes involved in the development of the embryo probably continue throughout the life of the individual. Sometimes the controlling mechanisms that maintain the differentiated state seem to be disrupted. The gradual loss of the ability to replace old cells with new ones is called *aging*.

Uncontrolled cell division results in cancerous growth. Much research is being devoted to learning the causes and potential treatments of cancer.

Guide Questions and Problems

1. What are the basic processes that occur during embryonic development? Give examples from your study of chick or human embryos to explain your response.
2. What is differentiation? Why is it significant?
3. What may be the relationship between gene action and the differentiation?
4. If an ice pack is kept on the back of a Himalayan rabbit for a period of time, the hair under the pack will grow in black, gradually replacing the white hair. What does this suggest about the influence of environment on gene expression?

Ideas for Independent Study

1. Are identical twins really identical in every way? You might use this question to begin studies of the effects of environment on development.

8-29.

2. Regeneration occurs widely among living things. For example, if a ray (arm) of a sea star is removed, a new ray develops, or regenerates. The ray has the same size, form, and function as the ray it replaces. In regeneration, all the processes of embryonic development are involved. Cell division, growth, differentiation, and organization take place. But the development involved in regeneration occurs at a later stage in the life of the organism. You may want to investigate processes of regeneration directly. Excellent test organisms include planarians, frog tadpoles, and plants such as geranium and coleus.

3. The lantana blossoms in Figure 8-29 are all growing on the same bush. Notice that some of the blossoms are pink and some are yellow. Propose a hypothesis to explain how the same plant can have flowers of two different colors. How might you test your hypothesis?

Related Readings

BOOKS

BSCS. *Animal Growth and Development.* A Collegiate Minicourse Program. Philadelphia: W. B. Saunders, 1976.

BSCS. *Biology in Three Dimensions.* Northbrook, Ill.: Hubbard Scientific, 1975.

Jenkins, M. M. *Embryos and How They Develop.* New York: Holiday House, 1975.

Milne, L. V. and M. Milne. *Growth and Age.* A BSCS Pamphlet. Boulder, Colo.: BSCS, 1964.

Moog, F. *Animal Growth and Development.* A BSCS Laboratory Block. Lexington, Mass.: D. C. Heath, 1968.

ARTICLES

Arehart-Treichel, J. "How Infections Threaten the Fetus and Newborn." *Science News,* April 12, 1975, page 242.

Etkin, W. "How a Tadpole Becomes a Frog." *Scientific American,* May 1966, page 76.

Hayflick, L. "Human Cells and Aging." *Scientific American,* March 1968, page 32.

Rutter, W. J. and N. K. Wessells. "Phases in Cell Differentiation." *Scientific American,* March 1969, page 36.

Taylor, T. G. "How an Eggshell is Made." *Scientific American,* March 1970, page 88.

9 Change Through Time–Evolution

Time. Some have said that time is what life is all about. The birth, aging, and death of an animal, and the changes of plant life with the seasons involve time. The continuity of genetic material from one generation to the next happens over time. In Chapter 8, you learned how development occurs in some animals. For some, such as the chick, the process requires only days or weeks. For others, such as the elephant, embryonic development requires almost two years.

But now let us think about even longer periods of time. Scientists estimate that the earth is perhaps six billion years old. The earliest life forms – bacteria and some one-celled plants – appeared three to four billion years ago. Yet the age of the dinosaurs was a relatively recent 200 million years ago. And the first truly human creature – *Homo erectus* – appeared a scant 1.3 million years ago (Figure 9-1). What has occurred during all this time is really the story of a different sort of development – the development of life on the earth.

9-1. An artist's idea of what an early human ancestor, *Homo erectus* may have looked like more than one million years ago. Modern people are called *Homo sapiens*—a different species from *Homo erectus*.

THE THEORY OF EVOLUTION

What does evolution mean to a biologist? It means that living things change. It means that the species of plants and animals that now live on the earth were not the first species of plants and animals. It also means that many species that once flourished no longer exist. They are **extinct.** We know of extinct species only by the fossil remains we find.

Evolution also means the field of inquiry in which scientists try to discover how change occurs. It is a field that is concerned with many questions: Why are there different species? How does a species change over time? How rapidly does evolutionary change take place? What controls the rate of evolutionary change? What is the past history of plants and animals living today? What plants and animals lived in the past but are now extinct?

The theory of evolution as we know and understand it today owes much to the work of Charles Darwin in the mid-nineteenth century. Although Darwin never labeled his hypotheses as "evolution," he was interested in how species change and how new species come about. His many years of work led to explanations that have proved to be valid. But Darwin was not the first to think that existing species might evolve into new ones. Ancient Greek philosophers, such as Democritus and Empedocles, correctly interpreted the importance of the fossils of sea animals embedded in rock high in the mountains. Linnaeus, who developed a system still used for animal and plant classification, thought at first that no new species had ever arisen. But after studying for many years, Linnaeus decided that new species do arise from hybrids between the original species. But Darwin was more persuasive than others for two reasons. First, he assembled a great mass of evidence, verified its accuracy, and presented it convincingly. And second, his hypothesis stated *how* change in organisms might take place—a contribution no one else had made.

Darwin's Hypothesis of Natural Selection

On November 24, 1859, the first edition of Darwin's *On the Origin of Species* was published. The book was so popular and so controversial that its first printing was sold in one day. Although the theory of evolution has undergone many changes since Darwin's time, Darwin's original thinking still serves as a convenient introduction to the subject. Here is his analysis:

First, there are many differences among the individuals of every species. In a population, or group, of these individuals, variations occur. Usually it is safe to say that no two individuals are exactly alike. Darwin knew or suspected that many of the individual differences could be inherited.

Second, the population size of all species tends to increase because of reproduction. One amoeba, for example, divides and produces two. These two divide, and the next generation numbers four. Then there will be 8, 16, 32, and so on.

Third, this increase in the size of populations cannot go unchecked. If it did, the number of individuals of any species would outgrow the food supply and the available living space.

Fourth, it is obvious that this huge increase seldom occurs in nature. The number of organisms in a species does not continue to increase over long periods of time. In fact, the sizes of many populations seem to remain nearly the same over time (Figure 9-2). How can this be explained? Observations of natural populations show that many individuals die before they are able to reproduce.

Why do some individuals die early, but not others? Darwin thought there must be a sort of "struggle for survival." The individuals of a species "compete" for food, light, water, places to live, and other things important for their survival. The "struggle" or "competition" may be either active or passive. That is, sometimes animals actually fight for food or the opportunity to mate. In other cases, there is no direct combat or competition. The first animal that happens to find a suitable living area may settle there, thus preventing the area from being used by others. In either case, *individuals with certain characteristics, or traits, will survive and produce offspring more often than individuals without them.*

Consider, for example, how the African cheetah (Figure 9-3) came to be such a fast funner. Cheetahs are hunters. They capture their food—mostly antelopes, gazelles, and birds—by first stalking near their prey. Then they run the prey down with a terrific burst of speed over a short distance. In any population of cheetahs, some can run faster than others. Those that run fastest are most successful in getting food. Those that are better at getting food also are more likely to survive.

But survival is not the whole story. If the characteristics that make an organism better able to survive in its environment are *inherited,* then those who survive are likely to pass on those characteristics to their offspring. For example, the surviving cheetah is likely to

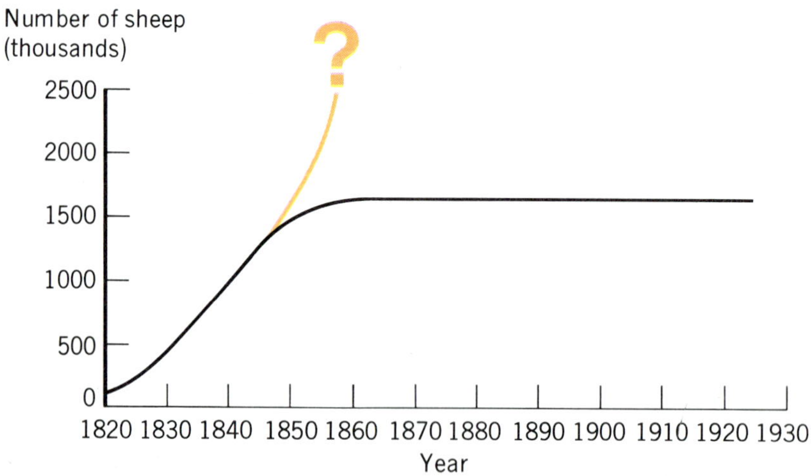

9-2. When sheep were originally introduced into Tasmania, the population grew rapidly for a while, but eventually leveled off. *Why did population growth not continue at its original rate?*

9-3. The cheetah, the fastest running animal on earth, can achieve speeds up to 120 kilometers per hour. *How might this ability have evolved?*

produce offspring with long, thin necks and powerful leg muscles, capable of great speed. Over many generations, then, one could expect an increase in the number of individuals that have these beneficial characteristics. The number with less beneficial characteristics would decrease. The organisms with the beneficial characteristics are likely to live longer and produce more offspring. Darwin called this process of survival and reproduction **natural selection.** Darwin thought that several factors were involved in natural selection:

1. The presence of variation among individuals in a population.
2. The hereditary basis of such variable characteristics.
3. The tendency of the size of populations to increase.
4. The "struggle for survival" (or competition for the needs of life).
5. A difference in the inherited characteristics that individuals pass on to succeeding generations.

Darwin took one additional step. The process of natural selection can explain how certain characteristics become more common in a population. But how do different *species* evolve? How, for example, can there be different species of maple trees (Figure 9-4)? Both are maples, but their characteristics are different. For Darwin, the an-

9-4. The sugar maple and the Rocky Mountain maple have very different characteristics. The formation of new species occurs over many generations when environments are different. The sugar maple **(left)** occurs in the forests of the eastern United States. The Rocky Mountain maple **(right)** grows in the mountains of Colorado, Wyoming, and other western states.

swer lay in the processes of natural selection. In different environments, different characteristics would be beneficial to the survival and reproduction of an organism. The bushy growth pattern of the Rocky Mountain maple would not be beneficial in another environment. Over time, populations of the same species that were isolated in different environments probably would evolve different characteristics. After many generations, completely new species could arise. This is an example of **divergence.**

Darwin's original hypothesis has grown and has been revised as more facts have been found. After a century of accumulating and testing data, Darwin's hypothesis has gained the status of theory. It has stood the test of time, repeated observation, and experimentation. You can gain some experience with the concept of natural selection by doing **Inquiry 9-1.**

INQUIRY 9-1 Natural Selection

In this inquiry, you will study the process of natural selection by working with a hypothetical population of organisms in a hypothetical environment. Members of the class will act as "predators," removing individuals of a species of "prey."

MATERIALS (per team)

sheet of newspaper
50 newspaper squares (1–1.5 cm²)
50 red paper squares (1–1.5 cm²)
50 white paper squares (1–1.5 cm²)

Procedure

Work in teams of four. One student (the keeper) sets up the environment before each round (generation). The other three in the group act as "predators." They remove prey from the environment before the prey can reproduce. All four students on a team collect and analyze data.

Lay a sheet of newspaper flat on a table or floor. This will be the environment where the prey species "lives." Take at least 50 each of the newspaper, white, and red paper squares (150 squares). Keep the three types separate, as each represents a different type of the *same* prey species. Some are brightly colored. The others are not. An example of such different populations is the species *Canis familiaris,* the common dog. Although dogs come in many different colors and sizes, they all belong to the same species. The paper squares represent individuals of different colors, but of the same species.

The keeper collects ten squares from each of the three prey populations. The keeper then mixes them and scatters them on the environment while the predators are not looking. Each predator may look at the environment *only* when it is her or his turn. When it is not your turn, simply close your eyes or turn your back until the keeper indicates that it is your turn. *Continue in order until each predator has removed five prey individuals.*

Shake off the individuals left on the environment and count these survivors according to their type. Enter the data for your group in the appropriate sections of a table like Figure 9-5. Place the data on the chalkboard also, so a class total can be reached. **1.** Does any population have more survivors than the others? **2.** Write a hypothesis that might explain this difference. **3.** Consider your hypothesis. If it is correct, what do you expect to happen to the num-

9-5. Sample table for recording team and class data

Generation		Paper Prey Species		
		Newspaper Individuals	*White Paper Individuals*	*Red Paper Individuals*
1	Team	10	10	10
	Class			
2	Team			
	Class			
3	Team			
	Class			
4	Team			
	Class			

ber of newspaper individuals by the end of the fourth generation? to the red paper individuals? to the white paper individuals?

The survivors will be allowed to "reproduce" before the next round begins. For each survivor, the keeper adds one individual of that same type. The next generation will then include survivors and offspring. This should bring the total prey number back up to 30. The keeper scatters these 30 individuals on the habitat.

Repeat the predation and reproduction procedures for three more generations. Calculate the change in the number of all three populations after each round. **4.** Does it take you a longer or shorter period of time to find one prey individual as you proceed through the generations? Give an explanation for this. **5.** How does the appearance of the surviving individuals compare with the environment? **6.** Is your hypothesis written in question 2 and your prediction in question 3 supported, or do they need to be revised? **7.** Were the red paper individuals suited or unsuited for this environment? Explain. **8.** Would you say this species *as a whole* is better adapted to its environment after several generations of selection by the predators? Explain. **9.** Now think of the "real" world. Is appearance the only characteristic which determines whether an individual plant or animal is suited to its environment? If so, explain. If not, give several other characteristics. **10.** In your own words, what is natural selection? What role does reproduction play in your definition?

Now you may test some of your own ideas about natural selection.

What would happen if there was a change in the environment, such as a change in color of the habitat? What would be the result if one type of paper square "reproduced" at a faster rate than the others?

DEVELOPMENT OF THE THEORY OF EVOLUTION

The clues that first led Darwin to the hypothesis of natural selection came from his own studies and experiences as well as those of others. He had to know something about the past history of the earth. He had to know how the individuals of a species vary. He had to know how species might change with time.

Geology laid one foundation for Darwin's work. In the early 1830s, Charles Lyell, an English geologist, published the *Principles of Geology: Being an Inquiry How Far the Former Changes of the Earth's Surface Are Referable To Causes Now in Operation.* Lyell was able to explain how the earth came to be as we now observe it. The mountains, valleys, deserts, rivers, lakes, and coastlines could have come about through the action of forces and natural conditions that still exist today. A river slowly carves a valley. Mountains are worn down to hills and finally plains.

Lyell described the evidence that led him to believe that such changes had been occurring through the ages. He thought these changes had been slowly altering the face of the earth. For the earth to change so slowly it had to be much older than was generally believed. This possibility fired Darwin's imagination. If the earth of today is so old and so changed, what was it like thousands of years ago? What kinds of life may have existed?

In 1831, when he was 22 years old, Darwin sailed as a ship's naturalist on H.M.S. *Beagle* (Figure 9-6). The purpose of the voygage was to chart the coasts of South America and some islands in the Pacific Ocean. Darwin studied the animal and plant life and observed geological characteristics of the regions he visited.

9-6. Charles Darwin and H.M.S. *Beagle,* on which he sailed in 1831 as the ship's naturalist. Observations that he made on this voyage suggested to him that a process of evolution occurs. The evidence that most impressed him was geographic variation of animal species and the close resemblance of fossil and living species.

9-7. The finches of the Galapagos Islands. These small islands are the home of numerous species of finches. There is much variation in the shape of the beaks and in the feeding habits among these different species. *How could we account for this diversity?*

9-8. COMPARISON OF RATES OF INCREASE

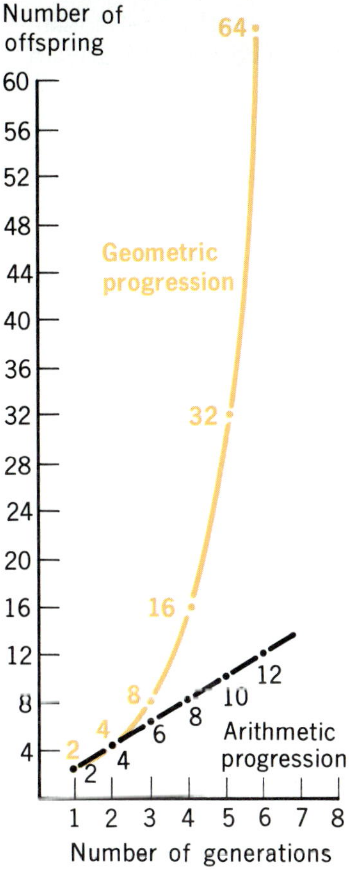

In South America, Darwin spent many hours ashore. He was impressed by the tremendous number and density of living things in the tropics. It seemed that any newly born individual had a small chance for survival. Later, the *Beagle* sailed to the Galápagos (gah·LAH·pah·gus) Islands 1,000 kilometers west of Equador in South America. Here Darwin made some important observations. He carefully studied the huge tortoises that lived on these islands. Much to his surprise, they were recognizably different on each island. Darwin also found finches on all the islands. All the species of finches were very similar, but they had beaks of different sizes and shapes (Figure 9-7). These different characteristics allowed the birds to feed on different kinds of food. How was one to explain all this diversity on a few small islands in the Pacific?

Darwin returned to England in 1836. Soon afterward, he read *An Essay on the Principle of Population* by the English political economist Thomas R. Malthus. Malthus argued that the rate of human reproduction was such that the population increases many times more rapidly than the available food supply. Malthus thought that the food supply increased arithmetically, but the population increased geometrically (Figure 9-8). This difference could only result in a struggle for food and, hence, for existence itself.

Although unaware of Mendel's studies, Darwin realized that there were hereditary differences among individuals of a species. He also knew that pigeon fanciers had developed many strange types of pigeons through careful selection. Plant breeders also had developed new varieties of plants by selecting and mating those with desirable characteristics. Were these clues to the work of nature? Did nature also select?

With these observations, Darwin had what he needed to formulate a hypothesis of evolution. He worked slowly and carefully. In 1842 he wrote, for himself, a brief 35-page sketch of this theory. Two years later he enlarged this into an essay of 230 pages. He showed this essay to his friends, but did not publish it. For the next 15 years, Darwin continued to collect facts to see whether they would support his ideas.

In 1858, Darwin received a letter from a fellow naturalist, Alfred Russel Wallace, who was on a collecting trip in what is now Indonesia. Wallace enclosed an essay he had written. He asked Darwin to read it and, if it seemed worthy, present it to the scientific community. In the essay, Darwin found, almost in his own words, the hypothesis of *the origin of the species by means of natural selection.* Darwin almost yielded to Wallace the honor of being the first person to announce the hypothesis. However, his friends arranged to present the two papers simultaneously. Darwin did so before the Linnean Society in London on July 1, 1858. Then he hastened to complete his book, *On the Origin of Species.*

Lamarck's View of Evolution

Over 50 years before Darwin's book was published, Jean-Baptiste Lamarck, a French naturalist, had offered a hypothesis to explain how evolution might occur. In an example often quoted, Lamarck tried to explain how giraffes had come to have such long necks. Figure 9-9a is based on Lamarck's hypothesis. Lamarck thought that giraffes had lived in areas where other animals were eating the grass. Therefore, the giraffes might have begun to eat the leaves of trees. Their necks and legs might have become longer from constantly stretching to reach the leaves. In time the changes had become inherited, or so Lamarck believed. Thus the environment would have *directly changed* the species.

Lamarck's idea was simple, attractive—and wrong. The problem was that Lamarck did not have any information about inheritance. He believed that characteristics could be acquired. He thought that certain characteristics developed by a parent would be passed on, through the reproductive cells, to the offspring. Today, we know

9-9. A comparison of Lamarckian and Darwinian hypotheses—the evolution of giraffes.

a Lamarck's hypothesis **Existing data do not support this hypothesis.**

Ancestral giraffes probably had short necks that were subjected to frequent stretching to enable the giraffes to reach the foliage of trees.

The offspring had longer necks that also were stretched frequently in the quest for food.

Eventually the continued stretching of the neck gave rise to modern giraffes.

b Darwin's hypothesis **Existing data support this hypothesis.**

Ancestral giraffes probably had necks that varied in length. The variations were hereditary.

Natural selection led to survival of longer-necked offspring.

Eventually only long-necked giraffes survived.

that this is not the case. A person may use black hair dye on naturally blonde hair. But the reproductive cells will not change. Black hair will not be passed on to the children. Here is another example. Ranchers have been cutting off the tails of sheep for many years. Yet baby lambs today are born with tails—just like thousands of generations of sheep before them. Of course we know that genes can be *expressed* differently in different environments. But the environment does not directly change the genes themselves—except in cases of mutations caused by radiation, chemicals, or drugs.

Figure 9-9b shows how Darwin's ideas differed from Lamarck's. Darwin suggested that individuals in a population who have certain characteristics are more likely to survive and reproduce than others. The genetic information for these characteristics is passed on to offspring and, over time, these characteristics become more frequent in the population. Thus, Darwin viewed the environment as a *selective* factor. The environment did not directly change the organisms. But the environment did have a role in determining which organisms would survive and reproduce.

Experiments on Natural Selection

Biologists have performed many experiments to test the hypothesis of evolution through natural selection. One example involves two color varieties—buff and grey—of deer mice. Under controlled conditions in a University of Michigan laboratory, these mice were exposed to the attacks of a barn owl. On one day, the floor was covered with pale soil, which matched the buff-colored mice. On the next day, the floor was covered with darker soil matching the gray mice. Piles of sticks provided places for the mice to hide. The room was kept nearly dark. Each day, four buff-colored and four gray mice were exposed to the owl for 15 minutes. On the pale soil the gray mice were conspicuous, but the buff-colored mice were not easy to see. On the dark soil the buff-colored mice were conspicuous but the gray mice were not easy to see. The test was run 44 times for each soil type. The owl caught nearly twice as many conspicuous mice as mice that were less obvious because of their coat color (107 to 62).

Experimental results like these relate to actual observations in nature. At White Sands National Monument in New Mexico the sand dunes are made of white gypsum crystals. White lizards, white grasshoppers, white beetles, and white mice live on these dunes. Outside the area, where the soil is reddish and darker, the animals of these same species are red-brown in color (Figure 9-10).

The light-colored animals that live on the white sand dunes are the product of mutation and natural selection going on for thou-

9-10. Mice that live in areas where the soil is light in color tend to be lighter than mice that live in places where the soil is dark. But there is always variation among members of a population.

sands of years. Any mutation that would give a paler coloration would be selected. Any mutation that would darken the animal would be eliminated. The white sand has no effect on the origin of the mutations. Mutations to dark or light occur by chance. But once mutations do occur, natural selection will determine what happens to them.

Most species of animals are relatively inconspicuous in their native environment. Conspicuous individuals are captured more often than inconspicuous ones. Some animals have a remarkable resemblance to their surroundings. The ptarmigan in Figure 9-11 is an example. The color and pattern of its feathers blend almost exactly with the plants, rocks, and soil of its environment. Furthermore, the color of the birds changes from summer to winter.

In recent years there have been many examples of selection in response to insecticides and drugs. You have probably heard that, when DDT was first used, it killed houseflies. Now, we find that not all houseflies are killed by DDT. How can we explain this?

During the first year of DDT use, nearly all flies sprayed were killed. However, a few flies, because of their genetic makeup, were not killed by DDT. This type of fly survived and reproduced. More resistant flies began to occur where DDT had been used. The insecticide became less effective. Eventually, wherever DDT was used for many years, only DDT-resistant flies were left. DDT does the selecting. It does not cause the genes of the flies to mutate to alleles that create resistance.

Other situations follow similar patterns. Some disease organisms are no longer killed by antibiotics. For example, certain diseases such as venereal diseases, have long been treated effectively with penicillin. Now strains of these organisms that are resistant to penicillin have developed.

Many experiments have been performed in the search for an understanding of how organisms become resistant. In one experiment, a culture of about 100 million bacterial cells was exposed to a relatively weak dose of penicillin. Fewer than ten bacterial cells survived. The offspring of these cells survived the same weak dose of penicillin. When the concentration of penicillin was doubled, nearly all these offspring were killed. A few survived, however. The survivors were isolated and grown. Their offspring were exposed to an even stronger dose of penicillin. This process was repeated five times. Finally, a strain of bacteria was obtained that could withstand a dose of penicillin 2,500 times as strong as the dose that killed most of the original bacteria.

Reports have been published of patients with bacterial infections that are resistant to antibiotics. These bacteria result from the same kind of action as that just described. Again we see the work of natural selection. As the environment changes—that is, as more antibiotics are added—the more resistant type of bacterium gains an advantage over the more susceptible.

9-11. Protective coloration in ptarmigans. Both birds are of the same species. The photograph at the left was taken early in summer. The bird's color is beginning to change. The photograph at the right was taken in winter. *Suggest hypotheses to account for the difference—that is, how the same individual changes with the season.*

THE EVIDENCE OF EVOLUTION

We can no longer speak strictly of "Darwinian evolution" because Darwin's original ideas have been modified and expanded greatly in our own century. We must, nevertheless, recognize the importance of Darwin's contributions. This work provided a framework for much research. Do you think we could have looked for answers to questions about evolution without having had Darwin's ideas to work from?

Evidence that supports the theory of evolution through natural selection was cited by Darwin and has accumulated since his time. In this section we will look at some modern evidence as well as some of the evidence put forth by Darwin.

Evidence from Inheritance

Darwin had very little understanding of heredity. Yet Darwin emphasized that only hereditary variations would have any meaning for the evolutionary process. Darwin was impressed by the success achieved by plant and animal breeders in selecting traits they thought were desirable.

Plant and animal breeders used **artificial selection** only on whatever variations they could see in the phenotypes. Despite this, they had been extremely successful in breeding special types of animals and plants. Darwin noted the similarities of artificial selection (by humans) and natural selection (by nature). He was especially impressed with the number of varieties of pigeons produced by artificial selection (Figure 9-12).

Probably no aspect of Darwin's theory has been more strengthened during the past century than our knowledge of inheritance. New alleles arise from mutations of genes. These mutations can be caused by many natural agents. Mutations may be produced by X rays, heat, some chemicals, or cosmic rays. These mutations nearly always have harmful effects. Probably this is because every species is already the product of many generations of natural selection. Thus, changing one gene without changing others is likely to produce a disorder or deformity.

Yet a new gene that produces a harmful effect in one environment may provide an advantage in another. Natural selection would preserve such a gene. Possessors of the new mutated gene would be better able to survive and to reproduce than individuals with the older gene. Then other genes that form a better system with the new one and genes that modify or cancel its harmful effects would be selected, too. Gradually, over many generations, the genotypes of the population would change.

9-12. Four varieties of pigeons. These varieties are the result of artificial selection carried out by pigeon breeders.

Mutations provide some of the "raw material" upon which natural selection operates. But they are not essential for evolution. Normal genetic variation within a population usually provides enough opportunities for evolution to proceed in new directions if the environment changes. And the changes can sometimes happen very rapidly. In 1937, a researcher at the Maryland Agricultural Experiment Station reported the following. A farmer seeded a pasture with ordinary plants of several kinds. Cattle were allowed to graze on half of the pasture. The other half was left for hay. Three years later, scientists from the research station got seeds from each half of the pasture. They planted the seeds in two experimental plots where all growing conditions were the same. They found that the seeds from the grazed half of the pasture produced plants that grew low and close to the ground. The seeds from the ungrazed half produced tall, upright plants. The conclusion was obvious: Cattle on the grazed half of the pasture had eaten the tall plants before the plants could produce seeds. Only the low growing plants had survived and reproduced.

Evidence from Geographic Distribution

During his voyage on the *Beagle,* Charles Darwin visited the Cape Verde (VURD) Islands. These islands lie near the equator off the western coast of Africa. They were formed by volcanic eruption. Darwin noticed the close resemblance of the plants and insects, birds, and other animals on the islands to those on the nearby shores of Africa. Several years later Darwin reached the Galapagos Islands. Although the climate and terrain of the two island groups were very similar, the plants and animals were totally different.

The plants and animals on each group of islands closely resembled those on the nearest mainland. It seemed reasonable to think that, in each case, ancestors of the island organisms had come from the mainland. On their island home, the organisms were isolated from similar species on the mainland. They would no longer be breeding and sharing genes with the mainland members of their species. The environment of each group of islands is very different from that of the closest mainland. Thus the species that survived in the new island environment would probably have characteristics different from those of their mainland ancestors. Eventually, through **isolation** and natural selection, new species would evolve on the islands.

Fossil Evidence of Evolution

The most important evidence for evolution is the thousands of fossils of animals and plants that scientists have found in rocks (Figure 9-13). Most of these organisms lived millions of years ago, often under conditions very different from those now existing.

A record of the past is kept in the earth's rocks. Two of the main classes of rocks are *igneous* (IG·nee·us) and *sedimentary.* Some ig-

9-13. Evidence from the fossil record is important in our efforts to learn about life in the past. The investigator shown in the photo on the left is unearthing in Nebraska the skull of a titanothere. The skeleton of the entire animal is shown on the right.

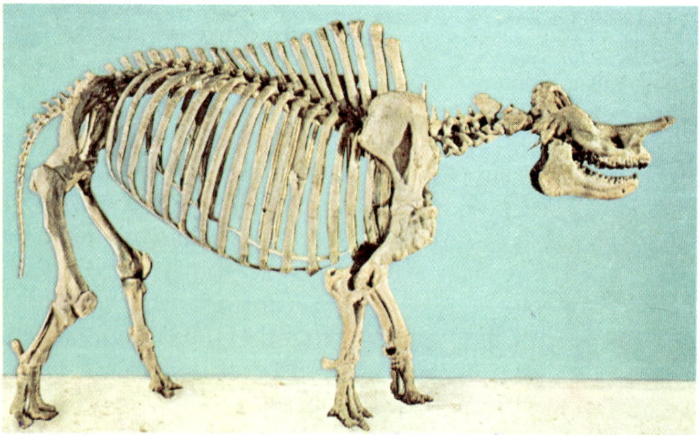

9-14. The stratification of rock layers in this mountainside was exposed when a highway was built.

neous rocks are formed when molten lava spews from volcanoes and cools. Sedimentary rocks (Figure 9-14) are formed when soil, sand, mud, pebbles, and rocks are deposited in layers, usually under water. Over long periods of time these sediments are changed by pressure into sedimentary rocks. As the rocks gradually change, changes also occur in the bodies of organisms buried in the sediments. Soft parts of bodies usually decay. But the hard parts—the bones or shell of the animal, or the cell walls of plants—often become preserved or fossilized (Figure 9-15). The preserved remains of these organisms are called **fossils.**

9-15. FOSSILIZATION

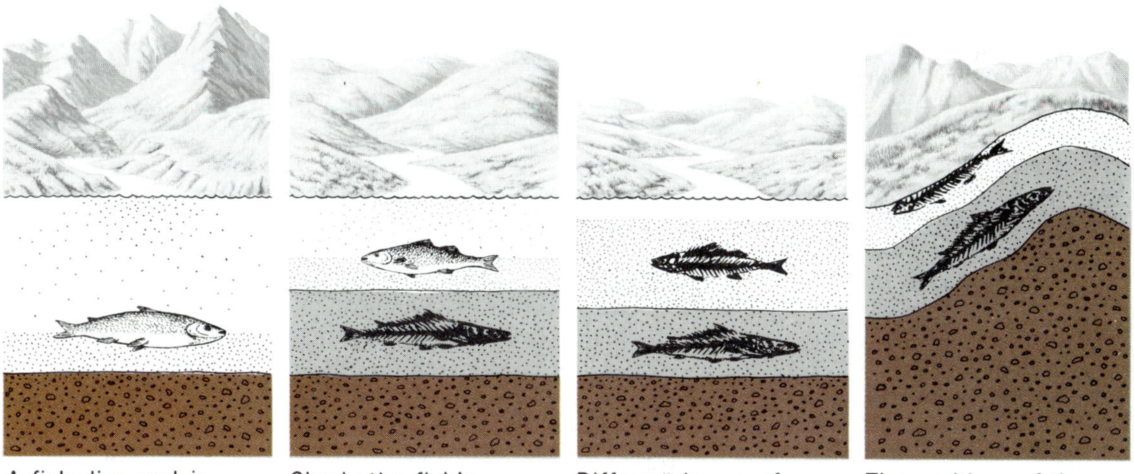

A fish dies and is covered by sediment washed down from the mountains.

Slowly the fish's remains are fossilized. Other fishes die and are covered.

Different layers of sedimentary rocks have fossils of different ages.

The positions of the sedimentary rocks and their fossils may be changed, and surface erosion may expose the fossils.

SCIENCE AT WORK: Life on Mars?

Scientists know that after spending years in research there is always the possibility that they may not find what they are looking for. Although something is always gained from any research, it may not be the hoped for result. Biochemists may develop a drug that they think will cure a disease. If the drug does not cure the disease, something has been learned, but the disease still has no cure.

In 1976, two spacecraft, Vikings I and II, landed on the planet Mars. Part of their mission was to perform a series of experiments designed to detect life on Mars or to add to our storehouse of knowledge about the origin of life here on earth. A gas-exchange experiment indicated that samples of Martian soil gave off substantial quantities of carbon dioxide and oxygen. Although it was possible that the carbon dioxide was produced by a biological oxidation process, it was more likely produced by some inorganic process. A second experiment was designed to determine whether or not Martian soil could convert carbon monoxide or carbon dioxide, labeled with the radioactive isotope carbon-14, into organic matter. A very weak synthesis did occur but nothing in comparison to that found in biologically active terrestial soils. The final experiment sought to detect the presence of metabolism by monitoring the evolution of gas containing carbon-14 from soil samples to which nutrients labeled with carbon-14 had been added. Some metabolism seems to have taken place but no data quantitatively reaching the kinds scientists get here on earth have been obtained.

The conclusion, as of this writing, is that the Viking landers are producing conflicting data as to life on Mars. The temptation is strong to say that the project has failed because life, as we know it on earth, does not seem to exist on Mars. But this is not the stand of the scientists. They hypothesize now that Mars may be in a pre-life state, perhaps with molecules as complex as DNA surviving on the Martian surface. If this is true, more stands to be learned about life's origins from the lack of life on Mars than from its presence. As one scientist said, "Life would gobble up its ancestral beginnings."

Relatively few animals and plants die in places where conditions are suitable for them to become fossils. This is especially true of those organisms that live on land. In addition, not all animals and plants have hard parts that can become fossilized. Nevertheless, scientists have found fossils of nearly all types of animals and plants. Even some very small and delicate creatures have been fossilized.

Look at Figure 9-15 again. Two ideas can be derived from it. First, each level of sedimentary rock contains the remains of organisms that lived at the same time the sediments were deposited. Thus, one can learn about the different animals and plants that lived at that time. The second idea is that the relative positions of the sedimentary rocks tell us about their *relative ages.* For example, in formations that have not been disturbed, the most recently formed layers will be on top and the oldest will be on the bottom.

The relative positions of the layers of sedimentary rocks can tell us which layer is older, but not *how much* older. Other methods must be used to determine the *absolute ages* of rocks. One method is to measure the accumulation of certain isotopes of lead or other elements. These are products of the breakdown of radioactive materials present when the rocks were formed. This is the most accurate way to measure the age of igneous rocks.

It is more difficult to measure the age of sedimentary rocks. They are often dated as being "no younger than" by dating an igneous rock that is found in a sedimentary layer. The igneous rock, of course, must be younger than the layer it penetrates. If we can date the igneous rock, we can give an approximate, minimum date to the sedimentary layer. It is sometimes possible to date two sedimentary layers, one just beneath the other. In such a case, a sharper definition of the age can be given.

Another radioactive method often used is based on the decay of carbon-14. Carbon-14 is a radioactive isotope of carbon. It occurs naturally in the atmosphere and is taken in by plants. Carbon-14 becomes a part of the plants' carbohydrates, fats, proteins, and nucleic acids. When animals eat the plants, carbon-14 becomes part of the animals' tissues. The carbon-14 dating technique can be used to estimate the age of organic remains fossilized less than 35,000 years ago. All of these methods, of course, involve a certain amount of error because they are based on the *probability* that certain natural events will occur.

With the use of such techniques, there have been many spectacular discoveries. For example, a fossil of a four-toed animal about the size of a large cat was found in layers some 60 million years old. From this animal a gradual line of descent, with continuing slight changes, can be traced to the modern horse (Figure 9-16). Similar evolutionary sequences have been established for the elephant, giraffe, camel, and many other organisms.

9-16. Five stages in the evolution of horses, as reconstructed from fossil evidence. Note the changes in size and proportions. Describe the changes in the leg and molar tooth.

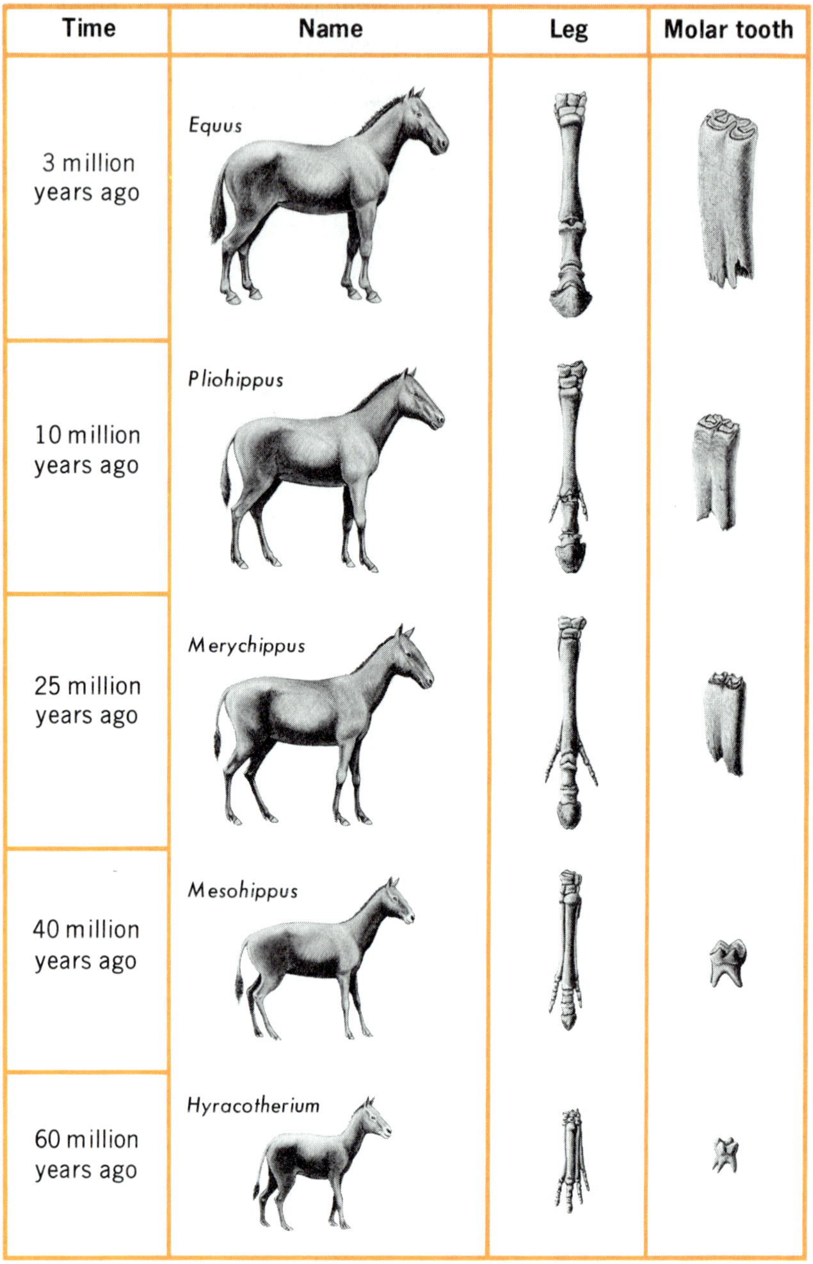

Evidence from Homology

Some biologists study comparative anatomy. That is, they compare the structures of organisms. Their studies show that individuals of the same species – or of larger groups such as all birds or reptiles –

resemble each other. For example, all birds have feathers, wings (though not all fly), and some kind of beak. If we assume that all birds evolved from a single ancestral type, we can see reasons for these similarities. The various descendants of the common ancestor became increasingly different in the course of evolution.

In many cases, we can even see similarities in the structure of very different organisms. The wing of a bat, the arm of a human, and the flipper of a whale look different. But they all have the same structural origin (Figure 9-17). The forelimbs of these three animals developed from the forelimbs of primitive reptiles. The presence of common patterns of structure is called **homology** (hoe·MOLL·oh·jee).

The processes of evolution tend to remodel existing structures. For example, fishes have gill arches that support the gills (the organs through which fish get dissolved oxygen from the water). In air-breathing land animals, gills are not necessary. But instead of disappearing, the gill arches have become modified into other organs. In humans, they have become the upper and lower jaws and the bones of the middle ear. Some of the muscles of the human face have evolved from the muscles of the gill arches.

Sometimes, however, an organ becomes reduced. It may even lose its function. Such organs are called **vestigial** (ves·TIJ·ih·al) **organs.** A good example is the human appendix. The appendix is a pouch that leads off from the large intestine. It has been removed from people without ill effects. In humans, it has no known function. In other animals, such as the rabbit, the appendix is well developed and is useful in digestion.

Humans are not the only organisms with vestigial organs. Whales and pythons have vestiges of hind limbs. Some birds, like the flightless kiwi and the Galapagos cormorant (Figure 9-18), have vestigial

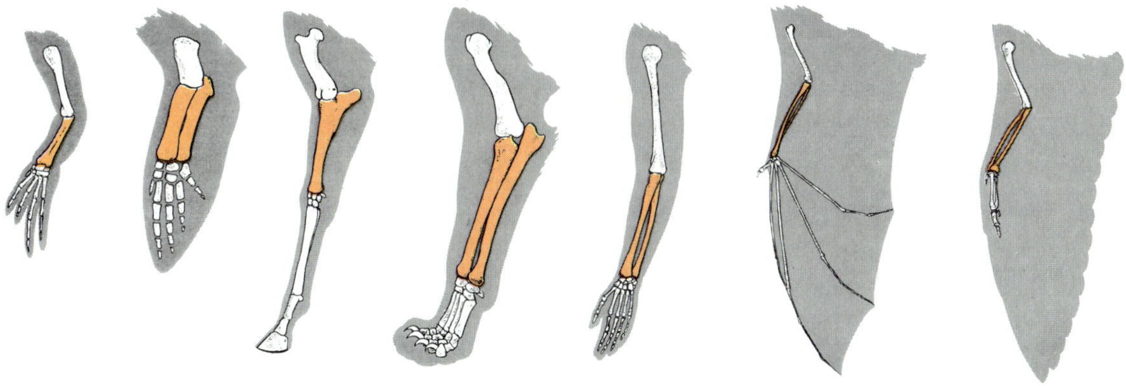

9-17. Bones in the forelimbs of seven animals. From left to right: the foreleg of a frog, the flipper of a whale, the forelegs of a horse and of a lion, the arm of a man, and the wings of a bat and of a bird.

9-18. Most species of cormorants are strong flyers with well developed wings **(left).** The flightless cormorant of the Galapagos, however, has poorly developed wings **(right).**

wings. These wings are organs that have no use but were neither lost nor modified into different structures.

From an examination of body structures, we can see evolution operating in two different ways. **Divergence** occurs as descendants of a common ancestral group become more different as time goes on. **Convergence** occurs when unrelated organisms begin to look more alike (Figure 9-19). How does convergence happen? The answer has to do with similar characteristics being selected in similar environments. Certain land animals took to the water millions of years ago. Although they are only distantly related, we can see many similarities in the structures of their bodies today. Seals and otters have hind legs that look like paddles. In whales and porpoises the hind legs are almost completely gone, but the tail has been modified into an extremely powerful swimming organ. Similarities in the structure and function of body parts come about not as a direct result of the environment (in this case, water) but because the environment acts as an agent of natural selection.

A similarity among the body plans of many organisms can be seen when comparative anatomy is studied. We also can see this unity when we study microscopic structures. Evidence of this unity is the uniformity of mitosis in all cellular organisms. Mitosis represents homology on the microscopic level. Another example is the similarity of meiosis and fertilization in all sexually reproducing organisms. Other physiological and biochemical similarities reveal the unity of body plan among living things. All organisms use nucleic acids. In most organisms the nucleic acid, DNA, is the means of inheritance.

9-19. Cacti and Euphorbs are not closely related plants, but their forms are very similar. The cactus on the left is a native of America. The euphorbs on the right grow in Africa. Both live in desert areas. *How might such convergent evolution have come about?*

And all organisms use ATP for energy transfer. Almost all plants use the same chlorophyll molecule for photosynthesis. Most striking of all is the universality of the genetic code. With few exceptions, the same sequences of bases in DNA code for the same amino acids.

How can we explain this unity and similarity in more than a million *living* species of animals and nearly half a million *living* species of plants? We have learned that all organisms *do* inherit characteristics from their ancestors. The evolution of all organisms from a common ancestor is one way of explaining unity of basic plan. Evolution also is a good explanation of the diversity we can see among similar organisms that live in different environments.

Evidence from Embryology

During the latter half of the nineteenth century, embryological development in animals was studied intensively. Many similarities among organisms were noted. Embryologists concluded that the embryonic development of an individual was much like the evolutionary history of the species. They suggested that one could trace the evolutionary history of a species by studying its embryonic development.

Today, embryonic resemblances are viewed with caution. Undeniably, similarities exist between embryos, as shown in Figure 9-20. But this does not mean that a human being passes through fish, amphibian, and reptile stages during embryonic development. What

9-20. COMPARISON OF EMBRYOS IN FOUR ANIMALS

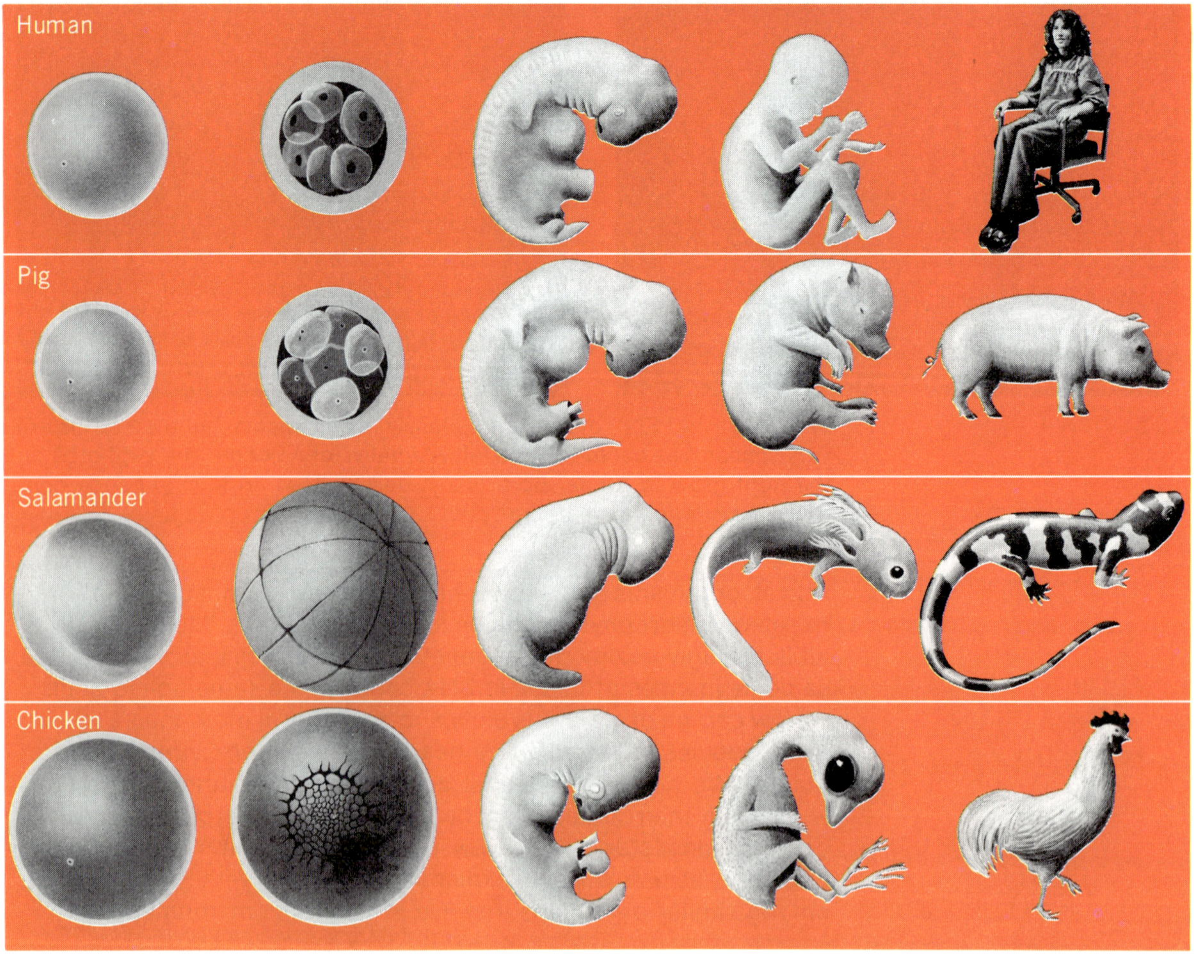

this statement does mean is that the same fundamental processes occur in the development of a great variety of structures. Similarities observed in embryonic development are most striking in the early stages.

Evidence from Biochemistry

Since Darwin's time, evidence supporting the theory of evolution has come from many sources. But in the last few decades, no area has been more active than biochemistry. Biochemists now can determine which amino acids make up a particular protein and in what order the amino acids occur.

The sequencing of the amino acids in proteins has led to comparisons of the same protein in different plants and animals. Cytochrome c, a chain of 104 amino acids, is one protein that has been studied extensively. Cytochrome c is important in processing and transferring energy in the cell. Nearly every living thing has cytochrome c. The substance is easy to purify, can be obtained in large quantities, and is a rather small protein. In studying these proteins in different organisms, scientists have found striking similarities in amino-acid sequences. How might such similarities be explained?

Chimpanzee	0	Bullfrog	18
Monkey	1	Fruit Fly	29
Horse	12	Bread Mold	43
Chicken	13	Box Elder Tree	46

9-21. This chart shows the number of differences in the sequence of amino acids in the cytochrome c of various organisms compared with the cytochrome c of humans. *How do such biochemical similarities and differences relate to evolution theory?*

Cytochrome c molecules from different organisms can be compared with human cytochrome c (Figure 9-21). The cytochrome c of chimpanzees is exactly the same as that of humans. The cytochrome c of bread mold or the box elder tree is very different from that of humans. The amino acids of cytochrome c in a donkey and a zebra are identical. Four of the amino acids in a chicken are different from those of a pigeon. The difference between the cytochrome c of a sunflower and a wheat plant is 15 amino acids. Thus, we see that closely related organisms have similar cytochrome c, while more distantly related organisms have different cytochrome c.

In Brief

Evolution is possible through the natural selection of inherited variations. Over time, a population may change slowly. Changes result when characteristics are selected that improve chances of survival in a particular environment. Smaller populations in isolated areas such as islands may evolve independently. What was once a single species may evolve into two or more different species. Or populations of unrelated organisms may evolve similar characteristics in similar environments.

Evidence originally supporting evolution theory has come from studies of genetics, embryology, anatomy, biochemistry, and the fossil record. And much work continues. So significant is the theory —not only for what it explains but also for the questions it stimulates—that it is indeed a cornerstone of modern biology.

Guide Questions and Problems

1. What evidence supports the following ideas?
   (a) Modifications caused by the environment, such as tanning of the skin by the sun, are not inherited.
   (b) Hereditary differences are of importance in determining whether or not an individual survives and leaves offspring.

2. What are the main differences between the Darwinian and the Lamarckian hypotheses of evolution?

3. How are artificial selection and natural selection alike? How are they different?

4. One of the most important fossils ever found was *Archaeopteryx,* an early bird with many characteristics of reptiles but with feathers like modern birds. Why is this fossil discovery important in evaluating Darwin's hypothesis of evolution by natural selection?

5. Anatomical and embryological similarities among animals today are striking. In what way are they evidence of evolutionary relationships?

6. Human activities, either through hunting or destruction of the environment, have eliminated many species of animals, such as dodo birds and passenger pigeons. Was this *natural* or *artificial* selection?

7. In what way does the theory of evolution make it easier for us to understand another theory in biology, the cell theory (Chapter 3)?

8. A basic principle of all the major sciences is this: Unless evidence can be discovered to the contrary, the basic laws that account for natural events today must also have applied to natural events in the past. Why is this principle important to the evaluation of evidence about the theory of evolution? Why is the principle a reasonable one?

Ideas for Independent Study

1. An area of biological investigation that is very active just now is the comparison of related organic compounds in a variety of different species. For example, living creatures have *proteases,* enzymes that break down proteins. Almost all higher plants have chlorophyll. Almost all animals with backbones have hemoglobin in the blood. To what extent are the proteases, chlorophylls, and hemoglobins similar from one species to another? This question is an introduction to the study of homology on the molecular level. If you wish to investigate this interesting field on your own, you can start with these two references in *Scientific American:* "Computer Analysis of Protein Evolution," by M. O. Dayhoff, July 1969, page 87; and "The Structure and History of an Ancient Protein," by R. E. Dickerson, April 1972, page 58.

2. People have greatly modified the characteristics of domestic species, such as cattle, horses, sheep, pigs, dogs, cats, corn, wheat, and orange and apple trees. Select one of these species and find out as much as you can about its origins and how it has changed under human influence. What problems are faced by domesticated species if they are forced to live in the wilds—that is, away from the human environment?

Related Readings

BOOKS

Appleman, P., ed. *Darwin.* New York: Norton, 1970.

Darwin, C. R. *The Origin of Species.* Garden City, N.Y.: Doubleday, 1960.

Kurten, B. *The Age of Mammals.* New York: Columbia University Press, 1974.

Moore, R. *Evolution.* New York: Time-Life, 1968.

Rhodes, F. H. T. *Evolution.* New York: Golden Western, 1974.

Volpe, E. P. *Understanding Evolution.* Deer Park, N.Y.: W. C. Brown, 1971.

Webb, S. D. *Evolution.* A BSCS Laboratory Block. Lexington, Mass.: D. C. Heath and Company, 1968.

ARTICLES

Bakker, R. T. "Dinosaur Renaissance." *Scientific American,* April 1975, page 58.

Barghorn, E. S. "The Oldest Fossils." *Scientific American,* May 1971, page 30.

Horowitz, N. H. "The Search for Life on Mars." *Scientific American,* November 1977, page 52.

10 Evolution at Work

Three conditions are necessary for evolution to occur. The first is *genetic variability*—members of a population are genetically different from one another. The second is *natural selection.* That is, the selection is determined by the nature of that population's environment. Differences in a population are sifted and sorted by natural selection. Some organisms survive and produce more offspring than others. In most places, for example, flies with wings are more likely to survive than wingless flies. But on a windswept island that has few plants to provide cover, flies and other types of insects may be more successful if they don't have wings. For example, on the Kerguélen Islands in the Indian Ocean, only wingless insects have survived.

The third condition is *reproductive isolation.* If a population cannot breed with other similar populations, mixing of genes does not occur. Reproductive isolation usually is a result of geographic isolation. On the Galápagos Islands, populations are isolated from their ancestral population on the mainland. Mountain ranges or large lakes and rivers also are barriers to reproduction. A few individuals of a population may migrate across a large river. They may survive and reproduce. These individuals will not be able to interbreed with the first population, however, because they are on the other side of the river.

But barriers to reproduction need not always be geographic. Populations may develop patterns of behavior that are reproductive barriers. Differences in courting behaviors, for example, may prevent mating among members of closely related neighboring populations. The time at which an organism reproduces also may be a method of isolation. Certain species of pines in California are one example. One type sheds pollen and fertilizes its cones in February or March, the other in April or May. These differences probably were selected in environments with different climates. But now, even where the pines grow side by side, the seasonal differences prevent the pines from interbreeding. Figure 10-1 shows how a factor, such as climate, can cause reproductive isolation.

VARIABILITY: THE RAW MATERIAL OF EVOLUTION

Genetic variability is caused mainly by two processes. One is mutation. The other is the recombination of genes.

A mutation is a change in the structure of a gene. Mutations are of several kinds. One sort is the simple substitution in a gene's DNA of one or more bases. The change of even one base in a DNA code may cause the production of a different amino acid. How can this happen? Remember that a sequence of three bases in DNA codes for a particular amino acid. Codes for amino acids are often quite similar. For example, the code in RNA for glutamic acid is -GAA-. The RNA code for valine is -GUA-. A change in a single base (from U in the middle to A, for example) will cause a different amino to be produced. As a result, the protein is changed. This kind of a mutation is responsible for sickle-cell anemia (Chapter 7).

Another kind of mutation occurs when a base is lost or added to the DNA chain. Suppose the DNA sequence for part of the hemoglobin molecule is -CCC-GAG-GAG-AAA- and a new base (T) is inserted after the first G. The sequence becomes -CCC-GTA-GGA-GAA-. The code to the right of the base with the inserted T now reads differently. The message is scrambled, and the polypeptide probably will be changed greatly from what it was. The new protein almost certainly will not work properly in the cell. Mutations such as these

10-1. REPRODUCTIVE ISOLATION

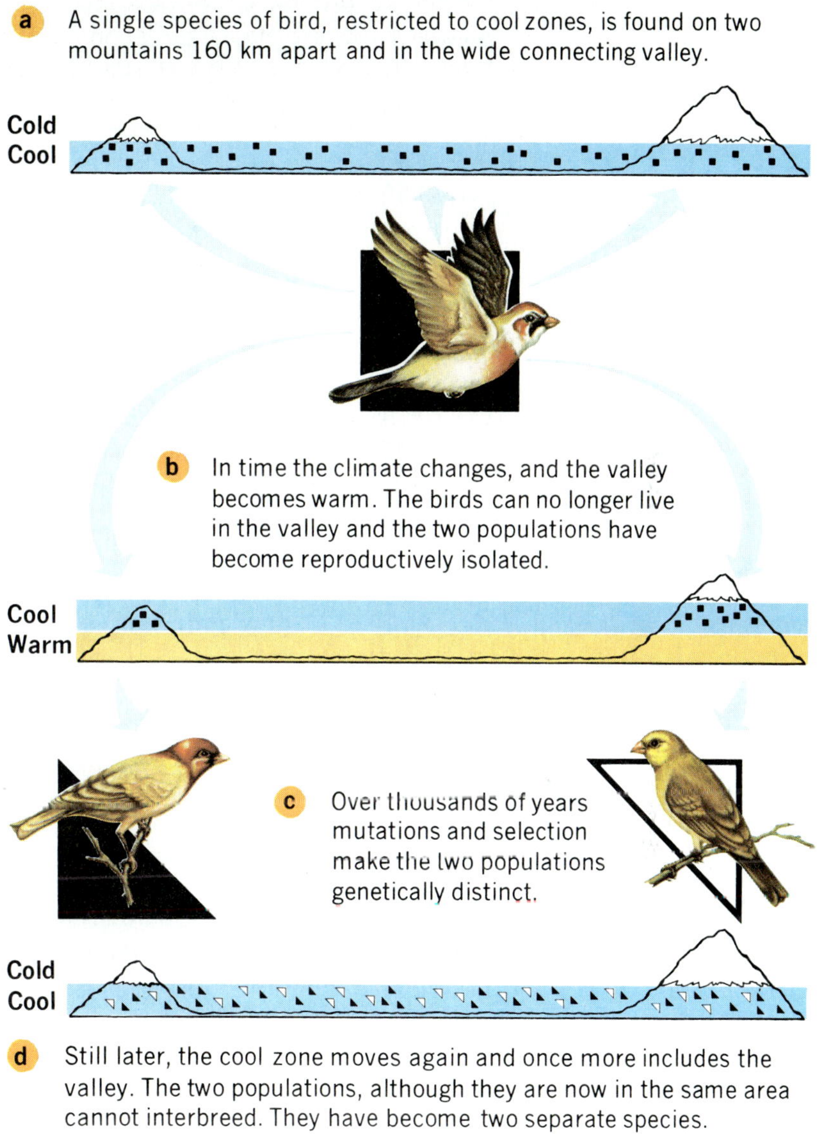

are nearly always fatal. Similar consequences follow if a single base is deleted. Figure 10-2 shows substitution, insertion, and deletion mutations.

What are the causes of mutation? Some mutations occur spontaneously. For some reason, a change occurs when a strand of DNA replicates. But certain conditions in the environment may increase the chance that mutations will occur. In 1927, Hermann J. Muller ex-

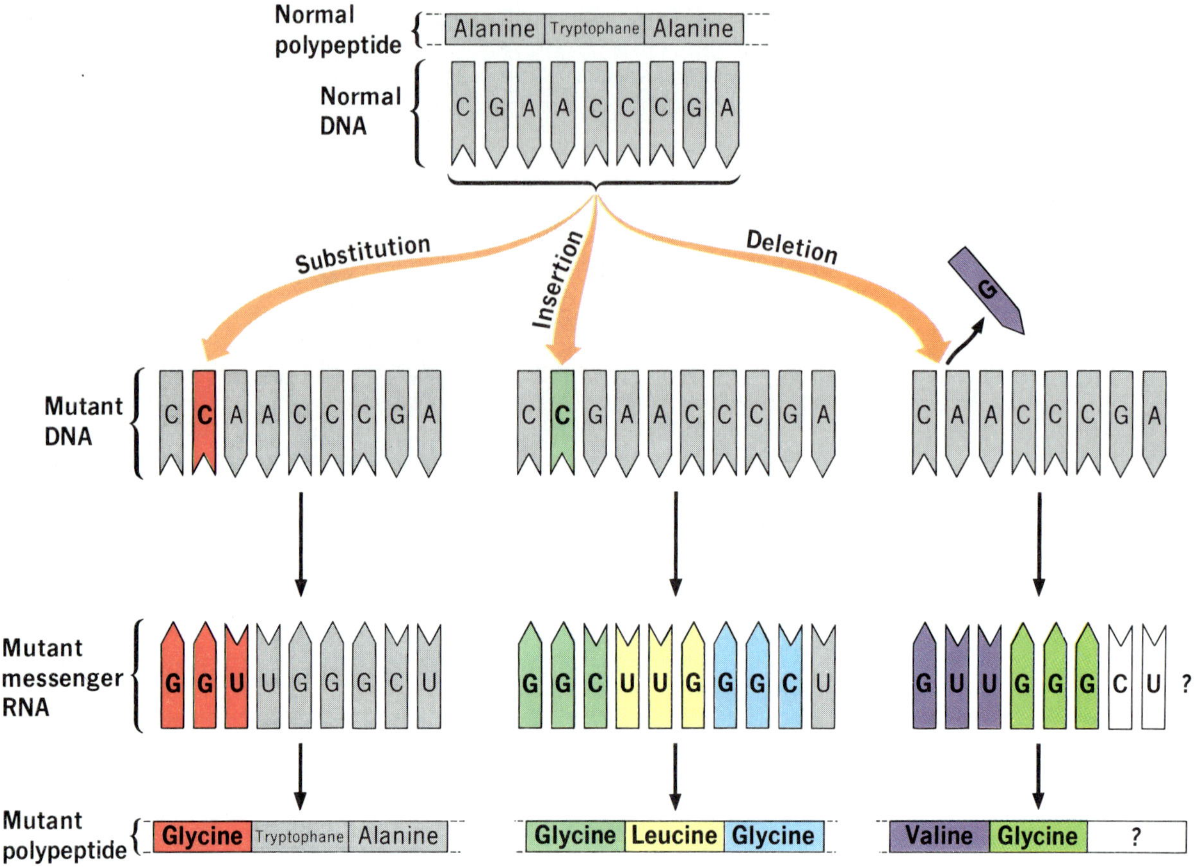

10-2. Basic types of DNA mutations. Note how the messenger RNA and the resulting polypeptide are affected.

posed fruit flies to X rays and discovered that the mutation rate in their genes was increased hundreds of times. In other experiments, an increase of 10°C in temperature nearly doubled the rate at which mutations occurred.

In the 1940s, Charlotte Auerbach's research with mustard gas showed the first *mutagenic* (mutation-producing) effect of a chemical substance. Since then, numerous substances have been tested for mutagenic properties. Mutagenic substances are found in automobile exhaust, smog, tobacco products, and in some foods. For example, caffeine in coffee and tea, although not proven mutagenic to humans, increases mutation frequencies in small animals.

Mutation and Evolution

We sometimes think of mutations as rare. And indeed they are if we consider only a single gene. How, then, can they be important in evolution? Mutations of a particular gene occur spontaneously only about once in 100,000 reproductive cells. If a population contains one million breeding individuals, then a certain mutation would be expected to occur at least a hundred times per generation. Thus mutation, even of a single gene, would occur often enough to be tested by natural selection.

Each individual in a population contains a great many different genes. Experts estimate that a human reproductive cell has somewhere between 100,000 and 2,000,000 different genes. If the mutation rate per gene is about one in 100,000 each sex cell would contain, on the average, at least one mutation. The diploid cell produced by the union of an egg and a sperm, therefore, would have two.

Every organism is already the product of thousands of generations of natural selection. Even a slight change in a single gene could make a difference in an individual's ability to survive and to reproduce. Yet, once in a great while, a mutation proves beneficial. The individual organisms that have this gene will survive and pass this gene to their offspring. Thus, the beneficial mutation will become more abundant in succeeding generations of the population.

A mutation that seems to be harmful may contribute to survival. For example, suppose a species of fish lives in a climate in which the temperature ranges from 35°C to 40°C (Figure 10-3, Time A). The

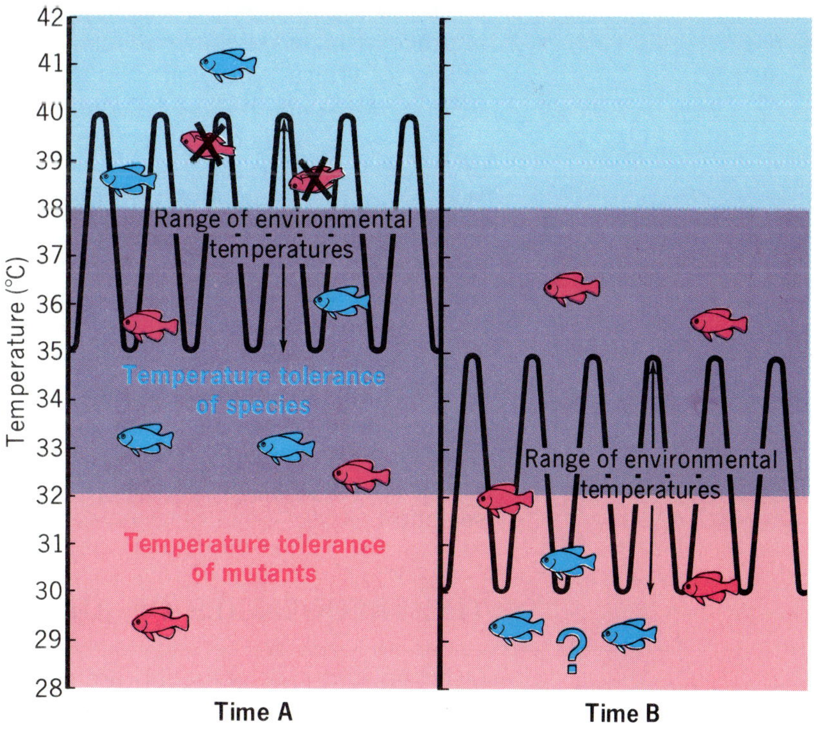

10-3. *What happens to the mutant (red) fish under the environmental conditions that exist at Time A? What happens when environmental conditions change (Time B)?*

normal individuals of this species can survive at temperatures from 32°C to 42°C. If a mutation shifted the range of tolerable temperatures in a few individuals to 28°C to 38°C, the mutant individuals could not survive when the temperature reached the hot part of the range. But if the climate also changed (Time B in Figure 10-3) and the temperature range was 30°C to 35°C, the new mutant type could live in it very well. What probably would happen to the former "normal" type?

The effect of any gene is only relative to the environment the population lives in. Furthermore, *most mutations are recessive.* Thus, a mutant gene will remain in a population for a long time, even though it might be harmful when homozygous. However, environmental conditions might change, and the mutant gene would be tested under the new conditions.

Recombinations of Genetic Material

Single mutations do not often provide for evolutionary change. Remember the giraffe (Chapter 9). A single mutant gene cannot produce all the bones, muscles, and blood vessels that make up that long neck. Actually, what is exposed to the action of natural selection is not a single gene, or even its effect, but the entire phenotype of an individual. Some genes suppress the effects of others. Some improve or increase these effects. More than just a mutation is exposed to natural selection. It is the combinations and recombinations of all genes that are selected. How is this brought about?

You have already discussed the independent assortment of pairs of chromosomes during meiosis. A human possesses 23 pairs of chromosomes. Even if each of these chromosome pairs contained only one pair of heterozygous genes, independent assortment would produce over eight million different sorts of sex cells. Thus, over 60 trillion (6×10^{13}) genetic combinations are possible in a child of one couple. Considering that there are *many* heterozygous pairs of genes, the actual number of possible combinations is enormous.

The variety of hereditary types that is possible in any species is virtually limitless because of genetic recombination during meiosis. New mutations continue to occur and form new genotypes that produce new phenotypes. These, in turn, are subjected to natural selection.

THE FREQUENCIES OF ALLELES IN POPULATIONS

Mutation and recombination are not evolution. They increase the variability in a population, but they do not cause changes in gene frequencies. Suppose we begin by considering a population in which a dominant allele **A** and its recessive allele **a** occur in a ratio of 50:50. Both alleles are equally adaptive. That is, the chance of survival and leaving offspring is the same for **AA, Aa,** and **aa** individuals. What will happen to the population in the succeeding generations?

The fact is that, with all members of the population surviving and reproducing equally well, the frequency of the two alleles, **A** and **a,** *will not change.* You can prove this to yourself by using a checkerboard as shown in Figure 10-4. If no difference occurs in rates of survival and reproduction, the ratio of dominant and recessive genes will be the same in each generation. For example, the ability to taste a chemical called PTC is inherited. To some people the substance tastes bitter. Others cannot taste it.

In one class in an American school, the PTC test showed 30 tasters of PTC and 10 nontasters. Each student took home a small amount of the PTC solution and tested his or her parents. Their results were as follows:

Parents	*Student Tasters*	*Student Nontasters*
taster × taster	20	2
taster × nontaster	10	5
nontaster × nontaster	0	3

From these data, can you determine which of the traits is dominant and which is recessive? Remember, when parents of the dominant phenotype are crossed, the recessive trait may appear among their offspring. But when both parents are of the recessive phenotype, all their children also should be of the recessive phenotype. In this test, only the nontaster × nontaster matings gave that sort of result. The nontaster trait must be recessive. Tasters may be either homozygous **(TT)** or heterozygous **(Tt).** Nontasters must be homozygous recessive **(tt).**

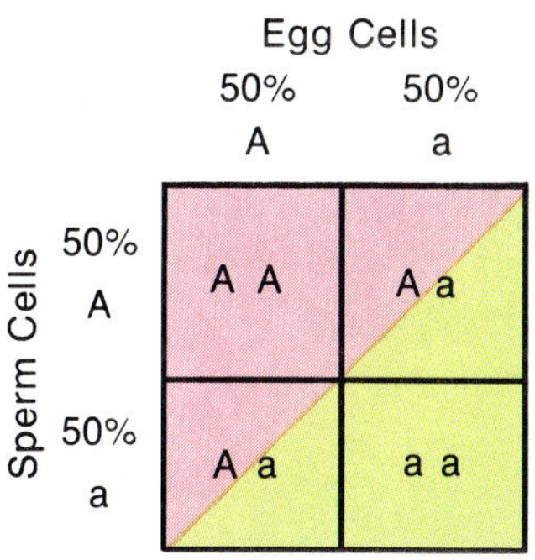

10-4. Of the 8 alleles in the first generation, 4 are **A** and 4 are **a.** The ratio is 50:50. *What will happen in the second and third generations?*

Ten of the 40 students were nontasters. That is 25 percent. We can use that information to figure the frequency of the two alleles—**T** and **t.** If 25 percent of the students in the class are nontasters, then the frequency of the genotype **tt** is 0.25. What two numbers multiplied together give 0.25? Multiplying 0.5 × 0.5 gives 0.25. If we want to obtain the frequency of the recessive allele **t** directly, all we have to do is take the square root of the frequency of **tt.** Thus $\sqrt{0.25}$ equals 0.5, the frequency of allele **t.** And since the frequency of both alleles must equal 1 (100 percent), then the frequency of **T** is 1 − 0.5, or 0.5.

Now let's consider a population in which 64 percent of the population are tasters and 36 percent are nontasters (Figure 10-5). What

10-5. A study of population genetics.

64%

Phenotype: PTC tasters

Genotypes: TT and Tt (but in unknown proportions)

36%

Phenotype: PTC nontasters

Genotype: tt (known)

Working directly with TT and Tt is not possible, since their relative proportion in 64 percent of the population is not known. Begin with the proportion of tt, known to be 36 percent. The frequency of gene t in the population is $\sqrt{.36}$, or 0.6. Therefore, 60 percent of the total number of genes (T and t) in the population must be t. and the remaining 40 percent T.

Sperm \ Eggs	0.6 t	0.4 T
0.6 t	0.36 tt	0.24 Tt
0.4 T	0.24 Tt	0.16 TT

Genotype proportions: 0.16 TT, 0.48 Tt, 0.36 tt

are the frequencies of genes? What is the genotype of the population?

To find the frequency of **t,** simply take the square root of 0.36 ($\sqrt{0.36} = 0.6$). The frequency of **t** is 0.6; hence the frequency of **T** is 0.4 ($1 - 0.6$). We can express this frequency of **T** alleles to **t** alleles as 0.6 to 0.4. Now we can determine the genotypes of the population by using the checkerboard, as in Figure 10-5.

The Hardy-Weinberg Principle

You might think that recessive genes slowly disappear from a population. If this were true, the percent of nontasters of PTC would change as time goes on. But in 1908, two scientists independently showed that this is not so. One of these was G. H. Hardy, an English mathematician; the other was W. Weinberg, a German physician.

According to the Hardy-Weinberg principle, the frequencies of genes and the genotype ratios remain constant in the absence of forces that can alter the frequencies of genes. The conditions that must hold for this principle to operate are:

1. Gene frequencies are the same in males and females.
2. There is no unequal mutation rate among the alleles. (That is, one allele does not mutate more frequently than the other.)
3. The population breeds at random with respect to the trait in question. (That is, mates are not chosen on the basis of the trait. For example, human beings mate at random with respect to PTC tasting.)
4. No migration occurs into or out of the population that would change the frequencies of the alleles.
5. The population is large enough to minimize any effects of chance.
6. No selection occurs that favors one allele over the other or one phenotype over the others.

Point 6 is very important. In order for the Hardy-Weinberg principle to hold true, there must be no natural selection of one type of allele. If such a pressure does exist, gene frequencies will change, although not as much as you might expect. Suppose that the recessive homozygous condition were lethal. The organism carrying two recessives would not survive to reproduce. Then frequency of the recessive gene in the population might approach zero. But there would still be recessive genes in the heterozygous individuals. Homozygous recessive individuals would still be possible in the next generation. Of course, mutations also occur. Recessive genes probably are never eliminated from a population.

The Hardy-Weinberg principle provides us with a way to calculate the frequencies of alleles in a population and gives us a base for measuring evolutionary change in populations. If the Hardy-Weinberg frequencies are maintained, the genes concerned cannot be changing in the population. If mutation, natural selection, migration, or chance are at work, then change will take place. The population will evolve.

HOW POPULATIONS CHANGE

Geneticists and other biologists who study populations have found four factors that make the frequencies of alleles in a population change (Figure 10-6). These factors are:

1. *Mutation.* If one allele mutates less frequently than the other, the more stable allele will gradually become more common. The more mutable allele will gradually become less common.

2. *Natural selection.* One allele may give its carrier a greater chance to reach adulthood and reproduce than the other allele does. Then the frequency of the advantageous allele will increase, and the frequency of the harmful allele will decrease. The environment acts as the selective agent.

3. *Emigration and immigration.* Suppose the emigrants from or the immigrants into a population have a different frequency of alleles

10-6. Factors that can change the frequency of genes in the gene pool.

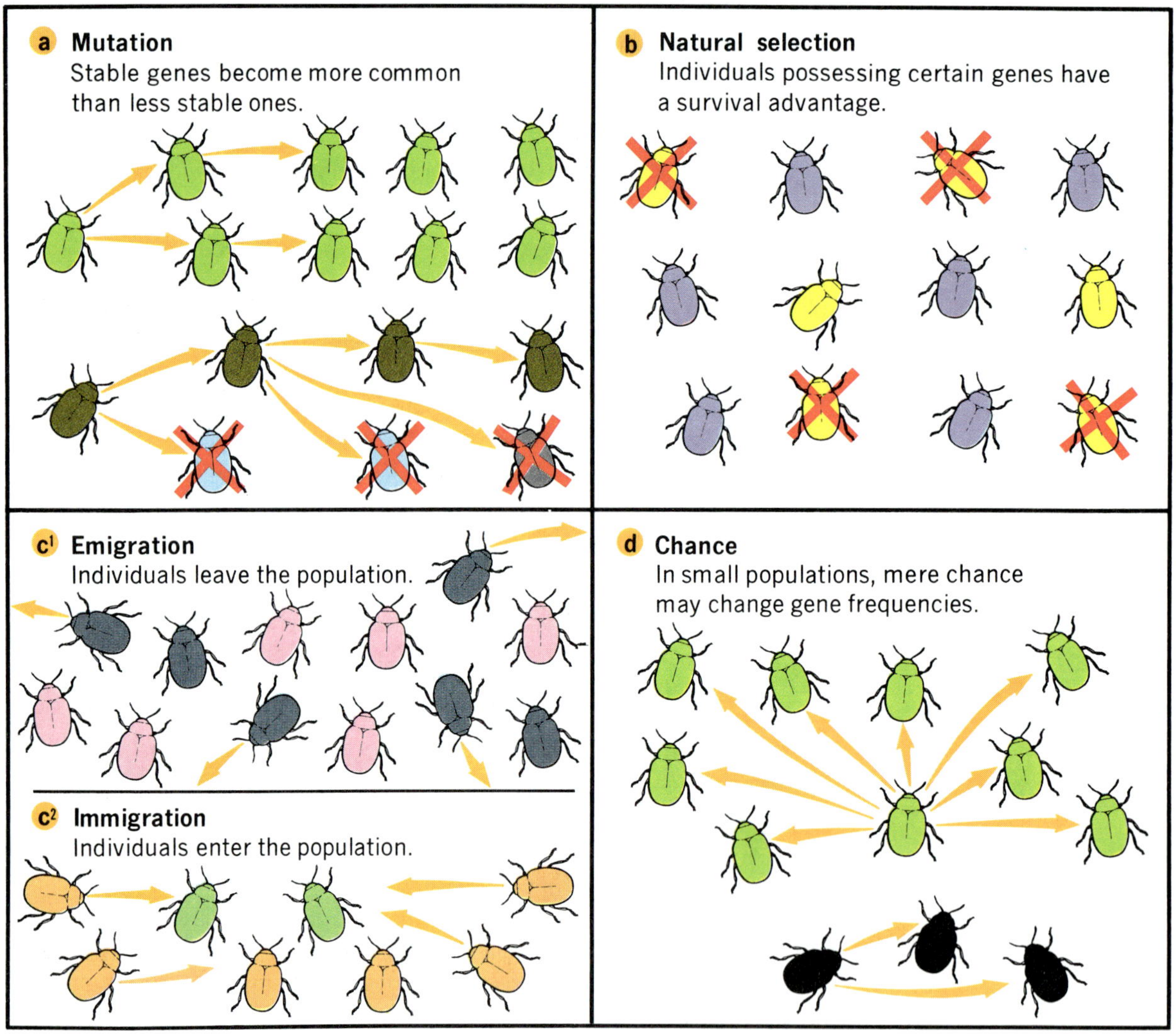

than the population. Then the frequency of alleles in the **gene pool**—all the genes in a population—will be changed.

4. *Chance.* When a population is very small and has few breeding individuals, chance alone may produce a change in the frequencies of the alleles.

You can gain direct experience with the Hardy-Weinberg principle and with factors that affect gene frequency by doing **Inquiry 10-1.**

INQUIRY 10-1 Population Genetics and Evolution

MATERIALS (per team)

can marked "gene pool"
can marked "eaten by predators"
red beans, at least 100
white beans, at least 100

Procedure (Day 1)

In the inquiry, beans are used to represent all the alleles for a particular trait in a population's gene pool. Red beans represent the dominant alleles **(A)** and white beans represent the recessive alleles **(a).**

Place 100 red beans and 100 white beans into the "gene pool" can. Without looking, one member of each team will pick 20 pairs of beans, at random, from the gene pool. Each pair of beans represents a diploid individual.

In this study, the genotypes of all individuals will be homozygous dominant (**AA,** two red beans), heterozygous (**Aa,** one red and one white bean), or homozygous recessive (**aa,** two white beans).

Copy Figure 10-7 and enter your data for the first trial of Generation 1. Record the number of **AA, Aa,** and **aa** individuals. Then determine the frequency of allele **A** and allele **a.** To do so, first count the number of red beans, representing **A** alleles, and the number of white beans, representing **a** alleles. Place these numbers in the proper column and also the total number of alleles (beans).

Now convert the number of each allele to a percentage. To determine the percent of **A** alleles, divide the number of **A** alleles by the total number of alleles. Similarly, find the percent of **a** alleles. In the last column, state these percents as a ratio to determine the gene frequency. **1.** The gene pool originally included 100 red beans and 100 white beans. Why have we assumed 25 **AA,** 50 **Aa,** and 25 **aa** in the original population? **2.** Why must the frequency of both genes added together always be 1?

Return all beans to the can and repeat the procedure twice more. Total the number of individuals for all three trials and compute the average gene frequency. **3.** How much difference is there between the gene frequencies of the first generation (average of all three trials) and those of the original population? **4.** Combine the data of all teams in the class and reconsider question 3. What happens to gene frequencies? Why?

Procedure (Day 2)

In this part of the inquiry, you will examine one of the evolutionary factors that may change gene frequencies in a population.

Begin again with 100 red beans and 100 white beans in the can. Select 20 pairs of beans as before and arrange the pairs in rows on the table. These pairs of beans represent 20 individuals with two alleles each.

Assume that the homozygous recessive individuals are white-colored moths, while those individuals with one or two dominant alleles are dark colored. Both types of moths live on trees with dark bark. Because the homozygous recessive individuals are not camouflaged (Figure 10-8a), they are found easily by predators and removed from the population. Therefore, these white moths cannot pass their genes on to the next generation.

Represent this selection process by removing all homozygous white individuals from your population of 20. Place these beans in the can marked "eaten by predators." **5.** You should not remove any of the heterozygous individuals. Why?

10-7. Data table for Inquiry 10-1 (Day 1)

	Genotype			Alleles			Percent of alleles		Gene Frequency (ratio of alleles)
	AA	*Aa*	*aa*	*A*	*a*	*total*	*A*	*a*	
Original population	25	50	25	100	100	200	50%	50%	0.5:0.5
Trial 1									
Trial 2									
Trial 3									
Average of three trials									

10-8 (a-b). There are two moths in each picture. *Which do you think is most likely to be eaten by predators?*

Copy Figure 10-9 and enter your data for Generation 1. (Be sure *not* to include the alleles "eaten by predators" in the number of alleles.) Remove and set aside the beans that are still in the gene-pool can. **6.** By removing all the beans left in the can, we are assuming that only the first generation offspring survive and reproduce. We do this as a matter of convenience only. What is wrong with this assumption?

Now suppose that the surviving individuals mate and reproduce. Count the white beans left in your population. For every white bean in your population, put two white beans in the can *plus* the one in your population. Do the same for the red beans. (In other words, the can will contain three times as many white beans and three times as

10-9. Data table for Inquiry 10-1 (Day 2)

	Genotype			**Alleles**			**Percent of alleles**		**Gene Frequency (ratio of alleles)**
	AA	*Aa*	*aa*	*A*	*a*	*total*	*A*	*a*	
Original population	25	50	25	100	100	200	50%	50%	0.5:0.5
Generation 1									
Generation 2									
Generation 3									
Generation 4									
Generation 5									

many red beans as you had in your first generation.) This tripling of the number of beans does not affect the *ratio* of red to white beans. It simply makes more beans available.

Each team will again draw 20 pairs of beans from the gene-pool can. These represent 20 diploid individuals of the second generation. Calculate gene frequencies, remove homozygous recessive individuals, and return the surviving individuals to the gene-pool can. Represent reproduction once again by tripling the number of each kind of bean put into the can. Continue until five generations are complete. **7.** What is the trend in the frequency of the recessive allele from the original to the fifth generation? **8.** What is the trend for the dominant allele? **9.** What effect do you conclude selection has on the gene frequencies of a population? **10.** Predict what would happen to the frequencies of alleles if there were a change in environment that caused the *light*-colored moths to be at an advantage (Figure 10-8b)? **11.** Combine data for all teams in the class. How do the gene frequencies calculated by your team compare with those computed by the entire class? What does this tell you about chance and population size with regard to gene frequency? **12.** Form a hypothesis about the relationship between gene frequency and selection in a population of organisms. **13.** How would you test this hypothesis?

Procedure (Day 3)

Natural selection, which you have already investigated, is only one of the four factors that may cause changes in gene frequencies. The other three are migration (either emigration or immigration), mutation, and chance (especially in very small populations). Using the beans as before, invent and test ways of demonstrating the effects of these factors on gene frequencies.

An Example of Evolution at Work

The Hardy-Weinberg principle shows that the frequencies of various alleles in a population do not change *of themselves*. However, over a long period of time changes usually do occur. Let's consider one example.

Two species of monkey flower are shown in Figure 10-10. One species is red. Its pollen-bearing structures are conspicuous *outside* the main part of the flower (Figure 10-10a). The other species is lavender, and the pollen-bearing parts are *inside* the flower (Figure 10-10b). Artificial crosses of the two species (Figure 10-10c) are possible. The hybrid is fertile, but only when artificially bred. The hybrid does not occur in nature. Why?

10-10. *How did the two species of monkey flower evolve? In this case, what were the main agents of natural selection?*

a The Scarlet monkey flower is pollinated by hummingbirds.

b The lavender Lewis monkey flower is pollinated by bumblebees.

c Hybrids can easily be produced artificially. However, the structure of the flower is such that neither bees nor hummingbirds can effect pollination.

The answer has to do with the animals that pollinate the flowers. To birds, red is the brightest part of the color spectrum. The flowers of the scarlet monkey flower are bright and conspicuous to hummingbirds. When the bird is sucking up the nectar at the bottom of the flower, its head feathers become coated with pollen. The pollen is borne on structures that protrude from the flower. When the hummingbird moves on, it carries pollen to the next flower. The lavender Lewis monkey flower is pollinated by bumblebees whose eyes are sensitive to violet and ultraviolet colors. The curved mouth parts of the bumblebee just fit the curve of the flower.

By artificial crosses in the garden, we can easily get all possible combinations of colors and flower shapes. But in nature we find only the original two combinations. Flowers with those particular combinations are pollinated most often, and they produce more seeds than do flowers with other combinations. We can say that mutations for pollen-bearing structures outside the flower are beneficial if combined with genes for red flowers. They are harmful, however, if combined with genes for purple flowers. But "beneficial" and "harmful" relate only to the presence or absence of hummingbirds or bumblebees.

Here is another example. In American zoos it has been easy to cross a polar bear with the giant Alaskan brown bear. The offspring (F_1) and their offspring in turn (F_2) are vigorous, fertile bears. However, they lack the adaptations of the polar bear for swimming and catching fish and seals in open water. The offspring do not have lots of fur and skin between their toes, and they lack the streamlined shape of the polar bear which helps it be such a good swimmer. The hybrid is not white, but has both brown and white hairs. On the other hand, the hybrid bears lack the size and color and habits that make the Alaskan bear so well suited to life on land. The Alaskan brown bear is a poor swimmer and gets its food by scooping salmon out of the shallows along the rivers. It also eats berries and small animals. The hybrid bears seem poorly adapted for such a life. They are in a sense misfits in nature, fit only for the zoo.

Human populations are also subject to the forces of evolution, as **Inquiry 10-2** will show.

INQUIRY 10-2 Sickle Cells and Selection

In 1910, James B. Herrick discovered sickle-shaped red blood cells in the blood of one of his patients. A short while after Herrick had written an account of this discovery, other doctors observed the same sickle-shaped cells. Today we know that the sickle-shaped red blood cells are due to a hemoglobin molecule that is slightly different from normal. This trait is determined genetically.

Sickle-cell anemia is diagnosed when a blood sample is placed in an atmosphere that has very little oxygen. The sickle-shaped deformity then occurs in the red blood cells. If the blood of the patient's parents is examined, less extreme deformities are found in the red blood cells of both parents. The parents usually have no anemia. However, some sickling of their red blood cells may occur if they go to high elevations or engage in prolonged physical exertion. This mild sickling without anemia is called sickle-cell trait.

Examine Figure 10-11. **1.** How many phenotypes are there in the family? What are they? **2.** If only one gene with two alleles (**S** representing the allele for normal hemoglobin and **S**′ representing the allele for sickle hemoglobin) is involved, how many genotypes might you expect? What are they? **3.** In the pedigree, what is the genotype of individuals 1 and 2?

Notice that individual 3 has very severe sickling. Many people with severe sickling die in infancy or childhood. **4.** Would you expect individual 3 to have genotype **SS, S′S′**, or **SS′?** **5.** What genotype would you expect in individuals 4 and 5? in individuals 6 through 11 and 13 through 15? **6.** If individual 16 marries a person with a genotype like her own, what is the probability of their children having sickle-cell anemia? sickle-cell trait? normal red blood cells?

In some populations of black Africans, the sickle-cell trait may be found in as many as 40 percent of the individuals. The gene is not found in most white populations. If 4/10 of the adult population is heterozygous **(SS′)** for sickle-cell anemia, the chance of a heterogyzote mating with a heterozygote is $4/10 \times 4/10 = 16/100$. Thus, approximately 1/6 of the matings would involve two heterozygotes. **7.** What fraction of the offspring in such a population is expected to be homozygous for the sickle gene? **8.** Would you expect individuals with sickle-cell trait **(SS′)** to leave, on the average, more or fewer surviving descendants than would individuals with normal red blood cells? **9.** Would the trend you cited in

10-11. A PEDIGREE FOR SICKLE-CELL ANEMIA

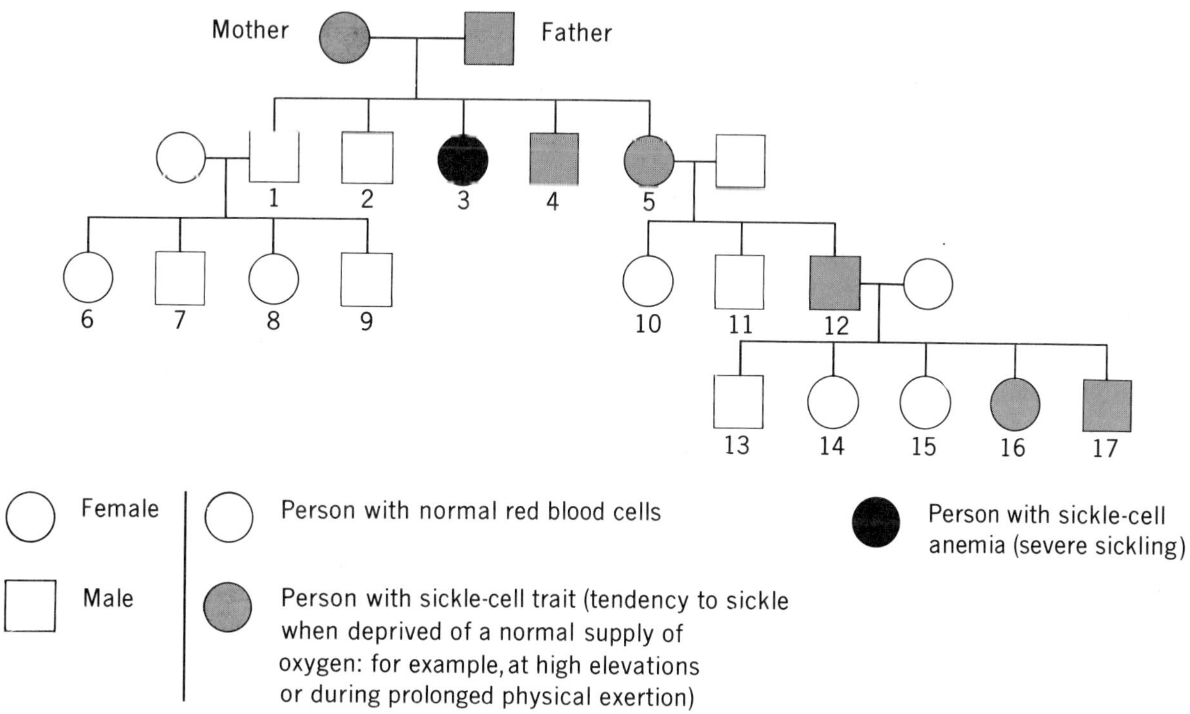

response to question 8 increase, reduce, or have no effect on the total number of genes for sickling in the population?

There is *no* evidence that the frequency of the gene for sickling is being reduced in the African populations. The question arises, how can the gene for sickling be maintained at such a high level while being selected against so strongly?

Place yourself in the position of the investigator. What hypotheses can you think of that might answer the question? Scientists have considered at least three hypotheses to account for the frequency of sickle hemoglobin in Africa. One involves the mutation rate of the gene. The second pertains to differences in fertility between people with the sickle-cell trait and those without. The third involves susceptibility to other diseases. **10.** Form a hypothesis involving mutation rate to explain the commonness of the sickle-cell allele in Africa. **11.** Form another hypothesis involving differences in fertility. **12.** Form another involving susceptibility to disease.

It has been estimated that the mutation rate necessary to replace the genes lost by selection would be about a hundred times as high as the known rate for other human genes. There is no direct evidence of such a high mutation rate for any gene in humans. **13.** Does this information support or weaken your first hypothesis?

There is evidence that individuals with the sickle-cell trait do not produce more children than nonsicklers do. **14.** Is your second hypothesis supported or weakened by this fact?

The third hypothesis is more difficult to test. The incidence of the trait has been plotted on maps. The gene is most common in a belt that extends across the central portion of Africa. Malaria is very common in this region. Recall briefly from Chapter 1 some characteristics of malaria, including its symptoms, how it is transmitted, and how its symptoms are produced. Individuals with the sickle-cell trait have hemoglobin that differs from normal hemoglobin. **15.** How might the sickle-cell trait influence the incidence of malaria?

This last question suggests a new hypothesis. To test this hypothesis, one investigator examined the blood of 290 children in an East African population. Both malaria and sickle-cell traits were common in the area. The results are tabulated below. **16.** What hypothesis do these data suggest?

	With Malaria	*Without Malaria*	*Total*
With sickle-cell trait	12 (27.9%)	31 (72.1%)	43 (100%)
Normal	113 (45.7%)	134 (54.3%)	247 (100%)

To test this hypothesis, 30 volunteers were inoculated with malaria parasites. The volunteers were adult males of about the same age and physical condition. None of the 30 had malaria parasites at the beginning of the experiment. Of the group, 15 had the sickle-cell trait and 15 had normal red blood cells. Two of the volunteers with sickle-cell trait and 14 of the volunteers without the trait developed malaria. When the observations were completed, the individuals who developed malaria were treated successfully to stop the disease.

17. How do these results support the hypothesis developed in question 12? **18.** How do the preceding observations show that, under certain conditions, the heterozygote may have a selective advantage over both the normal homozygotes and the homozygotes with sickle-cell anemia? **19.** How can a gene that is harmful under certain conditions be beneficial to a population under certain other conditions?

The incidence of sickle-cell trait in the early American black population (whose ancestors were African) is estimated to have been about 22 percent. In the United States, today, we have nearly eliminated the mosquito that transmits malaria. **20.** What survival advantage, if any, would the heterozygote have under these conditions? Recall that often an individual who is homozygous for the sickle-cell trait dies before reaching reproductive age. **21.** Predict what may be happening to the frequency of the allele for sickle cells in the United States? Explain. **22.** The incidence of sickle-cell trait in the present American black population is about 9 percent. How does this fact correlate with your interpretation in question 21?

THE ORIGIN OF SPECIES

When we speak of the evolution of new species, we are not referring only to the differences that occur. The domestic dog has evolved, largely through human selection. We now have many different breeds (Figure 10-12). We do not consider these breeds separate species because they usually can mate and produce fertile offspring. A new species must not interbreed with any other species to a significant extent, or its distinctness may disappear. It must be isolated reproductively.

In nature, the evolution of new species often occurs when part of a population becomes cut off from the main body of the population. Darwin noticed this in the Galápagos Islands. The original species of finches spread to many of the islands. Then in isolation on each island, new species evolved.

10-12. The differences among some of these domesticated dogs are far greater than differences between domesticated dogs and their wild relatives such as wolves and coyotes. *Why are the dogs considered to be of one species?*

Genetic Isolation

The same process can occur on continents as well and may give rise to a *ring of races.* For example, a salamander that is light and rather uniform in color lives in the coastal area of southern California. A darker race (Figure 10-13) of the same species lives farther north. In northern California a still darker, uniformly colored type ranges beyond the coastal regions into the interior of the state. In the interior, toward the southeast, the dark type is replaced by a mottled form. Farther south, in the Sierra Nevadas, across the hot, dry valley from the coastal habitat, this mottled form is replaced by a different mottled form. Finally the ring is completed by another type of boldly mottled salamander, which overlaps the range of the even-colored form in southern California.

Suppose we call these races, or subspecies, *a, b, c, d, e,* and *f,* in the order of the geographic distribution described above. Where they meet, *a* and *b* interbreed. Similarly, *b* and *c* interbreed. Also *c* interbreeds with *d, d* with *e,* and *e* with *f. Even though they overlap in territory, however,* a *and* f *do not interbreed—or do so only rarely.* If some natural catastrophe should destroy one of the subspecies *b* through *e,* the ring would break up into different species. They would be isolated genetically.

Even if two populations look alike, we assign them to different species if they do not breed with each other and produce vigorous, fertile offspring in nature. For instance, the coyote looks much like a mongrel dog. Hybrids between coyotes and dogs are rare. When such matings between the two species do take place, the male F_1 hybrids are sterile. Thus, dogs and coyotes can live side by side and remain distinct from each other.

Different courtship patterns also may isolate species. A male grouse, or woodcock, performs an elaborate dance in front of a female before mating takes place. This dance apparently stimulates her to release her egg cells. Since females are most strongly stimulated by the dance of males of their own species, there is little mating with males of other species.

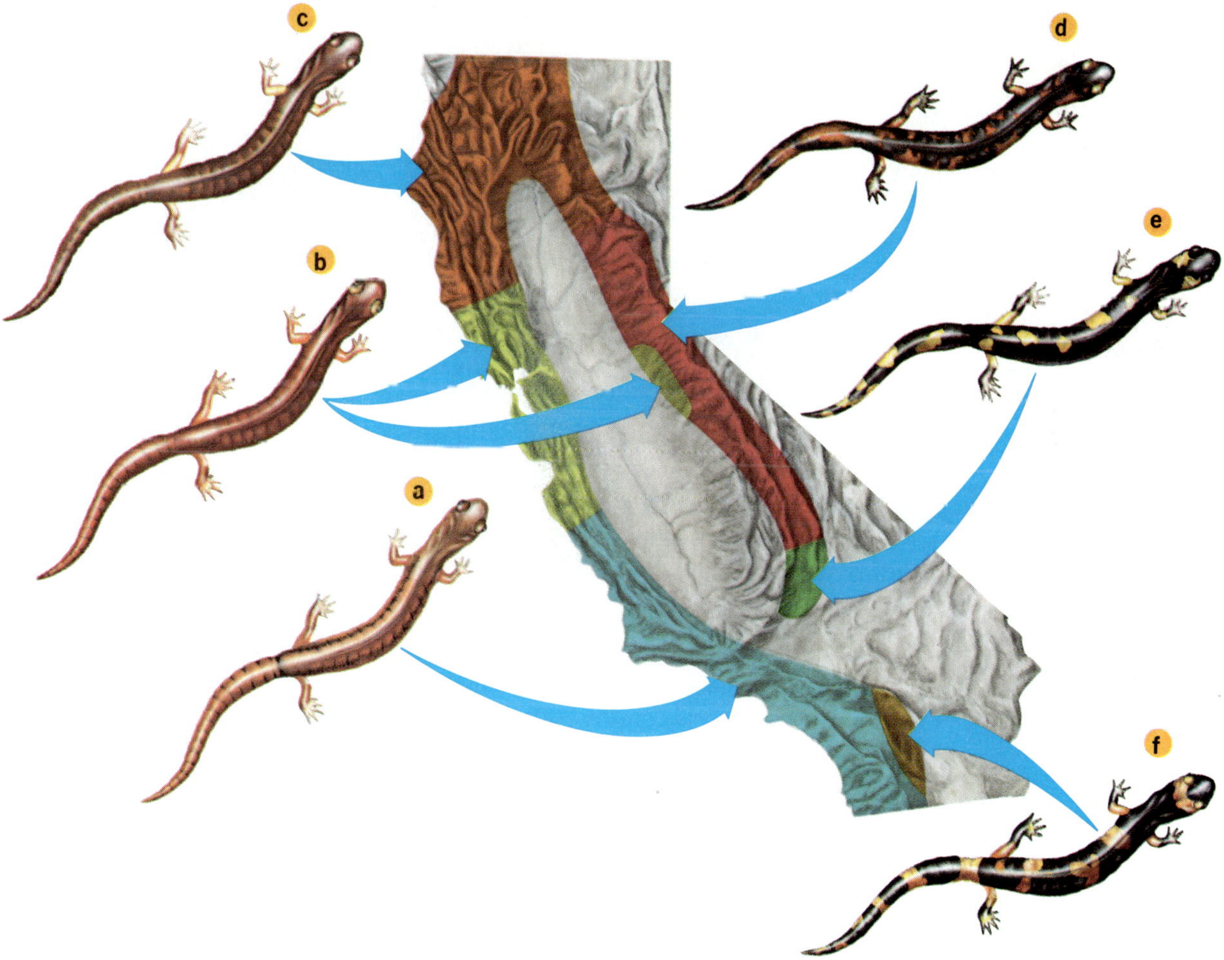

10-13. Geographic variation. In California, populations of these salamanders are distributed around the borders of the state but not the center.

SCIENCE AT WORK: A Living Fossil

Much of our knowledge of evolution is based on the discovery of thousands of fossils in all parts of the world. From bits and pieces of fossilized plants and animals, scientists have been able to reconstruct what early organisms probably looked like.

In 1938, something happened that gave scientists unexpected information about evolution. A fishing boat arrived at the port town of East London, South Africa. Courtenay Latimer, the curator of the local museum, went to the docks to search among the fish being unloaded. She was looking for interesting specimens for the museum's collection.

On this particular visit, she found a heavy, dirty, oily fish that she was not able to identify. The fish looked very much like the ancient fish that were the ancestors of modern amphibians such as frogs. But it couldn't be – the fossil record showed that these creatures had been extinct for 70 million years. Latimer sent sketches to several well-known ichthyologists (scientists who study fish). She was astounded, along with the whole scientific community, to receive the news: The coelacanth (seal·uh·kanth), lobe-finned ancestor to four-legged animals, was not extinct.

Thus, a museum curator from a small city in Africa had found a "living fossil." For a long time, it looked like this fish was to be one of a kind. Then, in 1952, a second coelacanth was caught near the Comoro Islands off the coast of East Africa. Since then, more than 80 coelacanths have been pulled from the Indian Ocean. None of them has lived.

The big prize being sought after today is a living coelacanth. Researchers would like to study its swimming motions, analyze its chromosomes, and get fresh tissues for a variety of studies. Aquariums around the world would like to exhibit a coelacanth. As the superintendent of the Steinhart Aquarium in San Francisco said, "It would be like having a dinosaur in your zoo."

Can mutation, recombination, selection, migration, chance, and barriers to cross-breeding explain the major trends of evolution? Can they explain, for example, the development of catlike and doglike animals from earlier ancestors, or the evolution of the horse from its primitive ancestors? The mechanisms that govern these major trends of evolution cannot be studied directly. They took place thousands or even millions of years ago. Nevertheless, a study of living populations and of fossils provides strong evidence that the same evolutionary forces that we see in operation today also have guided evolution in the past. One species may change gradually into another, quite different, species. Or one species may split into two or more species. The new species may continue to evolve. If this process operates, we would then classify the species in very different groups, like dogs and cats, even though they may have shared a common ancestry.

Polyploid Species

In most groups of animals and in many plants, the formation of new species is a long process sometimes requiring hundreds of thousands of years. But there is one way in which new species can originate almost instantly. A cross between two species may be *followed by doubling of the chromosome number of the hybrid.* In 1928, a Russian geneticist, Karpechenko (kar·PAYCH·en·ko), crossed a radish with a cabbage (Figure 10-14). The resulting plant was large and vigorous—but it was sterile. Karpechenko studied this plant and found that the chromosomes did not pair in meiosis. What was the reason for this? The nine chromosomes that came from the radish parent are not homologous to the nine that came from the cabbage parent.

Karpechenko found a way, however, to produce a fertile hybrid. Heat treatment caused meiosis to fail in some cases, and the number of chromosomes was not reduced. Thus, quite rarely, pollen grains and egg cells were produced that had 18 chromosomes (nine of the radish parent and nine of the cabbage parent). When sex cells of this type united, they produced a plant with 36 instead of 18 chromosomes. Since half of the chromosomes of this plant were from the radish and half from the cabbage, the plant was like the sterile hybrid. But when meiosis took place in the 36-chromosome plant, the 18 chromosomes from the radish paired with each other, and the 18 chromosomes from the cabbage paired. Thus, meiosis was normal, and each sex cell received 18 chromosomes. Offspring formed from the union of these sex cells are fertile. They breed true for an intermediate phenotype like that of the F_1 hybrid. Since the hybrid has more chromosomes than the plants that produced it—

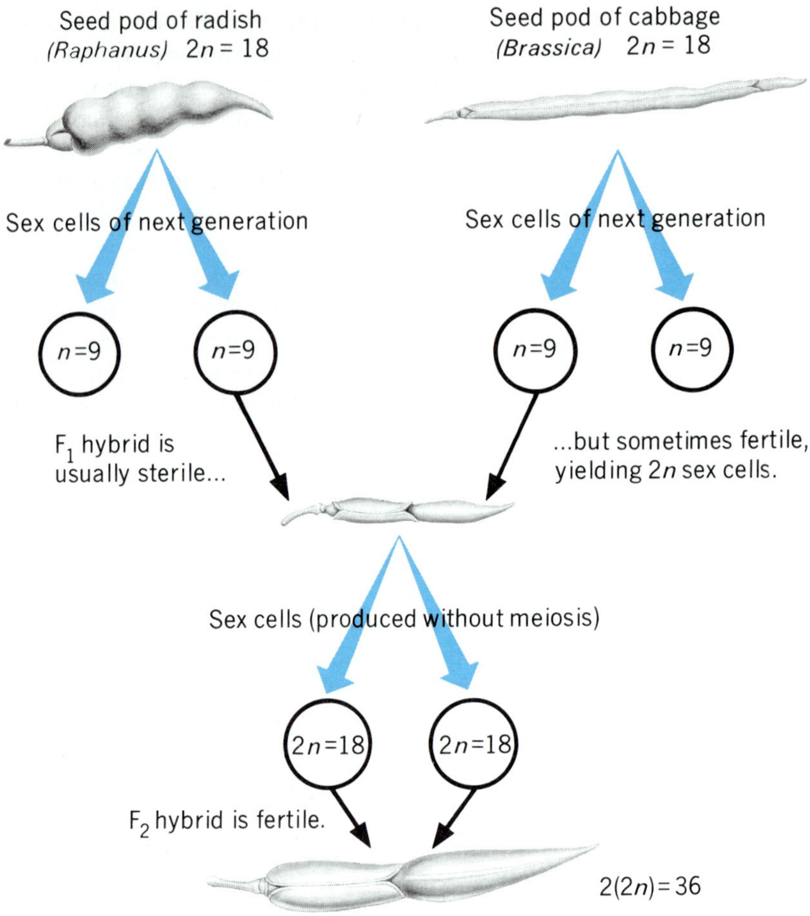

10-14. Karpechenko's cross of a radish with a cabbage. Meiosis in the F_1 hybrid was nearly always abnormal, producing defective sex cells. After heat treatment, meiosis sometimes failed altogether, and there was no reduction of chromosome number. The sex cells had 9 radish chromosomes and 9 cabbage chromosomes. When two such cells united, the F_2 had 18 radish chromosomes and 18 cabbage chromosomes. In this plant, meiosis was normal.

actually two diploid sets—it is said to be **polyploid** (POL·ih·ploid).

When the polyploid hybrid is crossed with either a radish or a cabbage, the offspring are sterile. The polyploid hybrid has all of the requirements of a distinct species. It looks different from other species, including its original parents. It is fertile and breeds true. And it forms sterile hybrids when crossed with any other species, including its own ancestors. Many wild species of plants are polyploids. Polyploids in animals are rare. Can you suggest why?

Karpechenko's polyploid species was the first species to be made artificially. Since then, many doubled hybrids have been made in various groups of higher plants. Scientists also learned that many well-known species are doubled hybrids that occurred in nature. Examples are cultivated wheat, cotton, and tobacco; garden flowers such as the marigold and larkspur; and wild species such as the blue flat iris and the Jerusalem artichoke.

Chromosome numbers also can be increased using chemicals. For example, *colchicine* (KOL·chih·seen) is a drug obtained from the yellow crocus. It is very effective in doubling the chromosome number of plant cells. Colchicine destroys the spindle on which the chromosomes normally separate into two groups. Biologists have produced many artificial polyploid species in this way.

Human Influence on Evolution

We generally think of evolution as occurring only in nature, over millions of years. But humans have been a force in the evolution of many species. We have altered natural selection by changing the environment. Long ago, people burned prairies or forests and plowed and planted fields. These activities upset the natural plant cover of the land and allowed weeds to grow rapidly along with the crops. Later, humans began to breed crop plants and animals for improved characteristics. The resulting populations were not better adapted to the natural environment; they were simply better suited to human needs.

One of the greatest successes in the development of human civilization occurred when people first noticed the superior size and yield of certain plants. Scientists now know that these are polyploid plants. When polyploid wheats and fruits occurred in the Old World and polyploid cottons and tobacco in the New World, people were able to select and breed them. Originally, corn was just a weedy grass. The Indians saw a few plants with superior qualities, such as keeping the seeds on the cob instead of shedding them when they were ripe. Gradually, over thousands of years, the weedy grass was cultured into maize. Rice, in the Orient, underwent a similar change. Along with wheat, these three grains feed most of the world's human populations today.

Genetic and evolutionary principles also have been applied to improve plants. One problem with cultivated tomato plants is wilting caused by the fungus *Fusarium* (few·SAY·rih·um). A gene for resistance to *Fusarium* wilt is present in a wild tomato species from Peru. This wild species, however, has poor fruit. A cross between the cultivated tomato and the wild one produced a hybrid resistant to *Fusarium*. Because the hybrid did not have the desirable taste of cultivated tomatoes, repeated crosses of the hybrid plants with cultivated plants were made in subsequent generations. As a result of introducing the gene for resistance to wilt into the best cultivated varieties in subsequent generations, a tasty *and* resistant tomato was obtained.

Early people recognized the importance of reproductive isolation. They kept the desired lines of domestic plants and animals separate by preventing the desirable types from breeding with less desirable types.

American cattle breeders now develop special traits in domestic cattle. The meat of shorthorn cattle is good, but their skin is thin and they are susceptible to certain diseases carried by insects. Brahman cattle, on the other hand, have thick skin that insects have difficulty biting through. But Brahman cattle do not produce very good meat. Breeders crossed American shorthorn cattle with Brahman cattle. From among the offspring, animals with the desired characteristics—good meat and thick skin—were selected. After repeated crosses, a strain that breeds true was developed. The result was a beautiful breed of cattle known as the Santa Gertrudis (ger·TRUE·dis), shown in Figure 10-15.

10-15. The Santa Gertrudis **(c)** was produced through artificial selection of the offspring of shorthorns **(a)** mated with Brahman **(b)** cattle. The Santa Gertrudis has the desirable characteristics of both the parent types.

a

b

c

Another example of how human activities have changed the course of natural selection is the peppered moth of England (Figure 10-8). When the Industrial Revolution began in the late eighteenth century, many factories were built. Soot and smoke from the factories gradually darkened tree trunks so they became gray or even black. By the middle of the nineteenth century, collectors of moths and butterflies began to notice a change in the color of the peppered moths. A previously rare type of peppered moth, one that was black, was becoming more and more abundant. Finally, this moth became common in the industrial areas of Britain. Elsewhere, the peppered moth still retained its light color.

In the twentieth century, a British scientist, H. B. D. Kettlewell, studied the peppered moth to learn why the solid black type was replacing the light type. The results of genetic crosses showed that the dark color is caused by one mutated gene. In the soot-darkened woods of England, birds ate more light-colored moths than dark ones. In areas away from the factories, the reverse was true.

In Brief

Mutation, recombination, selection, migration, and chance, combined with geographic isolation, are the principal mechanisms of evolution in animals and plants. Far from being of theoretical value only, a knowledge of the principles of evolution is helping us to develop better domestic animals and plants.

Guide Questions and Problems

1. Genes vary in mutation rates. What might be the result if a species of organisms had many genes that mutated often? Or, if many genes never mutated or did so at a very low rate?
2. What are some factors that can cause changes in the frequencies of genes in a population?
3. How may some gene mutations be harmful or beneficial, depending on special circumstances? Give an example to explain your response.
4. How might a population become divided into two or more reproductively isolated populations? What factors may determine whether such isolation eventually leads to a separate species?
5. The appearance and behavior of a house cat are very different from those of its wild relatives. Propose a scheme that explains the differences.

6. Why can we say that human beings are agents of "natural" selection? Give an example to illustrate your response.

7. Why is it not possible for us to study the evolution of species and larger groups "directly"? What do we mean by "indirect evidence"?

Ideas for Independent Study

1. Mutation rates are increased by exposure to high-energy radiations, such as X rays and nuclear radiation. Of course, people have always been exposed to some radiation (for instance, from naturally radioactive elements found in certain kinds of stone). In the last few decades, however, this exposure has increased greatly. X rays and radioactive materials are now used more often in industry, science, or weapons. Find out about new sources of radiation (for example, microwaves; medical and dental X-ray machines; commercial uses of X-ray machines; radioactive minerals, and byproducts of uranium mining and refining; and nuclear power plants). How might exposure to these sources of radiation affect human heredity? A place to start might be the chapter on "The Unseen Enemy: Infection and X-Rays" by Aubrey Milunsky, *Know Your Genes,* Boston: Houghton Mifflin, 1977, pages 113–120.

2. There are many practical applications of our knowledge about genetic change in plants and animals. An interesting case is the development of a variety of corn that has a special pattern of amino acids in its kernels. How has this variety been developed, and how has it affected the people who eat it? Dale D. Harpstead described this case in "High-Lysine Corn" in *Scientific American,* August 1971, page 34. Read this account and try to predict what long-range effects may result from such developments.

3. People have greatly modified the characteristics of various species of domestic animals and plants. Some of these are cattle, horses, sheep, pigs, dogs, cats, corn, wheat, and orange and apple trees. Select one of these species. Find out as much as you can about its origins. Try to determine how that species has changed under human influence.

What problems would be faced by humans and members of domesticated species if they were forced to live in the wild—away from the human environment? Look into specific cases where individuals or species have escaped or become isolated from humans. What changes have they shown as they adapted to life in the wild?

Related Readings

BOOKS

Gastonguay, P. R. *Evolution for Everyone.* Indianapolis: Pegasus, 1974.

Hamilton, T. H. *Process and Pattern in Evolution.* New York: Macmillan, 1967.

Kelly, M. G. and J. C. McGrath. *Evolution and Adaptation to the Environment.* Boston: Houghton Mifflin, 1975.

Lack, D. *Evolution Illustrated by Waterfowl.* New York: Harper & Row, 1974.

Olson, E. C. and J. Robinson. *Concepts of Evolution.* Columbus, Ohio: Charles E. Merrill, 1975.

Wallace, B. *Population Genetics.* A BSCS Pamphlet. Boulder, Co.: BSCS, 1964.

ARTICLES

Bakker, R. T. "Dinosaur Renaissance." *Scientific American,* April 1975, page 58.

Cavalli-Sforza, L. L. "The Genetics of Human Populations." *Scientific American,* September 1974, page 80.

Harpstead, D. D. "High-Lysine Corn." *Scientific American,* August 1971, page 34.

11 Human Evolution

The scientists and citizens who, in 1859, read Darwin's *On the Origin of Species* must have wondered, "Does this apply to me?" If other organisms have evolved, what about the human species? Are we the result of a long series of evolutionary changes? If so, what were our ancestors like? Where did they live? How did they live?

THE SEARCH FOR OUR ORIGINS

In 1859 there was no direct evidence for human evolution. It was known, however, that Europeans had not always lived as they did in the 1800s. The earliest known people in Europe used tools of stone, wood, and bone. They hunted for their food and lived in simple shelters. Later, people learned how to raise some plants for food.

Agriculture was developed. People learned to remove metals from the earth for tools, weapons, and ornaments. Obviously, there had been cultural change. Was biological evolution possible, as well?

Some evidence of human evolution came from comparisons of humans with other animals. Long before Darwin's time, biologists classified animals according to similarities. Humans are included, with apes and monkeys, in a large group called primates (PRY·mates). All the primates share certain characteristics. Darwin suggested that members of a group are similar because they evolved from a common ancestor. But, at that time, there was little evidence that supported or refuted this hypothesis. To find out what *did* happen in the past, we must search for the fossil remains of our ancestors.

Australopithecus

Many different fossil primates have been discovered in Africa. One of the first was named *Australopithecus* (aus·tray·loh·PITH·eh·kuss). Australopithecines lived in Africa perhaps five million years ago. These primates probably were not an evolutionary link between humans and their nonhuman ancestors. Except in the lower jaw and teeth, the skulls (Figure 11-1) of the australopithecines resemble those of modern apes. Australopithecines may have walked erect.

Recently, other fossil remains have been found in some of the same deposits as australopithecines. One skull, found in Kenya, is

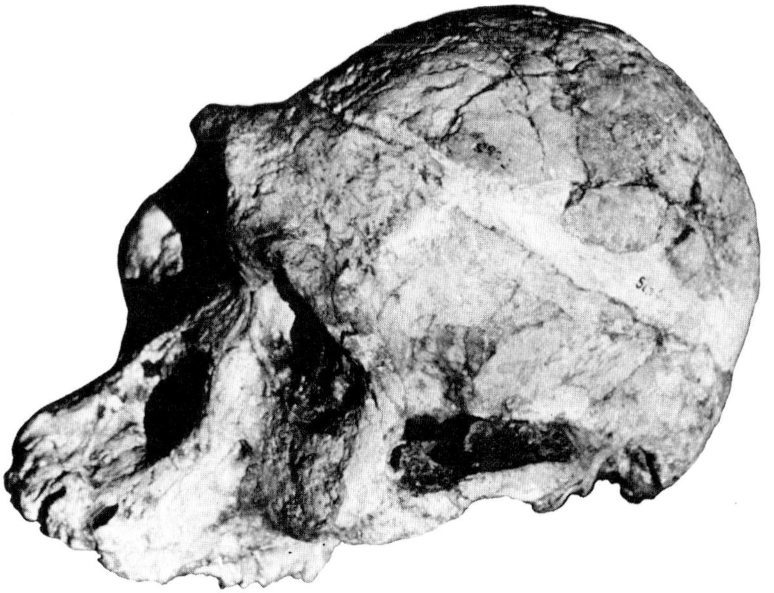

11-1. *Australopithecus.* From Sterkfontein, South Africa.

clearly human in size and is very ancient. Because primitive stone tools also were found in these deposits, scientists believe that humans and australopithecines lived at the same time and in the same areas. Perhaps they competed for food and shelter. The humans won. The australopithecines became extinct.

How long ago was this? It is not always easy to date sedimentary rock deposits. Some of the evidence, however, makes it possible to estimate that the first humans existed in East Africa about three million years ago.

The Discovery of Pithecanthropus

Several years after Darwin's death, a Dutch anatomist, Eugene Dubois (diu·BWAH), began to search for fossil remains of early humans. He started digging along the Solo River in Java. The deposits in this area are believed to be nearly one million years old. Dubois unearthed a small piece of a human jawbone, several teeth similar to those of apes, and part of a skull. A study of the skull suggested that the brain was too big to belong to an ape. But the brain was too small for any known human. Later Dubois found a fossil femur (thighbone) that was straight. Humans are the only erect primates, and their femurs are nearly straight. The fossil femur, however, might not have come from the same individual that the skull came from. Nevertheless, these fossils from Java seemed to be of a primate that lived long ago. Dubois named this fossil primate *Pithecanthropus erectus* (pith·eh·KAN·thro·pus ee·REK·tus). In recent years, biologists have reclassified the Java fossils as human. We now call this form *Homo erectus*. Remains of this extinct species have since been found in other parts of Asia and in Africa and Europe.

In 1924, the skull of a form commonly called "Peking man" (Figure 11-2) was discovered in China. Peking man also is classified as *Homo erectus*. During the next 12 years, parts of more than 40 similar individuals were dug up from a cave near Peking, China. A few crude flint tools had been found near the fossils in Java, but in China a whole array of chopping tools was found. These included rough stones chipped to an edge on one end, somewhat like a modern chisel. Charcoal was found nearby, indicating that this *Homo erectus* used fire.

In recent years, more than a hundred of these fossils have been found by Chinese scientists in the caves west of Peking. Every part of the body can now be fitted into place. The total picture is that of a type much like modern humans, except in the head and jaws. Large ridges of bone over the eyes, a massive jaw, and a receding chin are the most distinctive features. This type walked erect. Its body build

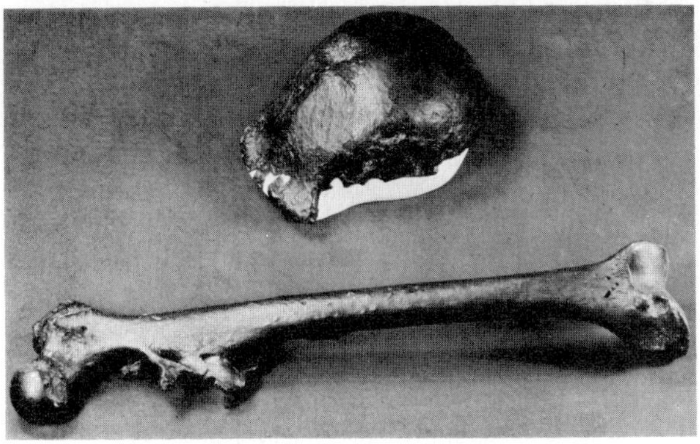

11-2. *Homo erectus.* The upper part of a skull and a thighbone **(top left)** are two of the specimens found in Java. The bust **(top right)** suggests the probable appearance. A skull **(left)** of *Homo erectus* from China.

was chunky and short. Individuals varied considerably, as might be expected. In general, the brain size was a bit smaller than that of people today.

The Neanderthal People

In one of the strange accidents of history, a fossil human was discovered just a few years before Darwin published *The Origin of Species.* In 1856, an unusual skull was discovered in the valley of the Neander River in Germany. It has come to be known as the Neanderthal skull (*tal* means "valley" in German). The skull was much like that of modern humans. The bones of the top of the skull, however, were very thick. The teeth were large, and there were heavy ridges over the eyes (Figure 11-3). Many years later, scientists determined that the skull belongs to a race that lived in Europe during the Ice Age, nearly 100,000 years ago.

The Neanderthals were not as ancient as *Homo erectus,* which existed at least 700,000 and possibly 1,000,000 years ago. The Neanderthals appeared only about 100,000 years ago. They differed from modern people, but not greatly. Skeletal remains show brain size was, on the average, about the same as ours. They were efficient makers and users of tools (Figure 11-4). Many students of human evolution include them in our own species, *Homo sapiens.*

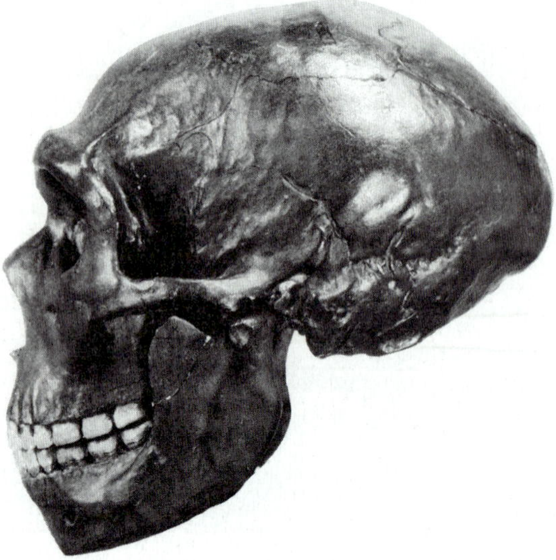

11-3. A fossil skull **(left)** and a reconstruction of what a Neanderthal probably looked like. Neanderthal remains and tools have been found in Europe, North Africa, Gibraltar, and the Near East. The skull shown here is an unusually large one (1,600 ml cranial capacity).

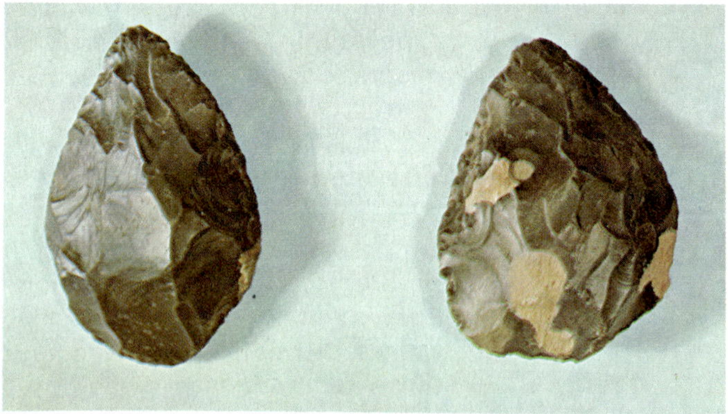

11-4. (Top) Neanderthal handaxes. **(Bottom)** Neanderthal hide-scrapers and pointed hunting implements.

SCIENCE AT WORK: Cooperating Scientists

Some people believe that scientific work is very secretive, that laboratories or field work stations are "off-limits" to everyone but the scientist. It is true that some types of government-sponsored projects and dangerous projects have tight security regulations. However, a great deal of scientific research depends on an atmosphere of openness. This is particularly true when the success of the investigation depends on cooperation between the scientists involved in the work.

A good example of this was the joint appearance of two archaeologists, Richard E. Leakey and Donald Carl Johanson at a news conference in 1977. Leakey announced the discovery, at the famous Koobi Fora site in Tanzania, Africa, of a complete *Homo erectus* skull that had been reliably dated at 1.5 to 1.8 million years old. Leakey announced the unearthing of two other highly significant fossils. One of these was a skull –2.5 to 3 million years old–from what may have been an earlier form of *Homo.* Johanson also had collected more than 150 fossilized bones from the Hadar region of Ethiopia. These were dated at 3.5 million years and had the characteristics of *Homo.*

This news conference was significant in several ways. It moved back *Homo erectus*, a major fossil on the road of human evolution, from 700,000 years old to possibly as much as three million years old. Two scientists, who had sometimes been seen as competitors in the search for humanity's history, cooperated in a major scientific effort. It also showed that a scientific discovery must be presented to the scientific community and to society as a whole. Until this happens, the discovery is not a part of the storehouse of scientific knowledge.

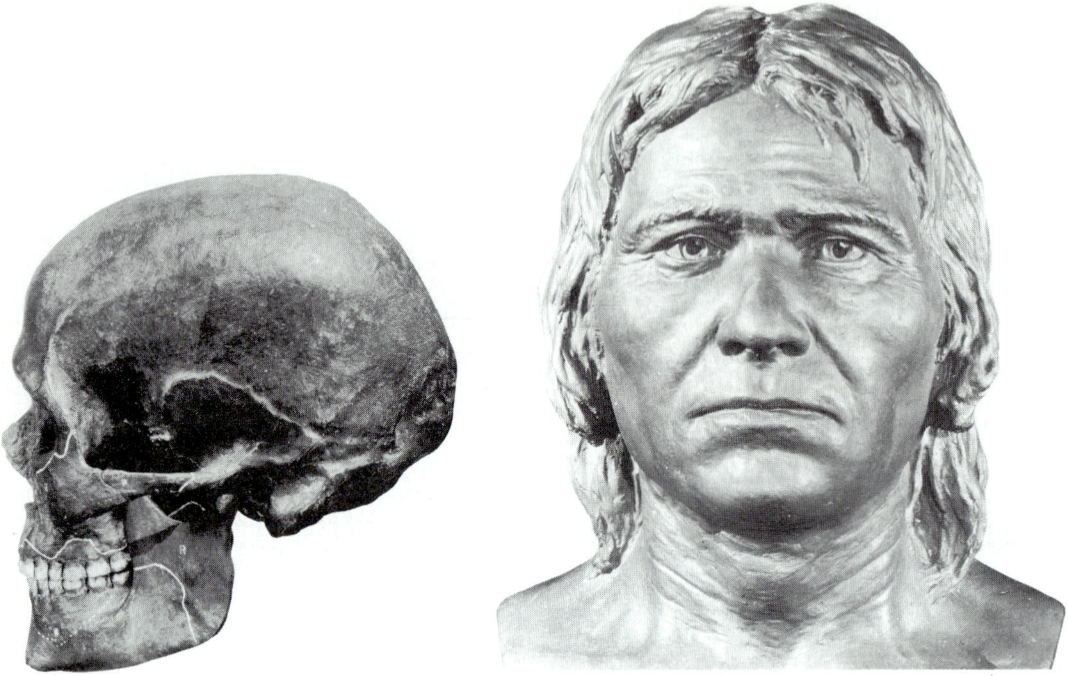

11-5. A skull of Cro-Magnon **(left)** and an attempt to portray appearance. The Cro-Magnons differed from the people of today only in minor respects. They made beautiful tools, were skillful hunters, and left us some remarkable paintings (Figure 11-6).

11-6. The dawn of painting. Caves in France and Spain contain examples of the oldest known paintings, done by Cro-Magnons as early as 20,000 years or more ago. This is from Lascaux in France. Many of the animals portrayed no longer are found in Europe. *How do you explain this?*

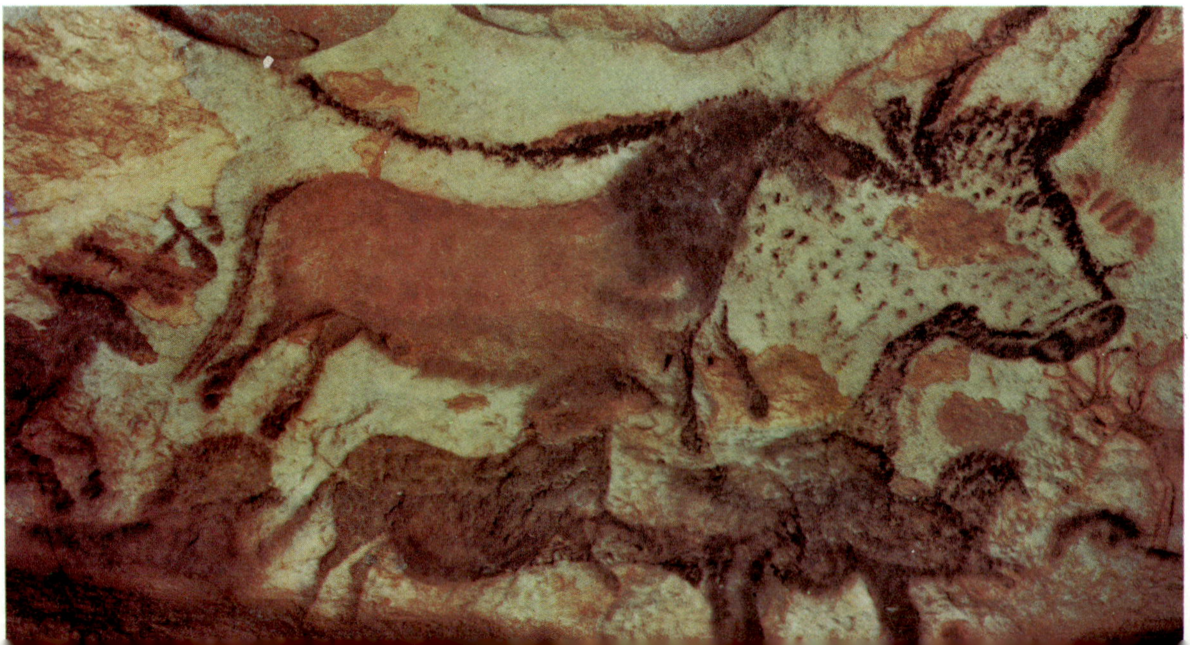

11-7. The search for evidence of our ancestors is going on all over the world. At a site in western Egypt, a team of scientists headed by Dr. Elwyn Simons of Yale is working in deposits that are about 40 million years old.

11-8. The lower jaw bone of an early monkey *Parapithecus.* A brush is being used to remove the sand from the very delicate specimen.

Cro-Magnon People

We do not know what happened to the Neanderthals. But we do know they disappeared before the end of the Ice Age. This was perhaps 35,000 years ago. Did the Neanderthals become extinct, or did they simply interbreed with the Cro-Magnons (crow·MAN·yons; Figure 11-5), who developed the peak of Stone Age culture?

The Cro-Magnons lived in a world inhabited by woolly mammoths, woolly rhinoceroses, cave bears, wolves, bison, reindeer, wild horses, and other large mammals. They hunted these animals, and they drew superb pictures of them. Cro-Magnon drawings have been found deep within caves in southern France and northern Spain and on the walls of cliffs in North Africa (Figure 11-6). The Cro-Magnons carved tools and ornaments from bone and ivory, and they made elegant arrow and spear points from finely chipped stone. They were strong, intelligent people of large stature, who made the best of their Ice Age climate and resources.

The fossils of human ancestors discovered so far are still rather few in number. Figures 11-7 through 11-10 show some of the scientists who are searching for our ancestors and what they are discovering. New finds and research continue to add to our knowledge of human evolution.

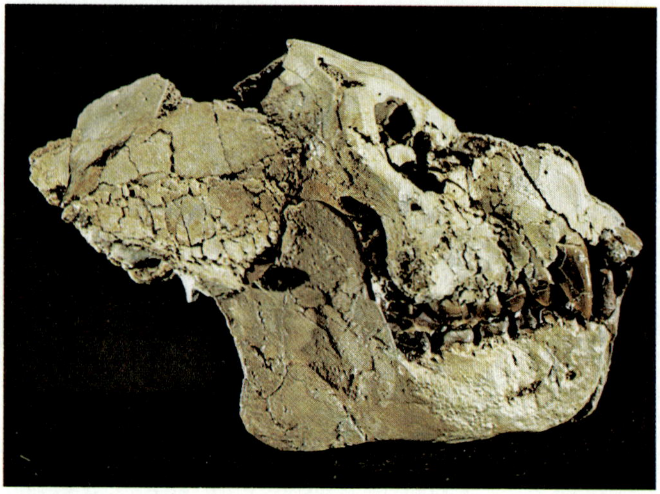

11-9. The cleaned and partially restored skull of an ape *Aegyptopithecus.*

11-10. Three lower jaws. Left and center are those of two Old World monkeys *Parapithecus* and *Apidium.* The jaw on the right is that of an Old World monkey that is alive today. *What similarities can you find?*

RECONSTRUCTING OUR PAST

No human beings lived 70 million years ago, but some primates did. These primitive primates did not look like apes and monkeys of today. They probably were small insect-eaters, much like today's tree shrew of Asia (Figure 11-11). From this ancestral form, we can trace three major lines of descent (Figure 11-12).

The first line is the Prosimians (proh·SIM·ee·uns). This group includes shrews, lemurs, and tarsiers. The ancient lemuroids probably had grasping feet and hands. We suspect that they had bushy

11-11. A tree shrew.

tails that were useful for balancing as they leaped from one branch to another. The skeletons of fossil lemurs are similar to those of lemurs living today (Figure 11-13).

Ancient tarsiers (TAR·see·erz) had large eyes and small snouts. These features suggest that vision was becoming more important than smell. The eyes "looked forward," as in humans, making three-dimensional vision possible. (How do such eyes compare with those of a dog, cat, or horse?) Despite these advantages, most tarsiers died out. Only a single modern species now exists in the East Indies (Figure 11-14).

The second line of descent was to the monkeys. New World monkeys have *prehensile* (pree·HEN·sil) tails—tails that can grasp. Old World monkeys have slender or stubby, nonprehensile tails (Figure 11-15).

The third major line of descent was to the apes and humans. The earliest fossil that is clearly an ape is *Aegyptopithecus* (ee·jip·toe·PITH·eh·kuss). *Aegyptopithecus* was found in deposits that are about 30 million years old. Another species called *Dryopithecus* (dry·oh·PITH·eh·kuss) first appeared about 20 million years ago. This species could be regarded as an ancestor of both humans and the modern apes. (Note the branching arrow in Figure

11-12. SUMMARY OF THE EVOLUTION OF PRIMATES

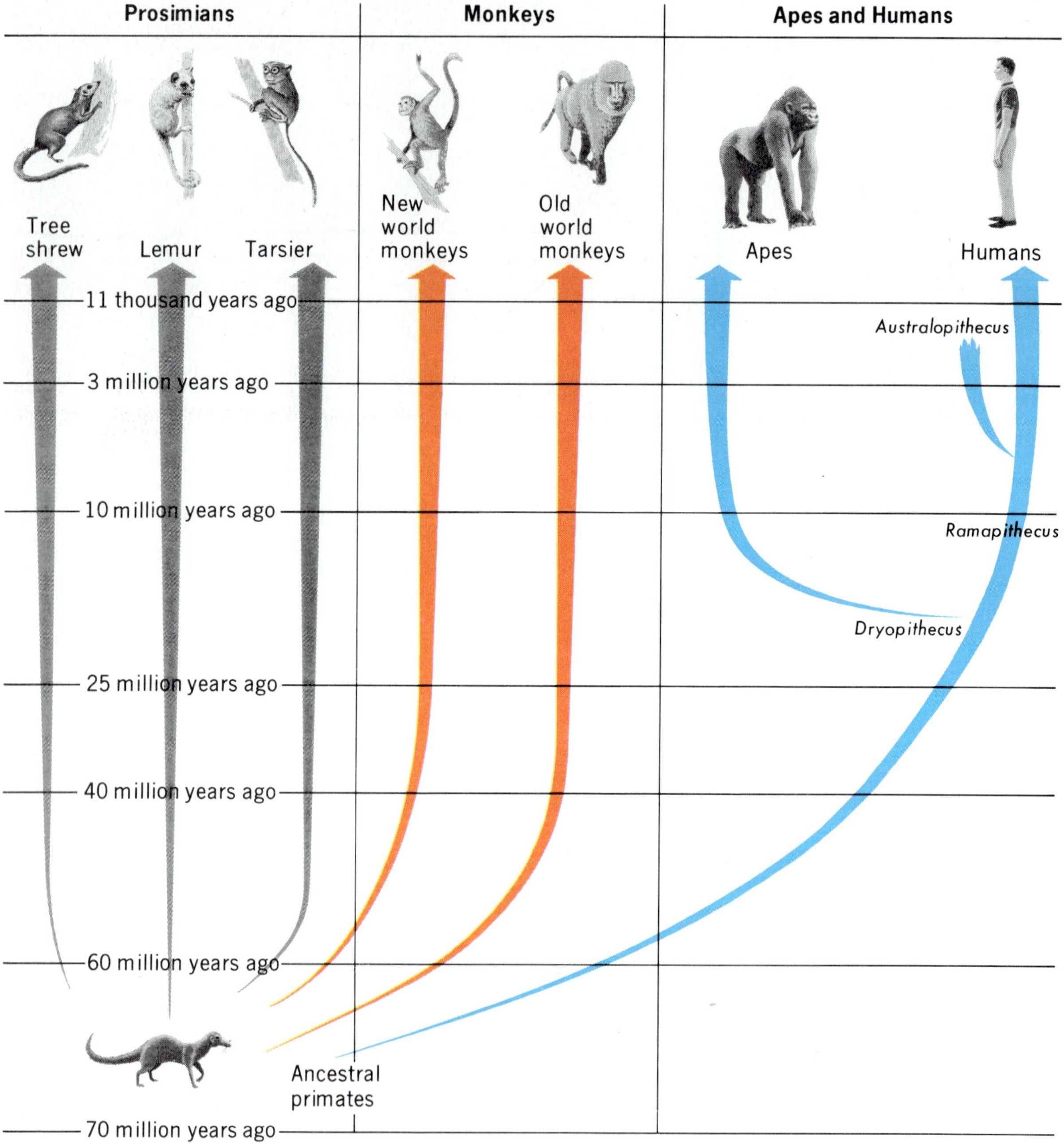

11-12.) *Ramapithecus* (ram·ah·PITH·eh·kuss), which lived from about 9 to 14 million years ago, is thought to be on the evolutionary path to humans. Thus, for the past 15 to 20 million years, the evolu-

11-13. A ruffed lemur.

11-14. A tarsier.

tion of humans and the apes has been entirely different. Before that it is presumed they shared a common ancestry.

A more complete understanding of our early evolution has been difficult to obtain because there are so few primate fossils. This seems to be a result of where and how primates lived. Most primates lived in the forested regions of the tropics. The conditions that lead to the formation of fossils rarely occur in tropical forests.

11-15. New and Old World monkeys. The squirrel monkey **(left)** is a New World monkey. The baboon **(right)** is an Old World monkey.

THE RACES OF HUMANS

Long before recorded history, humans had spread to all continents except Antarctica. In their isolated homelands, they evolved into the races we recognize today. All living humans can be classified, in a rough way, into such groups as the Negroid, Mongoloid, and Caucasoid races. Even scientists who specialize in the study of human culture do not agree on the number of human races. Some are satisfied with three or four; some would make the number 30 or 40. The features by which people might be so classified show tremendous variation within every population.

Thus, we can divide *Homo sapiens* into races on the basis of differences of many inheritable traits, such as the shape of the eye or the color of the skin. However, the different members of the human species are still much more alike than they are different. All have the same organs and similar physical characteristics. All human racial types are able to interbreed. All people living today are of one species.

Anthropologists study different human cultures. They agree that people who have been brought up in different cultures may think and behave differently. These differences are cultural, not genetic. If a small baby is moved from its place of birth to a different social environment, it develops a behavior like that of the society in which it is brought up. Many scientists therefore believe that differences in human behavior are most strongly influenced by the social environment, although these may be affected by both the physical environment and heredity.

CHANGING CULTURES

By 25,000 years ago, all people on the earth resembled one or another of the modern races of humans. If we could dress some of our ancestors in present-day clothes, we probably would not think of them as very different from ourselves.

Since then, the rate of cultural change has been very rapid. Even so, cultures have not changed at a steady rate. Early in human history, people lived as their ancestors had. Today, life styles change more rapidly. The rate of cultural change parallels the rate at which the human population has increased (Figure 11-16). In fact, most of the changes in our ways of life have occurred in the past 200 years.

Two hundred years ago, travel and communication were no faster than they were 2,000 years ago in ancient Greece and Rome. Food and clothing were made in much the same way, and health and freedom from disease were little better. Today, however, people can travel at the speed of sound and communicate at the speed of light.

11-16. WORLD POPULATION GROWTH

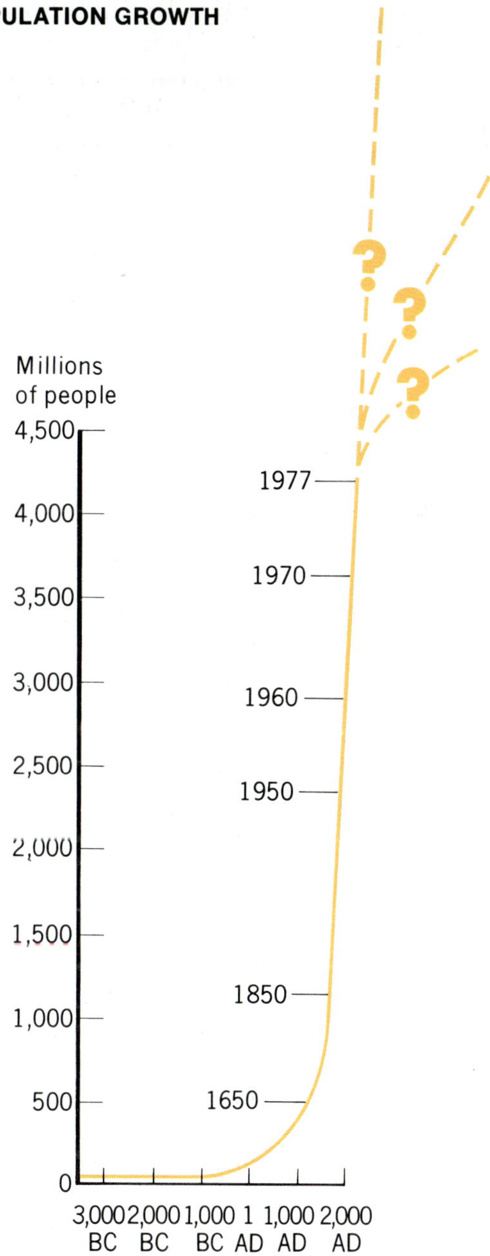

Many diseases have been eliminated. Nutrition has improved, following the discoveries of the body's need for vitamins in addition to proteins, carbohydrates, and fats. Remarkable increases in the food supply have been brought about by the use of new fertilizers and pesticides. Steam and, later, electricity brought light and power to homes and factories. Many other discoveries have revolutionized the way we live. The growth of science and technology are chiefly responsible for these changes in our lives.

How can we learn today about the lives of people in the past? The answer lies in interpretation of data obtained from the remains of past civilizations. You can gain experience with the sort of "detective" work that has to be done by doing **Inquiry 11-1.**

INQUIRY 11-1 Interpreting Evidence of Past Human Cultures

We cannot return to the times where people lived and worked thousands of years ago. As a result, we can never be sure that we understand all of the details of their cultures. Archaeologists search for clues or the remains of ancient people and civilizations. After these are found, dated, and studied, scientists formulate hypotheses to explain what they find or do not find. What emerges is a picture of life in a particular place many years ago. Although it is impossible to answer all questions, many reasonable and logical conclusions can be drawn. Some findings can be interpreted in a variety of ways, and archaeologists may disagree about which interpretation is correct. They try to find an explanation that is simple, yet fits the available facts.

In the inquiry, you will analyze some data found in archaeological digs. In nearly every case, several interpretations are possible. You and your classmates can suggest and share a variety of ideas. Think of as many possible interpretations as you can. In the final section of this inquiry, you will predict how archaeologists far in the distant future might interpret today's culture and life styles.

Part A. An Indian Cemetery in Newfoundland

One scientist investigated an Indian burial site in Newfoundland. The burial ground appeared to have been used between 4,000 and 5,000 years ago. The site is located at Port au Choix, a small village on the west coast of Newfoundland (Figure 11-17). Many skeletons, tools, and other artifacts have been dug up there. Answer each of the following questions, suggesting your own interpretations and possible explanations of the evidence. **1.** The graves were in very fine sand. One of the strips of sand was almost 1.6 kilometers long, varying from about 9 to 22 meters in width. The sandy strip lies

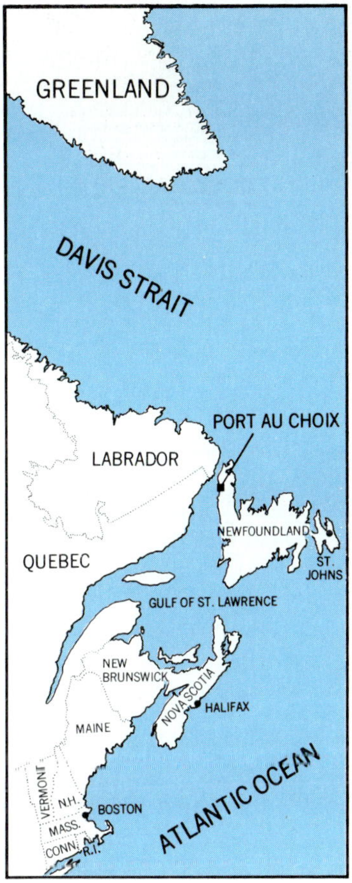

11-17.

about 6 meters above the high water mark of the shore of the ocean. What kind of an area do you think this might have been thousands of years ago when the Indians lived there? Why? **2.** What might account for the sandy strip being chosen as a prime burial site? **3.** Boulders or slabs of rock covered many graves. Why might this have been common practice among the Indians? **4.** The ages of 88 skeletons were estimated as follows:

Age	*Number of Skeletons*
newborn infants	12
less than 2 years old	15
2–6 years old	2
6–18	15
18–21	1
21–50	36
50 +	7

What do these data suggest about the lives of the indians?

5. The teeth of many of the adults were worn down. Often the enamel was so worn that the pulp was exposed. Some skeletons also had teeth missing, but there was no evidence of tooth decay. What might these findings suggest about the diet and life style of the Indians? **6.** What do you think might have been the use of the small objects shown in Figure 11-18a? What purpose might the holes have served? **7.** What do you think the objects shown in Figure 11-18b might have been used for? Why do you say this? **8.** What purpose might the objects in Figure 11-19 have served? Explain. **9.** What modern implement do the objects in Figure 11-20 resemble? What do you think their function might have been? **10.** A great number of woodworking tools were found in the burial site. Considering the location of the site (Figure 11-17), what do you think the Indians might have built from wood? **11.** What is your definition of the word "tool"?

11-18.

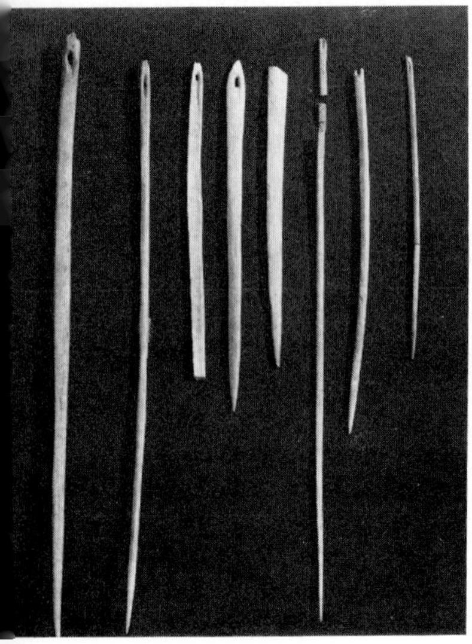

11-19.

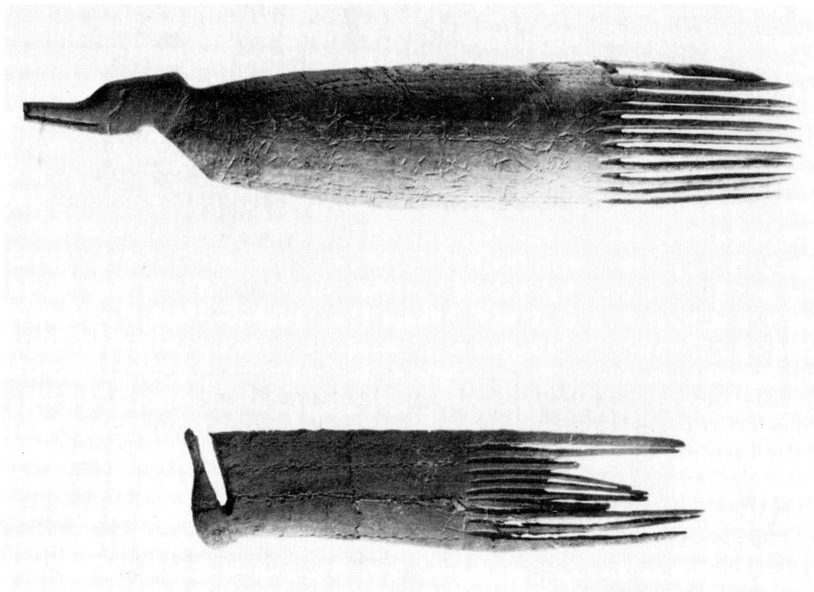

11-20.

Part B. Ancient People of Greece

Stone Age people inhabited Greece long before the dawn of Classical Greek civilization. Until recently, little was known of the culture of these people who lived in Greece between 5,000 and 20,000 years ago. Excavations in and around a cave on the east coast of Greece began in 1967. Since then, archaeologists have found much evidence that suggests changes in the culture and life styles of Stone Age Greeks. Consider each of the following findings and suggest how these facts might be interpreted. **12.** The oldest remains (20,000 years ago) in the cave are mostly the bones of a single species of horse and some tools made from flint (rock that forms sharp edges when it breaks). Newer remains (10,000 years ago) include bones of red deer, bison, horses, and a species of wild goat. These also include remains of wild plants such as vetch and lentils (pealike plants), shells of land snails, marine mollusks, and some small fish bones. These new remains were left at the time when a great Ice Age had come to an end. Suggest a hypothesis to explain the presence of all the remains as well as the greater abundance of newer remains. **13.** In still later finds, dated about 7250 B.C., very large fish bones were found—the fish might have been 100 kilograms or more. What does this finding suggest about a change in the life style of the people? **14.** At about the same time, tools made of obsidian, a kind of volcanic glass were found. The nearest source of obsidian is 150 kilometers away from the excavation site and across a body of water. How do you think these ancient people obtained obsidian? **15.** The site also yielded the oldest complete skeleton yet uncovered in Greece. A male of about 25 was buried in a shallow grave covered with stones. Certain bone abnormalities indicated that he may have been suffering from malaria. What does this suggest about the environment of the time? **16.** Remains from finds that are dated about 6000 B.C. are dominated by the bones of goats and sheep found in and around the cave. These animals were very different from animals found away from the cave. Also present are fossils of wheat and barley. Among the tools found are axes and millstones. What do these findings suggest about the life style of that time? **17.** The human graves at the site rarely included objects such as tools or jewelry. One infant was found near a marble vessel and a broken clay pot. A 40-year-old woman, who probably died about 4500 B.C., was buried with some bone tools and some obsidian blades. What might be the significance of these items found in the graves?

Part C. Excavating the Present

Imagine that you are an archaeologist living in the year 3000. A catastrophe has destroyed the written records of the twentieth century. You have only the remains that have been preserved over 1,000 years as clues to what life must have been like in the 1970s or 1980s. Imagine, furthermore, that you are excavating a site that was a school 1,000 years ago—the site of your school. **18.** What things do you think you would find? Make a list and then exchange it with a classmate. After exchanging lists, answer question 19. **19.** For each object, suggest several interpretations. (Remember, you know nothing about the time when the objects were left—the only clues available are the fossils.) **20.** Some materials such as stone and metal can last thousands of years. Other materials, such as paper and wood, decay rapidly. What important things would be absent in the archaeological dig that was once your school? What ideas about our society might you miss because the clues were not preserved? **21.** Consider questions 18 through 20 again. This time think about a different kind of site, such as a shopping center or business area, a residential street, a farm, or an industrial area. How would your findings and interpretations differ? **22.** Suppose you wanted to build a "time capsule" that would tell people in the far distant future about our lives today. What would you put in the capsule? Why? **23.** How would you define "culture"?

The Great Transition

By 25,000 years ago, people were making and using a variety of tools. And they talked and taught so that the younger generation learned from the experiences of the elders. They worked together and respected wise, intelligent leaders. They also could look into the future toward a better life. The fruits of these abilities have become the story of human progress.

Despite these advances, the people of 25,000 years ago were still at the mercy of their environment. For food, they hunted game and gathered wild fruits, seeds, nuts, and roots. If game animals were scarce, they starved. If a drought cut down the supply of wild food plants, again they faced the danger of starvation. They had to protect themselves from lions, tigers, cave bears, mastodons, and mammoths.

The Dawn of Agriculture

Once people learned to cultivate plants and domesticate animals, a settled life was possible. Archaeologists believe this happened between 10,000 and 7,000 years ago. With the remains of humans before 10,000 years ago, they have found only hunting tools, remains of wild plants, and tools for preparing these plants for food. With more recent remains, they have found hulls of grains related to modern wheat and barley. They also have found tools, such as sickles, which would be useful in harvesting grains.

Many plants and items that could tell us about early culture are fragile and rarely leave traces. You can guess how difficult it is to study the history of clothing. Yet there are some remains to study. A Siberian hunter of 30,000 years ago was found frozen in his grave. He was clothed in neatly sewn and tailored garments.

Can you imagine how much had to be learned before people could raise enough crops to support themselves? They had to learn which seeds could be saved and which would grow quickly into mature plants. People had to learn to clear the ground, and how to dig it up or plow it to provide a loose, well-aerated soil. They had to learn when to plant to avoid freezing temperatures or drought. They had to learn how to keep animal pests away from their crops. They also had to defend their crops from other tribes. They had to learn how to harvest. Finally, they had to learn how to store crops so there would be food until the next growing season. And, even without knowing about genetics, they learned to select the most desirable forms of food plants. The probable geographic origins of some of the modern domestic food plants we use today are shown in Figure 11-21.

11-21. GEOGRAPHIC ORIGINS OF SOME IMPORTANT DOMESTICATED PLANTS

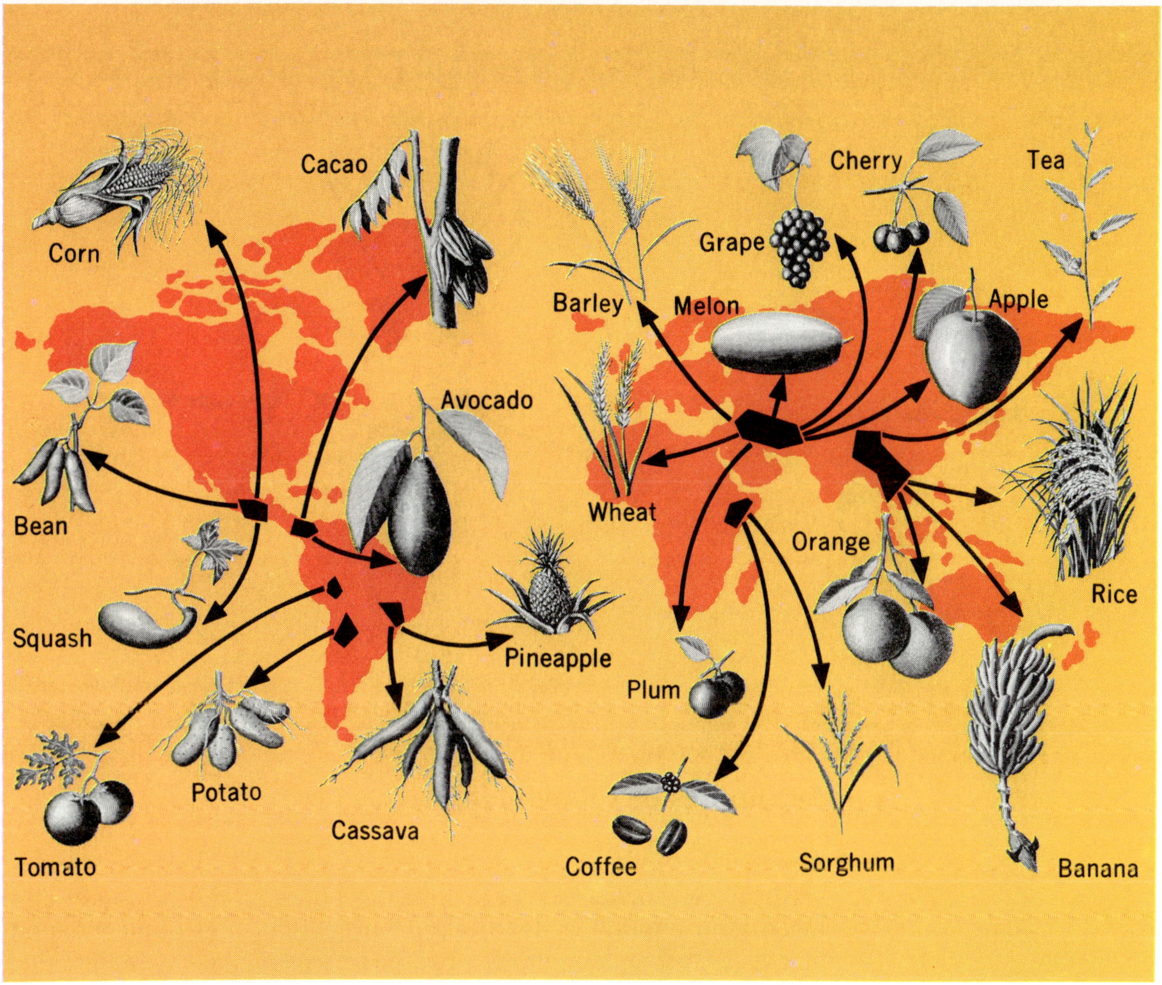

Domestication of Animals

Domesticating wild animals required skill and patience. Much more was involved than simply capturing the animals and keeping them fenced in. It is very unlikely that early people could have caught and handled adult wild beasts. They probably began by raising young animals. As these animals grew, they must often have become too unruly to handle. But some of them became tame enough to keep and breed in captivity. The principal animals raised today are cattle, hogs, sheep, goats, horses, dogs, cats, and chickens. All were domesticated by our prehistoric ancestors. The places of origin of domestic animals are shown in Figure 11-22.

11-22. GEOGRAPHIC ORIGINS OF SOME IMPORTANT DOMESTICATED ANIMALS

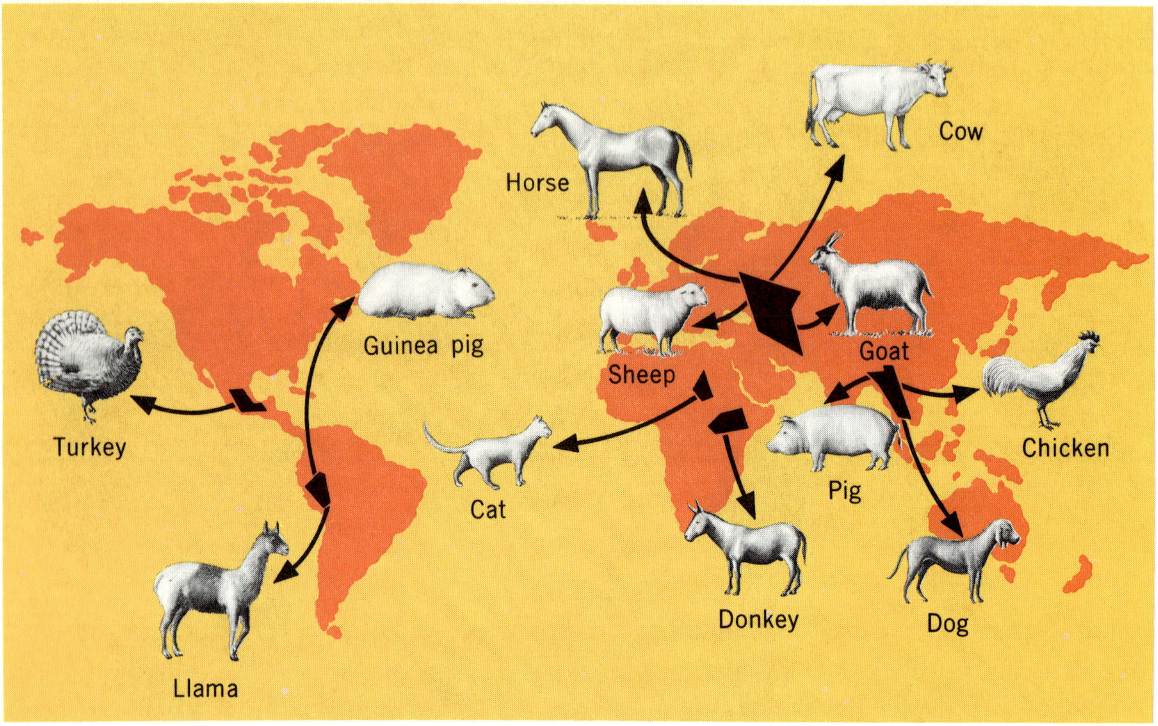

Beginnings of Village Life

Hunters had to keep moving to find enough game to keep themselves alive. How could these tribes of hunters have stayed in one place long enough to learn about agriculture and domesticating animals? Recent observations of the remains of ancient settlements suggest one possible answer. Some early tribes lived on the seashore or along lakes and rivers. Fishhooks are found in the remains of some settlements that existed before the dawn of agriculture. Some of the first people who settled down and lived in villages apparently relied on fish and shellfish for food. Once a tribe had discovered a good place to fish, they could stay there indefinitely.

You will be sure to recognize this as a hypothesis, not fact. The change from a nomadic to a settled life probably took place many times under different conditions. The earliest villages in both the New and Old Worlds were isolated settlements. Different plants and animals were domesticated at different locations. Some of the domestications probably took place at almost the same time in different places. For example, in the Middle East, parts of Africa, and the highlands of Mexico, the climate at one time was dry and cool.

These dry, cool areas were more suited to the cultivation of grains than to the raising of tropical roots. The wild grasses ancestral to wheat, barley, and corn must have grown very well in these areas.

Clothing

Next to food, one of the most important concerns of our early ancestors was clothing—except, perhaps, in the tropics. Other primates are hairy, even those living in the tropics. Early European people undoubtedly made clothing from animal skins. Cro-Magnon people may have made clothing as long ago as 35,000 years. These probably were made from the hides of animals, cut and neatly sewn together with rawhide thongs. Some Eskimos still make clothes from animal hides (Figure 11-23).

Clothing took an entirely new turn with the invention of weaving. The weaving of cloth is almost as old as agriculture. It started in Egypt and Mesopotamia (mess·uh·puh·TAY·mee·uh), where linen was woven from the fibers of flax plants. Wool and cotton fabrics

11-23. Eskimos wearing clothing made of animal skins.

were later inventions that involved the spinning of short fibers into long threads.

Was weaving invented quite independently in the Old and New Worlds? How did cotton come to be recognized on both continents as an ideal natural fiber? The Indians of Peru made cloth of wool, woven from the hair of domesticated llamas (LAH·muz) and alpacas (al·PAH·kuz). How and when did they learn to do this? No one knows, but one thing is sure. Inventiveness and ingenuity are not recent developments in the human species.

CULTURAL CHANGE AND THE FUTURE

The change from a Stone Age way of life to modern civilization took only a tiny fraction of the time required for humans to evolve physically from their early ancestors. The change was made possible by the things humans learned and discovered. Through language, both spoken and written, knowledge is transmitted from one generation to the next. At first, the rate of discovery was slow. In recent centuries, the rate of discovery has increased with great rapidity.

Changes in human culture have occurred independently of physical changes in the human body. Thus, we have a kind of **cultural inheritance** that is very different from biological inheritance. Biological inheritance is the transmission of traits from one generation to another. This is done through genes and chromosomes. Cultural inheritance is the transmission of beliefs, traditions, skills, and knowledge from one generation to another. Your culture is the product of the human beings who lived before you. Your culture also depends on influences from members of your own generation. It depends on friends, relatives, authors, teachers, scientists, artists, engineers, and all of the people who contribute to a society.

Your biological inheritance is determined by the genotypes of your parents and by the accidental events of meiosis and fertilization. We cannot choose our genes. In contrast, our cultural inheritance allows us some freedom of choice. We can choose which traditions we will follow. Or we can choose how knowledge will be passed on to our children.

Another important characteristic of cultural change becomes clear when we think of our relationship to the environment. Most other animals are affected by natural selection. Those individuals ill-suited to the environment are less likely to survive and reproduce. Humans are less affected by the environment. We often can alter the environment to fit our needs. Some animals become adapted to cold weather through the natural selection of alleles. Their bodies show an increase in the amount of fat or fur. Others become adapted through selection of alleles that cause them to go to sleep in a warm shelter during the winter. Humans have become adapted to cold in different ways. They learned how to make and wear warm clothes and to build and heat their houses. Birds and bats fly because of great changes that have evolved in their front limbs, backbones, and other structures. People fly by building machines that carry them through the air faster and farther than any bird.

A key element in all human cultures is cooperation. Cooperation is highly developed in many societies of animals. Ant hills and beehives are familiar examples. Ants probably cooperate instinctively. But humans have *learned* how to cooperate. Our laws, morals, and codes of ethics are rules that regulate our social, or cooperative, behavior. Our rules try to balance what is good for the group with what is good for the individual. As society has become more complex, the emphasis on cooperation also has become greater.

The last four centuries of our cultural heritage have been termed the period of the Scientific-Industrial Revolution. It began with the introduction by Galileo, Harvey, Redi and others of critical observations and experiments. There were two earlier periods of tremendous change in our history—the first when toolmaking began, and the second when agriculture was introduced.

What happens next seems to depend on whether the scientific knowledge, upon which our present technological civilization is based, will continue indefinitely or will come to a halt. What do you think? Will we soon know everything about nature and about ourselves? Will the era of exploration and discovery in science come to an end? At least we can say that at this moment no one can see the end of the scientific enterprise.

In Brief

Questions concerning the origin of the world and its inhabitants are some of the most profound that can be asked. These are questions that human beings have asked throughout history. Finally, a generally accepted scientific answer emerged. Charles Darwin, in the middle of the nineteenth century, formulated the concept of natural selection as the basis for evolution. This rapidly became the organizing concept of the biological sciences. Today, we look upon all species, including human beings, as the products of a long evolution, going back to the origin of life itself. What every species is and does is a reflection of its evolutionary past.

As important as biological evolution in humans is, we have remained the same species for many thousands of years. Thus, the significance of biological evolution seems dwarfed when compared to the explosion of cultural change, especially within the last few centuries. Human culture has shown several periods of rapid, drastic change. The first was the agricultural revolution; another was the industrial revolution.

The latest stage of cultural revolution results from the vast expansion of knowledge. In addition, we now communicate through the use of telephones, radio, and television, as well as communications satellites. Knowledge is growing so rapidly that we can compare a generation of cultural change in the past century with just a few years at the present time.

Guide Questions and Problems

1. What evidence supports the idea that humans have evolved?
2. Of what significance is the fact that hemoglobin in humans and in gorillas differs in only a few amino acids out of several hundred?
3. Why would the discovery of tools and their relationship to fossils of primitive primates be important in interpreting whether these primates were human?
4. Criticize or support the decision of scientists to concentrate the search for our ancestors in Africa and in other parts of the Old World rather than in the Americas.
5. Some who are unfamiliar with the theory or the evidence of evolution keep alive a misconception that scientists believe "humans have descended from monkeys or apes." All three groups of primates referred to by the statement are living primates rather than fossil ones. How would you clarify the evidence and correct the statement?
6. The abilities to use tools, to talk, and to teach and learn from one's experience and the experience of others are the basis for cultural change. What special characteristics of human evolution have led to these abilities?
7. What were some of the major problems that had to be solved before a settled community life could replace the nomadic ways of life?
8. Imagine that you are living long ago when village life was beginning. What use might you make of the wild ancestors of today's domestic animals—cattle, hogs, sheep, goats, horses, dogs, cats, and chickens?

Ideas for Independent Study

1. Cultural changes have made it possible for people to escape from some of their biological limitations. Conduct an independent study of how people have used their knowledge and technical skills to increase the food supply, to increase physical strength and powers of observation, to lengthen life, and in other ways. Think about your own life. How many of your activities and interests would be possible in a culture similar to that of people living 20,000 years ago?

2. A recent book by E. O. Wilson of Harvard University is *Sociobiology: The New Synthesis,* Harvard University Press, 1975. It traces the origins of cooperative behavior among animals. Compare the societies of ants and termites with the social organization of our early ancestors. What are the major differences? Compare social behavior in birds, lions, and baboons with early human social organization.

3. Observing monkeys and apes in the wild has given us new understanding of the role of teaching and learning in the life of animals. Two paperback books that might help you start the study of this topic are *In the Shadow of Man* by Jane van Lawick-Goodall, Houghton Mifflin Co., Boston, 1971; and *Primate Ethology,* edited by Desmond Morris, Anchor Books, Doubleday and Company, Inc., Garden City, N.Y., 1969. Another interesting related study is described in an article by Sheo Dan Singh, "Urban Monkeys," in *Scientific American,* July 1969, page 108.

4. Although the course of cultural change has affected all people to a certain extent, today different groups of people are at different stages of change. Search the current literature to find out what happens to people living in a hunting or agricultural society when they suddenly come in contact with a modern industrial society. How fast do changes set in? How do the changes affect the quality of life for individuals?

5. Today, people cultivate many plants and animals for food and as sources of fibers and other useful materials. Some of these organisms were domesticated by humans long before the dawn of history. Nevertheless, modern humans have learned to cultivate a great variety of microorganisms, insects, and plants. These are used as new food sources, in controlling disease, in scientific experiments, and for other purposes. What are some specific examples of these newly cultivated forms? What possibilities are there for further developments along these lines?

Related Readings

BOOKS

Cole, S. *Leakey's Luck: The Life of Louis Seymor Bazett Leakey, 1903–1972.* New York: Harcourt Brace Jovanovich, 1975.

Day, M. H. *Fossil Man.* New York: Grosset and Dunlap, 1970.

Howell, F. C. *Early Man.* New York: Time-Life, 1965.

Lerner, I. M. *Heredity, Evolution and Society.* San Francisco: Freeman, 1968.

Shapiro, H. L. *Peking Man.* New York: Simon and Schuster, 1975.

Human Variation and Origins, An Introduction to Human Biology and Evolution: Readings from Scientific American, San Francisco: Freeman 1967.

ARTICLES

Cavalli-Sporza, L. L. "Genetic Drift in an Italian Population." *Scientific American,* August 1969, page 30.

Martin, R. D. "Ascent of the Primates." *Natural History,* March 1975, page 52.

Oxnard, C. E. "The Place of the Australopithecines in Human Evolution: Grounds for Doubt?" *Nature,* December 4, 1975, page 389.

Simons, E. L. "Ramapithecus." *Scientific American,* May 1977, page 8.

Smith, J. M. "Molecular Evolution and the Age of Man." *Nature,* February 13, 1975, page 497.

Zihlman, A. and D. Cramer. "Human Locomotion." *Natural History,* January 1976, page 64.

PART THREE

DIVERSITY

12 Diversity in the Living World

As you read the chapters on genetics, development, and evolution, you may have gained some insight into the patterns and processes of the living world. In particular, you may now have a better idea of how organisms change through time. Natural selection acts on a genetically variable population. The characteristics selected adapt the organisms to their environment. These processes have operated for millions of years and continue to operate today.

Over two million species of animals and nearly 300,000 species of plants live on the earth today. It would be impossible for any person to know all about so many different kinds of organisms. In fact, it is not necessary to do so. Certain organisms are similar enough that they can be thought of as belonging to the same group. For example, a bullfrog and a green frog are different species, yet their ways of life have many features in common. Both spend most of their lives along the edges of ponds and lakes. Both eat insects. They digest their food, breathe, excrete, and reproduce in almost identical ways. The nuclei of their cells contain the same number of chromosomes. They respond in similar ways to heat or cold and to potential predators, such as snakes. Thus, if you knew about the life of a bullfrog, you also would have a good idea of the structure and function of a green frog. And, as a matter of fact, you could apply your knowledge to most kinds of frogs.

CLASSIFICATION

One way to make sense of the great diversity of organisms living today is to group them according to similarities. The characteristics selected may be general enough to include a great many different kinds of organisms. If very specific details are used for grouping, perhaps only one species will be included. Thus, a system of groups within groups can be formed and used. To see how such a category system might work, consider the example shown in Figure 12-1.

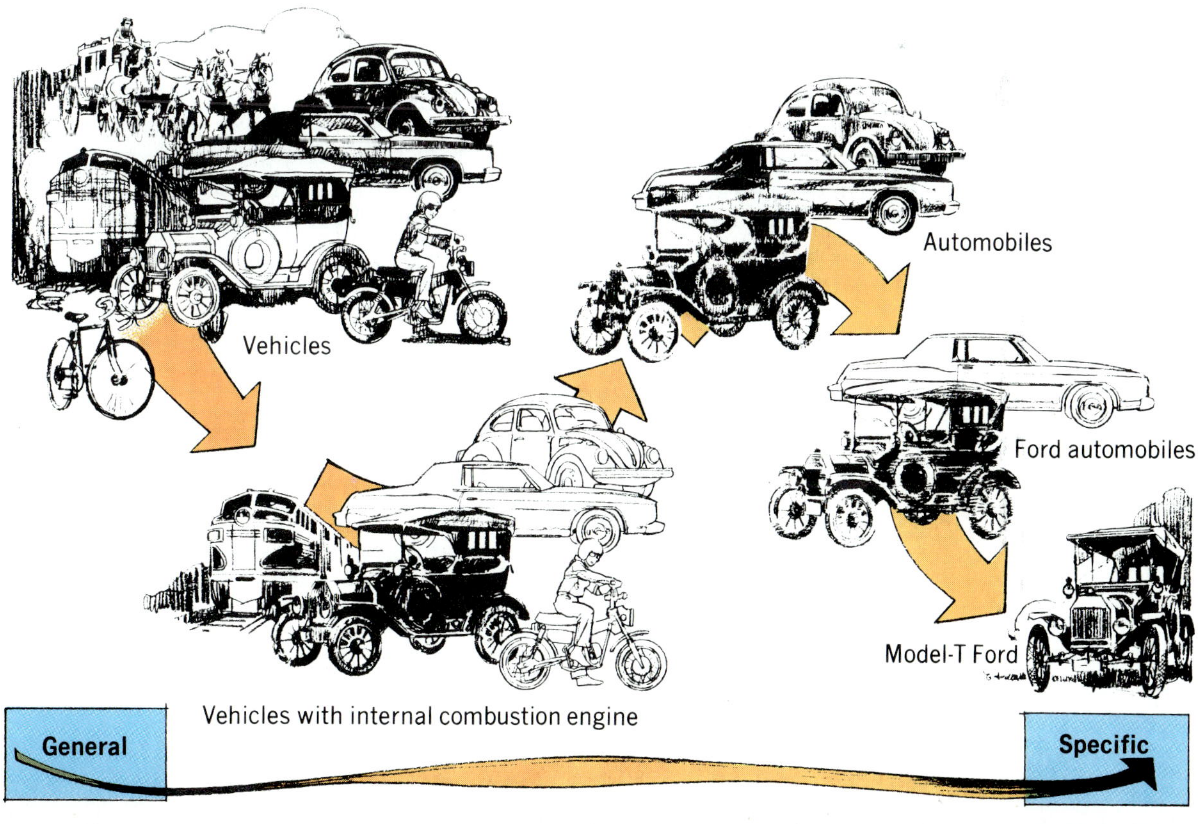

12-1. A classification system is based on shared characteristics. General characteristics allow for large groupings of very diverse things. More specific characteristics result in smaller groupings.

Suppose you are told that object X belongs to the group "vehicles." Even if you have never seen this particular X, you would be able to make some very general predictions about its structure and function. It would probably have wheels or runners, be used for carrying objects or people, and so on. If you were then told that X belongs to a more specific group, "vehicles with internal combustion engines," you could make more specific predictions. It would probably have spark plugs and pistons and use a fuel derived from petroleum. If you were told that X is an "automobile," you could make still more specific predictions. You might then learn that X is a "Ford automobile," and that the specific Ford being discussed was the model-T. Finally, you could classify X into the smallest group. This group would include only one kind of vehicle, the model-T.

A Classification Key

Classifying anything, whether vehicles or plants, is based on comparing characteristics. The more characteristics two things have in common, the more alike they are considered to be. Lists of characteristics are called *classification keys*. Keys become more complex as they move from very general characteristics to very specific ones.

The easiest kind of key is one that uses pairs of short phrases or, preferably, two single words. But the simpler the key, the greater is the chance of error in identification. A single word or phrase may not provide enough information to identify all the objects or organisms in the group. To see how a very simple key might work, consider the horses shown in Figure 12-2. Suppose you did not know the names used for their coat colors. You could use the simple key below to identify each kind, using color characteristics to name them. Study each of the five horses individually. Starting with number one on the key, answer yes or no to each characteristic. If the answer is no, move on to the next number until you come to the right identification. Note that each step in the key requires only one comparison. For example, is the coat solid or spotted? Note that each step eliminates one horse from further consideration.

1. Solid, reddish-brown: Sorrel
1a. Spotted or solid color(s) other than reddish-brown—go to No. 2.

2. Entire body tan or blonde: Palomino
2a. Spotted or of more than one color—go to No. 3.

3. Body solid tan or blonde; mane and tail dark brown: Buckskin
3a. Spotted—go to No. 4.

4. Spots large and uniformly distributed over entire body: Pinto
4a. Spots small and confined to saddle area of back: Appaloosa

12-2. *Can you use a simple key to determine the common names of these horses?*

Later in this chapter, you will be using some keys to classify plants and animals. For now, you can gain experience in making and using a key by doing **Inquiry 12-1.**

INQUIRY 12-1 Making and Using a Classification Key

MATERIALS (per team)

deck of playing cards
an assortment of pictures clipped from magazines

Procedure

Develop a key to classify the playing cards. Hint: You may find that you need several "sub-keys" that are specific for a particular group of cards, such as those of similar suit or color. Develop the key so that someone else could identify any single card using only the key. (Assume that the other person does not know the name of the card.)

After your keys are constructed, trade them with other teams. Select a single card from the deck and ask team members to pretend that they do not know the name of the card. See if the other team can accurately name the card using only your key. Try the same procedure using only the key prepared by another team. **1.** Did your key work for other students? Why or why not? **2.** Compare all the keys prepared by all teams in the class. How do they differ? **3.** In what ways are they the same? **4.** How might they be improved?

Develop and test a similar key for the assortment of magazine pictures. **5.** Did you find it easier or more difficult to prepare the key for the pictures? Why? **6.** What problems do you think biologists face in attempting to classify a wide variety of organisms?

Developing a Biological Classification System

Systems for classifying organisms have developed gradually over the past 300 years. In the late seventeenth century, an Englishman, John Ray (1627–1705), developed a clear concept of species. To Ray, a species consisted of offspring of similar parents. The concept has since been modified. We now look upon a species as a group of individuals that can breed with one another and produce fertile offspring. As a rule, individuals of one species do not breed with indi-

12-3. This liger is the result of an experimental cross between a lion and a tiger. Such breedings do not occur in nature.

viduals of other species under natural conditions. However, many species may crossbreed in a laboratory. Figure 12-3 shows one example.

In the eighteenth century, Swedish biologist Carolus Linnaeus (KAR·oh·lus lih·NAY·us) established the system that is used today for naming organisms. His system is known as **binomial nomenclature** (by·NOH·mee·ul NOH·men·klay·chur). Linnaeus gave each known species a name of two words. Both are often Latin words or Greek words with Latin endings. The first word of the name is the name of the **genus** (plural, *genera*) to which the organism belongs. For example, cats are grouped in the genus *Felis* (FEE·lis). The second word is descriptive. *Felis leo* is the name of the lion, one species of cat. The scientific name of the common house cat is *Felis domesticus* (doh·MES·tih·kus); of the tiger, *Felis tigris* (TI·gris). All belong to the same genus, but each is a different species.

Why are the scientific names of organisms usually based on Latin? When scientists began classifying organisms, Latin was an important language in science. Most scientists knew Latin as well as their native language, and could communicate in Latin with scientists of other countries. Today, scientists still use Latin names—even though Latin is no longer the international language of science.

But why does a robin, for example, require a scientific name? The "robin" in England is a very different species from the "robin" of North America. By using the Latin names that everyone accepts, scientists from England and North America can communicate more easily, avoiding the confusion that might arise from using the word "robin."

A species includes organisms of only one kind. A genus may include several species. Step by step, organisms can be classified

12-4. EXAMPLES OF ANIMAL CLASSIFICATION

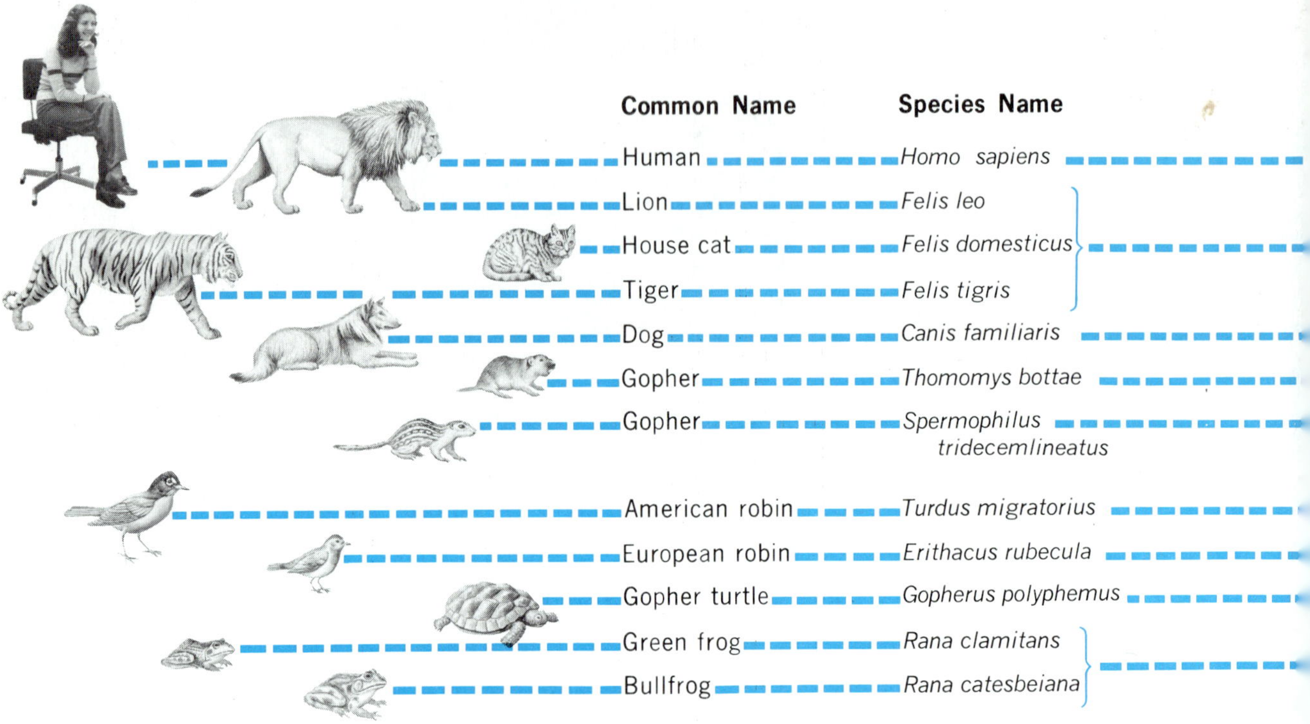

into progressively larger groups: family, order, class, phylum, and kingdom. The relationships among these various groups are illustrated in Figure 12-4.

Consider the examples shown in the figure. All humans are classified as species *Homo sapiens.* Humans also belong to larger groups that include other organisms. The family Hominidae (hoh·MIN·ih·dee) includes not only modern humans, but also extinct humans not of the genus *Homo.* The order Primates (pry·MATES), includes humans along with lemurs, monkeys, and apes.

The three cats—lion, house cat, and tiger—belong to the genus *Felis.* They also belong to the family Felidae (FEE·lih·dee). Generally a **family** includes related genera. (In Figure 12-4 this is shown only in the case of the two genera of robins.) The dog, on the other hand, belongs to the genus *Canis* (KAY·nis), which also includes wolves and coyotes. The dog also belongs to the family Canidae (KAN·ih·dee).

Dogs and cats, however, resemble one another in many ways. On the basis of their resemblances, the families Felidae and Canidae are put in a common **order,** Carnivora (kar·NIV·oh·rah). Generally an order includes *related* families.

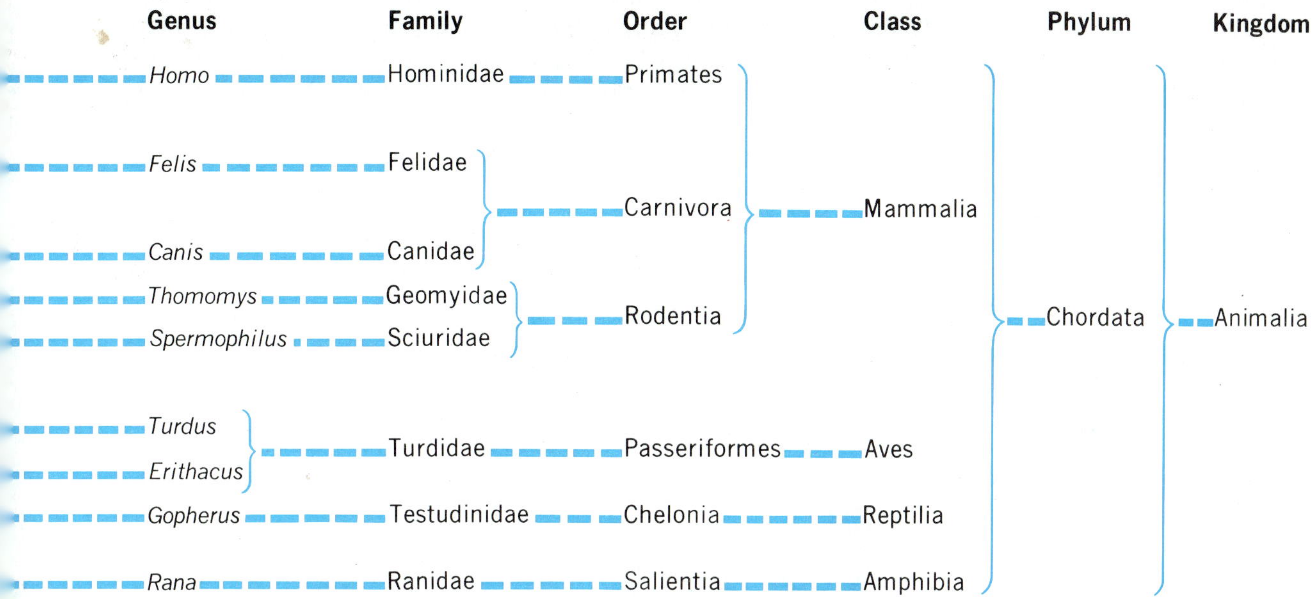

In a similar way, the two gophers are not enough alike to be included in the same genus, or even in the same family. However, they are similar enough to be classified together in the same order, Rodentia (roh·DEN·she·ah).

The first seven species shown in Figure 12-4 are different enough to be put in three different orders. Yet they are alike in many ways. All are covered with hair, they nurse their young with milk, and their red blood cells are without nuclei. Because of these and other resemblances they are combined in the class Mammalia (ma·MAY·lih·ah). A **class,** then, is composed of related orders.

What is the relationship of the class Mammalia to the other classes in the list? All members of these classes have backbones. Their internal organs also are alike in many ways. All have gills, or structures like gills, in their embryonic stages. Because of these general resemblances, these classes of animals are placed in the phylum Chordata (kor·DAY·tah). A **phylum,** then, includes similar classes.

Finally, the Chordata are grouped with many of the other animal phyla (not represented in Figure 12-4) that make up the Kingdom Animalia.

Homology, the Major Basis of Classification

All organisms classified in a particular group have many fundamental similarities in structure. It is not always easy to recognize these similarities. At first glance, the insects shown in Figure 12-5 seem to have little in common. We might wonder how a butterfly, a silverfish, and a louse could be grouped together. All are placed in the phylum Arthropoda, class Insecta because, among other things, they have an outside skeleton (or exoskeleton), six jointed legs, three basic body parts (head, thorax, and abdomen), and a single pair of antennae on the head.

Structures that are similar because of their common origin are said to be *homologous* (hoh·MOL·oh·gus). The homologous structures usually are the ones considered when organisms are grouped in a classification scheme.

There is a definite link between homology, as a basis of classification, and evolution. In searching for homologies, scientists are look-

12-5. *What do these organisms have in common?*

ing for clues about the relationships among organisms. For example, all animals with backbones have pancreas cells that make insulin. This is interpreted to mean that the insulin-secreting mechanism evolved in some remote ancestor of these animals. The cells that make insulin are very much alike in all these animals. It seems impossible that these cells could have evolved independently in each vertebrate species.

Clues about evolutionary relationships also can be found in protein analysis. For example, the cytochrome c of a pigeon and a turkey differ by four amino acids. The difference between a pigeon and a turtle is eight amino acids. Data of this nature suggest a close relationship. One might conclude that a pigeon is more closely related to a turkey than it is to a turtle. This conclusion is supported by morphological and fossil evidence.

Biologists have developed classification systems based on homologies of structure, function, and biochemistry. Such systems are always being revised as more is learned about organisms. The rest of this chapter shows several such systems. It also introduces some of the major groupings of organisms and shows examples for each group. In **Inquiry 12-2** you will use a classification key to identify some organisms.

INQUIRY 12-2 Animal Classification

Precise naming of animals is invaluable in all branches of biology. Reporting the proper scientific name of the organism with which you are working is essential. Other scientists can then repeat your experiments using exactly the same kind of organism. For example, identifying the organism as a crayfish is not enough. There are over 200 species and subspecies of crayfish in the United States alone.

In the earlier days of naming animals, the science was called *taxonomy* (tak·SAHN·oh·mee), which meant merely "the assigning of names." Today we speak of it as **systematics** (sis·teh·MAT·iks). Systematics involves not only assigning names, but also arranging animals in groups that show their evolutionary relationships to one another.

MATERIALS

preserved or mounted specimens of
- *animals from 10 major animal phyla*
- *animals from the 4 major arthropod classes*
- *animals from 7 chordate classes*

additional mounted and preserved specimens as desired and available

Procedure

Try to identify the phylum to which each of the animals belongs. First use the very simple key below, then the more detailed key that follows it. At each step of the key, determine if the animal has these characteristics. If not, you must go on until you find the animal properly described.

A few of the terms used in the keys may be unfamiliar to you. Figure 12-6 may help. Consult a dictionary or your teacher for any other words you don't understand.

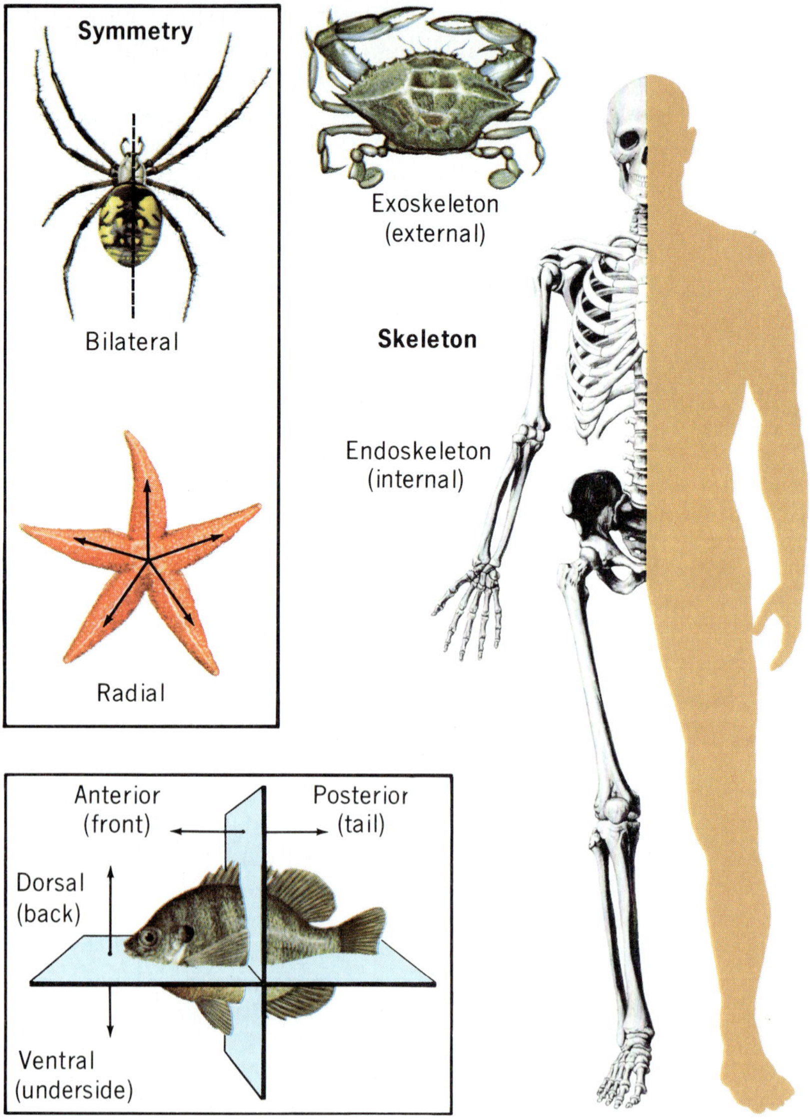

12-6. Some terms important for animal classification.

Key A: Simple Key

1. Microscopic: *Protozoa*
1a. Visible to the unaided eye—go on to No. 2

2. Holes: *Porifera*
2a. No holes—go on to No. 3

3. Body a two-layered sac with one opening: *Coelenterata*
3a. Body not saclike, or has two openings, or three or more cell layers—go on to No. 4

4. One opening to the digestive tract: *Platyhelminthes*
4a. Two openings to the digestive tract—go on to No. 5

5. Radially symmetrical: *Echinodermata*
5a. Bilaterally symmetrical—go on to No. 6

6. Nonsegmented worm: *Nematoda*
6a. Segmented or with shell—go on to No. 7

7. With a shell: *Mollusca*
7a. Segmented—go on to No. 8

8. Legless: *Annelida*
8a. With appendages—go on to No. 9

9. Exoskeleton: *Arthropoda*
9a. Endoskeleton: *Chordata*

The more you know about animals and their structures the more detailed a key you can use. The next key lists a few more characteristics than the one above. It takes into account a few more exceptions. This key will allow you to identify the forms more accurately. Using the same animals, classify them again with Key B.

1. Are the results the same? Account for any differences.

Key B: More Detailed Key

1. Usually microscopic in size; body usually a single cell. If more than one cell, each cell almost independent of the others: Phylum *Protozoa*
1a. Usually visible to the naked eye; multicellular; cells dependent on one another—go to No. 2

2. Body wall covered with holes; radially symmetrical or asymmetrical; adults immobile, fastened to one spot: Phylum *Porifera*
2a. Body wall without numerous openings; radially or bilaterally symmetrical; immobile or free-moving—go to No. 3

3. Saclike body of two cell layers; radially symmetrical; digestive cavity with one opening; tentacles frequently present: Phylum *Coelenterata*

3a. Body with three embryonic cell layers; bilaterally symmetrical, at least in larvae—go to No. 4

4. Digestive cavity with one opening; body flattened; mostly parasitic: Phylum *Platyhelminthes*

4a. Digestive cavity with two openings; body more rounded—go to No. 5

5. Adults radially symmetrical; body pattern usually in "5's"; all marine: Phylum *Echinodermata*

5a. Adults bilaterally symmetrical; marine, freshwater, or terrestrial—go to No. 6.

6. Wormlike; no shell, scales, or prominent sense organs; mostly parasitic; nonsegmented: Phylum *Nematoda*

6a. If wormlike, also with shell, scales, segmentation, or prominent sense organs; mostly free-living and nonwormlike—go to No. 7

7. Nonsegmented with shell; muscular foot; mostly slow-moving or living attached to one spot: Phylum *Mollusca*

7a. Segmented, mostly with no shell or muscular foot; mostly rapidly moving and free-living—go to No. 8

8. Legless worms; no prominent skeleton; segmentation obvious: Phylum *Annelida*

8a. Mostly with jointed appendages; obvious exoskeleton or endoskeleton; segmentation frequently not obvious externally—go to No. 9

9. Exoskeleton of chitin: Phylum *Arthropoda*

9a. Endoskeleton of cartilage or bone: Phylum *Chordata*

Keys are used to identify organisms not only as to phyla but for all systematic categories. You can find keys in many books for nearly every group of plants and animals. Classification keys involve a great deal of specialized terminology. This terminology is necessary for accuracy and brevity.

The next key will identify most members of four classes of arthropods. Using the specimens available to you, identify the selected arthropods according to class.

Key C: Classes of Arthropods

1. Wormlike adults with many body segments, and with one or two pairs of legs per segment: Class *Myriapoda* (mir·ee·OP·o·duh)

1a. Nonwormlike adults; fewer body segments, with either one pair of legs per segment or none—go to No. 2

2. Two pairs of antennae; marine or aquatic; gills: Class *Crustacea* (krus·TAY·she·uh)
2a. One pair of antennae or none; terrestrial or freshwater, tracheal tubes—go to No. 3

3. Four pairs of legs: Class *Arachnida* (uh·RAK·nih·duh)
3a. Three pairs of legs: Class *Insecta* (in·SEK·tuh)

Turn to the main classes of the phylum *Chordata*—to which humans belong. Use the next key to help you identify chordates according to class.

Key D: Classes of Chordates

1. Fins or no appendages; marine or aquatic fishes—go to No. 2
1a. Legs, wings, or flippers, but not fins as appendages; mostly terrestrial—go to No. 3

2. Slimy, scaleless skin; no paired appendages; reduced fins; no jaws; snakelike: Class *Cyclostomata* (sy·klo·STO·muh·tuh)
2a. Leathery skin with small scales; prominent fins; jaws; no bones: Class *Chondrichthyes* (kon·DRIK·thih·eez)
2b. Scaly skin with prominent scales; jaw bone present: Class *Osteichthyes* (OS·teh·thih·eez)

3. No scales, feathers, or hair: Class *Amphibia* (am·FIB·ee·uh)
3a. Scales, feathers, or hair—go to No. 4

4. Scales: Class *Reptilia* (rep·TIL·ee·uh)
4a. Feathers: Class *Aves* (AY·veez—birds)
4b. Hair: Class *Mammalia*

Use the following information to construct a key to the five classes of mollusks.

Class *Amphineura* (am·fih·NYOO·ruh): well-developed head or special sense organ; gills in mantle; dorsal skeleton composed of several separate plates (Figure 12-7a).

Class *Scaphopoda* (ska·FOP·o·duh): single tusklike shell, uncoiled and open at both ends; small tentacles, foot, and mantle cavity (Figure 12-7b).

12-7. Some mollusks for keying. (Continued on page 358.)

12-7. (continued)

Class *Pelecypoda* (pel·eh·SIP·o·duh): bilaterally symmetrical, with a shell of two halves (or valves). Many live attached in one place; feed by filtering microorganisms from the water; use gills as respiratory structures. Mostly marine but some freshwater; none terrestrial. A few swim fairly well. May reach 1.8 m in diameter (Figure 12-7c).

Class *Gastropoda* (gas·TROP·o·duh): single shell with one opening; shell commonly twisted in a spiral. Marine or terrestrial. Breathe with gills or lungs. Some lack shells; their internal organs, however, are always twisted as if in a twisted shell. Well-developed head and sense organs (Figure 12-7d).

Class *Cephalopoda* (sef·uh·LOP·o·duh): free-living, fast-moving forms, with large image-forming eyes, large and obvious tentacles, mostly with greatly reduced shells. Frequently move by "jet propulsion" as water is squirted from the mantle cavity (Figure 12-7e).

After you have constructed a key using this data, try it out with known members of each mollusk class.

2. As shown in this inquiry, what features seem to be used most often for classifying animals?

3. Do you think a suitable key might be made for the internal structures of these animals? Why or why not?

4. Suppose a key to the animal phyla was designed that was absolutely fool-proof and had references to all exceptions to the usual features used in classifying. How might you describe such a system?

5. Why wouldn't a key to orders of insects separate them on the basis of presence or absence of an exoskeleton? on the basis of numbers of legs?

6. How valuable would such a key be, based on the presence or absence of wings? on differences in body form?

7. Would it be useful to construct a key that identified species of animals on the basis of their behavior? their geographical distribution? their food preferences? their economic importance?

8. What would be wrong with classifying animals simply on the basis of color?

9. Butterflies, birds, and bats all have wings. Why aren't they classified together?

10. Cats, seals, and bears are classified in the same order of mammals (*Carnivora*). These animals do not look very much alike. Why are they considered members of the same order?

SCHEMES OF CLASSIFICATION

Like any other aspect of science, systems of classification are changing constantly. The systems change over the years as scientists learn more. Scientists may begin to understand a certain kind of organism better, or they may simply agree that one system seems more logical and easier to use than another. We are all familiar with two classification kingdoms—plants and animals. Most of the organisms we encounter every day seem to belong very definitely to one or the other. But in recent years, many organisms that do not fit into either the plant or the animal kingdom have been identified. The *Euglena* shown in Figure 12-8 is an example. It is green and carries on photosynthesis like a plant. But it swims about like an animal. To which kingdom does it belong?

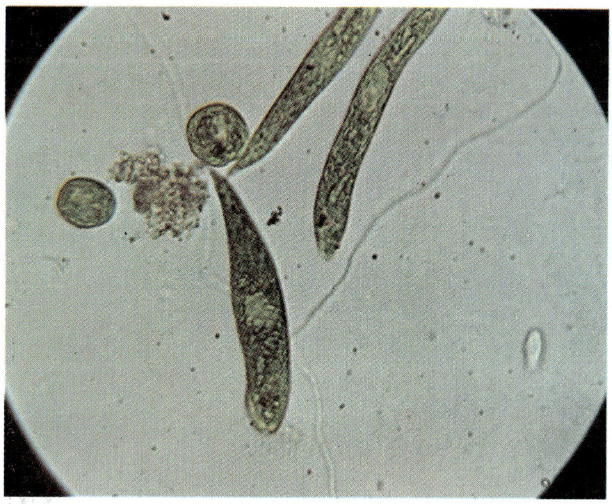

12-8. *Euglena*—plant or animal?

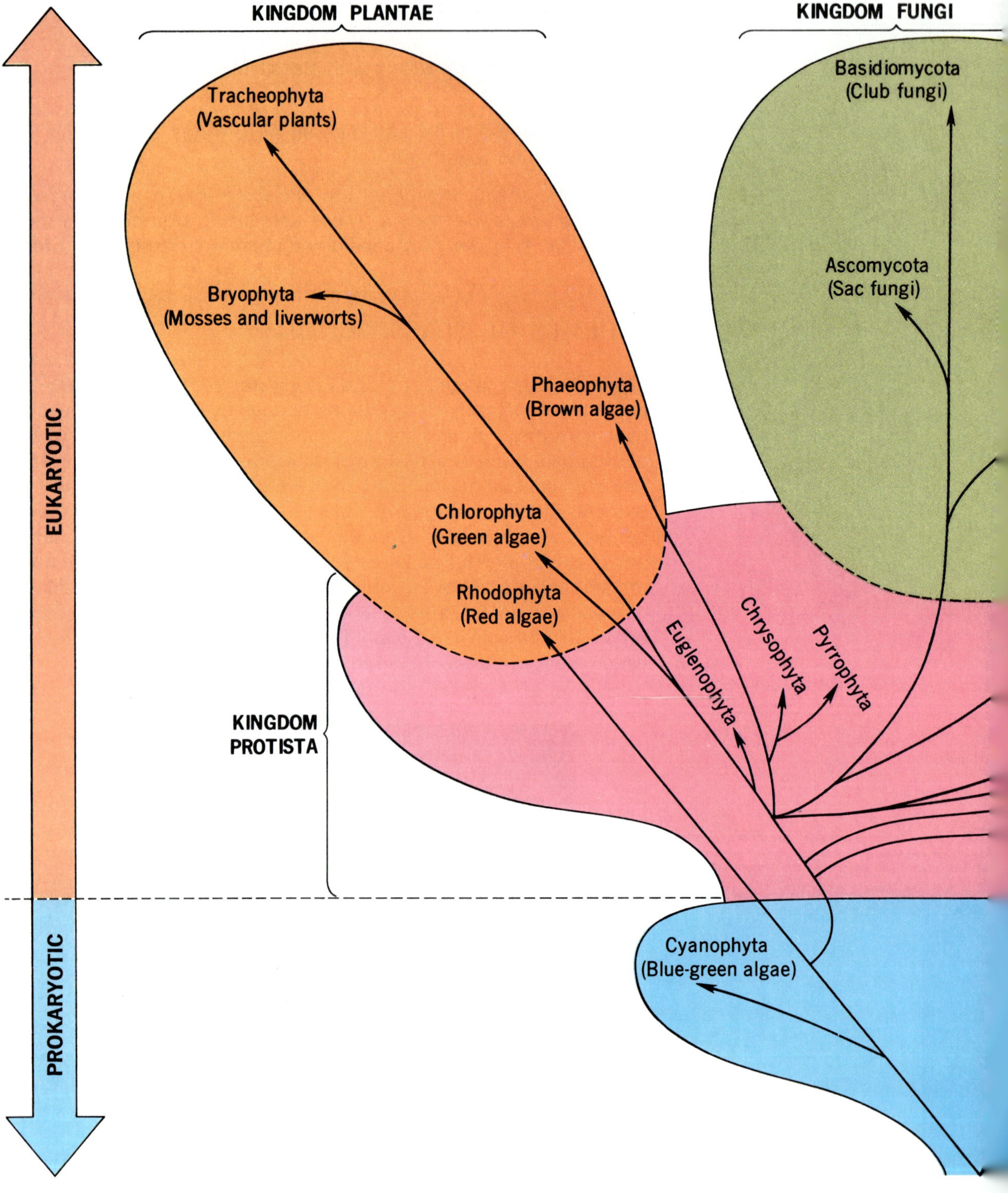
KINGDOM PLANTAE
KINGDOM FUNGI
Basidiomycota
(Club fungi)
Tracheophyta
(Vascular plants)
Ascomycota
(Sac fungi)
Bryophyta
(Mosses and liverworts)
Phaeophyta
(Brown algae)
EUKARYOTIC
Chlorophyta
(Green algae)
Rhodophyta
(Red algae)
Euglenophyta
Chrysophyta
Pyrrophyta
KINGDOM
PROTISTA
Cyanophyta
(Blue-green algae)
PROKARYOTIC

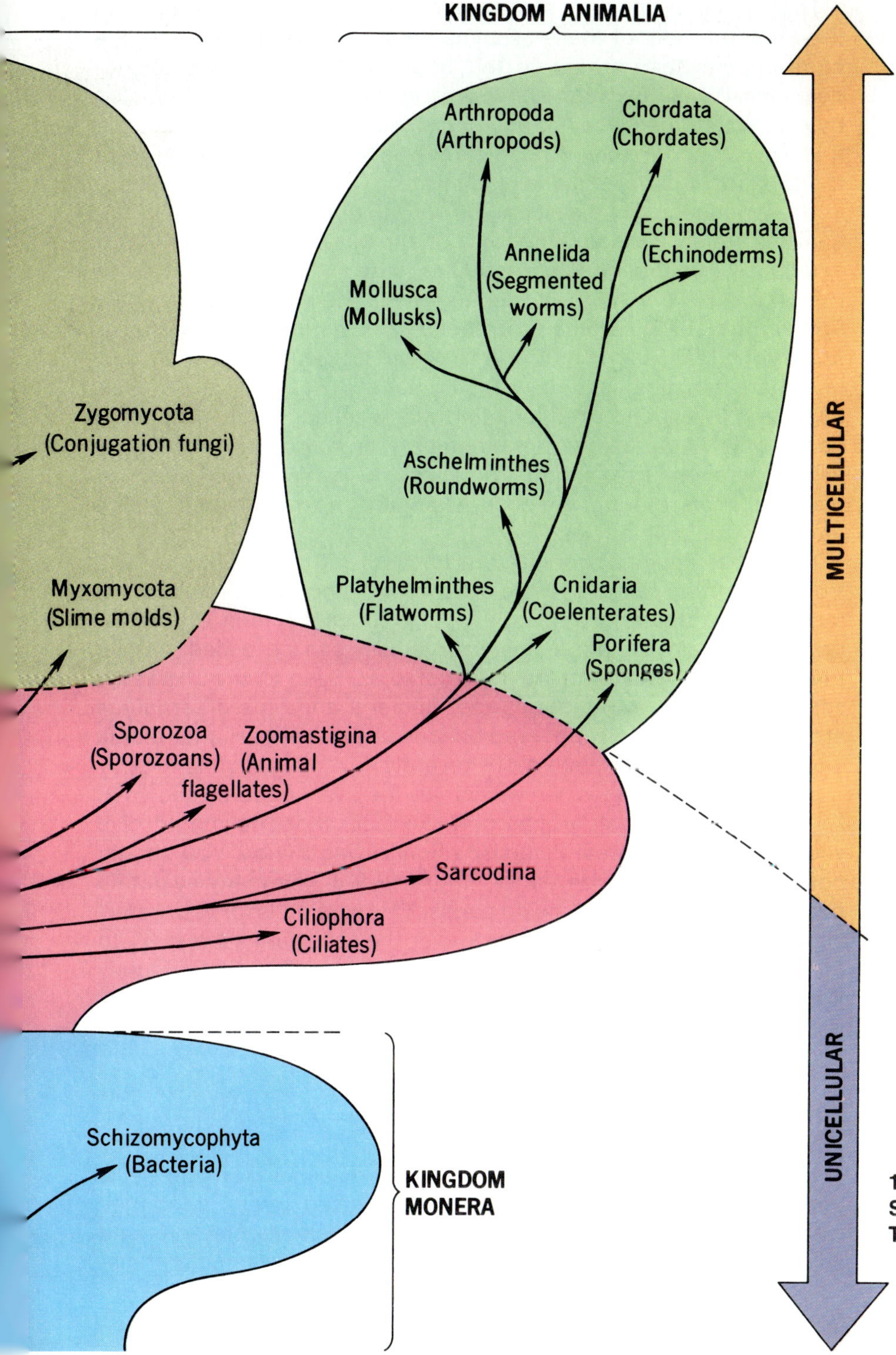

12-9. A FIVE KINGDOM SYSTEM OF CLASSIFICATION

There are classification systems that include two, three, four, and five kingdoms. All of these systems are in use, although some are more popular than others. In this book, we will use the five kingdom classification system shown in Figure 12-9. We have selected this system not because it is any better than the others, but because it is easy to understand. It also allows us to pay some attention to those exceptional or unusual organisms.

The logic behind the system shown in Figure 12-9 is simple. Organisms are divided into kingdoms on the basis of two sets of characteristics. The first characteristic has to do with whether there is a true nucleus—one with a nuclear membrane—within the cell. **Prokaryotic** (PRO·care·ee·ot·ik) **cells** lack a nuclear membrane. **Eukaryotic** (YOU·care·ee·ot·ik) **cells** have a true nucleus. The second characteristic has to do with whether the organisms are unicellular (single-celled) or multicellular (many-celled). Of the five kingdoms, only Monerans are prokaryotic. They also are unicellular. Protists are eukaryotic and unicellular. Organisms classified as plants, animals, or fungi are both eukaryotic and multicellular (some have only multiple nuclei).

Two other important points are implied in Figure 12-9. Moving from the bottom of the chart to the top, we encounter organisms of increasingly greater complexity. We think of bacteria and blue-green algae as "simple"—if any organism can be called simple. Plants, animals, and fungi are more complex. Second, the chart says something about evolutionary age. Bacteria and blue-green algae are very old. They first appeared billions of years ago. Higher plants and animals evolved much more recently.

In the following pages, we will survey the five kingdoms shown in Figure 12-9. The photographs and short descriptions should give you an idea of the great diversity among living things. You should not attempt to memorize either Figure 12-9 or the many scientific names you find in the following pages. This section is for your reference and enjoyment. As you look over the following survey, try to answer these questions: How are the organisms alike? How are they different? Why are members of the same phylum grouped together? What other grouping systems can you suggest?

KINGDOM MONERA

The Monera are the only organisms that have prokaryotic cells. Prokaryotic cells are those with no organized nuclei contained within a nuclear membrane. These cells also lack mitochondria and chloroplasts and have simple chromosomes (a single, circular molecule of DNA). This kingdom consists of the bacteria and the blue-green algae.

Phylum *Schizomycophyta* (Bacteria)

Bacteria are one-celled organisms with cell walls but no organized nuclei. All the specimens on this page are microscopic. The earliest fossils date to several billion years ago.

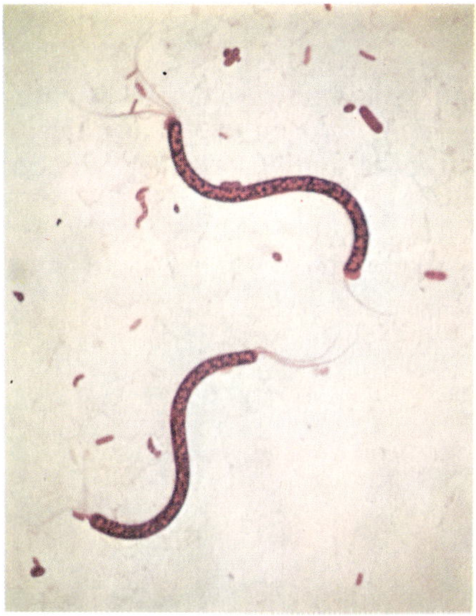

Spirilla

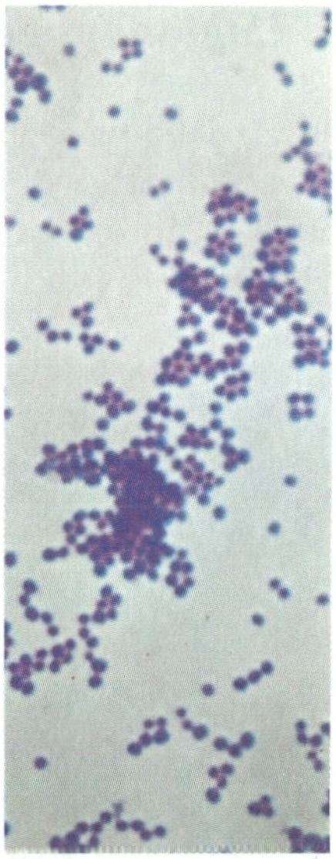

Cocci

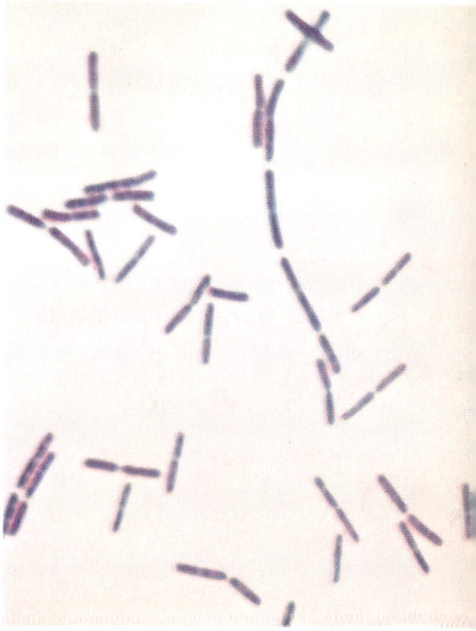

Bacilli

Phylum *Cyanophyta* (Blue-green algae)

Like the bacteria, blue-green algae are one-celled organisms that have cell walls but no organized nuclei. They have chlorophyll but no chloroplasts (chlorophyll-containing structures). Blue-green algae appear in the fossil record of over three billion years ago.

Oscillatoria, a freshwater blue-green alga

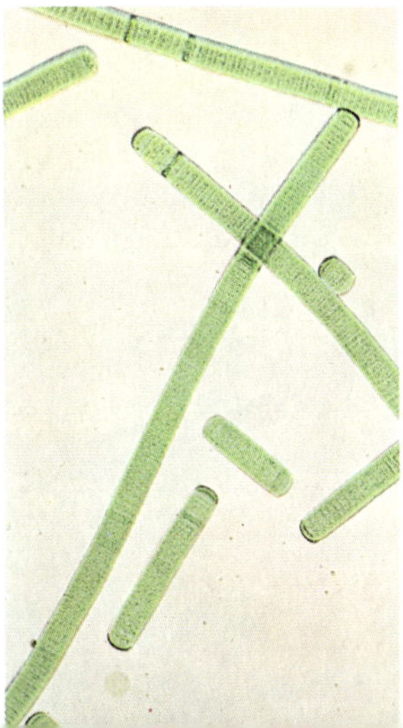

Fossil blue-green algae

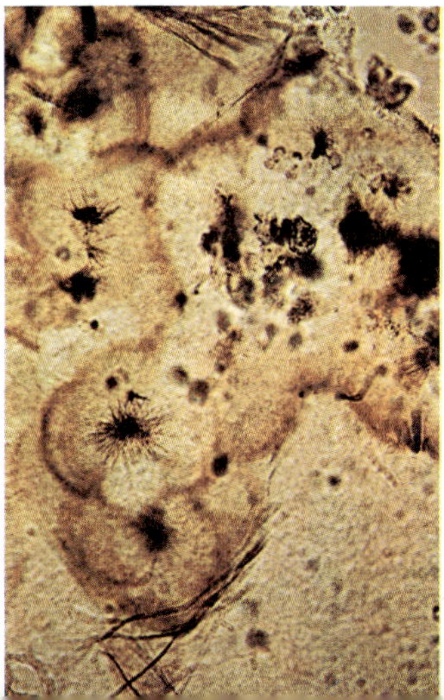

Nostoc, a freshwater blue-green alga

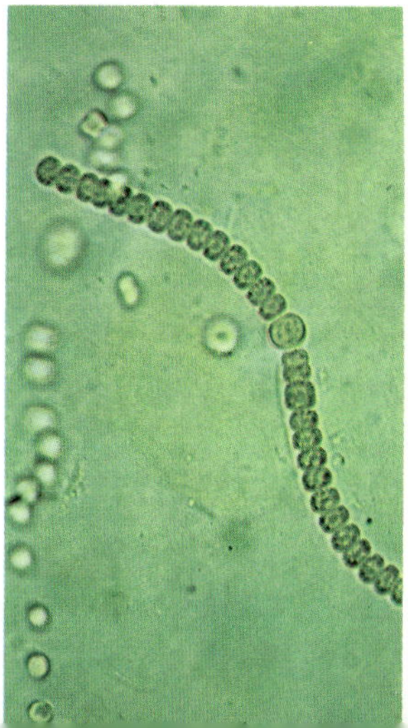

KINGDOM PROTISTA

The Protista are unicellular organisms. They have eukaryotic cells, complex chromosomes of DNA plus protein, and mitochondria.

Phylum *Euglenophyta*

These organisms have some plantlike and some animallike traits. The most common of the group, *Euglena* (page 359), has chloroplasts, but like an animal, it moves about and has a gullet through which it ingests its food.

Peranema. A member of the phylum Euglenophyta that lacks chlorophyll.

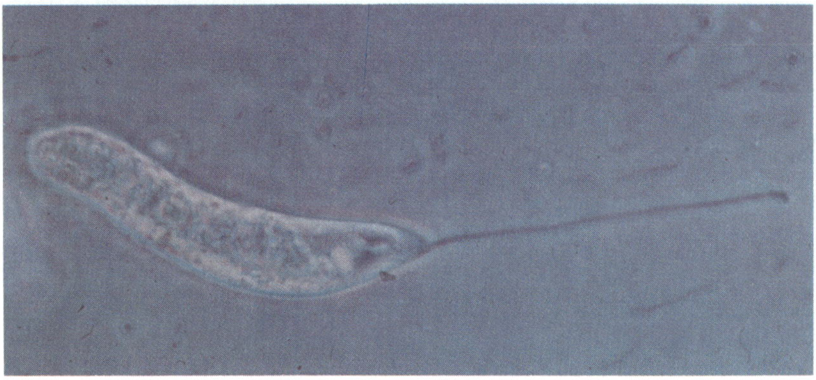

Phylum *Chrysophyta* (Golden algae)

Most of these organisms are unicellular, although a few are colonial (single cells grouped together). They are photosynthetic and store their food as oil or as a complex carbohydrate. Most of the species are diatoms—the principal component of plankton, the first link of most food chains in the water.

Diatoms are the most abundant members of this phylum.

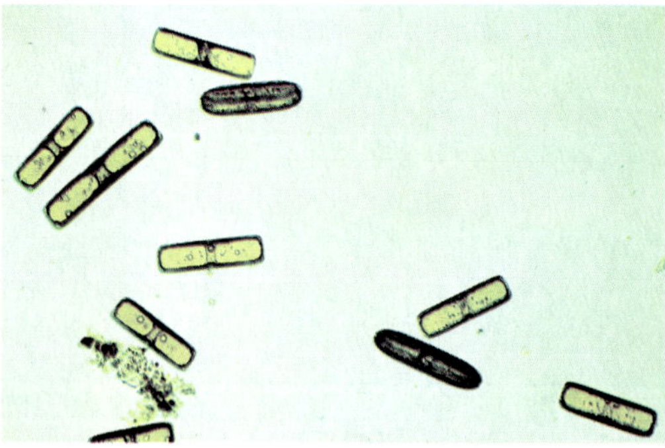

Phylum *Pyrrophyta* (Fire algae)

This group gets its common name "fire algae" from the fiery red color of many members. The most numerous species of this group are the dinoflagellates. These are the organisms responsible for the "red tides" that may kill large numbers of fish.

Peridinium

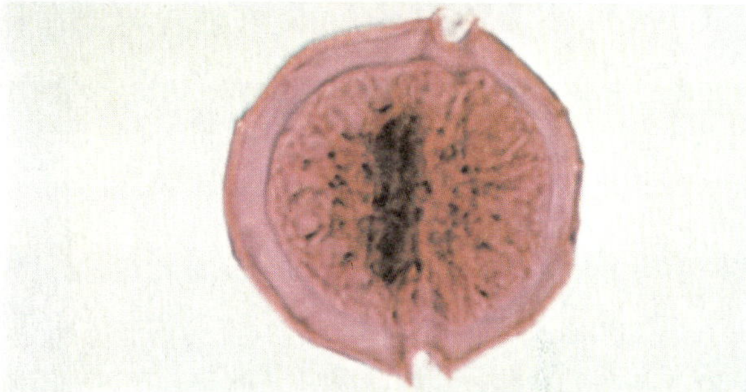

Carolina Biological Supply Company

Phylum *Sporozoa*

Members of this phylum are internal parasites living in blood cells, muscles, or internal organs of their hosts. The parasites are often carried by ticks, flies, or mosquitoes. One of the most familiar sporozoans is *Plasmodium vivax,* the organism that causes malaria (Chapter 1).

Phylum *Zoomastigina* (Animal flagellates)

Most organisms in this phylum have flagella but no outer cell wall. Most are free-living, but some are parasitic.

Noctiluca, marine flagellates

Trichonympha lives in the guts of termites and digests wood.

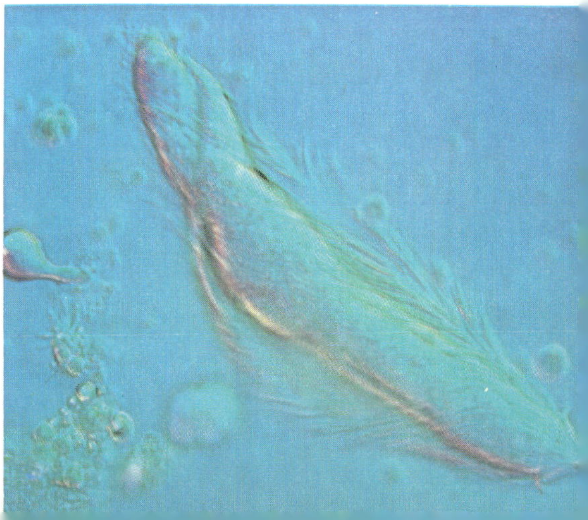

Phylum *Sarcodina*

Members of this phylum are amoebalike, with no wall outside their cell membranes. They generally move and feed by forming pseudopodia. They are found in both fresh water and salt water. Many Sarcodina, such as Foraminifera, secrete chalky shells around themselves.

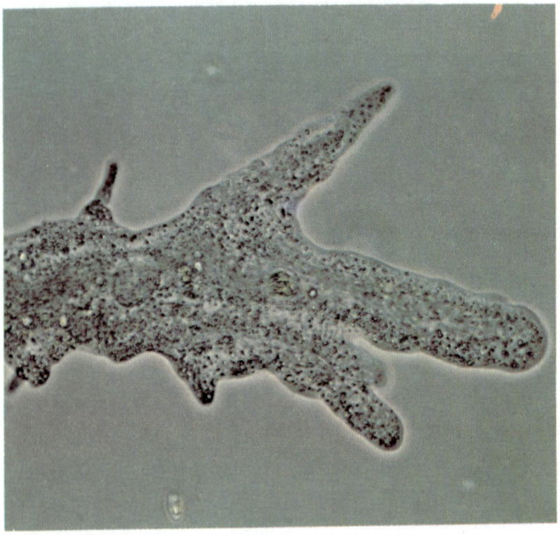

Amoeba changes shape by forming pseudopodia. In this way *Amoeba* moves about and finds food.

Foraminifera are marine creatures that secrete shells. They are abundant in the fossil record.

Phylum *Ciliophora* (Ciliates)

These organisms have cilia, tiny hairlike extensions, that are used as a way of moving about and, in some species, for food gathering. *Paramecium* is a member of this phylum.

Blepharisma, a freshwater ciliate

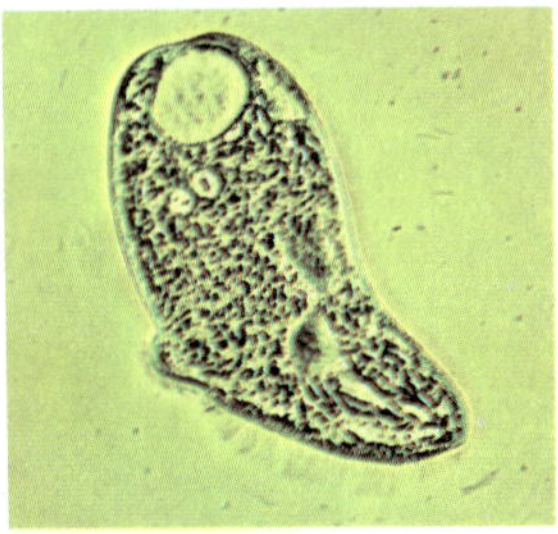

Paramecium is the most familiar and extensively studied ciliate.

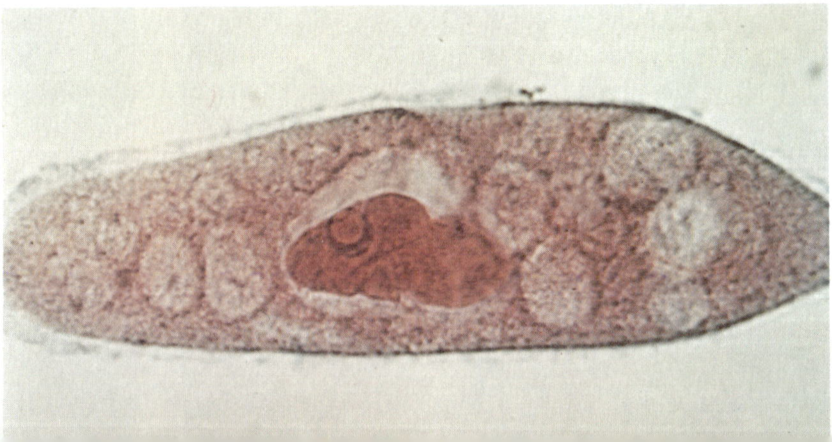

Carolina Biological Supply Company

KINGDOM PLANTAE

The Kingdom Plantae contains several hundred thousand species. These range in size from the microscopic to some of the largest organisms ever to have lived on earth. Animals rely on plants for food and oxygen.

Phylum *Rhodophyta* (Red algae)

These multicellular, red algae contain a red pigment in addition to chlorophyll. They generally live in the tide pools of warm seas. Red algae are a source of agar, which is used in media for growing bacteria in the laboratory.

Agardhiella, a red alga

Stenophycus, fossil red algae

Phylum *Chlorophyta* (Green algae)

The green algae have chloroplasts and a cell wall. They store starch. In these characteristics, they resemble green land plants. Green algae vary from one-celled forms to colonies made up of many cells. The cellular colonies of colonial green algae like *Volvox* may be thought of as the first step toward the more complex multicellular organisms.

Ischadites, a fossil green alga

Spirogyra, a freshwater green alga

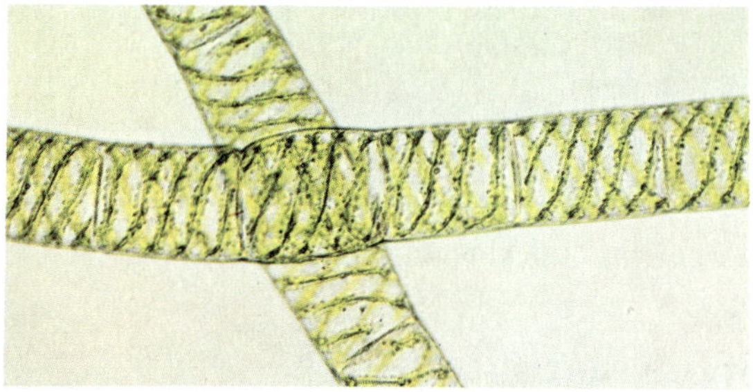

Phylum *Phaeophyta* (Brown algae)

Brown algae contain brown and yellow pigments in addition to chlorophyll. They are multicellular, with tissues that often resemble stems, leaves, and roots. All brown algae are ocean-dwellers.

Fucus

Kelp

Phylum *Bryophyta* (Mosses and liverworts)

The bryophytes are green land plants that lack a vascular system—true leaves, stems, and roots. Most are small plants living in moist environments.

Marchantia, a liverwort

Moss

Phylum *Tracheophyta* (Vascular plants)

Vascular plants have special tissues through which water and food materials move. That is, they have true leaves, stems, and roots.

Subphylum *Psilopsida*

These are the most primitive vascular plants known.

Psilotum

Equisetum, horsetails

Subphylum *Sphenopsida*

Horsetails and their relatives have cones at the tips of their reproductive branches. The horsetails once grew very large and were among the plants from which coal was formed about 400 million years ago.

Fossil *Sphenophyllum* branch. The stems bear whorls of small leaves at regular intervals.

Subphylum *Lycopsida*

This group dates from 430 million years ago. Some, such as *Lepidodendron*, were treelike. Small club mosses such as *Lycopodium* survive today.

Trunk cast of *Lepidodendron*

Lycopodium, a club moss

Subphylum *Pteropsida*

This major group of vascular plants includes all the ferns and seed plants.

Class *Felicineae* (Ferns). About 300 million years ago, ferns were the dominant plant species on earth. They are still found in abundance today in a variety of environments.

Ferns

The unfolding fronds of ferns are called "fiddleheads."

Class *Gymnospermae.* Ginkgoes, cycads, and conifers typify this class of plants. Most produce seeds in cones. Like the ferns, gymnosperms once were much more abundant than they are today.

Leaves of a ginkgo tree

Araucaria, the monkey puzzle tree

Microzamia, a cycad

A close-up of cycad cones, *Microzamia*

The world's largest organisms are the redwoods. At the base of the trees on the left is a car.

Branch of a *Pinus,* white pine tree

Bamboo

Class *Angiospermae* (Flowering plants). Today, these are the dominant land plants—vascular plants that flower and form fruits with seeds. The origin of the angiosperms and how their reproductive patterns evolved are a mystery. The oldest fossil angiosperms date from about 225 million years ago. There are two subclasses of angiosperms.

Subclass *Monocotyledoneae.* Monocots have slender leaves with parallel veins. Flower parts (petals and sepals) occur in multiples of three. The embryo has one seed leaf (cotyledon). Grasses and grain crops are monocots, as are some trees and decorative flowering plants.

Musa, banana plant

Lilium, lily

Narcissus

Common corn is a monocot.

Subclass *Dicotyledoneae.* The leaves of dicots usually have branching veins. Flower parts (petals and sepals) are usually in multiples of four or five. There are two seed leaves (cotyledons). All the common deciduous trees, most fruit and vegetable plants, and most decorative flowering plants are dicots.

Dicots are named for the two seed leaves, called cotyledons. The cotyledons (shown on the lower part of the stem) drop off as the plant matures.

Waratah, *Telopea*, a native of Australia

Cornus, bunchberry

Cactus in flower

Silver maple

KINGDOM FUNGI

The morel *Morchella* looks a little like a mushroom.

The Fungi lack chlorophyll and cannot carry on photosynthesis. Instead of being multicellular, most are multinucleate. Fungi are among the most diverse groups of organisms. They also are the most difficult to classify. Some biologists favor including them with plants or with protists, rather than classifying them as a separate kingdom.

Phylum *Ascomycota* (Sac fungi)

The sac fungi are great problems to agriculture and industry. Apple scab, brown rot of stone fruits, and mildews on grapes and roses are caused by sac fungi. On the other hand, such commercially important products as baker's and brewer's yeast are also sac fungi.

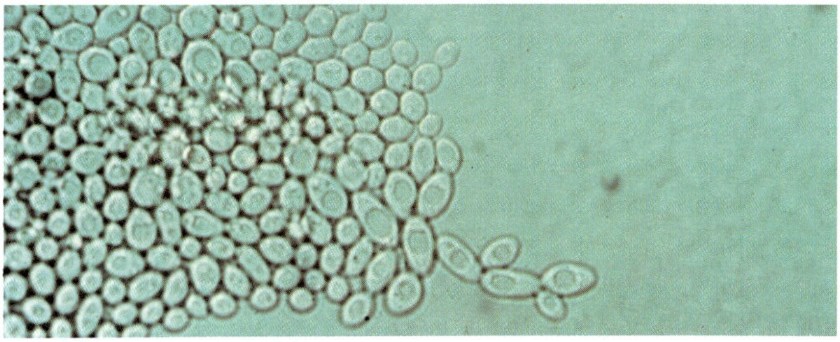

Saccharomyces is baker's yeast.

Can you see why the bracket fungus is so named?

Phylum *Basidiomycota* (Club fungi)

Included in this group are rusts, smuts, puffballs, mushrooms, and bracket fungi. The rusts and the smuts attack cereal grains, grasses, and trees.

Mushrooms

Phylum *Zygomycota* (Conjugation fungi)

Most members live in the soil, feeding on dead plant and animal matter. Some are parasites of plants, insects, or small animals living in soil.

The most familiar member of this group is *Rhizopus,* common bread mold.

Phylum *Myxomycota* (Slime molds)

The slime molds are not true fungi. Their life cycles have two distinct phases—an amoebalike stage with no cell walls (called the plasmodium) and a fruiting stage that produces thick-walled spores. Little is known of their fossil history.

A portion of the plasmodium of the slime mold *Physarum.*

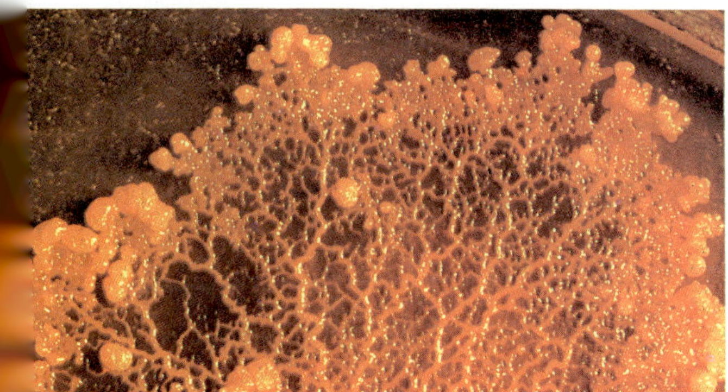

Fruiting bodies of *Physarum* growing on the surface of rotting wood.

KINGDOM ANIMALIA

Members of the Kingdom Animalia are the dominant life form on the earth today. From the sponges to the chordates, the animal kingdom shows great diversity. Among its members are the insects, the most numerous of all animal species.

Phylum *Porifera* (Sponges)

Sponges are the simplest of all animals. They have many cells but no highly specialized tissues or organs. Often they do not even have a definite shape. Sponges have been found in rocks that are more than 600 million years old.

A sponge

Two different kinds of sponges

Phylum *Cnidaria* (Coelenterates)

Coelenterates have a central digestive cavity and a nervous sytem of a primitive kind. Like the sponges, all are aquatic and most are marine. Though simple in anatomy, coelenterates are responsible for the largest structures produced by any living organism. In the warmer seas of the earth, corals and other coelenterates form enormous reefs that surround many islands. Besides coral, common members are the jellyfish, sea anemone, and Portuguese man-of-war. One well-known freshwater coelenterate is the *Hydra.*

Astroides, a coral

Madrepora, a coral

A sea anemone

Portuguese man-of-war

Phylum *Platyhelminthes* (Flatworms)

These flat-bodied worms are small, inconspicuous animals with digestive, nervous, excretory, and reproductive systems. Little is known of the fossils of these soft-bodied organisms. Some of the flatworms alive today—flukes and tapeworms—are parasites on animals or humans.

Dugesia, a freshwater planarian

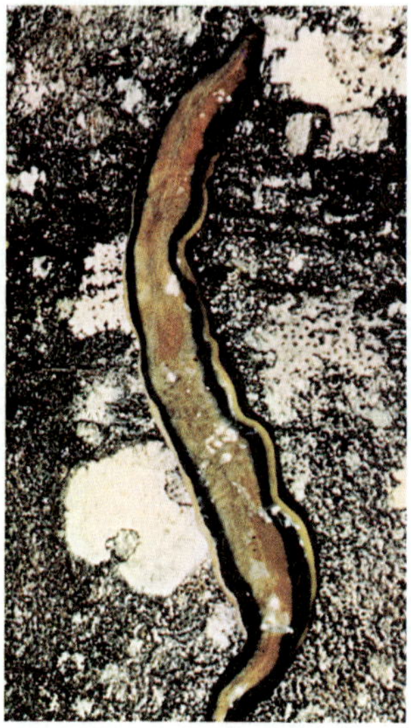

A land planarian from the tropics

Carolina Biological Supply Company

The tapeworm *Taenia* is parasitic.

Phylum *Aschelminthes* (Roundworms)

Roundworms are found in a variety of environments—in the soil, in seas, in fresh water, and as parasites on plants and animals.

Ascaris is an intestinal roundworm of the pig.

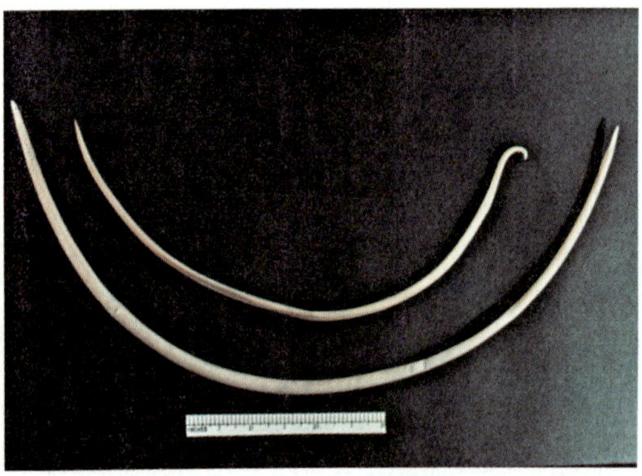

A roundworm, *Trichurus*, a parasite in humans.

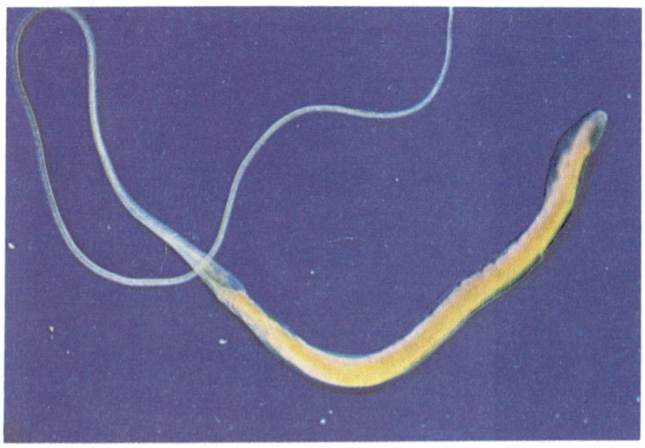

Phylum *Annelida* (Segmented worms)

These worms have segmented bodies that are much more complex than those of flatworms and roundworms. The annelid worms have a circulatory system and, in some species, special structures for respiration. Most annelids live in the ocean. Some swim freely. Others live in burrows or in tubes that they secrete. A few species of annelids live in fresh water or in moist soil.

The common earthworm is an annelid. Note the segments of the body.

Phyllodoce, a marine annelid.

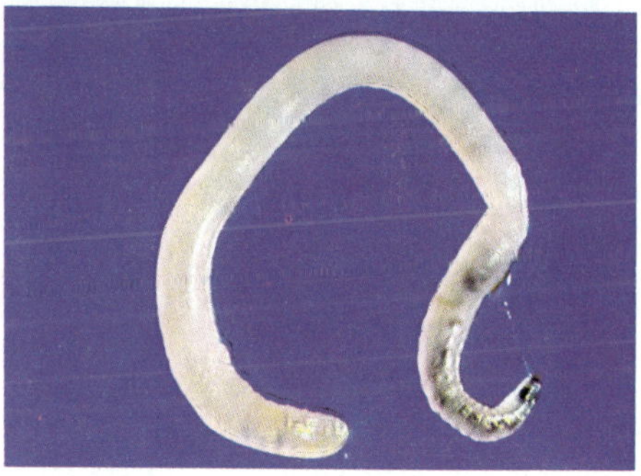

Enchytraeus, a land annelid.

Bispira, a marine annelid in its tube

Mytilus, marine mussels

Phylum *Mollusca*

The mollusks are much like the annelid worms in their general level of complexity, but the body is not segmented in most species. Nearly all mollusks have shells, which may be single, double, composed of several parts, or small and situated within the body. Mollusks live in the ocean, in fresh water, or on land.

Most species of mollusks, especially those with heavy shells, move slowly, if at all. The underside of the body is specialized as a muscular foot. A few mollusks, such as the squid, can swim rapidly. Many mollusks are used as food—clams, scallops, mussels, oysters, and snails.

Carolina Biological Supply Company

Helix aspersa, a garden snail

Ilex, a squid

Paradoxides, a fossil trilobite from over 500 million years ago.

Phylum *Arthropoda*

Arthropods are the most complex and most numerous of all animals without backbones (invertebrates). Many species have well-developed brains, complex sense organs, and complex behavior patterns. The body is enclosed in a tough exoskeleton, jointed in the legs and often in other parts of the body. Because arthropods are so numerous, varied, and important, we will look at some of the major classes.

Class *Trilobita.* All members of this class are extinct. Because trilobites were heavily armored, they were easily fossilized. Therefore, they form an important part of the fossil record.

Class *Arachnida.* All members of this class have twelve appendages. Eight of these are legs. Four are modified for feeding.

Garden spider

Scorpion

Harvestman

Class *Crustacea.* These are the dominant arthropods of the sea. These arthropods have two body regions and many appendages, including two pairs of antennae. Included in this group are shrimp, sowbugs, sand fleas, lobsters, crayfish, crabs, and barnacles.

Marine Studios, Marineland, FL

Peneus, the pink shrimp

Stone crab

Barnacles

Carolina Biological Supply Company

Scolopendra, centipede

Class *Chilopoda.* Centipedes, which capture and eat insects and other small animals, are the representative members of this group. They have one pair of legs on each body segment.

Class *Diplopoda.* Millipedes eat plants and dead animal material. They have two pairs of legs on each body segment.

Carolina Biological Supply Company

Harpaphe haydeniana, millipede

Class *Insecta.* By far these are the most numerous and diverse arthropods. There are one to three million species of insects. And the number of individual insects is staggering—about a billion insects for every person on earth. Insects differ from other arthropods. They have three basic body parts (head, thorax, abdomen), a single pair of antennae on the head, and three pairs of legs originating from the thorax.

Walking stick

Libellula saturata, dragonfly

Hippoboscidae, louse fly

Stenopelmatus intermedius. Jerusalem cricket

Pneumoridae, bladder grasshopper

Marine Studios, Marineland, FL

Astropecten, a sea star

Phylum *Echinodermata*

Echinoderms are marine animals with radially symmetrical bodies. They may be shaped like circles, spheres, cylinders, or stars. The body is usually divided into five parts or some multiple of five. Organ systems are at a low level of organization. The development of the embryo somewhat resembles the development of the embryo of the chordates. (Chordata is the phylum that includes animals with backbones.) For this reason, biologists believe that the echinoderms are the invertebrates most closely related to chordates.

Echinus, a sea urchin

Sea star

Sea urchin

Pumpkinseed fish

Phylum *Chordata*

This is the phylum that includes humans. Chordates possess, at one time or another in their lives, a dorsal nerve tube and a notochord (a stiff rod of tissue ventral to the nerve tube). They have paired gill slits—which have a respiratory function in the lower chordates. The oldest fossil chordates date from 500 million years ago. Most chordates are vertebrates—that is, they have a vertebral column, or backbone. We will consider a few of the classes of chordates.

Class *Osteichthyes* (Bony fishes). Fishes are the principal large-bodied animals of the sea and fresh water. They are the most primitive vertebrates. Their fossils date from 500 million years ago. Fishes have gills for respiration and most have an outer covering of scales.

Pristigenys alta, short bigeye

Selene vomer, lookdown

Fossil *Eryops*, a primitive amphibian from rocks more than 200 million years old.

Class *Amphibia*. Nearly 400 million years ago, some advanced species of fish seem to have evolved into amphibians. Fins gradually evolved into arms and legs. Two main groups of amphibians survive today—those with tails (the salamanders), and those without tails (frogs and froglike species). Both groups generally live in or near water for some or all of their life cycle.

A tree frog

Bufo, a toad

A horned lizard

Class *Reptilia.* Evolution away from a dependence on water, which began with the earliest amphibian, was completed in about 40 million years with the appearance of primitive reptiles. Unlike amphibians, whose embryos generally develop in water, reptiles have embryos that are either kept in the mother's body until they are born or develop in eggs laid on dry land. The outer covering of a reptilian body, usually scales, prevents excessive water loss in a land environment. The major groups of living reptiles are turtles, snakes, lizards, crocodiles, and alligators. Dinosaurs, at one time the earth's dominant animals, are now all extinct.

Reptiles are thought to be the ancestors of both the birds and the mammals.

Alligator

Fossil *Tyrannosaurus*, a huge meat-eating dinosaur of about 100 million years ago.

Atlantic loggerhead

Class *Aves*. Between 190 and 150 million years ago, one group of primitive reptiles evolved into birds. The earliest known bird was *Archaeopteryx*. It lived in Europe about 150 million years ago. It had many reptilian characteristics such as teeth and a long bony tail. But it also had feathers—the most unique characteristic of birds. Birds differ from lower chordates in having a nearly constant body temperature. But their embryos develop in eggs not unlike those of some reptiles. However, parent birds give their young prolonged care and feeding after the offspring hatch. Though there are birds with a limited ability to fly and even some flightless birds, sustained flight is a trait of most living birds.

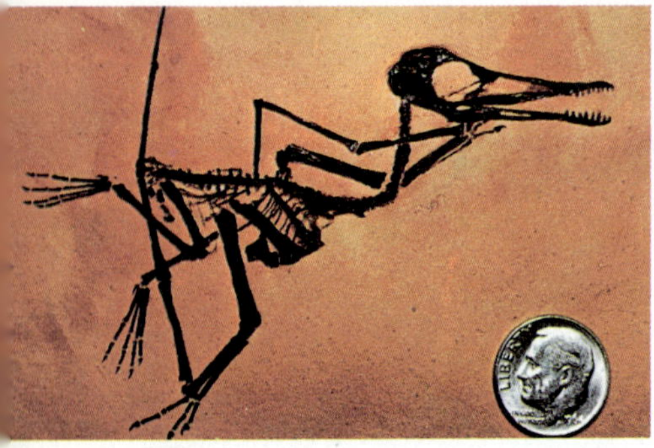

Fossil *Pterodactylus*, a tiny flying reptile

Penguins are among the few flightless birds.

Pelican

Fossil *Archaeopteryx*, the first known bird

Cyanocitta stelleri, Steller's jay

Gallus domesticus, rooster

Parrots

Class *Mammalia.* Between 190 and 65 million years ago, some rather primitive reptiles evolved into mammals. But the most primitive fossils fail to suggest the future importance of the mammals. Dinosaurs dominated the landscape of the time. Yet it was the mammals, covered with hair, and feeding their offspring with milk, that were to dominate the future. Mammals today are of three major groups. Monotremes lay eggs. Marsupials have pouches in which the young develop. Placentals have special internal structures that nourish the unborn young before birth.

Fossil *Lycaenops*, a mammal-like reptile from rocks about 250 million years old

Tachyglossus, the echidna, is a monotreme.

Phascolarctos cinerens adustus, koala bear, a marsupial

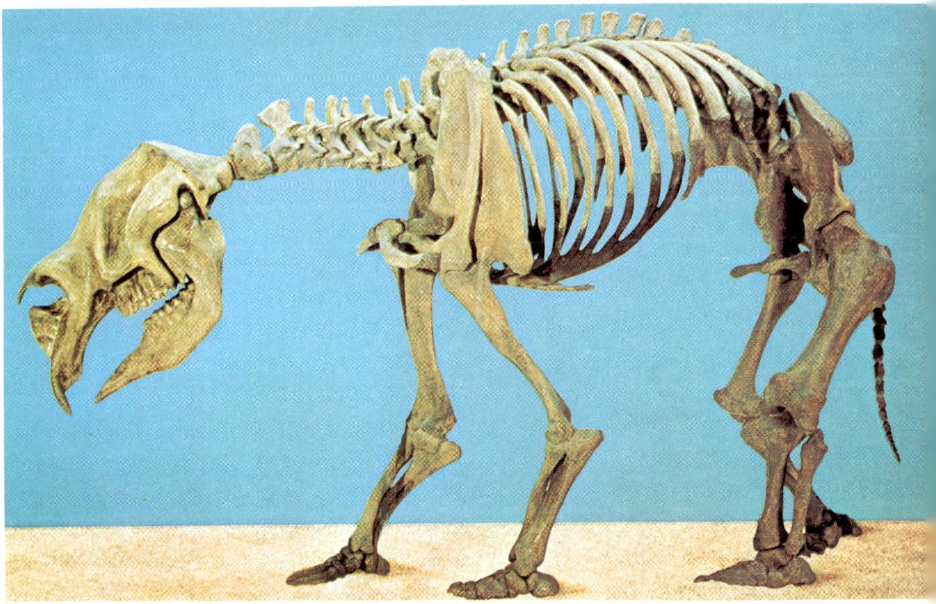

Diprotodon, a giant fossil marsupial that lived in Australia about 3 million years ago.

Tenrec, an insectivore

Most of the familiar mammals are placentals. The young are kept within the mother's body until they reach an advanced stage of development. During this period, the placenta supplies the embryo with nourishment from the mother. The oldest known placental mammals were very small. They were insectivores (insect-eaters). Some relatives of the earliest insectivores still survive.

Living placental mammals are placed in about 16 orders, beginning with the insectivores (shrews, moles, and hedgehogs). Bats, the only true flying mammals, make up another order.

Mole

Mexican vampire bat

An order of toothless mammals, or edentates, includes anteaters, armadillos, and sloths. The aardvark order has a single genus.

Giant anteater

Orycteropus, the aardvark

About 200 species are included in the order of primates. *Homo sapiens* is one of these species. There have been other human species, but they are now extinct. In many ways, primates seem most closely related to insectivores. Indeed, tree shrews seem so closely related to their shrewlike ancestors that many biologists disagree about whether they are primates.

The lemur group includes the true lemurs, lorises, galagos, and their relatives. They are among the most primitive primates alive today. Tarsiers also are primates.

Tree shrew

Tarsier

White-fronted lemur

There are four more advanced families of living primates: New World monkeys, Old World monkeys, the great apes, and human beings.

Douc Langur, an Old World monkey

Red howler, a New World monkey

Homo sapiens, human

The order of rodents is one of the most successful mammalian orders. Besides muskrats, beavers, and rats, it includes squirrels, chipmunks, gophers, mice, woodchucks, porcupines, and guinea pigs, among others. The incisor teeth of rodents grow throughout life, replacing portions worn away by gnawing and cutting hard materials.

Rabbits and hares are placed in a separate order of mammals—the lagomorphs. Rabbits and various rodents are widely used in biological and medical research.

Cottontail rabbit

Squirrel

Prairie dogs

Although mammals evolved on land and have been most successful there, the pressures of evolution seem to have sent some of them back to the seas. The cetaceans—whales, dolphins, and porpoises—live in the ocean. Their forelimbs are fins; the hindlimbs are but vestiges under the skin, which has almost no hair. Although they live in the ocean, cetaceans use their lungs for respiration.

Tursiops, a bottle-nosed dolphin

Marine Studios, Marineland, FL

The order of carnivores includes our most common pets, dogs and cats. Also included are bears, lions, tigers, and other cats. As their name implies, carnivores eat meat. Bears, raccoons, and some of the dogs are really omnivorous—they eat animals and plants.

Wolf

Bengal tiger

Lion

The ungulates are hoofed mammals. There are two major orders, which are not closely related. The odd-toed ungulates have one or three toes. The even-toed ungulates have two or four toes. The living horses and zebras, with one main toe, and the rhinoceros, with three toes, belong to the order of odd-toed ungulates. The tapir is usually included in this order even though it has three toes on the hind foot and four on the front foot.

The order of even-toed ungulates includes the elk, deer, llama, gazelles, hippopotamuses, pronghorns, camels, goats, and many others. Our principal food mammals—cattle, sheep, and pigs—belong to the order of even-toed ungulates.

Hippopotamus

Zebra

Bighorn sheep

Hereford cattle

The elephant order includes the African elephant and a fossil species of mastodon. The elephants are the largest mammals that live on land. If you have ever watched an elephant in the zoo, you will have observed that the trunk serves many of the same functions as our hand or arm. The tusks are greatly enlarged teeth of the upper jaw.

Mammut, a fossil mastodon

Loxodonta, an African elephant

LIFE THROUGH TIME

The classification of the organisms shown in this survey is based on evolutionary relationships. Through millions, even billions of years, life on earth has evolved from the tiniest and simplest forms into the tremendous complexity and variety we see today. Let's briefly trace what biologists think evolutionary history must have been like.

Before Life Began

Astronomers, geologists, and chemists are reasonably sure that the earth was formed billions of years ago, when a mass of extremely hot gases and dust particles condensed into a hot, molten (liquid) mass. At such high temperatures, atoms combine, separate, and recombine rapidly to form many kinds of molecules. As the molten earth slowly cooled, the heavier dust particles probably accumulated and formed its core. Lighter particles formed the surface layers. Water vapor in the gases condensed and filled low places on the surface with water, forming oceans. Rivers carried dissolved minerals into the oceans, and other substances settled out of the atmosphere into the water. From these elements, new chemical compounds were formed. Scientists know a lot about the behavior of chemical substances at high temperatures. Thus scientists feel certain that, at first, the earth's crust was very different from what it is now.

The primitive earth probably had an atmosphere of methane (CH_4), ammonia (NH_3), hydrogen sulfide (H_2S), hydrogen (H_2), and much water vapor (H_2O). These substances must have combined gradually into increasingly complex molecules. Finally, the complex molecules formed organized groups of molecules. These groups of molecules had a unique feature. They could organize the chemical substances around them into other groups of molecules like themselves.

What could these hypothetical molecules have been? Having no direct evidence, scientists can only guess. Since the molecules could self-replicate, or make copies of themselves, possibly they were a nucleic acid like DNA or RNA. Of all the molecules we know of today, only DNA and RNA can be replicated directly. Could these first replicating molecules have been a nucleic acid? Perhaps. And if they were, we would have an explanation of why DNA or RNA is the basis of inheritance in all organisms.

Is there really any evidence for such a hypothesis? In 1953, Stanley Miller published an article entitled "A Production of Amino Acids Under Possible Primitive Earth Conditions." In a special apparatus, he simulated the conditions believed to exist on the primitive earth. In Miller's apparatus, water was boiled continuously in a flask. The rest of the apparatus was filled with gases—methane, ammonia, hydrogen, and water vapor. These gases were thought to be present in the early atmosphere of earth.

Miller sent electrical discharges through the gases. The electrical discharges represented violent electrical storms that may have taken place around the primitive earth. Miller could have used ultraviolet radiation, since it probably was a common form of energy at that time. In any case, the main requirement was intense bursts of energy.

Miller allowed the experiment to continue for a week. By the end of that time, his colorless solution had turned red. He analyzed the solution and found that it contained a great variety of organic molecules. Some he could not identify with certainty. However, some of the molecules present were amino acids, the building blocks of proteins, which are the basic components of living matter.

In similar experiments, other investigators have obtained the bases adenine and guanine. Ribose and deoxyribose also have been produced. Thus, it is no longer so far-fetched to imagine a slow evolution of organic molecules in a primitive ocean.

Under laboratory conditions, scientists have observed the assembly of amino acids into microscopic spheres. These spheres are called *coacervates* (coh·AS·ur·vates). Coacervates also may be formed by mixing together various proteins and other materials. Coacervates have a membranelike outer boundary. These organized droplets apparently also grow. Careful experiments have ruled out the possibility that the growth is caused by intake of water. But coacervates have no nucleic acids. Thus, their growth processes cannot be the same as those occurring in living creatures.

One of the characteristics of a living thing is that it can reproduce its own kind using the raw materials of the environment. Modern living things require nucleic acids and proteins to reproduce. The primitive systems of living matter probably were formed and disintegrated on many occasions. But some systems survived, and from these, life on earth may have originated. Because coacervates have some of the properties of living things, such droplets might have been an important step in the origin of life.

In **Inquiry 12-3,** you can produce coacervates and study the conditions under which they form.

INQUIRY 12-3 Coacervate Formation

MATERIALS

gelatin dissolved in distilled water
gum arabic dissolved in distilled water
0.1 M hydrochloric acid
living amoebas (optional)
pipette
medicine dropper
test tubes with corks
microscope slides
cover glasses
microscope
pH test paper

Procedure

Mix together 5 ml of the gelatin solution and 3 ml of the gum arabic solution in a test tube. Gelatin is a protein. Gum arabic is a carbohydrate. Measure the pH of this mixture.

Place a drop of the liquid on a slide and observe it under low power of the microscope.

Carefully add the hydrochloric acid solution, drop by drop, to the test tube. After adding each drop, mix well. Then wait a few seconds to see if the mixture becomes cloudy. If the liquid in the tube remains clear, add another drop of acid. Continue adding acid a drop at a time until the mixture becomes cloudy. Then take another pH reading.

Carefully observe a drop of the liquid under the microscope. Look for structures resembling spherical droplets. Often the droplets are grouped together in a shape not unlike that of an amoeba. If you cannot see them, adjust the light and use high power. If you still do not observe coacervate droplets, repeat the procedure from the beginning, for you may have added acid too rapidly. When you are successful, record your observations and make sketches of the coacervate droplets.

Examine a slide of living amoebas and compare their structure and organization with the larger coacervates.

Add more hydrochloric acid to the test tube, a drop at a time. Mix after adding each drop. Measure the pH after every third drop. Continue until the solution becomes clear. Examine a drop under the microscope and measure the new pH. **1.** How do the materials that you used to make coacervates compare with those that might have been present in the ancient ocean? **2.** In what pH range did the coacervate droplets form? **3.** Did the pH change as you gradually added more acid to the solution? **4.** When hydrochloric acid was added beyond a certain point, the coacervates disappeared. What might you add to the test tube to make the coacervates reappear? **5.** How might the coacervate droplets be made more visible under the microscope?

Earliest Fossils

The earliest known organisms to be preserved as fossils were almost certainly cellular. They were probably similar to modern microorganisms, particularly blue-green algae and bacteria (Figure 12-10). However, soft-bodied organisms such as bacteria and algae rarely are fossilized. The organisms that fossilize well, such as the mollusks, echinoderms, arthropods, and vertebrates, all have hard parts such as bones or other substances. Yet these forms are not

12-10. A limestone reef several billion years old. Biologists believe the concentric layers of limestone were formed by the blue-green algae.

found in rocks formed more than 600 million years ago. It could be that any fossils contained in older rocks were destroyed, since the rocks were greatly altered by heat, pressure, and erosion.

About 600 million years ago, conditions on the earth must have changed. The rocks of this period contain evidence of a rich variety of animals and plants. All lived in water. The land masses must have been desolate places. But the oceans were full of life (Figure 12-11).

12-11. A model of life in the sea 600 million years ago.

On the ocean floor were sponges, coelenterates, and brachiopods (BRAY·kih·oh·podz)—shelled animals that resemble mollusks but belong to a separate phylum. Among the more active animals were certain arthropods, the trilobites (TRY·loh·bites). Some were 30 centimeters in length. They swam and crawled along the ocean floor. Eventually, they became extinct.

As millions of years passed, new kinds of animals and plants evolved. Great reefs were built by corals and algae. Along these reefs lived brachiopods, mollusks, and other animals. The nautiloids (NAW·tih·loidz) were especially common. These mollusks were the ancestors of today's squids and octopuses.

Green Plants Move onto the Land

About 500 million years ago, the earth's atmosphere changed. Green plants began to grow on the land. These green plants evolved into many different types. One group remained small like the algae of the ocean. They lived on the almost-bare rock surfaces where moisture is abundant at some seasons, but soil and mineral matter are scanty. This first group gave rise to the mosses and liverworts. A second group evolved an ever-increasing amount of conducting and supporting tissue and true leaves. This group evolved into the ferns and all other higher plants.

12-12. A reconstruction of a forest in the coal-forming age.

12-13. A model showing the earliest chordates, about 350 million years ago. At the bottom center are models of ostracaderms, early jawless fish. At the left are several species of more advanced fish. These fish had jaws and paired fins.

Land plants were not very common until about 350 million years ago. By that time, there were forests. The plants in these forests were not like those in our forests today. The trees were actually giant species of club mosses, horsetails, and ferns.

An even more spectacular development of land plants occurred up to about 265 million years ago. This is the time when the great coal deposits were formed. Coal beds are the fossilized remnants of plants that grew in swamps. The world in those days was largely tropical and subtropical, even in areas where it is now very cold, such as Greenland and Antarctica. The forests were jungles of lush vegetation. The treelike forms were related to the club mosses (Figure 12-12).

These spectacular examples of evolution among the plants were rivaled by those of the animals. Ancestral dragonflies with wing spans of 50 to 75 centimeters, giant cockroaches, scorpions, and other primitive insects lived in the forests. Many species of amphibians lived in swamps.

Animals Move onto the Land

Early chordates (Figure 12-13) were primitive fish called ostracoderms (os·TRAK·oh·dermz). The ostracoderms had a primitive vertebral column, so they were vertebrates as well as chordates. In

SCIENCE AT WORK: The Oldest Life?

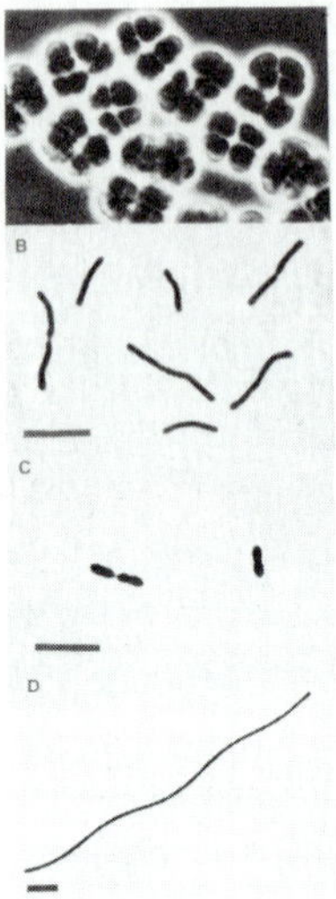

At first glance, characterizing, naming, and classifying organisms might appear very dull indeed. But this is seldom true. Taxonomy involves asking questions about all kinds of "unusual" living things.

One such group of organisms has been examined recently. The results of the investigations have shaken our traditional view of life in the past. The organisms are called methanogens (meth·AN·oh·jins). They look like bacteria. They survive only in warm environments that are free of oxygen—on the sea floor, in sewage, in the stomachs of cattle, and in the hot springs of Yellowstone National Park. They live by converting carbon dioxide and hydrogen into methane, the principal component of natural gas.

Scientists have known about methanogens for 25 years, but the organisms were hard to study in the laboratory. It is difficult to provide the right kind of culture environment. Finally in 1973, one scientist figured out a way to create the oxygen-free environment in which the organisms could grow. This development opened the door for more intensive study of the methanogens.

Scientists at the University of Illinois who analyzed the RNA of the methanogens were in for a few surprises. The RNA sequences were radically different from those of any organisms ever studied before. Biochemists in Germany found that the cell walls of the methanogens lack peptidoglycan, an essential component of the cell membrane of all conventional bacteria.

Some researchers have suggested that these organisms belong to a previously unrecognized category of life forms. The suggestion has stirred some controversies. Not all biologists agree that the evidence warrants new classification schemes.

Most do agree, however, that further study of the methanogens will yield some evidence about the early evolution of life. The conditions in which methanogens live are very much like those thought to exist during the first billion years of the earth's history. Perhaps the methanogens were the first organisms to evolve, even before bacteria. Methanogens may provide significant clues about how life could have evolved from the chemical "soup" of the ancient seas.

some, the body was covered by bony plates and scales. The ostracoderms lacked jaws and paired fins. Some of the ostracoderms were bottom feeders. They sucked mud containing food material into their mouths. There are no ostracoderms living today, but the lamprey and hagfish are close relatives.

More advanced fish with jaws were also present at this time. Some of these fish had paired fins. In the course of evolution, the paired fins of fish evolved into the paired legs of chordates. The chordates, however, lived on land. Examples of these chordates are the amphibians, reptiles, birds, and mammals.

The amphibians were the first chordates to become adapted to life on land. Lungs, which were already present in some fish, became the chief structures in which gases were exchanged between the environment and blood. The earliest amphibians looked like, and must have behaved like, fish (Figure 12-14). These animals probably lived for millions of years in the oceans before the terrestrial types evolved.

All living amphibians are much modified from their ancestors. Yet these amphibians still retain many features that show how the transition from water to land was made. Modern amphibians are quite numerous and successful. However, these amphibians do not survive in a great variety of environments. Amphibians must live close to water. Most amphibians still lay their eggs in water, and their young develop in the water.

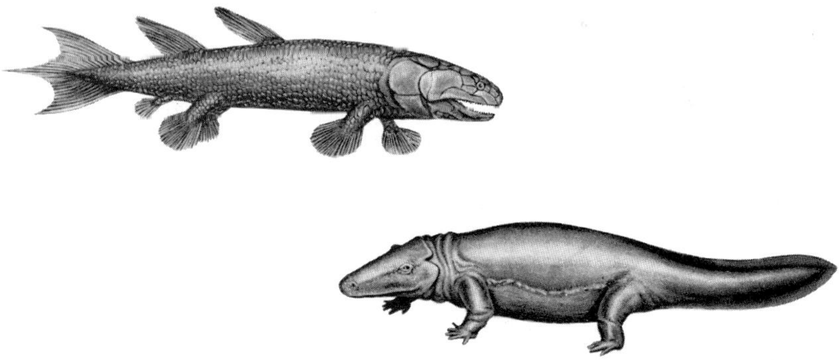

12-14. A lobe-finned fish **(left)** and a primitive amphibian **(right).**

About 250 million years ago, the environment began to change. Some parts of the world that had been warm became colder. In some areas where land became elevated, shallow seas were formed. Great mountain chains were pushed upward. The Appalachian Mountains are the remains of a mighty upheaval that occurred at this time. Many forms of marine invertebrates became extinct as their environment changed.

Changes on land increased the variety of environments. No longer were the continents low and covered with steamy jungles. There were hills and lowlands, swamps and deserts, and varied climates. Up to this time, the amphibians had dominated the swamps (Figure 12-15). Now the reptiles had a great advantage. They could lay their eggs on land, far away from swamps and streams. The age of reptiles had begun.

Many new kinds of reptiles evolved. These varied from small species to the giants among the dinosaurs (Figure 12-16). Some became flying reptiles, others evolved into aquatic forms. Some ate meat; others ate grass and other plants. Some lived on dry land; others lived primarily in swamps.

The plant world also was changing. The earlier forests of club mosses, horsetails, and ferns were now overshadowed by forests of cycads (SY·kads), ginkgoes (GING·kohz), and conifers. Representatives of these three groups are still living.

Between 185 and 135 million years ago, one group of reptiles gave rise to the birds. Most of these birds had feathers instead of scales

12-15. A scene in Texas about 250 million years ago.

12-16. Extinct reptiles (not drawn to scale). One was a flying reptile (*Pteranodon;* ter·AN·oh·don); two were swimmers (*Elasmosaurus;* ee·LAS·moh·sore·us and *Ophthalmosaurus;* ahf·THAL·moh·sore·us). The rest lived on land.

(Figure 12-17), but in many ways they were still basically reptilian. They had teeth, a long bony tail, with tail feathers set in like a fan. Only one living species of bird, the hoatzin of South America, has retained claws on its wings. And even in the hoatzin, the claws disappear by the second or third week of life.

12-17. An ancient landscape. The dinosaur is *Allosaurus* (al·oh·SAUR·us). One living and one dead ancestral bird, *Archaeopteryx* (ar·key·OP·ter·iks), are seen. Among the Jurassic plants are the cycads at the left and right foreground.

The reptiles also gave rise to another class of chordates, the mammals. For nearly 100 million years, mammals were small and scarce.

About 100 million years ago, the dinosaurs reached their evolutionary peak. Many new groups of dinosaurs developed. Each group was adapted to a different way of life. Then, while seemingly at the height of their development, these reptiles became extinct. The reason for their extinction is one of the puzzles of science. When reptiles became extinct, an inconspicuous group, the mammals, began to increase. Of course, the change was not sudden. It happened over millions of years.

AGE OF MAMMALS

About 65 million years ago, there was an explosive evolution of mammals. All of the major groups that we know today were present. Many others evolved and then became extinct (Figure 12-18). Some of the newly evolved mammals looked like the reptiles. The porpoise, for example, resembles some of the extinct aquatic reptiles in body shape and fish-eating habits. This similarity of form and way of life in different groups of organisms is a good example of convergence. In convergent evolution, animals with very different ancestors become more alike as they evolve in similar environments.

Mammals and birds seem to have arisen from different reptilian ancestors. But both mammals and birds have a far more effective protection against changes in environmental temperature than is found in any reptile. Mammals maintain a constant body temperature partly because they are covered with *hair.* In addition, most female mammals *retain their eggs in the body,* where embryonic de-

12-18. Ancient mammals. The large mammal on the right is *Uintatherium* (you·in·tah·THIR·ee·um). At the left in the foreground is a primitive carnivore, *Mesonyx* (MEZ·oh·nix). Behind the *Mesonyx* is an early kind of horse, *Hyracotherium* (hy·rak·oh·THIR·ee·um).

velopment takes place. Thus, mammals usually bring forth living young (in contrast to laying eggs). They care for the young after birth and feed them with milk secreted by the *mammary glands.* Many of the mammals living ten million years ago were not very different from those living today. Figure 12-19 shows some mammals that lived in North America.

Changes in environment influenced the evolution of animals and plants. Fossil evidence shows that the earliest angiosperms (flowering plants) probably lived in mountainous regions, while the conifers and cycadlike plants dominated the plains. The angiosperms did not become the dominant land plants until less than 70 million years ago.

In angiosperms, the methods of fertilization, seed dispersal, and establishment of seedlings are more effective than in gymnosperms. The pollen of gymnosperms is carried by wind or water. However, in most angiosperms, insects and birds move from one flower to another, carrying pollen on their bodies. In addition, the seeds of most gymnosperms are spread by wind and water. But birds and small mammals may transport the large seeds of angiosperms, such as acorns and nuts, for very long distances. Very often, the seeds of the angiosperms have a better chance to survive and grow in the new locations.

The fossil record of insects at this time is very imperfect. But there is some evidence that the bee family was evolving. Butterflies also appeared. These insects seem to have taken over the role of pollinating flowers from the more primitive beetles, which pollinate the most primitive flowering plants even today. Thus, beetles, bees, butterflies, birds, and early mammals must have been a strong selective influence in the evolution of flowering plants. By the end of the age of dinosaurs, about 70 million years ago, the plants were essentially like those of modern times.

In Brief

In this chapter, we have considered three major ideas. First, we looked at systems of classification — ways of naming organisms and grouping them in some easy-to-manage way. Then, we considered briefly some of the diverse living things that inhabit oceans, lakes, streams, mountains, and forests. Perhaps the picture survey gave you some appreciation for the incredible variety of organisms on earth — from the most primitive single-celled organisms to the most complex of the higher vertebrates. Last, we considered the evolutionary history of organisms. Using evidence from the fossil record, we attempted to recreate what the world may have been like millions and even billions of years ago.

12-19. Early mammals. Several of these species became extinct only within the past 10,000 years. Some were hunted by early humans. Probably some hunted early humans.

This chapter is the first in a unit we have called diversity—and perhaps you now see why. In the chapters that follow, we will consider diversity not only of structure but of function as well.

Guide Questions and Problems

1. Classification systems are based on comparisons of how organisms are alike and how they are different. Select any five (or more) organisms that are familiar to you and develop a classifica-

tion scheme that would reflect both their similarities and their differences. Do not use scientific names. Invent your own, and explain your reasons for classifying the organisms as you did.

2. The basis of classification is homology—usually of body structures and means of reproduction. Review the survey of the five kingdoms (pages 362–393) and select any two groups for comparison. Identify (a) the similarities that both groups have in common and (b) the differences that account for the two groups being separated.

3. In previous chapters, we have discussed viruses—particularly bacterial viruses called bacteriophage. Viruses are not included in either Figure 12-9 or in our survey of the five kingdoms. What reasons can you suggest to account for the omission of viruses from our classification scheme?

4. Why do we assume that plants evolved before animals? Why do we assume that plants lived on the land before animals did?

5. Reptiles evolved after amphibians, probably from amphibian ancestors. Give evidence that supports this statement.

Ideas for Independent Study

1. Many excellent keys are available for the identification of all kinds of organisms. Field guides prepared by R. T. Peterson are found in most libraries. Select an interesting group of organisms and go into the field with a key. Try to identify the organisms you find. Trees and birds usually are easy to see and identify.

2. No one knows why the dinosaurs became extinct, although a radical change in the environment certainly seems to have been a part of the answer. Investigate the age of dinosaurs. Consider the various hypotheses about the extinction of the dinosaurs. One of the newest and most interesting hypotheses about dinosaurs is that, unlike modern reptiles, they were "warm-blooded." Their body temperature was maintained at some constant level as in the birds and mammals of today. Investigate this idea further. A reference that will help you get started is "Dinosaur Renaissance," by R. T. Bakker, *Scientific American,* April 1975, page 58.

3. Over the history of the earth, the continents appear to have moved and shifted. The force of these movements has resulted in mountain building, formation and regression of inland seas, and other major changes in the geology and climate of the earth. The continents appear to move on giant plates—thus the emerging science called plate tectonics. Investigate plate tectonics. A good reference to get you started is "Plate Tectonics and the History of Life in the Oceans," by J. W. Valentine and E. M. Moores, *Scientific American,* April 1974, page 80; and "The Oldest Rocks and the Growth of Continents," by S. Moorbath, *Scientific American,* March 1977, page 92.

Related Readings

BOOKS

Carr, A. *The Reptiles.* New York: Time-Life, 1968.

Carrington, R. *The Mammals.* New York: Time-Life, 1963.

Hanson, E. D. *Animal Diversity.* Englewood Cliffs, N.J.: Prentice Hall, 1964.

Hutchins, R. E. *Insects and Their Young.* New York: Dodd, Mead, 1975.

Peterson, R. T. *The Birds.* New York: Time-Life, 1968.

Schaefer, J. *An American Bestiary.* Boston: Houghton Mifflin, 1975.

ARTICLES

Clarke, B. "Natural Selection—Causes of Biological Diversity." *Scientific American,* August 1975, page 50.

Gould, S. J. "The Five Kingdoms." *Natural History,* June-July 1976, page 30.

13 Diversity in Form and Function

Most of the surface of the earth—land, water, and air—is inhabited by living creatures. It is true that only a few organisms can live in the south polar regions, which are covered largely by ice. And only a few species of plants and animals can survive the hot noonday sun in the deserts of Death Valley (photo above). Yet organisms do survive in the most harsh environments, such as the depths of the ocean where sunlight never penetrates (photo above).

Although plants and animals inhabit large areas of the surface of the earth, they evidently thrive best where the physical environment provides for certain requirements. It must offer a reliable and sufficient supply of water. Oxygen, too, must be plentiful. In addition, plants must have supplies of carbon dioxide and nitrogen compounds from which they manufacture glucose, amino acids, and other organic compounds. Animals must have access to food of some kind.

In this chapter we will take a closer look at some of the ways that organisms carry out their life functions in a wide variety of environmental situations.

THE WORLD OF MICROORGANISMS

Among the variety of organisms on our planet, no group (except perhaps the insects) is so widespread as the bacteria. These single-celled organisms live in a variety of environments. Plants and animals require oxygen to live; some bacteria do not. The energy that supports most life comes from the sun, trapped in the carbohydrates manufactured by plants. Yet some bacteria obtain all the energy they need from inorganic chemicals found in rocks, soil, or water. Evolution usually is a very slow process, taking place over thousands, even millions, of years. Yet some bacteria reproduce so quickly and produce so many offspring that the characteristics of the population may change in a few days or weeks.

Bacteria are everywhere. They are in places we would little expect. Bacteria have been isolated from the ice packs of the Arctic and Antarctic, and from the hot, humid rain forests of the tropics. It would be impossible to name a part of the land or water in which bacteria are not found.

You can investigate the wide distribution of bacteria—as well as their diversity of form—by doing **Inquiry 13-1.**

INQUIRY 13-1 Microorganisms Everywhere

In this inquiry, you will culture and observe microorganisms that live all around us.

MATERIALS (per team)

dilute soil suspension
4 petri dishes containing sterile, nutrient agar
peppercorn infusion
crystal violet stain
microscope slides
forceps or tongs
dropper or pipette
compound microscope (with oil immersion objective—optional)
bunsen burner
incubator at 27°C (optional)
grease pencil or waterproof marker
2 sterile cotton swabs
immersion oil (optional)

Procedure (Day 1)

The petri dishes contain nutrient agar. The agar is a derivative of seaweed. It looks much like gelatin. Agar is a liquid at high temperature but becomes solid at 42°C and below. The nutrient agar contains salts and other chemicals that provide an ideal food and growing environment for many kinds of bacteria.

The nutrient agar in the plates is sterile. (Sterile means free of microorganisms.) Your teacher used sterile petri dishes and sterilized the culture medium. This was done by subjecting it to a very high temperature. This temperature could be obtained only under pressure. Your teacher used either a pressure cooker or a larger device called an autoclave.

Use very careful laboratory procedures when working with bacteria. Some bacteria cause serious diseases. Never touch with your hands any of the bacteria that grow on the plate. After completing your observations, seal the culture plates. Your teacher will dispose of them.

Before you begin, your teacher will instruct you in the proper methods of cleaning lab tables, handling petri dishes, and sealing the culture plates. Make sure you understand these instructions before you go on with this inquiry.

Now you are ready to inoculate your plate. (Inoculate means placing bacteria and other microorganisms on the culture medium.)

Label the bottom of unopened petri dishes with your name and the date. Label the plates with the numbers 1 through 4.

Expose Plate 1 to the air for 30 minutes. Then replace the top of the petri dish.

Use a sterile cotton swab to collect microrganisms from some surface. You might try the desk, drinking fountain, cafeteria table, door knob, or any other surface.

Rub the cotton swab carefully over the surface of Plate 2. Do not dig the swab into the agar. Swab the plate quickly and close it immediately.

Rub your fingers through your hair. Then touch your fingertips to the surface of the agar in Plate 3. Close the plate immediately after doing this.

Dip a sterile cotton swab into the soil suspension. Run the swab over the surface of the agar in Plate 4.

Turn the plates upside down and stack them in the place where your teacher indicates. If your plates are left at room temperature (about 22°C), it may take two or three days to see any real results. If you have an incubator set at 27°C, the bacteria will grow more quickly. Do not incubate the cultures at a temperature any higher than 27°C.

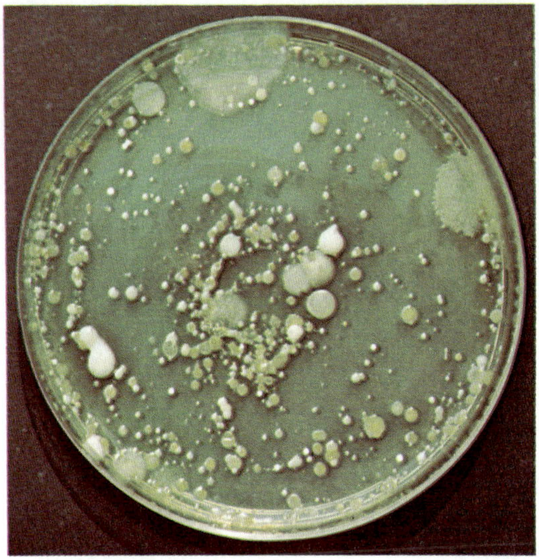

13-1. This plate was inoculated with a sample of soil. Note the fungus growing at the top center of the photograph.

Procedure (Day 2 or 3)

Examine each of your four plates. It is not necessary to remove the cover from the petri dish unless moisture has condensed on the surface and you cannot see through the glass. If you must remove the covers, *do not* touch anything on the surface of the agar. After you have made your observations, dispose of the cultures.

Record your observations and comparisons by answering the following questions: **1.** The areas of growth on the surface of the plate will appear roughly circular. Each of these areas is called a **colony.** A colony is composed of millions of bacteria—all the direct descendants of a single cell. Which of your four plates has the most colonies? Offer a hypothesis to explain why you might get more bacteria from this source than from the others. **2.** There are probably many different kinds of colonies on your plates. Identify three or four that look different. Describe the differences in terms of size, shape, and color. Also decide whether the colonies are smooth, rough, or glistening. Are the edges of the colonies smooth or jagged? **3.** You may have some colonies of fungi on your plates (Figure 13-1). How do the fungal colonies differ in appearance from the other colonies? **4.** How might microorganisms that cause disease be spread on surfaces such as the drinking fountain or the desk?

Now you are ready to observe some bacteria with a microscope. Bacteria are easier to see if they are stained. Follow this procedure:

Spreading the bacteria on the microscope slide:

a. Select a clean slide.
b. Use the dropper or pipette to obtain a drop of peppercorn infusion.
c. Place the drop of infusion on the slide.
d. Let the drop dry in the air to form a film on the slide.

Fixing the bacteria:

a. Holding the slide with forceps or tongs, quickly pass the slide two or three times through the flame of the bunsen burner. Keep the film side up. Do not hold the slide in the flame, as this will destroy the bacteria. Heating will fix the bacteria and make them stick to the slide.
b. Let the slide cool to room temperature.

Staining the bacteria:

a. Cover the film with two or three drops of crystal violet stain.
b. Leave the stain on the film for about 30 seconds.
c. Pour off the stain, and *gently* rinse the slide in a beaker of water.

Use low power of the microscope to focus on the stained bacteria. **5.** Describe what you see using low power.

Move the high-power objective lens into place. If necessary, refocus with the fine adjustment. **6.** What greater detail is seen with high power?

It is generally necessary to use a higher power microscope objective than the 43X on most student microscopes in order to study bacteria. For this purpose an oil immersion objective is used. It gives a much higher magnification. If you wish to study bacteria with this objective, it is first necessary to place a drop of immersion oil between the objective and the slide. **7.** What details do you see now that you did not see before?

Three general forms of bacteria exist (Figure 13-2). The cocci (COCK·seye; singular, coccus) are round. They may appear single,

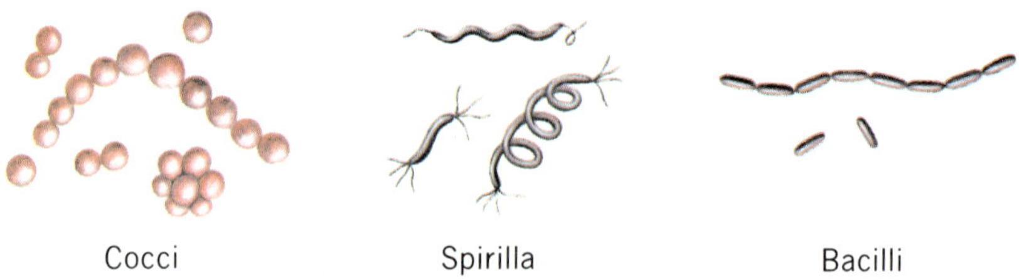

13-2. Three basic forms of bacteria: cocci, bacilli, spirilla.

in pairs, in clusters, or in chains. The rod-shaped bacilli (buh·SIL·eye; singular, bacillus) may appear long and slender, or short and thick. The spirilla (spuh·RIL·ah; singular, spirillum) are spiral or coil-shaped. **8.** Which of these forms do the organisms on your slide most closely resemble?

THE FUNCTIONING ORGANISM

So far, we have seen that organisms vary greatly in their appearance. We also have seen that organisms live in a variety of environments, as your work with **Inquiry 13-1** may have suggested. But no matter how different their environments, all organisms have certain functions in common. For the moment, let's focus our attention on animals. What functions must the body of an animal carry out in order to maintain life? Try to devise your own answers to that question. Then consider the following list. The animal must:

Obtain food composed of proteins, carbohydrates, fats, vitamins, and inorganic substances.

Digest food, that is, break it down into smaller molecules such as amino acids and simple sugars.

Transport the molecules of digested food through all parts of the body.

Assimilate the molecules of digested food to form structures of the cell.

Eliminate remains of undigested food.

Respire, transferring some of the energy in the molecules of digested food to ATP and similar molecules that serve as energy sources for cell processes.

Exchange gases with the environment—absorb oxygen and expel carbon dioxide.

Excrete liquid wastes, including excess water and nitrogen compounds.

Behave in such a way that it obtains food, avoids enemies, and defends itself against unfavorable conditions.

Reproduce, that is, produce offspring.

Grow and *develop.*

By closely examining how one animal carries on its life functions, you can better understand how other animals function. The unicellular organism *Paramecium* carries out all its life-sustaining activities and reproductive functions in one small cell. For this and other reasons, *Paramecium* is a convenient organism for a study of the animal way of life.

SCIENCE AT WORK: The Puzzle of the Viruses

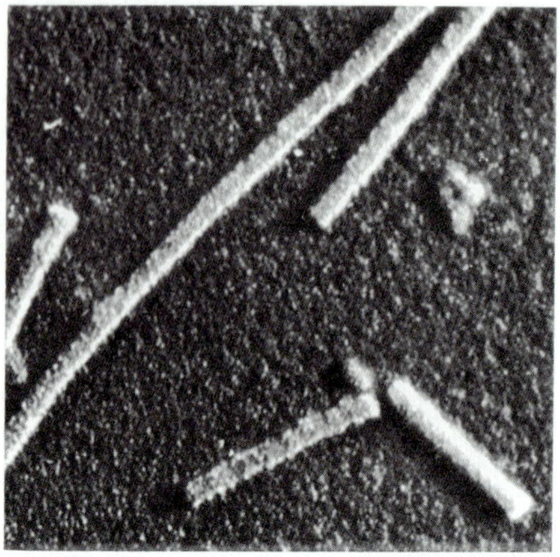

By the late 1800s, bacteriologists had demonstrated that many diseases of humans and other organisms were caused by bacteria. Some diseases puzzled bacteriologists, however, because they could find no organisms that were responsible for the symptoms. One such disease occurred in tobacco plants. The disease causes the leaves to wrinkle and become mottled. The mottled effect resembles a mosaic, so the disease was called *tobacco mosaic disease*.

The scientists learned that this disease could be transmitted from an infected tobacco plant to a healthy one. This was first demonstrated in 1892 by a Russian biologist, Dimitri Iwanowsky (duh·MEE·tree ee·vuh·NOF·skee). Iwanowsky extracted the juice from an infected tobacco plant. He strained the juice through a very fine filter made of porcelain to remove all bacteria from the juice. Next, Iwanowsky rubbed some of the filtrate on leaves of a healthy tobacco plant. The healthy plant soon showed symptoms of the disease (see photo above).

By 1900, similar disease-producing substances – viruses – had been discovered in many organisms. Viruses cause many of the major diseases of plants, animals, and humans – for example, influenza, poliomyelitis, and the common cold.

In 1935, W. M. Stanley, an American microbiologist, finally isolated the tobacco mosaic virus (TMV). Seen with the compound microscope, the isolated virus mass appeared as sliver-shaped crystals. The electron microscope showed them to be rods (see photo above). The virus was a nucleic acid linked to a protein. These closely resembled the nucleoproteins of chromosomes.

In the isolated, purified condition, removed from the cells of the host, TMV seemed quite dead. And the virus would not reproduce outside the tobacco plants. None was successful. The question was asked: "Are the viruses really living?" Stanley took a very small amount of purified TMV, dissolved it in water, and rubbed it on the leaves of healthy tobacco plants. These treated leaves soon developed the tobacco mosaic disease.

Many biologists have tried to get different kinds of viruses to grow and reproduce outside the living cells of their hosts. All attempts have failed. Viruses are true parasites. They require the enzyme systems in living cells of other organisms to carry on their life processes.

As you work through **Inquiries 13-2** through **13-6,** ask yourself, "What have I observed about the life of *Paramecium?*" "How do these observations of *Paramecium* build on my understanding of life in general?"

INQUIRY 13-2 Structure and Function in Paramecium

The relationship between structure and function is always significant, but not always obvious. Careful and detailed observations of living organisms—sometimes over a long period of time—are necessary to see which structures are performing the life functions. In this inquiry, you will identify the structures of the single-celled *Paramecium.* You will form hypotheses about the possible functions of these structures.

MATERIALS

Paramecium *culture*
methyl cellulose solution
iodine or methylene blue stain
medicine droppers (pipettes)
microscope slide
cover glass
compound microscope
colored pencil

Procedure

With a pipette containing several drops of methyl cellulose solution, make a small ring (about the diameter of a pencil) in the center of a clean slide. With a second pipette place one drop of *Paramecium* culture inside the ring. Add a cover glass. Examine the culture drop carefully with low power. Use high power when you wish to examine detail. **1.** What is the general shape of the animal? **2.** How does *Paramecium* move? (Movement will become slower as the methyl cellulose solution diffuses toward the center of the drop.)

In your notebook make a simple drawing of *Paramecium* as you see it through the microscope. Include all the parts of the animal that you observe.

Form hypotheses about the functions of the structures you observe. We suggest that you number your hypotheses according to the scheme that follows.

a. Structure(s) associated with how the animal moves.
b. Structure(s) associated with food intake.

c. Structure(s) associated with food digestion.
d. Structure(s) associated with elimination of the remains of undigested food.
e. Structure(s) associated with oxygen and carbon dioxide exchange.
f. Structure(s) associated with excretion of fluid wastes.
g. Structure(s) associated with removal of excess water.

Using these letters, label the structures in your diagram according to what you believe their functions to be. Obviously, you will not be sure you have labeled the structures correctly until you finish your studies of *Paramecium*. However, use your knowledge of living processes to help you form your hypotheses. Do not worry at this time about the names of the parts you have labeled.

After you have finished your observations of living *Paramecium*, add a drop of iodine or methylene blue stain to the slide. **3.** What structures in the *Paramecium* can you see now that you did not see before?

Add these structures to your diagram, using a colored pencil to distinguish them from the structures you saw in the living organism. **4.** What structures that you observed in the living organism are no longer visible? **5.** In testing your hypotheses, which gave you more information, the living organisms or the stained, dead ones? **6.** What functions of a living organism are not represented in your seven hypotheses?

Compare your sketches with Figure 13-3. **7.** What structures were you unable to observe? Why? **8.** What clues to function do

13-3. *Paramecium caudatum* is a useful organism for laboratory studies.

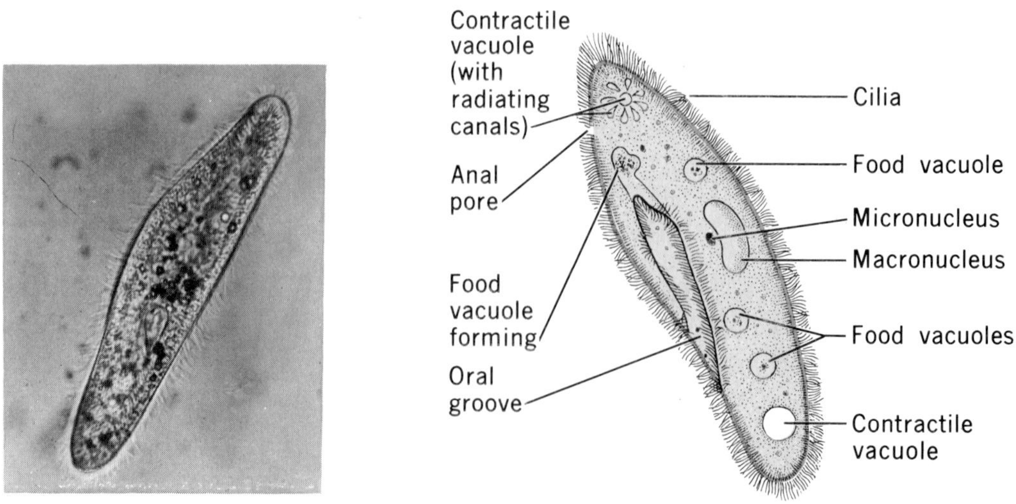

you find in the names of the structures shown in Figure 13-3? **9.** Reconsider your hypotheses. What could you do to test one or more of them? If possible test one of your hypotheses before going on to **Inquiry 13-3.**

INQUIRY 13-3 Locomotion of Paramecium

MATERIALS

Paramecium *culture*
methyl cellulose solution
carmine solution
microscope slide
cover glass
medicine droppers (pipettes)
compound microscope
cotton fibers

Procedure

Make a thin ring of methyl cellulose on a clean slide, as before. Then add a drop of *Paramecium* culture. Cover with a cover glass and observe with low power.

Before the methyl cellulose has diffused into the water, the animals will move very rapidly. When they begin to slow down, switch to high power to observe motion in detail. **1.** Which end of *Paramecium* is usually in front as the animal moves?

The front end of an animal is referred to as *anterior.* The back end is referred to as *posterior.* In your notebook, label an outline drawing of *Paramecium* according to what you believe to be its anterior and posterior ends. **2.** How do these ends differ in shape? **3.** On the basis of your observations, do you believe that the body wall of this animal is stiff or flexible? **4.** Can you identify a top (or *dorsal*) and a bottom (or *ventral*) side? **5.** Diagram the motion of *Paramecium* as it moves in one direction. Use arrows to show the directions of the motions of the body.

Gently raise the cover glass on your slide. Add a few cotton fibers to your drop of *Paramecium* culture. **6.** Describe what happens when *Paramecium* bumps into one of these fibers.

In your notebook, sketch a line to indicate a cotton fiber. Add arrows to show the direction of movement of an animal that has just bumped into a fiber. **7.** What do the reactions of *Paramecium* to an obstacle in its path suggest about its behavior?

Add a small drop of carmine solution to the drop of *Paramecium* culture on the slide. With high power, observe the surface of one of the animals. The hairlike structures you see are cilia. Notice carefully what happens to carmine particles that touch the surface of *Paramecium*. Adjust the light until you can see individual cilia. **8.** What do you believe is the function of the cilia? **9.** What does the pattern of movement of the carmine particles tell you about the activities of the cilia? **10.** On the basis of your observations in this inquiry, review and revise (if necessary) the hypothesis you formed in **Inquiry 13-2** of how *Paramecium* moves from one place to another. **11.** In what way has a study of movement in *Paramecium* contributed to your understanding of locomotion in all animals?

INQUIRY 13-4 Ingestion and Digestion in Paramecium

MATERIALS

Paramecium *culture*
yeast suspension containing Congo red powder
methyl cellulose solution
microscope slide
cover glass
medicine droppers (pipettes)
compound microscope

Procedure

Prepare a ring of methyl cellulose on a clean slide, as before. Place a small drop of *Paramecium* culture inside the ring. Add a drop of the yeast suspension. Cover and observe immediately (if possible, within eight to ten seconds) with high power.

Observe the ingestion and digestion of food particles in the *Paramecium*. The Congo red stain is red at pH 5 and blue at pH 3. It changes from red to blue in the vacuoles in about three minutes. Observations must be made rapidly. **1.** What color are the yeast cells? **2.** What is the source of their color? **3.** What happens to the yeast cells that are encountered by the *Paramecium*? **4.** Which parts of the *Paramecium* are first involved in the ingestion of the yeast cells?

Notice that the colored yeast cells become enclosed in a tiny droplet, a food vacuole. **5.** Where are the food vacuoles formed in *Paramecium?*

Carefully watch a food vacuole and note any changes in it. Within two or three minutes you should begin to notice a change in color in the food vacuole. **6.** Describe the change in color. How do you account for the change? **7.** As you continue to watch, describe any further color changes that occur in the food vacuoles of the *Paramecium.*

Continue to observe the food vacuoles. By means of circles representing food vacuoles and arrows representing their movement, diagram the path a food vacuole takes from its formation to its disappearance. **8.** How do the food vacuoles move through the cytoplasm?

Watch carefully to see how undigested food in a food vacuole is eliminated from *Paramecium.* **9.** Where in the body does this happen?

Refer to the hypotheses you formed in **Inquiry 13-2** regarding the structures in *Paramecium* that you thought responsible for food digestion and the discharge of undigested food. Were you correct? **10.** On the basis of the observations in this inquiry, summarize how you now believe food to be ingested (taken in) and digested by *Paramecium.*

In this inquiry it has not been possible to observe whether *Paramecium* is selective about its choice of food. **11.** What experiments could you perform that would test the food selectivity of *Paramecium?*

INQUIRY 13-5 Contractile Vacuoles in Paramecium

In *Paramecium* there are structures known as contractile vacuoles (Figure 13-3). Each contractile vacuole fills up with fluid, contracts (expelling its contents), and then seems to disappear. The vacuole becomes visible again, filling with fluid, and repeats this process. The canals leading into the vacuole give it a star-shaped appearance.

MATERIALS

Paramecium *culture*
methyl cellulose solution
microscope slide
cover glass
medicine droppers (pipettes)
compound microscope
Chinese ink

Procedure

Make a ring of methyl cellulose on a clean slide, as before. Then place a drop of *Paramecium* culture in its center. Add a cover glass and observe with high power until a contractile vacuole is located. **1.** How many contractile vacuoles are there in *Paramecium*? **2.** Where are they located? Draw them in place in an outline sketch. **3.** Do the contractile vacuoles move through the body as food vacuoles do, or do they stay in one place? **4.** Is the fluid within the enlarged vacuole clear or granular? **5.** If radiating canals are seen around a contractile vacuole, are they more evident when the vacuole is full or when it is empty? **6.** How many radiating canals apparently empty into a contractile vacuole? **7.** How many contractions does the vacuole make in one minute? (Measure this rate using a watch or clock with a second hand.) **8.** If you identified any contractile vacuoles in **Inquiry 13-2,** what hypothesis did you form about their function?

The techniques for testing your hypothesis about contractile vacuoles require equipment not available in your laboratory. However, you can refer to observations made by others who have investigated the question, "What is the function of a contractile vacuole?" Four of the observations follow.

OBSERVATION 1: Urea (an end product of nitrogen metabolism) has not been found in any great amount in the fluid of the contractile vacuole.

OBSERVATION 2: If vacuoles of freshwater *Paramecium* do not function, the body swells.

OBSERVATION 3: Injection of distilled water into the organism increases the rate of contraction of the vacuole. (More contractions per minute.)

OBSERVATION 4: Water intake with food in food vacuoles is only a fraction of the amount of water that is expelled by the contractile vacuole.

9. What effect does Observation 1 have on a hypothesis that the contractile vacuole is an excretory organ used for eliminating nitrogenous wastes? **10.** Considering Observations 2 and 3, what probably is the function of a contractile vacuole in *Paramecium?* **11.** List the ways in which water might enter the body of *Paramecium.* **12.** From Observation 4, what would you suggest is the major method of water entry? **13.** Make a hypothesis that will account for the function of a contractile vacuole. **14.** Suggest additional investigations that would tend to confirm or disprove this hypothesis.

INQUIRY 13-6 Reproduction in Paramecium

MATERIALS

Paramecium *culture*
mating types of Paramecium *(2 cultures)*
methyl cellulose solution
iodine stain
prepared demonstration slides of fission and conjugation in Paramecium
medicine droppers (pipettes)
microscope slide
cover glass
compound microscope

Procedure

Place a drop of *Paramecium* culture in a ring of methyl cellulose on a slide. Cover it with a cover glass. Locate one animal that seems to be constricted in the middle. Perhaps you can find several individuals showing different degrees of constriction. If so, add a drop of iodine to the slide. This will kill the animals and stain their nuclei. **1.** How does the nucleus of a constricted animal differ from the nucleus of a slipper-shaped one? **2.** What is the general shape of the nucleus or nuclei in the constricted animals?

Carefully examine the nucleus with high power. The large structure is called the *macronucleus* (MAK·roh·NEW·klee·us). If possible, locate a smaller, spherical structure that seems to be embedded in the surface of the macronucleus. This is the *micronucleus* (MY·kroh·NEW·klee·us). **3.** How many micronuclei can you see? Make a sketch of a constricted animal and add the nuclear structures to your drawing.

Shift back to low power. Look for animals that show several stages in the process of constriction (in the middle of the body) and separation. Make sketches that begin with what you regard as the earliest sign of the process. Continue your sketches until you have drawn *two* smaller, individual animals that appear to be held together only by a narrow band of cytoplasm. **4.** What will be the result of the completion of this constricting process? **5.** Which animal is the parent? **6.** The process you have observed is a kind of *asexual* reproduction. Literally, asexual means "without sex." Why is asexual a good word to use to describe this process?

In the *Paramecium* cultures or in prepared slides, locate stages in which two individuals seem stuck together side by side. Draw an outline of these individuals. **7.** How can you tell by observation whether what is occurring differs from the asexual reproduction you have observed?

Stain *Paramecium* as you did earlier. Notice the condition of the nucleus in each of the individuals joined together. **8.** How does it differ from what you observed before? **9.** On the basis of your observations, what seems to be occurring between the two joined individuals?

The process that you just observed involved the pairing or *conjugation* (kon·joo·GAY·shun) of two individuals. It is a process of reproduction. The process of conjugation begins when nuclear material is exchanged between the two mating types. It ends as the new individuals separate.

In 1937, Tracy M. Sonneborn (SON·neh·born) discovered that two individual *Paramecium* joined in this way are different—even though the differences cannot be seen. The differences are revealed by the behavior of the individuals. Not all individuals are able to conjugate with each other. There are different mating types. An individual of mating type I will conjugate with an individual of mating type II, but not with another type I individual. Type II individuals also will not conjugate with one another. **10.** Would you describe conjugation as an asexual or sexual process? Why?

In **Inquiry 13-2** you formulated some hypotheses about structures and their possible functions in *Paramecium.* You also made a drawing. As you worked on **Inquiries 13-3** through **13-6,** you revised your original drawing. **11.** On the basis of this series of inquiries, which of your hypotheses were supported? Which were not? **12.** Which of your hypotheses remain untested? How might you test them? **13.** On the basis of your work with *Paramecium,* do you think it is a simple or a complex animal? Why? **14.** Why does it serve as a good experimental animal for the study of structure and function?

In Brief

By studying bacteria and *Paramecium,* you may have gained some awareness of how organisms can be studied. You may now have some idea of how at least one single-celled organism, the *Paramecium,* carries out certain life functions. These functions—digestion, respiration, transportation of materials, excretion, coordination, and reproduction—are common to all animals.

The rest of our unit on diversity will explore in some detail how these functions are carried out in several different kinds of animals. Later, we also will consider some of the special structures and functions of green plants.

Guide Questions and Problems

1. Which will use the greater number of extracellular enzymes in meeting its nutritional requirements, a parasitic bacterium or one living in soil? Explain your answer.

2. The graphs in Figure 13-4 show possible patterns of growth in bacterial populations. Numbers of organisms are plotted on the vertical axis, time on the horizontal axis. Describe the pattern shown in each graph. Suggest a set of environmental conditions under which each growth pattern might occur.

3. Why must animals digest proteins, large carbohydrate molecules, and complex fat molecules before these food materials can be used?

4. Consider the following situation. A biologist observed that *Paramecium* living in a pond had a contractile vacuole that emptied four times every minute. In a 1 percent salt solution, *Paramecium* shrank, and the biologist could see no contractile vacuole. When placed in water without salts of any kind, *Paramecium* became swollen. The contractile vacuole then emptied ten times every minute. What explanation can you give for these observations?

5. If the food vacuoles in *Paramecium* are serving as digestion chambers, what types of enzymes would you expect to find in them?

6. What data would show whether *Paramecium*, like people and many other animals, uses oxygen and releases carbon dioxide? Plan an experiment that could answer this question.

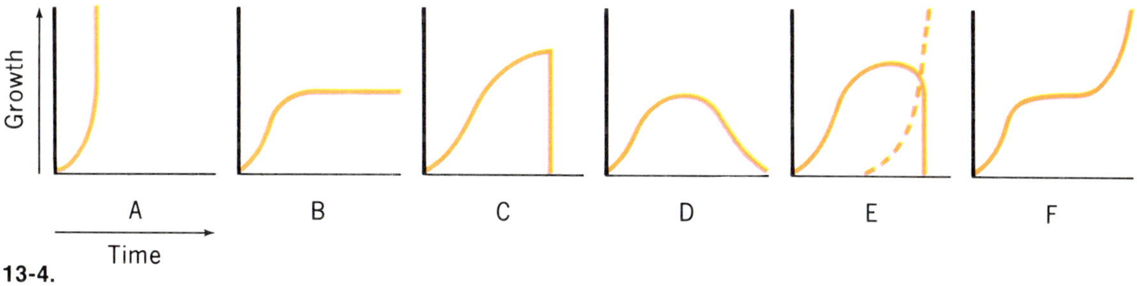

13-4.

Ideas for Independent Study

1. Look carefully at some of the environments in which living creatures can exist. Compare the environment of *Paramecium* with your own. For instance, how do the conditions vary in light and temperature? How available is water? What food sources are available? Select an environment very different from your own or that of *Paramecium* for intensive study. Wet tropical forests, arctic regions, and deserts are some obvious examples. For other situations, you might look at one of these articles in *Scientific American:* "Urban Trees" by T. S. Elias and H. S. Irwin, November 1976; "Kangaroos" by T. J. Dawson, August 1977; "Microbial Life in the Deep Sea" by H. W. Jannasch and C. O. Wirsen, June 1977; "Active Animals of the Deep-Sea Floor" by J. D. Isaacs and R. A. Schwartzlose, October 1975.

2. What are the possibilities that life exists on other celestial bodies? Find out what is known of conditions in our solar system and our universe. Is there enough oxygen, light, water, and so forth to support life as we know it? Many publications concerning the possibilities of life on other planets are available from the National Aeronautics and Space Administration. You can request a list of NASA booklets by writing the Education Office at the NASA Center serving your state. Your teacher or library may have a list of these offices.

Places to start in *Scientific American* are: "The Search for Life on Mars" by N. H. Horowitz, November 1977; "The Search for Extraterrestrial Intelligence" by C. Sagan and F. Drake, May 1975; "Mars" by J. B. Pollack, September 1975.

Related Readings

BOOKS

DeKruif, P. *Microbe Hunters.* New York: Harcourt Brace Jovanovich, 1966.

Dubos, R. and M. Pines. *Health and Disease.* New York: Time-Life, 1965.

Frobisher, M., R. D. Hinsdill, K. T. Crabtree, and C. Goodheart. *Fundamentals of Microbiology.* Philadelphia: W. B. Saunders, 1974.

Gabriel, M. and S. Fogel, eds. *Great Experiments in Biology.* Englewood Cliffs, N.J.: Prentice-Hall, 1955.

Lewin, R. *In Defense of the Body: An Introduction to the New Immunology.* Garden City, N.Y.: Anchor/Doubleday, 1974.

Reid, R. *Microbes and Men.* New York: E. P. Dutton, 1975.

Sussman, A. S. *Microbes: Their Growth, Nutrition, and Interaction.* BSCS Laboratory Block. Boston: D. C. Heath, 1964.

Tabor, W. and R. A. Tabe. *The Impact of Fungi on Man.* Boulder, Co.: BSCS, 1967.

ARTICLES

Batra, S. W. T. and L. R. Batra. "The Fungus Garden of Insects." *Scientific American,* November 1967, page 112.

Berg, H. C. "How Bacteria Swim." *Scientific American,* August 1975, page 36.

Smith, I. M. "Death from Staphylococci." *Scientific American,* February 1968, page 84.

Valentine, J. W. and E. M. Moore. "Plate Tectonics and the History of Life in the Oceans." *Scientific American,* April 1974, page 80.

14 Animal Structure and Function: Digestion

All animal cells have similar requirements, although these requirements differ in detail. Animals must have a supply of water, certain amino acids, simple sugars, fatty acids, vitamins, and oxygen. They also need many other organic and inorganic substances. They must be able to dispose of their waste products. These are, principally, carbon dioxide, water, and nitrogen compounds such as ammonia (NH_3), urea, or uric acid.

Single-celled organisms, such as *Paramecium*, capture and digest food, obtain oxygen, dispose of wastes, move, and reproduce. All of these activities are accomplished in a single cell (Figure 14-1a).

Most animal species, however, are made up of many cells. This introduces a problem. Most cells of multicellular animals are far from the external environment (Figure 14-1b). Unlike *Paramecium*, a cell

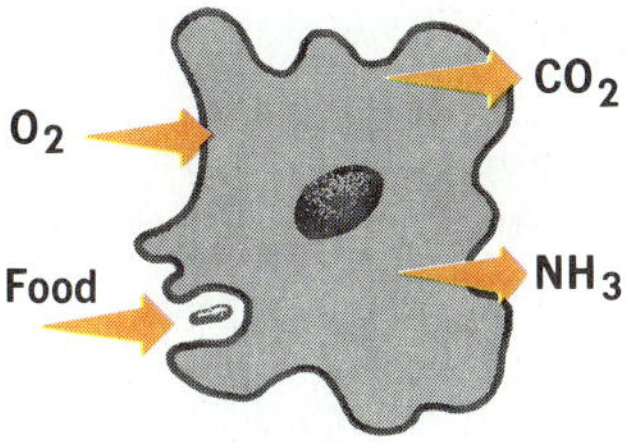

a One-celled organisms exchange materials directly with the outside environment.

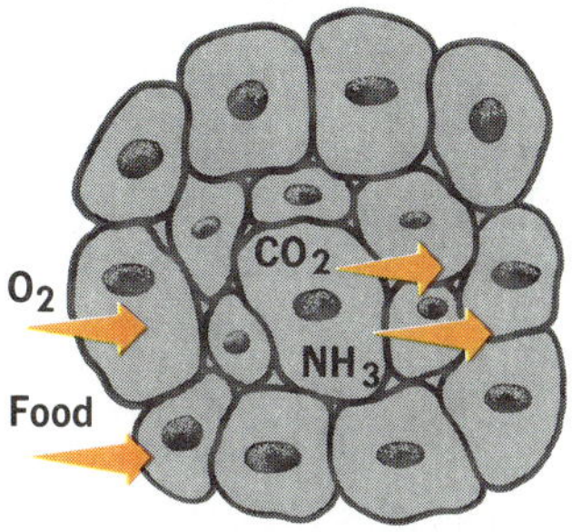

b A different situation exists in the cells of a multicellular organism. How can materials be exchanged?

14-1. (a) One-celled organisms exchange materials directly with the outside environment. **(b)** A different situation exists in the cells of a multicellular organism. *How are materials exchanged here?*

deep inside a large animal cannot take in food particles directly from the outside. Nor do the wastes from the cell go directly into the outside environment. The wastes go into the internal environment that surrounds this cell and all the other cells that are producing wastes.

In multicellular animals, cells have different functions—they become specialized. Different groups of specialized cells take in food and digest it. Other groups get rid of wastes. Still other specialized cells move the organism, provide support, or carry on reproduction. They receive stimuli and coordinate the various activities of the whole organism. Multicellular animals have transport systems within their bodies. These transport systems carry food and other materials to each cell and remove its wastes.

No matter how specialized a cell's functions become, the cell still has the same basic requirements for staying alive. Yet specialized cells cannot provide for all their own needs. Thus, systems of specialized cells have evolved in multicellular organisms. These systems are dependent on each other. Certain groups of cells, tissues, and organs perform one set of functions. Other groups perform still other functions. As a result, a multicellular organism meets all the needs of its individual cells.

Let's consider a human muscle cell. It is highly specialized for contraction. The muscle cell depends on cells of the digestive tract for its supply of digested food materials. It depends on cells of the lungs for oxygen from the external environment. It depends on the cells that form the blood vessels to carry digested food substances and oxygen to it. The cells of the blood vessels also carry away the muscle cell's waste products. Cells of the brain and nervous system coordinate all these complex activities, including the contraction of the muscle cell.

Every cell in the body, in its own specialized way, contributes to the functioning of the whole body. But we also can think of the whole body as functioning in a manner that keeps its cells alive.

METHODS OF DIGESTION

The substances in the food we eat that our cells require are water, minerals, inorganic salts, amino acids, simple sugars, vitamins, and fatty acids. These small molecules can pass across cell membranes and enter the cells of all animals. But amino acids, simple sugars, vitamins, and fatty acids are rare in the natural environment. Organisms almost never have these substances *directly* available to them. In nature, all except vitamins (and sometimes these, too) are usually parts of much larger molecules, such as proteins, carbohydrates, and fats. And these large molecules are contained in larger pieces of food. Through the process of digestion, animals break down large molecules into substances cells can use.

An organism first finds and ingests the food. The large food pieces are broken into smaller pieces and mixed with water. After these small food pieces are in solution, the large organic molecules are digested by enzymes into smaller molecules. These small molecules, which can pass easily across cell membranes, are picked up by the transport system and circulated to all cells of the body.

Amino acids from proteins, simple sugars from starch, and fatty acids and glycerol from fats are found in all living creatures. Every species uses these small organic molecules to synthesize its own specific cellular structures. For example, many people eat beef. The beef protein is digested to amino acids, which are used in the synthesis of specific human proteins.

Single-celled organisms such as *Paramecium* take small food particles into a food vacuole. Another freshwater Protist, *Amoeba proteus,* engulfs the food in extensions of the cytoplasm (Figure 14-2). Small plant cells, animals, and even bits of nonliving material are enclosed in food vacuoles. Enzymes enter the vacuoles from the cytoplasm and break down the large molecules of food. The smaller molecules that are formed, such as glucose and amino acids, cross the membrane of the food vacuole into the cytoplasm.

While all of this is happening, the food vacuoles are being moved around in the cytoplasm of the cell. Digested foods thus are trans-

14-2. Feeding and digestion in *Amoeba proteus.*

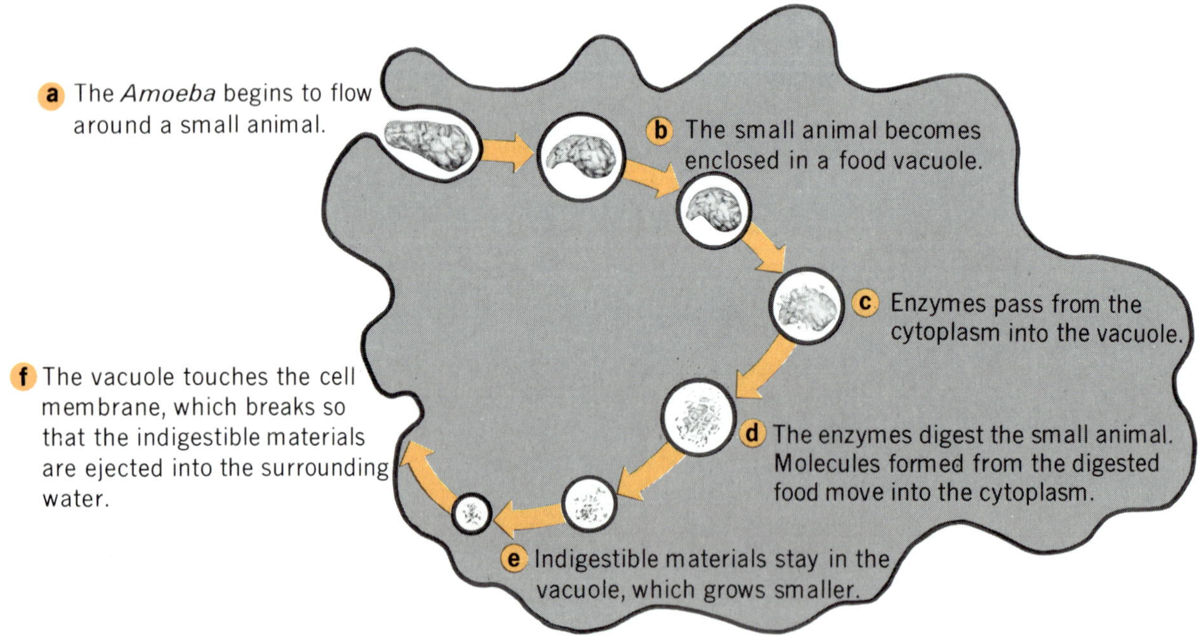

ported wherever they may be used. The problem of food transport is solved easily in one-celled organisms.

Gradually, the food vacuoles become smaller as food molecules leave them. Some indigestible materials remain, however. The food vacuoles move to the outer part of the cytoplasm until they touch the cell membrane. At this point, the membrane opens and the indigestible contents of the vacuole burst out into the surrounding water. Then the cell membrane quickly closes. An *Amoeba* may contain as many as 50 food vacuoles at one time, each with food in different stages of digestion.

Since this process occurs completely within a single cell, it is called **intracellular digestion**. In multicellular animals, food is taken in from the outside. Digestive enzymes are secreted from cells into the digestive tract where this food is digested. This process is called **extracellular digestion**.

Small coelenterates, such as sea anemones, corals, and *Hydra*, paralyze even smaller animals with stinging cells on their tentacles. The tentacles slowly move the stunned animal into the mouth. It is then ingested into the digestive cavity (Figure 14-3). The cells that line the digestive cavity secrete enzymes into the cavity. The enzymes begin the digestion of the large molecules in the body of the captured animal, or prey.

However, most digestion in *Hydra* takes place *not* in the digestive cavity but inside body cells, in the manner of *Amoeba*. The body of the prey is broken down gradually into smaller molecules in the digestive cavity. These molecules are taken into vacuoles formed in the cells that line the cavity. Enzymes inside the cells continue the process of digestion. Digestion in *Hydra* thus is both extracellular and intracellular. It occurs in the digestive cavity and within the cells, breaking down the food to amino acids, fatty acids and glycerol, vitamins, and simple sugars, as well as some inorganic substances. Those products of digestion formed in the digestive cavity are absorbed by the cells lining the cavity. Any undigested material from the body of the prey is ejected from the digestive cavity. Since there is only one opening, it serves *Hydra* both as a mouth to take in food and an **anus** to expel wastes (Figure 14-3).

A planarian (flatworm) also has a saclike digestive cavity. There is only a single opening, the **pharynx,** which serves as both mouth and anus. In the planarian, the opening is on a tubelike projection on the underside of the body, almost midway between the anterior and posterior ends (Figure 14-3).

Although a planarian can capture small, living, aquatic animals, most of its food consists of the bodies of dead, aquatic animals. Digestion is not very different from that in *Hydra*. Some digestion occurs in the digestive cavity. In addition, the cells lining the digestive cavity engulf and digest the broken-down food molecules.

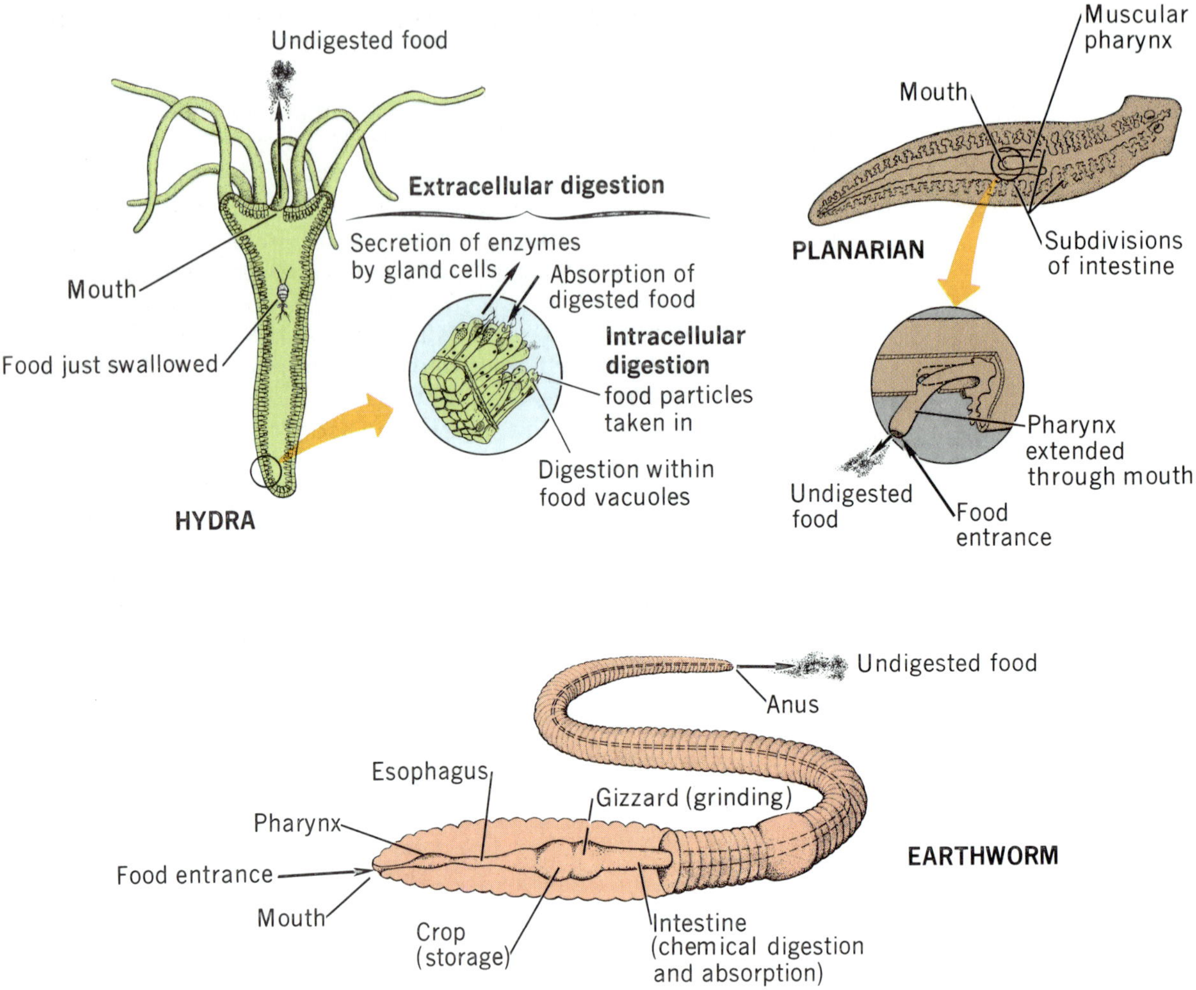

14-3. Digestive systems of three multicellular animals: *Hydra,* planarian, and earthworm.

Compared with *Hydra* and planarians, the digestive system of earthworms is much more highly specialized. Instead of being a cavity with a single opening, the earthworm's digestive system is a tube with two openings—a mouth at one end (Figure 14-3), an anus at the other.

Each region of the earthworm's digestive tract performs a specific function. The food (largely soil that contains decomposing organic matter) is ingested and moved rapidly to a storage organ, the crop. It is then ground and broken into smaller pieces in the gizzard. In the intestine, enzymes secreted from cells that line the intestinal wall digest the food. Cells of the intestinal wall then absorb the small molecules and distribute them to other cells by the transport system. Undigested material is pushed out of the animal through the anus. A one-way, tubular digestive system like that of the earthworm is found in most groups of complex animals.

THE HUMAN DIGESTIVE SYSTEM

In humans, as in earthworms and many other complex animals, the digestive system is a long tube—the **alimentary** (AL·uh·men·tuh·ree) **canal.** The alimentary canal processes food in an efficient, assembly-line manner (Figure 14-4). The digestive process begins in the mouth, where food is chewed into small particles. The

14-4. THE HUMAN DIGESTIVE SYSTEM

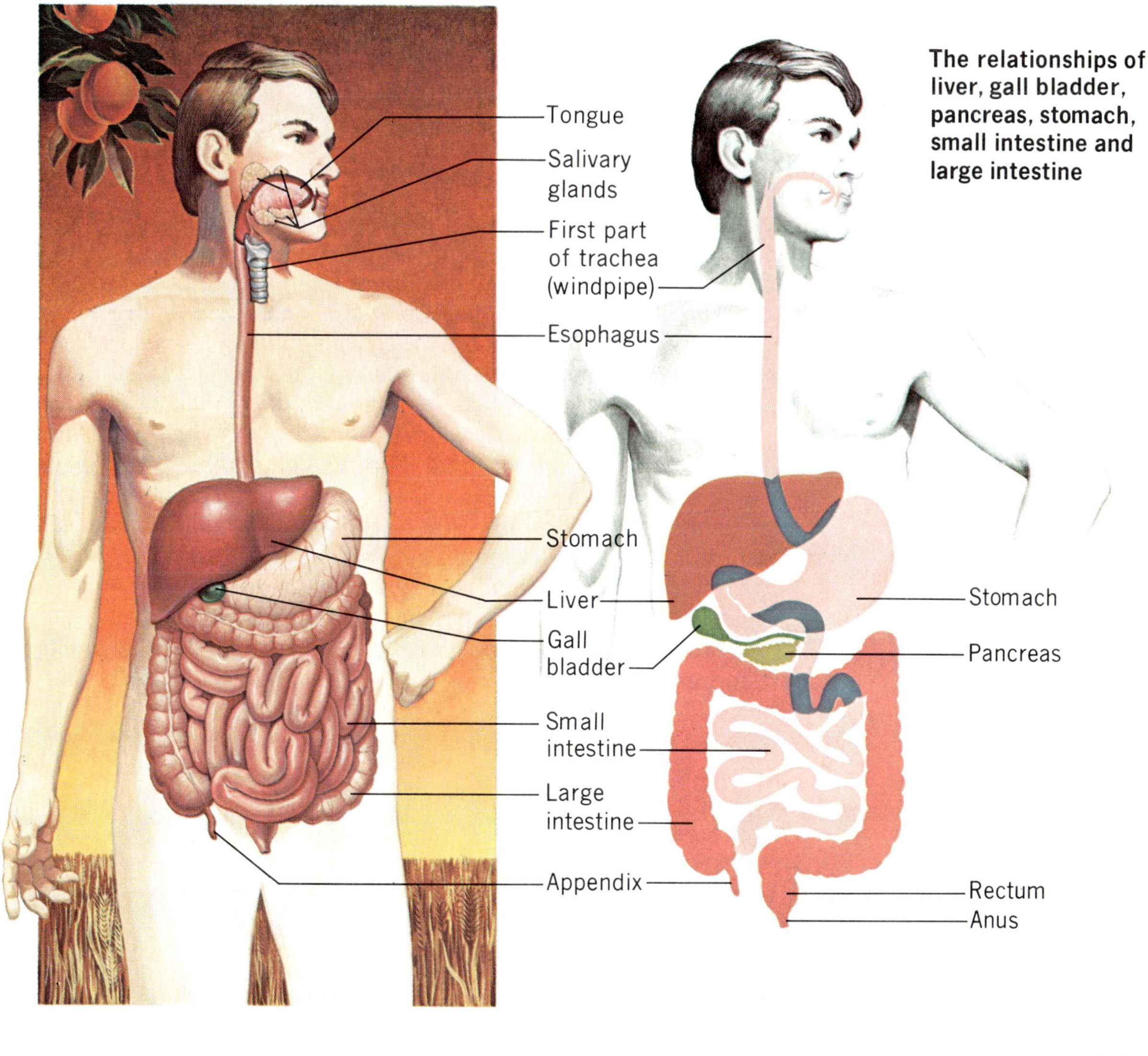

food particles move down the **esophagus** (ih·SAHF·uh·gus) to the stomach. Then the stomach carries out one sort of digestion. From the stomach, food is moved into the small intestine, where the final processes of digestion occur.

Several types of **glands** are associated with the digestive system. Glands are organs or tissues that form substances inside their cells and then empty them to the outside. The substances are called secretions. The most important glands of the digestive system are three pairs of salivary glands in the mouth, the pancreas, and the liver. All these glands connect with the alimentary canal by means of tubes, or ducts. The salivary glands and the pancreas secrete juices that contain digestive enzymes. The liver secretes **bile,** which also aids digestion but does not contain enzymes. Other glands are located in the lining of the digestive tube. These glands empty their secretions directly into the digestive tract.

Digestion in the Mouth

In the mouth, the teeth and tongue grind the food into small particles and mix it with **saliva** (suh·LYVE·uh), which moistens it. Saliva is produced by the salivary glands (Figure 14-4) and empties into the mouth through ducts. Saliva has two main functions. It adds mucus (MEW·kus) and water to the food. Mucus is a slippery secretion from cells and glands. It moistens and lubricates the food. The water dissolves some substances. Water and mucus ease the passage of food. The mixing of food with water in the mouth is very important for the digestion that occurs there. Enzymes digest food molecules more efficiently when the food and enzymes are in solution. Saliva contains an enzyme, amylase, that begins the conversion of starch to smaller molecules of a double sugar.

Food is chewed into small pieces and moistened in the mouth, and digestion of starches begins. When the food is swallowed, it passes through the esophagus to the stomach. Quick movement down the esophagus is achieved by a process called **peristalsis** (per·ih·STALL·sis; Figure 14-5). Peristalsis is the alternating contraction and relaxation of the muscular wall of the esophagus. This creates a wave of movement along the esophagus that pushes the food toward the stomach. Peristaltic waves move food through the entire alimentary canal.

In the stomach, the food is broken down by other enzymes. You can observe how one of these enzymes works by doing **Inquiry 14-1.** The principles of protein digestion are introduced in Inquiry 14-1. These same principles are applicable to this process wherever it is found in animals.

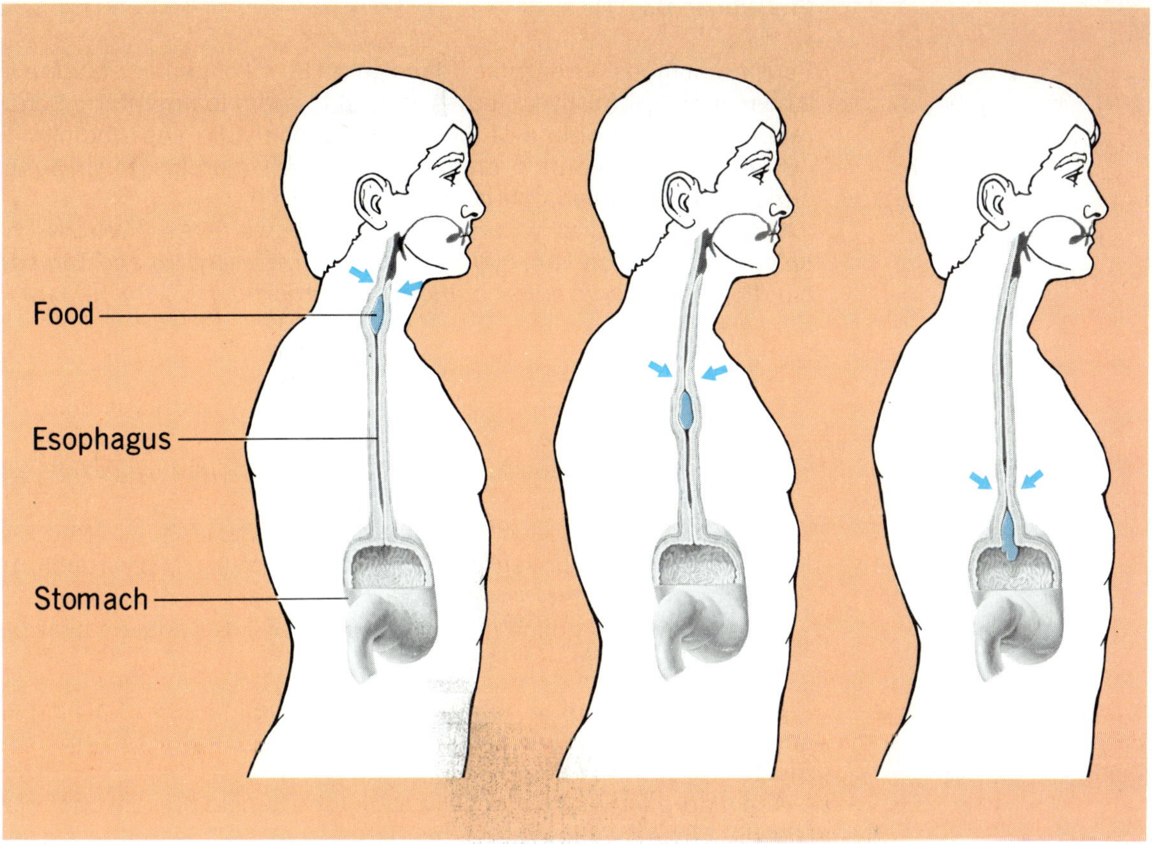

14-5. Movement of food through the alimentary canal. Peristalsis in the wall of the esophagus forces food through it and into the stomach.

INQUIRY 14-1 Protein Digestion

MATERIALS (per team)

0.5 pepsin solution
0.1 M hydrochloric acid
0.1 M sodium hydroxide
toluene
2 hard-boiled hen's eggs
6 test tubes
graduated cylinder, 100-ml
2 beakers, 100-ml or larger
hot plate or bunsen burner
tripod stand (if bunsen burner is used)
test-tube rack
knife
millimeter ruler
incubator
pH paper
forceps

Procedure (Day 1)

Pepsin is a digestive enzyme. Place about 50 ml of pepsin solution in a beaker. Boil it for five minutes. Cool it to room temperature. Separate the albumin of one of the eggs from the yolk. The albumin, or "white," is almost pure protein. Cut the solid albumin into six 5-mm cubes. You will need one cube for each test.

Label and prepare six test tubes as follows. (CAUTION: *Hydrochloric acid and sodium hydroxide can injure your skin and damage clothing. If you spill either of them, rinse thoroughly with cold, running water.*)

Tube 1: 15 ml of pepsin, albumin cube.
Tube 2: 15 ml of pepsin, 10 drops of 0.1 M hydrochloric acid, albumin cube.
Tube 3: 15 ml of pepsin, 10 drops of 0.1 M sodium hydroxide, albumin cube.
Tube 4: 15 ml of boiled pepsin, albumin cube.
Tube 5: 15 ml of boiled pepsin, 10 drops of 0.1 M hydrochloric acid, albumin cube.
Tube 6: 15 ml of boiled pepsin, 10 drops of 0.1 M sodium hydroxide, albumin cube.

Measure the pH of all tubes using very small pieces of pH paper and forceps. **1.** What are the differences in the pH of the test tubes, if any?

Add a drop or two of toluene to each tube to help prevent bacterial growth. Place the tubes in an incubator set at 37°C. Describe the appearance of the tubes.

Procedure (Day 2)

Examine the six tubes after 24 hours. If necessary, continue your observations for another 24 hours. **2.** After 24 hours, what changes, if any, have occurred in the test tubes? **3.** What is the effect of boiling on the action of pepsin? On what observations do you base your answer? **4.** Test tubes 1, 2, and 3 differed only in that Tube 2 contained hydrochloric acid, Tube 3 contained sodium hydroxide, and Tube 1 contained neither. What was the purpose of this difference? **5.** On the basis of your observations, which test tube had the best conditions for the digestion of egg albumin by pepsin? **6.** What conditions do you predict might be found in the stomach, the organ of digestion? **7.** In **Inquiry 2-3,** you saw that a low pH had a negative effect on the action of the enzyme, catalase. How do your results in this experiment compare with that one? What does this suggest to you about the activity of enzymes and their environment?

Procedure (Day 3)

Work in teams of three or four to design a simple experiment to determine what other factors may affect pepsin activity. For example, would the activity of pepsin change with a change in temperature? How might the presence of other foods affect pepsin activity? What effect does chopping, chewing, or grinding the egg albumin have on the rate of digestion?

Write down the hypothesis you decide to test. Outline your procedure, the equipment and materials you will need, and a time schedule for your experiment. Show your instructor your experimental design so that any problems can be cleared up before you proceed.

Carry out your experiment and report to the class what you have done and what you have learned about digestion.

Digestion in the Stomach

Considerably more digestion occurs in the stomach than in the mouth. Glands in the lining of the stomach secrete gastric juice, which contains pepsin, a protein-digesting enzyme, and hydrochloric acid. The chemical work of the stomach is aided by its muscular activity. Its walls contract and relax. These movements mix gastric juice with the food.

What causes gastric juice to be secreted? What prepares the stomach for the arrival of food? Sometimes even the sight or thought of tempting food causes increased gastric secretion. Food in the mouth stimulates nerves in the cheeks and tongue. These nerves pass messages to the brain, which sends messages to nerves in the wall of the stomach. Here, nerves stimulate the gastric glands. As a result, there is some gastric juice in the stomach when the food arrives.

When food touches the lining of the stomach, more gastric juice is secreted. But food alone is not enough to stimulate a maximum secretion of gastric juice. If you eat only foods with little protein in them, the stomach does not secrete much gastric juice. On the other hand, the presence of protein, which is digested by the enzymes in gastric juice, results in an abundant secretion.

Biologists know quite a bit about how proteins stimulate secretion of gastric juice. The large protein molecules are broken down into smaller molecules by the gastric juice that is present when the food reaches the stomach. These smaller molecules stimulate some of the cells lining the stomach. They cause these cells to secrete a substance called *gastrin*. Gastrin enters the blood system. The blood carries it to all parts of the body, including the stomach. In the stomach, gastrin stimulates cells to secrete more gastric juice. Gastrin

SCIENCE AT WORK: Alexis St. Martin's Stomach

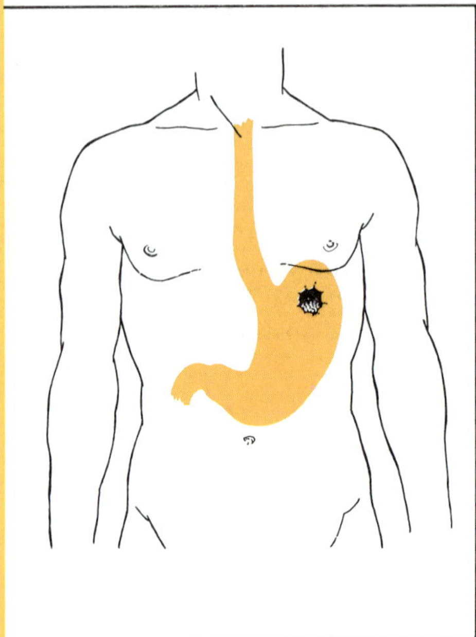

A basic part of any scientific research is observation. Unless observations are accurate and complete, very little data of any value is obtained. Today we have many machines and instruments that extend our powers of observation.

But this wasn't always so. At one time, the human body was a real puzzle to investigators. Certain internal sounds, such as the beating of the heart and gurglings and rumblings from the stomach, could be heard. But there was no direct way of finding out what was really going on inside the human body.

In 1822, in Michigan, the strangest of circumstances occurred. This situation allowed William Beaumont, a physician in the U. S. Army, to see into a human body and make some important observations. A badly wounded man, Alexis St. Martin, was brought to Beaumont. The accidental discharge of a shotgun had blown a large hole in St. Martin's left side. Beaumont patched up his patient and eventually the wound healed. But it never closed properly. The wall of the stomach healed by growing to the skin and muscles of the body wall. As a result, there was a permanent hole in the left side of the body that led to the interior of the stomach.

Beaumont realized that this was a wonderful opportunity to study the function of the stomach. St. Martin cooperated, and the interior of his stomach was studied by Beaumont for 11 years. The doctor made many observations. St. Martin's stomach secreted a fluid. When the fluid was put on meat, the meat was digested. The stomach moved and churned when food was in it. Beaumont kept very careful and complete records of his observations. They laid the groundwork for later research that could be undertaken as the tools of science became more efficient.

has no effect on other parts of the body. If there is protein in the stomach, gastrin is secreted and causes a large amount of gastric juice to be secreted. If only fats and carbohydrates are present, only small amounts of gastrin are produced.

Gastrin is a **hormone** (HOR·mohn). As such, it belongs to a class of complex organic molecules that are secreted by cells and are carried by the circulatory system. They have very specific effects on other cells—often in different parts of the body from where they are secreted. For this reason, they are often called chemical messengers.

Protein digestion is an example of the controlled changes that occur in response to the needs of the body. Since these controlled changes are so important, it will be useful to introduce a special word for such control. The word is **homeostasis** (ho·me·oh·STAY·sis). Homeostasis is the regulation of body processes. It works in such a way that conditions in the body's internal environment are maintained within the best range for cellular functions. In the case of digestion, homeostasis means that the secretion of gastric juices is regulated according to how much it is needed for digesting proteins.

In the stomach, digestion of proteins has begun but is not completed. Starch digestion is temporarily halted. Pepsin, which digests protein molecules, works best in the acidic medium provided by hydrochloric acid. But the digestion of starch by the enzymes in saliva occurs in a basic (alkaline) medium. These enzymes are inactivated by the acidic pH of the stomach. After food enters the stomach from the mouth, starch digestion stops. It does not begin again until the food reaches the small intestine.

Digestion in the Small Intestine

In adult humans, the small intestine is more than six meters long. Within the intestinal cavity, peristalsis of the walls of the tube mixes the food with secretions. These secretions come chiefly from the pancreas, liver, and walls of the intestine itself. Enzymes in these secretions complete almost all of the digestive processes.

The secretions also contain salts that adjust the pH of the digestive tract contents back toward a slightly basic pH. This is important, for the enzymes that work in the small intestine function best at a basic pH. This is in contrast to the stomach enzymes, which function best at an acidic pH.

The pancreas (Figure 14-4, page 433) is a large gland in the abdominal cavity just below the stomach. Its secretion, pancreatic juice, reaches the intestine through small ducts that connect the pancreas to the first part of the small intestine. Pancreatic juice does not flow continuously, but it is secreted rapidly when food enters the intestine. When digestion is not taking place in the intestine, the secretion of the fluid slows.

Pancreatic juice contains enzymes that act on the three main classes of food. One of its enzymes, amylase, breaks down starch to double sugars. Another enzyme, lipase, breaks down fats to fatty acids and glycerol. Pancreatic juice contains trypsin and other enzymes that partially break down proteins. These enzymes break some of the bonds of the polypeptide chains.

Intestinal juice is secreted by the cells of the inner lining of the small intestine. Enzymes from the intestinal glands break down double sugars. The double sugars are changed to simple sugars that the body cells can use. Other enzymes complete the breakdown of the polypeptide segments to amino acids. Thus, many different kinds of enzymes are necessary to carry out all the complex chemical processes of digestion.

Bile is secreted by the liver. It contains no enzymes, but it has an important role in the digestion of fats. Fats do not dissolve in water. Therefore, they are not easily broken down by digestive enzymes, which must be dissolved in water to become active. Bile contains bile salts, which break up fats into small globules. This increases the surface area of the fats and speeds up enzyme action.

Absorption in the Small Intestine

The cells of the intestinal lining absorb the molecules formed by digestion. Then the amino acids, simple sugars, some vitamins, salts, and water enter the bloodstream. Some other vitamins, fatty acids, and glycerol go into other, less familiar transport vessels called the **lymphatics** (lim·FAT·iks), which you will read about in Chapter 16.

The process of absorption takes time, as the volume of food to be absorbed and transported is large. The inner surface of the intestine is wrinkled into thousands of small "fingers" called **villi** (VIL·eye; singular, villus), which project into the intestinal cavity (Figure 14-6). This vastly increases the available surface and makes possible fast absorption of food molecules.

Most of the food molecules that enter the bloodstream from the small intestine are carried to the liver. The liver removes excess glucose from the blood and stores it as glycogen. When necessary, the liver converts glycogen back to glucose to maintain the proper level of glucose in the blood. The liver also stores vitamins and iron. These are added to the blood as the body needs them. Amino acids are stored until body cells need them for synthesis of enzymes, hormones, or cell structures. Some excess amino acids are converted

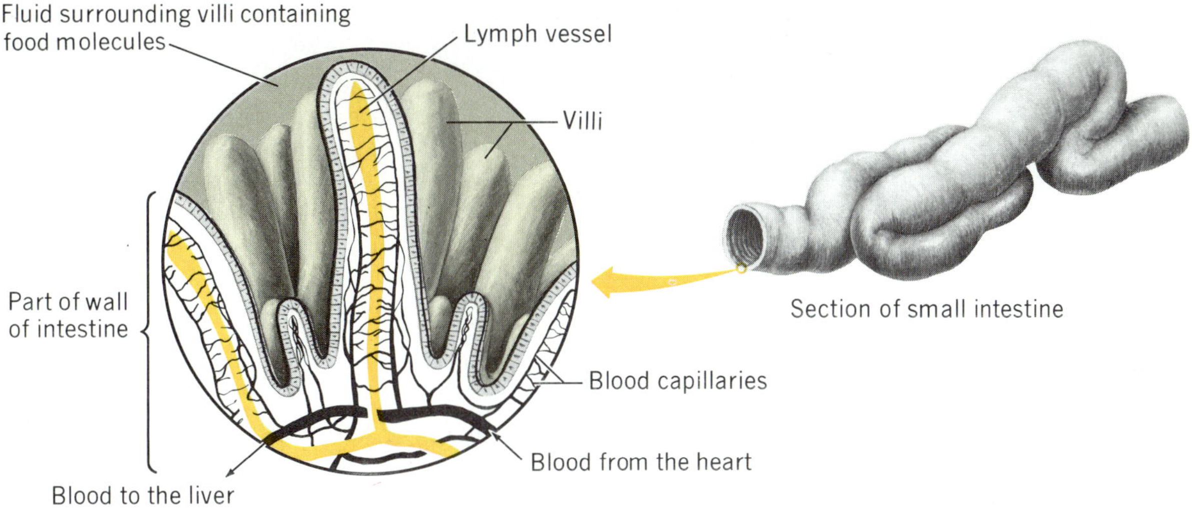

14-6. A drawing of part of the small intestine, with a section enlarged to show the villi of the inner wall. Each villus contains a branch of a lymph vessel and a network of blood capillaries.

to glucose, glycogen, or fat by a series of chemical reactions. These can be used as energy sources.

Where the small intestine enters the large intestine there is a sac. Off this is a little fingerlike projection called the appendix (Figure 14-4, page 433). Sometimes material becomes trapped in it and infection, called appendicitis, results.

Absorption in the Large Intestine

Material that cannot be digested passes from the small intestine to the large intestine. This material contains a large amount of water. One of the main functions of the large intestine is to absorb the water and return it to the bloodstream.

Much of the material in the large intestine contains bacteria and worn-out cells shed from the walls of the digestive tract. The bacteria in our intestines are necessary for our health. They synthesize some of the vitamins required for our normal body functions. The vitamins are absorbed into the bloodstream and are carried to the cells of the body.

The body periodically discards the small amount of water and solid matter that remains in the large intestine after absorption is complete. These discarded materials are the **feces** (FEE·seez). A large part, but not all, of the fecal material is material that our bodies are not able to digest. Feces pass through the rectum and out of the body through the anus (Figure 14-4).

In Brief

Large protein molecules are broken down into amino acids by the protein-digesting enzymes in gastric, pancreatic, and intestinal juices. Carbohydrates are acted on by starch-digesting enzymes in saliva, pancreatic juice, and intestinal juice, producing simple sugars. Fats are digested into fatty acids and glycerol by enzymes in pancreatic juice and intestinal juice.

Digestion begins with a tremendous number of different sorts of molecules. It ends with comparatively few. There are about 20 kinds of amino acids, a few kinds of simple sugars, fatty acids, water, dissolved salts, and small amounts of special compounds, including vitamins.

These end products of digestion are small molecules that can pass through cell membranes by diffusion or active transport. Some can be used by cells for energy. Others are the kinds from which a cell makes its own specific structures.

Research has shown that all phases of the complex process of digestion—enzyme activity, movement of materials along the alimentary canal, secretion, and absorption—are coordinated. Nerves, hormones, and the activities of cells of the digestive tract play important parts in this coordination. The net result is the smooth, efficient operation of the food-supply system on which human life depends.

Guide Questions and Problems

1. What role does digestion play in the life of animals?
2. What is the advantage, if any, to chewing food rather than swallowing it whole?
3. Why is the action of bile in fat digestion not considered an enzymatic action?
4. Sometimes the stomach is damaged by disease and must be removed. A person can live without a stomach on a carefully controlled diet. Why is this possible? Would it be possible to live if all of the small intestine was removed? Would it be possible to live without the large intestine?
5. What specialized features of the small intestine account for the efficient absorption of digested food?
6. How is it possible for a giraffe, drinking water with its head down, to move the water upward to its stomach?

7. Carbon dioxide, water, ammonia, and indigestible materials may be thought of as waste products. Some biologists, however, hesitate to call them wastes. What arguments would you suppose they use in supporting their views?

8. After major surgery, patients are sometimes "fed through the veins" during the early period of recovery. The food is placed directly into the blood in the veins. It receives no treatment in the digestive system. What must be the nature of this food?

Ideas for Independent Study

1. People living in different parts of the world have very different diets. Find out as much as you can about some of these diets. Find out what problems they may offer in processes of digestion and absorption. For example, milk is not equally as "good" a food for all people. Look into questions of why this is so.

2. Microorganisms grow in vast numbers in the digestive tract of humans and other animals. There are many interesting problems to investigate in connection with these microorganisms. Here are a few ideas about places to start:

(a) How do antibiotics, given as medicines to control infections, change the patterns of growth of microorganisms in the digestive system? What effects do these changes have on health?

(b) What effects do single-celled animals in the human digestive tract have on health?

(c) How are observations of bacteria from the digestive system used to trace pollution of lakes and streams?

(d) What are the differences in patterns of nutrition of grazing animals (such as sheep and cows) and humans? Find out how microorganisms play a part in these differences.

(e) Termites live on a diet of wood as long as large populations of a certain ciliated protozoan exist in their digestive tracts. Without these protozoans, a termite literally starves to death. What kind of relationship between the termite and the protozoan would explain this?

3. Imagine that you are an engineer in the biomedical field and wish to design an artificial digestive system. You hope to help people whose systems have been affected by injury or disease. What major functions would you have to build into your artificial digestive system?

Related Readings

BOOKS

Asimov, I. *How Did We Find Out About Vitamins?* New York: Walker & Co., 1974.

Nourse, A. E. *The Body.* New York: Time-Life, 1970.

Schmidt-Nielsen, K. *Animal Physiology.* Englewood Cliffs, N.J.: Prentice-Hall, 1975.

Sebrell, W. H. and J. J. Haggerty. *Food and Nutrition.* New York: Time-Life, 1967.

ARTICLES

Davenport, H. W. "Why the Stomach Does Not Digest Itself." *Scientific American,* January 1972, page 86.

Kermode, G. O. "Food Additives." *Scientific American,* March 1972, page 15.

Young, V. R. and N. S. Scrimshaw. "The Physiology of Starvation." *Scientific American,* October 1971, page 14.

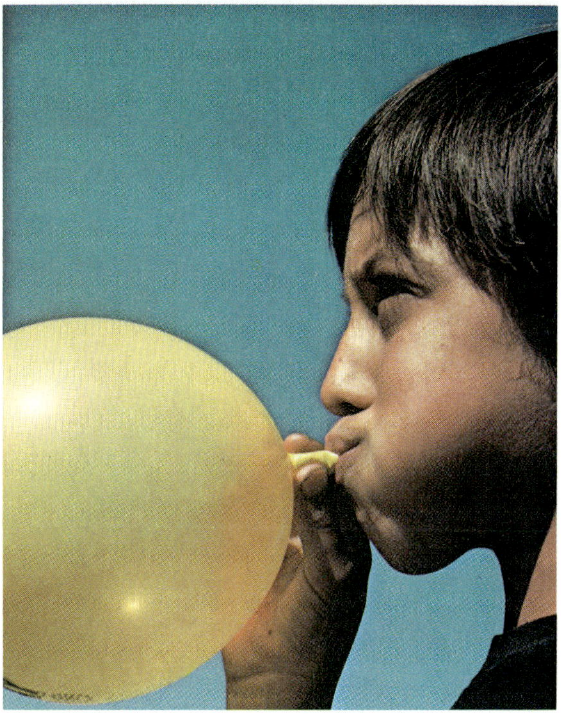

15 Animal Structure and Function: Respiration

With a few exceptions, most living cells require oxygen to obtain energy from food molecules. In obtaining energy, carbon dioxide is produced as a waste product. Therefore, cells must have some means of obtaining oxygen and of getting rid of excess carbon dioxide. These processes are part of the total chemical functioning of a living organism. We refer to the total chemical activity of a cell as metabolism.

Respiration includes all the events concerned with getting oxygen to the cell and disposing of carbon dioxide. It also includes all the chemical processes of a cell that are involved in the transfer of energy from glucose to ATP. *Breathing* includes all the processes involved in pumping air in and out of our lungs. Thus, breathing is part of respiration, but the two are not the same.

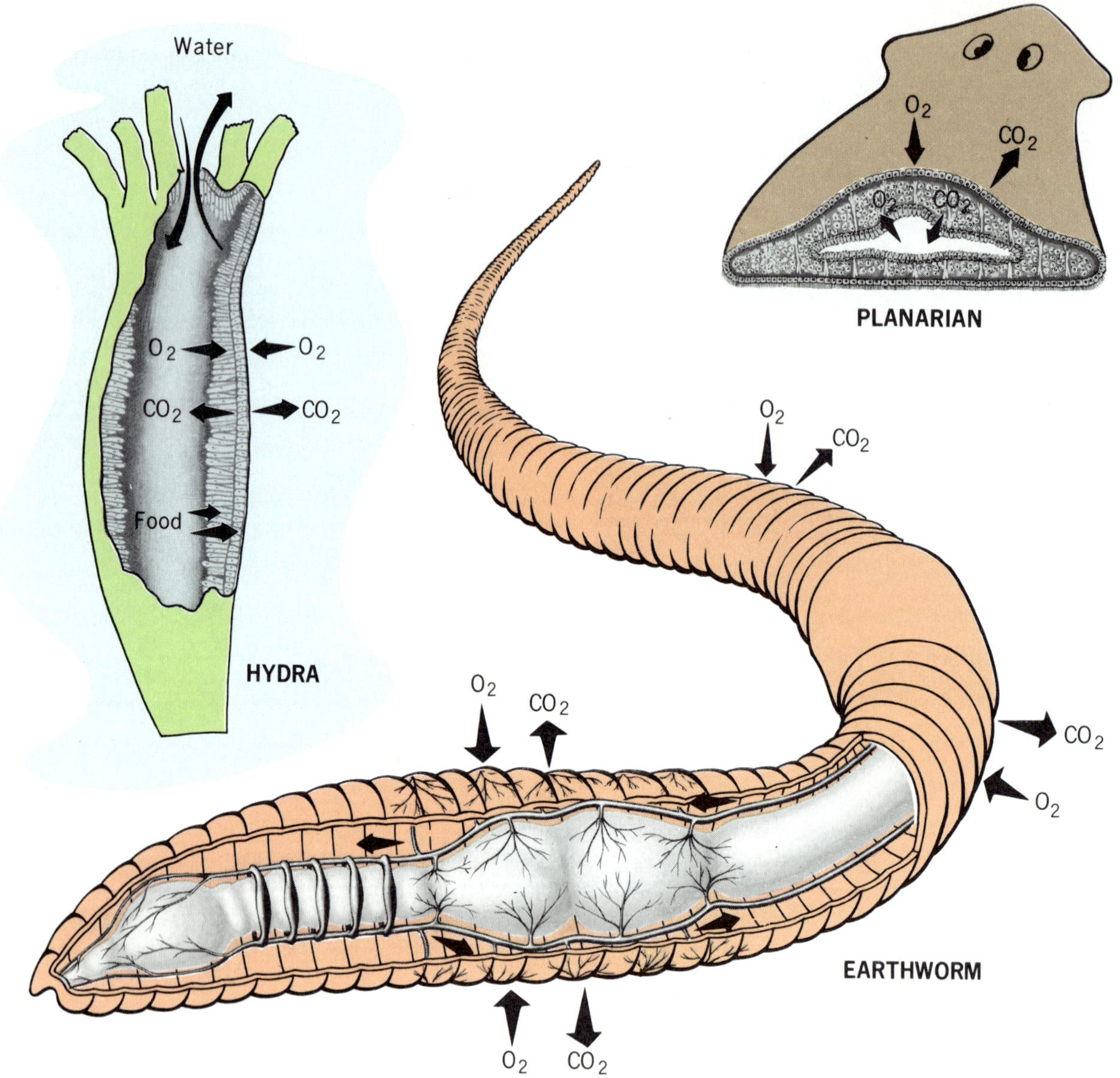

15-1. Respiration in *Hydra*, planarian, and earthworm.

PATTERNS OF RESPIRATION

Respiration in single-celled organisms, such as *Amoeba* and *Paramecium*, is relatively simple. Oxygen molecules dissolved in water diffuse across the cell membrane and into the cytoplasm. Diffusion and the movements of the cytoplasm carry oxygen to the mitochondria. In the mitochondria, oxygen is used in cell respiration, a process which produces carbon dioxide. Diffusion and movements of the cytoplasm carry the carbon dioxide produced during respiration to the cell membrane. The carbon dioxide diffuses across the cell membrane into the water.

In small aquatic organisms like *Hydra* and planarians, there are no organs specialized solely for respiration (Figure 15–1). The cells of both organisms are in thin layers. Oxygen diffuses into their bodies across the membranes of the outermost cells. In the case of *Hydra*, oxygen also diffuses from the water in the digestive cavity into the adjacent cells. Carbon dioxide leaves *Hydra* and planarians by simple diffusion. Most of the individual cells of these animals exchange oxygen and carbon dioxide directly with the surrounding liquid much as single-celled organisms do.

Larger and more complex organisms cannot rely solely on diffusion to supply cells with oxygen and to remove carbon dioxide. The oxygen that diffuses in would move too slowly to be used by any except the outermost layers of cells. The cells in the interior would die from the lack of oxygen.

A transport system containing a fluid (blood) forms the link between the inner cells of the animal and the outer respiratory surfaces. In the earthworm, for example, the skin is richly supplied with microscopic blood vessels, or **capillaries** (KAP·ih·ler·eez), of the circulatory system (Figure 15-1). Oxygen diffuses across the outer cells of the skin and the walls of the capillaries into the blood. The blood then moves around the body carrying oxygen to all the cells. Carbon dioxide is carried away from the cells by the blood to the skin where these molecules diffuse out of the animal.

Larger amounts of oxygen and carbon dioxide can be transported in liquid of the blood in various ways. One of these is the combination of oxygen with hemoglobin found in red blood cells. About five milliliters of oxygen can dissolve in a liter of water (at room temperature). The same amount of human blood at the same temperature will hold 250 milliliters of oxygen in chemical combination with hemoglobin. Thus, human blood can carry about 50 times as much oxygen as can water.

Many animals, especially larger ones, have specialized surfaces for the exchange of oxygen and carbon dioxide. In general, these surfaces are thin and moist. They are made up of folds or pockets, which greatly increase the surface area through which oxygen and carbon dioxide can diffuse. Many aquatic animals have folded respiratory surfaces called **gills** (Figure 15-2). Since there is no problem of drying out, these gills are on the outside of the animal. They are protected by a covering called the operculum (oh·PERK·yoo·lum). Water containing oxygen constantly passes over the surface of the gills. The oxygen diffuses across the gill membrane and into the blood of the circulatory system. Carbon dioxide moves from the blood into the water where it is carried away.

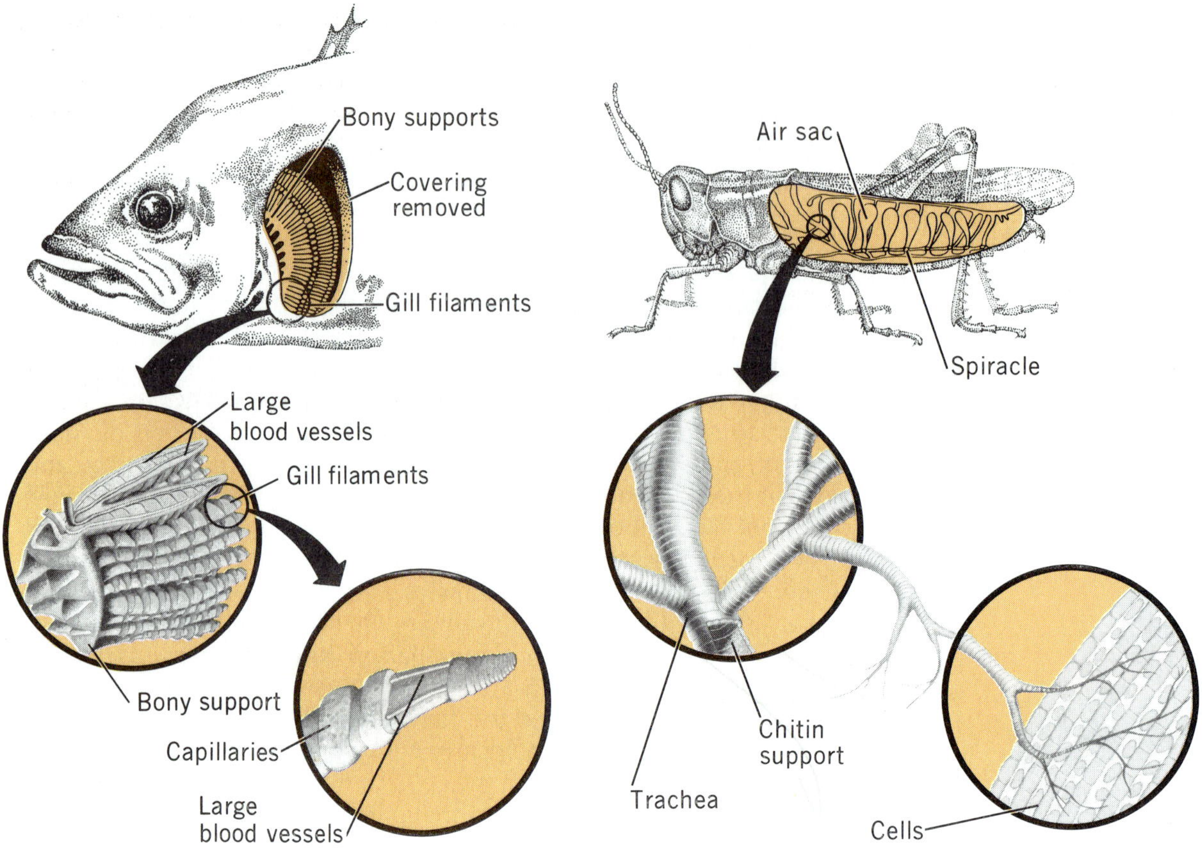

15-2. Gills are made of thin filaments richly supplied with blood vessels. Each filament is made of delicate plates containing many capillaries. In insects, oxygen and carbon dioxide are exchanged through the branching tracheae.

In animals that live on land, the air dries the delicate respiratory surfaces. Land animals, therefore, have different ways of carrying oxygen and carbon dioxide to and from the cells.

Insects have **tracheae** (TRAKE·ee·ee; singular, trachea), or tracheal tubes (Figure 15-2). The grasshopper has ten pairs of openings on the sides of its body. These openings, called **spiracles** (SPY·ruh·culs), lead to a series of tracheal tubes that extend throughout the body. Air is pumped through these tubes by contraction and expansion of the grasshopper's abdomen. When the abdomen expands, the front four pairs of spiracles are open. Air enters them. Then, as the abdomen contracts, the front spiracles close and the posterior six pairs open. Air enters these spiracles. This forces the air through the tubes and eventually out of the body.

The tracheal tubes branch repeatedly and reach all parts of the body. The inner ends contain a fluid. Oxygen dissolves in the fluid and then diffuses into the cells. Carbon dioxide diffuses out of the cells and is removed as air moves through the tracheal tubes. Notice that the tracheal tubes carry oxygen and carbon dioxide directly to and from the cells. The blood transport system in insects does not move oxygen.

The ancestors of insects and of other arthropods may have had a circulatory system with hemoglobin. The ancestors of arthropods are believed to have been similar to annelids. As the primitive arthropods evolved, however, hemoglobin may have become less of a survival advantage for land-dwelling arthropods.

HUMAN RESPIRATION

In humans, the exchange of gases with the environment occurs through the linings of the lungs, which are folded into millions of tiny microscopic air sacs (Figure 15-3). The total surface of these air

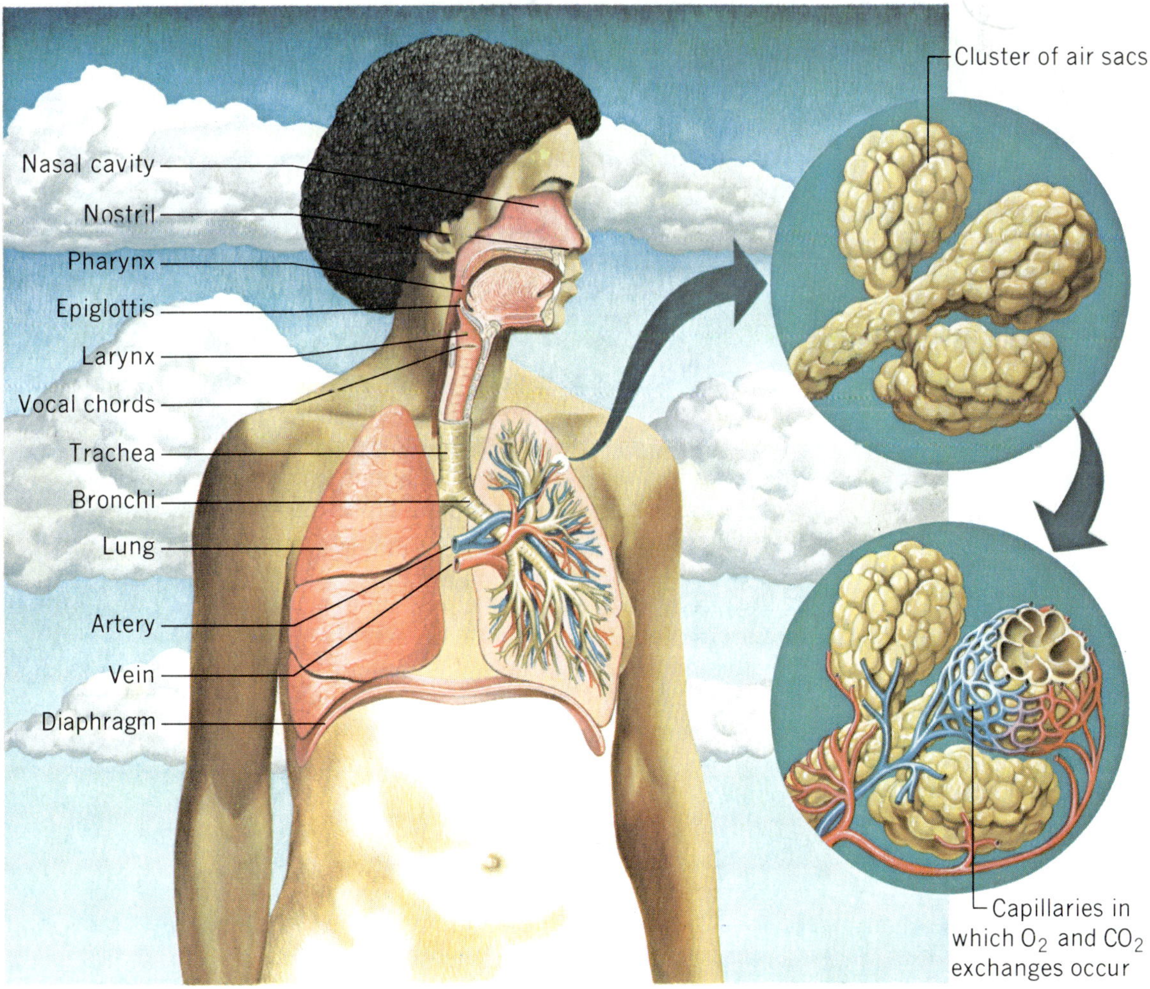

15-3. The parts of the human respiratory system involved in breathing. The lung has been cut away to expose the branching system of bronchial tubes. Part of the lung has been enlarged to show air sacs and their relation to capillaries. Millions of air sacs in each lung give the tissue a spongelike appearance.

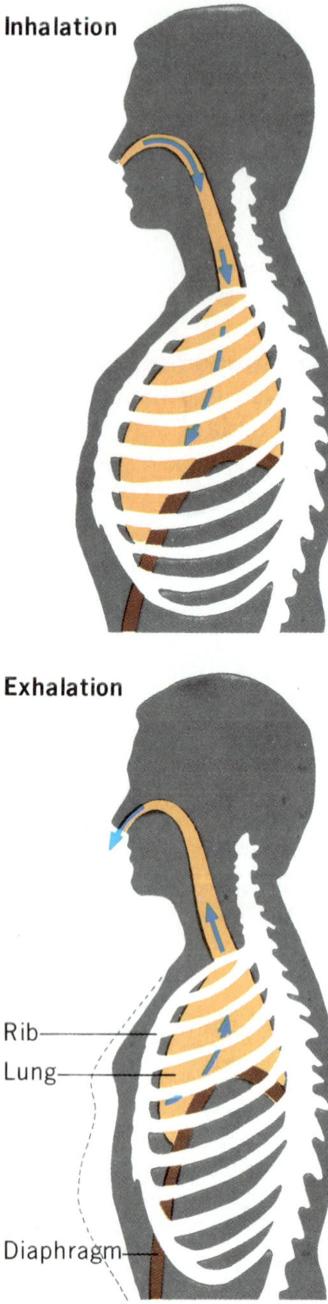

15-4. Movements of breathing in humans. Notice how chest size decreases and the diaphragm relaxes during exhalation.

sacs is enormous. If flattened out, they would cover an area of about 93 square meters. How does this compare with the area of a football field?

Diffusion of oxygen and carbon dioxide between the air sacs and the circulatory system occurs easily and rapidly. Each air sac is richly supplied with capillaries. Thus, the blood is very close to the air in the lungs. Blood flows continuously through the vast capillary network of the lungs. Red blood cells pick up oxygen and carry it through the arteries to the rest of the body cells. Blood flowing through capillaries in the tissues of the body picks up carbon dioxide and carries it through the veins to the lungs. The carbon dioxide diffuses from the capillaries into the air sacs.

The air inside the air sacs must also be exchanged with that in the environment. The exchange could not be made by diffusion alone. It would be impossible to move oxygen and carbon dioxide fast enough between the inside of the lungs and the outside environment. Actually, the air is pumped into and out of the lungs as we breathe.

Breathing is the process of inhaling and exhaling. The lungs themselves neither draw in air nor push it out. Instead, the chest wall and the **diaphragm** (DY·ah·fram) act as a large pump. The chest wall is made up of ribs with muscles between them (Figure 15-4). When we inhale, the muscles contract, pulling the ribs upward and outward. The diaphragm is a dome-shaped sheet of muscle that extends across the body below the lungs. When the diaphragm contracts it flattens out.

The movement of the chest wall and the flattening of the diaphragm increase the size of the chest cavity. When the volume of the cavity increases, the internal air pressure decreases. Then, air rushes into the lungs from the outside. This is inhalation. The reverse action then occurs. The chest wall is lowered and moves inward, and the diaphragm relaxes and takes on its dome shape. These changes reduce the size of the chest cavity and increase the pressure around the lungs. The elastic tissue of the lungs returns to its original shape, and the air is squeezed out of the body. This is exhalation.

As air moves into the nasal passage, dust particles are filtered out. The air is moistened by water vapor from the tissue lining the passage. The temperature of the air is adjusted to that of the body. The air then passes to the back part of the throat, down the trachea, through the **bronchi** (BRONG·ky), and into the small air sacs of the lungs. Food is prevented from entering the air passage by a flaplike valve, the **epiglottis.**

How does the amount of carbon dioxide exhaled vary with changes in the amount of exercise you do? In **Inquiry 15-1** you will find out.

SCIENCE AT WORK: SIDS

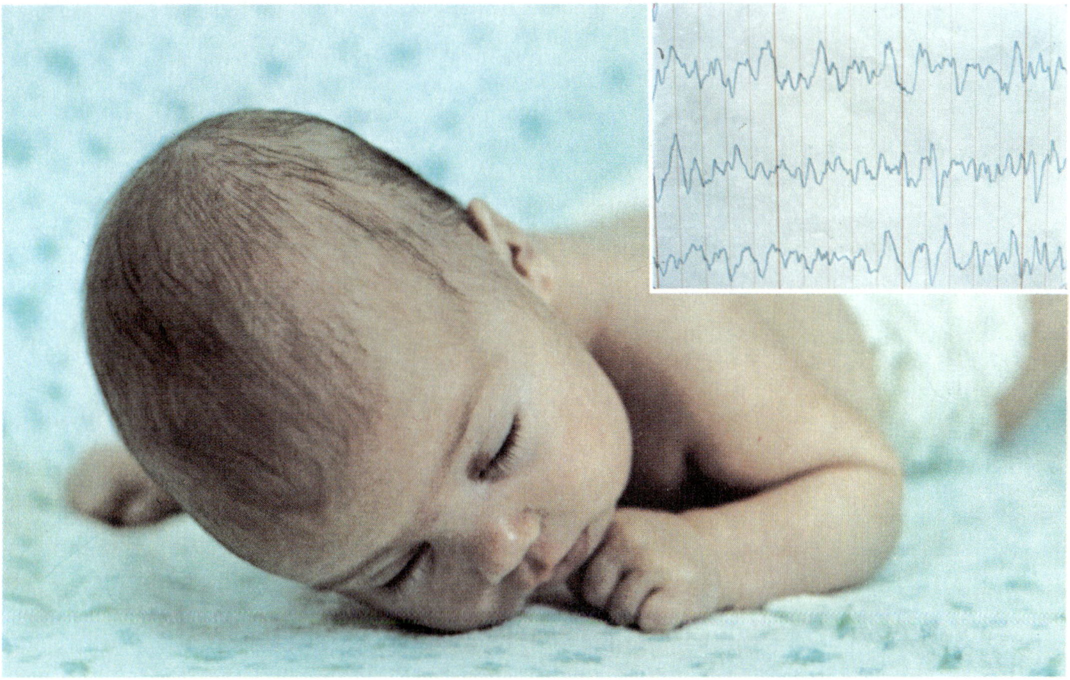

An area of research being given a great deal of attention by the scientists in the Pediatrics Department at Georgetown University in Washington, D.C. is that of sudden infant death syndrome (SIDS), sometimes called crib death. SIDS affects about one in 350 babies but strikes about one in 70 siblings (brothers or sisters) of crib-death victims. Because of that fact, the Georgetown University team recommends home monitors for all SIDS siblings and also for babies subject to sleep apnea (temporary cessation of breathing).

Sleep apnea is thought to be one of the causes of SIDS. Even if it does not cause death it can be threatening to the normal mental development of the child. Brain damage may result from a lack of oxygen during the apnea. A monitoring device would make a beeping noise to alert parents when a baby has stopped breathing for over 30 seconds.

A scientist in New Jersey has invented a mechanism that detects and corrects apnea immediately after it occurs. He developed a respiratory stimulator. This machine immediately stimulates breathing by suddenly inflating a small pneumatic mattress under the upper half of the infant's body. The mattress inflates after the device detects apnea. Its quick lift causes a sudden loss of equilibrium, and the startled infant begins breathing normally. It is similar to experiences you may have had with fast starting elevators. The sudden rise or fall of the elevator usually causes you to gasp for breath.

Only time and the continued use of this apparatus will tell if it can reduce the incidence of the sudden infant death syndrome.

INQUIRY 15-1 Carbon Dioxide and Human Respiration

MATERIALS (per team)

1% phenolphthalein solution	*rubber stopper to fit flask*
0.4% sodium hydroxide	*medicine dropper*
graduated cylinder, 100-ml	*straws*
graduated cylinder, 10-ml	*piece of white paper*
2 flasks, 250-ml	*stop watch (or watch with second hand)*

Procedure (Day 1)

Measure 100 ml of tap water in a graduated cylinder. Transfer it to a flask. Add 5 drops of phenolphthalein. Swirl the flask to mix the materials. Is there any change in the color of the water? If it is pink, you can assume there is very little or no carbon dioxide in the water.

If there is no color change, the tap water must have some carbon dioxide in it. To find out how much, add sodium hydroxide a drop at a time. It will help if you place the cylinder on a piece of white paper. Count the number of drops and stop when the water turns slightly pink. Swirl the water constantly while you are adding the sodium hydroxide.

When you have a pink color, stopper the flask. Save this sample of water and use it as your *standard* of pinkness.

If you added sodium hydroxide, determine how much you added. Using the same dropper, add an equal number of drops of water to an empty 10-ml graduated cylinder. Read from the scale of the cylinder how many milliliters your drops contain. Divide the number of drops by the number of milliliters. This will tell you how many drops from your dropper make 1 ml.

Prepare a second flask with water and phenolphthalein as you did the first flask. Add sodium hydroxide until you have the same pinkness as your standard. (This should be the same number of drops of sodium hydroxide.)

Designate one member of the team to be the experimental subject. Have the subject sit very quietly for one minute and then blow gently and steadily into the water through a straw or glass tube for ten seconds. **1.** What changes do you observe in the color of the solution?

After blowing for ten seconds, add sodium hydroxide drop by drop. Count how many drops it takes to cause the same pinkness as that of your standard. Convert the number of drops to milliliters by dividing by the drops-per-milliliter figure you obtained earlier. Multiply the number of milliliters by ten. This product is equal to the number of micromoles of carbon dioxide in the solution. **2.** How many micromoles of carbon dioxide did you exhale in ten seconds after resting for one minute? **3.** Do you think this number is an accurate measurement of the carbon dioxide you exhale in ten seconds?

Empty and rinse out the second flask. Prepare it with water and phenolphthalein again. Now, the subject should exercise mildly for one minute by running in place. Immediately, he or she should blow into the water sample for ten seconds. **4.** How many micromoles of carbon dioxide did you exhale in ten seconds after one minute of mild exercise?

Prepare the flask again and repeat the above procedure. But this time the exercise should be more vigorous (jumping jacks, running outdoors on the track field, and so forth). **5.** How many micromoles of carbon dioxide did you exhale in ten seconds after one minute of vigorous exercise?

Record the number of micromoles of carbon dioxide given off in each of the three experimental conditions on the chalkboard. Compare your results with the rest of the class. **6.** What is the average of each of the three experimental conditions? **7.** How is the production of carbon dioxide related to activity? **8.** What factors other than exercise might play a part in determining the amount of carbon dioxide given off?

Coordination of Breathing Rates

At this moment you are probably breathing at a rate of 16 to 20 times a minute. You are probably taking shallow breaths. Now let us suppose that you begin a game of tennis or walk upstairs rapidly. Much more carbon dioxide will be produced in your muscle cells than when you are at rest.

From your past experience and your observations in **Inquiry 15-1,** you can describe what happens as the rate of carbon dioxide formation increases. More carbon dioxide moves from your cells, across the cell membranes, and into your blood. More carbon dioxide passes from the blood capillaries across the lung membranes into your lungs. Your rate of breathing increases greatly. Your heart beats faster and blood is forced around your body more rapidly. The result of these various changes is the rapid removal of carbon dioxide from your muscle cells.

Measurements of the concentration of carbon dioxide in the blood show that it increases somewhat during exercise. But it does not go as high as we might expect from the great increase in carbon dioxide production in muscles. We can explain this in terms of the increased breathing rate. Increased breathing speeds the removal of carbon dioxide from the blood. From many other experiments, we can draw a related conclusion. Whenever the concentration of carbon dioxide in the blood rises, breathing tends to increase. Moreover, breathing rate is decreased if the concentration of carbon dioxide in the blood decreases below normal levels. But how does the concentration of carbon dioxide in the blood cause a change in breathing rate?

Experiments have shown that there are many cells in a part of the brain called the *breathing center.* These cells are connected by nerves with the muscles responsible for the breathing movements. Nerve messages are sent out of the breathing center to the breathing muscles. These rhythmic bursts produce the alternating muscle contractions and relaxations that cause inhalation and exhalation. When the concentration of carbon dioxide in the blood increases, the messages from the breathing center change. The rate of breathing movements increases, and the amount of carbon dioxide in the blood decreases. Through the coordination of blood flow and breathing, excess carbon dioxide is eliminated from the body.

This is not the full story of how breathing rate is regulated. Specialized nerve endings in blood vessels, with connections to the breathing center, can detect changes in concentrations of both carbon dioxide and oxygen in the blood. They also take part in controlling the activity of the breathing center. Many other kinds of nerve messages affect the rate of breathing. Can you control your breathing at will, even while all these other coordinating mechanisms are operating? You may want to design and carry out an inquiry to answer this question.

The control of respiration rate is another excellent example of homeostasis—the regulation of body processes and of the internal environment. In respiration, homeostasis means that breathing is regulated by the amount of carbon dioxide produced by the body cells.

Transport of Oxygen

Air contains about 21 percent oxygen. In the lungs, the air is in close contact with capillaries in the lung membranes. Oxygen molecules diffuse from the lung cavity. They move across the capillary walls and into the blood. Here, the oxygen molecules combine with hemoglobin in red blood cells.

Hemoglobin is a very large protein molecule. In spite of its size and complexity, it has been studied so carefully that we know its structure better than that of most other protein molecules. Each hemoglobin molecule contains four iron atoms. These account for the special function of hemoglobin. Oxygen molecules from air can combine loosely with the iron atoms. One molecule of oxygen, O_2, combines with each of the four iron atoms in a hemoglobin molecule. The combination of oxygen and hemoglobin is called *oxyhemoglobin* (ok·sih·HEE·mo·glow·bin). We can express the reaction as

$$Hb + O_2 \longrightarrow HbO_2$$

(The symbols Hb and HbO_2 are abbreviations for hemoglobin and oxyhemoglobin, respectively. They are not true chemical symbols.) There are some important things to learn about this reaction. The interactions of Hb and O_2 are very sensitive to the surroundings. If the pressure of oxygen is high, most of the Hb will combine with O_2. If, on the other hand, the pressure of oxygen is low, most of the HbO_2 present will break down into Hb and O_2. Thus, the reaction can be reversed very easily:

$$Hb + O_2 \leftrightharpoons HbO_2$$

Not only can hemoglobin combine with oxygen, but oxyhemoglobin can break down into hemoglobin and oxygen.

If hemoglobin is exposed to air at sea level, nearly every molecule combines with oxygen to form oxyhemoglobin (Figure 15-5). At 13 kilometers above sea level, the pressure of oxygen is much lower—

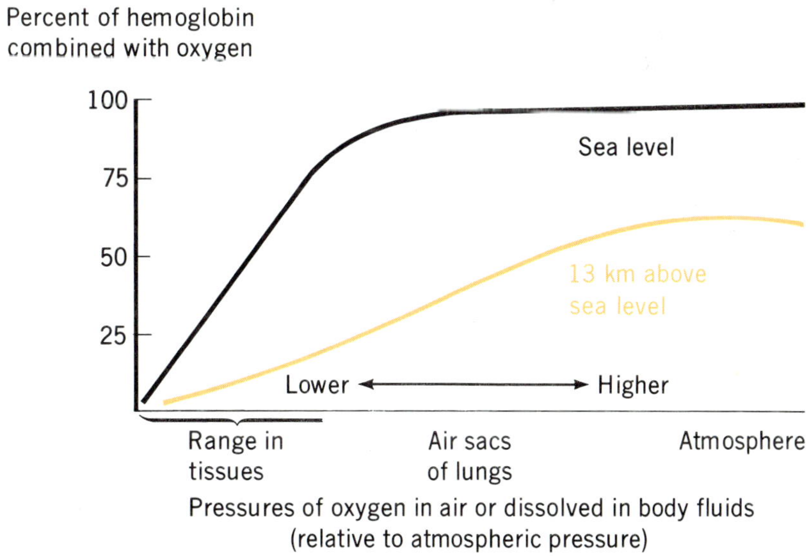

15-5. Differences in oxygen concentration in the atmosphere and in body tissues.

about one fifth as great as at sea level. Under these conditions of reduced oxygen pressure, only about half as many molecules of oxygen are combined with hemoglobin. In fact, human life is impossible at such an altitude without an extra supply of oxygen. Such an extra supply of oxygen is built into modern aircraft. These have pressurized cabins that maintain a more normal air supply, regardless of altitude.

The air next to the membranes in our lungs is not the same as the surrounding air. When we are at rest, or exercising only a little, our breathing exchanges only a small part of the air in the lungs. The concentration of oxygen in the lung cavity is about two thirds of that in fresh air. Nevertheless, at sea level, this is enough for about 95 percent of the hemoglobin molecules to form oxyhemoglobin.

In the capillaries of the tissues, red blood cells meet a very different environment. The tissue cells are continually using oxygen. Hence, the concentration of oxygen is quite low here. As the blood reaches these tissues, oxyhemoglobin molecules break down and form hemoglobin and oxygen. The oxygen then enters the cells by diffusion.

Before we leave this discussion of oxygen and hemoglobin, there is an important problem to point out. Some other gases besides oxygen can combine with hemoglobin. One of these is carbon monoxide. Carbon monoxide attaches so strongly to hemoglobin that it actually displaces oxygen. In other words, it prevents oxygen from combining normally with hemoglobin. Carbon monoxide often is produced in burning. It is present in automobile exhaust, tobacco smoke, and in many other kinds of air pollution. In terms of adequate supply of oxygen to the body cells, what dangers can result from air pollution?

Transport of Carbon Dioxide

In the reactions that occur within cells in our bodies, an excess of carbon dioxide is produced. The excess carbon dioxide diffuses into the capillaries. Carbon dioxide is carried in the blood in several forms. Most is carried in the fluid of the blood as the biocarbonate ion. The bicarbonate ion is formed as a result of a reaction between carbon dioxide and water. Smaller amounts of carbon dioxide dissolve in the blood or combine with hemoglobin and other molecules.

Carbon dioxide is carried by blood in much larger amounts than would be possible if it were simply dissolved in the fluid of the blood. Special enzymes in red blood cells speed the formation and breakdown of bicarbonate ions. When blood reaches the lungs, carbon dioxide is released very quickly into the air sacs. Thus, the blood serves as an excellent carrier of the respiratory gases, oxygen and carbon dioxide.

In Brief

Before going further, we should pause and take stock. What have we learned so far about life processes in humans and other animals? Reactions are constantly going on in body cells. Some of the most important involve the breakdown of organic compounds and exchange of energy. Two end products of these reactions are carbon dioxide and water. There must be a way to provide a constant supply of the molecules being used and a way to dispose of those being produced.

The digestive system supplies food molecules that are transported to all cells either directly by diffusion or by the circulatory system. In the cells of most animals, a supply of oxygen is necessary for complete breakdown of glucose molecules. In some animals, such as *Amoeba, Hydra,* and planarians, oxygen is obtained by diffusion. In other animals, the combined functioning of the respiratory and circulatory systems make oxygen available to the cells.

A large amount of carbon dioxide is produced as a waste product. In small animals such as *Hydra* and planarians, carbon dioxide diffuses from the organism directly into the external environment. In larger, more complex animals, such as earthworms and humans, the elimination of excess carbon dioxide is more complex. Carbon dioxide diffuses from cells and is removed by the circulatory and respiratory systems.

Guide Questions and Problems

1. What properties do most respiratory surfaces have in common?
2. Animals preserve homeostasis, within narrow limits, by regulation of intake and loss of substances. How is this principle illustrated in respiration in animals?
3. The same exertion at sea level and at 3,600 meters above sea level produces different breathing rates and different degrees of fatigue. What factors are operating to cause the differences?
4. How is blood able to take on oxygen in the lungs and release it to the tissue cells?
5. Aquatic mammals breathe with lungs, as you do. What special adaptations might they have that permit them to remain underwater longer than you can?

6. You plan to climb Mt. Everest. Your equipment includes a supply of oxygen you will need when you reach high altitudes. A friend suggests that this equipment is a waste of space for the following reasons: The air in the mountains is pure—no smog or dirt. There are very few other organisms to use the air. It has nearly the same percentage of oxygen as in the lowlands. How would you answer?
7. What is "artificial respiration"? Why may this be a life-saving procedure?

Ideas for Independent Study

1. Technology has been very useful in correcting disturbed patterns of respiration in disease. Look into the design of oxygen tents, iron lungs, and high-pressure chambers. Find out what problems they can solve, as well as difficulties in their use.

2. Many problems concerning respiration come up as soon as people leave the surface of the earth. They must leave the pressure and composition of air to which they are adapted. Among the more difficult problems are lack of oxygen and major changes in air pressure. This can result in conditions such as "the bends" that deep-sea divers may get. The bends are the result of the sudden formation of bubbles of nitrogen gas in the blood and body tissues.

Investigate this topic from several points of view. How have animals adapted to life at high altitudes and under water? What technical methods can be used to adjust the environment for deep-sea divers and for people in high-flying aircraft or space vehicles?

3. You may find it interesting to look into techniques of respiratory control in yoga and in skills and arts such as diving, running, glass blowing, playing wind instruments, and singing.

Related Readings

BOOKS

Ochsner, A. *Smoking: Your Choice Between Life and Death.* New York: Simon and Schuster, 1970.

Overmire, T. G. *Homeostatic Regulation.* A BSCS Pamphlet. Boulder, Co.: BSCS, 1963.

Weart, E. L. *The Story of Your Respiratory System.* New York: Coward-McCann, 1964.

The Royal College of Physicians. *Smoking or Health.* New York: Beekman Publishing, 1977.

ARTICLES

Hock, R. J. "The Physiology of High Altitude." *Scientific American,* February 1970, page 52.

Hong, S. K. and H. Rahn. "The Diving Women of Korea and Japan." *Scientific American,* May 1967, page 34.

Schmidt-Nielsen, K. "How Birds Breathe." *Scientific American,* December 1971, page 72.

Shannon, D. C. "Impaired Regulation of Alveolar Ventilation and the Sudden Infant Death Syndrome." *Science,* July 22, 1977, page 367.

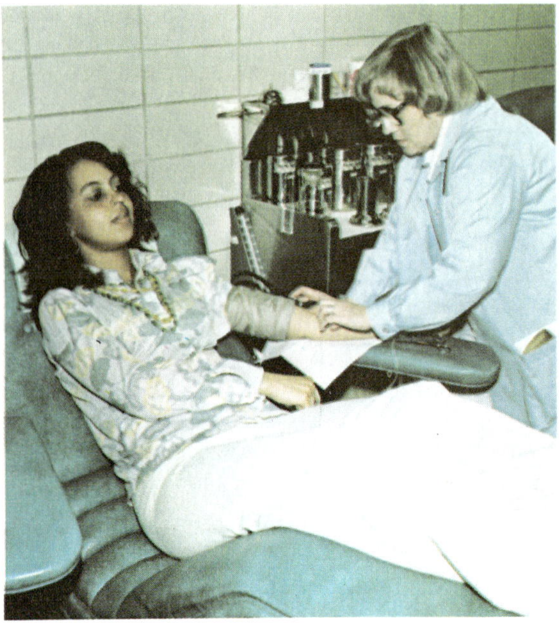

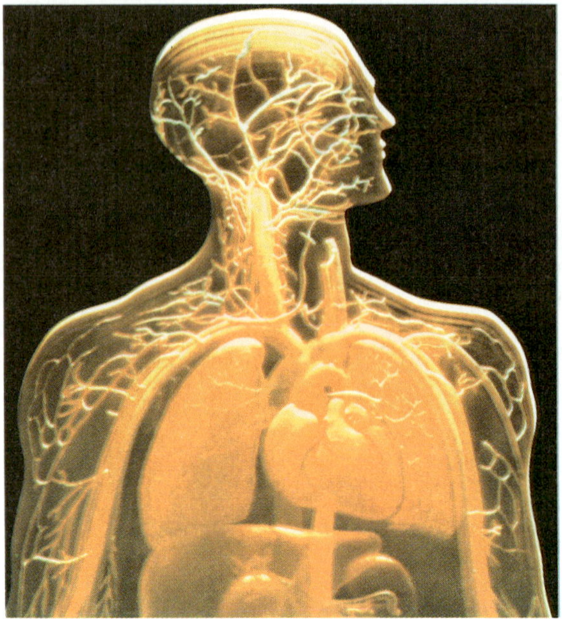

16 Animal Structure and Function: Transportation

One way in which molecules move is by diffusion. Molecules and ions of substances dissolved in body fluid or in a cell's cytoplasm are never at rest. They bombard and are bombarded by other particles. This results in movement and distribution of dissolved substances.

Diffusion, however, is a very slow process. It is fast enough to carry some particles around inside very small cells or areas of cells. But diffusion cannot move materials quickly enough in most cells and especially in the bodies of large organisms. Large organisms require additional means for transporting materials.

METHODS OF TRANSPORTATION

Even single-celled organisms, such as *Amoeba* or *Paramecium*, do not depend solely on diffusion to transport materials throughout their cytoplasm. Movements of the cytoplasm and food vacuoles help distribute materials throughout the cell. In many cells, structures such as microtubules and the endoplasmic reticulum (Chapter 3) also are important in moving molecules from one part of the cell to another.

Many small, multicellular organisms, such as *Hydra* or a planarian, have no specialized transport system. Most of the cells are at the surface of the organism or very close to it. As you learned in the previous chapters, digested food molecules are absorbed by the cells lining the saclike digestive cavity. Some of the molecules resulting from digestion diffuse from the inner cell layer to the cells of the outer layers. The cells of the outer layers, therefore, depend entirely on the inner cells for their organic food substances. The cells making up the cylindrical, hollow body wall of *Hydra* exchange oxygen and carbon dioxide directly with the watery environment.

The body of a planarian is very flat, as shown in Figures 14-3 (page 432) and 15-1 (page 446). Every cell is close to the outside of the body, where it easily absorbs oxygen and eliminates carbon dioxide. The distribution of digested food materials is taken care of largely by the highly branched digestive sac itself. This sac occupies much of the interior of the animal. No part of the body is far from some branch of the digestive sac.

In many groups of more complex, multicellular animals, there is a specialized system that transports materials throughout the body. The **circulatory system** of the earthworm is similar in its basic structure to that of humans and many other complex animals. The essential features of a circulatory system are shown in Figure 16-1. The circulatory system of the earthworm consists of a series of tubes,

16-1. The circulatory system of an earthworm. *Why does the earthworm need a more specialized system of transportation than the Hydra?*

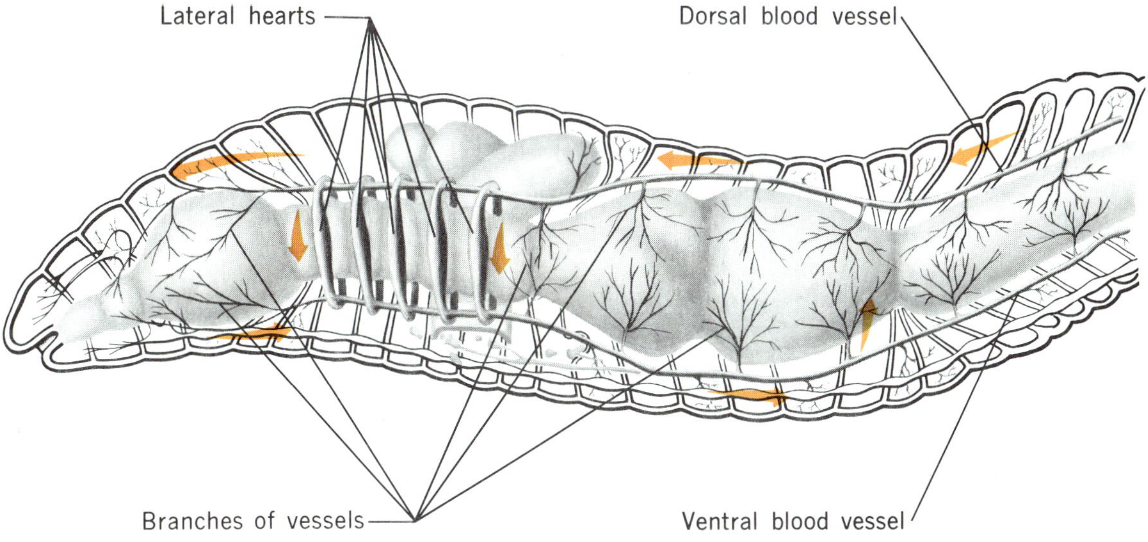

the blood vessels, which contain a fluid, blood. Blood is circulated to and from the cells throughout the body in these blood vessels. In the skin, the blood picks up oxygen and loses carbon dioxide. In the intestine, digested foods enter the blood and are delivered to body cells everywhere. Wastes from body cells are transported by the blood to excretory organs and, in the case of carbon dioxide, to the skin.

Some blood vessels have walls that pulsate (lateral hearts). This pulsating moves the blood. Blood circulates through the earthworm's body in the dorsal and ventral blood vessels. These vessels branch and go to all parts of the body. The ones that go to the digestive tract are indicated in the diagram. Eventually, the smaller vessels branch into capillaries.

The microscopic capillaries are so numerous that every cell is next to, or very near, a capillary. The branches of the vessels that enter the walls of the digestive tract break up into many capillaries. These capillaries have very thin walls. Molecules move quite rapidly across them. The end products of digestion, such as amino acids and glucose, enter the cells lining the intestine. Then they enter the capillaries. Once in the bloodstream, these end products of digestion are distributed to all parts of the body.

The capillaries play a unique role in the circulatory system. Only these vessels have walls thin enough for molecules to diffuse through them. Consequently, all the important exchanges that occur between blood and the environment, or between blood and cells, occur through capillaries. This is true for all animals in which capillaries are part of the circulatory system.

The blood of the earthworm is always circulated in many enclosed tubes—blood vessels. This type of circulatory system is known as a

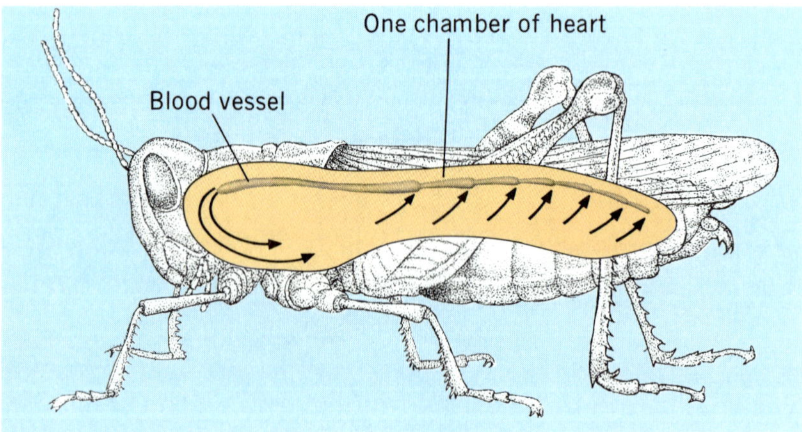

16-2. The circulatory system of a grasshopper. *Is this system less efficient than the circulatory systems of earthworms and humans?*

closed circulatory system. Humans have this kind of circulatory system.

The circulatory system of insects is built on quite a different plan from the systems of earthworms and humans. Insects have an **open circulatory system.** That is, the blood is not always enclosed in tubes. Much of the time it flows in large spaces among the tissues of the body. Does this system seem more or less efficient than the closed circulatory system?

For example, the grasshopper has a small heart located in the upper portion of its body (Figure 16-2). The grasshopper's heart consists of several chambers. The heart pumps blood out toward the head in a single vessel. This vessel ends in the head region, where blood flows out of the vessel and into spaces among the body tissues. The blood moves back through the body cavities. It carries the products of digestion to all parts of the body. Eventually, the blood returns to the heart to be pumped forward once again. Oxygen and carbon dioxide are carried directly to the cells through a separate system of tracheal tubes (Chapter 15).

In the laboratory, you can study the activities of the heart in an invertebrate animal as you do **Inquiry 16-1.**

INQUIRY 16-1 A Living Invertebrate Heart

In this inquiry, you will study the working heart of a living water flea, *Daphnia.* This organism is commonly found in freshwater ponds and lakes. It is so transparent that, when placed under the low power of a microscope, its beating heart can be seen (Figure 16-3). You will be able to study a number of environmental effects on the action of the heart. *Daphnia* can also be used for determining the effects of drugs on heartbeat.

MATERIALS (per team)

Day 1

Daphnia
medicine dropper or pipette
microscope slide
thermometer, at least 0°–40°C
beaker or wide-mouth jar, 400-ml or larger
beaker, 50-ml
bunsen burner
tripod
watch with sweep second hand
graph paper
stereoscopic dissecting microscope
petroleum jelly
toothpicks

16-3. The internal structure of *Daphnia.*

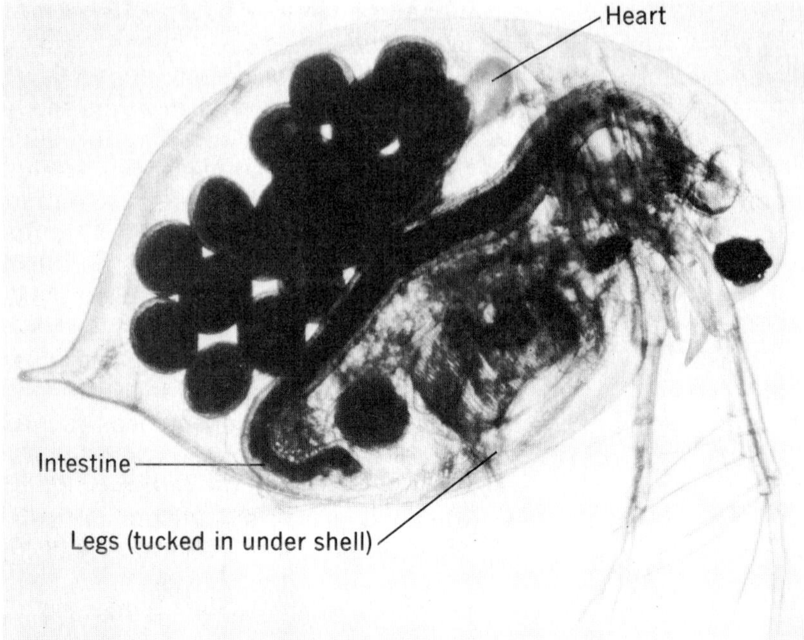

Day 2

Daphnia
microscope slide
medicine dropper or pipette
5% ethyl alcohol
10% ethyl alcohol
stereoscopic dissecting microscope
graph paper
petroleum jelly

Procedure (Day 1)

Each laboratory team will work with *Daphnia* at one of the following temperatures: 0°, 12°, 24°, or 36°C.

Prepare a water bath at the temperature you plan to use. Partially fill a large beaker with water and heat it, or add ice, to bring it to the desired temperature. Place a 50-ml beaker containing a few *Daphnia* and water into the larger beaker. Swirl the contents of the small beaker to maintain a constant temperature throughout. Wait at least ten minutes before you measure the heart rate. The temperature of the water bath should not vary more than 1° or 2°C while *Daphnia* are in the water bath or during the testing period.

To determine the rate of the heartbeat, place a clean glass slide in the larger beaker for one minute (to bring its temperature to that of the *Daphnia*). Remove the slide, dry it quickly, and put a small dab of petroleum jelly in its center, using a toothpick. Use a medicine dropper to remove a *Daphnia* from the smaller beaker, taking as little water as possible. Place the *Daphnia* on the jelly very carefully. With a clean toothpick, move the *Daphnia* into a position for easier observation.

Add a few drops of water from the smaller beaker. Do not add a cover slip. Place the slide on the stage of the microscope and bring the heart into focus. You may have to move the *Daphnia* to do this. Count the heartbeats for 15 seconds. Multiply by 4 to get the number per minute. Perform this procedure quickly so that the temperature changes as little as possible. The heartbeats are usually very fast. The following method will help you to count these beats: Place a pencil dot on a piece of paper for each beat of the heart. The heartbeat is rhythmical. If you keep in time with it you can approximate the beats rather well. When you have finished, count the dots.

Repeat this procedure with a fresh *Daphnia* at least three times. Calculate the average of your three trials. Record this average in your notebook. Combine the class data on the chalkboard in a table such as the one in Figure 16-4. When all the data have been reported, record the completed table in your own data book.

Temperature	*Average heart rate per minute*	*Initials of team members*

16-4.

Make a graph of the data. Plot the rate of heartbeat along the vertical axis and temperature along the horizontal axis. **1.** What is the effect of temperature on the rate of heartbeat? **2.** At what temperature is the heart rate the slowest? the fastest? **3.** Predict what would happen to the heart rate if you continued to increase the temperature in 5°C intervals. **4.** Does your own heartbeat change in the same manner as that of *Daphnia* in different temperatures? Explain. **5.** Your graph shows the experimental heart rates taken by different students using different *Daphnia*. What might be a more reliable way to run this temperature study?

Procedure (Day 2)

A number of substances, known as depressants, slow the activity of some nerves. Other substances, known as stimulants, have an opposite effect. Does alcohol have a depressing or stimulating effect on the heartbeat of *Daphnia*? Recall what you know about *Daphnia* and about the effects of alcohol. Consider these along with the results you obtained from Day 1. Write a hypothesis about the effects of alcohol on the heartbeat of *Daphnia*. Using any of the materials listed for Day 2, design a simple experiment to test your hypothesis. List the materials you will need and describe the procedure for the experiment. Do you have a control for your experiment? Check with your instructor to see if there are any problems. Then proceed. Obtain your data, analyze it, and draw a conclusion about your experiment. Does your hypothesis have to be modified? Discuss your experiment and the results you obtained with your classmates.

Human Transport System

In humans, transport of fluids is by either the **blood system** or the **lymphatic system.** These systems work together. We will look at the blood system first.

The blood system consists of the heart and a series of blood vessels. The heart is a muscular organ. Its contractions force blood into the **arteries.** Arteries branch repeatedly, becoming progressively smaller. They carry blood to all parts of the body. The blood then enters the capillaries. Because of their very thin walls, the exchange of fluids and dissolved molecules between the blood and the cells occurs in the capillaries.

Blood flows from capillaries into **veins,** which join one another and eventually carry the blood back to the heart. Just as arteries can be thought of as distributing vessels, veins are collecting vessels. Blood moves constantly from the heart through the arteries, to the capillaries, through the veins, and to the heart again. You can imagine the path of the blood as a circle. Thus, we speak of the circulation of the blood.

16-5. William Harvey.

The Discovery of Blood Circulation

The fact that blood circulates in the body was demonstrated by William Harvey and Marcello Malpighi (mal·PEE·gee) in the seventeenth century. In order to understand their achievement, you should consider some of the hypotheses about blood that were accepted in the sixteenth and early seventeenth centuries. These hypotheses dated from ancient times.

Aristotle believed that blood was formed in the liver. From there it would flow to the heart, then through the veins to all parts of the body. Arteries were not thought to function in the distribution of blood. They were thought to contain air. About 500 years after Aristotle's time, Galen (A.D. 130–200), a Greek physician and author of important books on medicine, found that arteries actually contained blood.

In the sixteenth century, people understood neither the function of blood nor how it moved. Generally, it was believed that blood flowed in and out of the heart in the same vessels.

William Harvey, an English physician and scientist of the seventeenth century, studied the hearts of embryos of many vertebrate animals, as well as mollusks and water fleas. He also examined the hearts of living animals. As a result of these observations, Harvey came to realize that the heart is a pump. By measuring its change in size as it contracted, and the rate at which it contracted, Harvey discovered that the heart pumps out a large quantity of blood. It seemed impossible that so much blood could be produced by the liver as Aristotle had suggested. From the data, Harvey made quite a brilliant deduction: The same blood must flow out of the heart and then return to it. That is, the blood must circulate. The blood pumped by the heart into arteries must then return to the heart through veins.

Harvey tested this hypothesis in many ways. One study demonstrated the presence of valves in many veins (Figure 16-5). These valves are placed in such a way that blood can flow in veins in only one direction – toward the heart. You can repeat this demonstration, with a partner, by following the directions shown in Figure 16-6. Place a tourniquet above your partner's elbow. The large knots that develop are formed by valves (Figure 16-6a). If you press on one part of a vein and then, with another finger, streak the blood upward beyond the next valve, that part of the vein will remain empty (Figure 16-6b). Apply a finger of the other hand to the distended part of the vein above the valve and press downward. You find that you cannot force the blood back through the valve (Figure 16-6c).

Harvey studied the veins of dead animals and found that he could push a probe easily in one direction. But often it would not move in the other direction. Study the structure of these valves shown in Figure 16-7. In which direction would you expect blood to flow more easily: **a** ⟶ **b** or **b** ⟶ **a**? How could you test your hypothesis? Valves are always directed in one way in veins – with the direction of "easy movement" toward the heart. How does this observation serve to refute Aristotle's hypothesis of blood flow?

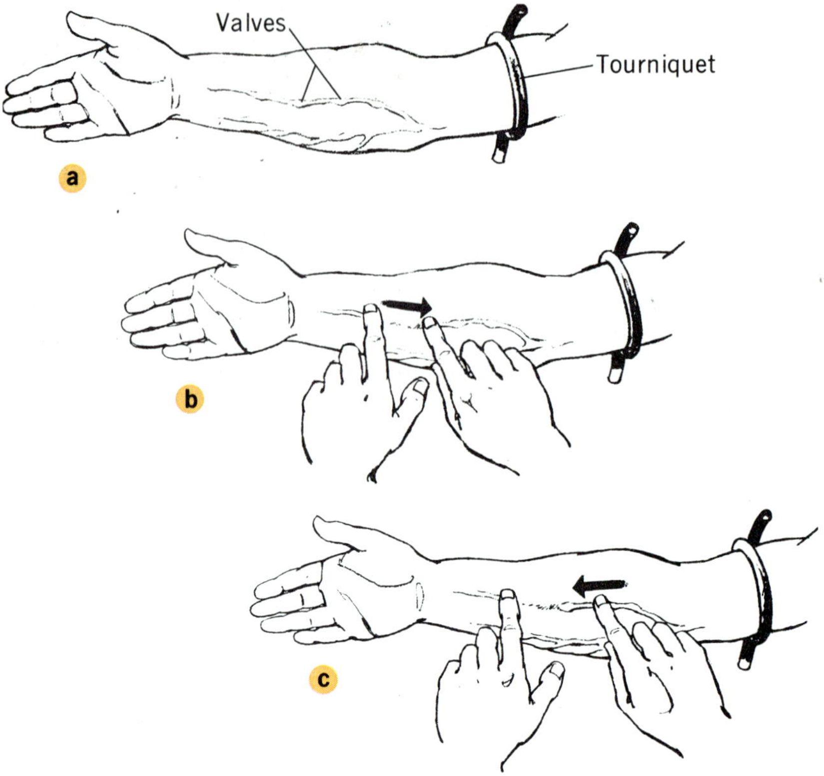

16-6. William Harvey demonstrated the presence of valves in many veins. These valves work in such a way that blood can flow in one direction only—toward the heart.

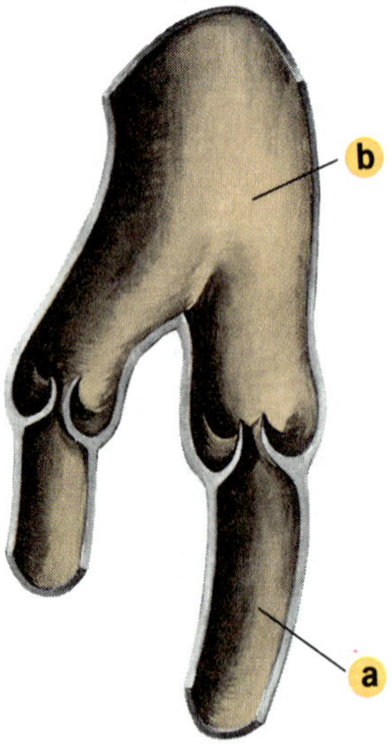

16-7. The valves in veins. A short section of a branching vein, shown in longitudinal section.

Although Harvey was able to give evidence that blood circulates, there were a few questions about circulation that could not yet be answered. Harvey thought that connections must exist between the arteries and the veins. But he was unable to find them. The instrument that could have made it possible to see these connections—a microscope—was not available.

So Harvey presented observations and arguments that blood circulates. The hypothesis was not proven. But Harvey's work was a major accomplishment among the painfully slow and difficult steps taken by scientists to try to understand how the human body functions. Only a few years after Harvey died, the Italian biologist Marcello Malpighi was able to describe the structures that complete the path of blood circulation. Malpighi was able to do this because microscopes had become available.

You may observe these structures for yourself by completing Inquiry 16-2.

INQUIRY 16-2 Observations on Circulation

MATERIALS

goldfish in a bowl or an aquarium
dip net
half of a petri dish
2 wads of absorbent cotton (1 thin and 1 thick)
2 halves of a microscope slide
medicine dropper (pipette)
compound microscope or stereoscopic dissecting microscope

Procedure

Soak a *thin* wad of cotton in water and spread it toward one end of the bottom of a petri dish. At the other end place a clean half-slide. Soak a thick wad of cotton in water and have it ready for the next step.

Using a net, remove a fish from the water. Place the fish in the petri dish with its head and body on the moist cotton. Its tail should lie on the half-slide.

Now place the thick, soaked wad of cotton over the body of the fish. Place another half-slide over the tail. The setup should look like Figure 16-8.

If the fish flips its tail out, as it may, just put it back between the glass slides. *Remember to keep the cotton wet.*

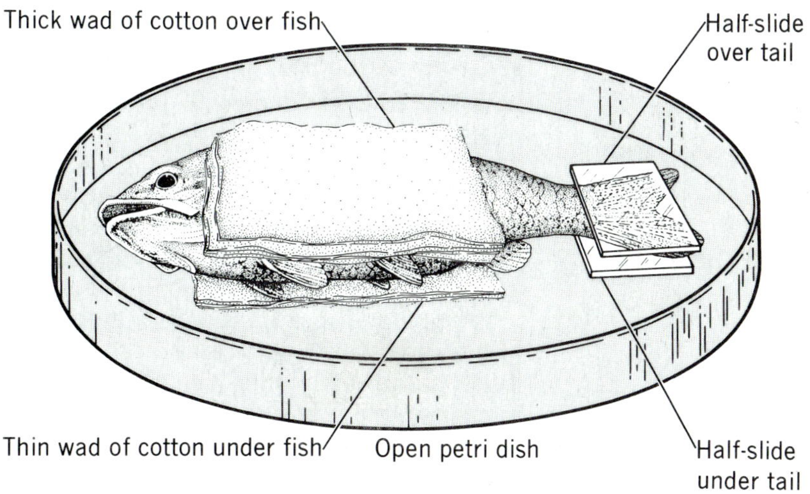

16-8. Preparation for observing circulation in a living fish.

Remove the clips from the stage of your microscope. Place the petri dish on the stage so that the fish's tail is over the hole in the stage.

Focus on the tail. Then move the dish around until you find a part of the tail in which you clearly see flowing blood. The smallest of the blood vessels you can see here are capillaries. Look for a small artery at a point where it divides. The two forks of this division are the capillaries. **1.** Does blood flow more slowly or more rapidly in capillaries than in the arteries?

The small objects moving through the capillaries are red blood cells. **2.** What is their shape? **3.** How many red blood cells can pass side by side through the capillaries? **4.** Formulate a hypothesis to account for any advantage in having red blood cells pass through the capillaries as they do.

Trace a capillary in the direction of the blood flow until it joins with another to form a slightly larger vessel, a tiny vein. **5.** Does the blood move more rapidly or less rapidly after it flows from the capillaries into a vein? **6.** Why is it an advantage that the capillaries have thin walls? Would you expect much exchange between red blood cells and the body cells in arteries or veins? Why or why not?

The Heart

The human heart is a double pump. It has four chambers that act as the pumps (Figure 16-9). The two lower chambers are the **ventricles** (VEN·trih·kulz). The left ventricle pumps blood to all parts of the body except the lungs. This blood supplies the cells with digested food substances and oxygen. It also collects carbon dioxide and nitrogenous wastes from the cells. Then the blood returns to the right upper chamber of the heart, the right **atrium** (AY·tree·um; plural, atria). From here, the blood passes to the right ventricle, where it is pumped to the lungs (Figure 16-9) through the **pulmonary** (PUL·mun·nehr·ee) **artery.**

In the lungs, blood passes through lung capillaries. As it does so, carbon dioxide is given off and oxygen is taken in. The blood is returned through the pulmonary veins to the left atrium, then the left ventricle, and is pumped into the **aorta** (ay·OR·tah) to be carried around the body once again.

The blood always flows through the chambers of the heart in one direction. The direction of flow is controlled by valves (Figure 16-9). There are valves between the right atrium and ventricle as well as between the left atrium and ventricle. When the ventricles begin to contract, these valves close. As a result, blood passes into the pulmonary artery and into the aorta, not back into the atria.

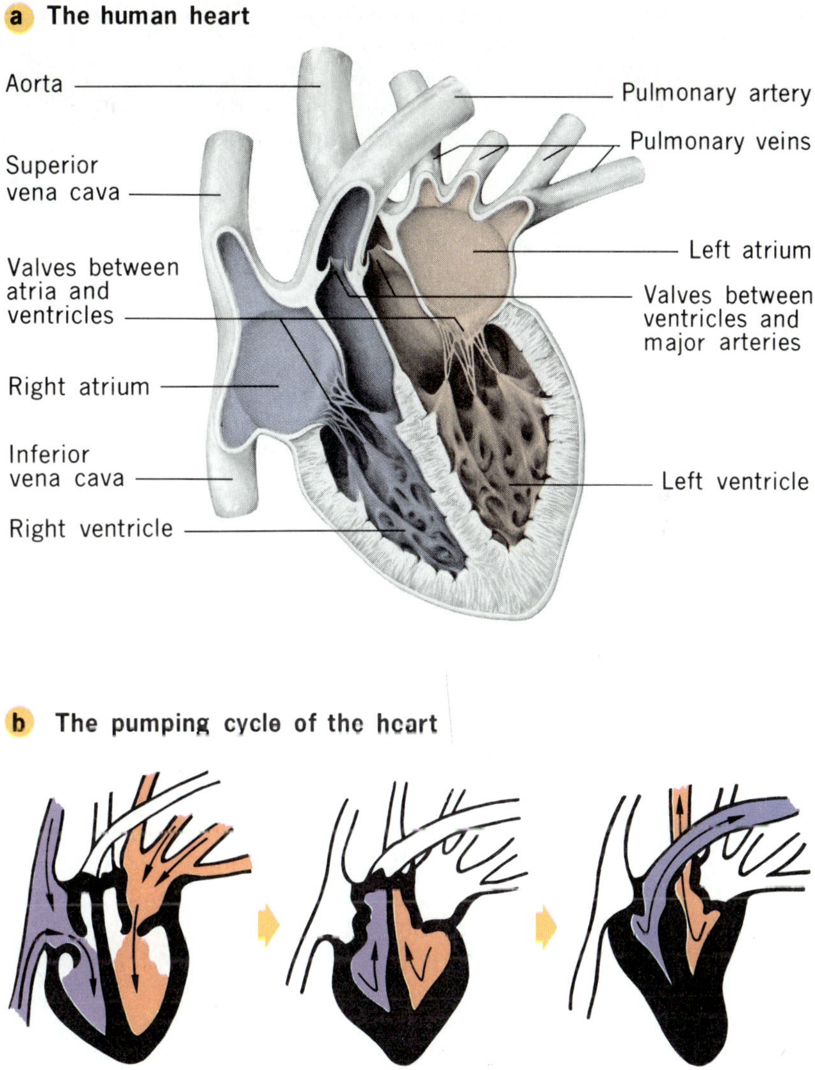

16-9. The human heart. **(a)** The parts of the heart. **(b)** Right ventricle contracts—blood passes into the pulmonary artery. Left ventricle contracts—blood passes into the aorta.

There also are valves between the left ventricle and the aorta and between the right ventricle and the pulmonary artery. After the ventricles have contracted and forced blood into the aorta and pulmonary artery, they relax. If there were nothing to hold blood in the aorta and the pulmonary artery, it would flow back into the relaxed ventricles. However, the valves close and prevent backflow.

Your heart is about the size of your fist. This muscular organ must continue to contract as long as you live. Its activity is staggering.

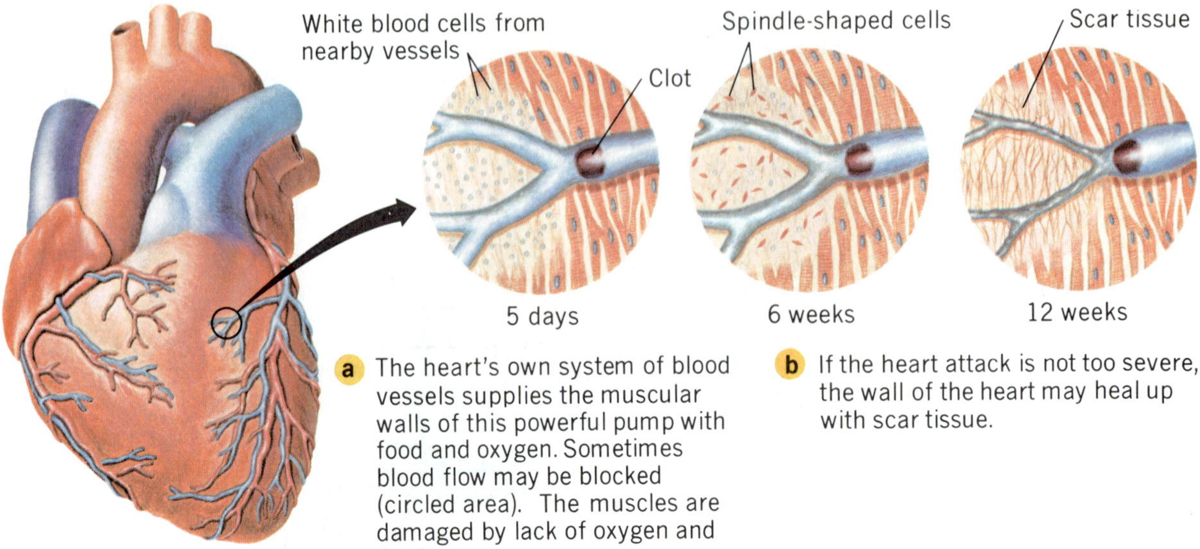

a The heart's own system of blood vessels supplies the muscular walls of this powerful pump with food and oxygen. Sometimes blood flow may be blocked (circled area). The muscles are damaged by lack of oxygen and foods.

b If the heart attack is not too severe, the wall of the heart may heal up with scar tissue.

16-10. The heart attack and possibilities for recovery.

Check your own heart rate (in beats per minute). Then estimate how many times your heart has contracted since you were born.

Often the heart carries on its work even following the damage of a "heart attack." Such failures in function are of numerous types, one of the more common of which is illustrated in Figure 16-10a. Coronary arteries supply blood to segments of heart muscle. If one of these arteries is blocked by a blood clot, muscle fibers in that segment die. If the damaged region is small, repair processes start. White blood cells enter the site and consume the dead tissue (Figure 16-10b).

Another type of cell, which is spindle-shaped, also enters the damaged area. These cells begin producing strong connecting fibers that form scar tissue, which does not contract. After a few months, the area is repaired. The scar tissue draws the edges of healthy muscle tissue closer together. Muscle function, however, is never quite the same as before. If the damage is too great, the heart ceases to function and the person dies. Heart disease is a major cause of death in the United States. A doctor often can prevent a heart attack by increasing the oxygen in the blood and keeping the blood circulating.

Arteries, Capillaries, and Veins

Blood is driven out of the ventricles by their contractions. It enters the aorta and the pulmonary artery under a great deal of pressure.

The repeated contractions of the ventricles maintain the pressure that continues to force more blood into the aorta and the pulmonary artery (Figure 16-11). The pressure also is maintained by the elastic walls of these arteries. The walls are stretched with each heartbeat by the outward flow of blood. They return to their original size be-

16-11. THE HUMAN CIRCULATORY SYSTEM

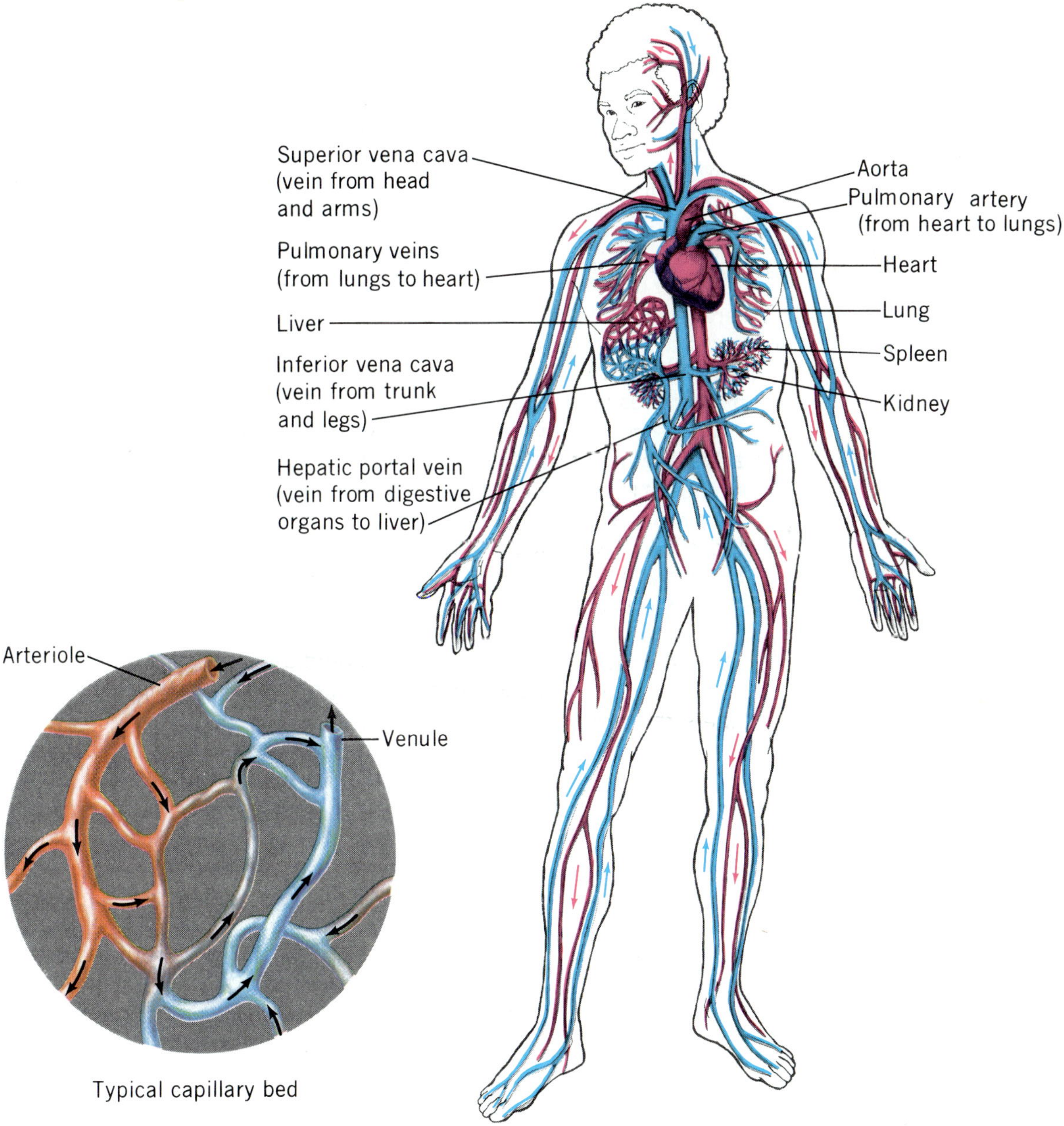

Typical capillary bed

tween heartbeats. The pressure is great enough to send blood to all parts of the body through the arteries and to return it to the heart through the veins.

If we compare the artery leading to a particular body organ with the vein that receives blood from that organ, we will see three main differences. First, the artery has a smaller inside diameter and a thicker wall than the vein. The walls of an artery must withstand greater blood pressures than those of a vein. Second, the artery's elasticity helps change the pulsing flow of blood, as it leaves the heart, into the steadier flow seen in the small arteries. Third, veins have valves but arteries do not.

As blood is pumped out of the heart through the aorta, more blood enters the heart through two large veins. The smaller veins that carry blood from the head and arms join above the heart. These form one large vein, the **superior vena cava.** Veins carrying blood from the lower body join to form the **inferior vena cava.** Both of these veins empty into the right side of the heart (Figure 16-11).

Individual capillaries, which are microscopic in size, are the most numerous blood vessels in the body. They are networks of tiny arteries called **arterioles** and tiny veins called **venules** (Figure 16-11). It is through the capillaries that the actual exchange of materials takes place between blood and individual cells (Figure 16-12). There is a constant flow of material between the capillaries and the fluid that surrounds and bathes the cells.

As blood enters the capillaries, water and dissolved substances from the blood move out by filtration through the membranes. Most of the protein molecules and blood cells stay in the capillaries. Oxygen diffuses from blood cells to tissue cells. Amino acids, glucose, and waste substances like urea also move across capillary walls by diffusion. The cells of our bodies live in an internal liquid environment that maintains the conditions necessary for proper cell function.

The amount of blood reaching any organ at any given time is precisely controlled. Tiny muscles in the walls of arteries open into each branching group of capillaries. These muscles regulate blood flow through the capillaries. For example, during strenuous exercise, such as running or swimming, a great deal of blood is in the capillaries and other small vessels in the muscles. Very little blood is in the capillaries of the digestive system. (Does this suggest to you why it is not wise to exercise a lot immediately after a meal?)

Control systems also operate to maintain a fairly steady blood flow to important organs like the brain and the heart itself. These controls operate in spite of great changes in the pattern of blood flow. This is fortunate, indeed, for even a brief interference in the blood flow to these organs may be harmful or may even cause death.

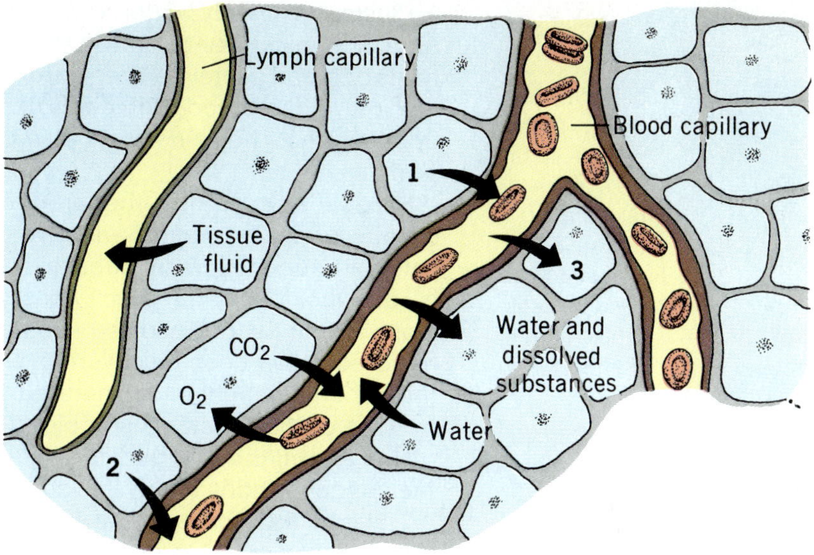

16-12. The exhange of materials at a capillary. *What do you think is happening at points 1, 2, and 3?*

The transport system, under the control of nerves, also helps maintain a constant body temperature in humans and other warm-blooded animals. When the body (and the weather) is warm, the arteries supply more blood to the capillaries just under the skin. (What evidence can you give to support this?) Body heat radiates out of the blood through the skin. When the outside temperature is low, the body is in danger of losing too much heat. In this case, the small arteries under the skin are partially shut off, and skin capillaries receive far less blood. (What evidence can you give that supports this?)

Blood flows from capillaries into small veins. From these, blood flows into larger veins that return it to the heart. The major veins have valves that keep blood from flowing backward, away from the heart. The flow in veins is dependent on many factors, including the body's position with respect to gravity, movement of muscles, and breathing. People sometimes faint when they stand still too long. This occurs because blood is no longer being returned to the heart fast enough to maintain a steady flow to the brain.

Blood

Blood is a complex substance. About 55 percent of it is a fluid, the **plasma.** The other 45 percent consists of various cells and cell products (Figure 16-13).

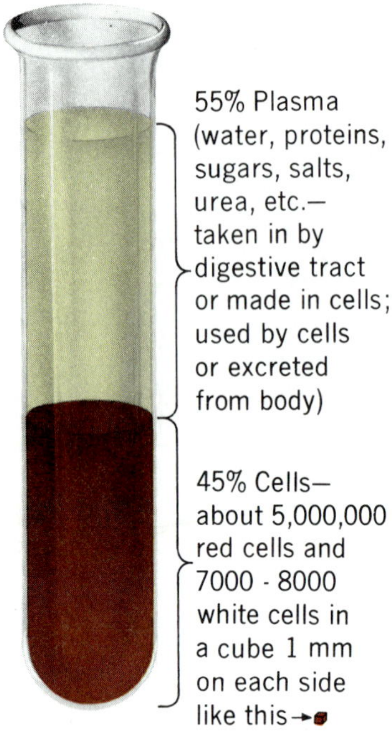

16-13. We can separate a blood sample in a centrifuge. The cells and molecules are present in concentrations that remain nearly constant.

Plasma is mainly water with a large number of inorganic and organic substances dissolved in it. There are, for example, several types of blood proteins. These proteins have special functions, such as clotting the blood. Some blood proteins are hormones. The antibodies that help destroy disease organisms also are blood proteins.

Plasma also carries the products of digestion—amino acids, simple sugars, fatty acids and glycerol, and other materials used by cells. Nitrogenous waste substances, such as urea and some of the carbon dioxide produced by cells, are found in plasma. We could make a long list of dissolved substances in human blood plasma—hundreds have been discovered. Yet we are far from being able to give a complete chemical picture of this remarkable fluid. Concentrations of the substances in blood plasma vary slightly depending on the activities of the body's cells. But elaborate mechanisms keep the concentrations nearly constant.

Three kinds of cells or cell-like structures—**red blood cells, white blood cells,** and **blood platelets**—are present in blood (Figure 16-14). The red blood cells are most numerous. Each cubic millimeter of blood contains about five million of them. Red blood cells are produced in bone marrow—special tissue in the center of some bones. As each red cell matures, it undergoes a most unusual change. Its nucleus is pushed out of the cell. This leaves an incomplete cell that can survive for only a few months. But this cell is packed full of hemoglobin, the protein that carries oxygen.

There are several kinds of white blood cells. Some are produced in bone marrow. Others are produced in special structures of the lymphatic system, the **lymph nodes,** and in the spleen. A cubic millimeter of blood usually contains 7,000 to 8,000 of them. Each

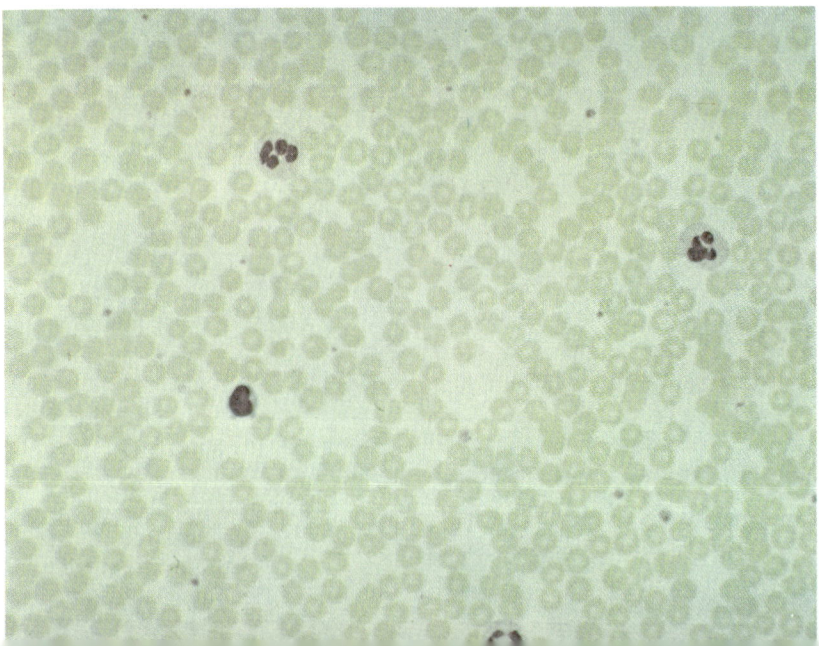

16-14. A stained smear of human blood as seen under the compound microscope. The pale, rounded structures with light centers are red blood cells. The larger cells with dark nuclei are white blood cells. Small, irregular objects are the blood platelets.

SCIENCE AT WORK: Serendipity

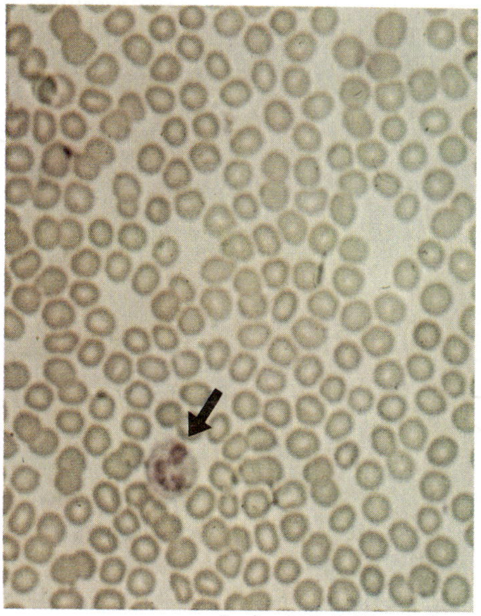

Carolina Biological Supply House

Serendipity, the aptitude for discovering valuable or agreeable things accidentally, frequently plays an important role in science. Dr. Alexander Fleming discovered the antibiotic, penicillin, when one of his experimental bacterial cultures became contaminated with a mold. He put the culture aside. Several days later, Fleming noticed that all the bacteria in the area of the mold had disappeared. Further experimentation led to the discovery of penicillin. Fleming had not been looking for such results in the original research, but was able to see that he had come upon something important quite by accident.

Another example of serendipity and good science occurred recently to several scientists at Georgetown University in Washington, D.C. These scientists were working with two hormones involved in the body's processing of sugars and insulin. These hormones will not bind as well to fat and liver cells of diabetics and potential diabetics as to those of normal individuals. This knowledge could lead to a test for diabetes. But removing fat or liver tissue from a patient for testing was not easy. In further research, the Georgetown University team found another indicator. They learned that the two hormones also failed to bind to the white blood cells (shown stained in the photo above) of a person with a diabetic condition.

Using the white-blood-cell system, the team looked at adult diabetics. They also tested individuals with a history of diabetes in their families. And they tested a normal control group. All the diabetics tested, and nine of ten of those with diabetes in their families, showed deficiencies in hormone binding to white blood cells.

This test may offer a way of predicting whether adults whose genetic histories indicate a possibility of diabetes will actually develop it. These people could then change to diets that might help prevent this serious disease.

white blood cell has a nucleus. The chief function of white blood cells is to protect the body against bacteria. Some of these cells move their cell membrane around the bacteria, take them into the cytoplasm, and destroy them. The pus that forms in an infected wound is largely a mass of white blood cells that have died destroying bacteria. Other white blood cells produce antibodies. White blood cells, then, combat infection by destroying bacteria and by forming antibodies.

Blood platelets are much smaller than either red or white blood cells. They play a key role in the clotting of blood. This is the main process that stops blood from flowing out of damaged blood vessels.

Blood Clotting

Much investigation has gone into questions of why and how blood clots. People have known for centuries that clotting is important in stopping blood flow from a wound. But they have not always understood the mechanism of clotting. Some have suggested, for example, that the exposure of blood to air initiates clotting. But such is not the case. Under certain conditions, blood exposed to air will remain unclotted for hours.

What actually happens is another example of the precise way in which the body maintains itself. The mechanism that controls clotting is not one process; it is a series of processes. When you cut your finger, the damaged tissues release a group of substances called thromboplastins (THRAHM·boh·plas·tins). Disintegrating blood platelets also release thromboplastins. The thromboplastins start a chain of events that involve other substances in the blood. These substances are fibrinogen (fy·BRIN·uh·gin), calcium ions, and prothrombin (pro·THRAHM·bin). In the presence of calcium ions and thromboplastins, prothrombin is converted to thrombin. Blood will not clot unless thrombin is present. Then, in the presence of thrombin, fibrinogen changes to fibrin. As the name suggests, fibrin is the material that forms a fibrous network that eventually becomes the clot.

Why platelets disintegrate at the site of a wound is unknown. How damaged tissues produce and release thromboplastins also is unknown. What is clear is the effectiveness of these processes. So long as there is no wound, and therefore no thromboplastins, fibrinogen remains in the blood in its original state. It will not, under ordinary circumstances, be changed to fibrin. But the release of thromboplastins from disintegrating blood platelets and from damaged tissues sets the mechanism in operation.

16-15. THE HUMAN LYMPHATIC SYSTEM

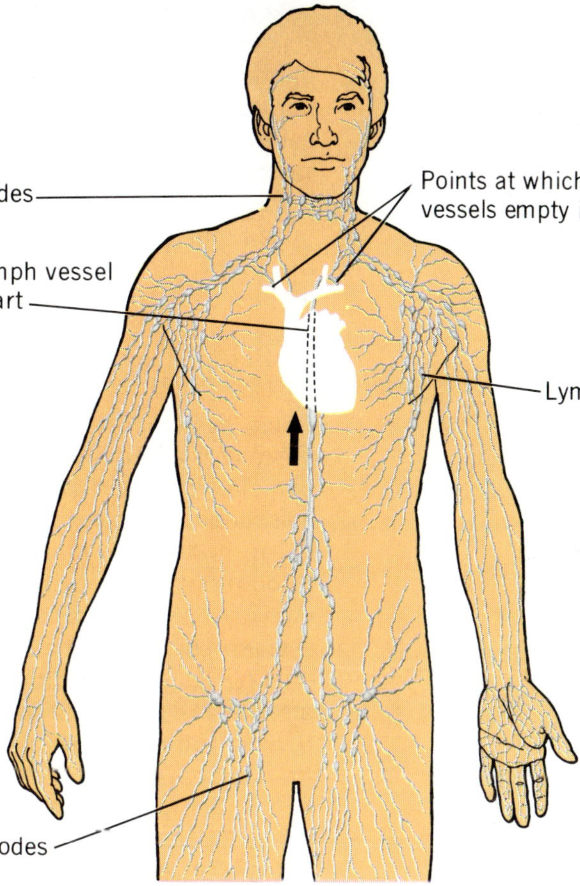

Lymphatic System

Some blood plasma constantly passes through capillary walls and enters the spaces between the cells as tissue fluid. Figure 16-12 shows this process. Most of the blood cells and some of the proteins of the plasma remain in the bloodstream. Some white blood cells, however, may push out between the cells of the capillary walls.

The exchange of molecules takes place between the cells and the tissue fluid surrounding them. Then some of the fluid returns to the blood capillaries; some enters the lymph vessels. Fluid traveling in lymph vessels is called **lymph**. Lymph flow depends on muscle contractions, breathing movements, and the presence of valves in lymph vessels. In the digestive tract, many small branches of lymph vessels extend into the villi, the fingerlike projections in the wall of the intestine. Most digested fat molecules are absorbed from the intestine into these lymphatic branches. Lymph vessels eventually connect with large blood veins near the heart (Figure 16-15).

Located along the lymph vessels are lymph nodes, which filter harmful substances and bacteria from the lymph. There are vast numbers of white blood cells in lymph nodes. Many white blood cells are produced here by cell division. Lymph nodes are found at various places in the body, such as around the major organs, in the limbs, under the arms, and in the neck. During infections, the lymph nodes may swell as they remove bacteria and other particles from the lymph.

In Brief

Some system of transporting materials throughout the body is present in all highly developed animals. Even the earthworm has a complex circulatory system with distributing and collecting vessels, capillaries, and specialized vessels that pump the blood.

The human circulatory system transports chemical substances throughout the body. Everything that individual cells receive and eliminate is moved by blood and lymph, the fluids of the circulatory system. The blood is pumped by the ventricles of the heart through the arteries—the distributing vessels of the blood system. From the arteries the blood enters the capillaries. Some of the plasma containing nutrients and oxygen leaves the capillaries and passes into the spaces around the cells as tissue fluid. Here, materials are exchanged with the cells.

The tissue fluid surrounding the cells now contains carbon dioxide and other waste materials. Most of the tissue fluid reenters the blood capillaries and then enters the veins. Some of the tissue fluid enters lymph vessels and eventually returns to the bloodstream through large veins near the heart. Finally, blood and lymph enter the heart from the veins.

Guide Questions and Problems

1. The more cells an animal has, the more dependent it is on a specialized transport system. Why is this?

2. If you were told that a certain animal has no special transport system, what deductions could you make about the animal without obtaining additional data?

3. What are the differences in structure among veins, arteries, and capillaries? How are these differences related to the functions of each kind of vessel?

4. Valves in veins of the legs sometimes fail to stop backflow of blood. What could be the effects of this failure?
5. What is the relationship between blood plasma and lymph?
6. Arteriosclerosis (ar·TER·ee·oh·skla·ROH·sis) is a condition also called "hardening of the arteries." One of its effects is reduced elasticity of the arteries. What circulatory problems might arise from this effect on the vessels?
7. In the development of a human infant, the wall between the two atria of the heart sometimes fails to close completely. Compare this condition with a heart having only one atrium and one ventricle. What would happen if blood from the lungs and from the rest of the body mixed in the heart? Why do you suppose the afflicted babies are called "blue babies"?

Ideas for Independent Study

1. Because the heart is often damaged by disease, much work has been done on designing artificial hearts. What problems must be overcome in order to design an effective artificial heart? Compare the difficulties of making an artificial heart with making artificial capillaries.

2. Astronauts and cosmonauts who travel in space have many problems to overcome, including the effects of increased gravity (during acceleration) and decreased gravity (apparent weightlessness). Changes in gravity have very major effects on circulation, especially on blood flow through the veins. If you are interested in the general problems of humans in space (how their bodies must adjust to conditions of space travel), this would be a good place to start an investigation on your own.

3. Capillaries are often damaged by bruises, cuts, and burns. (Even a lot of exercise can result in tearing of muscle capillaries.) What happens to the injured capillaries and to the cells depending on those capillaries for transport? During the course of the school year, make a list of minor accidents that could affect capillaries in your body. The repair of damaged capillary areas is one of many processes that regulate the internal environment of cells.

4. Long before people understood how blood circulates, they recognized the connections between parts of the body involved in circulation and important aspects of life. People have even developed religious or magical meanings for circulatory factors such as blood flow, heart, and pulse rate. If you are interested in novels and poetry, classics, psychology or anthropology, you may want to see how such ideas are expressed in literature.

Related Readings

BOOKS

Denton, J. G. *New Parts for Old: The Age of Organ Transplants*. Palisade, N.J.: Franklin Publishing, 1974.

Gordon, A. S. *Blood Cell Physiology*. A BSCS Pamphlet. Boulder, Co.: BSCS, 1963.

ARTICLES

Adolph, E. F. "The Heart's Pacemaker." *Scientific American*, March 1967, page 32.

Dudrick, S. J. and J. E. Rhoads. "Total Intravenous Feeding." *Scientific American*, May 1972, page 73.

17 Animal Structure and Function: Excretion

Animal cells produce waste products in the process of metabolism. The chief excretory products are carbon dioxide, water, and nitrogen-containing compounds. The nitrogen compounds result from the metabolism of amino acids. Carbon dioxide is removed through the respiratory surfaces. Other waste molecules are removed by specialized excretory organs.

When the concentration of any substance is higher than that needed by the cell to carry on its life processes, the excess of that substance becomes a waste product. For example, animal cells use small amounts of carbon dioxide. But much more carbon dioxide is formed during cell respiration than can be used by the cells. The excess is a waste product that is removed by the respiratory system. Another example is water. If more water enters a cell than the cell can use, the excess is removed as a waste product. Even necessary substances like vitamin C may accumulate in excess amounts. These also are excreted until normal concentrations are reached.

Excretion, then, is not merely getting rid of waste molecules. It is a process that constantly regulates and controls the concentration of chemicals within the fluids of the body. An optimum internal environment is maintained for normal functioning of the cells. The excretory organs, therefore, are important in maintaining homeostasis.

METHODS OF EXCRETION

Many single-celled organisms, such as *Amoeba* and *Paramecium*, live in water. These organisms also produce water as a waste product. The cell membrane of *Amoeba* is permeable to water, but the concentration of water in its cytoplasm is lower than the concentration in the surrounding pond. Because of this difference, water constantly enters the body of *Amoeba* more rapidly than it moves out. We considered this process of water entry by osmosis in **Inquiry 3-3**. If no other processes were involved, the body of the *Amoeba* would swell and burst. In *Amoeba*, however, water is removed by a contractile vacuole (Figure 17-1), which regulates and maintains optimum water balance for normal cell function.

The chief nitrogenous waste product in most single-celled organisms is ammonia (NH_3), which is very poisonous to cells when

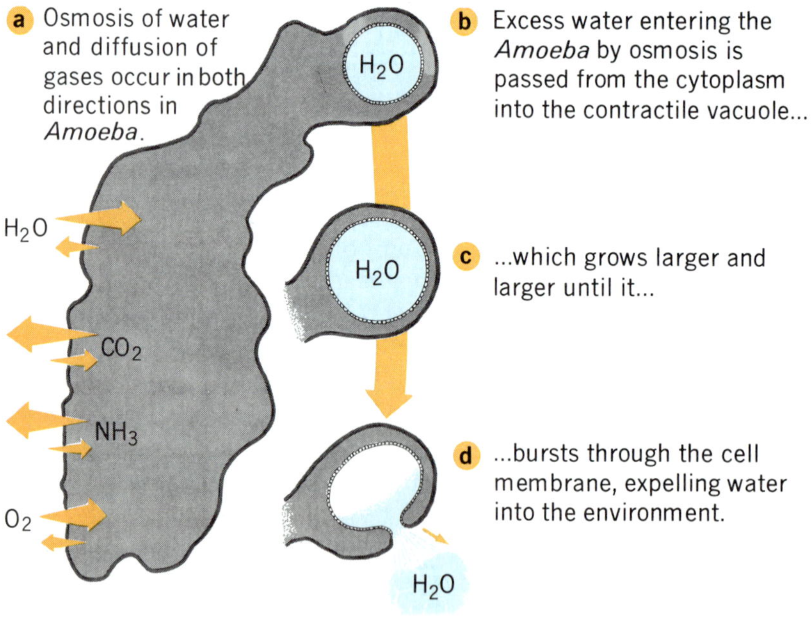

17-1. Like all other animals, *Amoeba proteus* exchanges substances between the cell and the outside environment. Oxygen, carbon dioxide, and ammonia move by diffusion. Water moves into the cell by osmosis and is removed by the contractile vacuole. The net exchanges are indicated by the heavier arrows.

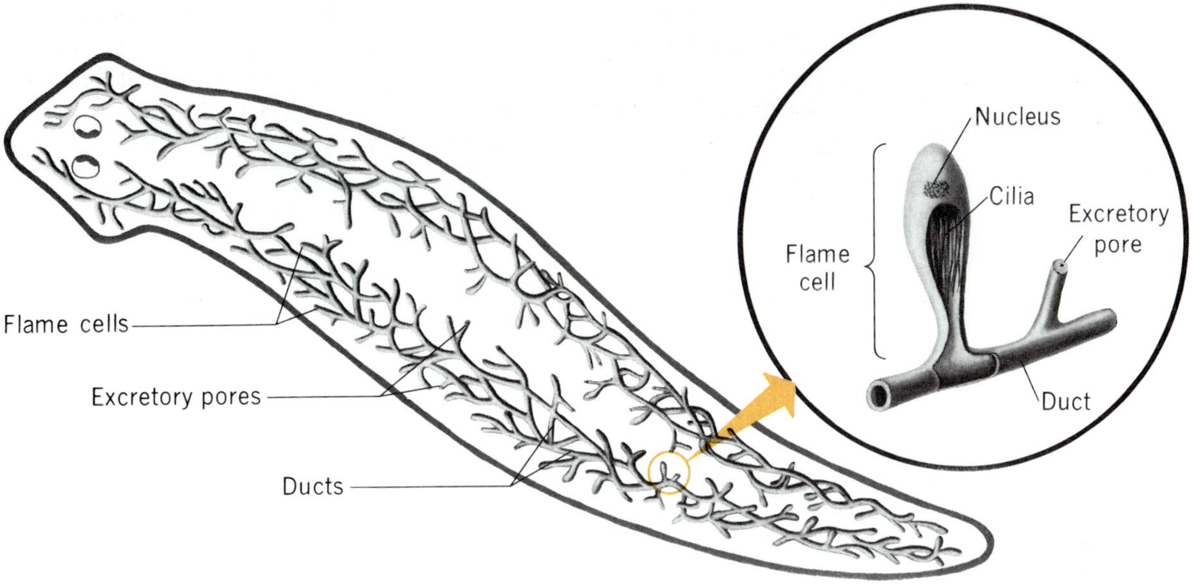

17-2. The excretory system of a planarian. A flame cell and an excretory pore are shown enlarged. Wastes and excess water removed by flame cells are eliminated through pores in the body wall.

present in large concentrations. But ammonia is a small molecule. It readily diffuses through the cell membrane before damage can be done (Figure 17-1).

In planarians, the excretion of carbon dioxide and much of the ammonia is simply a matter of diffusion. These molecules dissolve in water and form other molecules. They are carried away in the water of the environment.

A planarian does have an excretory system. This system consists of a series of branching tubes, or ducts. The ducts end in *flame cells* (Figure 17-2). A tuft of cilia projects into the space within each flame cell. (In a living animal, the movement of the cilia reminded early observers of a flickering candle flame—hence the name "flame cell.") Flame cells remove liquid wastes from the body and pass them into the ducts. The ducts carry the wastes to the excretory pores on the surface of the body.

In planarians, as in *Amoeba* and other small organisms that live in fresh water, water constantly diffuses into the cells and must be removed. This is accomplished in planarians largely by the flame cells of the excretory system.

An earthworm has excretory tubes (Figure 17-3), two for almost every segment of its body. The inner end of each tube is in the body cavity. The body cavity contains a fluid. Within this fluid many waste substances collect—among them ammonia and **urea**, the chief ni-

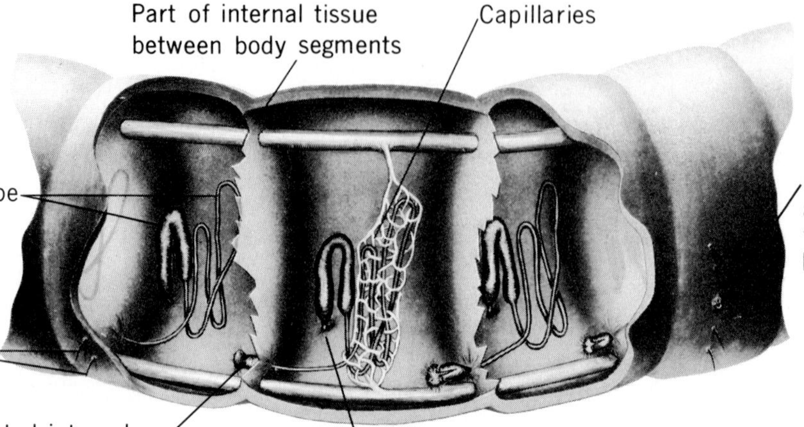

17-3. Excretory tubes in the earthworm. The near side of the body, including the alimentary canal, is cut away. Most of the segments of the body have a pair of excretory tubes, one on either side. Capillaries from the circulatory system surround the excretory tubes.

trogen-containing wastes formed by the earthworm. These and other waste materials are removed by the excretory tubes, which have openings encircled with cilia. The cilia propel fluid into each excretory tube. Each excretory tube originates in one segment and ends as an opening in the body wall of the segment to the rear. Contractions of the muscular wall of the excretory tube then force the fluid out of the body.

Capillaries of the circulatory system are closely associated with the excretory tubes. Some waste materials from the capillaries pass into the excretory tubes and other materials are reabsorbed from the tubes back into the capillaries. As a result of this exchange, the fluid excreted to the outside is quite different in composition from the fluid that originally entered the tube.

Nitrogenous wastes are excreted by various animals in different molecular forms—ammonia, urea, and uric acid. Ammonia, which is poisonous to cells, is produced only in organisms that also excrete large quantities of water. Many aquatic animals excrete ammonia (Figure 17-4).

Urea is formed from ammonia, but it is less harmful to cells. Urea can remain in the body in fairly high concentrations without damaging the cells. Water is necessary for the excretion of urea. Many animals that live on land and have adequate supplies of water excrete urea (Figure 17-4). In humans, at least a liter of water per day is required to carry away urea produced by the cells.

		WASTE PRODUCTS AS A PERCENTAGE OF TOTAL NITROGENOUS WASTES		
ENVIRONMENT	ORGANISM	**Ammonia**	**Urea**	**Uric Acid**
Aquatic	Protozoa *(Amoeba)*	98+		
	Planaria	100		
	Coelenterates *(Hydra)*	52	4	
	Squid	67	2	2
	Aquatic insects	90		
	Fish (freshwater)	74	9	
Aquatic Changing to Terrestrial (amphibians)	Frog (tadpole)	78	20	
	Frog (adult)	8	80	
Terrestrial	Mammal (humans)	4	84	2
	Terrestrial insects	0	trace	92
	Python (snake)	9		89
	Bird	3	10	87

17-4. Forms of nitrogenous waste excreted by organisms in relation to their environment.

The formation of **uric acid**, rather than ammonia or urea, is an evolutionary adaptation in animals that live on land where very little water is available. Uric acid is almost completely harmless to cells and can be excreted with small amounts of water. In fact, insects produce solid uric acid crystals in their excretory tubes.

The grasshopper disposes of uric acid and other waste products of cell metabolism through a specialized excretory system. This is made up of very long, fine tubes often called **Malpighian** (mal·PEE·gee·uhn) **tubules** (Figure 17-5). (They are named after Marcello Malpighi, the scientist who first described them.) One end of each tube opens into the digestive canal. The other end floats freely in the large blood cavities. The Malpighian tubules extract waste products from the blood and pass them into the digestive canal. In the digestive canal, these waste products are mixed with the feces. Almost all water is reabsorbed and returned to the blood. Eventually, the wastes are eliminated through the anus as very dry material containing crystals of uric acid.

17-5. The excretory system of a grasshopper.

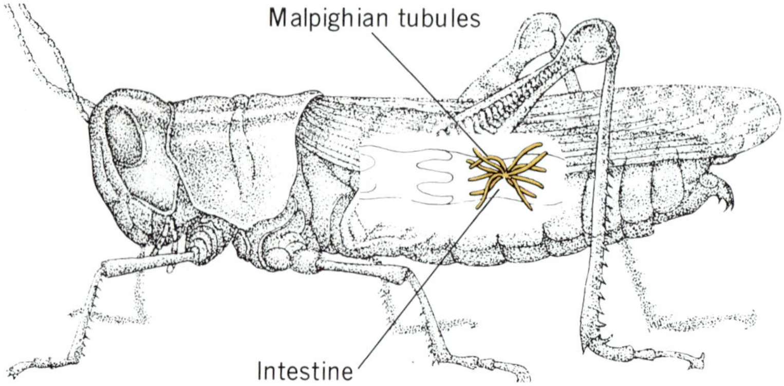

Excretion in Humans

In humans, carbon dioxide is excreted by the lungs. Some water, salts, and urea are excreted through the skin as perspiration. A few substances also are taken up from the blood by the liver. These are passed, as part of bile, from the liver into the intestine and are excreted to the external environment. But by far the most important

17-6. THE HUMAN EXCRETORY SYSTEM

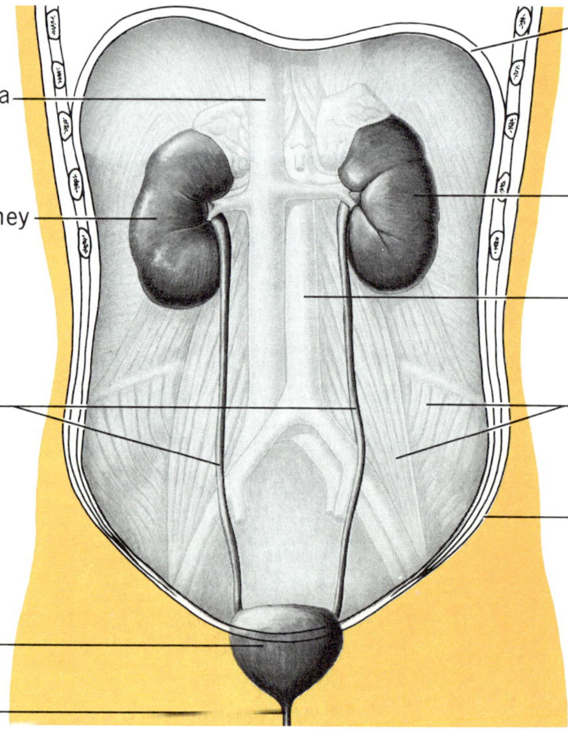

(The structures on the upper ends of the kidneys are the adrenal glands. They are not part of the excretory system.)

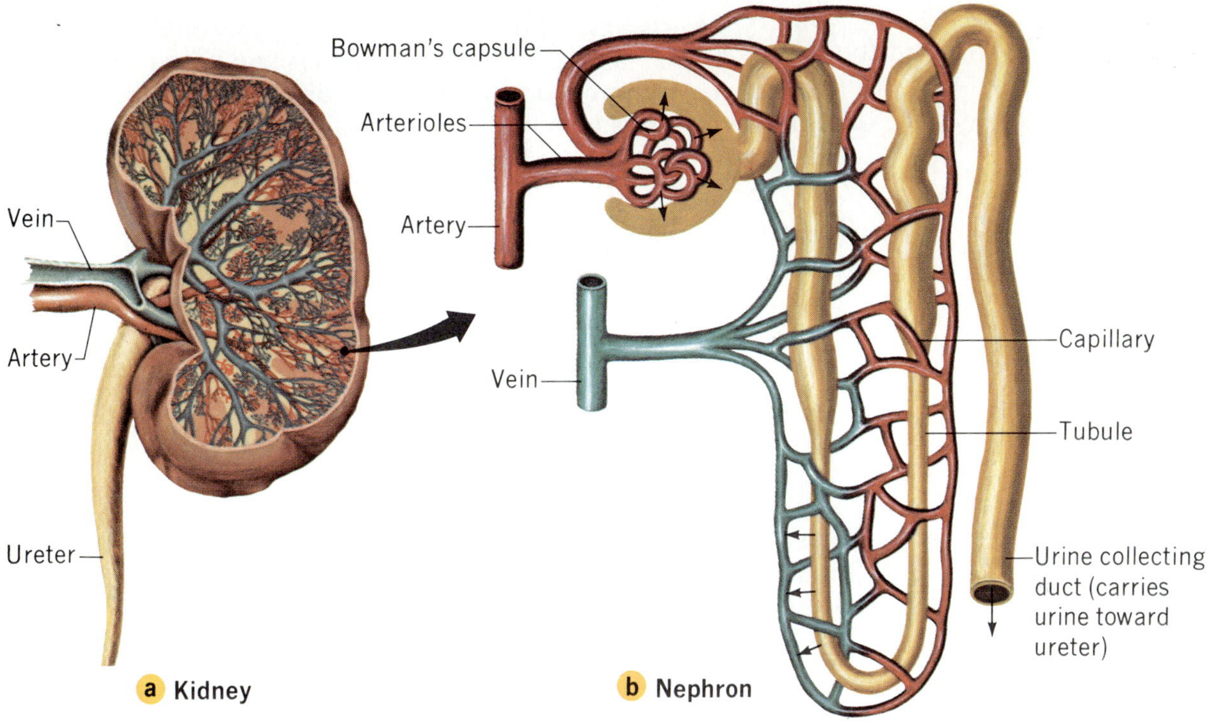

17-7. A kidney in longitudinal section **(a)** with an enlargement of a single nephron **(b).** *What substances are transferred from the blood to the Bowman's capsule?*

excretory structures are the **kidneys** and certain structures associated with them.

A human has two kidneys. One is located on each side of the spine, just above the hipbone (Figure 17-6). A duct, the **ureter** (YOUR·it·er), goes from each kidney to the **bladder**. The bladder opens to the body surface through the **urethra** (you·REE·thrah). **Urine**, the final product of kidney function, flows through the urethra to the outside.

Each kidney is made up of about a million tiny tubes, the **nephrons** (NEFF·ronz; Figure 17-7 a and b). The nephrons remove waste materials from the blood. Each nephron is long and coiled. One end opens into a duct that collects urine. The other end is closed, but it is expanded into a cup-shaped structure called **Bowman's capsule**. This structure fits tightly over a mass of blood capillaries. There are many other capillaries, too, surrounding each tubule along its twisting path to the ureter.

It has taken years of work to understand even the general outline of human kidney activity. Many tests have been carried out. By comparing the contents of various substances in blood plasma and urine, biologists have learned that the two are very different. Blood plasma always has a very constant composition. Urine varies depending on body conditions. Some examples of the results of measurements made in healthy people at rest are shown in Figure 17-8. Review the data in this figure.

17-8. Average concentrations of substances in plasma and urine

Substance	*% in plasma*	*% in urine*
Water	92.0	95.0
Proteins	7.0	0
Glucose	0.1	0
Sodium ions	0.32	0.35
Nitrogen-containing waste products:		
Ammonia	0.0001	0.04
Urea	0.03	2.0
Uric acid	0.002	0.05

1. What is the ratio of concentration in plasma to concentration in urine for each of these substances? (Divide the value for plasma by the value for urine.)
2. Of the substances shown, which are more concentrated in urine than in blood?
3. Try to think of a simple explanation of how urine could be formed from plasma.

Did you notice how different the composition of urine is from that of the plasma? For instance, there are no glucose or protein molecules in urine. Urea molecules are 60 times more concentrated in urine than in plasma. Apparently, the kidney removes some molecules from the plasma and retains others. As you can see, the kidneys regulate the composition of body fluids.

But how do the kidneys function? How does the selective excretion of some of the molecules occur? These questions could be answered only after a series of complex experiments was performed by A. N. Richards. In the 1920s, Richards began to study frogs and salamanders. These animals have nephrons that are relatively large (1/10 mm) and easy to work with. Richards compared the concentrations of substances, such as protein and glucose, in blood plasma with their concentrations in fluid collected from the end of a nephron (Figure 17-9a). Figure 17-9b shows some data from these studies. How do the data support the hypothesis that fluid in the

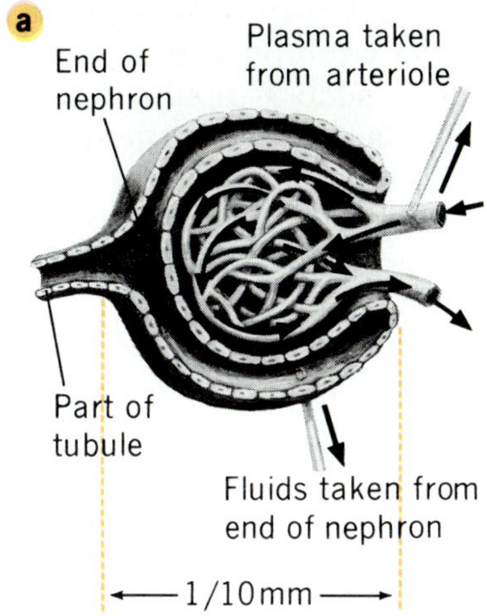

b

Substance	Percent of value in blood plasma
Blood cells	0
Proteins	0
Glucose	100
Urea	100

17-9. Richards' experiment **(a).** Concentrations of substances in fluids collected from the end of nephrons **(b).**

nephron is first formed by filtration of plasma fluid through the capillary membranes? Now look at the concentrations of glucose and urea. How do these concentrations (relative to plasma) compare with their concentrations in urine? (Refer again to Figure 17-8.)

From these and other experiments, biologists gradually gained a more complete understanding of how the kidneys function. As blood passes through capillaries at the end of the nephron (Figure 17-7b), some plasma fluid is forced into the nephron tube by the blood pressure. Red and white blood cells and most blood proteins do not cross capillary membranes into the nephron. But through the process of filtration most of the smaller molecules do (Figure 17-10a). As the fluid moves along the nephron, cells remove glucose and other substances that are useful to the organism. These substances are taken up by the cells of the nephron and reabsorbed into the bloodstream (Figure 17-10b). The nephron cells also take up substances from the blood in the capillaries around the tubules. These substances then are passed into the nephron (Figure 17-10c). Thus, two-way traffic of molecules occurs across the nephron wall. Molecules that cells can use move from the tubules into the blood capillaries, and waste molecules move from the blood capillaries into the tubules. The final product in the tubules is urine.

Glucose and many other substances are reabsorbed from the nephron and restored to the blood even though their concentrations are higher in the blood than in the tubule. Something other than simple diffusion must take place. The process involves active transport, which uses energy. Amino acids, fatty acids, vitamins, hormones, and many salts are removed from the tubule and returned to the blood by active transport. Most of the water is reabsorbed in the lower part of the nephron and is returned to the blood. Certain waste substances, such as urea and uric acid, are greatly concentrated in the tubule. These remain in the urine. Thus, the nephron cells are constantly recycling important molecules. They also discard useless, harmful, or excess molecules. It is no wonder that one biologist, H. W. Smith, has called the kidneys "the master chemists" of the body fluids.

Kidney activity varies, depending on the needs of the body at a particular moment. The activity of the kidneys is closely controlled by hormones and other factors. If you drink a large amount of water, it passes across the lining of your intestine and enters your blood. The excess water is removed rapidly by the nephrons. Similarly, if you eat a large amount of salt, this also enters the blood. The excess is removed by the nephrons.

The nephrons keep the composition of the blood constant. In addition, the lungs remove carbon dioxide, and sweat glands in the skin remove water and dissolved salts. A few other organs, including

17-10. MAJOR STEPS IN FORMATION OF URINE

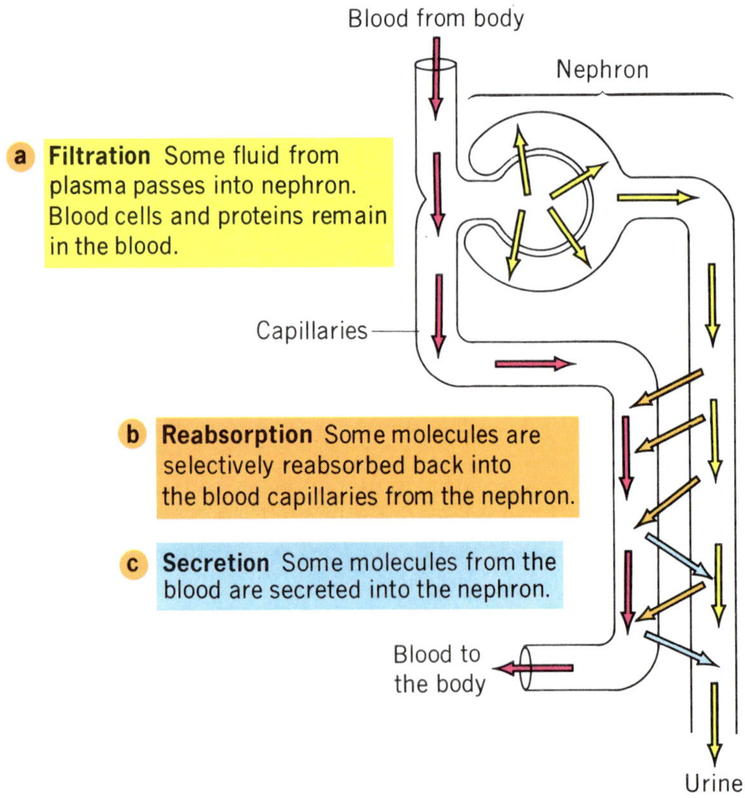

SCIENCE AT WORK: Science and Technology

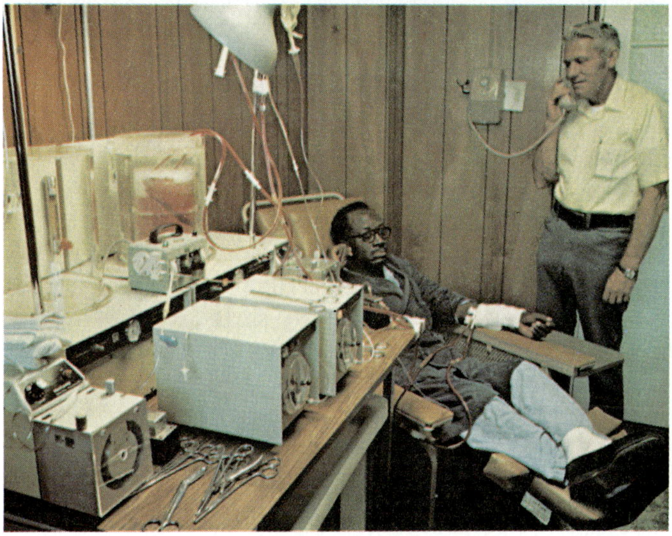

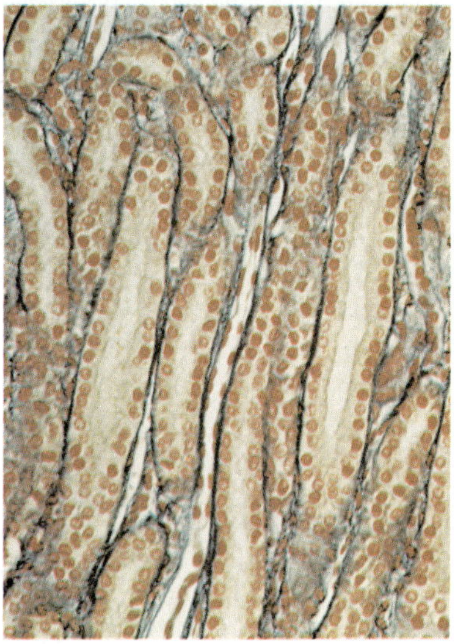

One definition of technology is "applied science." Many technologists are concerned with putting the discoveries of science to use in ways that are beneficial to people. Science and technology are not independent. They work together toward a common goal – improving the quality of life for all people.

After many years of study, biologists have developed a rather complete understanding of how the human kidney functions. As you know, the kidney acts as a filter. It removes impurities and excess materials from the blood. In doing so, it regulates the chemical composition of the body fluids.

However, kidneys do not always function properly. If the condition cannot be remedied, the chemistry of the body becomes so unbalanced that prolonged illness and eventual death result. In some cases, kidney transplants have been successful. But few people will donate a kidney. Another problem is that the tissue of a donated kidney must be very similar to the patient's kidney tissue. Those who need a new kidney must sometimes wait for months or even years.

Technologists have developed a machine that makes this long wait possible. Dr. Willem Johan Kolff developed an artificial kidney. He made the first simple model in his native Netherlands during the 1940s. Since that time, technologists have redesigned and improved the "kidney machine." Today it is saving countless lives.

Although many improvements in the design of the artificial kidney have been made over the years, the basic design is still the same. Blood from the patient is passed through a long membranous tube. The tube is immersed in a solution containing a low concentration of salts. Excess salts and other wastes, the molecules normally removed by the kidneys, leave the blood through tiny pores in the membrane. Larger materials, blood proteins and blood cells, remain in the blood and are returned to the body. Regular treatments help keep the body chemistry in balance. Rarely, however, is this treatment a cure. The machine helps to keep the patient alive, but the ultimate hope is still a kidney transplant.

the liver, play minor parts in excretion. All of these organs are important in the regulation of the internal environment, so that cells can function normally.

PROBLEMS OF HEAT PRODUCTION

In addition to chemical waste materials, heat is constantly formed in living processes. This heat also is a "waste product" of metabolism. Animals must get rid of it in one way or another.

In small animals like *Paramecium*, heat is lost easily through the body surface to the surrounding water. Bigger animals have a large cell mass as compared with their body surface. They must have other ways of getting rid of excess heat. Some of the ways in which heat is removed from these animals are already familiar to you. Blood circulation plays a major part. It carries heat from inside the body to the skin, where the blood vessels expand and diffuse the heat to the surface. There, the heat is lost to the environment. Heat also is lost when perspiration evaporates from the skin. If the respiratory rate increases, the exhaled air also removes heat.

In birds and many mammals, the lungs and mouth also play a part in heat removal.The dog shown at the beginning of this chapter is panting on a hot day. The panting helps cool its body. In addition, the body covering of feathers or fur changes with changing demands for heat loss. That is why dogs, cats, and some other animals tend to shed their coats in spring and summer.

You can study changes in processes of heat loss in an inquiry of your own design. For instance, you might measure skin temperature and patterns of perspiring when you are quiet and rested, and then measure them after vigorous exercise. How does skin temperature compare with mouth temperature? Your observations may lead to further questions, such as these: How long do changes in skin temperature and sweating continue after exercise? What is the cellular origin of the excess heat produced in exercise? What would happen if this heat were simply stored in the body?

Heat also can be conserved or produced in greater amounts. Such is the case in birds and mammals that have a constant body temperature. Body temperature does not change with environmental temperatures in these animals as it does in fishes and reptiles.

The part of the brain known as the **hypothalamus** (hi·poh·THAL·uh·mus) is responsible for maintaining the body temperature within a fairly constant range. Shivering is one means of increasing body temperature in cold environments. The muscle movements in shivering generate heat. Also, blood vessels near the skin may contract, allowing less blood to flow to the surface.

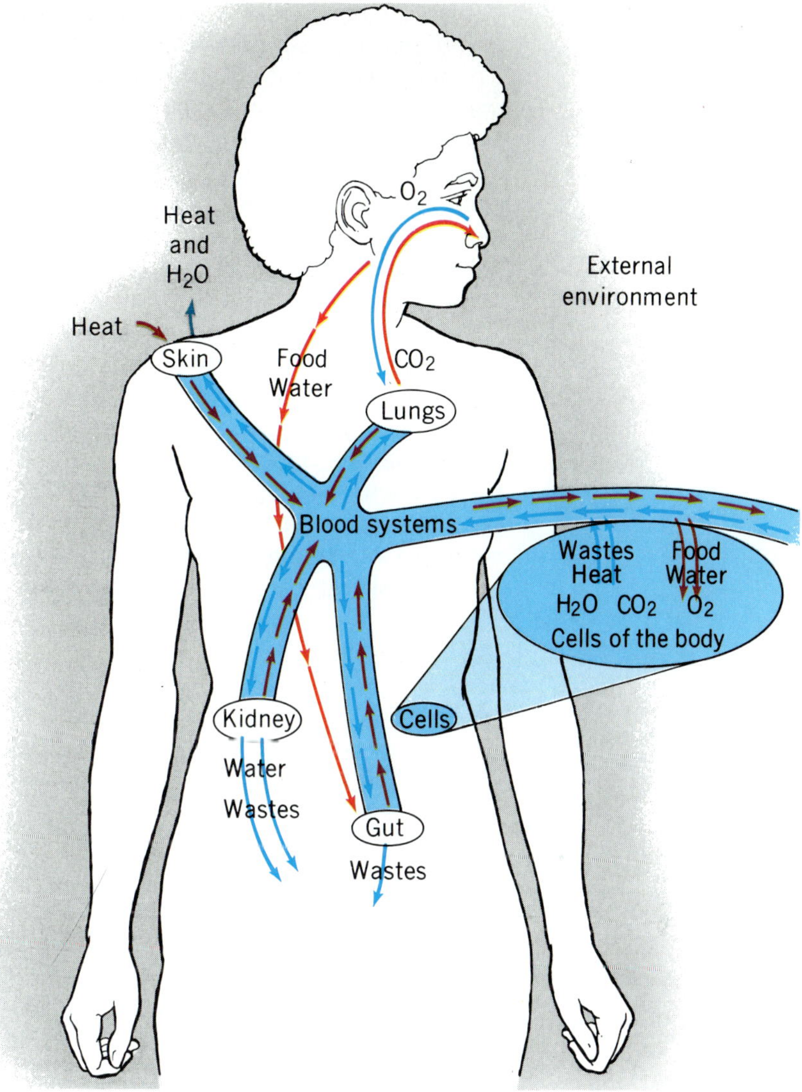

17-11. Exchanges that take place between internal and external environments.

A Review of Homeostasis

Homeostatic regulation of the internal environment is complex and far-reaching. It involves not only chemical compounds but also energy balance.

In complex animals, groups of cells are specialized to carry out specific functions. Tissues, organs, and organ systems perform certain activities that maintain homeostasis in the organism (Figure 17-11). The digestive system processes food into small molecules that are absorbed easily across membranes. The respiratory system transfers molecules of oxygen and carbon dioxide between the external environment and the capillaries in the lungs.

The transport system rapidly distributes nutrients and oxygen to the fluids that surround the cells. This system also carries wastes away from the cells. The excretory system constantly regulates the concentration of molecules within the transport system. It retains useable molecules at precise concentrations and removes waste molecules. It also regulates the water level of the internal fluids.

Temperature regulation also is critical. If the blood temperature in a human rises by as little as 6°C, the change may be fatal. Groups of specialized cells add chemical substances to and remove them from the blood. These are found in the digestive, respiratory, and excretory systems. Heat exchange is regulated as well. All of the control processes are specialized and require energy.

There is one important process missing in Figure 17-11. There is no means for coordinating all the activities. Yet we have mentioned control as a necessary part of each system. All multicellular animals have two interrelated means of coordination – nerves and hormones – which we will discuss in the next chapter.

In Brief

The life of a multicellular individual depends on the life of its cells. The organism as a whole obtains food and oxygen. But these are used by individual cells. The organism also removes the wastes from the individual cells. Single-celled organisms, such as *Paramecium*, are able to operate normally even when the external environment changes. Excess water is removed by contractile vacuoles and nitrogenous wastes diffuse out of the cell.

In humans, the kidneys are the most important organs that remove wastes (except for the functions of the lungs in removing carbon dioxide). They are the "master chemists" that control the composition of the blood. Exchanges occur between blood capillaries and the nephrons of the kidney. Through filtration, reabsorption, and secretion, waste substances are removed from the blood while vital substances are retained.

Guide Questions and Problems

1. The closed end of a nephron allows most of the materials of the blood, except proteins and blood cells, to enter the cavity of the nephron. How does the body recover the necessary sugars,

amino acids, vitamins, minerals, and so forth, that are filtered through the nephron?

2. In this and preceding chapters you learned how necessary homeostasis is in maintaining life. With this idea in mind, can you suggest why blood and urine analyses are valuable tools to a physician?

3. During physical exercise, your body's heat production increases. What homeostatic mechanisms are involved in getting rid of the excess heat? How do these mechanisms function?

4. When you have lost a lot of body heat, homeostatic mechanisms cause heat production to increase. They also cause heat loss to decrease. What are some of these mechanisms and how do they function?

5. It is quite fair to speak of all the processes that rid the body of excess heat and chemical substances as "pollution controls" for the internal environment. How do these processes affect the external environment?

Ideas for Independent Study

1. Find out as much as you can about how urine composition can be used to answer questions about health and disease of the human body. Investigate how urine composition can give clues to the functions of various organs.

For example, the presence of plasma proteins in urine often means that there is disease present in the kidney itself. Can you guess, even without further investigation, why that might be so?

Physicians often check for the presence of abnormal amounts of glucose, amino acids, and related compounds. These may indicate that a patient has diabetes mellitus or phenylketonuria. Or they could mean that there are disturbances in the chemistry of body cells in general and in their regulation. Drugs and poisons often are excreted in the urine. These can be detected there by sensitive chemical tests.

2. People in different parts of the world have developed clothing, diets, housing, and ways of life that help them meet problems of heat production and loss. Investigate some of these ideas in more detail. For example, compare the life patterns of people living in the tropics with those living in arctic regions.

Related Readings

BOOKS

Hamburger, Jean, et al. *Structure and Function of the Kidney.* Philadelphia: Saunders, 1971.

Moffat, D. B. *The Control of Water Balance by the Kidney.* Oxford Biology Reader. London: Oxford University Press, 1971.

ARTICLES

Dempsey, D. "Transplants Are Common; Now It's the Organs That Have Become Rare." *The New York Times Magazine,* October 13, 1974.

Subak-Sharpe, G. "What Your Blood and Urine Tell About You." *Reader's Digest,* October 1975.

18 Animal Structure and Function: Coordination

Any organism has a system that coordinates its various functions. This is true even of single-celled forms such as *Paramecium*. In laboratory experiments, *Paramecium* responds to changes in temperature or in concentrations of various chemicals, and to electric currents. The response usually involves movement. Movement, in turn, depends on coordinated motion of the cilia.

Coordination is a response to external as well as internal conditions. Organisms react to various conditions in the world around them. The activities of an organism often change in response to some change in the environment.

Complex animals are made up of many different kinds of specialized cells working together. They could not survive if their parts did not work in this coordinated way. Most multicellular animals have two related systems of coordination: the **nervous system** and the

endocrine (EN·doe·krin) **system.** These two systems, working together, control body processes. Many multicellular animals also have a **skeletal system** and a **muscular system** that allow them to respond to changes in the environment. Any such change that invokes a response from the organism is called a **stimulus.**

NERVOUS SYSTEMS

Many organisms have a nervous system – cells, or systems of cells, that enable them to detect and respond to internal and external stimuli. The nervous system of *Hydra* looks like a net of threads that extends throughout the animal (Figure 18-1). *Hydra* has no brain or any other center of nervous coordination. In *Hydra*, nerve impulses move in all directions from a point of stimulation.

Hydra has special cells in the outer cell layer that are sensitive to stimuli like chemicals or pressure. These cells stimulate nerve cells, which then carry nerve impulses to other parts of the body. But the nerve net in *Hydra* functions primarily to produce local contractions. *Hydra* has no specialized muscle cells, but most of its cells can contract. *Hydra's* movements are very limited, but it can bend and shorten its body and move its tentacles.

Planarians have a very simple form of a **central nervous system;** that is, many nerve cells are concentrated in some part of the body (Figure 18-2). The concentration of nerve cells in the anterior end of

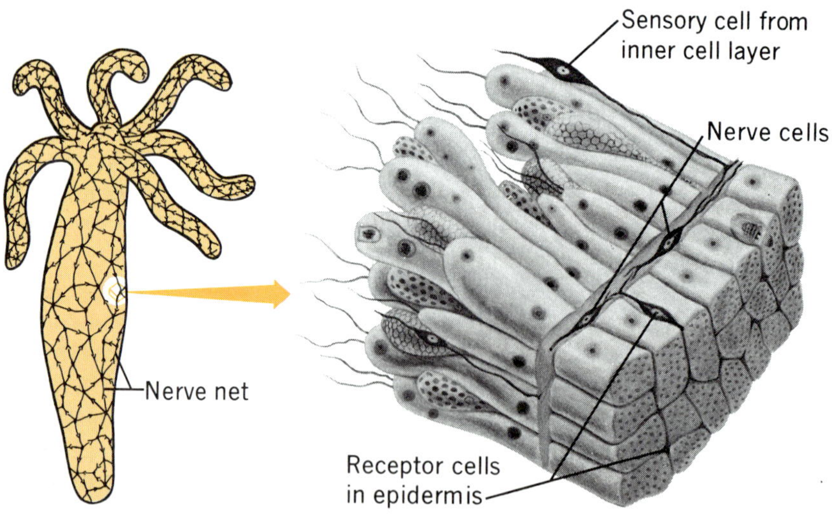

18-1. The nervous system of *Hydra.* The relationship of nerve cells to other specialized cells is shown in the enlarged drawing.

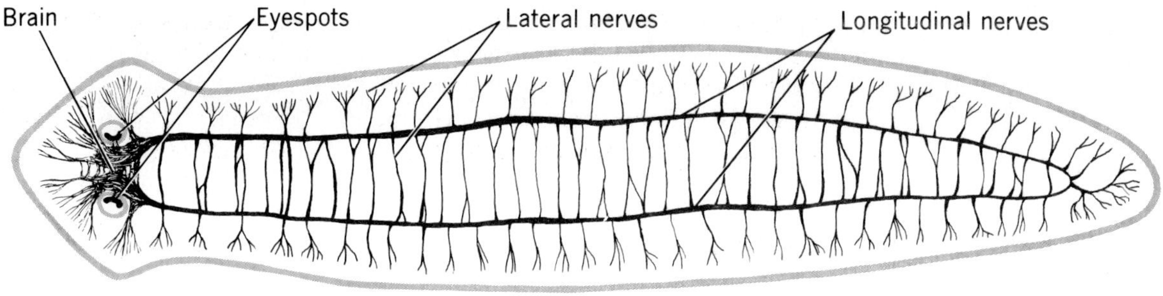

18-2. The nervous system of a planarian.

an organism is called a brain. Bundles of nerve cells make up two longitudinal nerves that reach from one end of a planarian to the other. These two nerves are connected by lateral nerves, which extend across the planarian's body.

In a planarian's head, a group of special cells is connected directly to the brain. These cells are sensitive to light, pressure, touch, and chemical stimuli. The central nervous system contains nerve cells that transfer impulses from these sensitive cells to nerve cells that control muscles. Removing the brain of a planarian reduces its ability to move. The action of the cilia is not disturbed, but the muscular movement necessary for crawling stops completely.

Nerve Function

In humans and some other animals, the central nervous system is more complex (Figure 18-3). The central nervous system includes the brain and the spinal cord. The brain is composed of several different parts. The larger, upper part of the brain is called the **cerebrum** (SER·eh·brum). Specialized areas of the cerebrum control sight, speech, hearing, motor activities, and other functions. The **cerebellum** (ser·eh·BEL·um) lies under the back part of the cerebrum. It functions in controlling balance and coordinating movement. The **medulla** is the part of the brain that regulates vital physical activities such as breathing. The brain connects with the spinal cord. The spinal cord is actually a centralized cluster of nerves.

Even simple multicellular animals have structures that receive stimuli and respond to them. **Receptors** are structures that receive stimuli. Receptors may be nerve cells, or they may be organs that are specialized for detecting stimuli. Most animals, including humans, have many receptors, including those for touch, taste, smell, temperature, light, and sound. **Effectors** are structures that carry out responses, such as muscle cells and gland cells.

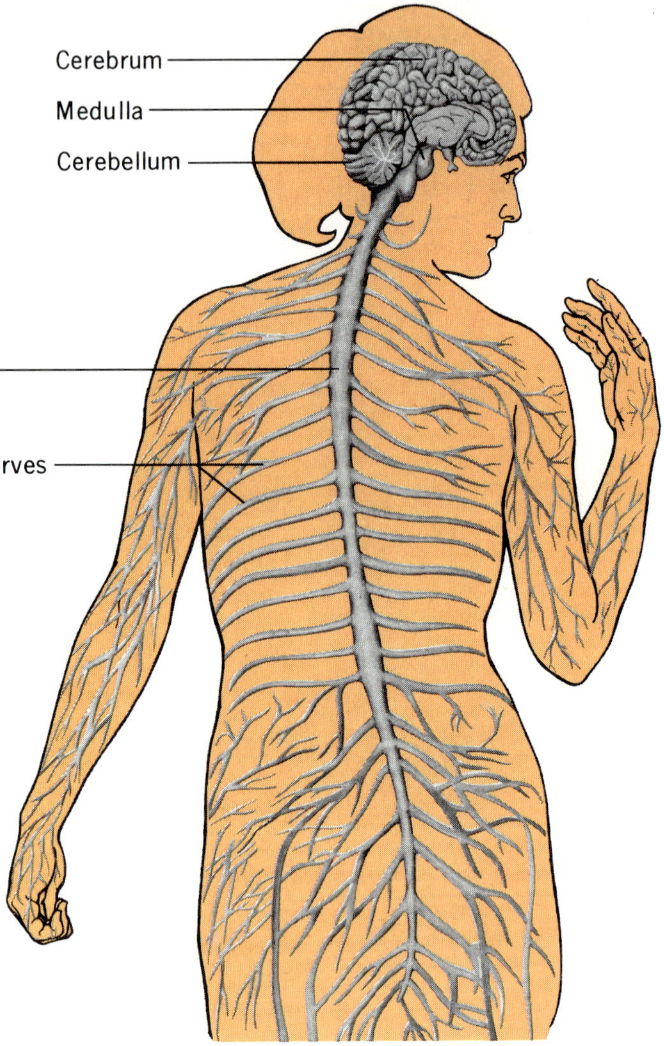

18-3. The human brain and spinal cord are connected to all parts of the body by a network of nerves.

The unit of structure and function in the nervous system is a nerve cell, or **neuron.** Neurons consist of three main parts: the **dendrites,** the **cell body,** and the **axon** (Figure 18-4). The dendrites and axon extend from the cell body. They sometimes are referred to as neuron fibers. Dendrites carry nerve impulses toward the cell body. The axon carries impulses away from the cell body. Most neurons are fairly short, but some are very long. Some of the longest in the human body reach from the fingers or toes to the spinal cord.

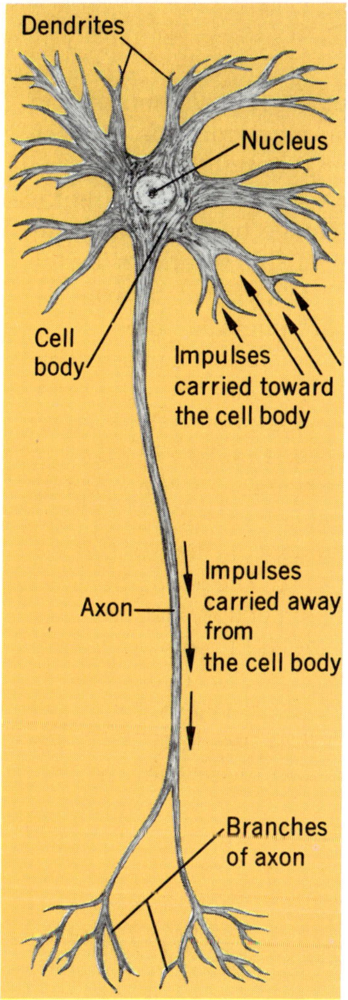

18-4. In this typical neuron, note that the axon always carries impulses away from the cell body.

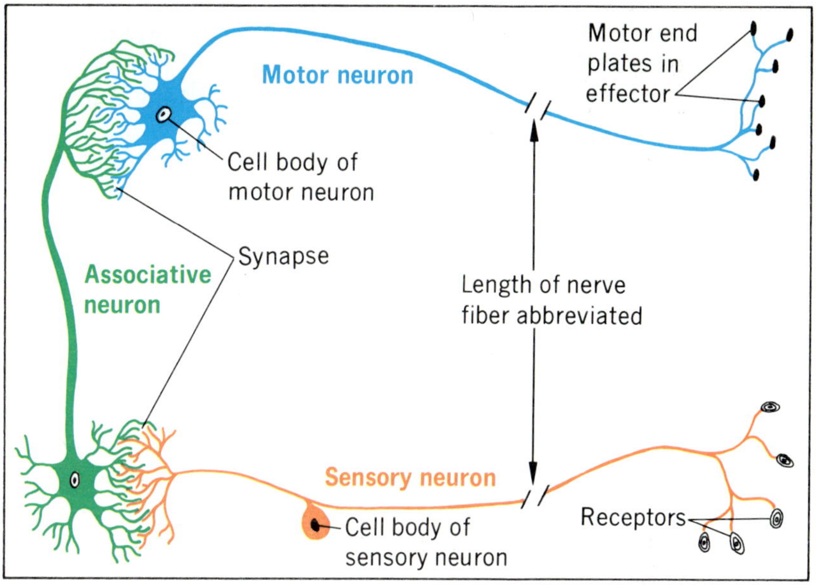

18-5. Kinds of neurons.

The nervous system is made up of different kinds of neurons (Figure 18-5). **Sensory neurons** carry impulses from receptors to the central nervous system. **Associative neurons** in the central nervous system transmit impulses to **motor neurons.** The motor neurons carry impulses from the central nervous system to **motor end plates** in the effectors. Thus, one kind of nerve cell picks up the stimulus and another kind carries the message that produces the organism's response.

Neurons transmit **nerve impulses.** A nerve impulse is a wave of changes traveling along the nerve cell membrane. These changes travel in one direction through the neuron. The impulse begins at the dendrites of one neuron and travels rapidly to its axon. From the axon of this first neuron, the impulse is transmitted to the dendrites of the next neuron.

But what exactly is a nerve impulse? We know that it is partly electrical, a flow of charges as in a conducting wire. But it is also chemical. Changes in the concentration of certain ions inside and outside the nerve cell membrane cause the movement of a "current" along the nerve.

Neurons are not attached to each other. A gap exists between the axon of one neuron and the dendrites of another. This gap is called the **synapse** (SIN·aps). The axon of one neuron releases a chemical that diffuses across the synapse and produces a new nerve impulse in the dendrites of the next neuron. A nerve impulse moves only in one direction through each neuron and across the synapse.

The Central Nervous System

The central nervous system is made up of associative neurons. In humans, the spinal cord and brain contain many millions of these nerve cells. Many branching outgrowths connect associative neurons in the brain or spinal column with the sensory and motor neurons in other parts of the body. Associative neurons carry impulses from sensory neurons, leading from the receptors to the motor neurons, which stimulate effectors. Let's look at an example of how receptors, neurons, and effectors are interrelated.

Suppose you are walking without shoes and step on a sharp rock. You rapidly lift your foot (Figure 18-6) and you feel pain. What happens in the nervous system to account for these events? The skin of the foot contains many receptors. Some are stimulated by strong pressure. Others respond to any intense stimulation. When you step on the sharp rock, the receptors are stimulated.

The receptors start nerve impulses in the sensory neurons of the foot, which extend up the leg to the spinal cord. These impulses are transmitted across the synapses between the sensory neurons and associative neurons in the spinal cord (Figure 18-6). Some of the associative neurons also have synapses with motor neurons. The nerve impulses are transmitted across the synapses to the motor neurons that extend to the leg muscles. When the nerve impulses reach the muscles, the muscles contract and move the leg. This simple pathway from receptor to sensory neuron, to associative neuron, to motor neuron, to effector, is a **reflex arc.**

At the same time other associative neurons in the spinal cord send impulses to the brain. It is not until these impulses reach the brain that you feel pain. You probably have moved your foot by reflex action by the time your brain becomes aware of the pain. Perhaps you will hold the injured foot with your hands. For you to do this, your brain must send nerve impulses along the neurons of the spinal cord and out to the muscles of arms and legs.

Sensory Neurons and Receptors

Sensory neurons carry impulses from all parts of the body to your spinal cord. They carry impulses from your ears, eyes, taste buds, nose, and skin, as well as from all the internal parts of your body. You are aware of some of these messages. Even with your eyes closed, you know the position of your arms—whether they are outstretched or at your sides. Sensory neurons associated with muscles constantly carry to the brain impulses that indicate the position and degree of contraction of the muscles. Other sensory neurons function without your awareness. These bring to your brain or spinal

18-6. A REFLEX ARC

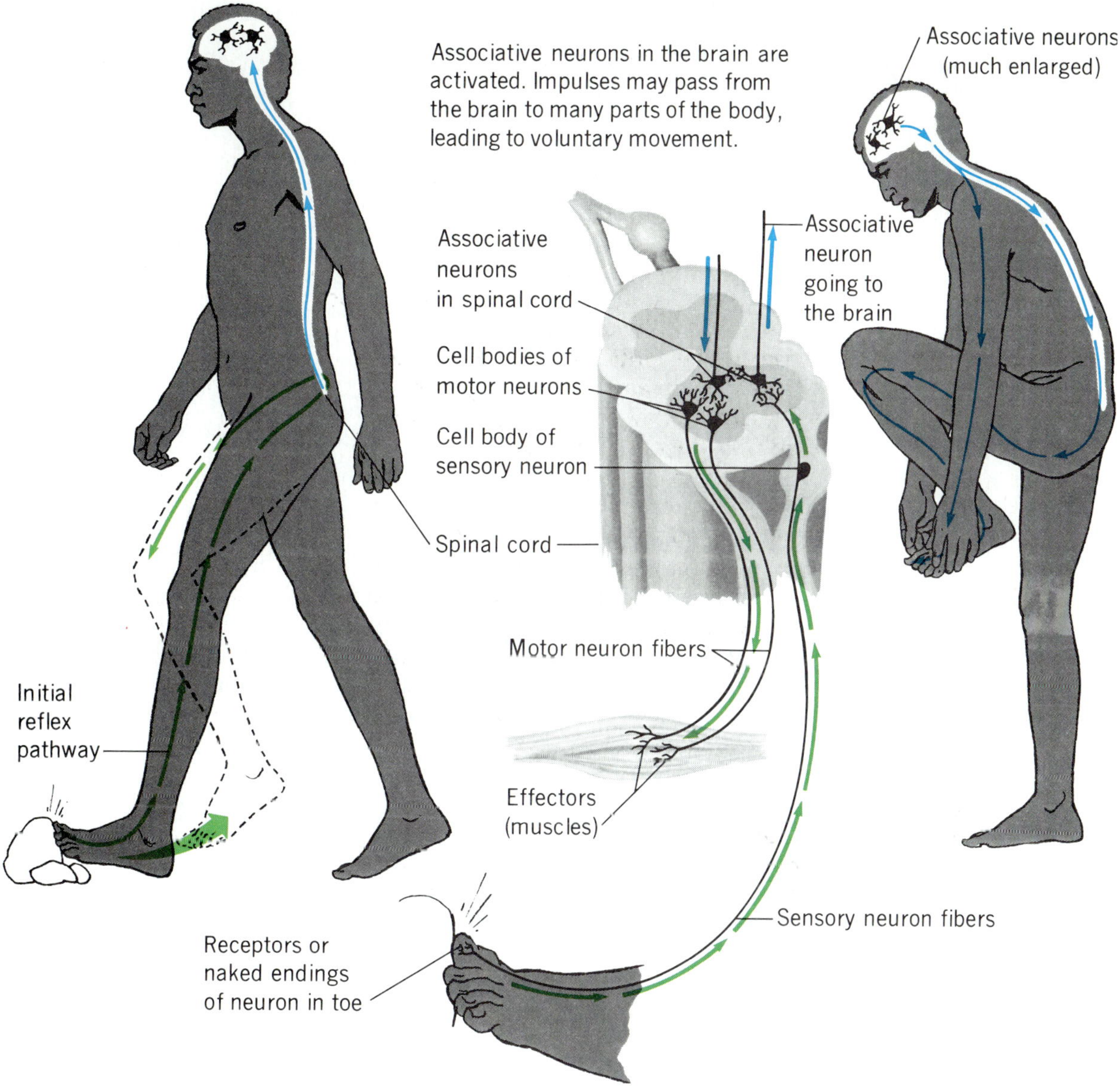

cord information from the heart, liver, digestive tract, blood vessels, kidneys, and other internal organs.

Some information about the external world is received through specialized receptors – the **sense organs.** Each sense organ is specialized to receive a specific type of stimulus. Our ears are stimulated by air vibrations. Our eyes are stimulated by light of specific wavelengths, and our taste buds and organs of smell are stimulated by chemical substances.

Our sense organs, however, receive only a small portion of the stimuli present in the environment. For example, the colors of light that stimulate our eyes vary from violet to red. These and other colors differ in their wavelengths. Violet has a very short wavelength. Red has a very long wavelength. The colors between violet and red are only a small portion of the total range of wavelengths that actually occur. We cannot see the waves shorter than violet (the ultraviolet, for example). Waves that are longer than red are infrared and radio waves, which we also cannot see. We know of the existence of such waves only when we use special instruments to detect them. For example, thousands of broadcasting stations throughout the world transmit radiation that is not detected by your sense organs. Yet this radiation can be detected, amplified, and converted to sound by a radio.

At this point, you should carry out **Inquiry 18-1,** in which you will observe the functions of some of your own sensory receptors.

INQUIRY 18-1 Sensory Reception in Humans

How do we know about the world in which we live? We receive information through receptors. The activities of these various receptors, coordinated by the brain, are called the senses: touch, sight, hearing, smell, and taste, for example.

Work in pairs for this experiment. One partner (the experimenter) will apply the stimulus (signal to the receptors). The other (the subject) will indicate what sensation is produced.

MATERIALS (per team)

1-3 cm^3 cubes of apple, onion, and potato
solutions of:
- *10% sodium chloride*
- *5% sucrose*
- *1% acetic acid*
- *0.1% quinine sulfate*

metric ruler
cotton swabs (2 per student)
2 flat-headed, straight pins
large cork or rubber eraser
rectangles (2.5 × 8 cm) of stiff construction paper, various colors
sheet of white paper
pen or pencil

Procedure (Day 1)

Begin by considering your sense of touch. Suppose you were to guess the number of touch receptors on the underside of your wrist, the back of your neck, or the tips of your fingers. **1.** Where would you think you would find the most receptors per square centimeter of surface area? Why?

Test your prediction using a large cork. Push two straight pins into the cork 2.5 cm apart. Push them the same distance into the cork.

The experimenter will gently touch the inside of the subject's wrist in several places with the flat pinheads. Each time, the subject will report how many pins are felt. The experimenter will then move the pins closer together on the cork and test the wrist again. Record the distance between the pins each time. Continue until the subject can feel only one pin. Record the distance between the pins.

Now do the same on fingertips and the back of the neck. **2.** When the pins were close together, why didn't you feel both pinheads? **3.** Suppose the pinheads are 1 cm apart when you can feel only one of them touching your wrist. And this distance is 2 cm when the back of your neck is tested. Can you conclude that there are more or fewer touch receptors in the wrist than in the neck? Explain your answer. **4.** Which area actually had the greatest number of touch receptors? the fewest?

To test your peripheral (side) vision, use rectangles of several colors. The subject will stare straight ahead at some object in the distance. The experimenter will take one colored rectangle and move it slowly forward from an area to the right of and slightly behind the subject's right ear. The rectangle should be about 50 cm from the head at all times. When the subject first sees the rectangle, he or she will tell the experimenter to stop, holding the paper at that point. The subject will try to identify the color of the paper. If the response is correct, repeat the procedure using another color. If the color given is incorrect, the experimenter will continue to move the rectangle forward until the subject is certain of the color. **5.** Is there any difference between the point at which the rectangle is first seen and the point at which the color is described accurately? **6.** Does this imply that all of the receptors in the eye are the same? Explain.

Now you can test some aspects of color vision. Draw a small dot in the center of a sheet of white paper. Hold Figure 18-7 about 20 cm away from your eyes and stare at the dot in its center for 20 to 30 seconds. Immediately shift your eyes to the dot you made on the white paper. Stare at the dot for 10 to 15 seconds. **7.** Describe what you saw. **8.** Suggest a hypothesis to explain what you saw in terms of the receptors in the eye.

18-7.

Now investigate your senses of taste and smell. The experimenter will tell the subject, "Close your eyes and hold your nose." The experimeter then will place the pieces of onion, apple, or potato—one at a time—on the subject's tongue. The subject will try to identify each piece as it is placed on the tongue. **9.** How accurately were the pieces identified by taste alone?

Repeat the tests. This time the subject will not hold her or his nose. **10.** What do the results indicate about the relationship between the senses of smell and taste? **11.** How do these signals received by your tongue and nose differ from the signals that your fingers received?

The tongue has receptors sensitive to salt, sour, bitter, and sweet tastes. You can find out if these receptors are located uniformly over the tongue. Draw a map of the tongue on a piece of paper. With cotton swabs, the experimenter will apply one of the solutions to one area on the subject's tongue. Record whether or not the substance can be tasted. Continue to apply the solution to other spots on the tongue until the entire surface has been tested. Record the results each time the swab is applied.

Test the other solutions in the same way. Use a code letter for each solution to indicate on the tongue map places where that solution was tasted. Rinse the mouth with water before testing a different solution. **12.** Is the entire surface of the tongue sensitive to each of the four chemicals tested? Where is the tongue most sensitive to salt? sugar? bitter? sour? **13.** Now, think back to what you learned about smell and taste and explain what usually happens to your sense of taste when you have a severe cold. **14.** What can you now say about the location, specificity, number, sensitivity, and independence of sense receptors? **15.** What other senses do you have besides those you studied in this inquiry? Is it possible to learn anything about the world without the use of one or more of your senses?

Procedure (Day 2)

In the first part of the inquiry, you learned about the many receptors in your body. How are these receptors related to the actions of our bodies? How do we react once our receptors receive a stimulus? How long does it take for an effector to respond? The only equipment you will need is a meter stick.

The subject will sit down with one arm resting on the desk so that the hand is extended past the edge of the desk. The subject will curl the last three fingers of the extended hand into a fist, leaving the thumb and forefinger free. The thumb and forefinger should be parallel to the ground and 3 to 4 cm apart.

The experimenter will stand, holding the bottom of the meter stick between the finger and thumb of one hand. The small numbers of the meter stick should be at the bottom.

Without a warning, the experimenter will drop the stick. The subject will try to catch it with just the thumb and forefinger of the extended hand. The longer it takes the subject to react, the farther up the stick he or she will catch it. Record the point on the meter stick at which the subject caught it.

Repeat the test five times, recording this distance each time. Determine the average for the five trials. **16.** What happens in other parts of your body before you react and catch the meter stick? **17.** What receptors are involved?

Repeat the exercise while the subject is counting by 3s to 99. Average the time (distance on the meter stick) it takes the subject to catch the meter stick while counting. **18.** What effect does thinking have on the reaction time?

Repeat the procedure while the subject's eyes are closed. The experimenter will snap his or her fingers to signal when the stick is released. Repeat five times and calculate the average. **19.** Compare this average with the average in the first step of this experiment. In which situation was reaction time shortest? **20.** What can you conclude about the reactions to the different signals?

What other things might influence the reaction time involved in catching the meter stick? Design a simple experiment to test the reaction time if the signal is a light touch on the back of the wrist. **21.** Are there changes in the environment you cannot detect? Explain and give an example.

Motor Neurons and Effectors

From the brain and spinal cord, nerve impulses are carried to motor neurons. The motor neurons then carry these impulses to motor end plates, where the axon comes in contact with the cells of the effector. The effectors of the body are muscles and glands, which carry out specific activities. Muscles contract, and an organism moves about in its environment. Muscle contractions also occur in organs within the body. Glands are stimulated to secrete hormones.

The regulation of when, and how much, these effectors act is under the control of the nervous system. Stimuli are received by sense organs, and nervous impulses travel along sensory neurons to the brain. The brain analyzes these impulses. It then sends nerve impulses along motor neurons to the effectors that will carry out the proper response.

Studying Motor Neurons

As is so often the case in science, the answer to one question may come from trying to answer an entirely different question. This was certainly the case in one investigation into how neurons function.

All vertebrate hearts beat in a rhythmic way. Nerves extend from the central nervous system to the heart. One of these is the **vagus nerve.** When nerve impulses pass from the brain along the vagus nerve, the heart beats more slowly. Nerve impulses passing along another nerve—the **accelerator nerve**—speed up the rate of heart contraction.

When scientists removed a frog's heart from its body and the heart continued to beat, they began to look for an internal mechanism in the frog's heart that stimulates contraction. This research revealed a special area in the heart called the **pacemaker.** The pacemaker sends out rhythmic impulses to the rest of the heart, causing the heart muscles to contract.

But how do the vagus and accelerator nerves regulate the rate of contraction of the heart? In 1921, German biologist Otto Loewi (LUH·vee) carried out an important experiment. Loewi removed the heart and vagus nerves from two frogs. Loewi then stimulated the vagus nerve of one frog and watched its heart slow down and stop. Then he placed some fluid from the stopped heart into the other heart, which was beating normally. Almost at once, the second heart stopped just as if Loewi had stimulated its vagus nerve. Loewi concluded that a chemical substance must be released by the heart when its vagus nerve is stimulated. Many later experiments confirmed Loewi's hypothesis. The chemical was identified as an organic compound, acetylcholine (as·uh·til·KOH·leen).

The accelerator nerve to the heart increases its contraction rate. Its action was found to be associated with a chemical substance that is different from acetylcholine. This substance, epinephrine (ep·ih·NEF·rin; also called adrenaline), stimulates the pacemaker, causing it to send out impulses more rapidly.

Both of these chemical substances are associated with the function of many other nerves. They play a part in transmission of nerve impulses to muscle and gland cells throughout the body. Acetylcholine and other transmitter substances also are involved in the stimulation of one neuron by another across a synapse.

There are many puzzles still unsolved about the roles of chemical substances in nerve functions. Loewi and his frog hearts started us on a fascinating trail that probably will lead to many important answers.

Brain Functions

Although many animals communicate with one another, humans have the ability to make known their thoughts by means of a complex spoken language. And only humans preserve an understanding of themselves and the world about them in written records. But the human nervous system is similar to that of other mammals. Even primitive vertebrates have receptors, neurons, brains, and effectors with the same general pattern as ours.

As in other vertebrates, the brain and spinal cord are the coordinating centers of the human nervous system. Nerve impulses from all neurons are the same; they are electrochemical changes along the cell surface. How these impulses are interpreted, however, depends on which part of the brain receives them. If you were to touch a neuron connected with the eye, you would have the sensation of light. If you were to touch a neuron from the ear, you would have the sensation of sound.

Sensory neurons from each type of sense organ connect with specific areas in the cerebrum of the brain (Figure 18-8). When nerve impulses are received from the sensory neurons, the brain becomes aware of some change in the external or internal environment. Other regions of the cerebrum initiate movement of the body. Nerve impulses from these regions cause contraction of specific muscles bringing about motion of arms, legs, and head. Other regions control the movement of eye muscles and muscles of speech.

The specific regions where complex processes such as thinking, planning, learning, and remembering take place have not been located precisely. There are specific parts of the brain that seem to be concerned with sleep, rage, simple pleasure, and sexual feelings. These have been discovered in experiments done on animals, using

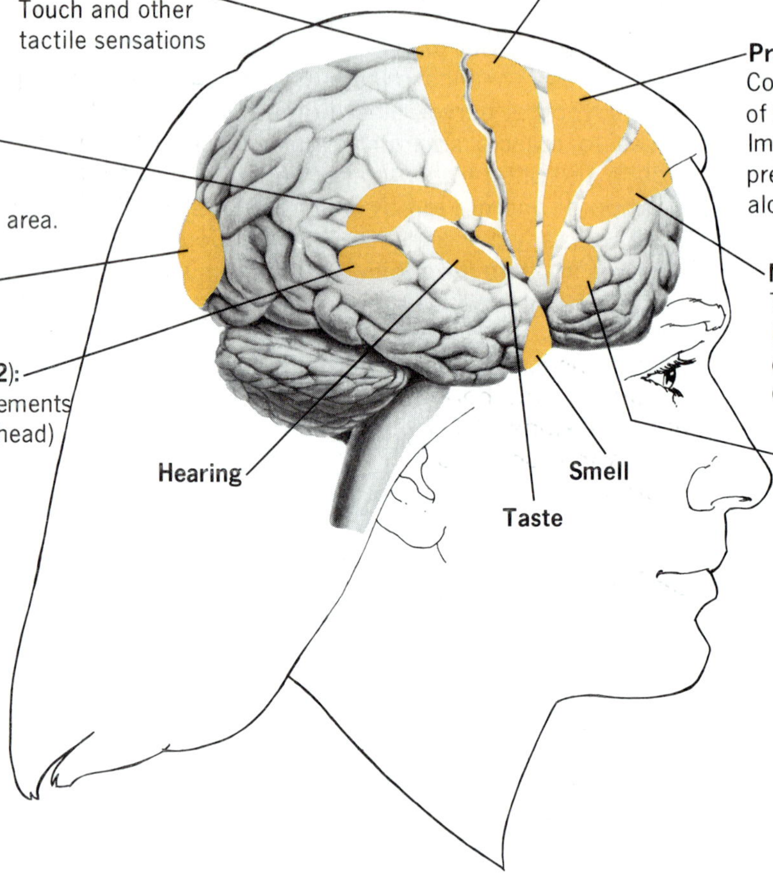

18-8. Some functional areas of the human cerebrum.

electrodes that can be placed in an animal's brain without damage or pain to the animal. The electrodes stimulate a given region and the response is observed.

Delicate operations to remove diseased parts of the human brain also have revealed the locations of centers of certain activities. But very little is known about the specific brain regions involved in consciousness or the more complex functions of the brain. Some evidence does suggest that the right and left sides of the brain are specialized. In right-handed people, the left side seems to carry out most logical and verbal activities. The right side seems to be specialized in intuition, creativity, and spatial relationships. Research is being carried out to determine the exact nature of these specializations. Scientists also are investigating the processes of communication between the two sides of the brain.

HUMAN ENDOCRINE SYSTEM

Loewi's observations of frogs' hearts and vagus nerves, as well as the studies conducted by other biologists, helped break down the idea that functions of the nervous and endocrine systems are independent of each other. At one time, biologists thought the nervous system worked only because of the electrochemical messages carried by neurons. The endocrine system, on the other hand, was believed to work only by secreting chemical substances, the hormones. Nerve impulses traveled as much as 100 meters a second, but hormones circulated less rapidly in the blood. At first, biologists saw little in common between these two coordinating systems.

Later, however, this point of view changed when biologists discovered that nervous control of effectors depends on acetylcholine and other transmitter substances. One of these transmitter substances is identical to the hormone epinephrine, secreted by glands located near the kidneys. This is not too surprising, for the part of these glands that secretes epinephrine is actually a modified group of neurons.

Endocrine Glands

Hormones may be secreted by individual cells, but often the cells are clustered together in the form of glands. The locations of the endocrine glands of humans are shown in Figure 18-9. The hormone produced by a gland is secreted directly into the blood, producing a specific effect when it reaches certain cells of the body.

The **thyroid** (THI·roid) **gland** is in the throat. It secretes thyroxin (thi·RŌK·sin), which regulates the rate of chemical activities of cells. The **parathyroid** (par·uh·THI·roid) **glands** are embedded in the back of the thyroid gland. The hormone secreted by these glands functions in the control of calcium.

The **thymus gland,** located below the thyroid, produces a hormone, thymosin, that affects the production of antibodies by the lymph system. An **adrenal** (A·DREEN·uhl) **gland** is located on top of each kidney. The adrenal cortex, the outer part of each gland, secretes several hormones, one of which is cortisone. These hormones are important in maintaining blood-sugar levels, regulating sodium-potassium metabolism, and in stimulating certain male sexual characteristics. The adrenal medulla is in the center of each adrenal gland. It secretes epinephrine (adrenaline) and a related substance, norepinephrine. Epinephrine stimulates heart action and causes blood vessels to constrict. Both epinephrine and norepinephrine play a part in the body's use of carbohydrates. Epinephrine also helps the body meet emergency situations.

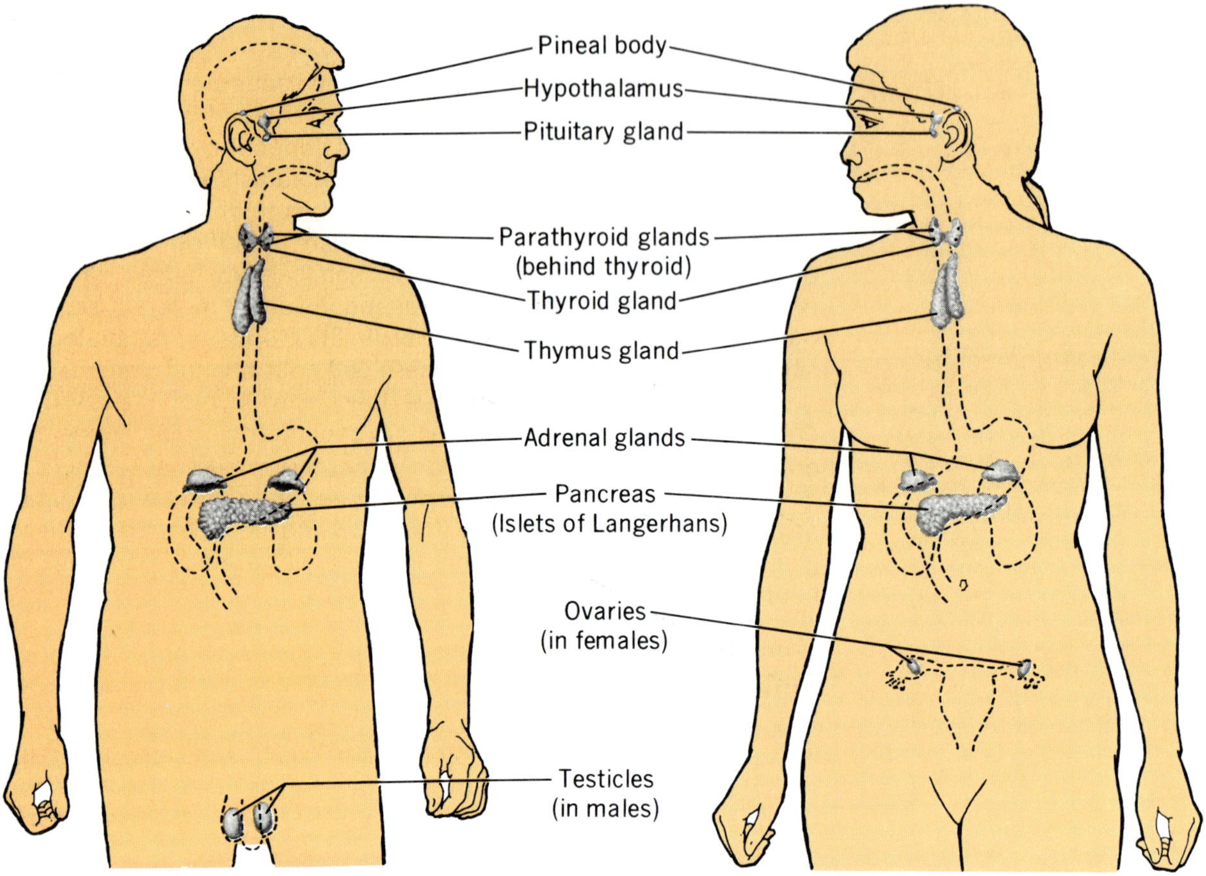

18-9. Some of the endocrine glands in the human body. The dotted lines show some major organs such as the stomach and kidneys.

The pancreas has many functions. In particular, it produces enzymes important in digestion. In addition, the **islets of Langerhans** (eye·lets of LAHNG·er·hahns), which are located in the pancreas, secrete two hormones. One hormone is insulin and the other is glucagon (GLEWK·ah·gon). These hormones have important roles in controlling the way glucose and some other organic compounds are used in the body.

The **testicles** (TES·tuh·kulz) are the male reproductive glands in humans. In addition to producing sperm cells, they secrete a hormone, testosterone (tes·TOS·ter·ohn). Testosterone is secreted in early adolescence and influences the maturing of the male. The **ovaries** (OH·vah·reez) of the female secrete several hormones that influence the maturing of the female. These hormones function in the prenatal development of a human infant and in control of the female reproductive cycle.

The **pituitary** (pih·TOO·ih·tehr·ee) **gland** works in close association with cells of the brain and is under the brain's control. It secretes a number of hormones. Some of the pituitary hormones affect other endocrine glands. For instance, the pituitary secretes thyroid-stimulating hormone, which controls the secretion of thyroxin by the thyroid gland. Some pituitary hormones control the rate of bone growth, secretion of milk by the mammary glands, and development of ovaries and testicles. Others control the activity of the adrenal cortex, removal of water from the nephrons of the kidneys, and contraction of the muscles of the uterus.

The action of the pituitary is, in turn, influenced by the products of other glands. The hypothalamus and the pineal (PIE·knee·uhl) body in the brain both secrete hormones that regulate the functions of the pituitary.

Regulation of Hormone Secretion

Hormones are secreted from endocrine glands at a closely controlled rate (Figure 18-10). A good example of how glandular secretions are regulated is the interaction between the thyroid and the pituitary. Special cells in the hypothalamus just above the pituitary are

18-10. The feedback mechanism between the pituitary and the thyroid glands. This mechanism controls the level of thyroxin in the blood. To trace the story, begin at **a** in each drawing and follow the sequence of letters.

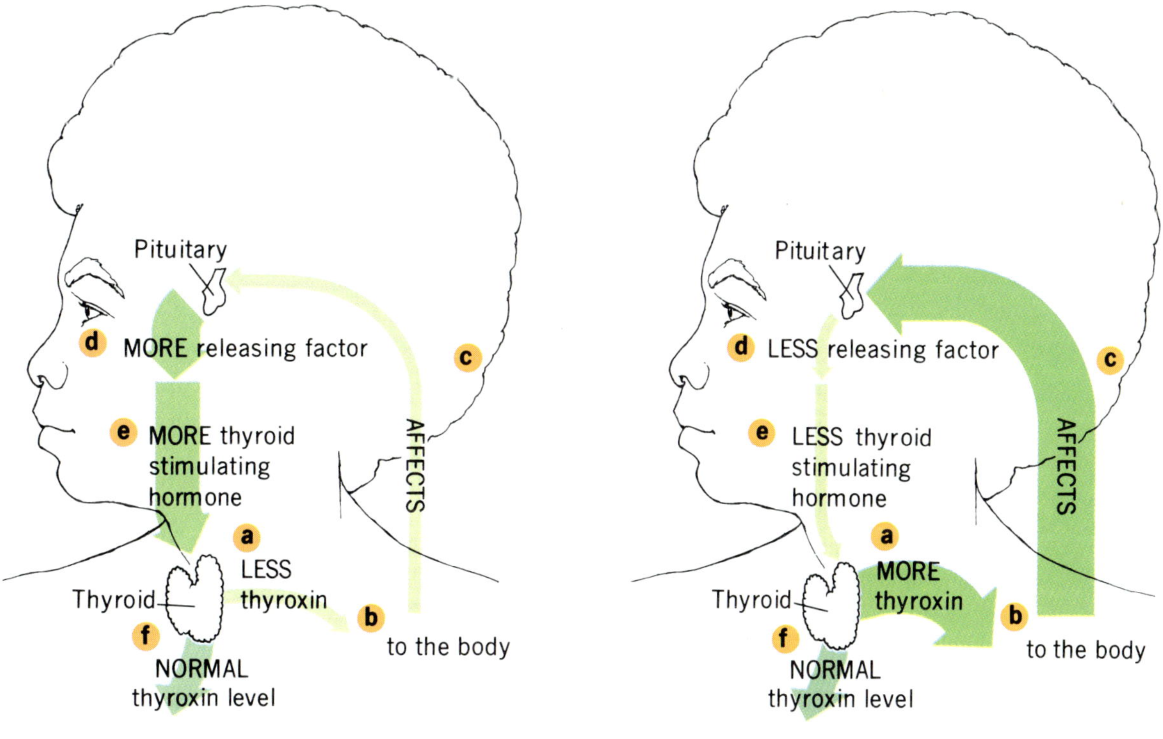

sensitive to the changing concentration of thyroxin in the blood. When the concentration of thyroxin is low, the cells of the hypothalamus secrete a hormone called a **releasing factor.** This hormone travels in the blood to the thyroid gland and stimulates its cells to secrete thyroxin. If the thyroxin concentration in the blood is high, the pituitary secretes less of the releasing factor, and the thyroid does not produce as much thyroxin.

Regulation of hormone secretion is an example of a *feedback mechanism.* That is, the amount of hormone secreted by the gland is controlled by the hormone itself. Feedback is an important part of regulation in homeostasis. Let's take the thyroid as an example. Since an increase in thyroxin in the blood decreases the rate of thyroxin production in the thyroid, the feedback in this case is negative. For some years, engineers have used the principle of negative feedback. A thermostat is a good example of a mechanism of negative feedback. When the room gets warm enough, the furnace shuts off.

The Study of Hormones—Diabetes Mellitus

The blood of humans contains only a trace of glucose—about 0.07 to 0.12 percent. This trace, however, is vital for the life of all the cells. If the amount of glucose drops to about 0.03 percent, convulsions and death will result. On the other hand, problems occur if the glucose level rises too high. In the disease diabetes mellitus, the concentration of glucose in the blood may reach a level above 0.14 percent. Unless this rising blood sugar level is controlled, the disease becomes very serious. The control of glucose concentration, then, is very important.

Diabetes mellitus has been recognized as a specific disease since the days of the Romans, but nothing was learned about its cause until the late nineteenth century. In 1889, observations by two German physiologists suddenly opened up the way to the experimental study of diabetes. These scientists were using dogs to study the pancreas. They began by removing the pancreases from some of the dogs. They thought that the enzymes in gastric and intestinal juices were sufficient to allow the dogs to digest their food and remain in good health. But all the dogs died shortly after their pancreases were removed.

Before these dogs without pancreases died, the scientists observed something unusual about their urine. Chemical analysis showed that it contained a large amount of glucose. The scientists realized that the dogs without pancreases had diabetes mellitus.

The pancreas is composed mainly of two types of cells. One type, by far the more abundant, was thought to secrete pancreatic juice.

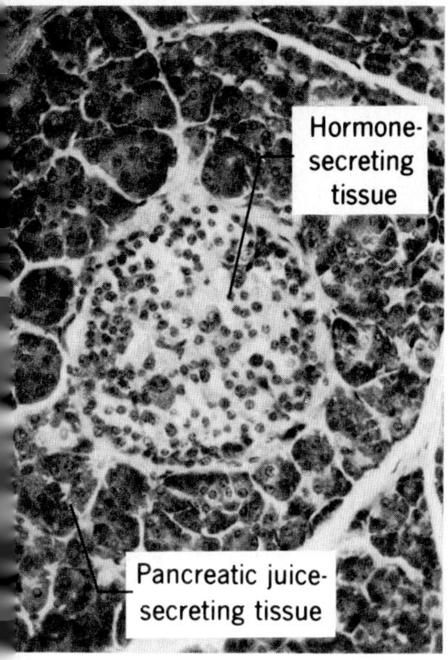

18-11. A photograph of a microscopic section of human pancreas. The light, circular area contains cells of the islets of Langerhans. The darker, surrounding cells secrete pancreatic juice.

The other type of cell forms in little clusters, called the islets of Langerhans. These are shown in Figure 18-11. The islet cells are not connected to the pancreatic ducts, which carry pancreatic juice to the small intestine. Their only contact with the rest of the body is by way of the bloodstream. The islet cells secrete a hormone that regulates the amount of glucose in the blood. The hormone is insulin.

Biologists have been able to isolate insulin in pure form. It is a protein molecule, differing slightly in different species, yet very similar in all mammals. Insulin promotes the use of glucose in the body's cells. As a result, glucose does not remain concentrated in the blood.

Diabetes can be caused by many hormonal disturbances. In the pancreases of some humans who have died of diabetes, the cells that secrete pancreatic juice are normal. However, the cells of the islets of Langerhans are definitely abnormal. But no matter what the specific cause, diabetes can only be treated, not cured, by insulin. A diabetic person must control the level of glucose in the blood. This is done by eating a carefully regulated diet. In addition, some diabetics must receive specific amounts of insulin or other drugs at regular times.

A detailed study of special extracts of the pancreas led to the discovery that the islets of Langerhans also secrete the hormone glucagon. The effects of glucagon are almost opposite those of insulin. Glucagon tends to raise the amount of sugar in the blood, while insulin lowers it. Glucagon and insulin, then, take part in the homeostatic mechanism that controls the body's use of carbohydrates.

Other hormones also are involved in this homeostatic mechanism. The control of the concentration of glucose in the bloodstream is a complex process. Body cells take up glucose rapidly from the bloodstream, and glucose is added to the bloodstream from the intestine and liver. In normal conditions, the level of blood sugar is closely controlled. Part of the control comes from the balanced secretions of insulin, glucagon, adrenaline, cortisone, a pituitary hormone, and the releasing factors of the brain.

MUSCLES AND BONES

In our brief discussion of coordination, we must not overlook perhaps the two most obvious systems—the skeletal and the muscular. These two systems are so much a part of us that we seldom think about their structure and function. But our muscles and bones are really the structures that allow us to respond to stimuli—both from the internal and from the external environment.

Skeletal Systems

There are two kinds of skeletal systems. One is on the outside of the animal and is called an **exoskeleton.** Insects have exoskeletons. Their bodies are supported by hard structures that form an external covering. Humans and many other kinds of animals have **endoskeletons**—hard structures of support inside the body. Either kind of skeleton supports the organism, aids in movement, and often acts as protection. For example, the bones of the human head protect the brain.

Animals with backbones are the vertebrates. This group is named after the small jointed sections of backbone, called **vertebrae** (VUR·tuh·bree), that protect the nerve cord. If you touch the middle area of your lower neck, you can feel some of the vertebrae. The backbone is flexible, rather than rigid, because of these interlocking vertebrae.

The supporting skeleton of humans and many other vertebrates is made of bone. Bones consist of living bone cells surrounded by hard, nonliving material. The dense material around the bone cells is secreted by the bone cells themselves. Nearly half of the bone consists of calcium compounds that give the bones strength and hardness. A series of canals runs among the bone cells. These canals connect the bone cells to each other and to blood vessels.

Normal bone structure is the result of a balanced exchange of materials between the bone tissue and the body's internal environment. A balanced level of minerals is maintained in bone by the action of hormones and by vitamins and minerals supplied in the diet. The thyroid and parathyroid glands affect bone formation. The skin produces hormones that help regulate the levels of calcium and phosphate in the blood and bones. The growth of the skeleton is controlled by a pituitary hormone.

Sometimes, things can go wrong in the control of bone composition or growth. During childhood, too much or too little of the pituitary growth hormone can retard or speed up growth. Then, too, poor nutrition can lead to problems in bone development and strength. Bones that lack calcium as a result of starvation are soft, spongy, and easily broken.

The Human Muscular System

Muscles as well as bones play an important part in supporting the body. Muscles also allow us to move. They are responsible for some movements that we cannot see or control. The movements of the stomach and blood vessels are the result of the action of smooth muscles within our body. Such movements usually are considered

SCIENCE AT WORK: Biofeedback

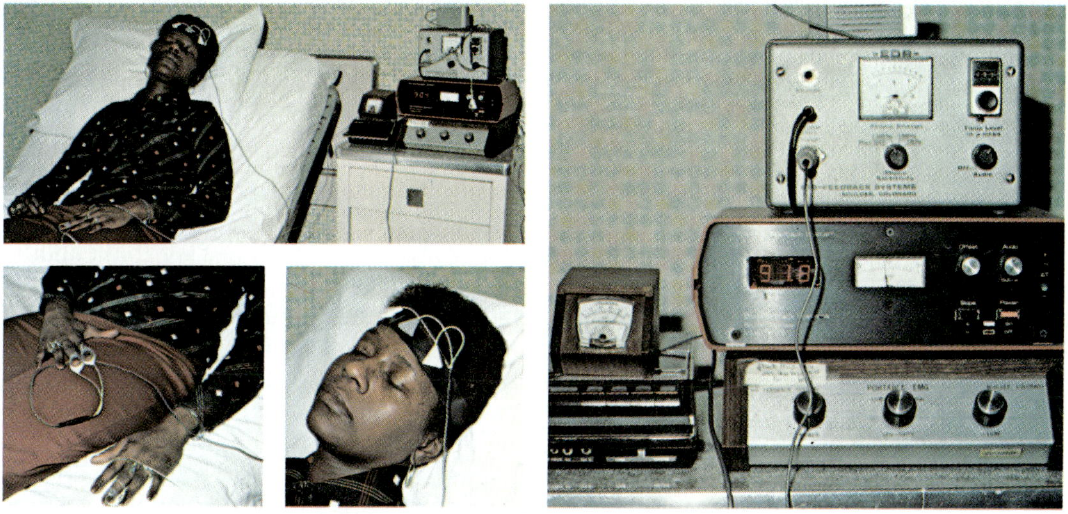

The anatomists who studied the human nervous system during the nineteenth and early twentieth centuries firmly established that it was composed of two distinct parts. One part is a voluntary nervous system by which the human mind consciously controls the movement of arms, legs, jaws, and posture. There also is a part that controls internal organs. Blood vessels, heart, stomach, glands, and many other parts of the body function without any conscious control by the human mind. This involuntary system of control is the autonomic nervous system.

As has been true throughout the history of science, new knowledge is now causing researchers to take another look at the human nervous system. It may be that we have more control over some of our bodily functions than was previously thought.

Research centers around the country are working on the following hypothesis: Given information about how any internal physiological system is operating, the human mind can learn to control the activity of that system. This activity may be heartbeat, blood pressure, or secretion of digestive juices. It does not seem to matter which function it is, as long as information about how it is behaving is made available.

At the Menninger Foundation in Topeka, Kansas, scientists observed that a woman's migraine headaches disappeared when a surge of blood flow to her hands increased. In experiments, migraine sufferers were taught to increase the blood flow to parts of the body other than the head. Eighty percent learned to control their headaches.

This ability of the mind to control body activities that were once thought to be involuntary is termed biofeedback. This phenomenon occurs when biological functions are monitored and information is "fed back" to people who then attempt to control a particular body function. Experiments on biofeedback are still going on. While some studies indicate that these techniques have promise for the relief of migraine headaches or high blood pressure, other work shows little benefit to be derived. It will take more research to find out if people can truly learn to control some of the biological processes once thought to be completely involuntary.

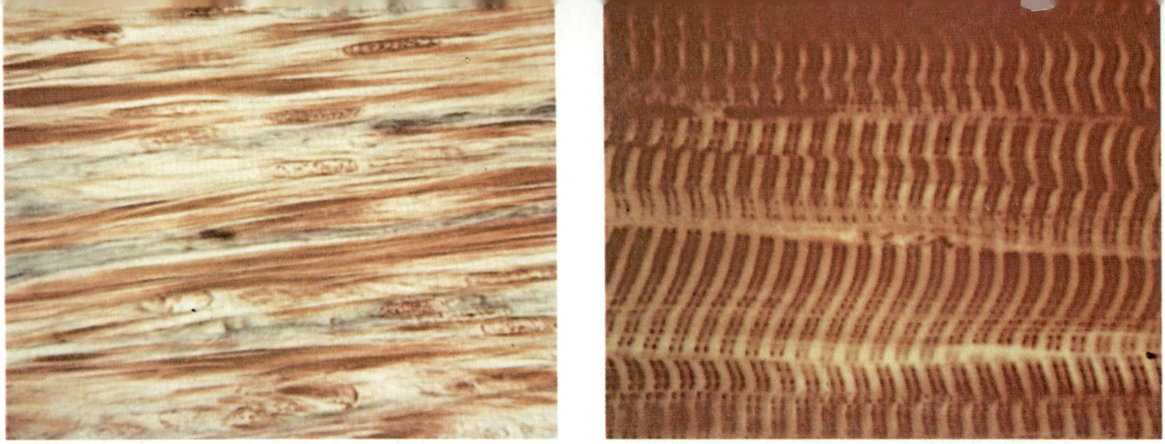

18-12. Compare the structure of smooth muscle **(left)** with striated muscle **(right).** *What differences can you identify?*

involuntary. Our voluntary movements, the ones we think about, are brought about through the contraction of striated muscles.

Striated muscle differs from smooth muscle in many ways. First, striated muscle contracts much more rapidly than smooth muscle. Second, nervous control of striated muscle generally is voluntary. Conscious messages from the brain bring about an increase or decrease in the rate and degree of muscle contraction. These types of muscle also differ in their structure. Smooth muscle is composed of individual cells with a single, centrally located nucleus. Compare the cells shown in Figure 18-12. Clearly defined individual cells cannot be seen in striated muscle. When this muscle is formed, repeated division of a single cell produces a long, continuous fiber. There do not seem to be any separations between the cells. Each fiber, therefore, contains many nuclei.

Some vertebrates, including humans, have another type of striated muscle. The muscle is called cardiac muscle, because it is found in the heart. Cardiac muscle is striated, yet it is involuntary.

The structure of striated muscle accounts for its ability to contract strongly and quickly. Muscles are actually bundles of many fibers. Each fiber is composed of smaller fibrils made of still tinier filaments. Some filaments are thick; others are thin. The arrangement of fibers causes the striated appearance shown in Figure 18-12.

The contraction of a muscle is, like all other body processes, a chemical process. Evidence indicates that muscle proteins contract in the presence of ATP, the energy storage molecule already discussed in Chapter 3.

In Brief

Coordination in an organism involves receiving information and responding to it. The information—stimuli—may be of many kinds. For instance, the odor of a skunk, the hot air near a flame, and the noise of an automobile horn are kinds of stimuli. The concentration of carbon dioxide in the blood, the presence of food in the stomach, or glucose in the blood also are stimuli. These various stimuli can lead to a variety of responses. Holding your nose, moving away from the flame, and covering your ears are responses. So are breathing more rapidly, the release of gastrin, and the release of insulin.

Most multicellular animals have specialized structures that detect, transmit, coordinate, and respond to stimuli. Information about the stimulus is transmitted by nerve impulses in sensory neurons. The central nervous system coordinates the stimulus with an appropriate response. The response occurs through action of effectors (muscles and glands). The skeletal and muscular systems are responsible for movement, as well as support of the organism. The ultimate result of coordination is survival—maintaining normal homeostatic conditions for the life of cells.

Guide Questions and Problems

1. What basic function do receptors have? How might this function be related to the fact that most animals have a head end? What relationship do you think might exist between the kinds of receptors an animal has and where it lives?
2. What are the major parts of a neuron in a human? Describe their functions.
3. What evidence supports the idea that nerve cell impulses stimulate other nerve cells, muscles, and glands by chemical means?
4. What functional differences are there among sensory, motor, and associative neurons?
5. Suppose you have just placed your hand on a hot stove. Describe the events that would follow and relate them to your nervous system.
6. If all neurons transmit nerve impulses in the same way, how can light and sound result in such distinctly different sensations?
7. Suppose you wanted to know more about the function of a particular gland suspected to be an endocrine gland. What might you do in an investigation into its function?
8. Eating a lot of candy can cause a temporary increase in blood glucose concentration. What control mechanisms bring the blood glucose concentration back to its normal value?
9. The photographs at the beginning of this chapter show people engaged in some very challenging activities. Explain how these activities involve coordination of the nervous, endocrine, skeletal, and muscular systems.

Ideas for Independent Study

1. Nerve and gland activity of animals is often rhythmic. The cycles may last for seconds or fractions of seconds, for days or parts of days, or months or years. Have you observed any cycles of activity in yourself? Record some of your body functions (such as heart rate, breathing rate, temperature, hunger, alertness, etc.) for several days. Then try to detect evidence of your own biological rhythms.

2. Our understanding of the world is completely dependent on our senses. Probably no two people have quite the same sensory impressions in the same environmental conditions. Investigate some of the following:

(a) Young children can hear sound of more shrill quality (shorter wavelength) than can teenagers and adults. Can you recall any incidents from your own experience that could be accounted for by the changing patterns of sensitivity of your hearing?

(b) What would the world seem like if you were able to "see" or "hear" radio waves?

(c) Astronauts have reported that they "see" flashes of light even with their eyes closed. What could account for this?

(d) Some people believe that a few gifted individuals can communicate through the use of "extrasensory perception" using senses other than taste, smell, sight, temperature, hearing, touch, sense of equilibrium, and sense of body position. What evidence exists for extrasensory perception? Design experiments that might test whether or not it exists.

3. Do you think we know everything that we need to know about coordination? Actually, the more we learn, the more unsolved problems we encounter. You may want to research how the hormone calcitonin was discovered. Although scientists have been studying this hormone, many questions are still unanswered. The article "Calcitonin," by H. Rasmussen and M. M. Pechet, *Scientific American,* October 1970, page 42, will help you.

Related Readings

BOOKS

Klemm, W. R. *Science, the Brain, and Our Future.* Indianapolis: Pegasus, 1972.

Le Baron, R. *Hormones: A Delicate Balance.* Indianapolis: Pegasus, 1972.

Silverstein, A. and V. B. Silverstein. *Epilepsy.* Philadelphia: Lippincott, 1975.

Smythe, R. H. *Vision in the Animal World.* New York: St. Martin's, 1975.

Stevens, S. S. and F. Warshofsky. *Sound and Hearing.* New York: Time-Life, 1969.

Wilson, J. R. *The Mind.* New York: Time-Life, 1969.

ARTICLES

Favreau, O. E. and M. C. Corballis. "Negative Aftereffects in Visual Perception." *Scientific American,* December 1976, page 42.

Heimer, L. "Pathways in the Brain." *Scientific American,* July 1971, page 48.

Lester, H. A. "The Response to Acetyl
Scientific American, February 1977, page

McEwen, B. S. "Interactions betwe
mones and Nerve Tissue." *Scientific A*
July 1976, page 48.

19 Animal Structure and Function: Reproduction

Every species of organism can reproduce new individuals of that species. In multicellular animals, methods of reproduction are varied and some are quite complex. In one respect, reproduction differs from all other functions of animals—it is not necessary for the well-being of the individual. An individual cannot survive after a respiratory or digestive system has been removed. But any reproductive organ—even the whole system—can be removed. With proper treatment, the individual can continue in good health after such an operation.

Although reproduction contributes nothing to the survival of the individual, it is very important to the survival of a species or a population. If the reproductive systems in *all* members of a species were to stop functioning, the existing individuals would live out their lives and die. The species would die with them. Reproduction, therefore, is the mechanism that produces new generations and maintains a species.

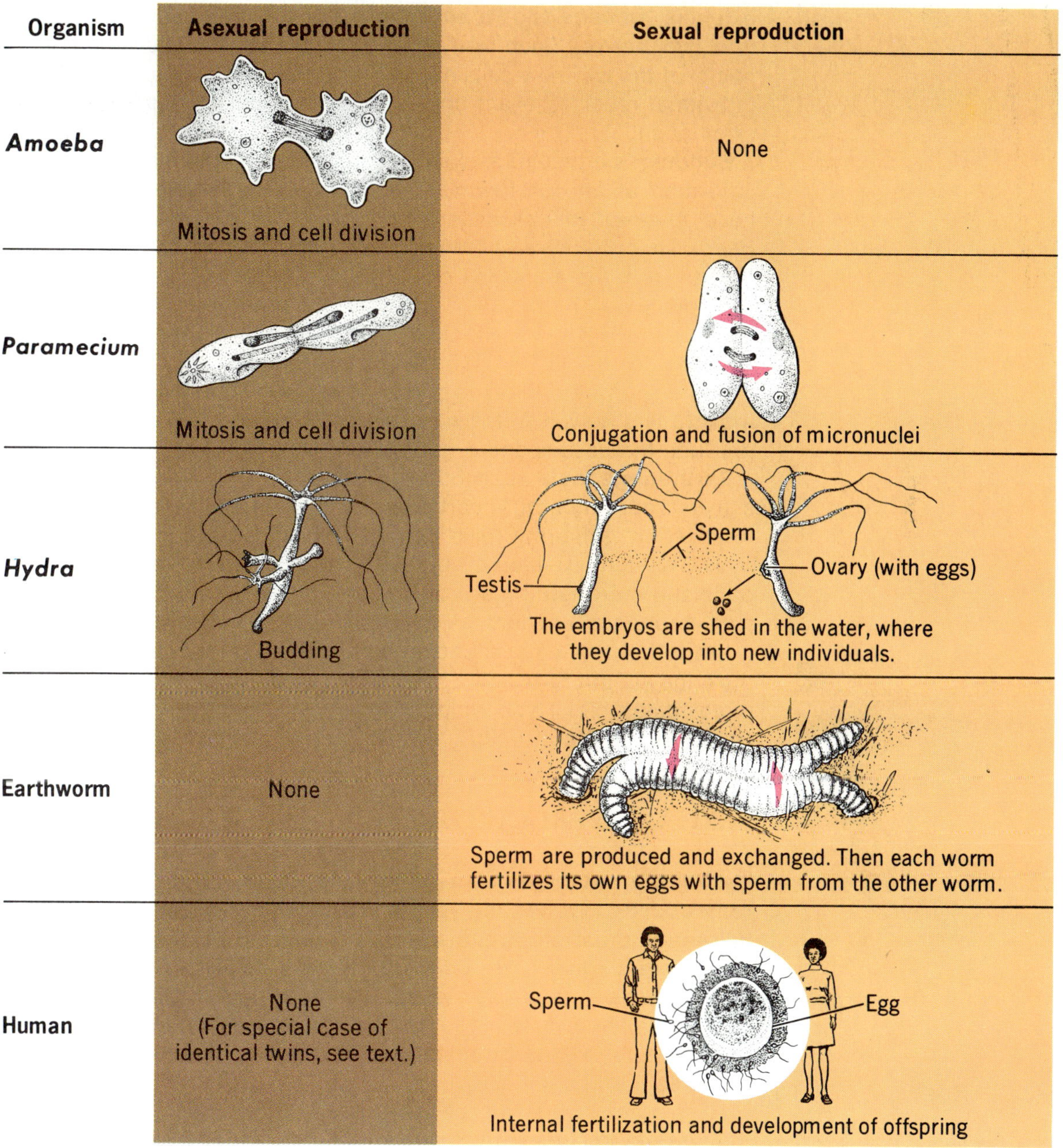

19-1. Examples of asexual and sexual mechanisms of reproduction.

ASEXUAL REPRODUCTION

In many animals, reproduction is asexual (Figure 19-1). **Asexual reproduction** involves neither meiosis nor the fusion of reproductive cells. There is only one parent involved in asexual reproduction. For example, *Amoeba* and *Paramecium* reproduce asexually by dividing into two halves. Each half then becomes a new individual. The nuclei of the two new cells divide by mitosis, not meiosis. Each new individual receives exact copies of all the genetic material of the original cell.

Budding is a type of asexual reproduction in which a new animal is formed as an outgrowth, or bud, on an older animal. You may have observed this process of budding in your laboratory study of *Hydra*. In coelenterates such as *Hydra*, some cells form an outgrowth from the animal's body. This bud eventually separates as a new individual (Figure 19-1). In other coelenterates such as corals, the new buds remain attached. The individual corals thus form a colony.

Some worms such as the planarian reproduce asexually by dividing into two halves. Mitosis and differentiation of cells in each half produce the tissues and organs of the missing half.

Parthenogenesis (par·thuh·noh·JEN·uh·sis) is another type of asexual reproduction. In parthenogenesis, the egg develops without fertilization. Offspring form from diploid eggs produced by the female. The eggs develop by mitosis rather than by meiosis. Thus, the eggs and the offspring are diploid. In species in which this type of parthenogenesis occurs, all the individuals are females.

In another type of parthenogenesis, meiosis does occur, but again the eggs are not fertilized. The unfertilized eggs develop into monoploid males. This type of parthenogenesis occurs in bees, ants, wasps, and other insects.

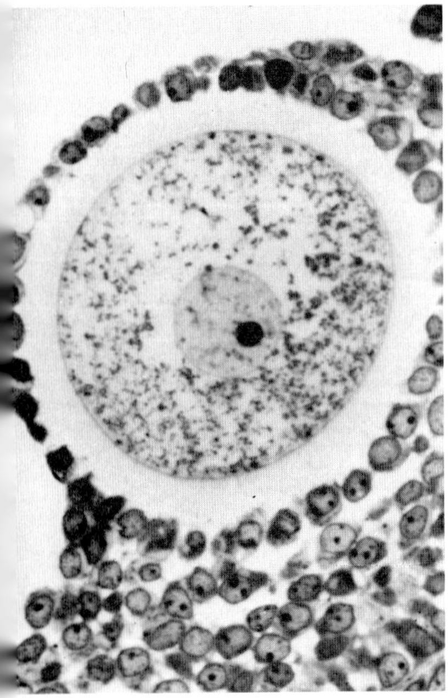

19-2. A human ovum as seen with a compound microscope.

SEXUAL REPRODUCTION

Sexual reproduction occurs in many multicellular animals. Sexual reproduction involves meiosis, which produces nuclei that contain half the number of chromosomes that were present in the cells of each parent. During meiosis, the nucleus of a cell that has the diploid number of chromosomes divides (Chapter 4). The result is two cells that have the monoploid number of chromosomes. Reproductive cells with a monoploid chromosome number are called **gametes** (GAM·eets).

Most animals that reproduce sexually have specialized organs, the **gonads** (GO·nadz), which produce two kinds of gametes. The ovaries, in females, produce eggs, or **ova** (OH·vuh; singular, ovum). The testes, in males, produce **sperm**. (In humans, the testes are sometimes called testicles.) Fertilization occurs when a monoploid sperm and monoploid ovum unite and form a diploid cell—the zygote. The zygote has the same diploid number of chromosomes as other members of its species.

Sexual reproduction insures a variety of genetic types in the population. Each parent produces gametes that contain a variety of genes. Ova and sperm combine at fertilization, producing many variations in the offspring. Some of these variations increase the chance that the offspring will survive and reproduce. Individuals that are less fit may not survive long enough to reproduce. Thus, sexual reproduction is of great importance in evolution.

Mechanisms of sexual reproduction are as variable as the organisms themselves (Figure 19-1). For example, *Paramecium* sometimes undergoes a process called conjugation (con·jew·GAY·shun). Certain species of *Paramecium* have, in addition to a large macronucleus, one or more tiny micronuclei. Under certain conditions, two *Paramecium* come together and attach at the area of the oral groove. The micronuclei undergo meiosis. All but two of the resulting nuclei disintegrate. Usually, the macronucleus also disintegrates. Each individual now contains two nuclei. The two individuals exchange one micronucleus. In each individual, the exchanged micronucleus fuses with the one that remained, producing a new diploid nucleus in each individual. This nucleus then divides, and one of the new nuclei develops into a macronucleus. Because genetic material is exchanged between two *Paramecium,* we say that sexual reproduction has occurred.

Like *Paramecium, Hydra* can reproduce either sexually or asexually, but the mechanisms are different. Occasionally, *Hydra* stops budding. Certain body cells then differentiate into ovaries or testes. These organs may develop on different individuals. Sometimes a single individual will develop both ovaries and testes. Sperm are released into the water. They enter the ovaries and fertilize the ova. Later, the embryos are released into the water, where they develop into adult *Hydra*.

Earthworms have both male and female reproductive organs and form both sperm and ova. But they never self-fertilize. When mating occurs sperm are exchanged between two individuals.

Although reproduction in humans is sexual, the formation of identical twins has some similarities with asexual reproduction (Figure 19-1). Identical twins develop from one zygote. At some time during development, the dividing cells separate into two masses. Each mass continues to develop separately, producing two embryos. Identical triplets, quadruplets, and quintuplets form in a similar way. In fraternal twins, two different eggs are fertilized by two different sperm. The twins that result are no more alike than brothers and sisters born at different times.

In humans and many other animals, fertilization and development occur inside the body of the female. Gametes from the two sexes are quite different. The ovum is usually round or oval and has no means of independent movement (Figure 19-2). Ova of different species

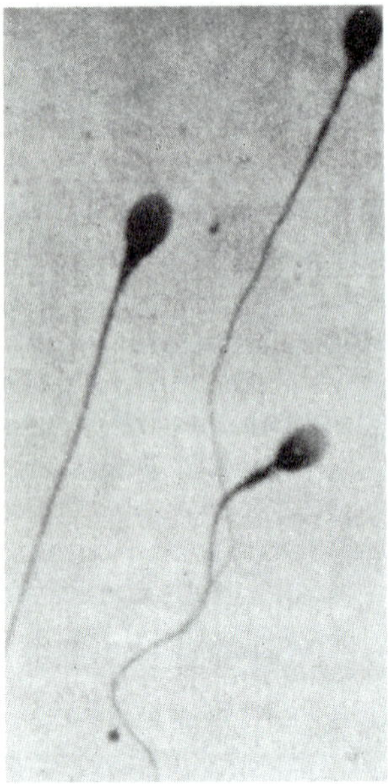

19-3. These three sperm cells have been enlarged several times more than the ovum in Figure 19-2.

19-4. Secondary sex characteristics in four species. *Describe how the females differ from the males of each species.*

contain varying amounts of yolk. The yolk is food for the developing zygote. Sperm generally are very small. In most cases, a sperm has a long, slender flagellum (Figure 19-3). The beating of the flagellum moves the sperm.

Males and females are identified by the type of gametes they produce—sperm by a male, ova by a female. Testes and ovaries are called primary sex characteristics. In many species, other differences occur between males and females. These are known as secondary sex characteristics (Figure 19-4).

Frequently, secondary sex characteristics develop in response to hormones produced by the testes or ovaries. You can investigate the effects of one such hormone on developing chicks in **Inquiry 19-1.** You will be able to observe how the male sex hormone, testosterone, causes development of secondary sex characteristics.

INQUIRY 19-1 Effects of Reproductive Hormones

The male hormone, testosterone, is produced in the testes. It affects development of the testes, the primary sex characteristics. It also affects development of secondary sex characteristics. Some examples of secondary characteristics in male chickens are the color of feathers, comb development, and behavior patterns. Is it possible to observe these effects after administering testosterone to young male chickens?

MATERIALS (per team)

4 or more 1-day-old male chicks
housing and feed for chicks
alcohol solution of 4 different dyes
oil
testosterone (in oil), 5 ml each of 2 concentrations:
10 mg per ml of oil
50 mg per ml of oil
70% alcohol solution
hypodermic syringe, 1-ml
hypodermic needle, No. 25
millimeter ruler
balance
cotton swabs
cotton

Procedure (Day 1)

1. Predict how you think the young chicks will look and act after testosterone is injected.

Conduct this study carefully, so you will not harm the chicks. Divide the class into at least two teams, A and B. These teams will exchange data at the end of each day.

Divide your team's chicks into four groups (Groups I, II, III, and IV), with one or more birds in each group. Mark each chick by carefully swabbing some dye on the top of its head or on the back between the wings. Use four different colors. Key the colors to the treatments you will use.

For Group I, you will inject testosterone subcutaneously (under the skin). For Group II, you will apply the hormone on the comb. Group III, a control, will receive an injection of plain oil. For Group IV, also a control, you will apply plain oil on the comb.

Team A will use the 10-mg testosterone solution. Team B will use the 50-mg solution. Both teams will measure the effects of the treatment.

Note: In order to use as few chicks as possible, your teacher may choose to share one experimental setup among all classes. If this is the case, another class may inject the chicks. But all students will be able to collect and analyze the data.

Comb development can be determined by taking three measurements: the length of the comb, its height at the highest point, and its greatest width at the base. For this last measurement, hold the ruler directly above the comb and look down at each side of the base for an estimate. Multiply the three measurements by one another to obtain the comb index: $L \times W \times H$. The comb index is about equal to the volume of the tissue in the comb.

Determine the comb index for all chicks before treatment, including the controls. For each group, add the individual indexes. Then divide by the number of birds in the group. Record this *average index* on a chart similar to the one shown in Figure 19-5.

19-5. Sample table for recording data

Team ______ *Concentration of Testosterone Used* ______________

	Group I (injection)	**Group II** (on the comb)	**Group III** (control)	**Group IV** (control)
Day 1 Average weight				
Observations of behavior				
AVERAGE COMB INDEX				
Day 2 Average weight				
Observations of behavior				
Day 3 Average weight				
Observations of behavior				
Day 4 Average weight				
Observations of behavior				
Day 5 Average weight				
Observations of behavior				
AVERAGE COMB INDEX				

Group I Chicks. Inject 0.1 ml of testosterone solution into each chick. Make the injections subcutaneously near the base of the neck or in the crop region at the base of the front of the neck. Clean this area and the hypodermic needle with alcohol before making the injection. Massage the area gently after the injection. This will distribute the fluid under the skin. Repeat the injection for five consecutive days. Find the mass in grams of each chick daily.

Group II Chicks. Using a cotton swab, gently rub the comb of each chick with 0.1 ml of the testosterone solution. Try not to let the solution run onto the feathers. One person can hold the chick while another applies the solution. Repeat the treatment for five consecutive days. Find the mass in grams of each chick daily.

Groups III and IV Chicks. These groups are controls. Treat Group III exactly as you did Group I, but use plain oil instead of oil that contains testosterone. Treat Group IV as you did Group II, but for this group use plain oil.

Each day, watch the behavior of the four groups of birds. Look for fighting, crowing, aggression, or dominance. Record your observations and describe the behavior of the chicks.

After five days of treatment, compute the comb index for all birds, including the controls. Record all your data. Complete your data sheet by exchanging records with the other team. You also may wish to pool data with other classes to increase the sample size. **2.** Why were there two controls? **3.** Suppose a member of your team proposed that chicks that received no treatment at all could serve as controls. What would be wrong with such a suggestion? **4.** From your data, describe the effects of different concentrations of testosterone on comb development when applied subcutaneously, and when applied directly to the comb. **5.** If injections and external applications were carried on indefinitely, which chicks do you think would be the first to crow? **6.** Why were young chicks used instead of roosters? **7.** Was the mass of the chicks affected by the hormone injection? **8.** Capons are roosters that had their testes removed early in life. What do you think would be the effect of injecting testosterone into these older birds? Suggest an experiment to test your hypothesis. **9.** Summarize the effects of testosterone on the chicks' behavior and sexual development.

Fertilization

Sperm are motile cells. They are propelled through a liquid environment by their flagella. In some species of animals that live in water, fertilization occurs when ova and sperm are released into the water. This external fertilization occurs in such animals as frogs and many kinds of fish. After fertilization, the embryos develop in the water.

Internal fertilization occurs in many other species. The sperm fertilize the ova within the female's body. This method occurs in some aquatic animals—many crustaceans and species of fish, including guppies and some sharks. In land animals, internal fertilization is very important. Sperm can survive only in a fluid. The liquid environment within the males, where the sperm form, and in the females, where they swim, increases the chances for fertilization and survival of the zygote.

Also important in internal fertilization is timing. In most cases, the sperm and ova live only a short time. The ova must reach the proper stage of development and be released from the ovaries at about the same time that sperm are present in the female system. In many mammals, timing of fertilization is controlled by hormones.

In many insects, sperm are deposited in the reproductive system of the female. They are stored in a special pouch and are not released until the ova are laid. In bees, for example, sperm may be stored for several years.

Development of the Embryo

After fertilization occurs, the developing embryo is fed and protected in a suitable environment. The ova of many organisms contain yolk, which nourishes the developing embryos. In many cases, the proper temperature for development is assured because ova are produced and fertilized only in a certain season of the year. In other cases, such as that of birds, the developing offspring are kept warm and protected by the parent or parents.

Many fish and amphibians, such as frogs, release sperm and ova into the water. Each ovum is surrounded by layers of jelly, which the sperm can penetrate. These layers will protect the developing embryo. The ovum also contains enough yolk for the embryo's early development. Oxygen diffuses into the eggs from the surrounding water. Waste materials diffuse out of the eggs and are carried away by the water.

In birds and many reptiles, a different kind of reproductive system has evolved. Fertilization is internal. Then a protective shell forms around the ovum. The shell keeps the embryo from being crushed and protects it from drying out. The fertilized, protected ovum is deposited on land. Membranes inside the shell aid in respiration, provide a liquid environment, store nitrogenous wastes, and nourish the embryo.

In some animals with internal fertilization, the embryo develops within the body of the female. In some of these animals, the embryos develop within embryonic membranes, using the yolk that was contained in the ova. This kind of internal development occurs in some sharks and reptiles. In humans and other mammals, however, the mother's body provides nourishment to the embryo.

REPRODUCTION IN MAMMALS

In mammals, the group of animals to which we belong, the females have glands that secrete milk. These are called **mammary glands.** The milk feeds the young after birth, before they can obtain food on their own. This class of animals was named Mammalia because of this characteristic.

The duckbill platypus of Australia is a member of the most primitive group of living mammals. In these mammals, a shell forms around a fertilized ovum. The ova are deposited in a nest and incubated, or kept warm. When the embryos hatch, glands on the female's underside secrete milk. The young lick the milk off the fur.

The American opossum and many mammals of Australia, such as the kangaroo, belong to another group of mammals, the marsupials. In marsupials, fertilization and early development occur within the female's body. The small ova have only enough yolk to provide for the development of very immature young. The tiny, almost helpless offspring emerge from the female's body into a pouch on her abdomen. Here, they are sheltered and kept warm while their development continues. They also find food, for the female's mammary glands open into the pouch. The offspring live on the female's milk until their development is complete.

Humans belong to the group of placental mammals. This group is named for the placenta. The placenta is a specialized organ that develops within the mother's body and provides food, oxygen, and waste removal for the developing embryo. Fertilization is internal. Early development of the young occurs in the female's body. We will study human reproduction as an example of reproduction in placental mammals.

The process begins, as with other animals that reproduce sexually, with the sperm of the male and the ova of the female. In the male human, the testes are called testicles (Figure 19-6). The two testicles are located in an outpocketing of the body wall called the **scrotum** (SKROH·tum). Sperm are produced in large numbers (hundreds of millions) in the testicles. As the sperm pass through the sperm duct, fluid is added to them forming **semen** (SEE·muhn). The fluid is manufactured by the seminal vesicles, the prostate gland, and the Cowper's glands. The development of the testicles is regulated by male sex hormones. Usually during the early teens, the male's voice becomes lower in pitch and hair grows on the face. These events are controlled by male sex hormones secreted by the

SCIENCE AT WORK: Teamwork

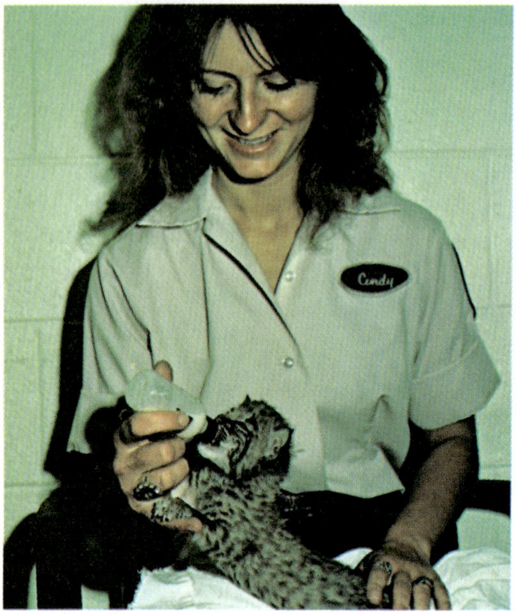

To many people, zoos are strictly places of entertainment. They visit zoos to laugh at the antics of the monkeys, to be terrified (safely) by the roaring of lions and tigers, or to exchange stares with the massive gorilla.

Not many people recognize that zoos also are places where various types of animal research take place. One such area of research is breeding in captivity—the controlled breeding of animals that normally breed only in the wild.

One of the major hazards to animals living in the wild is the destruction of their habitats. In the future, we can expect that there will be even more destruction. And when habitats are destroyed, species of animals that are so totally dependent on those habitats may become extinct. Captive breeding in zoos may help to preserve these species of animals.

Captive breeding involves the time and talents of many people working as a team. Among them are zoologists who have studied the physiology, particularly the reproductive physiology, of the animals involved. Animal keepers maintain an environment and provide food that will encourage breeding and support a pregnant female. Veterinarians at the zoos understand and can treat medical problems that might develop during the pregnancy and birth processes. Human nurses care for the newborn if the mother cannot.

This team approach to a biological problem is characteristic of the way scientists work. Rarely, if ever, are problem-solving efforts the work of one individual. Frequently, one or two people seem to get all the credit for developing an important scientific concept. But these individuals are usually the first to say that they were only a part of a larger team and that all team members played an important role in the successful outcome of their research.

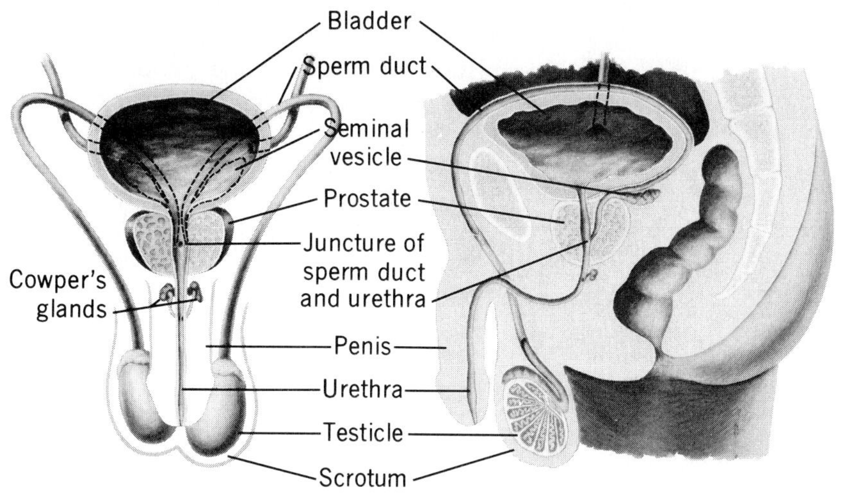

19-6. Simplified diagrams of the reproductive system of a human male. Left is a cut-away view from the front. On the right is a side view, as if cut along a midline between the left and right sides of the body.

testicles. When these events occur, the production of sperm by the testicles has begun.

In the female, the two ovaries are located on either side of the female's body (Figure 19-7). Ova develop in tiny structures in the ovary called **follicles** (FOL·ih·kulz). As each follicle grows, it develops a cavity that is filled with fluid. Each follicle contains a single ovum. When the ovum matures, the follicle breaks open (Figure 19-8). After the egg is released, the follicle fills up with cells, and a mass of blood capillaries forms. This structure is called the **corpus**

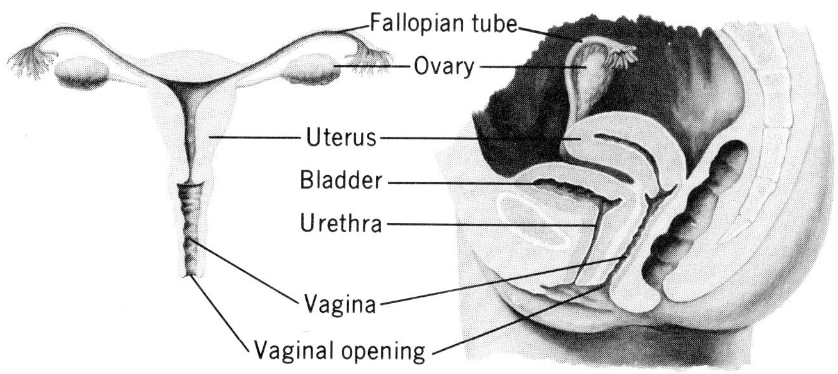

19-7. Simplified diagrams of the reproductive system of a human female. The views are the same as those used in Figure 19-6.

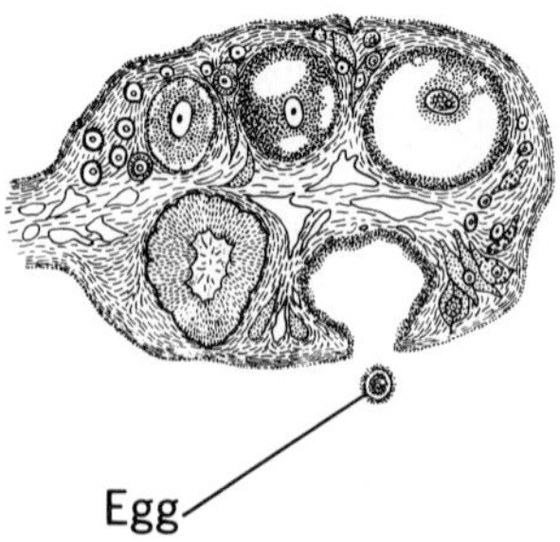

19-8. A section of an ovary showing the maturation and release of an ovum and the formation of the corpus luteum.

luteum (KOR·pus loo·tee·um; Latin, meaning yellow body). The corpus luteum secretes female sex hormones. The tiny ovum, which is about the size of the period at the end of this sentence, bursts from the ovary. This release of the ovum is called **ovulation** (ah·vyoo·LAY·shun). Generally, the ovum enters the **Fallopian tube,** a duct that extends from the ovary to the muscular, thick-walled **uterus** (YOU·ter·us).

The development of secondary sex characteristics in females is also regulated by sex hormones. Between the ages of 10 and 16, the shape of the body changes and the breasts enlarge. These and other changes usually precede the onset of **menstruation** (men·stroo·AY·shun). Menstruation is a release of blood, mucus, and cellular debris formed by the breakdown of the soft tissues lining the uterus. These pass to the outside through the **vagina** (vuh·JY·nuh), the canal that leads from the uterus to the outside of the body. Once menstruation has begun, the female is producing ova and is capable of having a child.

Pregnancy

When sperm from the male are deposited in the vagina, they swim through the uterus and into one of the two Fallopian tubes. Contractions of the vagina and the uterus also move the sperm to the tube where fertilization usually occurs. The nucleus of one sperm enters the ovum and fuses with its nucleus. If no sperm reach the ripe ovum, fertilization does not occur, and the ovum disintegrates.

Mitosis and cell division begin as the fertilized ovum moves down the Fallopian tube. The embryo that develops attaches to the soft tissues of the uterus. The inner wall of the uterus is now thick, soft, and moist with fluid. Its blood supply is greatly increased. Once attached, the embryo sinks into the soft, inner uterine wall.

Certain cells of the embryo then develop into membranes that help to nourish, protect, and support it. Figure 19-9 shows these membranes. The chorion, the outermost membrane, forms first. Tiny, fingerlike projections grow from its surface into the soft tissues of the uterus. Gradually, small pools of rapidly moving blood form around these projections. The tissues of the chorion and the adjacent parts of the uterine tissue make up the placenta.

Another embryonic membrane, the amnion, grows around the embryo itself. The cavity within the amnion becomes filled with fluid. This fluid-filled cavity keeps the embryo moist and protects it from bumps and jolts. Another membrane that forms is the allantois. It originates from the digestive canal of the embryo.

The **umbilical** (um·BIL·ih·kul) **cord** connects the embryo and the placenta. The umbilical cord contains blood vessels that carry

19-9. The membranes that surround the developing embryo.

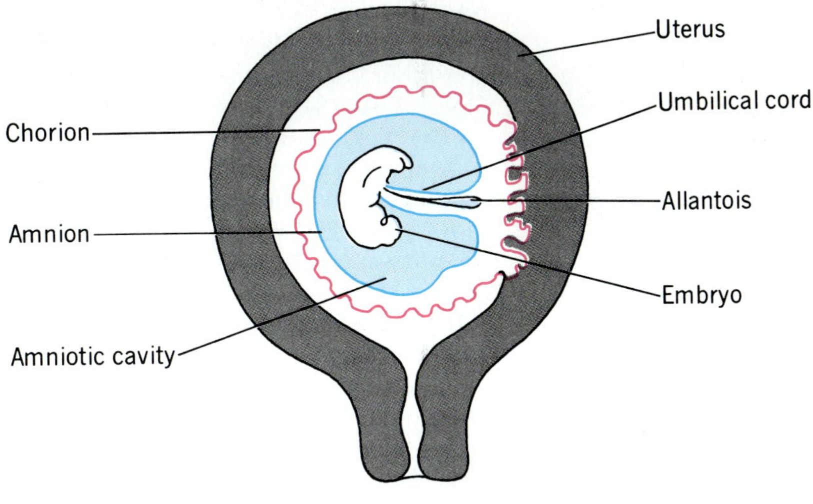

nourishment to the embryo and remove the embryo's wastes. Normally, there is no direct blood flow between the mother and embryo. Their separate blood vessels come close together, however, in the placenta (Figure 19-10). Oxygen and nutrients diffuse from the mother's blood, across the membranes of the placenta, into the em-

19-10. Materials are exchanged between mother and embryo across the placenta.

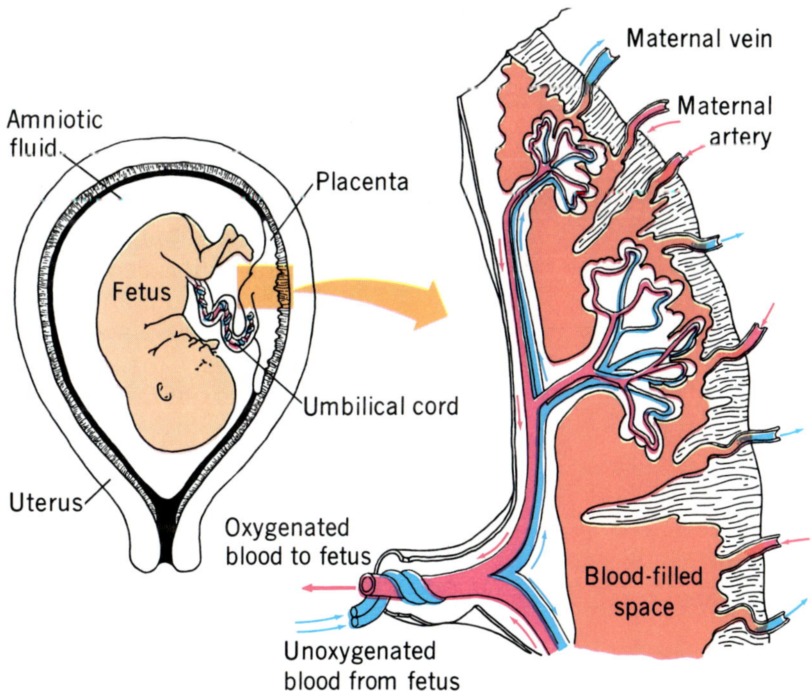

bryo's blood. Alcohol and drugs taken by the mother also can cross the placenta and enter the embryo's blood. Carbon dioxide and other wastes diffuse from the embryo's blood into the mother's blood. These waste products are carried to the mother's lungs and kidneys for removal.

Pregnancy in the human female lasts, on the average, nine months, or 280 days. It may vary in individual women, however, by as much as two or more weeks from this average. The average length of pregnancy also varies by species. It is about 63 days for the domestic cat and dog, 330 days for the horse, 280 days for the cow, and 20 to 22 days for the rat and mouse.

Birth

As pregnancy progresses, the fetus, which is the name we give the embryo after the twelfth week of development, grows. As this happens, the uterus increases in diameter. Usually, at about the ninth month after fertilization, the fetus turns so that its head is toward the opening of the uterus into the vagina.

We still do not know exactly how the birth mechanism is triggered. But changing concentrations of hormones in the blood and the changing condition in the muscles of the uterine wall are thought to be factors. Childbirth begins when the muscles of the uterus repeatedly contract and then relax (Figure 19-11). This series of contractions is called *labor*. At first, muscular activity of the uterus is just strong enough to move the baby slowly toward the vagina. At this stage, the amnion usually breaks, and its fluid contents are released.

As the contractions of the muscles become stronger and more frequent, the baby is pushed through the vagina and into the outer world. The umbilical cord is tied off and cut. (The small piece of cord remaining attached to the baby shrivels and falls off within a few days. The navel marks the place where it once entered the body.) After the birth of the baby, the muscular contractions continue until they push out the placenta, which is commonly called the "afterbirth."

During the last part of pregnancy, a watery, lymphlike fluid called colostrum (koh·LOS·trum) collects in the mammary glands. For the first few days after the baby is born, the mammary glands secrete only colostrum. This fluid is high in protein and contains many antibodies that protect the baby against infections. After a few days, milk, which is rich in food materials, is secreted. If the baby does not feed from the mammary glands of its mother, the secretion of milk soon stops.

19-11. Childbirth **(a)** Position before birth.

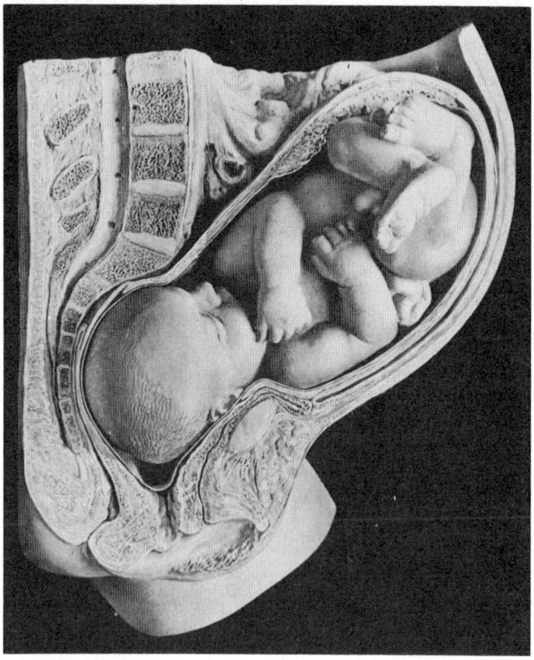

(b) Birth is beginning.

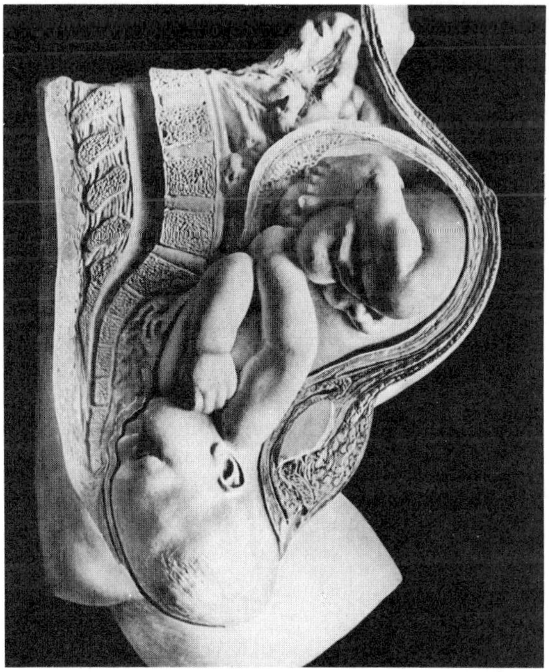

(c) The head emerges first.

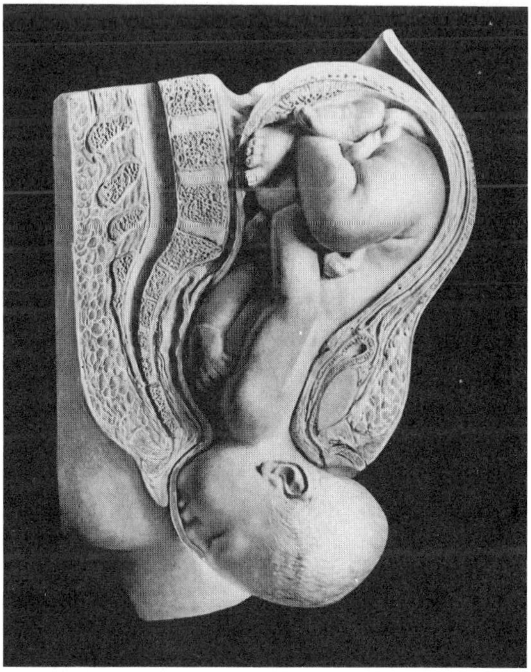

(d) The body and placenta follow soon after.

REPRODUCTIVE CYCLES

The periodic repeating of changes in structures and function of the female reproductive system is a complex cycle. The cycle has been studied in laboratory animals. A female laboratory rat repeatedly goes through a short period of increased desire to mate, which has been called **estrus** (ES·trus), or heat. Ovulation generally occurs during this period of estrus. The whole pattern of changes, occurring from one period of estrus to the next, is called the **estrous cycle.** Rats, mice, dogs, sheep, and many other mammals also have estrous cycles. These cycles vary in length. The laboratory rat has an average estrous cycle of five days. The period of estrus occurs only twice a year in female dogs.

The reproductive cycle of the human female (and of some other primates) is called the **menstrual** (MEN·stroo·al) **cycle.** The human menstrual cycle takes about 28 days, although much variation occurs among individuals. Ovulation usually takes place about 14 days after the beginning of the last menstrual period. This also varies. As the cycle begins, several egg follicles enlarge in an ovary, although only one usually matures. If more than one egg matures, the result is fraternal twins or multiple births. Hormones, called estrogens (ES·tro·jens), are released from the follicle. These hormones stimulate growth of the soft tissues of the uterus. They increase cell division and blood supply in the uterus and cause tissue fluid to accumulate.

After ovulation, the corpus luteum is formed, estrogens and another hormone, progesterone (proh·JES·tur·ohn), act on the uterus. Progesterone stimulates the thickening of the uterine wall and the development of glands and blood vessels. The soft tissues soon form a spongy layer suitable for the attachment of a fertilized ovum. These changes in the uterus must take place before a fertilized ovum can become attached and develop normally.

If fertilization and attachment of the embryo do not occur, menstruation will start. Normally, the onset of menstruation is closely associated with a decrease in the production of estrogens and progesterone. After menstruation, the cycle begins again.

In the United States, human females usually start to menstruate in their early teens, although there is great variation. Generally sometime during a woman's forties or fifties, the menstrual cycle ceases. This is referred to as the **menopause** (MEN·oh·pauz).

The Role of the Ovaries

The ovaries are responsible for timing the cycle of development and the release of the inner wall of the uterus during menstruation. This was shown in experiments with young female rats. The reproductive cycle had not started in these rats.

Experiment 1. The rat was anesthetized. Its body wall was opened, and the ovaries were removed. The rat recovered completely, but never had a reproductive cycle. What hypothesis could explain these results?

Experiment 2. The body wall of this rat was opened and closed, but the ovaries were not removed. The reproductive cycle began at the same age as in rats that had not had the operation. How does this observation help test the hypothesis you made based on observations in Experiment 1? What is such an experiment called?

Experiment 3. A rat was operated on exactly as in Experiment 1 and was allowed to recover completely. The reproductive cycle did not occur. The animal then was injected with fluid containing substances from the ovaries of mature rats. The reproductive cycle began. What hypothesis could account for these observations?

Experiment 4. The same procedures were carried out on this rat as on the rat in Experiment 3, except that the injected fluid was a salt solution. The rat never had a reproductive cycle. Why is this experiment important in testing hypotheses from Experiment 3?

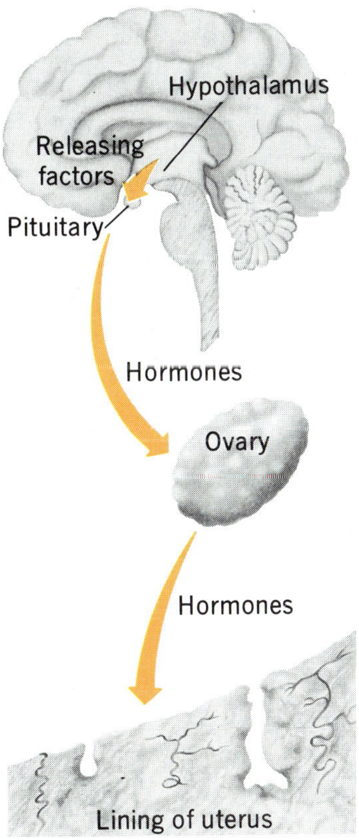

19-12. Hormones from the hypothalamus, pituitary gland, and ovary control the female reproductive cycle.

As a result of further experiments, we now can be reasonably sure that changes in the reproductive cycle result from the actions of chemicals produced by the ovaries. For example, the cells of the follicle in which the ovum matures secrete estrogens. They are responsible for the thickening and growth of blood vessels of the uterine wall. After the follicle bursts and releases the ovum, the corpus luteum forms. The corpus luteum secretes estrogens and progesterone into the blood, causing further development of the uterine wall. Since the ovaries secrete hormones, they can be considered endocrine glands.

The Role of Pituitary Hormones

Ovarian hormones regulate the uterus, but what regulates the ovaries? What controls the growth of an ovum and a follicle and the formation of a corpus luteum? What maintains the development of the uterus during pregnancy? We cannot answer these questions completely, but experiments have given us some clues.

Hormones released from the pituitary gland influence the ovaries. We also know that the pituitary gland is controlled by hormones called releasing factors. Releasing factors are secreted by the hypothalamus. They stimulate secretion of hormones from the anterior pituitary. These hormones influence the ovaries, which in turn secrete hormones that affect the uterus as shown in Figure 19-12.

Two hormones released from the pituitary affect the reproductive system of humans. One is called the follicle-stimulating hormone (FSH). This hormone stimulates the follicles to grow, but does not affect the uterus.

The second pituitary hormone is called luteinizing (LOO·tee·ih·ny·zing) hormone (LH). It brings about very little change in the ovaries if the follicles are small. If the follicles are well-developed, however, LH stimulates ovulation. Each follicle becomes a corpus luteum. LH also stimulates the corpus luteum to secrete mainly progesterone. Thus, the releasing factors from the hypothalamus begin a chain of secretions from the glands. This series of events causes development of the uterus.

But what regulates the secretion of releasing factors from the hypothalamus? They appear to be sensitive to concentrations of estrogens and progesterone in the blood. The hypothalamus is regulated by the hormones that it causes the pituitary to produce. You may recognize this as an example of feedback control. Figure 19-13 summarizes these feedback control mechanisms. Figure 19-14 shows in detail the changes that occur during the menstrual cycle.

19-13. Hormonal mechanisms controlling the menstrual cycle.

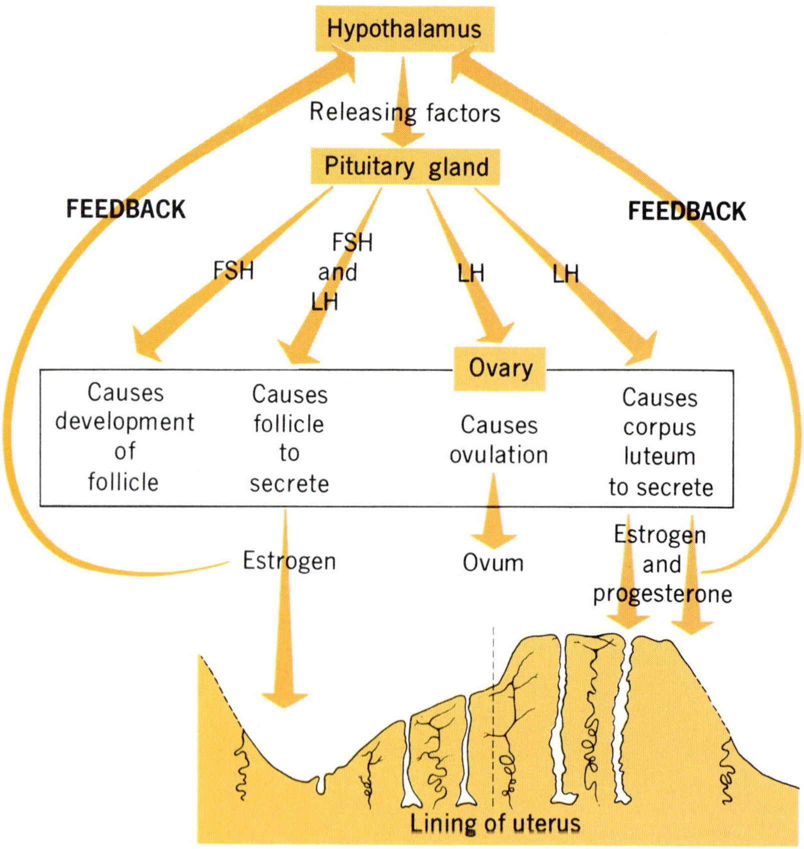

19-14. Hormonal control of the female reproductive cycle.

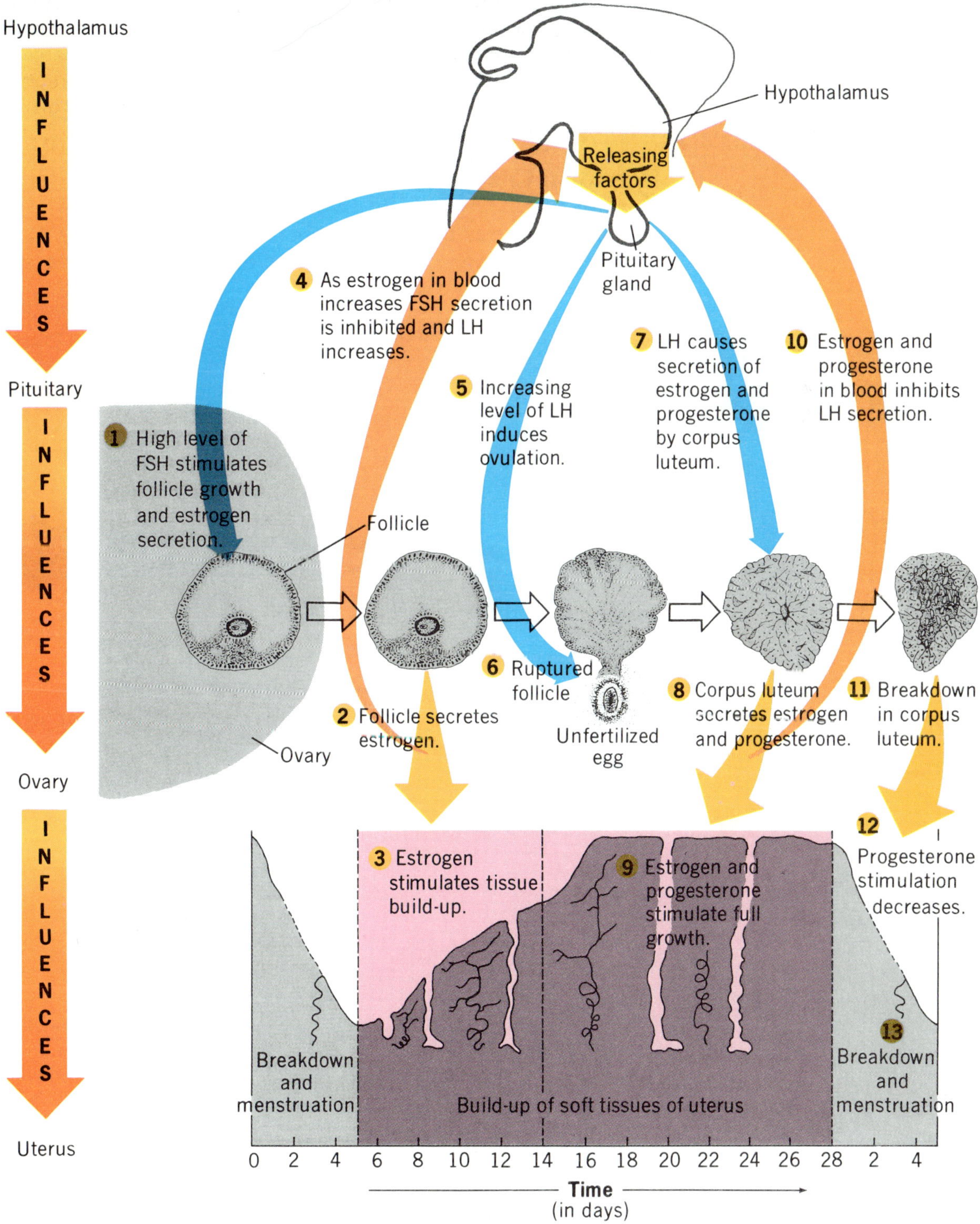

The same pituitary hormones that regulate female reproduction also control reproductive processes in males. In the males, LH and FSH control secretion of the sex hormones necessary for production of sperm and for the normal structure, activities, and behavior of the male. Thus, in both sexes reproductive function is precisely controlled, chiefly through the regulating role of the pituitary gland.

Hormonal Control by the Placenta

Figure 19-15 shows how the condition of the uterine wall is maintained during pregnancy. There is good evidence that the placenta

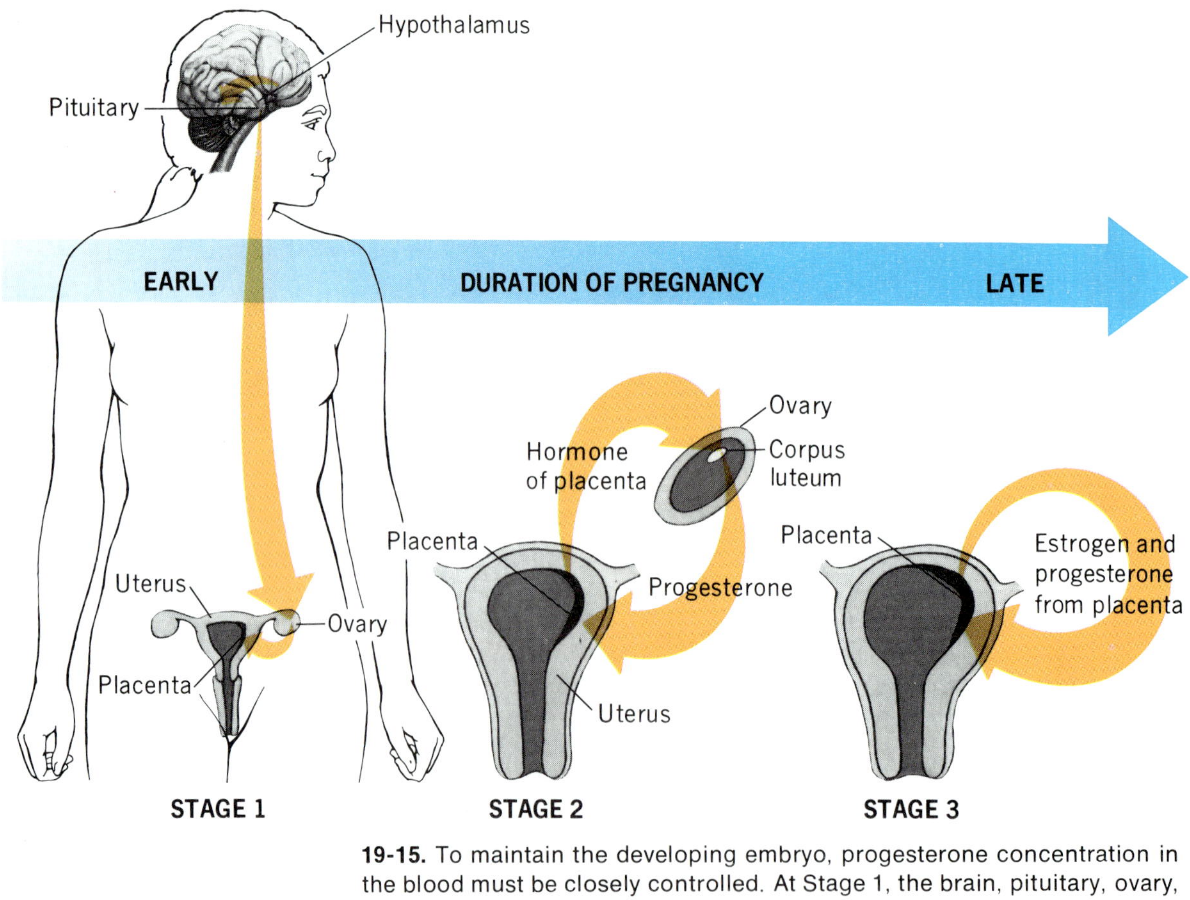

19-15. To maintain the developing embryo, progesterone concentration in the blood must be closely controlled. At Stage 1, the brain, pituitary, ovary, and placenta interact. At Stage 2, the placenta secretes a hormone stimulating the release of progesterone from the corpus luteum. At Stage 3, the placenta begins to secrete estrogen and progesterone, so that the presence of the ovary becomes unnecessary.

in some mammals (rats, mice, rabbits, monkeys, horses, humans, and a few others) secretes hormones. These hormones function somewhat like those of the pituitary. They differ, however, in structure and function and are not the same in all mammals. In the human embryo, the chorion produces a hormone that acts somewhat like LH. During pregnancy, this placental hormone stimulates the corpus luteum to secrete progesterone. The placenta itself, however, gradually develops the ability to secrete progesterone and estrogens.

If the ovaries of female rats, mice, opossums, or rabbits are removed during pregnancy (except, perhaps, at the very end), pregnancy stops. The young usually die. In other mammals such as guinea pigs, the ovaries can be removed in the latter part of pregnancy without serious effect. The pregnancy usually continues, and the young are born at the normal time. Apparently, the ability of the placenta to produce hormones similar to those of the ovaries and the pituitary has evolved only in certain mammals. Thus, when the ovaries are removed from humans, monkeys, and some other mammals, the placenta takes over and carries the pregnancy to completion. With the ovaries removed, however, the reproductive cycle will not occur again.

In Brief

Two main patterns of reproduction have evolved in animals—asexual and sexual reproduction. Asexual reproduction, in some species, is the process of one animal dividing and becoming two identical individuals. In other cases, a bud forms on the parent organism. The bud then separates and becomes a new individual. When parthenogenesis occurs, eggs are formed that develop—unfertilized—into diploid organisms. In another type of parthenogenesis, meiosis occurs when the eggs are formed. Not all the eggs are fertilized, but all develop. The unfertilized, monoploid eggs become males.

Sexual reproduction begins with the production of monoploid gametes—sperm and ova. In some species, both kinds of gametes are produced in the same individual. Fertilization may occur in the same individual, or it may involve the exchange of sperm between two animals. Occasionally, meiosis occurs in single-celled animals. When this happens, cell division results in two cells with monoploid nuclei. These cells combine and form a new diploid individual.

In the form of sexual reproduction that occurs in humans and many of the higher animals, sperm are formed in the males and ova are formed in the females. Fertilization and development of the offspring occur within the female's body.

Many systems have evolved in different types of animals that insure fertilization and successful development of the young. One of these is the reproductive cycle, which varies in different animals. In many animals, these cycles are closely regulated by hormones.

Guide Questions and Problems

1. Evolution is thought to occur more rapidly in animals and plants that reproduce sexually than in those that reproduce only by asexual methods. What is the biological basis for this idea?
2. As a general rule, ova are considerably different in form and size from sperm. How are these differences related to the role each plays in the reproductive process?
3. Adult female mammals produce only a few offspring. An oyster may give rise to millions of young oysters. To what can you relate this difference in reproductive capacity?
4. In rabbits, ovulation takes place only after mating. How does this pattern of ovulation compare with that in human females? Can you suggest what might control ovulation in the rabbit?
5. Suppose that a previously undetected gland is found in sheep. A biologist suspects the gland may be important in regulating the reproductive cycle in female sheep. What kind of experiments might shed light on this hypothesis?
6. In your study of the endocrine glands in Chapter 18, you learned that the pituitary gland produces thyroid-stimulating hormone (TSH). This stimulates the production of thyroxin by the thyroid gland. As the thyroxin collects in the blood, it inhibits the further production of TSH by the pituitary gland. The inhibition lessens as the thyroxin is used up in the body. What similar situation exists in the human reproductive cycle? What is such a mechanism called?

19-16. Expanding numbers of humans place demands on earth's finite resources.

Ideas for Independent Study

1. Different animal species have widely differing patterns of reproduction. Find out something about this range of reproductive processes. What correlation exists between environmental conditions and the chances for survival of offspring. What factors threaten the survival and development of the zygote in different environments? (Compare, for instance, warm ocean water with dry land in arctic or tropical regions.) Some unusual adaptations are found in mouth-breeding fish, penguins, and kangaroos. Compare these with the patterns in other species such as the primates.

2. Some species have remarkably effective ways of insuring the survival of offspring. Humans are outstandingly successful in this respect. Imagine a situation in which humans have a survival rate of 50 percent, with an average of four children for each pair of parents. What would be the number of living descendants of a single pair of parents after ten generations? Suppose that the survival rate is increased to 75 percent. How would this affect the number of living descendants ten generations later? What factors might increase the survival rate of young in any given species of mammal?

Related Readings

BOOKS

Goldstein, B. *Human Sexuality.* New York: McGraw-Hill, 1976.

Harrison, R. J. *Reproduction in Man.* New York: Norton, 1967.

Keller, D. E. *Sex and the Single Cell.* Indianapolis: Pegasus, 1972.

ARTICLES

Arehart-Treichel, J. "Human Reproduction and Aging." *Science News,* November 6, 1970, page 297.

Segal, S. "The Physiology of Human Reproduction." *Scientific American,* September 1974, page 52.

20 Photosynthesis and Plants

All living cells release energy, and all use energy in synthesizing new molecules. But in the long run, all living cells—except for a few bacteria—depend on the sun for their energy. Only certain specialized cells of green plants and a few bacteria, however, can trap solar energy in the synthesis of carbon compounds. This food-making process is called **photosynthesis.**

EARLY EXPERIMENTS ON PHOTOSYNTHESIS

In 1772, Joseph Priestley conducted some important experiments. In one of these, Priestley placed a shoot of a mint plant in a container of water. He inverted a glass jar over the shoot so that air could not enter. The mint plant remained alive for several months. In a similar experiment, a burning candle quickly went out when covered with a jar. After the candle had burned out, Priestley added a shoot of mint under the same jar. After a few days, the candle, when lighted, continued to burn for a short time. A mouse also survived under a jar with a mint plant. Apparently, the plant had restored something to the air that kept the flame and the mouse alive.

For several years, other investigators tried to verify Priestley's experiments without success. In 1779, however, Jan Ingenhousz discovered that plants restore the air only when exposed to light. Ingenhousz also learned that only the green tissues of plants are involved in this process. Soon the effects of Priestley's "restored air" were traced to the release of oxygen by plants. Next, scientists found that plants growing in light absorb carbon dioxide, which is used in the formation of carbon compounds. In 1804, Nicolas de Saussure showed that plant growth results from the intake of both carbon dioxide and water.

In 1845, Julius Robert Mayer identified the main events in photosynthesis as the absorption of light energy and the changing of light energy into chemical energy. The chemical energy is then stored in compounds that cells form from water and carbon dioxide. Oxygen is released into the air from the plant.

Chloroplasts

Botanists had observed that oxygen is produced only near certain microscopic structures in photosynthesizing cells. These structures are called **chloroplasts.** Chloroplasts can be seen with almost any microscope. But knowledge of the internal structure of a chloroplast has come from observations with the electron microscope (Figure 20-1). Many chloroplasts contain small, disk-shaped bodies called grana. Each of the grana is made up of a number of flat plates called lamellae. Lamellae are composed of layers of chlorophyll, protein, and lipid molecules. **Chlorophyll** is a green pigment that absorbs light energy. The layered structure allows maximum exposure of chlorophyll molecules to light. It also brings chlorophyll in contact with the protein layers, which contain the enzymes required for photosynthesis.

Chlorophylls

Five different kinds of chlorophylls have been identified as *a, b, c, d,* and *e.* Chlorophyll *a* is believed to be present in all photosynthetic plants. The others are thought to occur in various combinations in different plants, along with chlorophyll *a.* Chlorophyll molecules are

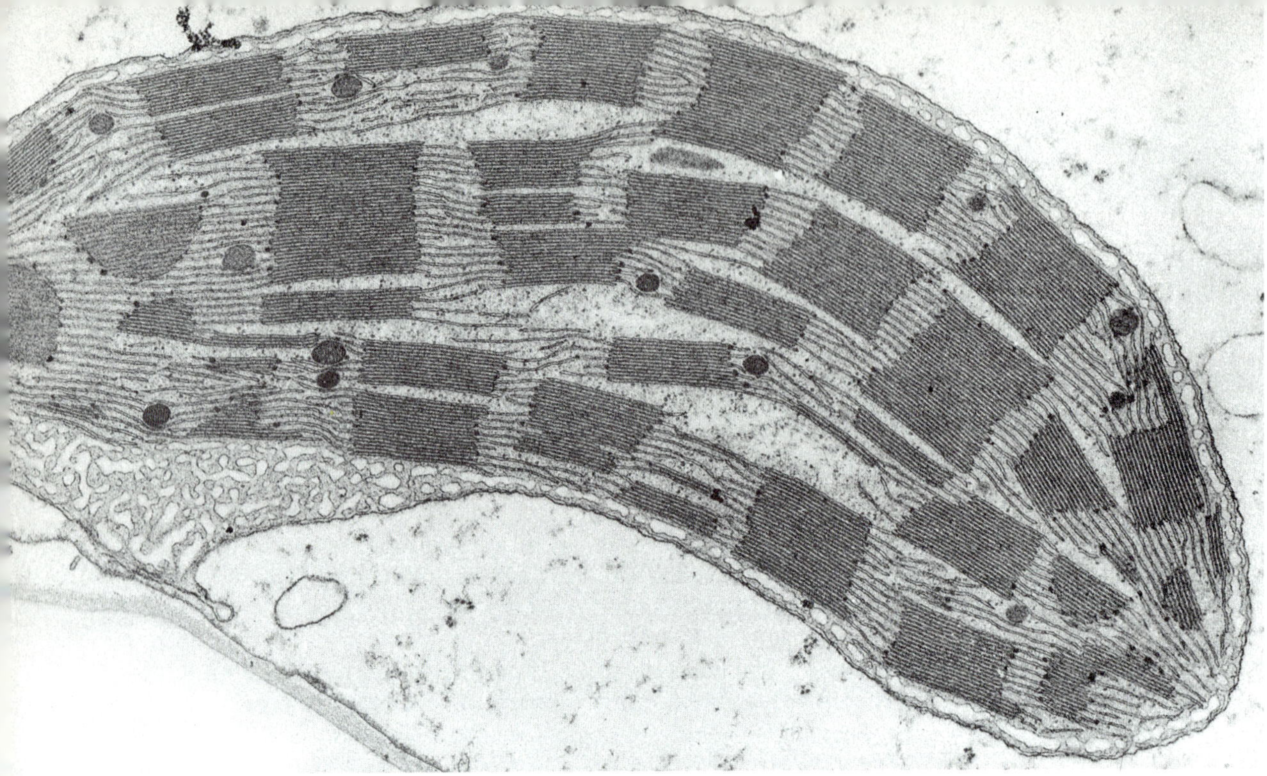

20-1. An electron micrograph of a chloroplast (magnification 26,000 ×). The structures somewhat like stacked coins are the grana. The "coins" in each granum are the lamellae.

quite complex. The formula for chlorophyll *a*, for example, is $C_{55}H_{72}O_5N_4Mg$. Although the structural formulas of the chlorophyll molecules have been worked out, little is yet understood about how plants produce them. We do know that, in most cases, they are produced inside chloroplasts and that very few are formed in the absence of light. Chlorophyll is the most common plant pigment, but others also are present. In **Inquiry 20-1,** you will use a process called chromatography (krom·uh·TOG·ruh·fee) to separate the pigments of a leaf. In **Inquiry 20-2,** you will learn more about the roles of chlorophyll and light in photosynthesis.

INQUIRY 20-1 Separation of Leaf Pigments

How does a biochemist know that the color of leaves is the result of a mixture of several kinds of pigments? Many methods are used to separate cell substances from the leaves and from each other.

In the late nineteenth century, a Russian chemist discovered the principle of chromatography. At first, this method was used only to separate pigments. By the late 1930s, however, chromatography was also being used to separate colorless substances. In the original technique, separation was done on paper. Now it can be done on other materials as well.

MATERIALS (per team)

spinach leaves
acetone
developing solution (8% acetone, 92% petroleum ether)
several strips of filter or chromatography paper
fine sand
test tube, 25 × 200 mm
cork to fit test tube
test tube, 18 × 150 mm
pipette, with a very fine tip
mortar and pestle
test-tube rack
forceps
funnel
funnel support
cheesecloth, 10 × 10 cm²
scissors
glass-marking crayon
cleansing tissue
2 pencils
paper clip

Procedure

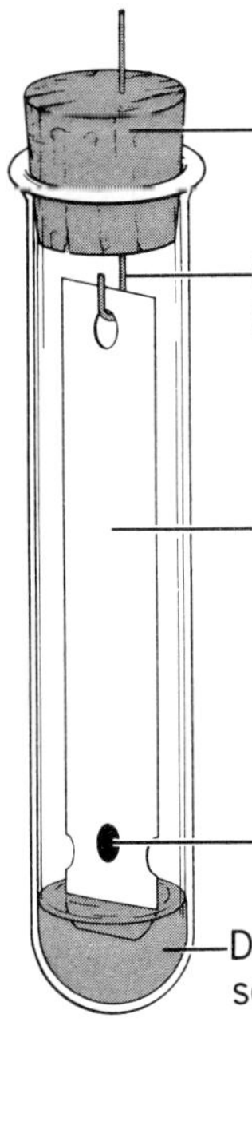

(CAUTION: *Because ether is being used, at no time should there be an open flame in the room.*) Using the larger test tube, assemble the apparatus as shown. But do not yet add the developing solution and pigments. Handle the paper carefully with the forceps. Even a small amount of oil from your fingers will affect the results.

Adjust the paper in the tube so it does not touch the sides or bottom. Mark on the test tube the level of the *lower* end of the notch.

Remove the paper strip from the hook. Then pour developing solution into the test tube to a depth of 5 mm *below* the mark that you made.

Place the cork with the hook attached (but without the strip of paper) in the test tube. Place the tube in an upright position in a rack. Support the strip of paper across two pencils so that it does not touch the table.

Put four or five spinach leaves, a little sand, and 10 ml of acetone into a mortar. Grind thoroughly. The acetone now contains extracted pigments.

Place a layer of cheesecloth in a funnel. Add a layer of cleansing tissue.

Pour the acetone into the funnel. Collect the filtrate in the smaller test tube. **1.** What is the color of the filtrate? **2.** Is there any evidence that more than one pigment is dissolved in the acetone?

Using a fine-pointed pipette, place a drop of the pigment extract on the paper between the notches. Allow it to dry. Add another drop in the same place, and allow it to dry. Repeat until you have placed at least four drops on the paper—one on top of another.

When the final drop has dried, remove the cork from the large test tube and hang the strip on the hook.

Insert the cork with the paper strip attached into the test tube. *Do not allow the pigment spot to touch the developing solution.* Adjust the length of the hook to avoid this. Be sure the cork is tight.

Watch the developing solution rise. When it *almost* reaches the hook, remove the cork from the tube. Take the paper off the hook with the forceps and recork the tube. Hold the paper until it dries.

Examine the chromatogram. **3.** How many bands of color can you see? **4.** How many bands might be made up of chlorophylls? **5.** What other colors can you see in the chromatogram? **6.** Why were you unable to see these colors in the leaf? **7.** Do you think that all the leaf pigments were soluble in the acetone? Why or why not? **8.** Suggest a hypothesis to explain the change of color that often occurs when a leaf dies.

Now consider the process by which the pigments were separated. **9.** From what point did all the pigments start as the developing solution began to rise? **10.** When did all the pigments start to move, and when did they all stop? **11.** In what characteristic, then, must the pigments have differed?

INQUIRY 20-2 Light and Leaves

Leaves contain pigments that absorb energy from the sun in photosynthesis. Starch is one of the compounds stored in the leaves. What happens when sunlight falls on a leaf? Is all of the light used by the leaf? What effect does the amount of light available have on the amount of starch produced?

MATERIALS (per team)

geranium plants, with lightproof screens
chlorophyll solution from Inquiry 20-1
200 ml of 95% ethyl alcohol
iodine solution
beaker, 600-ml
beaker, 500-ml
2 petri dishes
small, flat-sided bottle
prism
diffraction grating
hot plate
tongs
sheet of white paper

Procedure (Day 1)

Hold the prism so that a beam of light (either sunlight or artificial) passes through it and falls on a sheet of white paper. **1.** What has happened to the beam of light?

Hold the diffraction grating in front of a strong beam of light. Adjust the angle by moving the grating back and forth. **2.** What do you see?

Fill the flat-sided bottle with chlorophyll solution. Place the bottle in front of an aperture leading to a strong light source. Use the grating to examine a beam of light that has passed through the bottle. **3.** How does the light pattern produced by natural light compare with the pattern produced by light that has passed through the chlorophyll solution? **4.** What is the significance of this change?

The geranium plants you will use were kept in the dark for 36 hours. Then lightproof "screens" were attached to parts of the leaves. These screens are opaque paper or photographic negatives. They keep out light from the part of a leaf that they cover. Immediately following the dark period, the geraniums, with the screens still attached, were exposed to light for 12 hours.

Place the 600-ml beaker, half-filled with water, on the hot plate. Put 200 ml of 95 percent ethanol in the 500-ml beaker. Set this beaker inside the 600-ml water bath. Allow the water to boil gently, then turn off the hot plate. (CAUTION: *Heat alcohol only in a water bath on a hot plate, never over an open flame. Water will extinguish an alcohol flame, so have a container of water ready.*)

Pick two leaves from a plant, one with a screen and one without. Remove the screen. Also break off one stem so you can distinguish one leaf from the other.

Using tongs, put both leaves in the boiling water bath (outside the alcohol beaker) for one or two minutes. Then put the leaves into the hot alcohol for several minutes, until the color of the leaves becomes whitish.

Remove the leaves and blot them on paper towels. Put each leaf into a petri dish. Flood both leaves with iodine solution. **5.** What can you conclude? **6.** On the basis of your observations, describe the effect of light on the process of photosynthesis that occurs in a leaf.

BIOCHEMISTRY OF PHOTOSYNTHESIS

In many experiments, biologists measured the amounts of carbon dioxide (CO_2) and water (H_2O) taken in by chlorophyll-bearing cells that were exposed to light. They also measured the amounts of oxygen (O_2) and glucose ($C_6H_{12}O_6$) that were formed.

We can summarize the photosynthetic process as follows:

$$\underset{\text{carbon dioxide}}{6\ CO_2} + \underset{\text{water}}{6\ H_2O} \xrightarrow[\text{enzymes}]{\text{chlorophyll, light energy}} \underset{\text{glucose}}{C_6H_{12}O_6} + \underset{\text{oxygen}}{6\ O_2}$$

By labeling the oxygen in carbon dioxide and in water with isotopes, scientists learned that the oxygen released by plants is obtained from water.

Chemical reactions occur so rapidly in photosynthesis that identification of the intermediate steps seemed impossible to early investigators. By 1905, however, biologists had discovered that two distinct sets of reactions occur during photosynthesis. One set—the "**light reactions**"—occurs only while a chlorophyll-bearing cell is exposed to light. This one is followed by a set for which light is *not* required—the "**dark reactions.**"

The Light Reactions

Experimental evidence indicates that, within chloroplasts, light energy is changed to the chemical energy of ATP in two different ways (Figure 20-2a). The first is a cyclic set of reactions. Light energy is absorbed by a molecule of chlorophyll. This causes an electron that carries much energy to be expelled from the molecule. This high-energy electron is passed along a series of electron carriers. As the electron moves along this series, its energy is transferred to enzyme systems. These systems catalyze the change of ADP to ATP. Before the electron—now with low energy—returns to a chlorophyll molecule, two or more molecules of ATP have been formed.

The second way in which light energy is changed to chemical energy is a noncyclic set of reactions (Figure 20-2b). One of the other pigments present in chloroplasts—as yet unidentified—absorbs light and expells a high-energy electron. Its energy goes into the formation of an ATP molecule from ADP. At the same time, light energy is absorbed by chlorophyll *a* molecules. Each of these molecules expels another high-energy electron. The first electron—minus its high energy—goes into the chlorophyll. Two of the high-energy electrons from two chlorophyll *a* molecules join two hydrogen ions (H^+). The hydrogen ions were formed by the ionization of two water molecules and a molecule of NADP (oxidized nicotinamide adenine dinucleotide phosphate). The result of this activity is another kind of energy-rich molecule, $NADPH_2$.

20-2. LIGHT REACTIONS

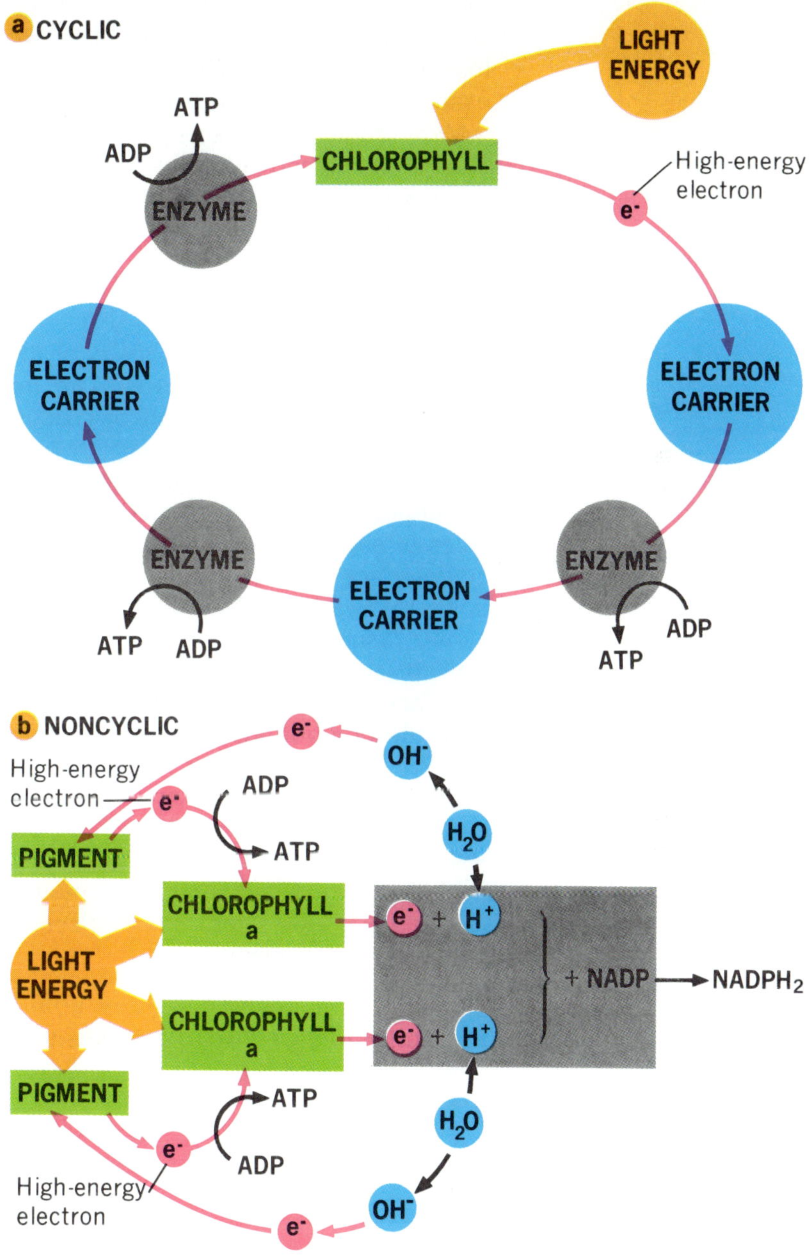

Formation of hydrogen ions from water also provides OH^- ions. Each OH^- ion furnishes an electron to replace the electron that light energy removed from the molecule of the unidentified pigment. Each OH^- ion then becomes an OH *radical*—a chemical term for atoms that are neither an independent molecule nor an ion. The OH radicals combine. This combination results in the formation of molecules of water and oxygen. The oxygen is given off by photosynthesizing cells. The oxygen may pass out of the plant or be used in the plants own respiratory processes.

Many details of these light reactions are still uncertain. Future biochemical discoveries may change our understanding of them in many ways.

The Dark Reactions

Scientists have worked out many of the more than 20 reactions involved in the dark reactions. The result of this phase of photosynthesis is the formation of glucose molecules which contain chemical energy.

The carbon source in the dark reactions is carbon dioxide from the atmosphere. Carbon dioxide is combined with complex molecules already in the cells, and, in a series of steps, molecules of glucose are formed. The glucose molecules may then be broken down into various other materials needed by the plant. Glucose is stored, usually as starch, until it is needed for respiration or for growth.

To accomplish all this, a supply of energy is required. It comes from the compounds formed during the light reactions—ATP and $NADPH_2$. As the energy is released from ATP in the dark reactions, ADP is formed. Likewise, as energy is released from $NADPH_2$, NADP is formed. The ADP and NADP are converted back into ATP and $NADPH_2$ during the light reactions. ATP and $NADPH_2$ thus act as carriers of energy from sunlight to multicarbon compounds. Look at Figure 20-3. This figure summarizes the relationship between the light and dark reactions.

20-3. SUMMARY OF PHOTOSYNTHESIS

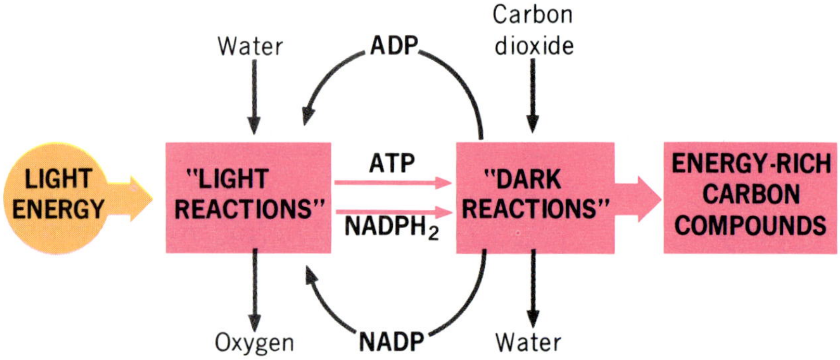

SCIENCE AT WORK: More Food for the World

In a recent report on the efficiency of agricultural production, agriculture was defined as the business of collecting solar energy and storing it as food energy in plant and animal products. Right now, in most plants, the process of photosynthesis is only about 1 to 3 percent efficient. Scientists believe that photosynthesis could be about 13 percent efficient. Finding ways of raising the efficiency of photosynthesis to that 13 percent level is an active area of agricultural research.

One reason for inefficiency in photosynthesis is that about half of the carbon dioxide taken in by most crop plants is lost. This waste carbon dioxide comes from the breakdown of glycolate. Glycolate is a compound formed in making protein in the leaves.

Some plants grow very rapidly. Examples of such plants are corn and sugar cane. These plants do not form glycolate in the process of making proteins.

Scientists are trying to eliminate glycolate metabolism from other food plants. If this can be done, these plants may grow faster and produce larger crops. Scientists are now working on two possible approaches. They are trying to develop a spray that would inhibit glycolate metabolism. Another idea is breeding new strains of crops that lack the glycolate gene.

Either of these two approaches could result in crops that produce food much more efficiently than present crops do. In this way, the agricultural industry could increase the food supply for the expanding human population.

LEAVES

At some time in the history of land plants, flat, green structures evolved that were exposed to the sun's rays. In general, botanists consider these structures—leaves—as organs, since they are composed of a number of tissues and perform some general function in the life of the plant.

Examine the leaves on any one plant. No two are exactly alike. Shapes and sizes may vary somewhat among plants of the same species. These differences will depend on the ages of the plants, the amount of light received, and other factors. Yet, for any particular species, the leaf shape is usually distinctive and constant enough to be useful in identification (Figure 20-4). Whatever the leaf form, however, there is no doubt that the primary function of most leaves is photosynthesis.

20-4. (a) *Hibiscus rosa sinensis;* **(b)** *Tupidanthus,* umbrella plant; **(c)** *Coleus blumii;* **(d)** *Codiaeum pictum,* croton.

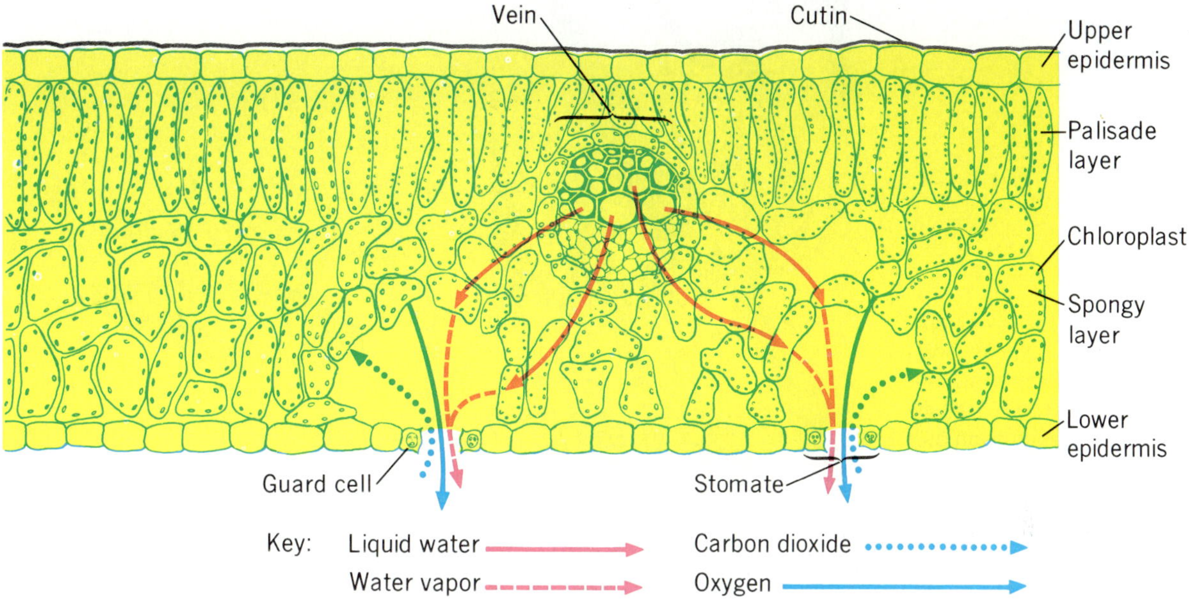

20-5. Part of a leaf as it might appear in a cross section under the microscope. The arrows show the movement of carbon dioxide, oxygen, and water into or out of the leaf.

Figure 20-5 shows a leaf blade in cross section. Several different kinds of cells can be seen in the figure. The **mesophyll** cells (palisade and spongy tissues) contain most of the chloroplasts. Most of the photosynthetic activity takes place in the closely packed palisade layers of mesophyll.

The raw materials of photosynthesis are water and carbon dioxide. Plants obtain carbon dioxide from the air and water from the soil. The amount of carbon dioxide in the air is fairly constant, but the amount within chlorophyll-bearing cells does vary. Carbon dioxide, therefore, is either given off or taken in by cells. An exchange of gases between inner leaf cells and the air occurs through slitlike pores, **stomates**. Stomates occur in the outer layers of leaf tissue, the **epidermis** (ep·ih·DER·mis). Usually, most of the stomates are found on the under surface of the leaf. The stomates lead into the spaces of the mesophyll.

The leaf veins are tubular structures through which liquids flow. Water comes to the leaf from the roots by way of veins. Air enters the stomates and moves among mesophyll cells. Although water molecules are present in air as water vapor, they are never as abundant in air as they are in the leaf. Thus, the plant loses water that diffuses into the air. This loss is called **transpiration**. Each stomate is surrounded by a pair of specialized cells, **guard cells** (Figure 20-6). The

20-6. **20-6.** Part of the lower epidermis of a leaf, as seen through a compound microscope. The narrow opening between each pair of guard cells is a stomate. *How do the guard cells function in conserving water in the leaf? in obtaining carbon dioxide?*

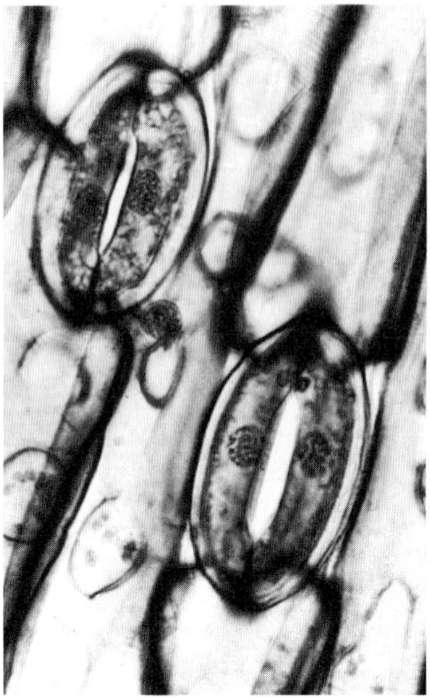

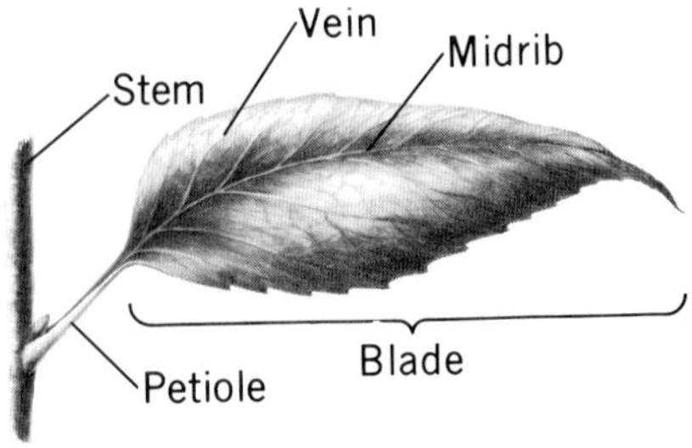

20-7. The external anatomy of a leaf. Leaves vary greatly in form. Whatever their differences, most leaves contain chlorophyll.

inner walls of the guard cells are thick. When their water content and thus their volume increase, the cells bend outward and the stomate opens. When the water content decreases, the stomate closes—but never completely. In this way, the stomates at least partly control the loss of water through transpiration.

In most land plants, the stomates are open during daylight hours, when the water content of the air is relatively low. They are closed at night, when the water content is usually higher. Experiments have shown, however, that the action of guard cells seems to be associated more with the concentration of carbon dioxide in the cells than with control of water loss. But botanists still have much to learn about how stomates work.

Regardless of the action of guard cells, land plants lose much water through the stomates of their leaves. But they also lose some water directly through their epidermal cells. This loss is reduced by the **cutin** (CUE·tin), composed mostly of a waxy material, that covers the outer surface of the epidermal cells. The cutin may be very thin, as on lettuce leaves, or rather thick, as on pine leaves.

The leaves of many flowering plants—a bean plant, for example—have two main parts. One is a stalk, or **petiole** (PET·ee·ohl), which attaches the leaf to a branch. The other is the thin, flattened **blade** (Figure 20-7). If you hold the blade of a bean leaf up to a bright light, you will see a network of veins extending from a **midrib.** The midrib is a continuation of the petiole into the blade. The veins are the transport system of the leaf. They contain special cells through which water moves into the leaves. Sugars, made in photosynthesis, also travel from the leaf to the stem and roots through the veins.

You will observe leaves and their internal structures as you conduct **Inquiry 20-3**.

INQUIRY 20-3 Leaf Structure and Function

MATERIALS (per team)

variety of mature plants
prepared cross section of a leaf
compound microscope

Procedure (Day 1)

Look at the various plants and the different kinds of leaves. **1.** What might be the advantage of a flat, thin leaf blade to the photosynthetic capacity of a plant? **2.** How are the leaves arranged on the stems of different plants? **3.** Relate this arrangement to the photosynthetic capacity of the plant.

In order to understand better the relationship between structure and function in a leaf, you will have to look inside it.

Examine the cross section of a leaf with high power. Consider its structure in relation to light and carbon dioxide absorption and to water supply. Find the chloroplasts in the cells of the leaf. Make a diagram of the chloroplasts of a few typical cells. **4.** What shape are the chloroplasts? **5.** Which leaf cells contain the chloroplasts? Compare the number of chloroplasts in the cells of the palisade layer with the number in cells of the spongy layer. **6.** How might this distribution of chloroplasts be advantageous in photosynthesis?

In the center tissues of the leaves, you will find veins of different sizes. Find a cross section of a small vein. The vein is surrounded by a sheath of cells that also are photosynthetic. Notice some of the empty cells with thick walls in the upper part of the sectioned vein. **7.** What might be the function of these cells?

Now take a close look at the covering layers of the leaf—the upper epidermis and the lower epidermis. These are single layers of cells. The outer surfaces of the epidermal cells are covered by cutin. You may not be able to see this thin yellow layer; it may have been removed when the slide was made. **8.** In what indirect way might the cutin affect the efficiency of photosynthesis?

Locate a stomate and its guard cells. Locate the spongy tissue just above the lower epidermis.

9. How does the structure of the spongy tissue differ from that of the other leaf tissues? **10.** Note how the stomates are located in relation to the spongy tissue. What do your observations suggest about the possible function of the spongy tissue? **11.** What functions of a leaf have been demonstrated in this inquiry and with what structures is each function associated?

STEMS

The stem of a plant connects roots and leaves. It carries water and minerals from the roots to the leaves and returns the products of photosynthesis from the leaves to the roots. Leaves and flowers grow from the stem, and food is stored in its tissues. There are two types of stems.

A woody stem is hard, tough, and strong. Its strength comes from many cells that have stiff, thick walls. Even when these cells dry out, their walls remain strong. The huge stems of trees, which we call trunks, are so strong that they are used in construction.

In a herbaceous (her·BAY·shus) stem, the tissues are rather soft and easily crushed. Tomato, geranium, and milkweed stems are examples. Plants that have herbaceous stems usually do not grow very tall. Although the cell walls give some support, if the cells do not get enough water the stems may wilt, bend, or even collapse.

Conduction in Stems

Two kinds of materials are transported in stems. Usually, water and mineral salts are transported upward from the roots, through the stem, into the leaves. In the summer, sugars produced in photosynthesis move in the opposite direction. Early in the spring, however, food that has been stored in the roots and stem begins to move upward. In the case of sugar maples, we say, "The sap is rising," and we "tap" these trees to remove the sap and make maple sugar.

What structures of the stem make these movements of material possible? Let us first consider woody stems. If we cut across the trunk of a large tree, the exposed surface will be more or less circular in outline (Figure 20-8a). The outermost layer is the bark. The inner parts of the trunk are wood. A thin layer of cells, the **cambium**, separates the bark from the wood. The cambium is the growth

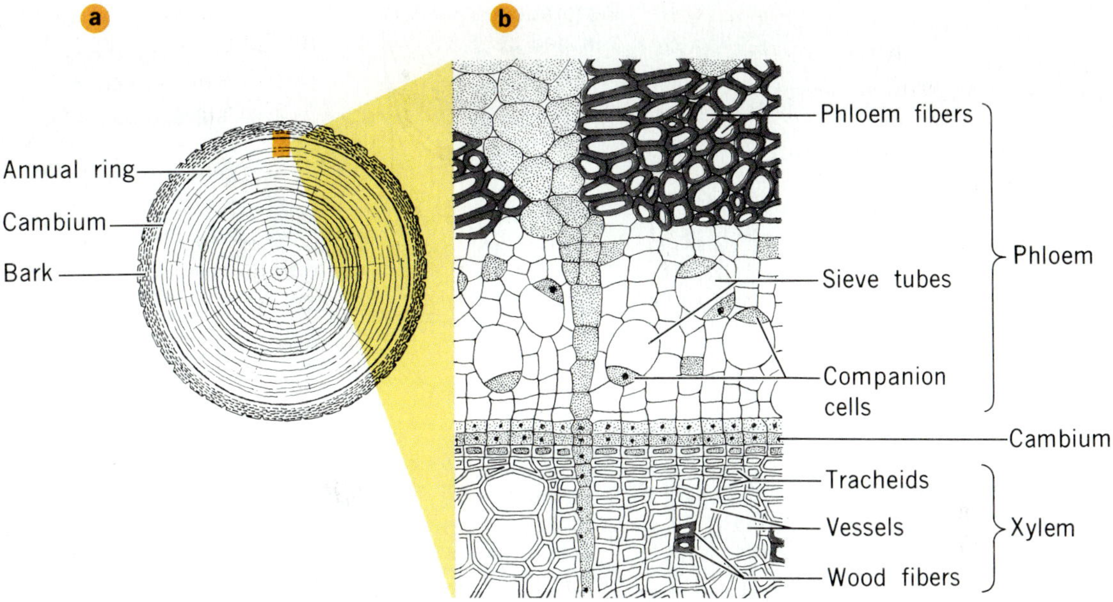

20-8. (a) A drawing of a cross section through a woody stem. This type of stem is characteristic of oaks, maples, elms, and many other trees. **(b)** An enlargement of part of the cross section showing the relationship between the vascular cambium and the phloem and xylem.

layer from which new tissues arise. The new tissues increase the diameter of the stem. If we use fine sandpaper to smooth the cut surface of the trunk and then wet it, we can see many layers of wood, like a series of rings one inside the other. The rings are usually the new growth of each successive growing season.

This much we can see with the unaided eye. To learn more, we must use a microscope to examine tissues from the inner bark, cambium, and the outer regions of the wood (Figure 20-8b).

Two types of tissue are especially important to the functioning of the plant. The first is the wood tissue, or **xylem** (ZY·lem), made up of cells that have strong, resistant walls. Xylem conducts water and dissolved minerals up from the roots to all parts of the plant. Two of the structures in the xylem, tracheids (TRAY·kee·ids) and vessels, are shown in more detail in Figure 20-9a and b. Conduction of materials in the tracheids takes place through pits in the cell walls. The vessel cells are situated on top of one another in a vessel. The ends of the vessel cells are open. Both of these structures are really dead cells. Their nuclei and cytoplasm have disappeared. The other cells in the xylem include a large number of fibers. These are long, narrow cells that have particularly thick walls. They contribute greatly to the strength of the wood.

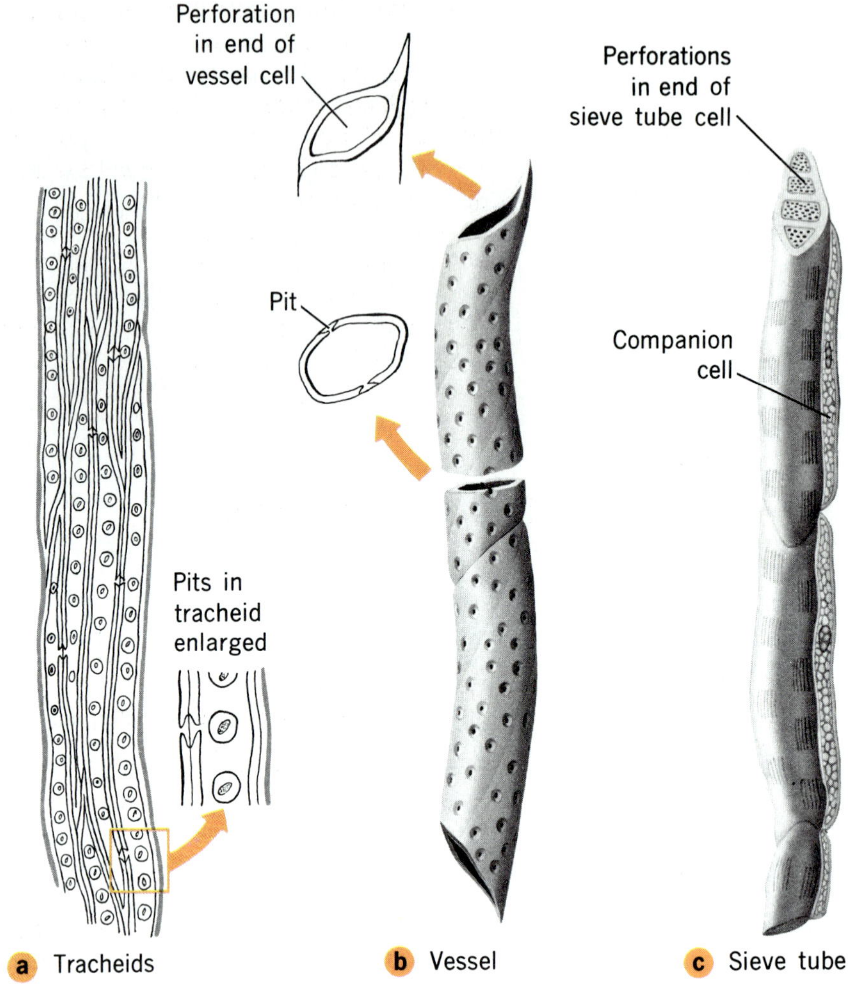

20-9. Some conducting cells of the stem – xylem and phloem. **(a)** The group of tracheids shown here is in longitudinal section (stem cut lengthwise). Conduction in these cells is facilitated by pits in the cell walls. *What is another function of tracheids besides conduction?* **(b)** Two vessel cells are shown from one side. Notice how the cells sit on top of one another in the makeup of the vessel. **(c)** Conduction in phloem cells occurs along the sieve tube. Companion cells, each with a nucleus, lie close to the sieve tubes and may aid in regulating their function.

If you look again at Figure 20-8, you will see another specialized area, the **phloem** (FLOW·em), close to the cambium. The phloem is made up of **sieve tubes**. Phloem is responsible for the movement of food (glucose) from the leaves, where it is made, to all other parts of the plant. These structures have an unusual pattern of development. The sieve tube cells develop one above the other in a young stem. As they grow, small perforations appear in their end walls, which overlap. Thus, cytoplasm is continuous from one cell to the next. In the final stages of development, the nuclei of the sieve cells disintegrate. The cytoplasm, however, remains alive as long as the sieve tube continues to function. A companion cell that does have a nucleus lies close to each sieve tube cell. The companion cells may aid in regulating the function of the sieve tube cells. Figure 20-9c shows a sieve tube. Nutrients pass along the phloem from the leaves to the roots, where they are stored, or to flowers and fruits, where they are used in development.

The structures already discussed occur in woody stems. In **Inquiry 20-4** you will make direct observations of a herbaceous stem.

INQUIRY 20-4 Herbaceous Stems

20-10. The shoot of a flowering plant. A *shoot* is the stem with attached leaves.

MATERIALS (per team)

young bean plant
prepared slides of cross section of a herbaceous stem
colored pencils (red, blue, brown, green)
compound microscope

Procedure (Day 1)

Observe the stem of the bean plant. Notice how the leaves are attached to the stem. The site of attachment is called the node (Figure 20-10). **1.** Does the shape of the stem differ along its length? **2.** What color is the stem? **3.** Judging from its color, what might be one function of a herbaceous stem? **4.** Should this be listed as a function of all stems? Why or why not?

Using low power, study a cross section through the internode of a stem. Compare your slide with Figure 20-11 to become familiar with the general regions of the stem. Like a leaf, a herbaceous stem is covered by an epidermis. **5.** How many cells thick is the epidermis?

The most noticeable feature of the cross section is a ring of *vascular bundles*. Each bundle may be wedge-shaped with the narrow end of the wedge directed toward the center of the stem. The center of the stem is made up of large, thin-walled cells. This is the *pith*. Between the outer edges of the vascular bundles and the epidermis you will find the *cortex*.

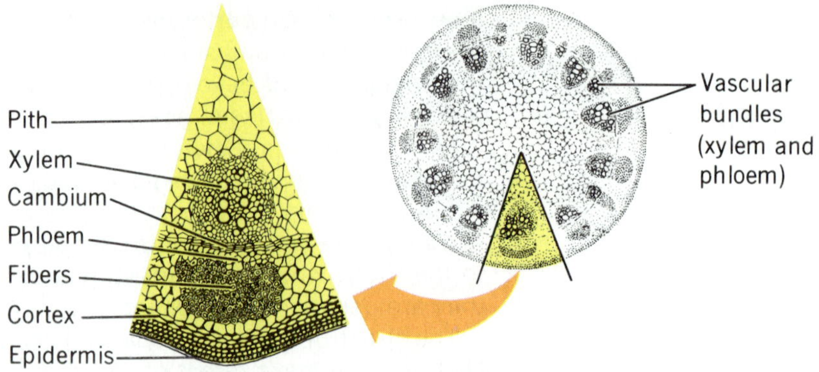

20-11. A cross section of the internode of a sunflower stem.

Draw a circle about 15 cm in diameter. In the circle, outline the vascular bundles as they appear in your slide. Do not draw in any cells.

Using high power, look in the cortex and vascular bundles for thick-walled, empty cells. Show their location in red on your diagram. **6.** Suggest a possible function for these cells. **7.** What hypothesis can you form about a possible relationship between these cells and the substance we know as wood? **8.** Suggest a way to confirm your hypothesis.

The leaves need water for photosynthesis. Water is absorbed by the roots and is conducted through the xylem of the stem to the leaves. Use a blue pencil to color the xylem in your drawing. **9.** In what structures of the stem did you look for xylem cells? **10.** What other functions might these cells have? **11.** What relation do these cells have to those previously noted in your stem drawing?

Many different carbohydrates are manufactured by green leaves. Some of this food is used by the leaves. Some is transported to other cells of the plant where it is either used or stored. These materials are conducted through the phloem. With a brown pencil, color that part of your diagram that represents the phloem. **12.** Where in the stem is the phloem tissue?

Excess food manufactured by leaves is stored elsewhere in the plant. For example, in a potato plant food is stored in the stem—but this stem is underground. The "potato" is a highly specialized portion of the stem. **13.** In what form is the food stored in the potato?

Using high power, carefully examine the stem tissues to find cells that store food. Color the storage cells in the diagram green. **14.** How is the presence of stored food indicated in these tissues? **15.** What region of the stem is devoted primarily to storage? **16.** Summarize the functions of the stem you observed in this inquiry.

Growth Patterns in Stems

The stems of woody plants grow in height and in diameter each year. How does growth take place? The cambium is the thin layer of cells between the phloem and the xylem. These cells divide rapidly. They build up new xylem tissue on one side of the cambium and phloem tissue on the other. The rate at which cambium cells divide depends on environmental conditions. For instance, during winter or dry spells, few new cells are formed. When growth begins again in the spring, cells in the cambium divide rapidly.

The first xylem cells are larger than those produced later in the summer. This difference causes the ringlike pattern that is typical of woody stems of plants growing in temperate climates. Since a new ring of xylem tissue is produced by the cambium during each growing season, the rings are called *annual rings*. By counting the number of annual rings in the stump of a tree, we can tell quite accurately the age of the tree at the time it was cut. Annual rings are not as evident in the wood of tropical plants. How do you explain this?

ROOTS

Before we consider the structure and function of roots in detail, you should make your own observations by doing **Inquiry 20-5**.

INQUIRY 20-5 The Structure and Function of Roots

On the basis of what you already know about green plants, suggest hypotheses for the function of roots. In this inquiry, you will test some of your hypotheses.

MATERIALS (per team)

radish seedlings germinating in a petri dish
grass seedlings germinating on water
root system of a grass plant
bunch of carrots
forceps
stereoscopic dissecting microscope
prepared slide of a root cross section
compound microscope
cover glass
colored pencils (blue, brown, green)

Procedure (Day 1)

A plant would soon die without water. Plants, therefore, must absorb water. **1.** When we water a plant, what part absorbs the water?

Recall your study of the absorbing surfaces of animal digestive tracts (Chapter 14). **2.** What would you expect one of the characteristics of water-absorbing structures of a plant to be?

Some roots store more food than others. Look at the root system of the grass plant and the carrot. **3.** Which probably contains more stored food?

Look at the radish seedlings in the petri dish. (Do not remove the cover from the dish.) Use the dissecting microscope and look through the cover of the dish at the fuzzy mass of tiny structures around the root. These structures are called root hairs. **4.** On what part of the root are the root hairs longest? **5.** What probably happens to the root hairs if a plant is pulled out of the soil? **6.** Suggest some things that seem to be missing from the carrot but are present in the grass. Why are they missing?

You can see that root hairs are delicate structures. However, it is hard to tell if they are made of one cell or of many cells. With a forceps, pick up a young grass seedling. Mount the seedling in a drop of water on a clean glass slide, without a cover glass. With the low power of your compound microscope, find the root tip. (You can see that the root differs from the young shoot, which is green.) Look along the root for root hairs. **7.** Is a root hair made up of one or many cells? **8.** What relationship do root hairs have to the epidermis of the young root? (To answer this question, you may have to put a cover glass on the specimen and examine it with high power.) **9.** Are root hair cells living or dead? **10.** How does the function of the root epidermis compare with the function of the stem and leaf epidermis? **11.** What important process is involved in the absorption of water by root hair cells?

Observe with low power the prepared slide of a cross section of a mature root. Find the epidermis, cortex, and pith. Find the xylem and phloem in the root (Figure 20-12). Draw a circle to represent the root cross section. Using the blue pencil, color in the location of the xylem on your diagram. Using the brown pencil, color in the location of the phloem. **12.** How does the position of xylem in the root compare with its position in the stem? **13.** Where is the phloem in relation to the xylem?

Examine the cells and tissues of the root with high power, looking for stored food. Color the region of food storage on your diagram with a green pencil. **14.** What region of the root contains the greatest amount of stored food? **15.** In what form is the food stored in the root?

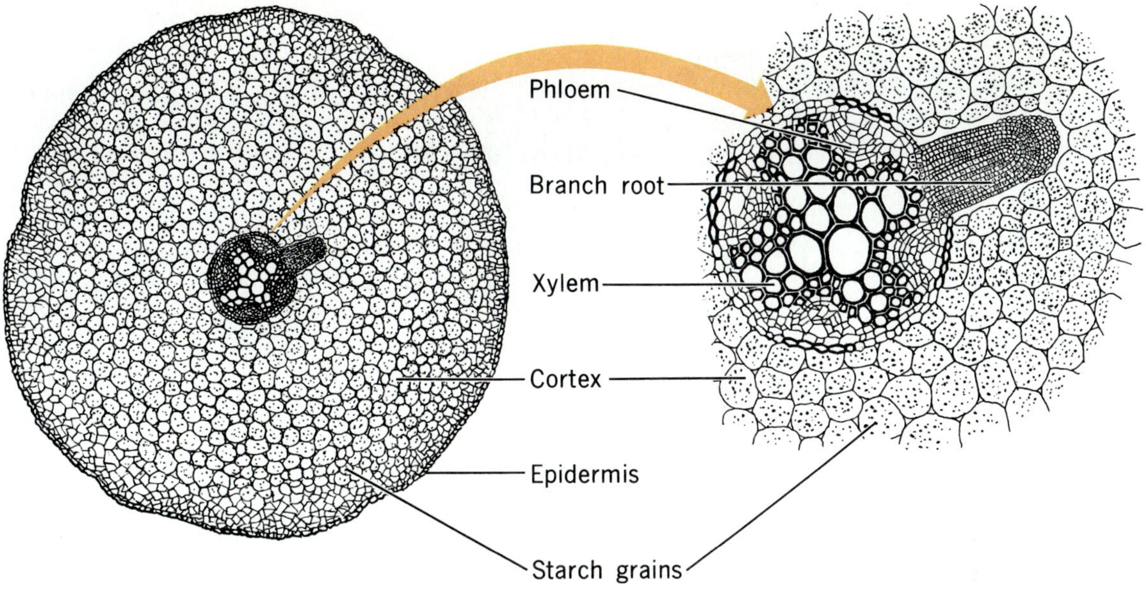

20-12. An enlarged cross section of a buttercup root with the vascular core and a young branch root further enlarged at the right. Compare this arrangement of tissues with that of the stem. *On the basis of the vascular structure, which do you think would be the most effective supporting organ — the root or the stem?*

Conduction and Absorption in Roots

A small rye plant has a root system made up of about 14 million primary roots and branches with a total length of about 600 kilometers. The total surface area of its root system is more than 600 square meters.

The roots of most plants are not green and they grow below ground. Root systems absorb water and minerals from the soil, anchor the plant into the ground, conduct materials, and store food.

The main conducting and strengthening tissues of the root are grouped together at its center, forming a rod-shaped core. This core extends the full length of the root. This central core of tissue is easy to see, in cross section, under a microscope (Figure 20-12). The cells immediately behind the root tip (Figure 20-13) elongate, pushing the root tip through the soil. The root tip is covered with a root cap, a layer of loosely attached, dead or dying cells. These cells protect the root tip from injury as it is forced through the soil. As root cap cells fall off, they are replaced by dividing cells of the root tip. A little farther back from the tip are the many delicate root hairs that give the tip a fuzzy appearance.

The root hairs grow out into spaces between soil particles. The particles in a typical, well-watered soil are surrounded by a thin

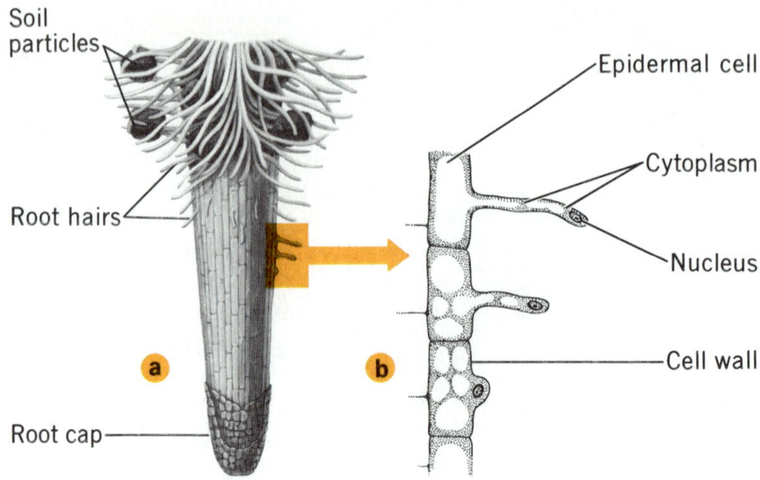

20-13. (a) External features of a root. The root cap protects the growing tip from injury. **(b)** Root hairs on epidermal cells above the root tip add greatly to the total cell surface directly exposed to soil and water. *How does this benefit the plant?*

layer of water. The water contains dissolved salts. In soils that are not completely saturated with water, many air spaces occur between the soil particles. The thin, pliable root hairs come into direct contact with the soil water and the gases in these spaces. The thin walls of the root hairs are permeable to gases, water, and minerals. Oxygen in the air spaces or dissolved in soil water diffuses into the root hairs.

Minerals are used to form many compounds necessary for the continuing life of the plant. In addition to carbon, hydrogen, and oxygen, plants need sulfur, phosphorus, calcium, magnesium, iron, and potassium.

Traces of other elements such as copper, zinc, boron, manganese, and molybdenum also must be available. These and other trace elements may be necessary parts of enzymes or coenzymes, without which the plant would die.

MOVEMENT OF MATERIALS IN PLANTS

As far as we know, the tallest living tree is a California redwood, 112 meters high. The leaves at the top of this tree are supplied with water that is absorbed through millions of root hairs buried in the ground more than 100 meters below. Huge amounts of water move into, through, and out of this tremendous organism. Where does the energy to move the water come from?

Water from the soil that is in contact with the root surfaces moves up the xylem vessels to the leaves at the top. What force supplies the pull at the upper end of the column? Water evaporates from the cells of the spongy and palisade layers of the leaf and passes out through the stomates. As you recall, this process is called transpiration. The effect of transpiration is to pull the column of water in the xylem of the root and stem continuously up to the leaf. As transpiration occurs, the concentration of soluble materials in the leaf cells increases. The difference in salt concentration causes water in the stem xylem to be pulled into the leaf and water in root xylem to move into the stem. Thus, soil water moves into the root, up the stem, and evaporates from the leaves. But it appears that the upward pull produced by transpiration is only part of the story. Under certain conditions, roots have enough pressure to exert an upward push on the water. Also, cells of the root and stem seem to "soak up" water, like a wick or blotting paper.

Sugars produced by photosynthesis move from the leaf, through the sieve tubes of the phloem, to other parts of the plant. We are not certain how this movement occurs. We do know that it may be many thousands of times more rapid than could occur by simple diffusion. The direction of movement usually is from the leaf, down the stem, and into the root, although movement in the opposite direction also occurs. In many cases the direction of movement seems to contradict the laws of diffusion. In sugar beets, for example, the sugars are moved downward and stored in the enlarged roots, where the concentration of sugars is already far higher than in other parts of the plant. For this sugar transport to take place, energy is required. Although the exact mechanism is unknown, it is clear that active transport must be involved. Recall from Chapter 3 that such processes often take part in moving materials across cell membranes.

STORAGE OF MATERIALS IN PLANTS

In a desert, rain may fall only a few times each year. Yet a cactus plant must have water continuously if it is to survive, grow, and reproduce. The roots of a cactus usually cover a wide area around the plant but extend only a short distance below the soil surface. Water absorbed by the roots is transported back to the main body of the plant, which actually is a very thick stem. Here the water is stored for long periods and is slowly released, as needed, for photosynthesis.

Food that is not needed at once for growth and repair is stored in the plant. In a corn plant, for example, about 25 percent of the food manufactured is stored. Foods are stored in the form of carbohydrates, proteins, fats, and oils. The stored food in white potatoes is in the form of starch. When you eat onions, you obtain food in the form of sucrose, which has been stored in the thickened leaves of the bulb. Roots that have large amounts of stored sugars or starches include sugar beets, sweet potatoes, and turnips.

Proteins, fats, and oils, along with carbohydrates, usually are stored in seeds and fruits. People have relied on these parts of plants for much of their food for thousands of years. It is possible to classify various civilizations by their dependence on particular kinds of cereal grains (wheat, rice, corn, oats, and barley, for example). There are the rice civilizations of the Far East, the wheat civilizations of the lands bordering the Mediterranean Sea, and the corn civilizations of the New World.

THE IMPORTANCE OF ALGAE

a

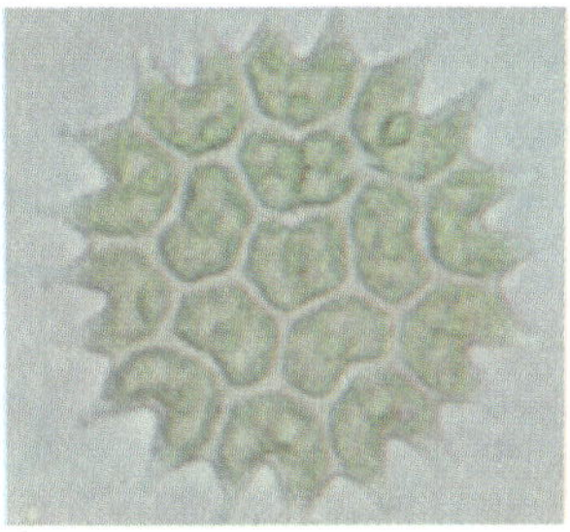

b

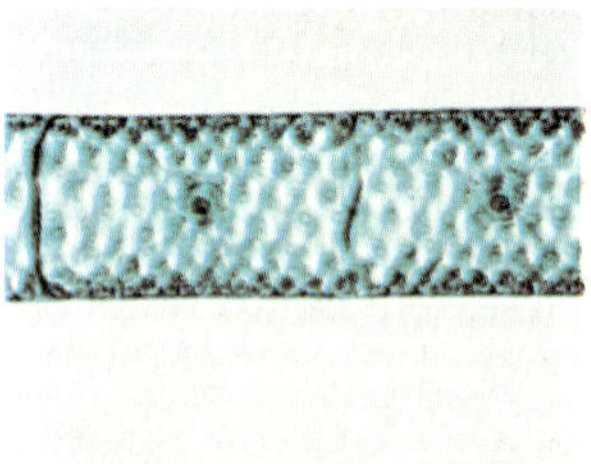

Carolina Biological Supply Company

c

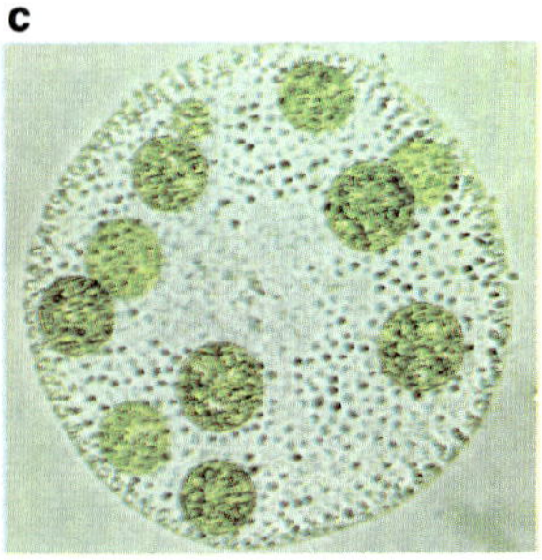

20-14. Microscopic algae. **(a)** *Pediastrum,* **(b)** *Spirogyra,* **(c)** *Volvox.*

Throughout this chapter, you have studied large, multicellular, vascular plants—the trees, crops, and flowers that we see everyday in farm, forest, and florist shop. But some of the most important plants are those we seldom see or notice. These are the algae (Figures 20-14 and 20-15). These photosynthetic plants live in water, where they supply most of the oxygen available to aquatic animals, from protozoans to fishes. This is why we include green plants in an aquarium with tropical fish.

Algae also are essential food producers. Like other green plants, the algae make their food in photosynthesis. Protozoans and other animals then eat the algae. Other small animals eat the protozoans. Larger animals may eat the smaller ones. In harvesting fish from the sea, people also rely indirectly on the algae for food. Clearly, without algae there would be no life in the seas, just as there could be no life on land without green plants.

Not all algae are green. In some, the pigment involved in photosynthesis is red, golden-brown, or blue-green (Figure 20-16). Some of the brown algae, the kelps, are among the longest plants in the world. They grow in the seas of northern coastal regions, in the deeper waters beyond the tide zone. Here they may grow to be 60 or more meters in length. They contain less available energy per unit of weight than corn, wheat, rice, and other grains. However, they are high in mineral content and certain essential elements, such as iodine, potassium, and nitrogen. These elements make brown algae useful as fertilizers and, in some cases, as food.

Substances obtained from brown and red algae are widely used. One of these substances is the agar used in preparing culture media in the laboratory. Another substance is algin (AL·gin). This is used in the manufacture of ice cream to give it a smooth texture. Algin also is used in shaving cream and shampoos. Many foods, especially in-

20-15. *Ulva,* or sea lettuce, is a common green alga of the tidal zone near the shore.

stant pudding, pie fillings, preserves, and candies, contain these substances obtained from marine algae.

In Brief

Photosynthesis is the process in which plants—mostly the green plants—use the energy of light in synthesizing food molecules. Photosynthesizing cells release oxygen.

In vascular plants, most photosynthetic activity takes place in the leaves. Some of the food products of photosynthesis are used in the leaves. Others are moved through the phloem of the plant stem and root to be used in other plant parts or to be stored. Water and minerals taken in by the roots reach the leaves through the xylem cells. As the water is lost from the leaves by transpiration, more water moves up through the xylem of the stem and root.

Algae are the photosynthesizing organisms in aquatic environments. Many algae are green, red, or golden-brown. In addition to providing oxygen and food for aquatic animals, these valuable plants are useful to people in many ways.

Guide Questions and Problems

1. If the world was without animals, would green plants eventually use up all the carbon dioxide and thus die? Explain.

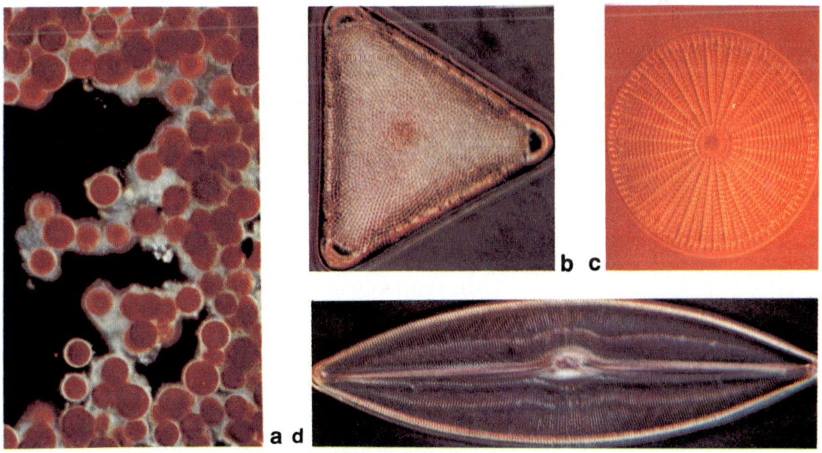

20-16. Not all algae are green. **(a)** *Porphyridium,* a red alga. Note that the many cells form a colony. Golden-brown algae **(b), (c),** and **(d),** are also known as diatoms. These have cell walls that contain silicon, the main component of glass. Their cell walls usually have beautiful geometric patterns.

2. Without oxygen, we could not live. Carbon dioxide is a waste product. Yet for photosynthesizing cells, carbon dioxide is needed and oxygen is released. Explain the relationships involved here.

3. What happens to light energy in a chloroplast?

4. How is a plant structurally adapted to get the materials needed to carry on photosynthesis and to move essential manufactured materials to other parts of the plant?

5. Solutions of minerals are sold in stores as "plant food" to be mixed with water and applied to the soil. Why is the name "plant food" incorrect?

6. The water in small ponds that have many microscopic green algae is often depleted of oxygen during certain periods in a day. At other periods the water is saturated with oxygen. What might be the cause of this fluctuation? When would you expect the carbon dioxide concentration to be greatest in such ponds? What effects would these fluctuations have on animals living in the ponds?

7. What would happen to the rate of water movement in a plant on a day that is rainy and cool? on a day so hot and dry that the plant wilts?

Ideas for Independent Study

1. Look more deeply into other aspects of photosynthesis. What are some unsolved problems in this field today? Would it be useful to be able to carry out this process in the laboratory or factory? Why or why not? What sorts of problems will have to be solved in order to make possible artificial photosynthesis?

2. You may find it interesting to observe directly the patterns and timing of conduction of materials through plant stems and roots. If so, we recommend that you begin with celery stalks. Trace the ascent of dye in the stem. Look at longitudinal as well as cross sections.

3. You can extend this study to a great variety of plant stems and roots, both herbaceous and woody. You also can vary environmental conditions. For example, try changing temperature, using controlled exposure to light and to darkness, reducing the number of leaves on the shoot or treating the leaves to prevent evaporation of water, damaging the leaves (mechanically or with heat), and so forth.

Related Readings

BOOKS

Galston, A. W. *The Life of the Green Plant.* Englewood Cliffs, N.J.: Prentice-Hall, 1970.

Gaffron, H. *Photosynthesis.* A BSCS Pamphlet. Boulder, Co.: BSCS, 1965.

Rutter, A. J. *Transpiration.* Oxford Biology Reader. London: Oxford University Press, 1972.

Wendt, F. *The Plants.* New York: Time-Life, 1963.

Wohlrabe, R. A. *Exploring the World of Leaves.* New York: Crowell, 1976.

ARTICLES

Albersheim, P. "Walls of Growing Plant Cells." *Scientific American,* April 1975, page 80.

Epstein, E. "Roots." *Scientific American,* May 1973, page 48.

21 Plant Growth and Reproduction

Plants provide the oxygen we breathe. They also are important food sources. Increased knowledge of plant structures and functions may help us improve the world's food supply.

The plants that are familiar to most people are rooted in place. They cannot move to a warmer climate or to a location that has more light or better soil. They must be able to survive where they start their lives.

PLANTS RESPOND TO THEIR ENVIRONMENT

Though unable to move around, plants can respond to the environment in many other ways. Plants respond to gravity. They also respond to light, or the lack of it, and to different concentrations of water, minerals, and nutrients in the soil. They even respond to rocks and other barriers that hinder root growth.

When a plant's response to a stimulus is a change in the direction of growth in one of its parts, the response is called a **tropism** (TROH·piz·m). If the response is toward the stimulus, the tropism is positive. If the response is away from the stimulus, the tropism is negative. You can observe some of these tropisms by doing **Inquiry 21-1.**

INQUIRY 21-1 Plant Responses to Environment

MATERIALS (per team)

Part A

4 soaked corn grains
6 petri dishes, about 150 × 20 mm
blotting paper or filter paper
cotton
modeling clay
masking tape
waterproof ink marker

Part B

soaked radish seeds
4 flower pots, 8 cm in diameter
4 boxes
blue and red cellophane
3 electric lamps

Part C

FILM: *Phototropism*

Procedure (Part A)

Is it necessary to plant a seed or grain right side up, so that the root will grow down and the stem will grow up? Seeds and grains often are planted by machine. The machine makes a narrow furrow, drops in the seeds, then covers them. What happens if a seed lands upside down?

Place four soaked corn grains on the bottom of an empty petri dish. Arrange them as shown in Figure 21-1, with their pointed ends directed toward the center of the dish.

Cut a piece of blotting paper to fit tightly inside the bottom half of a petri dish. Place the paper over the corn grains in the dish. Pack the rest of the dish tightly with cotton. When the dish is covered and set up on edge, the corn grains must stay in place. When you are sure the grains will stay in place, open the dish and wet the blotting paper. **1.** Why was it necessary to soak the grains before placing them in the dish? **2.** Why are the grains placed next to wet blotting paper?

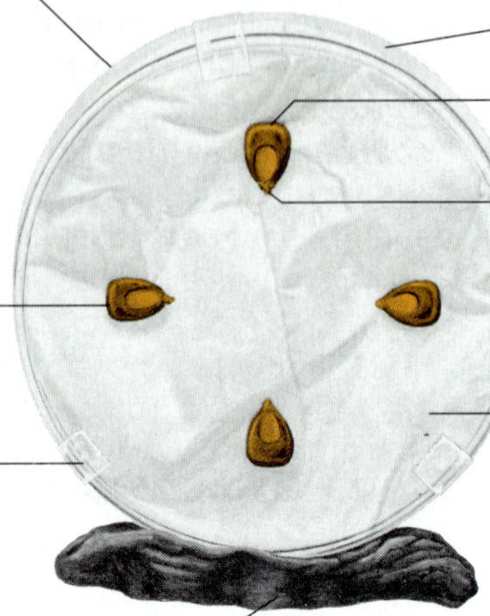

21-1. One technique by which you can investigate the orientation of shoots and roots in germinating corn grains.

Close the dish and seal it with strips of tape. Make a support for the dish out of modeling clay. Hold the dish on edge with one grain at the top. Press the lower edge of the dish into the clay. Mark the top of the dish with the word "top."

Observe the grains after three days. They should have begun to grow, or germinate. **3.** In what directions do the root and shoot grow out of each grain? **4.** To what factor in the environment is the root responding? To what factor is the shoot responding? **5.** What type of response is shown by the root? What type of response is shown by the shoot?

Procedure (Part B)

Plants respond to light in their environment. If you carry out the following experiments you will discover some of these responses.

Prepare four boxes as follows:

Box 1: This box should be lighttight. Seal the flaps, if necessary.
Box 2: Cut a window in one side of the box. Leave the window open, but seal the flaps. Place a lamp near the window.
Box 3: Cut a window and cover it with red cellophane. Seal the flaps. Place a lamp by the window.
Box 4: Cut a window and cover it with blue cellophane. Seal the flaps. Place a lamp by the window.

Plant 20 radish seeds in each pot. Water them. Place one of the pots in each box. Keep the lamps lighted constantly.

After five days, observe the radish seedlings. Do not take the pots out of the boxes or move them until your observations are completed. Compare the plant in Box 1 with the other plants. **6.** What is the apparent difference in these plants? **7.** What sort of response do the plants exhibit to light?

Compare the plants in Boxes 2, 3, and 4 with each other. Also, compare the size and color of the plants in Boxes 3 and 4. **8.** Which ones show the most obvious change in direction of growth? **9.** What, if any, difference is there in the responses to blue and red light? **10.** How does the effect of blue light on stem growth compare with that of red light? **11.** What hypothesis might account for the plants' responses to blue and red light?

Procedure (Part C)

Your teacher may be able to arrange for you to view and discuss the BSCS Single Topic Inquiry Film *Phototropism.* This film presents for your interpretation experimental evidence concerning the mechanisms of plant responses to light.

GROWTH SUBSTANCES

After many years of experimentation, biologists have isolated some of the active growth substances in plants. Collectively, they are called **auxins** (AUK·sins). Chemical analysis of auxins has shown their structure. One especially important auxin is called by its chemical name, indoleacetic acid (IAA). IAA is responsible for a number of plant growth responses. It is involved in the development of fruit. It can cause unpollinated flowers to develop into seedless fruits. IAA also is involved in the process of cell division, in the differentiation of xylem cells, and in many other natural processes.

Fruits may drop before they are fully formed or ripe. This early drop is a problem in orchards. Early drop can be controlled by spraying the developing fruits with another auxin.

Synthetic auxins have been produced and found to have properties similar to IAA. In some cases, these are much more active than IAA. Among these are indolebutyric acid (IBA), which is used to stimulate root development in cuttings. Alpha naphthalene acetic acid (NAA) is used to cause fruit to form in unpollinated flowers such as hot-house tomatoes. It also controls early fruit drop and regulates time of flowering in the pineapple. Other synthetic auxins are used to kill plants or to kill their foliage. When used properly, 2, 4-D can control some kinds of weeds. The substance is so poisonous to humans and animals, however, that its use is undesirable.

SCIENCE AT WORK: A Pest Becomes a Friend

Water hyacinth is a hardy, vascular plant with large leaves, thick stems, and violet blossoms. It thrives in warm waters and is a serious problem in Florida. Here, the hyacinths grow in such numbers that they may completely block waterways to all but the largest boats. This overpopulation is partly due to the fact that the manatee, commonly known as the sea cow, is bordering on extinction. Manatees live on vegetation consisting mainly of the water hyacinth. Now with only an estimated 600 manatees left in Florida waters, the water hyacinth grows with almost no interference from animal feeders. Millions of dollars are spent to eradicate the hyacinth.

Recently the National Aeronautics and Space Administration (NASA) has found that water hyacinths can be used as a natural "water treatment system." By absorbing pollutants through its roots, the hyacinth extracts things like mercury, lead, strontium-90, and complex carcinogenic compounds. NASA operates a water hyacinth system to remove chemical wastes on the NASA base in St. Louis, Missouri. The system consists of seven interconnected ponds teeming with water hyacinths. Water entering the pond is foul but water flowing from the last pond is clear and odorless, containing fewer chemicals than artificially treated drinking water. After a month or so, chemically saturated hyacinth plants are harvested. One hyacinth can breed 65,000 offspring a year so the remaining hyacinths quickly reproduce new ones.

NASA is now carrying on research which could lead to water hyacinths being used aboard spacecraft to consume crew wastes, take carbon dioxide from the air, and provide oxygen. The chemically saturated plants could then be harvested, dried, cleansed of impurities, and converted into food.

Another group of growth-controlling substances used in agriculture are the **gibberellins** (jib·er·EL·inz). Gibberellins were obtained originally from a fungus that infects rice plants in Japan. Gibberellins stimulate stem elongation in many different plants. They also stimulate production of new phloem cells by the cambium and have some effects on flowering.

REPRODUCTION IN PLANTS

Many kinds of plants and animals produce offspring by asexual as well as by sexual reproduction. Sexual reproduction involves the formation of eggs and sperm and the union of an egg and a sperm, forming a new individual. In asexual reproduction, a new individual is formed in many ways—but never by the union of egg and sperm. Let's consider some examples of asexual reproduction in flowering plants.

Asexual Reproduction

Each spring, farmers prepare "seed" potatoes for planting. They cut each potato into several pieces. Each piece includes one or more "eyes," which are actually bud clusters. When the pieces of potato are planted, the eyes develop into leafy shoots. The shoots obtain water, starch, and other nutrients from the pieces of potato. Soon roots appear at the low end of each shoot. Before long, the young potato plants are well established (Figure 21-2). In this way, quite a number of new plants can be grown from one potato.

The potato plant, banana plant, and other examples shown in Figure 21-2 have two things in common. First, single, mature plants produce new plants that are genetically identical to the parent. Second, the new individuals come from plant parts produced solely by mitosis and cell division. This type of asexual reproduction is called **vegetative propagation.**

Using a desirable variety of "seed" potatoes, a grower will get new potato plants that are almost identical to the desired type. Seeds produced by potato flowers would produce great variations in size and quality.

Sexual Reproduction

We do not have time to study sexual reproduction in all the plant groups. We will, however, examine a familiar, flowering plant. We will then contrast its reproduction with that of green algae, mosses, and ferns.

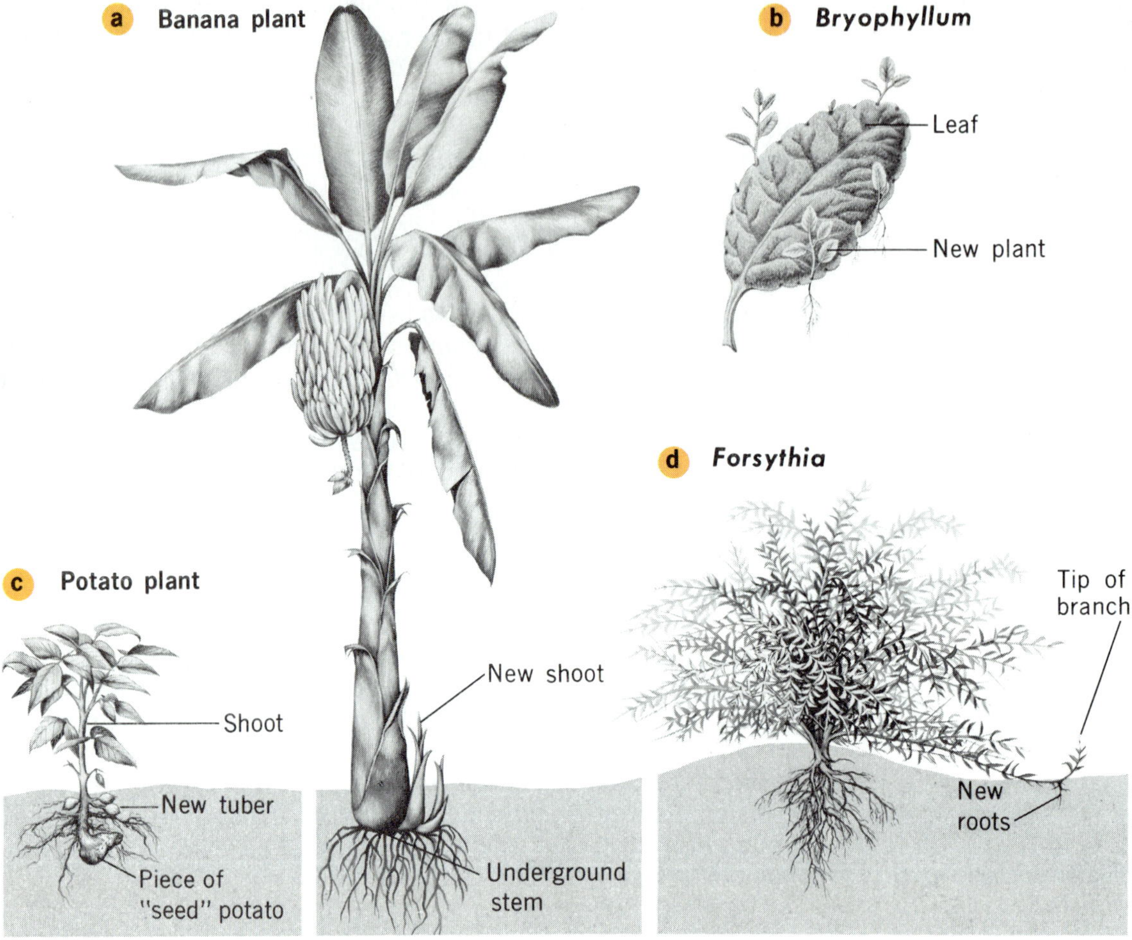

21-2. Asexual reproduction in four flowering plants: **(a)** New shoots produced at the base of a banana plant. **(b)** New plants on *Bryophyllum* leaves. The tiny plants drop off and grow. **(c)** Potatoes grown from pieces of potatoes. **(d)** *Forsythia* plants take root where branches touch the soil.

With few exceptions, all flowering plants produce seeds. The flowers are the reproductive structures of the plants. The outermost part of a flower usually is made up of green, leaflike structures. These parts, called **sepals** (SEE·pulz), cover the flower bud when it is young. Petals form inside the sepals. In some flowers – lilies, for example – the sepals look just like petals. Both sepals and petals are attached to the enlarged end of a stem.

Inside the petals are three to ten stalks, or **filaments,** each of which ends in a little sac. The small sacs are **anthers,** in which thousands of tiny pollen grains are produced. Each mature pollen grain contains two sperm nuclei. Each anther, with its slender filament, is a **stamen** (STAY·men) – the male part of a flower. The number of stamens varies in different plants.

The thick stalk in the center of the flower is the female part, called a **pistil.** The pistil is composed of an ovary (the enlarged part at its

21-3. The parts of a flower. The stamen is the male part. The pistil is the female part.

base), an elongated **style,** and a **stigma** at the end of the style. The parts of a typical flower are shown in Figure 21-3. (Note that, in some species, the stamen and the pistil are on *different* flowers.)

The process of sexual reproduction in a typical flowering plant is shown in Figure 21-4. **Pollination** is the transfer of pollen from anther to stigma. In some species, pollen is carried from the anther of one plant to the stigma of another by insects or the wind. This is cross-pollination. In other species, the flowers are self-pollinating (Figure 21-5). After pollination, each pollen grain forms a tube that penetrates the stigma. The tube grows down through the style of the pistil, and the two nuclei travel down the tube.

The ovary contains one or more immature seeds. (Only one is shown in Figure 21-4.) Each immature seed contains several nuclei. One of these is the egg nucleus. Another is called the fusion nucleus.

21-4. Pollination and seed formation. The endosperm in the seed **(d)** provides food for the developing embryo.

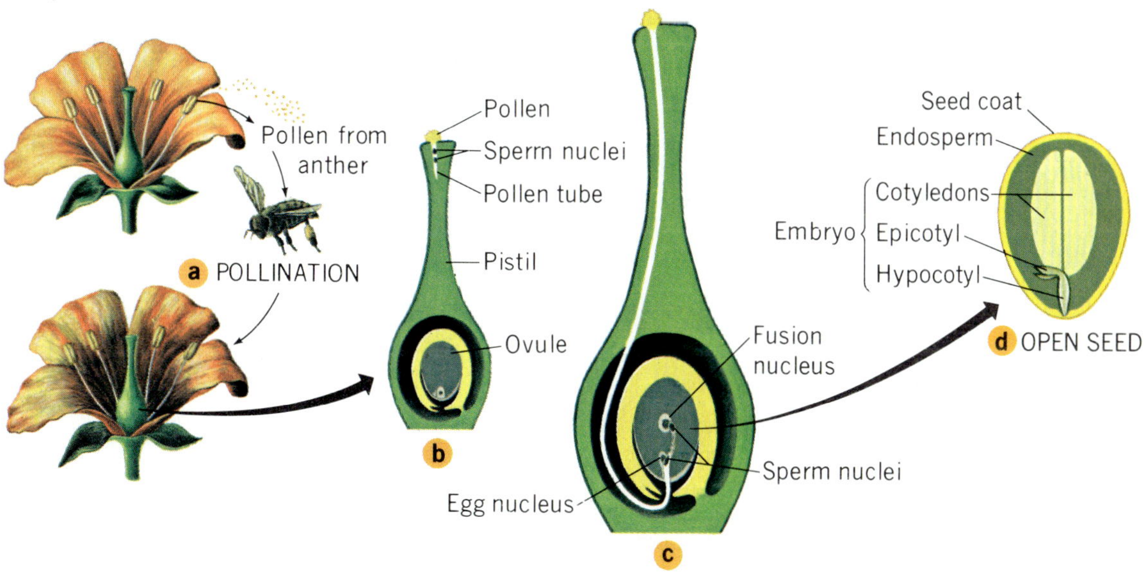

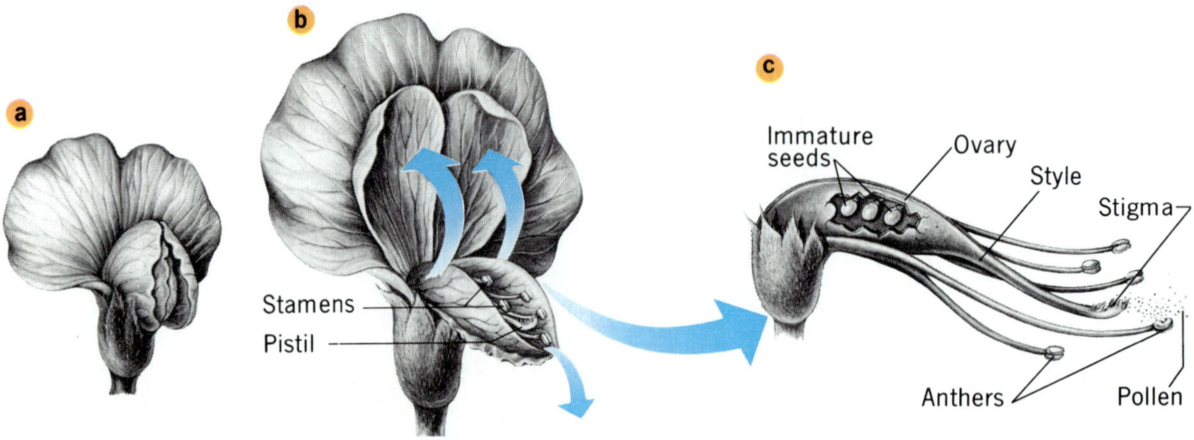

21-5. The flower and the fruit of the garden pea. **(a)** The flower is normally self-pollinating because its stamens and pistil are enclosed by modified petals that usually prevent cross-pollination. **(b)** The petals are shown opened. **(c)** An enlarged view of the reproductive structures of the flower.

One of the two sperm nuclei combines with the egg nucleus, forming a zygote. The other sperm nucleus combines with the fusion nucleus. In most plants, the combination of sperm and fusion nucleus begins the development of the **endosperm,** which stores food that is used by the embryo during its development.

After fertilization, the seed, containing embryo and endosperm, grows rapidly. A seed coat forms as the seed reaches full size. The seed then becomes dormant. The pistil, sometimes along with other parts of the flower, forms a fruit. Later, if conditions of moisture and temperature are right, the seed will germinate and begin to grow. You will examine the parts of a seed in **Inquiry 21-2.**

INQUIRY 21-2 From Seed to Seedling

MATERIALS

Parts A and B

soaked bean seeds
bean seeds, germinated 1, 2, 3, and 10 days
stereoscopic dissecting microscope or hand lens
dilute iodine solution
sharp razor blade

Part C

soaked, germinating corn grains
petri dish with starch agar
petri dish with plain agar
petri dish with starch agar and germinating corn grains
petri dish with starch agar and corn grains killed in FAA
iodine solution
sharp razor blade

Procedure (Part A)

What's in a seed? Look at some of the external features of a bean seed. Notice that the seed is covered by a tough, leathery coat. Look along one edge of the seed and find a scar. This scar marks the place where the seed was attached to the parent plant. **1.** What do you think is the function of the seed coat? **2.** What would you guess to be the function of a connection between a parent plant and a developing seed?

Remove the seed coat. Inside the seed coat you will find the embryo. Two fleshy halves called **cotyledons** (kot·uh·LEE·dunz) make up part of the embryo. These two halves also are found in peanuts, peas, and many other kinds of seeds.

Cut a little sliver from one of the cotyledons. Test the sliver with iodine solution. **3.** What would you deduce is the primary function of the cotyledons? **4.** What was the original source of the substance of the cotyledons?

Find the little plant attached to one end of the cotyledons. This is the rest of the embryo. Take a closer look at it with a hand lens or dissecting microscope. You will see that this part of the embryo has two miniature leaves and a root. The small leaves, plus a tiny shoot tip, make up the **epicotyl** (EP·ih·kot·ul) of the embryo. The root portion is the **hypocotyl** (HY·poh·kot·ul).

Procedure (Part B)

Look at bean seedlings that are 1, 2, and 3 days old, grown either in soil or in water. Then compare a ten-day-old seedling with the three-day-old seedling. **5.** What part of the plant becomes established first? **6.** What part of the embryo gives rise to the root of a bean plant? **7.** Where are the first leaves of the seedling? **8.** What part of the embryo produces the first leaves? **9.** What has happened to the cotyledons in the ten-day-old seedling? **10.** Where is the seed coat in this plant? **11.** Which part or parts of the embryo developed into the stem? **12.** How are the first two tiny leaves arranged on the stem?

Procedure (Part C)

Cut a soaked, germinated corn grain lengthwise with the razor blade. Test the cut surfaces with iodine solution. **13.** What food is present, according to the results of this test? **14.** What other foods might be present in corn that are not demonstrated by the test?

Test the starch agar and the plain agar with iodine solution. **15.** What difference do you observe?

On one petri dish, two or three corn grains have started to germinate in starch agar. Each grain was cut lengthwise and the cut surfaces were placed on the agar for about two days. Another petri dish contains starch agar and cut corn grains that have been killed. Remove both the germinating and the killed corn grains from the starch agar.

Cover the surface of the starch agar in each dish with iodine solution. After a few seconds, pour off the excess. **16.** Suggest hypotheses that might account for the appearance of the areas that were covered by the germinating corn grains. **17.** What kind of food would you expect to find in the clear areas? **18.** What do you observe in the petri dish that contained germinated but killed corn grains? Would dry corn grains work as well as moist, germinating ones? Explain.

If time allows, devise an experiment that might give evidence about your hypotheses in question 16.

Growth

You could hold a hundred seeds of the giant California redwood in the palm of your hand. Each seed, in favorable conditions, may develop into the largest living thing the world has ever known (Figure 21-6). Each of these tiny seeds contains genetic information that can result in a huge plant with hundreds of branches bearing millions of leaves. Extending into the soil from the base of the massive trunk are large woody roots, which branch and rebranch. These finally end in many millions of slender, pointed root tips.

All of the tissues of the plant embryo—the epicotyl, hypocotyl, and cotyledons—are made up of cells. These cells originate from repeated divisions of the zygote. When the seed matures and becomes dry, the cell divisions stop. When the seed is supplied with water, however, the cell divisions begin again. As each new cell grows to the size of the parent cell, the total volume of the cells increases. A full-grown plant has a volume millions of times greater than that of the original, fertilized egg cell from which it was formed. Thus, growth is the result of an increase in both the number and volume of the cells.

A mature plant, however, is not simply a collection of millions of identical cells. Rather, it is composed of many different types of cells. There are tracheids and companion cells, cells of the spongy layers of the leaf, and many more. So the development of the fertilized egg involves more than an increase in cell number and volume. It also involves differentiation into different kinds of cells and tissues. The process of differentiation, which was discussed in Chapter 8, is similar in both plants and animals.

21-6. The giant redwood trees in California are the largest living things in the world.

Alternation of Generations

Some of the marine algae, known as seaweeds, may be very large. *Ulva* is a common form of alga found in tidal pools of both the Atlantic and Pacific coasts (Figure 20-15, page 573). *Ulva* does not have leaves, stems, roots, or flowers.

Ulva has a very interesting *life cycle.* A life cycle is the sequence of stages from fertilization through growth and maturity to the production of eggs and sperm by the new organisms. Understanding the life cycle of *Ulva* will help you to understand the life cycle of some of the higher plants.

If you were to obtain fresh *Ulva* from a tidal pool, you might be able to observe part of its life cycle. Looking at fragments of the living green plants with the microscope, you would see that all the cells are not the same in structure. Some would be in the process of releasing tiny, motile cells into the water. Almost as soon as these cells are deposited in the water, they start swimming about. Some of the cells eventually pair and fuse with one another. This is fertilization—the pairing of gametes in sexual reproduction that results in a zygote.

The zygotes of *Ulva,* like those of most other marine plants, start to grow after a very short time. The rate of growth may be slow, but the zygote does grow into a new plant. The new plant looks just like the specimens observed in tidal pools.

With some luck and patience, you might be able to see the reproductive cells this generation forms. Except for the number of flagella, the new reproductive cells (Figure 21-7) look very much like the gametes formed by the original plant. However, the new reproductive cells do not behave like gametes. The new reproductive cells do not fuse in pairs.

These reproductive cells are not gametes. Rather, they are **spores.** Spores are asexual reproductive cells. In *Ulva,* the spores are produced by meiosis. But they do not fuse in fertilization as gametes do. Instead, the spores lose their flagella and begin to divide and grow. They grow into new plants that look exactly like those that grew from the zygotes.

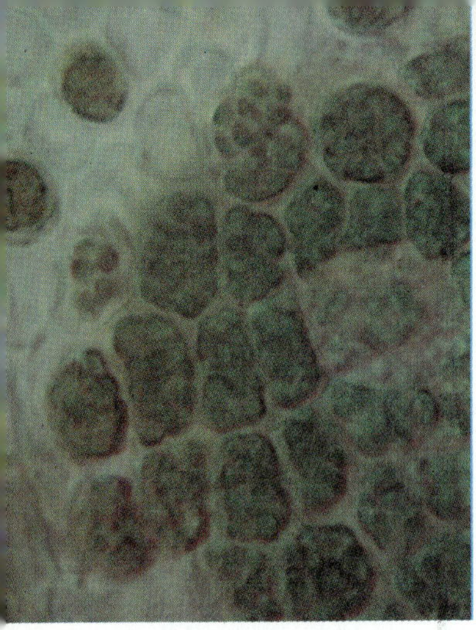

21-7. Spores of *Ulva*.

We see that there are *two* generations of plants in this life cycle. Although the plants look alike, one generation produces only gametes. The gametes fuse in pairs, and a zygote results. When the zygote grows, it gives rise to the spore-producing plant. The spores that are produced grow into plants that produce more gametes. The cycle is complete. It is a life cycle in which there is **alternation of generations** (Figure 21-8).

An alternation of generations occurs in those plants with two multicellular stages. One of the generations produces gametes and the other produces spores. A multicellular, gamete-producing generation, or **gametophyte** (ga·MEE·toe·fite), alternates with a multicellular, spore-producing generation, or **sporophyte** (SPOH·roh·fite), in the life cycle of the plant.

The two generations differ in the number of chromosomes in their cells. Study the life cycle of *Ulva* (Figure 21-8), starting with the

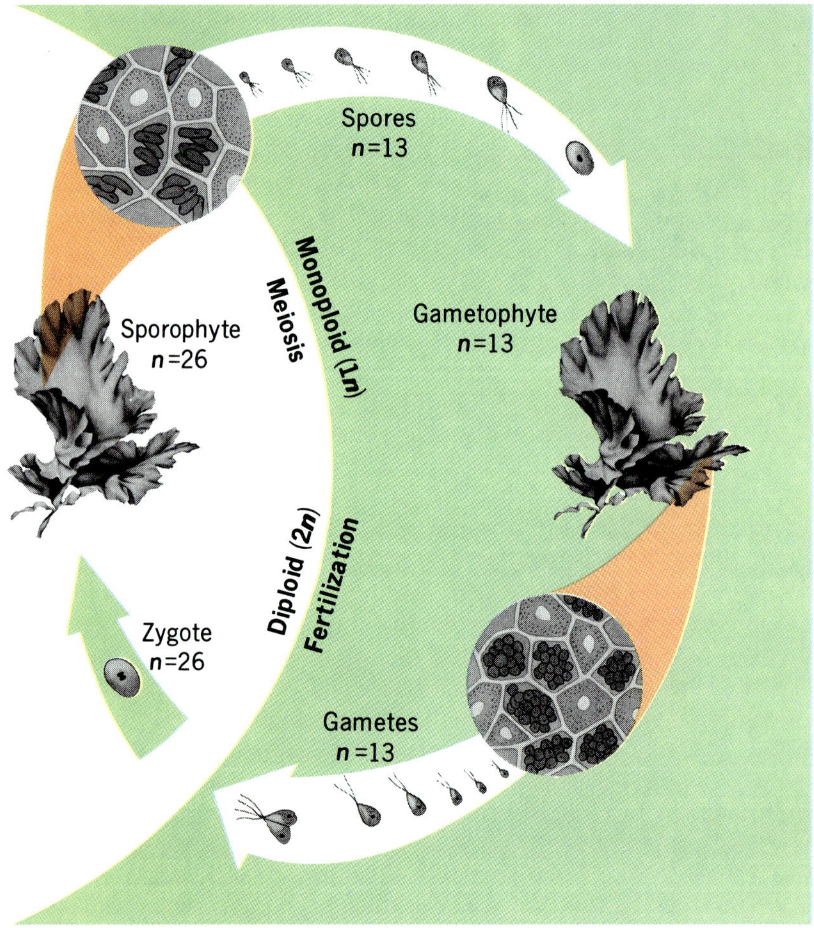

21-8. The life cycle of *Ulva* involves alternation of generations.

gametes and fertilization. Each gamete of *Ulva* has a nucleus containing 13 chromosomes. This is the monoploid number. Two gametes fuse, forming a zygote. The zygote has the diploid chromosome number of 26. The zygote grows by mitotic divisions. Thus, each cell of the new plant is diploid.

Certain reproductive cells of this diploid plant (the sporophyte) divide by meiosis forming monoploid spores. The monoploid spores grow by mitotic divisions into monoploid plants (gametophytes). Some cells of these plants produce monoploid gametes. Now we have returned to the beginning of the cycle.

HOW PLANTS ARE ADAPTED TO LAND

A very long time ago, all green plants lived in the ocean. Then, in the gradual, continuing process of evolution, these plants gave rise to plants that lived on land. What characteristics must have been selected that enabled plants living in the ocean to become plants that could live on land? First, new structures that permitted water absorption probably evolved. Land plants are not bathed in water but must absorb it from the ground. Special structures that allow the plant to hold water also evolved. Methods also evolved for absorbing carbon dioxide, needed for photosynthesis, from air rather than from water. We do not know exactly how these adaptations occurred. You can form some hypotheses, however, as you examine the structures of the mosses and their relatives, the liverworts.

Mosses and Liverworts

The mosses belong to a major group of green plants, the mosses and liverworts. Simple adaptations allow this group to live on land (Figure 21-9). You may have seen moss plants growing in clumps and forming mats on logs and on the forest floor. Liverworts may be found on rocks at the edge of a woodland stream. These plants grow best in damp, shaded environments. These plants, like *Ulva,* have a life cycle with distinctive alternation of generations. They have a

21-9. (a) Moss growing on rock; **(b)** male liverworts and **(c)** female liverworts of the genus *Marchantia.* These primitive green land plants are frequently found growing on rocks and soil of moist ravines.

a b c

sporophyte generation that produces spores followed by a gametophyte generation that produces gametes (Figure 21-10).

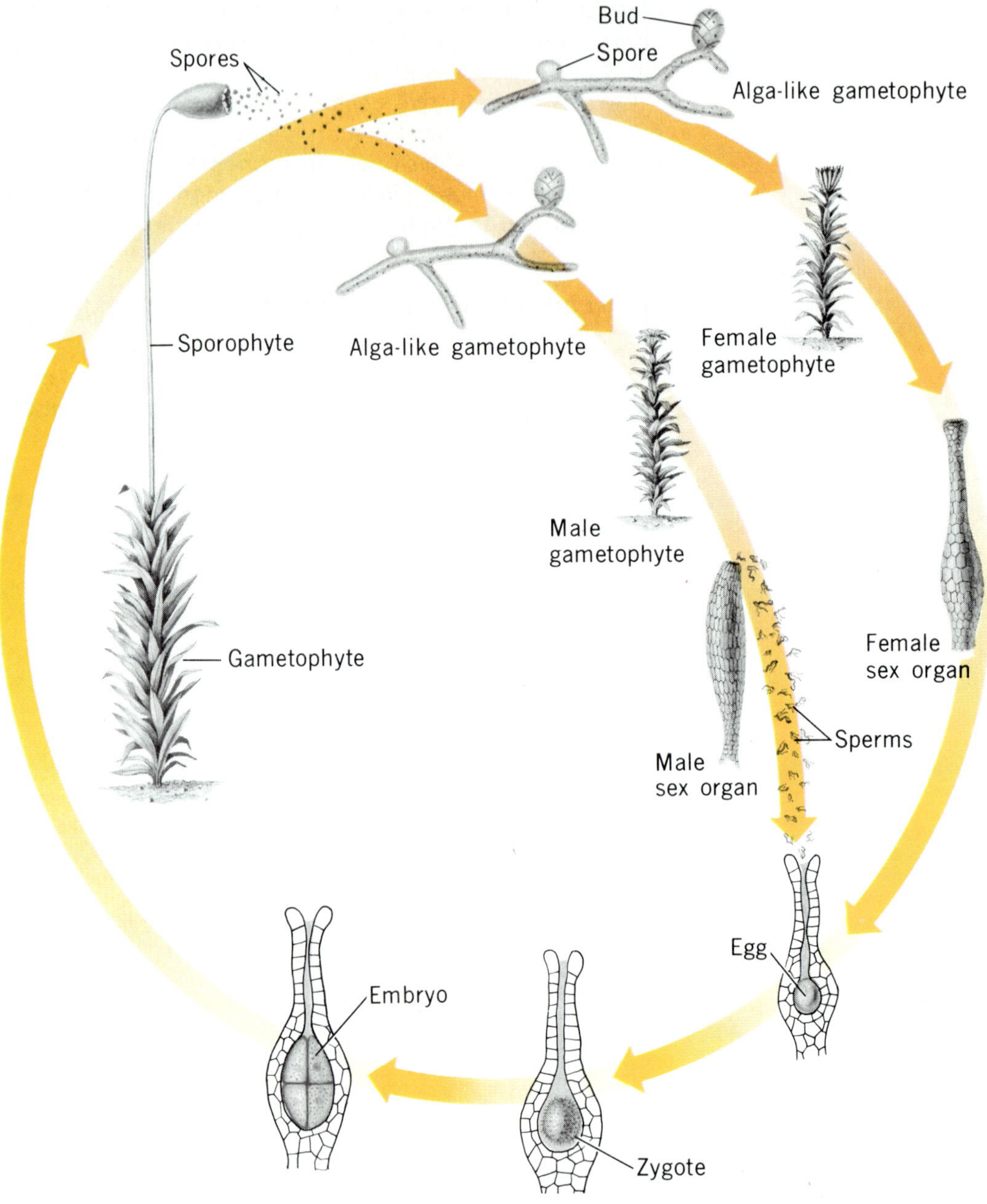

21-10. Compare the life cycle of a moss with that of *Ulva*, an alga (Figure 21-8). *What differences between the moss life cycle and that of Ulva may be the result of the land habitat of the moss? What evidence do you see that mosses may have originated from an aquatic ancestor?*

21-11. Rhizoids on the undersurface of *Marchantia*.

In *Ulva,* the two generations look the same although they differ in chromosome numbers. These generations in mosses differ, however, not only in chromosome numbers but also in the appearance of two generations.

Almost any clump of moss plants that you find is composed of many gametophytes all growing close together. Sporophytes grow out of the tops of the gametophytes. The gametophytes produce many sporophytes, which look like pins sticking out of the clump of moss plants.

If you look at the underside of the liverwort, *Marchantia* (Figure 21-11), you will see structures called **rhizoids** (RYE·zoidz). Rhizoids also occur in other mosses and liverworts. These structures are long, slender extensions of cells on the underside of the plant. The rhizoids may increase the plant's ability to absorb water from the soil.

In **Inquiry 21-3,** you will examine the differences between the sporophyte and the gametophyte generations in mosses.

INQUIRY 21-3 Mosses, Primitive Land Plants

MATERIALS (per team)

moss plants with sporophytes
moss plants with antheridia
moss plants with archegonia

Procedure

Take a moss plant with a sporophyte attached. Pull the filamentous stalk of the sporophyte out of the leafy shoot of the gametophyte. The two generations are now separated.

The sporophyte part of the moss has a smooth, leafless stem ending in a little capsule. With a dissecting needle, break open the capsule of the sporophyte. Pour its contents into a drop of water on a slide. Cover the slide with a cover glass. **1.** What are the structures you observe with the microscope? **2.** In the wild, how are these structures scattered from the mature plant?

Once released from the spore capsule, the spores can survive for several months. In time, these tiny cells will produce either male or female gametophytes. You may find Figure 21-10 helpful as you observe living moss plants. This figure illustrates the appearance of the different generations.

The spores of most mosses germinate readily on damp soil. They produce a filament that looks more like a branching green alga than a part of a moss plant. Next, a leafy shoot of the gametophyte grows from the filament. The leafy shoot is photosynthetic. **3.** How does this leafy shoot obtain water and minerals for growth?

The gametophyte has reproductive organs and produces gametes. Let us take a look at the sex organs at the upper end of the leafy shoot. Hold a male or female plant so that the tip is between your thumb and forefinger. Squeeze the tip and at the same time roll it between your thumb and forefinger. Submerge the tip of the leafy shoot in a drop of water on a microscope slide. Tease out the fragments from the tip with a dissecting needle. Some of the fragments will be the sex organs.

Put a cover glass over the fragments from each plant. Observe the preparations with the low power of your microscope. The male sex organs *(antheridia)* are simple, saclike structures. They produce many sperm cells. The female sex organs *(archegonia)* are flask-shaped and have long, twisted necks. An egg is formed within the swollen base of the archegonium. **4.** How does a sperm reach the egg? **5.** Where is the zygote formed? **6.** As the zygote develops, what does it become? **7.** In alternation of generations in a moss, which is the dominant, independent generation? Which is the less conspicuous generation?

Obtaining and Conserving Water

All mosses and liverworts are multicellular organisms. Does this multicellular organization have any advantages in a land environment? Mosses and liverworts are composed of hundreds of thou-

21-12. (a) When rains come liverworts rapidly become active and begin to grow. They will form a tiny green mat. **(b)** A short time later, the earth has dried. A few other plants have begun to sprout but the liverworts have curled up and look almost like the soil.

sands of cells. But only a small percentage of cells have surfaces directly exposed to the drying effects of the atmosphere. In addition, a layer of cutin on the surface of some liverworts and mosses reduces the rate of evaporation from surface cells.

Some mosses and liverworts that grow on exposed surfaces of rocks and trees have still another remarkable characteristic. Under certain conditions, they become so dry that they are almost brittle (Figure 21-12). Yet, within a short time after becoming wet from rain or other water, they turn green again and go on with their normal life processes.

Reproduction

You will recall from your study of moss plants in **Inquiry 21-3** that the gametophyte generation forms sperm and eggs. Similar patterns are found in all mosses and liverworts. The sperm have flagellae and can swim to the egg. This means that a film of water must be present for the sperm to swim in. A sperm swims to the egg and fertilizes it. A zygote is formed and an embryo develops inside the female reproductive structure. The coverings formed by the female organism protect the egg, the zygote, and the growing embryo from drying out and from mechanical injury.

This adaptation is of great importance in the success of mosses and liverworts in living on land. As a matter of fact, similar adaptations occur in all green land plants. In all, sperm move to the eggs, and zygotes form inside the female reproductive structures.

In mosses and liverworts, the two generations are quite different in appearance. The sporophyte grows attached to the gametophyte and is partially dependent on it. The sporophyte is relatively inconspicuous. This is not true of most other land plants, however. In the adaptation of aquatic plants to life on land, the sporophyte became the dominant generation. The gametophyte, on the other hand, became greatly reduced in structure. The sporophyte is the dominant stage in most familiar flowering plants.

In **Inquiry 21-4,** you will look more closely at ferns. This group is less complex than flowering plants but more advanced than mosses and liverworts.

INQUIRY 21-4 The Fern, a Primitive Vascular Plant

MATERIALS (per team)

whole fern plants
fertile fern fronds
living fern gametophytes
slide of cross section of fern stem
slide of fern gametophytes
microscope slides
cover glasses
compound microscope
stereoscopic dissecting microscope

Procedure (Part A)

Examine the fern plant. Notice the horizontal underground stem *(rhizome)*, the roots, and the leaves *(fronds)*. This is the sporophyte stage of the fern life cycle. **1.** What evidence is needed to determine whether a plant is a sporophyte or a gametophyte? **2.** What shape are the leaves before they are fully developed? **3.** How does the mature sporophyte of the fern compare with the sporophyte of a moss plant? Explain your answer in terms of complexity, length of life, and method of nutrition in the two kinds of plants.

Observe the upper and lower surfaces of the fronds. Notice that some have small, brown, dotlike or elongated structures on them. Scrape some of this brown material onto a slide. Prepare it for examination with the microscope.

The small, stalked structures are spore cases (Figure 21-13). Notice the row of thick-walled cells across the top and back of the spore case. Some of the spore cases probably will have been broken by handling. The spores will be scattered on the slide. Other spores can be observed still in the cases. **4.** What part of the moss

21-13. A cross section of a fern leaf, showing spore cases.

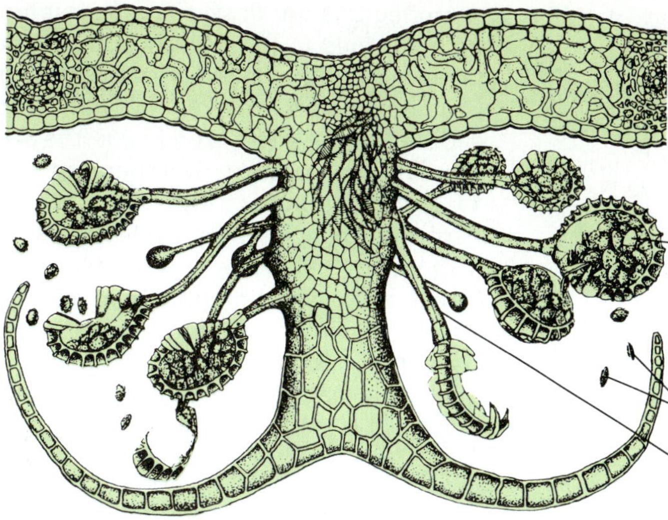

sporophyte is similar in function to the spore cases of ferns?

Study a demonstration slide showing fern stems in cross section. Find the vascular tissue—xylem and phloem. It may be in the form of a "C" or a ring of bundles within the stem. Some cells of the vascular tissues are empty, thick-walled cells. These are the cells of xylem. The phloem consists of thin-walled cells forming a layer around the xylem. **5.** How do these conducting tissues compare in structure and function with the conducting tissues that you studied in flowering plants?

21-14. Ant's-eye view of a fern gametophyte. The gametophyte is only a few millimeters wide. The undersurface is shown. The rootlike rhizoids are visible, but the male and female reproductive structures are not.

Procedure (Part B)

Put a gametophyte on a glass slide. Examine it first with the dissecting microscope. Look at the lower surface (Figure 21-14). Gametophytes are formed when spores settle from the air in a moist place. **6.** What structures appear to be responsible for water absorption? **7.** In what other plants have you seen these structures?

In addition to the absorbing structures, sex organs also are located on the lower surface. To see these structures, use the low power of your compound microscope. Antheridia are dome-shaped structures. These generally develop before the archegonia. Look for antheridia near the margins of the gametophyte. Crush an antheridium in a drop of water on a clean glass slide. Look for sperm in the water. They are quite small. **8.** On the basis of your observations, how do you think the sperm reach the egg cells in a fern plant? **9.** How does this compare with what happens in an alga and in a moss?

Archegonia are flask-shaped structures. The enlarged lower portion is buried in the tissues of the gametophyte. This portion contains an egg. The neck of the structure projects above the surface. If you cannot find the archegonium on the living plant, look at the slide. The egg is fertilized inside the archegonium. **10.** What generation, in alternation of generations, begins with a fertilized egg? **11.** Is it monoploid or diploid? **12.** Which generation is the dominant one in a fern life cycle? **13.** Which generation is photosynthetic? **14.** Why do ferns normally not live in dry environments even though their sporophytes have vascular tissue? **15.** Where would you look for fern gametophytes growing in the woods? Explain. **16.** How does the gametophyte of a fern compare with that of a moss in food-getting, length of life, and function? **17.** Is the gametophyte monoploid or diploid?

The Importance of Vascular Tissue in Land Plants

Xylem and phloem (Figure 21-15) are found in all green land plants except mosses and liverworts. These fluid-conducting tissues are called vascular tissues. Vascular tissue has two functions—conduction and support. Xylem is composed mostly of dead cells whose walls may be perforated, permitting rapid movement of water. The direction of movement generally is upward, from the roots to the leaves. The walls of the dead xylem cells usually are thick. They help support the plant above ground.

Phloem tissue is made up of both living and dead cells. Living cells of phloem conduct the materials manufactured by the photosynthetic leaves. Here the direction of movement is downward toward the stem and roots, where the food is stored. The dead, thick-walled cells of the phloem aid in support. You can think of the vascular tissue as a rapid transit system for movement of materials throughout the plant's body. Could this be one major reason land plants with xylem and phloem are large and those lacking these tissues are small?

In Brief

Plants respond to environmental factors such as gravity, light, and the quality of soil. Some natural and synthetic growth substances can be used to produce improved strains of plants.

Plants reproduce both asexually and sexually. In flowering plants, where sexual reproduction is most developed, a seed produced by a sperm and egg develops into a new plant.

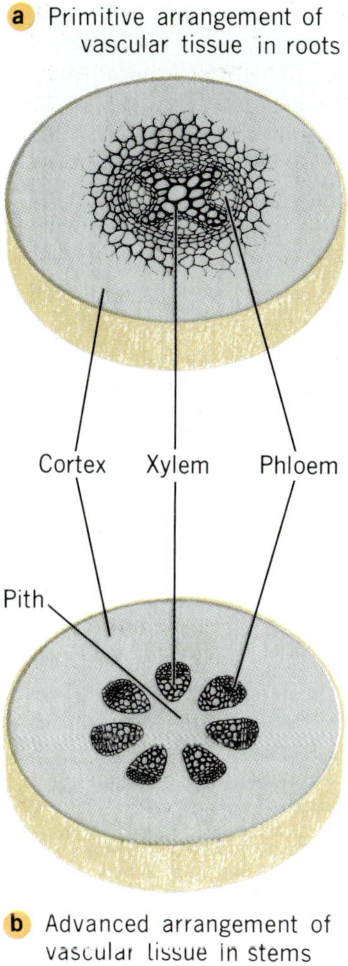

21-15. Arrangements of vascular tissue. **(a)** Roots of most vascular plants, and stems and branches of the more primitive vascular plants, have a central core of xylem, with phloem around it. **(b)** Stems of more advanced vascular plants usually have a central pith, surrounded by a ring of vascular tissue (often divided into bundles).

Scientists have evidence that green land plants evolved from green plants that lived in the ocean. Algae are examples of aquatic green plants. Mosses and liverworts are green land plants that have some similarities to algae. Alternation of generations occurs in mosses and liverworts. Ferns also have alternation of generations but are more complex than mosses and liverworts. Ferns contain vascular tissue similar to that found in flowering plants.

Understanding what makes plants grow and how they reproduce has practical implications. For example, it may help us get the most food value from our crops.

Guide Questions and Problems

1. How is vegetative propagation important to the survival of plant species that reproduce in this way? How is vegetative propagation important to humans in the cultivation of plants?
2. What are some of the practical uses of auxins and synthetic growth substances as they apply to farming and the raising of fruits and vegetables?
3. In what ways does light affect a green plant?
4. What part of a plant is represented by the following foods: apples, string beans, corn, cauliflower, beets, rice, lettuce?
5. Fruit production may be reduced in orchards near fields where DDT or other insecticides are used, even though the insecticides may not reach the fruit trees. How might this be explained?
6. What evidence can you cite to show that green algae are the most likely ancestors of green land plants?
7. Much evidence indicates that plants and animals originated in the ancient seas of the earth. From this place of origin, they colonized the dry land. Why was it necessary that green plants colonize the land before animals could do so?
8. How might you explain the fact that flowering plants have become the dominant form of plant life on land?
9. The small tree (Figure 21-16, left) is unusual in that its right part seems different from the left. You can see that the color is deeper and the branches seem to hang down rather than grow erect. On closer examination (Figure 21-16, right), you see that the small branches from the left side have tiny leaves. The branches from the right have peculiar seedlike structures. What observations would you make to explain why one part of the tree seems so unlike the other?

21-16.

Ideas for Independent Study

1. In many plants, the leaves show rhythmic movements during the cycles of day and night. Often their leaves spread out horizontally in the daylight and move into a more vertical position at night. This response occurs in bean plants and the sensitive plant *Mimosa* (Figure 21-17). It also occurs in common house plants called prayer plants (belonging to the related groups *Maranta* and *Calathea*). If you have a place to keep plants, you might want to study these 24-hour cycles of movement.

2. You can obtain many plants, or their seeds, from florists, seed companies, nurseries, and supermarkets. You might want to grow and study some plants that have special adaptations. For example, air plants, such as orchids, need not grow in soil. Some plants, such as the Venus flytrap, are carnivorous. Some—cacti, for example—are adapted to very dry conditions. *Bryophyllum* and the autumn crocus *Colchicum* have unusual growth patterns.

You might want to work with different conditions of light, temperature, soil, and so forth. Hydroponics, the technique of growing plants in water without soil, is worth looking into. You also can study plants in a botanical garden or a nearby park or vacant lot. How do they reproduce? How well are they adapted to such environmental factors as light, humidity, and temperature?

21-17. *Mimosa.* **(Left)** Leaves are extended in the daytime. **(Right)** At night, the leaves are closed.

3. Plants always have played a large part in the cultural life of people. Flowers, seeds, fruits, leaves, whole plants, and other botanical subjects can be found in decorative arts and designs since the earliest times. Myths, folk medicine, poetry, and literature contain endless references to plants. Investigate these sources. You might make some surprising discoveries about art and literature and add to your understanding of plants.

Related Readings

BOOKS

Bold, H. C. *The Plant Kingdom,* 3rd ed. Englewood Cliffs, N.J.: Prentice-Hall, 1970.

Jacobs, W. P. and C. E. LaMotte. *Regulation in Plants by Hormones—A Study in Experimental Design.* A BSCS Laboratory Block. Lexington, Mass.: D. C. Heath, 1964.

Lee, A. E. *Plant Growth and Development.* A BSCS Laboratory Block. Lexington, Mass.: D. C. Heath, 1963.

Milne, L. J. and M. Milne. *Because of a Flower.* New York: Atheneum, 1975.

Page, N. M. and R. E. Weaver, Jr. *Wild Plants in the City.* New York: Quadrangle/New York Times, 1975.

Salisbury, F. B. *The Biology of Flowering.* New York: Natural History Press, 1971.

ARTICLES

Batra, S. W. T. and L. R. Batra. "The Fungus Gardens of Insects." *Scientific American*, November 1967, page 112.

Fritts, H. C. "Tree Rings and Climate." *Scientific American*, May 1972, page 92.

Galston, A. W. "Bios: New Ways to Increase Man's Food." *Natural History*, October 1973, page 28.

Galston, A. W. "The Limits of Plant Power." *Natural History*, April 1975, page 22.

PART FOUR

INTERACTION

22 Animal Behavior

It was a sunny, summer morning. Bees buzzed around the flowers in the garden. A pair of white butterflies fluttered over the fence. They darted about, flying closely together, up and down, back and forth above the flowers. A caterpillar crawled up the stem of one of the plants.

Nearby, on the lawn, a bird hopped about, occasionally pecking at the ground. Once, as it did so, it pulled its head up suddenly. An earthworm hung from its beak. Quickly, the bird flew up to a tree branch on which a half-grown bird was perched. The young bird chirped loudly, opening and closing its beak. At one instant when the beak was open, the adult bird dropped the worm into the young bird's mouth. Then the adult bird flew back to the lawn and started hopping about again.

Suddenly, the bird flew up into the tree again and began chirping loudly. At the same time, a dog that had been sleeping under a tree awoke, pricked up its ears, and gave a low growl. The young bird on the branch became completely quiet and motionless. Only the caterpillar, the bees, and the butterflies seemed undisturbed. They continued crawling or flying around the flowers.

What caused the activity of the birds and the dog to change so suddenly while the insects did not seem to be disturbed? A large, gray cat had crept along a fence and into the garden. The cat's eyes were fixed on the young bird. The dog's eyes were fixed on the cat.

Does this description surprise you? It does not involve any of the biological principles you have studied so far. The **behavior** of organisms is, however, a most important part of biology. In fact, certain patterns of behavior are absolutely essential for the life of every animal. Certain behaviors tend to increase an animal's chances of surviving and reproducing. Why should this be so?

WHAT IS BEHAVIOR?

Nearly everything we have discussed thus far has involved the internal biology of individual organisms. When we considered food, we were concerned mainly with its chemical nature, digestive systems, and so on. We took for granted that the food was already present, to be eaten and digested. You know full well that much more is involved. Most animals have to search for food. Sometimes they find it; sometimes they do not.

Animal behavior is the sum of everything animals do—flying, walking, sitting, sleeping, eating, mating, rearing young, and so forth. A behavior may be very simple or extremely complex. Animal behavior is a difficult field to study. Yet it is one of the most fascinating areas of biological science.

Behavior is any activity that occurs in response to a stimulus or stimuli. For example, an internal stimulus, such as hunger, will cause food-seeking behavior. Of course, different kinds of animals seek different foods (Figure 22-1). And they also have different methods of obtaining their food. These aspects of behavior are determined mostly by the individual's physiology. The food must be something the body can use and something the individual is physically capable of obtaining.

Some behaviors occur in response to the presence of other animals—of the same species or of different species. Mating behavior and flight from a predator are examples. Some of the simplest behaviors occur in response to environmental conditions, such as gravity, light, and climate.

Innate Behavior

Many behaviors occur very predictably under certain circumstances. They seem largely independent of an animal's past experience. We refer to such behaviors as **innate,** or inborn. For example,

22-1. Food-seeking behaviors vary greatly among different animals. *What do you think is going on in each of these pictures?*

an animal often moves or positions its body in relation to a factor in the environment. These simple behaviors are called **taxes** (TAK·seez; singular, taxis—TAK·sis). The tendency to move toward a stimulus—light, for example—is a positive taxis. Moving away from the stimulus is a negative taxis. A tendency to move toward light (positive phototaxis) may bring an animal closer to a food source. In some cases, a negative phototaxis keeps an organism away from possible predators.

Many animals orient themselves in an up-down position in response to gravity. A taxis in response to gravity may assist an animal, like an insect or crab, in "keeping its feet on the ground." Taxes have been observed in a wide range of animals. However, taxes are difficult to detect in humans, probably because they are overshadowed by other, more complex behaviors.

Another simple and predictable behavior observed in many animals is a reflex. Like taxes, reflexes occur in response to specific environmental stimuli. But reflexes, unlike taxes, are responses of only one part of the animal's body. These reflexes often involve only a few neurons.

In addition to the simple responses of taxes and reflexes, other, more complex patterns of behavior also appear to be innate. Some of these more complex behaviors involve a series of activities, or "chain of events," in which each step is the stimulus for the next step. One such complex behavior pattern occurs in a small freshwater fish, the three-spined stickleback. The series of activities involved in courtship, mating, and fertilization of the eggs are described in Figure 22-2.

The behavior of sticklebacks depends on specific stimuli, as suggested in Figure 22-2. The male stickleback in the figure has a red belly. This is a color change that occurs only at a certain time of the year. The male builds a nest on the bottom of a stream. He then tunnels through it. The male will defend the nest against other male fish that approach it.

When a female fish approaches the site of the nest, the male makes dancing movements. This behavior occurs in response to the visual stimulus of the curved, gray belly of the female. The female is attracted by the dancing movements and the red belly of the male. She follows the male as he swims toward the nest. The female enters the nest and lays eggs. When she leaves, the male enters the nest and fertilizes the eggs. Later, he will guard the nest, protecting the young and aiding their development. The responses of the male and female sticklebacks to the stimuli result in the complex, predictable series of activities that make up this mating behavior.

You can learn much about the behavior of animals by observing them carefully. You also can make observations in a controlled laboratory situation, as outlined in the next four inquiries.

22-2. COURTSHIP AND REPRODUCTIVE BEHAVIOR OF STICKLEBACKS

a Male (right) threatens other males.

b Male's red underside attracts female.

c Male starts zig-zag dance toward nest; female follows male.

INQUIRY 22-1 Behavior of Paramecium

Remember *Paramecium*? Even this animal, with no brain or nervous system, has distinctive behaviors.

MATERIALS (per team)

Part A

Paramecium *culture*
2 long vials or test tubes
stopper for one of the test tubes
depression slide or small dish
medicine droppers (pipettes)
test tube rack
compound or stereoscopic dissecting microscope
black paper or metal foil
light source

Part B

Paramecium *culture*
dilute hydrochloric acid
microscope slide
cover glass
cotton thread
other acids, bases, salts, sugars (optional)

d Male guides female into nest. **e** Male taps female near tail; female lays eggs. **f** Male bites female; female leaves nest; male enters nest and fertilizes eggs.

Procedure (Part A)

Fill two test tubes or vials 1/4 full with a *Paramecium* culture. Place one test tube upright in a rack so that it receives uniform light from all directions.

Stopper the second test tube. Cover the top half of it with dark paper or metal foil. Lay the tube on its side. Illuminate the bottom half evenly.

Now observe the upright tube. In a rich culture of *Paramecium,* a concentration of the animals looks like a white cloud. Examine the tube for any concentration of the animals. **1.** What evidence do you find that the *Paramecium* are, or are not, moving toward one part of the tube?

In this controlled situation, the tendency of the animals to concentrate in one part of the tube suggests a response to gravity—*geotaxis* (jee·oh·TAK·sis). **2.** Is the response to gravity positive, negative, or neither? **3.** What other factors might account for a concentration in this test tube? **4.** What additional inquiries would you devise to check all variables, including gravity?

Observe the partially covered test tube placed on its side. **5.** Has the *Paramecium* culture concentrated at one end of the tube? **6.** Is the phototaxis of *Paramecium* positive, negative, or not definitely either? **7.** Why could this investigation not have been done with the tube in an upright position? **8.** How might this response to light be an advantage to *Paramecium?* **9.** How do these responses of *Paramecium* compare with plant responses (tropisms) you studied in Chapter 21?

Procedure (Part B)

A response to a chemical is called *chemotaxis*. The response of a *Paramecium* to food is chemotaxis.

Place a drop of *Paramecium* culture on a flat slide. Dip a small piece of thread into the dilute hydrochloric acid solution. Place the thread across the middle of the drop. Add a cover glass. **10.** Does the acid cause a positive or negative response in the *Paramecium* culture, or neither?

If time allows, determine the responses of *Paramecium* to other acids, bases, salts, sugar, and so forth. **11.** Summarize your observations of the behavior of *Paramecium*. **12.** What can you say about the predictability of the responses of *Paramecium* to different stimuli?

Thigmotaxis (thig·moh·TAK·sis) is a response to touch. If time allows, devise a simple investigation to determine whether *Paramecium* or another organism responds to touch.

INQUIRY 22-2 Behavior of Fruit Flies

MATERIALS (per team)

test tube containing several fruit flies (Drosophila)
lamp
light bulbs of different colors or white light bulbs and colored cellophane

Procedure

Hold the test tube of fruit flies horizontally about 30 cm from a glowing light bulb. **1.** How do the flies move in relation to the light?

Slowly turn the tube of flies vertically and then horizontally, or at an angle. Do not jar or tap the tube or otherwise disturb the flies. **2.** Why not?

Change the position of the tube. Hold it 100 cm from the light. Note the flies' response. **3.** How does it compare to the response at 30 cm?

Move the tube toward the light and then away from it. **4.** At what distance can you best observe the flies' response to the light? **5.** What is the minimum distance? the maximum distance? **6.** What do these observations indicate about fruit fly behavior in relation to the brightness or intensity of the light?

Using colored light bulbs, try your procedures again. (If you use white light bulbs and colored cellophane, hold the cellophane in front of the bulb. Do not wrap the cellophane around the bulb; it might burn.) **7.** How do the flies react to light of different colors? **8.** How does the position of the test tube affect the direction in which the flies move in response to light?

In addition to giving off light, the electric bulbs also give off heat. **9.** What effect might the heat have on your observations?

Devise a simple experiment that will test the reaction of the flies to heat or to cold. **10.** How does the direction of movement of the flies change as they react to different temperatures?

INQUIRY 22-3 Reflexes in Humans

Reflex actions occur without conscious effort when certain nerves are stimulated. We are born with some of these reflexes. We may acquire others later in life. Innate reflexes are not dependent on past experience or training, as are the acquired ones.

Following are a few investigations you can carry out in the classroom. Work in pairs, with one student as a subject, the other as an observer.

The Knee Reflex. The subject sits on the desk top or on a chair, with one leg crossed over the other so that the foot is off the floor. The observer uses the edge of the open hand to strike the subject's knee gently but quickly just below the kneecap. If the knee is struck in the proper place, a reflex action occurs. Compare all students in the classroom for the knee reflex. **1.** What is the response? **2.** How does it vary with the intensity of the stimulus? **3.** Do you think this response is innate or acquired? Why? How could you be sure? **4.** Can an individual consciously suppress the reflex?

The Eye Reflex. The subject sits on the desk or in a chair with the observer directly in front. The subject looks directly at the observer. The observer then suddenly raises both hands and claps them in front of the subject's face. (Be careful not to come too close!) **5.** What is the subject's reaction? **6.** Is this an innate or acquired response? **7.** If you repeat the clapping several times, can the subject control the response?

The Pupil Reflex. The subject sits on the desk. The observer watches the pupils of the subject's eyes. With both eyes open, the subject covers the right eye with the right hand, blocking out as much light as possible. After a minute, the subject uncovers the eye.

The observer compares the sizes of the pupils. **8.** What has happened to either or both pupils? **9.** Could the subject feel the change taking place?

The subject now covers both eyes and the investigation is repeated. **10.** Compare the response of both eyes to that of the right eye in the first test. **11.** What, if any, similarity is there between the reflexes observed in the human being and the responses of flies and *Paramecium* to various stimuli?

OBSERVING BEHAVIOR

It is impossible to imagine a time, no matter how long ago, when people did not make observations of animal behavior. Yet the science of behavior is still at an early stage. To develop any scientific field fully, we must organize observations, form hypotheses, and design experiments to test them.

The branch of biology concerned with the comparative study of behavior in animals is termed **ethology** (eh·THOL·uh·jee). **Psychology** is also the study of behavior. Often, though not always, psychological studies are concerned with humans. Knowing the terms ethology and psychology will help you locate sources of information about behavior.

Today, many methods are used for studying behavior. An important one is observing animals in their natural environments. Activities of animals also are recorded with sensitive measuring devices. Scientists can observe an animal's responses to a change in living conditions in specially designed cages. Surgical operations, drugs, and other techniques also enable us to learn more about animal behavior.

There are two main problems in studying animal behavior. The first is to be sure that the observations are accurate. The second is to make a useful interpretation of the observations.

Observations

It is difficult to make enough observations on which to base conclusions about animal behavior. For instance, what can we conclude about the bees in the garden described at the beginning of the chapter? Very little, except that bees can buzz around flowers. It would take far more study before we could say anything meaningful about the behavior of bees.

Often biologists must place animals in controlled conditions, in a laboratory or other restricted environment, to observe their behavior. But the biologists' conclusions must be tentative until they find

22-3. Marsh marigolds. **(Left)** Photographed on film that records light much as humans see it, and **(right)** on film that records ultraviolet light as bees see it.

out whether the behavior also occurs in the animal's natural environment.

Another problem of observation is that we usually see and measure only a small fraction of the total behavior of an animal. We can watch a bird fly, for example. But we do not see the changes that take place in its nerves, endocrine glands, respiration, circulation, and so forth, before and during flight. Yet all these systems influence the bird's behavior.

Finally, we have the problem of not knowing exactly what information an organism gets from its environment. That is, we cannot experience its **perception** of its surroundings (Figure 22-3). For example, can you tell whether you and your best friend see or hear *exactly* the same things? Do you see and hear the same things even when you are watching the same movie?

In general, birds have much sharper vision than people. Dogs react to smells and sounds of which people are entirely unaware. No doubt, the garden looked very different to the bird than it would have looked to you. The dog's perception of the garden probably was very different from that of the bird. As for the bees, the caterpillar, and the butterflies, they did not seem to respond at all to the cat in the garden. From the data in the story, could you conclude that these insects do not have the sensory equipment needed to sense the cat's arrival?

Interpretations

You may suggest that the insects did see the cat, but did not change their behavior because they did not "feel threatened." This is an illustration of the dangerous tendency to interpret animal behavior in human terms—*anthropomorphism.* We are tempted to explain animal behavior in terms of human emotions. Yet we have no evidence that animals either possess emotions or respond in emotional ways. The problem with this approach is that it gives quick, easy answers, thus blocking further scientific inquiry. Such explanations are dead ends that keep us from seeking more precise evidence of causes.

Sometimes, when we discuss behavior, we are tempted to interpret an organism's actions in terms of human plan or purpose. This pitfall in interpretation is called *teleology.* We might say that a worm burrows into an apple "to find food and shelter." A bird digs its beak into the same apple "to find the worm." This sort of explanation is not useful in biology. We are not obtaining enough information because we are not asking questions that we can answer by observation and experiment.

We can make meaningful observations and comparisons of animal behavior only if we learn most of the factors involved. This is not easy to do in the study of animal behavior.

INQUIRY 22-4 Behavior of a Caterpillar

MATERIALS (per team)

live caterpillar
plant material this caterpillar eats
plant material this caterpillar does not eat
paper
red, yellow, and blue food coloring

Procedure

Draw a 15-cm circle on a piece of paper. Place four pieces of plant material at opposite points around the circle. Use one piece of a plant the caterpillar normally eats. The other three pieces should be plant materials the caterpillar normally does not eat. Place the caterpillar in the center of the circle. **1.** How does the caterpillar move in response to the food? **2.** What factors in this inquiry tend to complicate the results and make them less accurate?

Change the distances of the food materials from the center of the circle. In this way you can determine the distance at which the caterpillar is able to locate food. Rotate the positions of the various food materials during the observations. **3.** What is the point of rearranging the foods?

Prepare a chart. List the four food items across the top. List the distance from the center of the circle down the left side. Record the number of times the caterpillar moves to each food source and how far it moves.

When the caterpillar locates the food in at least 80 percent of the trials, assume this to be the best distance for the caterpillar's response. **4.** What is this distance? **5.** In nature, what do you think would be the effect of moving a caterpillar far away from its food source?

How do you suppose a caterpillar locates its food – by odor, color, or shape? Design experiments to try to answer these questions. **6.** On the basis of your observations, how does this particular caterpillar locate its food? **7.** Have these experiments adequately demonstrated how this particular caterpillar locates its food? Why, or why not?

BEHAVIOR AND PHYSIOLOGY

How can we account for the predictable behaviors you observed in *Paramecium*, fruit flies, caterpillars, and humans? Behavior depends in part on the kind of information an animal *can* and *does* receive. It also depends in part on the kinds of responses an animal can make.

Recall the illustration of the simple reflex response of the foot to a painful stimulus (Figure 18-6, page 505). Sensory receptors in the skin send nerve impulses to associative neurons in the spinal cord. These associative neurons synapse with motor neurons. The impulses travel along the motor neurons and cause muscle cells in the leg to contract. These contractions move the leg – and thus the foot – away from the painful stimulus. This entire response can be described in terms of the chain: sensory stimulus→central association→effector response.

A different stimulus may result in a different response carried out by the same motor neurons. The muscles may contract so that the legs are held straight, rather than bending or moving. Different associative neurons would be involved. By knowing many such reflex pathways, we often can predict the response of an animal to a particular stimulus.

Let us consider again courting and mating behaviors in the stickleback (Figure 22-2). Here, many more stimuli are involved than in a simple reflex such as the response of the foot to a painful stimulus. Many more effectors also are activated. You might guess that much more complex central coordination is involved. This is undoubtedly true. Visual stimuli representing color (of the male) or form (of the female) must stimulate muscle activities. This is "danc-

22-4. Within one or two weeks, a leopard frog (*Rana pipiens*) changes from a tailed tadpole, with structure and behavior suited for life in water, to a tailless adult, with structure and behavior suited for life on land.

ing" in the male and "following" in the female. These events involve specific pathways in each fish's spinal cord and brain. But they also depend on endocrine regulation. They do not occur unless sex hormones are present in certain levels.

Many behavior patterns can be accounted for partially in terms of neurons and endocrine glands. Biologists have studied the central nervous system in a variety of animals. They have learned that specific regions of the central nervous system are involved in patterns of behavior. Such behavior patterns include drinking, reproductive behavior, fighting, and escape responses. The details are far from clear, and there are still many unanswered questions. For example, why does one pattern of behavior sometimes dominate the animal's activities as a whole? For example, an animal may be interested only in food at one time and only in mating at another time.

Different kinds of behavior also occur in animals at various stages of their development. Certain animals—frogs, for example—have lives that are divided sharply into different developmental stages (Figure 22-4). The behaviors of the tadpole include swimming, breathing through gills, and feeding on plant material. At the tadpole stage, there is no reproductive behavior. When the tadpole becomes an adult frog, it shows remarkable changes in its behavior. Many of these changes result from physiological changes. For example, unlike the tadpole, the adult frog has legs. Now it can hop on land or swim in water. But it swims in a manner quite different from that of a tadpole. As an adult, it feeds on other animals, mainly insects. It breathes with lungs rather than gills. And an adult frog also shows reproductive behavior.

What happens as a tadpole changes to an adult frog? Cellular and structural changes in the animal's body take place, along with behavioral changes. We could answer in general terms that the tadpole's nervous and endocrine systems are the basis for its tadpole behavior. By the same token, the adult frog's nervous and endocrine systems are the basis for its adult frog behavior. This, however, is not a satisfactory answer. Unfortunately, it is almost the only one we can yet give to this puzzling and fascinating question.

Behavior is closely regulated by endocrine glands as well as by the nervous system. It is not surprising, therefore, that changing levels of endocrine secretion in an individual's lifetime may be related to changing patterns of behavior. Young mammals show particularly abrupt changes in behavior at the time the reproductive system begins to function. Quite apart from the specific reproductive behavior, the general behavior of the animal also changes markedly. For centuries, farmers have recognized that infant cattle and horses are quite gentle. There are only slight differences in behavior patterns in males and females at this stage. When they mature, male horses and cattle often become more active and aggressive than females of the same species.

Do people experience similar changes as they grow from childhood through adolescence to adulthood? Think about your own life and the lives of relatives and friends you have known for several years. What trends of changing behavior do you recognize in humans? Can you relate any of these changes to physiological changes?

You might compare the overall behavior of students in the most elementary class in your school with the behavior of those in the most advanced class. As you do this, however, you will immediately encounter a problem that faces all students of human behavior. To what extent are any consistent changes the result of "growing up"—physical maturation, as in the tadpole becoming a frog? To what extent are these changes based on experience, education, and pres-

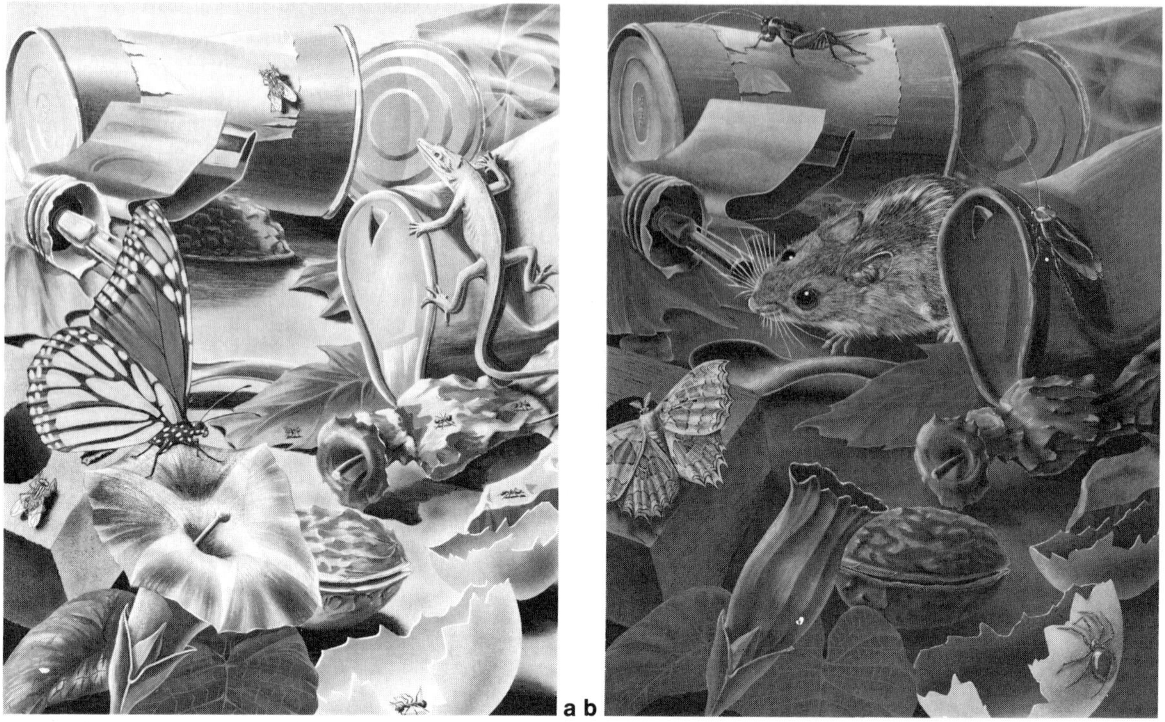

22-5. Almost all animals show some kind of rhythmic activity cycle. Two totally different groups of animals are found at a trash heap. **(a)** One group is active only during the day. **(b)** Another group is active only at night.

sure from society? These are the types of questions that complicate the study of human behavior.

PATTERNS OF BEHAVIOR

As we look over the entire range of behavior patterns in animals, some general concepts emerge. In comparing more advanced with less advanced animals, we find that more advanced forms have more complex behaviors. Simple taxes and reflexes seem to be less important in more advanced animals than they are in simpler animals. Among arthropods and some mollusks and vertebrates, elaborate innate behaviors often are found. In some higher mollusks and in vertebrates in particular, behavior can be modified by experience. In mammals, especially humans, remarkable brain processes like insight and reasoning occur.

Animals have many patterns of behavior that are closely related to time, space, place, and interaction with other organisms.

Behavior Related to Time—Cycles

Many daily activities of animals do not occur continuously. Instead, such activities as sleeping and food seeking seem to occur at certain times each day. Often the behavior is correlated with cyclic events in the environment. Many birds sing mostly at sunrise. Most moths fly at night and rest in the daytime. Most butterflies behave in just the opposite way. Some cyclic behaviors, however, happen without regard to changes of light and dark. Almost without exception, these timed behaviors are important to survival.

Many animals are inactive during the daytime. When it is light, they would be detected easily by predators. These animals seek food and are active at night (Figure 22-5). Almost all these behaviors occur in approximately 24-hour cycles. Such cycles are called **circadian rhythms.** Circadian rhythms are innate. However, they are influenced by light-dark cycles. The innate mechanism is reset each day by stimuli from the environment. In most organisms, circadian rhythms are readjusted to fit any new light-dark cycles within a few days.

Obviously, time is a major factor in regulating behavior in people. You may not feel your best when the school day begins. But some time during the day, you have a peak of alertness, muscular activity, nerve sensitivity, body temperature, and so forth. Similarly, you doubtless operate at a minimum in many of these respects about 12 hours later. Such timing is often based on cycles that are roughly 24 hours in length. You know this to be true from your own cycles of activity and sleeping.

Suppose you flew by jet to another country where the local time is six hours earlier than at home. If it were 8:00 A.M. at home, it would be 2:00 A.M. there. You might easily feel lively late at night (your body's morning) and very tired at 6:00 P.M. (your body's midnight). The experience would convince you that your body processes run by "clocks" that are not reset as easily as is your watch or alarm clock. It might take several days for your body to adjust to the local time. Such abrupt changes in local time are probably important factors in "jet lag."

Cycles also occur over longer periods of time. They seem to be adapted to favorable environmental conditions. For example, reproductive activities of some plants and animals occur at specific times of the year. Animals are attracted to locations where there is a rich food supply. They leave these locations or enter into an inactive state during unfavorably hot, cold, or dry seasons of the year. Often behaviors are closely correlated with the behaviors of other organisms. For example, a particular species of insect matures when certain plants bloom. As these insects use the flowers' nectar for food, they pollinate the flowers.

22-6. Many different animal species exhibit territorial behavior.

What human activities can you recognize as timed? Would it be possible to determine in each case which processes show timing because of social control and which represent actual body-timed cycles?

Behavior Related to Space—Territoriality

Some animals behave as if a particular area is exclusively their own. This is territorial behavior. The animals drive away other members of their own species that come into the territory (Figure 22-6). Only a few animals drive away members of other species as well.

This type of behavior has been studied carefully in many species of birds. Typically, male birds of a given species establish nesting sites some distance from one another. The borders of each territory are defended by the males. The territorial behavior of the male includes a characteristic song and a display of aggressive behaviors. The male and female birds can look for food around the nest site without interference from other birds of the same species.

Behavior Related to Place—Migration

Many species of vertebrates and some insects travel long distances during their lives. Birds often nest and rear their young during warm seasons far from the equator. As winter approaches, they move closer to the equator, spending a few months there. Then they travel north again as spring comes to the regions more distant from the equator. They may migrate many thousands of kilometers. Some birds return to the same place year after year. Often the young migrate before the adults. How do they find the way? Some mammals and insects have similar migration patterns.

SCIENCE AT WORK: Success After 40 Years

The work of science – research – is usually a team effort and proceeds over a long period of time. A prime example of this is the discovery a Canadian zoologist and his wife made after 40 years of searching. They studied the migration of a colorful butterfly, the monarch. The monarch is a very familiar summer sight in the eastern, mid-western, and a good portion of the western United States. But in winter the monarch butterfly vanishes from the colder regions. Where does it go?

A common method of studying the migration patterns of animals involves the tagging of the animals with some type of identification. Researchers hope that someone will then see the animal and tag during or after the migration and report back to the person who tagged the animals. Tiny labels carrying identifying letters and numbers and the words "Send to Zoology University Toronto Canada" were attached to the wings of thousands of migrating monarchs across the continent. Tagged specimens were returned from all over the United States and from Mexico.

In 1952 volunteers were asked to assist in the tagging and reporting program. Twelve people responded, launching the Insect Migration Association. By 1971 the group numbered 600 with thousands more taking part in the program over the years.

Much was learned about the life of the monarch because of the help of all these volunteers. Not until 1975, however, did one of the volunteers discover the long sought goal. In an area of the Sierra Madre Mountains in central Mexico, two nearly equal concentrations of monarch butterflies were found. Millions of the butterflies covered the trees and the ground and filled the skies in these areas. A tag was found on one of these butterflies. It had been marked months earlier in Minnesota.

Ten thousand of the newly discovered butterflies were tagged. Two months later, two of the butterflies tagged in Mexico were recovered in northern Texas. The prime wintering grounds of the monarch had been found. Forty years of work and the efforts of thousands of people working together had made this research effort a success.

Some species of fish have still another interesting type of migratory behavior. They hatch from eggs laid in freshwater lakes or streams. Then they move downstream and out to sea, where they spend most of their adult lives. When they are ready to reproduce, these fish migrate back to shore, then upstream to freshwater streams or lakes. There they deposit eggs and sperm, and there the next generation of young fish hatches. Certain species of fish (salmon, for example) will return to the same stream where they were hatched (Figure 22-7). Other species (eels, for instance) live in fresh water but migrate to the ocean to reproduce.

Much research has gone into the question of how animals navigate during long migrations. Apparently, they sometimes use cues like the position of the sun or stars or the character of light from the sky. Landmarks—a mountain, lake, or patch of forest—may help some animals traveling over land. But this certainly would not help a small bird in its flight from Hawaii to Alaska. Recognition of place by chemical senses may play some part, at least in fish migration. All these possibilities and many others are being investigated.

Behavior Related to Other Organisms—Communication

Sensory receptors give an animal valuable information about events within itself and about its environment. These receptors also inform the animal about the presence and activities of other living things. Communication among organisms has great biological significance. The information passed from one individual to another of the same species may concern food, danger, or readiness for mating. Even single-celled organisms sometimes exchange information. Biologists have evidence that slime molds can communicate by the use of specific chemical substances.

The chemical substances that are involved in sexual behavior and in many other behaviors are called **pheromones.** Pheromones have been found in many animals, including mammals. Recent discoveries indicate that some human behavior may occur partly in response to chemical signals.

Fire ants communicate with one another about the position of food. A foraging ant finds food. It returns to the colony, periodically touching the ground with its sting, which extends rearward from its abdomen. As it does this, a pheromone is released from a gland in the ant's abdomen. Other ants quickly travel out to the food, taking the path of the foraging ant's return. As each ant returns with food, it leaves a similar trail.

To confirm these observations, a biologist dipped a stick into an extract from one of the glands and drew a curving line from a colony of ants to a distant spot. The ants followed the path traced by the

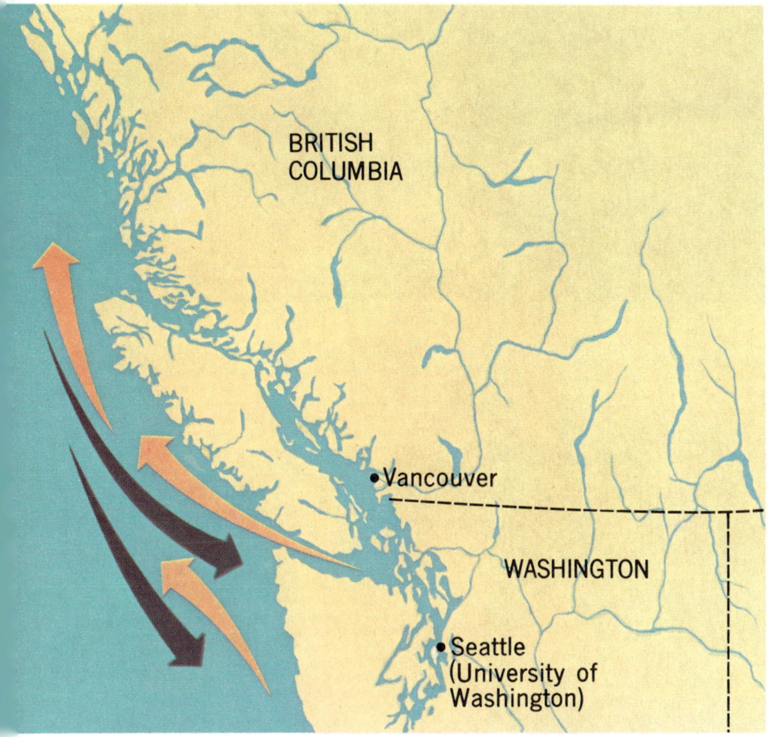

a b

22-7. Migratory behavior in salmon. **(a)** Salmon breed in fresh water in the northwest. Young salmon migrate to sea (orange arrows). After two to five years, they return to breed in the lakes and streams where they are hatched (gray arrows). **(b)** An adult male Chinook salmon swims upstream to his home breeding pond at the University of Washington.

stick. Further observations showed that the returning ants stopped leaving a trail when the food supply was gone.

Animals use other sensory information in their responses to species other than their own. A dramatic illustration is the behavior of bats in hunting for flying insects, especially moths. Bats are active at night. They utter high-pitched sounds. The sounds are reflected from objects even as small as the moths the bats hunt. This sonar system (echo location) allows bats to detect moths in the dark.

These moths also have sensory receptors that can detect the sound waves emitted by the bats. When a moth picks up bat sounds, it flies erratically, which must often help it escape. Some moths can emit sounds that are of nearly the same high pitch as the bat sounds. The moth sounds may cause the bat some difficulty in locating the moth. These moths are "jamming" the bat's sonar.

Scientists have investigated various methods of communication among animals of the same species. Many animals seek food in a relatively wide area, yet they always return to a particular "home base." Ants and bees have nests or hives where reproducing individuals, their young, and the food supplies are kept and protected. Other members of the community collect food and bring it back to the nest or hive.

a A foraging honeybee returns to the hive after finding a nectar source. Other bees soon leave the hive and fly to the same source. How does the first bee inform other bees about the food source?

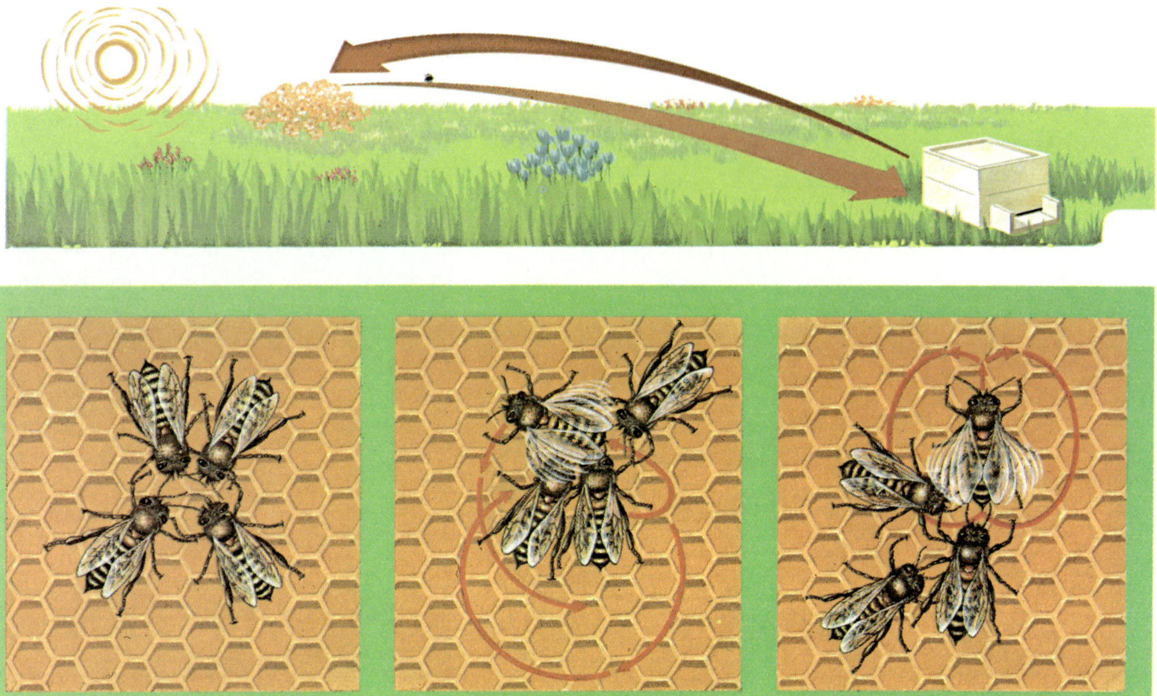

b Karl von Frisch found that the returning bee gives nectar to other bees, then begins a dance on the vertical face of the hive.

c If the dance is a round dance, other bees search for plants nearby.

d If the dance is a figure 8 with wagging of the abdomen, bees search at a distance.

22-8. How honeybees communicate.

When a foraging honeybee returns to the hive after finding a nectar source, other bees soon leave the hive and fly to the same source. The first bee does not go along as a guide. How does it inform the other bees where it obtained the nectar?

In one famous study of food-seeking behavior in honeybees, Karl von Frisch found that the returning bee gives nectar to other bees and then begins a dance on the vertical face of the hive (Figure 22-8). Other bees follow the dance with their antennae against the dancer's abdomen. A "round dance" seems to mean that the nectar is near the hive. If the dance is a "figure 8" with wagging of the abdomen, the other bees search farther from the hive. How far? The rate of movement in the dance seems to be involved. In what direction? The direction of the middle line of the figure 8 indicates the direction of the sun at the food source. In addition, sounds issuing from the thorax of the dancing bee have a pattern correlated with distance to the nectar source. Are there other factors involved?

Recent discoveries show that bees release special chemical substances that signal danger and other environmental information. Thus, in bees, there is evidence that light, touch, sound, and chemical stimuli are important in communication. How many of these play a part in human communication?

LEARNING

Up to this point in our discussion of behavior, you may have formed an impression that patterns of behavior are quite fixed. You might think that an animal's responses to environmental stimuli depend only on its endocrine system and its body structure. This is true of many behavioral patterns. It is equally true, however, that some behavioral patterns in animals may be modified by experience—by **learning.** Learning may increase an organism's chances of surviving and reproducing.

Although learning occurs in all animals that have organized nervous systems, the complexity of the learning varies. Often learning takes place in connection with some motivation. A hungry rat, for example, can learn to follow a complicated path if it always finds food at the end. If the same rat is kept well fed, it is not likely to learn the path. Hunger, thirst, sex drive, the drive to care for the young, and the tendency to avoid pain may motivate learning in complex animals.

Learning is an area in which human beings excel, for our capacity to learn is very great. We are aided by abilities that are present in some other animals but are developed in us to a unique degree. For example, our ability to reason and to use symbols aids us in learning.

Conditioning

An example of learning is the **conditioning** of simple responses. Dogs normally secrete saliva when they see or smell food, just as people do. The Russian biologist, Ivan Pavlov (PAV·lof), found that he could condition a dog to secrete saliva when no food was present. Pavlov rang a bell at the same time, or just before, food was given to the dog. Pavlov repeated this combination of stimuli—bell-ringing and food—many times. Eventually, the dog secreted saliva when the bell was rung, even when food was not present. The secreting of saliva at the ringing of the bell was a conditioned response. Other stimuli—a flash of light, for instance—worked just as well as the ringing of the bell.

Conditioning can occur in many kinds of animals—ranging from simple ones, like planarians, to complex organisms, like mollusks, birds, and mammals. You almost certainly have conditioned responses yourself. A good example is the increased rate of flow of saliva in response to the sound (or even the thought!) of words like *lemon, vinegar,* and *dentist's drill.*

Habituation

Another simple kind of learning is **habituation** (huh·BICH·oo·ay·shun). At first, an animal may respond to a particular stimulus in a particular way. If the stimulus is repeated at frequent intervals, however, the animal's behavior may change. For example, a fly becomes entangled in a spider's web. The spider will crawl quickly to the place in its web that moves. If you touch the web with a glass rod, the spider will rush over to the moving area. If you do this several times, however, the spider will become less responsive. It becomes habituated to the stimulus. Finally, the spider will not respond at all until it has been left undisturbed for some time. How would the spider then respond to a fly caught in its web? You can discover the answer for yourself.

Imprinting

An unusual type of learning is **imprinting.** This behavior has been observed in young animals and has been studied especially in young

22-9. A clear example of imprinting in Canada geese.

birds. Imprinting is an innate response to a stimulus that is learned. This *object of imprinting* becomes fixed in the memory and behavior patterns of the young birds (Figure 22-9).

We can describe the process of imprinting best by giving an example. Normally, geese hatch from eggs in the presence of the mother. Soon after hatching, the baby geese, or goslings, begin to follow their mother about. The mother leads the goslings to food and protects them from dangers.

The Austrian biologist Konrad Lorenz made a remarkable discovery. He experimented with goslings that were hatched in an incubator. These goslings had never seen an adult goose. The first thing they saw after hatching was Konrad Lorenz walking along slowly and making clucking sounds. The goslings followed Lorenz. Later, the goslings paid no attention to their real mother but continued to follow the biologist.

Additional experiments showed that geese, ducks, and other birds respond in similar ways during the first few hours of life. They will follow a variety of objects that move slowly and make sounds resembling those of the mother bird. Imprinting has been observed in a variety of animals. Usually, it occurs only in the earliest hours after hatching or birth. In ducklings, for instance, imprinting occurs only during the first 30 hours or so after the birds hatch.

Trial-and-Error Learning

Learning often depends on repetition of an experience. Here is an illustration of this kind of **trial-and-error learning.** A biologist places a hungry snail in a special container shaped like a "T." This container is called a T-maze. If the snail turns down the proper arm of the T-maze, it is given a reward of food. The snail starts from the base of the T and crawls down one or the other of the arms at the other end. In this situation, it has a 50-50 chance of getting the food reward.

The biologist allows the snail to become hungry again and then puts it back in the T-maze. The snail may still have only a 50-50 chance of turning toward the food. But if the test is repeated many times, the snail gradually learns, by trial-and-error, to turn right each time. After as many as 60 trials, it has nearly a 100 percent chance of taking the right turn and getting food. In this experiment, how could you make sure that the snail is not following its own scent in reaching the food?

Rats have more complex nervous systems than snails have. They can master complex mazes in a relatively small number of trials. Primates, porpoises, whales, and other higher mammals learn even faster. They also can learn extremely complex behaviors.

22-10. Peony, a chimpanzee, uses symbols to communicate with Debby, a human. *What is Peony's message?*

Reasoning

Conditioning, trial-and-error learning, and imprinting all play parts in modifying the behavior of many animals. In higher mammals, and in humans in particular, other special processes are important in adjusting behavior. They include reasoning and the ability to use symbols (Figure 22-10).

Consider the case of a hungry bird or mammal that is placed in a cage with one square cup and two round cups. Under the square cup is food. (The food is wrapped in plastic to conceal the odor.) The animal will find the food, sooner or later. It almost certainly could be trained—with more or less difficulty—to associate "square cup" with the reward of food. It also could be trained always to look for food in the cup on the right, or on the left, or in the center.

But suppose the animal is placed in a series of situations as shown in Figure 22-11. What is the clue that the animal must recognize before it receives the reward? Can you discover it before reading further?

A rat cannot learn to find the food in every test in this situation. A rhesus monkey may need 400 to 1,500 trials before it has a 90 percent chance of finding the food on the first try. You can judge for yourself whether this would be a difficult task for humans. (Try it on friends who are not in your class.) The animal must always go to the kind of cup of which there is only one. It requires a certain capacity to recognize an odd (different) member of a group. It also requires

22-11. *What pattern must the rat recognize to find the food?*

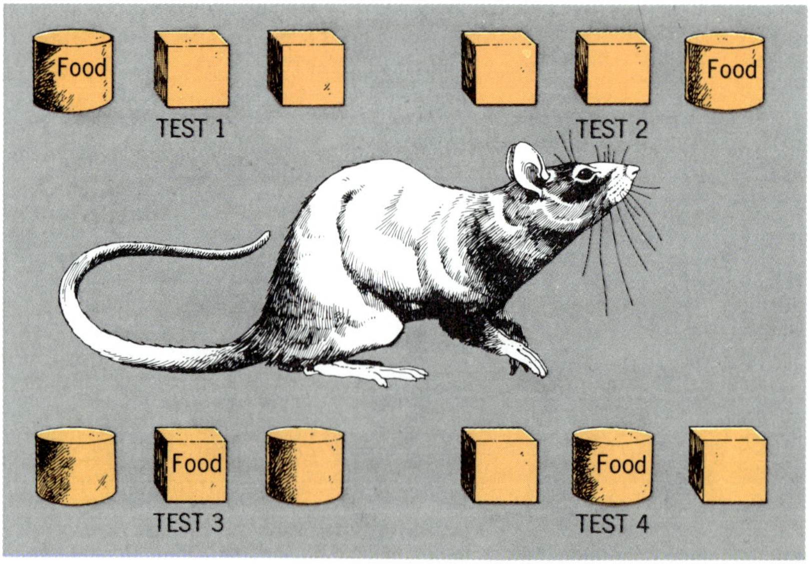

the ability to reason that food will be associated only with whatever there is one of.

Is Learning Permanent?

Apparently, learning is not always permanent. If a conditioned response is well established but is not tested again for a long time, it may disappear. Some behavioral changes are made more easily, and some last longer than others. They differ in other ways, too. For instance, imprinting can occur only very early in an individual's life. Learning in some animals may take place after only a few repetitions of a stimulus. In many other animals, learning may require many repetitions.

SOCIAL BEHAVIOR

Perhaps the examples of behavior we have been discussing will give you a basis for drawing some conclusions about interactions among animals—their social behavior. These interactions vary in complexity. Among the less complex interactions are encounters between two animals, for example, in trying to obtain the same food (Figure 22-12). Other interactions between two animals include predator-

22-12. This kill was made by a lion, who was the first to feed on the carcass. A hyena **(left)** and two jackals **(right)** now compete for their portions. Two vultures in the background wait their turn; they are last in this feeding order.

22-13. A predator-prey relationship exists between bears and fishes.

prey relationships (Figure 22-13) and mating behavior (Figure 22-14).

Often, many individuals of a given species come together in a relatively small space. When this happens, some sort of social interaction occurs. Sometimes it is no more than the chance meeting of many individuals at feeding places or in areas offering protection.

22-14. Many birds have elaborate mating displays. The male peacock **(left)** spreads his fantastic tail. The male frigate bird **(right)** has a vivid red throat pouch. The display of these secondary sex characteristics is necessary for normal mating.

22-15. Elephants frequently form separate male and female herds. The leader of this herd is an old female. The young are tended by the females. Generally, the male and female herds do not associate, except during breeding periods.

Behavior patterns in which great numbers of individuals travel together are seen quite often in insects, fish, and birds. Social organization at a rather simple level is seen in groups of birds or mammals simply living side by side.

At the other end of the scale are highly elaborate social behaviors in some insects and primates. Many animals interact over a fairly long time, even for a lifetime. Some species of birds, for example, pair for life.

In flocks of chickens and some other birds, a pattern of dominance and submission, the "peck order," is established. Usually, one member is at the top of the peck order. At the opposite end of the scale is a bird that does not resist if pecked by any other member of the group. Between the two extremes, a rather consistent order of "who pecks whom" is established. Each bird recognizes other members of the flock, as well as the position of each bird in the peck order.

Mammals that live in herds and other groups have similar social orders based on dominance and submission of the individuals. In some animal societies, a leader, often an old and experienced male, dominates and guides the behavior of the group. In other societies, separate groups with their own leaders are formed (Figure 22-15).

22-16. An ant community consists of hundreds or thousands of individuals. There are usually several types of individuals that differ in structure and behavior pattern. Each has its specialized function in the colony.

Another elaborate example is the social organization of insects such as bees and ants. In one ant species, many individuals live together in a single community (Figure 22-16). Each individual is specialized in body structure and behavior. Generally, a single, mature female (the queen) founds the community. She lays eggs that hatch into separate classes, or castes, of ants. Most of the ants are sterile females. They include workers that seek food and bring it back to the nest. Other workers also feed and care for the queen and the young. Sometimes there is a special caste of fighters that have powerful jaws and poison glands. A few males are hatched. These may be retained in the community to fertilize the queen. Certain individuals may serve merely to store reserves of food in their bodies. In some cases, entirely unrelated organisms, such as fungi or aphids, may live in the ant community. These organisms provide food for the ants and are cared for by the workers.

Such an elaborate social system requires a variety of mechanisms. These range from variations in body size and function to differences in the behaviors of individuals. Touch, chemical stimuli, and hormonal control play major parts in establishing and maintaining the special relationships among members of the group.

EVOLUTION OF BEHAVIOR

It is useful to look at behavior in the same way we look at structures and processes within living cells—in terms of evolution. Behavior also has evolved. After all, behavior involves the responses of an or-

22-17. The deer is well adapted to fleeing from a predator. It has keen senses of sight, hearing, and smell. It also has long legs, light body, and powerful muscles.

ganism to stimuli in its environment. Any animal living today is descended from animals of the past that succeeded in reproducing. That is, today's animals inherited their structures, cell processes, and innate behaviors. They, in turn, may pass these structures, processes, and behaviors on to their offspring. Their chance of doing so is much greater if their patterns of behavior are well suited to the conditions of their environment.

For example, the deer (Figure 22-17) has keen senses of sight, hearing, and smell. It is easily aroused by sensory stimuli. The deer can quickly sense the presence of a mountain lion. Its long slim legs, light body, and powerful muscles are advantageous in fleeing from such a predator.

A behavior that is advantageous to a species does not always help the *individual* to survive, however. A good illustration of this can be found in aphids. Aphids are small insects that suck juices from plant leaves and stems. In some species of aphids, the adults that emerge from the larval form may or may not have wings. The wingless forms cannot travel. They feed on whatever plants they happen to be on when they emerge. As soon as the winged forms emerge as adults, they fly upwards into the air, probably in response to light. As they move upward, the aphids are likely to be caught by the wind and carried some distance.

If we are considering the whole aphid population, this behavior appears to be highly beneficial. It insures that the newly emerged insects will be spread over a relatively wide area. The two kinds of aphids will not compete for a limited food supply in a small area. For a winged individual, however, the pattern may or may not be advan-

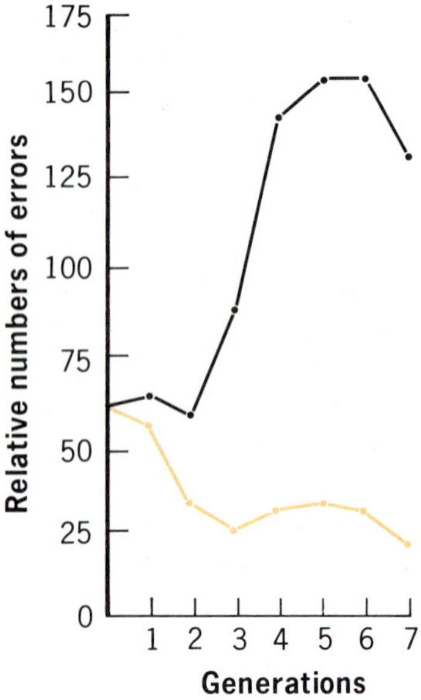

National Society for the Study of Education

22-18. Results of selective breeding of rats for seven generations. The original parents had the same ability in trial-and-error learning. Some of the offspring could learn more rapidly than others. These were mated. The better learners from these matings were mated. The yellow line shows the gradual decrease in errors. The black line shows the results of selecting offspring that made the most errors.

tageous. The insect may be dropped by the wind on a juicy leaf where it can feed. On the other hand, it may be dropped into a lake, where it will drown.

Inheritance plays a great role in determining the behavior patterns of each species. An isolated tadpole shows its characteristic behaviors even if it never sees another tadpole or frog. The goslings became imprinted to Lorenz as the object of the following behavior. In their general behavior, however, they were always geese.

Variations in the ability to learn also may be inherited. Thus, a rat that learns a maze quickly and one that learns it slowly may have inherited different traits. Selective breeding of rats can produce strains that show marked differences in the learning ability of individuals. This was demonstrated in an experiment. A pair of rats that had about the same ability to learn were mated. Among their offspring, some could learn more rapidly than others. These rats were mated. The slower rats also were mated with each other. From seven generations of such matings, fast and slow rats were selected and mated. The yellow line in Figure 22-18 shows the results of this experiment.

As you continue your study of behavior (for you cannot help making observations of animal behavior throughout your life), two ideas will be useful. You will find it helpful to analyze each behavior in terms of its survival advantage for the species. You also should think about its possible effects on the life of the individual organism.

DRUGS AND BEHAVIOR

Drugs are chemical substances, usually produced by green plants and fungi, that some people take by mouth or by injection. These chemicals change nervous function and behavior. Aspirin is a drug that relieves pain. Caffeine is a mild stimulant found in coffee, tea, cocoa, and cola drinks.

Many drugs may have far more serious effects. Among these are alcohol, amphetamines, barbiturates, marijuana (cannabis), cocaine, LSD and related compounds, and opiates such as morphine and heroin. These drugs change the balance of activity in the brain. This imbalance often leads to disturbed thought, perception, and behavior. Vision, hearing, and other senses may be distorted. Drugs may hamper the ability to reason. Profound feelings of well-being, happiness, anxiety, or fear may occur.

For centuries, people have used some drugs in small quantities (especially alcohol and marijuana) without obvious harm. All drugs, however, including aspirin and caffeine, are harmful if taken in excess. And some drugs have devastating effects even in small quantities. Alcohol injures the liver, nerve cells, and other structures.

Chromosome damage may be caused by some LSD preparations.

Besides these effects, there is the danger that people who take drugs in excess will develop drug dependence. In some cases, the dependence appears to be psychological. That is, a person has a strong desire to continue taking the drug. In many instances, drug dependence is physical. As a result of taking some drugs over a long period of time, certain physiological changes occur in the body. Among the drugs that can cause physical dependence are alcohol, barbiturates, and opiates. If the individual who has grown dependent on one of these drugs is suddenly forced to stop, she or he suffers intense pain and anxiety and may have convulsions and hallucinations. At present, we do not know exactly what causes drug dependence.

The dangers of damage to cells and organs and the development of drug dependence are only two of the problems associated with drug use. Drug-dependent individuals may not eat properly and may become badly undernourished. They may lose interest in normal life activities and may avoid their families and friends. Disturbed brain function may lead a person on drugs to do hazardous things. Swings of mood from elation to deep depression may result in injury or suicide. Because drugs are expensive, the person who is drug-dependent may steal or commit other crimes to get money.

For some people, the pleasures—or release from pain, anxiety, misery, or boredom—experienced with drugs seem so attractive that the threat of harm is ignored. People who become drug dependent may have thought they would be able to control themselves, or they may not have known the dangers.

Is there anything that can be done about excessive drug use? We might discuss the problems more openly and honestly. We could look for cures for people who have already become physiologically dependent on drugs. We should try to learn more about human drives, desires, moods, and perceptions. We need to understand the problems and difficulties that cause people to seek relief in drugs. And in making our own decisions about the use of drugs, we need to understand thoroughly the dangers involved.

In Brief

In this chapter, we have talked about various kinds of behavior—from simple to complex. Some behavior is unchanging and predictable. Other behavior can be modified by conditioning and more complex learning. It is difficult to study and interpret behavior. Yet it is interesting to see how much one can learn in such studies. Perhaps the inquiries in this chapter gave you some understanding of the present state of the study of behavior. You may have gained some idea of why scientists study behavior with excitement, much argument, and a keen sense of the importance of the work they are doing.

In part, a behavior exhibited by an animal is the result of that animal's genetic makeup, which also determines its structure and function. Behavior is closely regulated by the nervous system and endocrine glands. Some animals behave differently during various stages of their physical development.

Many patterns of behavior are innate. Examples are cyclic behavior, territoriality, and migration. These, however, may be influenced and modified by the animal's environment. Animals communicate with members of the same species and with other species by means of sensory stimuli. Sensory information may concern food, danger, or readiness to mate.

Biologists have conducted many studies in their efforts to find out how behavior can be modified by experience. Simple forms of learning include conditioning and habituation. An intriguing example of learning is imprinting. More involved methods of learning are found in more highly developed animals. Humans have a great capacity for reasoning and the ability to use symbols to communicate ideas.

Appropriate behavior patterns will contribute to an animal's survival. For example, a deer that responds to the presence of a mountain lion by fleeing probably will survive and reproduce. Sometimes, however, an adaptation that promotes survival of a species is not advantageous for all individuals of that species.

Humans have learned that drugs produce unusual effects on human behavior. Many studies have been made. Much information is available. However, we are far from knowing the full effects of drugs and the reasons for their indiscriminant use.

Now that you have a start on understanding animal behavior, you can continue this study on your own. You might start with observations of a caterpillar on a lettuce leaf, a bird with young in the nest, or your own pet dog or cat. Then turn to current publications for possible explanations of what you observed. Animals can become far more interesting as you begin to understand more about their behavior.

Guide Questions and Problems

1. Criticize the statement, "Squirrels bury acorns so they will have food in the winter." Give at least two additional examples of poor explanations of animal behavior.

2. Many insects seem attracted to lights on summer evenings. How would you determine if this is a positive phototaxis?

3. What problems might you encounter if you set out to observe the behavior of dogs using your pet dog as the experimental animal?

4. Suggest some factors that might change your behavior from drowsy to alert in a biology class.

5. Certain cyclic behaviors occur on an annual basis, such as migration and mating behavior. What might be the survival value of these behaviors? How can their existence be accounted for?

6. Social order exists in many animal groups where one animal becomes the group leader. What significance might this have for the survival of the group? What might be some unfavorable effects of this type of organization?

7. Peck order in chickens may change when submissive chickens are given injections of testosterone. These chickens then become more aggressive and move up in the peck order. What other events might affect an established peck order? Design experiments to test your ideas.

8. Male and female mockingbirds look alike. How might sex differences be communicated during mating season?

Ideas for Independent Study

1. Look into aspects of human behavior and customs involving chemical stimuli. Some examples are the use of incense, perfumes, spices and other food seasonings (onions, garlic). How are these used to affect mood, concentration, attitude, enjoyment? How do people react to unpleasant odors and tastes, such as those of polluted air, garbage, or animals? Can you find evidence in literature, history, and mythology that chemical stimulation may have been important in human life long before biologists knew about pheromones?

2. Study animals at home or in the school laboratory. Observe the behavior of mammals such as mice, hamsters, and gerbils; birds such as parakeets and weaver finches; or fish, including *Lebistes* (guppies) and *Tilapia*. Before starting on a study of this sort, check on housing, feeding, and general care of the species you will observe. Also design procedures for making meaningful observations and records of the animals' behavior.

3. What conditioned responses can you identify as part of your own behavior? Analyze your behavior and keep track of the conditioned responses you make throughout the day.

22-19.

4. What can we learn from the study of primates such as monkeys and apes? This is an active field of study. Many people believe that it has direct bearing on our understanding of human behavior. You may want to look into it. Observe animals in zoos and read about studies carried out under natural conditions. You might begin your investigation by reading *In the Shadow of Man*, by Jane Van Lawick-Goodall, Houghton Mifflin, Boston, 1971.

5. Beavers build dams like the one in Figure 22-19. Suggest hypotheses for the function of the dam in the beaver's life. How does a medium-sized animal, the beaver, perform such a complex engineering feat? What effects might the dam have on the living and nonliving things in the vicinity? Search your library for information on beavers. If beavers live near your home, you also can observe them directly.

Related Readings

BOOKS

Fincher, J. *Human Intelligence.* New York: Putnam's, 1976.

Follansbee, H. *Animal Behavior.* A BSCS Laboratory Block. Lexington, Mass.: D. C. Heath, 1965.

Kevles, B. *Watching the Wild Apes: The Primate Studies of Goodall, Fossey, and Goldikes.* New York: Dutton, 1976.

Langone, J. *Bombed, Bugged, Smashed, or . . . Sober: A Book About Alcohol.* Boston: Little, Brown, 1976.

Lee, E. E. *Alcohol—Proof of What?* New York: Messner, 1976.

Meyerriecks, A. J. *Courtship in Animals.* A BSCS Pamphlet. Boulder, Colo.: BSCS, 1962.

Tinberger, N. *Animal Behavior.* New York: Time-Life, 1965.

Weinswig, M. H. *Use and Misuse of Drugs Subject to Abuse.* Indianapolis: Pegasus, 1973.

Animal Behavior: Readings from Scientific American. San Francisco: Freeman, 1975.

ARTICLES

d'Aulaire, E. and O. d'Aulaire. "Ape That Talks With People." *Reader's Digest,* October 1975, page 94.

de Koenigsberg, V. C. "Between Man and Beast —Genetic-Behavior Connection." *Science Digest,* July 1975, page 36.

Egbert, A. L. and M. M. Lugue. "Among Alaska's Brown Bears." *National Geographic,* September 1975, page 428.

Grimspoon, L. "Marihuana." *Scientific American,* December 1969, page 19.

Shaw, E. "Fish in Schools." *Natural History,* October 1975, page 40.

Sturm, Shirley C. "New Insights Into Baboon Behavior. *National Geographic,* May 1975, page 672.

Topoff, H. "Ants On the March." *Natural History,* December 1975, page 60.

23 The Environments of Life

Spread out over the earth's surface, between the solid rocky crust of the inner earth and the upper reaches of the atmosphere, extends the world of life—a film of living matter. It is made up of grass, shrubs, and trees, of microorganisms, of worms, fish, rabbits, and wolves. This thin layer of life, along with the air, soil, water, and other nonliving matter that surrounds it, is called the **biosphere** (BY·oh·sfeer).

Organisms exist that live in almost every conceivable environment. There are algae that grow only in melting snow. Many animals and plants spend their entire lives in lightless caves. Some algae and bacteria live in hot springs where temperatures are near the boiling point. That these **habitats**—places where organisms live—are populated is evidence of genetic variability and natural selection.

Life is rich and varied wherever conditions are favorable. In the surface waters of the oceans and in the tropical rain forests, we find many kinds of green plants and animals. On the other hand, arctic regions, deserts, and the dark depths of the oceans are areas of great stress. Only a few species of animals and plants survive there. Between these two extremes, we find many variations.

The total of all the organisms and their interactions with each other and with their nonliving environment is called an **ecosystem.** Thus we can speak of a forest ecosystem as the interaction of all the living things in a particular forest with each other and with their nonliving environment. We can also speak of a drop of pond water as an ecosystem. Can you see how and why?

The green plants in all ecosystems produce their own food. They are the food source, directly or indirectly, for all animals. Thus, the green plants of an ecosystem are **producers.** The animals are **consumers.** Animals that eat plants are **herbivores.** Examples are leafhoppers, aphids, grasshoppers, moths and butterflies, and most mice and deer. Animals that feed on plants also are called primary consumers. Some beetles, spiders, centipedes, salamanders, and shrews are **carnivores**—they eat other animals. These carnivores also are called secondary consumers. Other animal species eat both plants and animals. These are the **omnivores.** Organisms such as mites, springtails, isopods, millipedes, snails, earthworms, crickets, and fungi feed on dead organic material. These are the **decomposers.** Decomposers break down the waste products and dead bodies of organisms into molecular form.

Let us look at the major ecosystems of the world and some of the organisms that live in each.

LIFE IN THE OCEANS

As we live on land, we may forget that the most extensive area of our planet is the oceans (Figure 23-1). The oceans cover about 70 percent of the earth's surface. The average depth of the ocean is roughly 400 meters. The chemical composition of ocean water is similar all over the world. Temperatures are nearly always between 0° and 30°C, and in many areas the range is much smaller.

Life in the oceans depends on the presence of light. Where there is light, plants carry on photosynthesis, grow, and reproduce. The plants are eaten by some animals living in the oceans. These, in turn, are prey for other animals. The mass of living matter in the oceans is far greater than that on land. But most of this life is not obvious to the unaided eye. If you were to examine the surface waters of the North Atlantic, the chances are that you would see only water.

23-1. Waves dash against rocks along the shore of the ocean.

Yet a liter of this water might contain 500,000 bacteria, 1,000,000 microscopic plants, and 150 microscopic animals.

Many of these tiny animals move by means of cilia, flagella, or legs of some sort. For the most part, however, the organisms in the surface waters are at the mercy of wind and waves. These drifters are known as **plankton.** When seen with the microscope, plankton exhibit a truly wondrous variety of forms and colors (Figure 23-2).

The producers in the plankton are single-celled plants. The most abundant are diatoms. These algae have an outer wall composed largely of silica, which has almost the same chemical makeup as glass. The chlorophyll in diatoms is similar to that of green land

23-2. Microscopic life in surface waters of the open sea. **(Left)** Diatoms (many are shown) and dinoflagellates (see arrow) are the world's leaders in food production. They account for perhaps 85 percent of the world's photosynthesis. **(Right)** The primary consumers of the sea include these minute crustaceans.

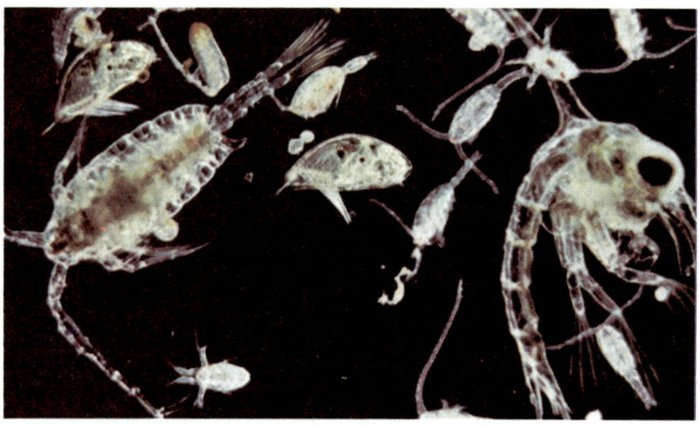

plants. Dinoflagellates (dy·noh·FLAJ·eh·laytes) also are abundant. Most of these protists contain chlorophyll.

Diatoms and dinoflagellates, plus a few other microscopic algae and some of the larger floating seaweeds, support all other life in the open ocean. Being photosynthetic, marine algae can live only where there is light. It is not surprising, therefore, to find them only in the surface layers of the water. They are nearly always in the top 100 meters. Among the consumers in the oceans are protozoans, crustaceans, fish larvae, and members of nearly every other group of animals (Figure 23-2). The animals that eat these small consumers range in size up to the whales, sharks, and giant squids (Figure 23-3). And, of course, there are many decomposers, primarily bacteria, in the ocean waters.

In the depths of the oceans, living conditions may be quite different. Sunlight does not penetrate the deep waters. There can be no photosynthesis, no local producers. All life in these dark depths depends on food that drifts down from above.

Life in waters near the shore is greatly affected by the land. Rivers carry organic and inorganic materials from the land to the oceans. Organisms living at the edge of the ocean endure greater environmental extremes than those living in the open ocean. Water temperatures also vary more here than in the open ocean. The salt content of the water is another variable. It decreases as rivers empty fresh water into the ocean. Runoff of rain water from the land also dilutes the ocean's salt water near the shore. The rise and fall of the tides also may affect shore organisms.

Along the shore, many organisms live attached to rocks or pilings. Each type of organism occupies a fairly constant zone in relation to the tides (Figure 23-4). A few live in the zone above the level of high tide. These are affected only by the ocean spray. Those of the tidal zone live between the levels of high and low tide. The organisms of these two zones must withstand periods of drying when the tide is low. The subtidal zone almost always is covered by water, and it supports a different group of organisms. Similar zones occur along the coasts of South Africa, Maine, and California.

The environment of the open ocean changes very little. The temperatures are more uniform than those of the land masses at the same latitudes. There is plenty of oxygen, except in the depths of the ocean. In the oceans, of course, water is abundant. This essential substance is not always so readily available on land.

LIFE ON LAND

Water diffuses easily in and out of marine organisms. But organisms living on land must conserve their internal water supply. Ter-

23-3. FOOD WEB IN THE OCEANS

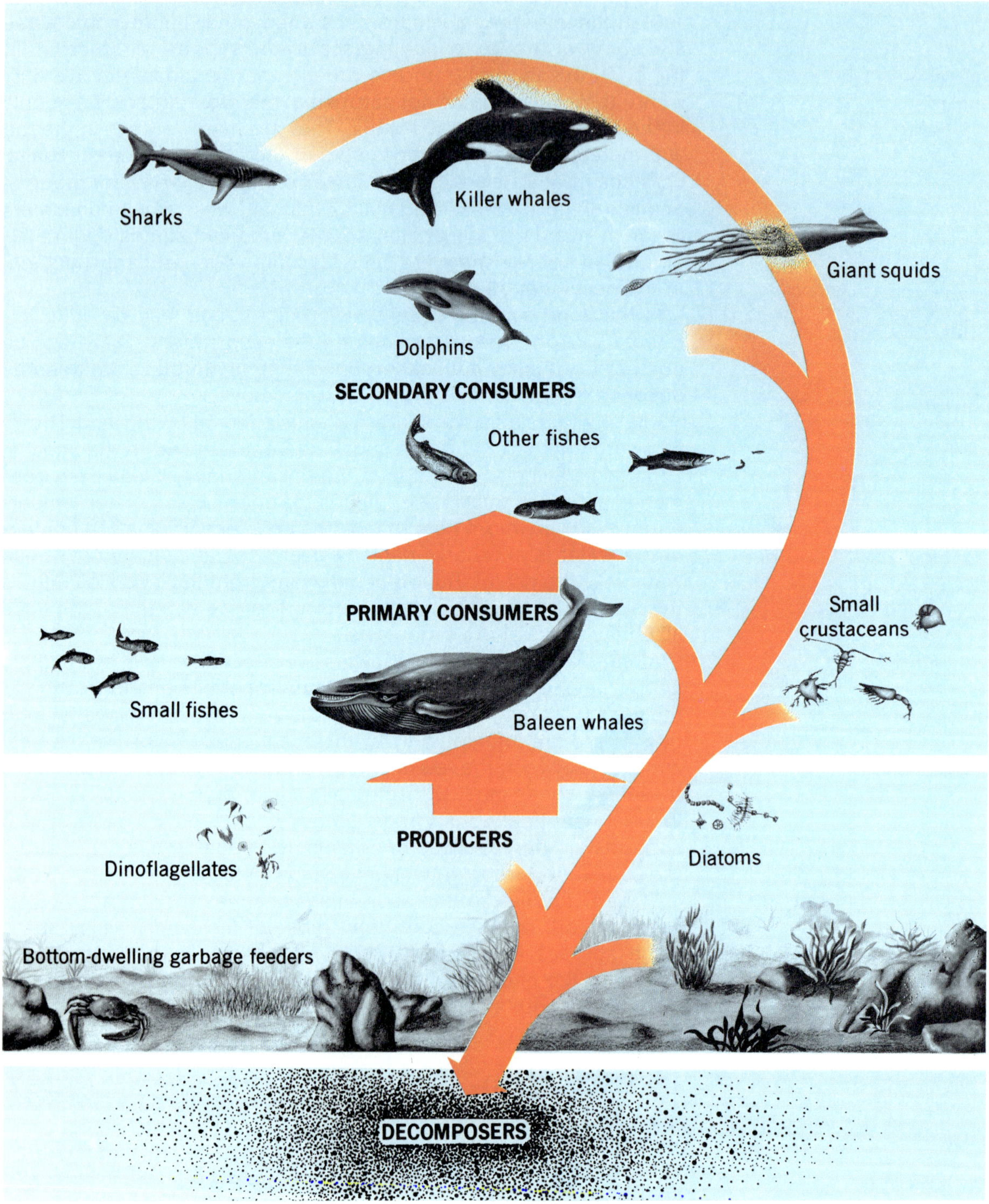

restrial plants and animals have special structures that conserve water. Although the skin of an animal gives off some water, for the most part it serves as a waterproof covering. Respiratory surfaces, where water may be lost by evaporation, are internal. Recall the lungs of chordates, tracheal tubes in insects, and air spaces behind the stomates in plants. In all these cases, narrow passageways limit contact with the drying air.

In different land areas we find different amounts of water—ranging from wet marsh lands to arid deserts. Soils vary tremendously in the amounts of minerals and dead organic material they contain. Oxygen and carbon dioxide are the only substances necessary for life that remain nearly constant. Oxygen and carbon dioxide are part of a continuous and ever-moving air mass.

Temperatures vary greatly on land, not only from one place to another but also with the seasons. The surface temperatures in some deserts may fall below 0°C in winter and rise above 50°C in summer.

23-4. A SHORELINE COMMUNITY

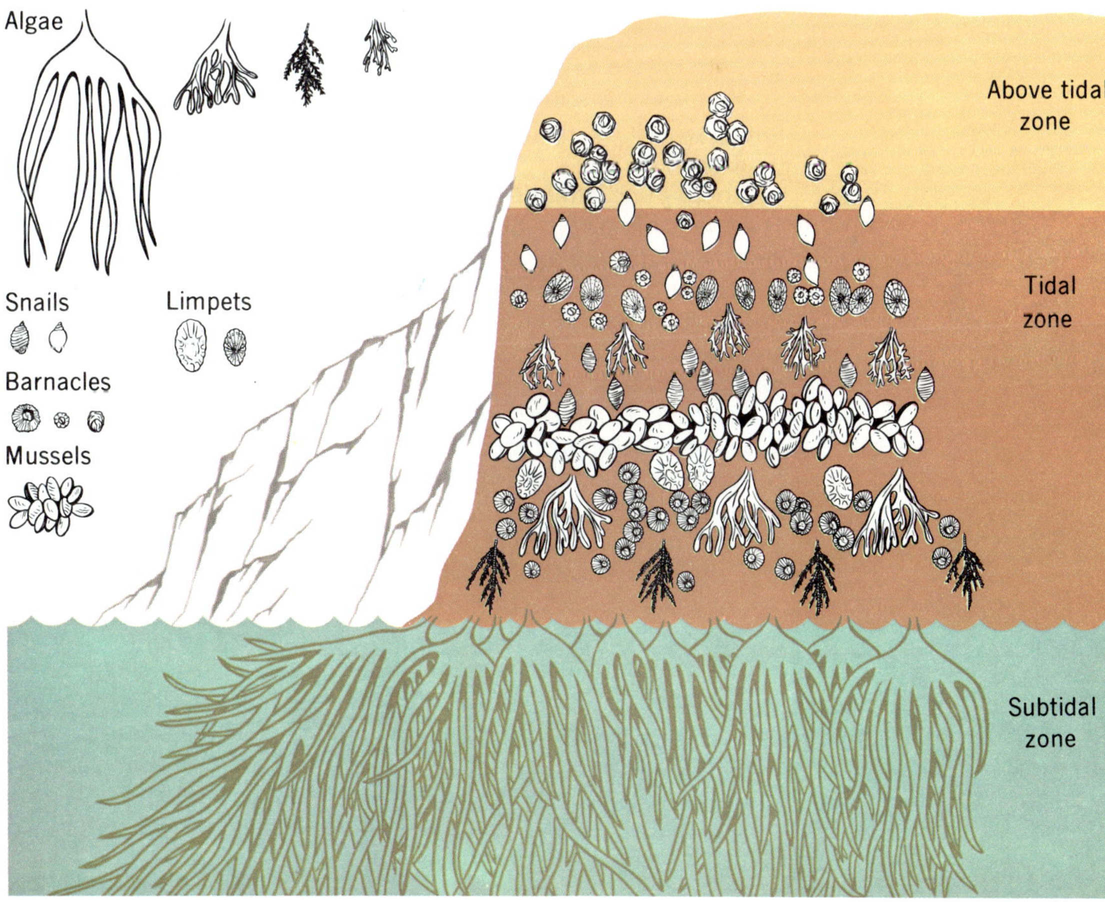

23-5. An astronaut must carry an earthlike environment when working on the moon.

No organism, in its active state, can withstand all of the environmental differences encountered in all land ecosystems. Humans are the most widely distributed species. We have learned to build shelters and to dress appropriately in extreme heat, cold, wetness, or dryness. We can even devise a costume that allows us to live briefly where the environment would not support life as we know it (Figure 23-5).

The soil and climate of a land ecosystem are important factors. They determine what green plants will grow. The plants, in turn, determine what animals can exist. In some places, the climate and soil will support a forest ecosystem. In others, they can support only a desert ecosystem. For the most part, animals that can live in a forest are quite different from those that can survive in a desert or any other ecosystem.

On the land masses of the earth there are five major types of ecosystems: forest, tundra, grassland, savanna, and desert. An ecosystem includes all **biotic** (living) things within its boundaries. But it is more than just a collection of plants and animals. Each organism interacts with other organisms and with the **abiotic** (nonliving) environment in which it lives.

You will study a land ecosystem in **Inquiries 23-1** and **23-2.** You probably will want to gather data for both inquiries at the same time. You and your teacher must plan this inquiry to fit an ecosystem that is convenient to your school. You may not have much choice, but let us look at some possibilities.

A forest provides opportunities to collect abundant data but is difficult to study as a whole. A prairie is almost as complex but is somewhat easier to study. Pastures and cultivated areas, such as cornfields, are rather simple to study.

Many schools have lawns with trees and shrubs. Here there may be fewer kinds of organisms than in the country, but you can be more thorough in your study. You also can study vacant lots and spaces between buildings. Even the cracks in pavement, gutters, and the area around playground trees often contain surprisingly diverse communities of organisms.

No biologist can identify every organism she or he sees. There are two ways to deal with this problem. One is to identify kinds only by general group names—for example, "trees," "spiders," "grass," "beetles," "turtles." These may be sufficient for developing many ideas about interrelationships. Another method is to collect a *specimen*—a sample individual or, in the case of large plants, a characteristic part such as a leaf. Mark the specimen with a letter (A, B, C, . . .). Whenever you need to refer to that organism, refer to its letter. You may be able to identify your specimen, after you return to the laboratory, by looking it up. Or you may show it to an expert.

The best observers take great care to preserve the ecosystem they are studying. Especially in cities, the pressure of people on natural ecosystems can be very great. A few careless biology classes, studying an ecosystem year after year, can become a significant factor in endangering the survival of a species.

INQUIRY 23-1 Producers in an Ecosystem

MATERIALS (per team)

plastic clothesline, 12 m, marked at 0.5 m intervals
strong twine
8 stakes, 18 to 25 cm long
hammer
millimeter ruler, 6 cm long
notebook and pencil
graph paper
right triangle
plastic bags
rubber bands

Procedure (Part A)

Teamwork will be necessary to make this survey. Plan the work carefully and take full notes.

Selecting and Plotting the Site. Choose an area with several trees, perhaps a lawn or a forest. Sites for studying different organisms should be of different sizes. These can be set up as in Figure 23-6. Square or rectangular study areas are called *quadrats.*

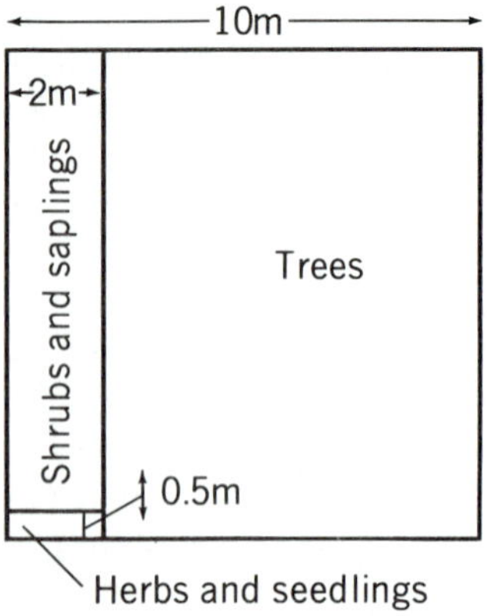

23-6. Plan of quadrats for use in the study of a forested community.

Drive a stake into the ground. Measure off 10 meters (m) in any direction from it, using the marked clothesline. Drive another stake at this point.

Stretch twine from the first to the second stake, tying it to each. Make a right angle at the second stake for the second side of your quadrat. Use the right triangle to get your right angle.

At 10 m from the second stake, drive a third stake and connect it with twine to the second. Continue around the square. Divide the large quadrat into smaller quadrats as shown in Figure 23-6. If the forest is thick, your quadrat may not be precisely square, because trees and bushes stand in the way. Be as accurate as possible.

Procedure (Part B)

Gathering Data. It is convenient to divide plants into three groups: trees, shrubs and saplings, and herbaceous plants and seedlings. In many areas, the ground is covered by litter (fallen leaves, stems, and so on) and certain flat plants, such as mosses and lichens.

Trees over 5 cm in diameter (about 16 cm in circumference) should be studied in the main quadrat (Figure 23-6). Observe trees that form the *canopy*—the forest top, which receives direct sunlight. Trees below this layer are called the *subcanopy.* Are the trees in your quadrat deciduous or evergreen? Deciduous trees lose all their leaves each year when unfavorable conditions begin. The leaves of evergreen trees remain on the trees for several years, and only a few are lost at one time.

Shrubs are low woody plants between 0.5 and 3 m tall. A sapling is a young tree that is 2.5 to 4 cm in diameter and 1.3 m high. Herbaceous plants are nonwoody plants that die down to the ground by winter time. These include grasses, grains, and small flowering plants. A seedling is a very young tree with a stem less than 1 cm in diameter.

Stand back and look at your study site. If necessary, walk around the outer boundary. Use the following symbols to sketch a profile of the ecosystem as you understand it.

Layers	*Plant types*	
Canopy: ♀	Woody vegetation	
Subcanopy: ⇧	Deciduous:	D
Shrubs and saplings: ()	Evergreen (coniferous):	E
Herbaceous plants	Herbaceous vegetation	
and seedlings: △	Grass:	G
Litter and flat plants: ⬭	Nongrass:	N

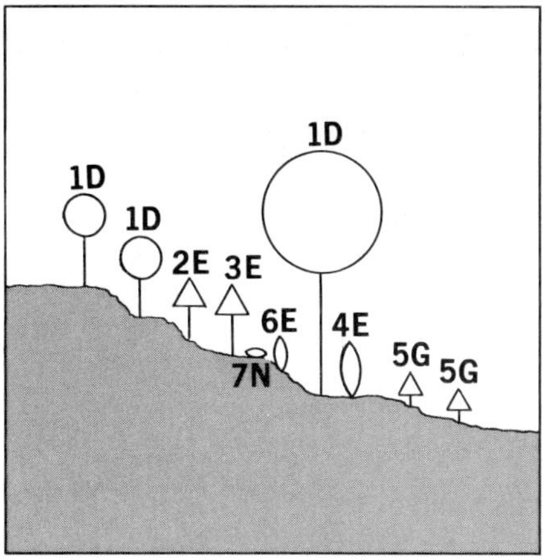

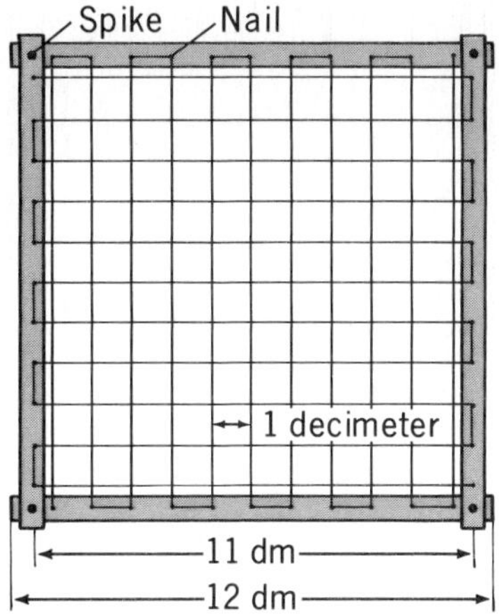

23-7. Example of a profile showing layers of vegetation.

23-8. Frame for a quadrat in a nonforested community.

Indicate with numbers the various plant types that are recognizably different. For example, in Figure 23-7, the first plant observed was a deciduous (D) canopy tree (1D). Three of these were found on the study area. The second species observed (labeled 2) was an evergreen (E) subcanopy species. Species 3 was also an evergreen subcanopy species, but recognizably different from species 2.

Once your profile of the site is made, you are ready for a closer look at its vegetation. Different teams should gather data on trees, shrubs and saplings, herbaceous plants and seedlings, and litter.

Abundance refers to the number of individuals of one kind within a given area. Count the kinds of trees in the canopy and subcanopy layers in the main quadrat. Count shrubs and saplings in one of the smaller quadrats. Also count herbaceous plants and seedlings in a small quadrat.

Special problems may arise. In a lawn, for example, there is no need to count blades of grass. However, a count of the "weeds" might be worthwhile, especially if comparisons are made between well-trodden areas and protected ones. A frame (Figure 23-8) may be useful for this work.

Make a table to record your counts. Across the top, list the kinds of plants in your quadrat. For example, if you are counting shrubs and seedlings you may list evergreen shrubs, rosebushes, raspberry bushes, other bushes, and seedlings. Under these headings record your count of each type of plant.

Teams that will study litter and flat plants should collect samples of litter in plastic bags. Gloves may be worn if you wish. Take samples from the ground around trees, shrubs, and herbaceous plants in a small quadrat. Secure the bag with a rubber band and mark it to indicate where it was gathered. You will study these samples later in the laboratory. At the study site, examine a sample 15-cm square of litter in place. Carefully pick the sample apart. Record the layers and composition of each layer. Also try to determine its physical condition. These observations should include state of decay, moisture level (damp, wet, dry), odor, location as to shade, and nearness to larger plants.

Procedure (Part C)

Analyzing the Data in the Laboratory. All the teams that counted the same kind of vegetation should now add their data. Then calculate the average number of each kind of plant to determine its abundance. Finally, make a list of all the plants on the study site in order of abundance.

To determine *density,* divide the total number of individuals of each kind by the area of the study site in square meters:

$$\text{Density} = \frac{\text{Total number of each species}}{\text{Total sample area (in m}^2\text{)}}$$

List the plants in the order of their densities.

The teams that are analyzing the litter and flat plants should also calculate the abundance and the density of each species (or kind) in the samples they took from the study area.

On the chalkboard, make a brief class summary of the important features of plants in the ecosystem you studied. Include such factors as profile, abundance, density, and litter. Record this information in your notebook.

Ecosystems are usually named according to the most obvious or important species of plants. For example, a deciduous forest might be called a beech-maple forest if these trees are dominant. Dominance is sometimes, but not always, determined on the basis of density. **1.** After you have carefully analyzed the data, determine the one or two species that appear to be most abundant. What will you name this ecosystem? **2.** How many layers of vegetation did you find in the entire study area? **3.** List the important species in each of these layers.

Some of the plants are in a layer because they are limited genetically to that maximum height. Others are young individuals that will some day reach higher in the ecosystem. **4.** Review your data to determine how many of these young individuals were among the plants in each layer.

Next, try to learn as much as you can about the animals that depend on the plants you have just studied.

INQUIRY 23-2 Consumers in an Ecosystem

In **Inquiry 23-1,** you studied the plants in an ecosystem and their relationships to one another. While it is convenient to separate the animals and plants of an ecosystem for study, we know that they interact with one another in many ways. As you conduct this inquiry on animals and their relationships, keep in mind the plants that are part of the same ecosystem.

For this field study, you will collect soil samples and specimens of invertebrates on the soil surface. You will collect flying insects and those crawling on, or attached to, plants. You also will make observations of vertebrates. Different teams should collect different samples or specimens.

Conduct your animal studies at the same site you used in **Inquiry 23-1.** In this way, you can study the relationships of all the organisms of the community. If you do not have a site plotted for study, refer to "Selecting and Plotting the Site" in **Inquiry 23-1.**

MATERIALS (per team)

In the field:

notebook and pencil
2 meter sticks
plastic bags
rubber bands
ink marker
frozen orange juice can, 5-cm diameter
sweep net
several specimen bottles containing 5-8% formalin or other preservative (may be 70% ethyl alcohol)
1 or 2 insect-killing bottles (ethyl acetate)

In the laboratory:

funnel
steel wool
gooseneck lamp with 25-watt bulb
compound or stereoscopic dissecting microscope
small vial
5-8% formalin solution (or 70% ethyl alcohol)
ring stand and clamp

Procedure (Part A)

Field Study. Collect soil and leaf litter samples from which you can extract tiny animals. Because there are so many of these organisms and their rate of reproduction is so great, you may collect and kill them for study. Collect the layer of undecayed litter from a small area. Seal this material in a small plastic bag. Label each bag with your team name and the quadrat from which it was taken.

From the cleared area, cut a soil sample 10 cm deep, using the orange juice can. Seal this sample in a plastic bag and label it. Take at least two soil samples from each main quadrat.

Look for invertebrate animals on the soil surface and among the litter of a small quadrat. Collect a specimen of each kind. Place it in a specimen bottle. Mark each bottle with an ink marker, recording the date, location, and team. Look carefully for spider burrows, insect nests, and other evidence of the presence of invertebrates.

Many animals can be found only on plants. Examine the plants in a small quadrat. Look on both sides of leaves, on stems, on flower buds, and on growing tips. Pick off a leaf, stem, or bud that has animals on it. Drop the entire specimen into a specimen container with preservative. Label with the date, location, plant type, and team number.

Collect flying insects with a sweep net. Walk in a straight line across the main quadrat. Make a low sweep with the net with every step you take. Brush the vegetation gently as you make your sweep. Quickly close the net. Transfer the insects to an insect-killing bottle. Label the bottle with the type of collection, date, area cover, and team number. If you take an additional sample, place it in a different bottle.

Amphibians and Reptiles. The vertebrates of an ecosystem are more difficult to find and study. As you walk through the main quadrat, turn over rocks and logs to find some amphibians and reptiles. Return logs and rocks to their original positions. Do not ruin the habitat for the organisms that live there. List the amphibians and reptiles you observe in order of abundance.

Birds. Describe or identify any birds that appear in the quadrat you are studying. Record the data you collect in a chart. Calculate the abundance and density of birds.

You may want to extend this part of the inquiry by setting up bird blinds to observe nests on weekends and after school. Binoculars are necessary for observation and identification, preferably 6 × 30, 7 × 35, or 7 × 50 power. Records of bird songs are a good aid to identification. Flight patterns, food sources, locations, and behavior patterns also are useful clues to help you identify the birds you observe.

Mammals. Sometimes you can observe mammals and other vertebrates directly. More often, you will need to look for traces of mammals—tracks, remains, or homes. Listen for animal sounds. It is difficult to determine abundance for mammals because they move about a lot. Record your estimate of the numbers and species of mammals in order of abundance. Indicate the density of each species in the study area.

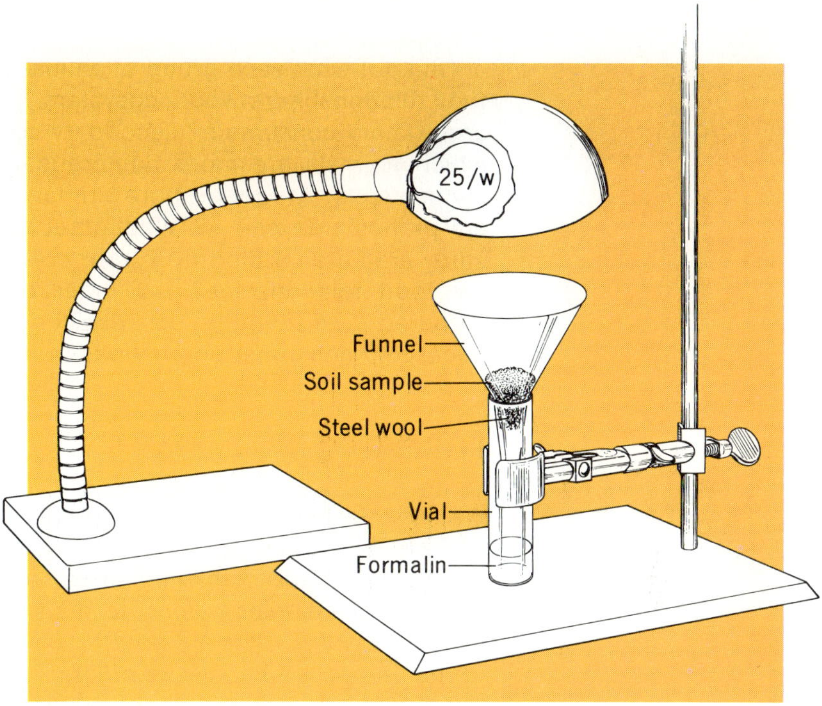

23-9. Apparatus for collecting animals in the soil.

Procedure (Part B)

Laboratory Analysis. To remove the animals from the soil samples, set up the apparatus shown in Figure 23-9. The neck of the funnel should be suspended into a vial of 5 to 8 percent formalin. The steel wool in the bottom of the funnel will hold back the litter and soil. Set the lamp to shine into the funnel from several centimeters above it.

The organisms in the samples will move away from the light and heat and fall down into the preservative. This process requires some time. Wait a week before you examine the animals in the preservative with the microscope. Record the species by name or by drawings. Record the number of individuals of each species in a column opposite the soil specimen from which it came. Total the number of each species found in all samples.

Sort the specimens of invertebrates you found on the soil surface and on plants into phyla and, when possible, classes. Also attempt to sort the arthropods into classes. Record the number of individuals of each group. Total the number of each group in all samples. List these in order of abundance. From your data, compute the density.

Relate the groups of animals you observed to the types of vegetation in this ecosystem. **1.** Which layer of the ecosystem has the greatest abundance of individual organisms?

Find out what each group of animal you found eats. Diagram the food relationships of your ecosystem. **2.** How many of the animals are primary consumers? secondary consumers? **3.** In your data, which animals are more numerous—primary or secondary consumers? **4.** Are there more primary or secondary consumers in any particular layer? **5.** What other animals known to occur in the study area are missing from your collections? How do they fit into the food relationships? **6.** After considering all the data you gathered in this inquiry, describe the interrelationships of the producer and consumer populations in this ecosystem.

FOREST ECOSYSTEMS

Wherever enough moisture is available and temperatures are not too low, the land supports forests. We have destroyed many of the forests that once blanketed broad areas of the earth, but some remain. Forest trees belong to many different species. Some are gymnosperms; others are angiosperms.

The photosynthetic tissue of large trees is far above the ground. Smaller plants grow in the shade of the forest floor. Some of these are shade-tolerant species that can grow in dim light. Some are young plants of the species that make up the forest. Most of the young trees soon die because there is not enough light for them. If one of the larger trees is removed, the smaller trees nearby begin to grow rapidly. Smaller or weaker individuals become shaded and die. After years have passed, probably only a single tree will stand in the place of the one removed long before.

Temperate Deciduous Forests

In the mid-latitudes of both the northern and southern hemispheres, forests of trees that lose their leaves in the fall—deciduous trees—thrive. For centuries prior to the coming of European settlers, American Indians depended on the deciduous forest ecosystem of North America (Figure 23-10) for their food and shelter. They hunted deer, bears, bison, squirrels, and wild turkeys. They harvested wild fruits, seeds, and roots. Their agriculture was restricted to a few small plots near their villages. At that time, the Indians were a part of the deciduous forest ecosystem. They did little to alter it.

All this changed with the appearance in North America of many European colonists. With axes and plows, the settlers cleared the land for cultivation. Gradually, the amount of forested land in the eastern United States declined. Today only small isolated and protected segments of this ecosystem remain.

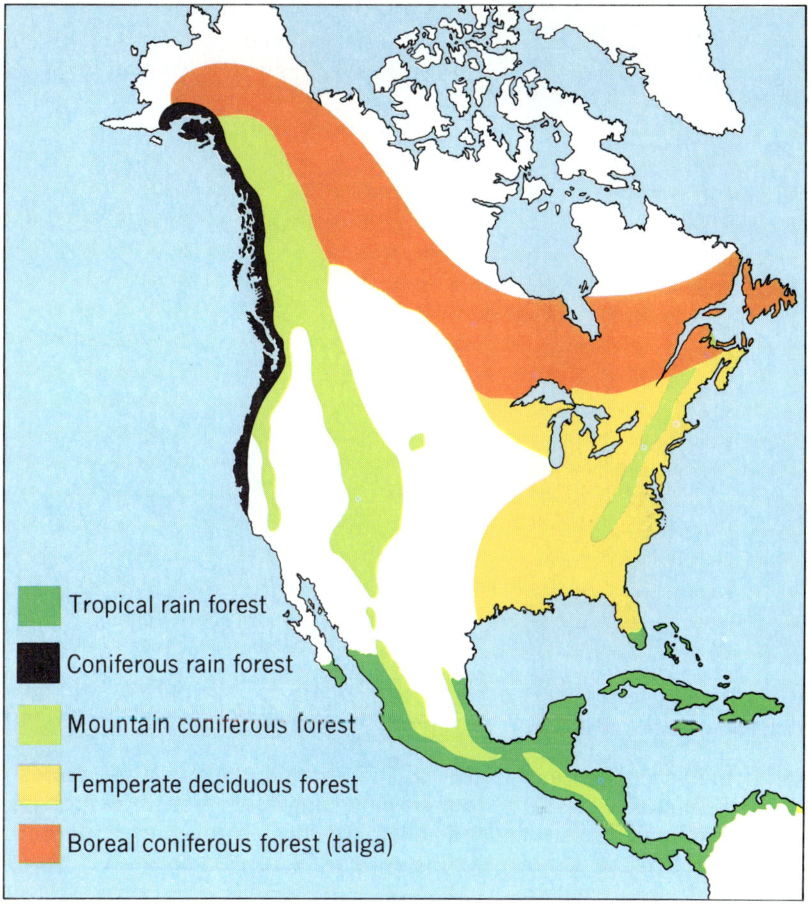

23-10. The distribution of forests in North and Central America before land was cleared for cultivation. The white areas are grasslands, deserts, tundra, and glaciers.

Some deciduous forest ecosystems in this area are still preserved. The most extensive are in the Great Smoky Mountains National Park of North Carolina and Tennessee. Here we can get some idea of what the original forest was like. Buckeye, basswood, beech, tulip trees, and many others make up the forest of the Smoky Mountains. Farther north in New England, we are more likely to see beech and maple trees (Figure 23-11). At the southern end of the Appalachian Mountains, oaks and hickories are common. All of these deciduous trees are hardwoods. In some regions, however, conifers, especially hemlocks, may be present.

The vegetation of the forest tends to be in layers, not unlike the zones of the tidal zone. In a deciduous forest (Figure 23-12), the

23-11. This maple grove in Vermont illustrates one of the dominant species of trees in climax deciduous forests of the eastern United States. *What are some other climax trees of this region?*

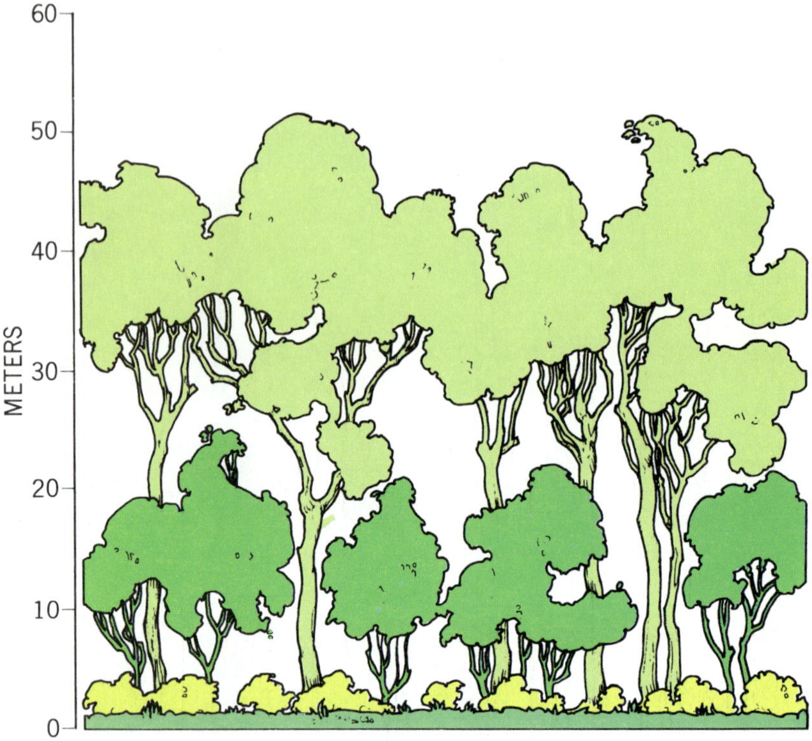

23-12. The layering effect in a deciduous forest. The upper two levels are composed of deciduous trees. The third level is composed mainly of shrubs, and the fourth layer is occupied by herbaceous plants that produce flowers in the spring (Figure 23-13).

upper level is composed of trees that form a canopy 25 to 60 meters above the ground. A second layer of trees reaches heights of about 6 to 16 meters. A third layer contains deciduous and evergreen shrubs 0.3 to 2.5 meters tall.

Herbaceous plants, such as hepaticas, Dutchman's breeches, wild ginger, bloodroot, May apple, and others compose the fourth layer, along with tree seedlings, on the forest floor (Figure 23-13). Mixed with these and on fallen logs and rocks is a fifth layer, mosses and lichens. The layering effect is due mostly to interactions between organisms and the available light and humidity. Down through the layers, light intensity decreases while humidity increases.

Coniferous Forests

Anyone who has traveled through the Sierra Nevadas of California or the Rocky Mountains is aware that, in ascending these moun-

23-13. Some of the flowering plants commonly found in deciduous forests include: spring beauty; trout-lily; toad shade; rue anemone; common blue violet.

tains, we go through several climatic zones in a short time. For example, when traveling westward toward the Rocky Mountains of Arizona, we come up out of the Painted Desert into the foothills. The area is dotted with rather scrubby-looking piñon pines and junipers. This open coniferous forest extends upward from about 1,500 meters through a rather dry zone to about 1,800 or 2,100 meters. Continuing upward to about 2,500 meters, we see a change in the kinds of conifers, reflecting an increase in moisture. Here are ponderosa pines, especially on the north-facing slopes where less drying takes place.

A little higher, around 2,800 meters, and again more prominently on north-facing slopes, Douglas firs and limber pines are abundant. The presence of Douglas fir is a good indication of increased moisture. Above 2,800 meters, the Douglas firs merge with spirelike Englemann spruce. In early June, melting snow covers parts of the forest floor in this zone. The temperatures are fairly cool as flowering plants appear (Figure 23-14). The spruce forest extends to the timberline zone (Figures 23-15 and 23-16) at about 3,500 meters, where wind and low temperatures are limiting factors.

The coniferous forest of the mountains changes in character from one zone to the next. However, not all coniferous forests are this diverse. In the far north, the *boreal* (BORE·ih·al) coniferous forest (Figure 23-10) is much more uniform. It is composed of spruce, pines, and firs. In parts of Canada and Russia, the vast northern spruce forests are known as the *taiga* (TY·guh). The climate in these forests is severe and similar to the climate at about 2,800 meters on a mountain. This far north, however, summers are short and less rain falls. The winters are more severe than in the spruce forest of a mountain. Very few trees other than spruce are adapted to this kind of physical environment. One large mammal that is very much a part of the boreal forest ecosystem is the moose.

23-14. Flowering plants of higher altitudes endure cool temperatures during their growth seasons. Snow buttercups; sunflowers; Indian paint brush; mountain bluebells.

23-15. Timberline in the Rocky Mountain National Park in Colorado. *What are some of the environmental factors characteristic of the high elevations that seem to limit tree growth and determine a timberline?*

23-16. These pine trees have been damaged by the winds. This effect is called flagging.

Tropical Rain Forests

Great forests also occur in many parts of central Africa, southern Asia, and the tropical parts of the Americas. The trees in these tropical forests usually have broad leaves. The leaves are shed irregularly rather than seasonally as in a temperate-zone deciduous forest. The tropical forest is always green. It never takes on the barren, wintery appearance of a deciduous forest in the temperate zone. Many tropical forests are called rain forests because rainfall is plentiful. Here temperature and humidity are always high. Frequently, the trees have such a dense mass of leaves that little light reaches the forest floor. Only the most shade-tolerant species can survive at the lower levels of this forest.

The variety of life in a tropical rain forest staggers the imagination. In such a forest, you might search a long time to find two trees of the same species. For every few dozen species of insects in a temperate deciduous forest, there are hundreds or thousands in a tropical rain forest.

The richness and diversity of life in a tropical rain forest is due to many factors. The environment has been stable for a very long time. The tropics were almost unaffected by the glaciers that covered much of Canada, the northeastern United States, and Europe as recently as 10,000 years ago.

In the warm, moist rain forests, plants grow where they could not grow in the temperate zone. Because there is no danger of drying

23-17. Epiphytes in the tropics. These plants grow on the branches of trees in the sunlight. They obtain water from the air.

23-18. Epiphytes growing in the coniferous rain forest in the state of Washington. The epiphytes are chiefly mosses and liverworts.

out, epiphytes (EP·uh·fites) cover many trees in the rain forest (Figure 23-17). These plants obtain moisture chiefly from the air. Many species of epiphytes, including orchids, grow in the tropics. The pineapple and fern families also occur in larger numbers.

A great variety of animals can live in such a highly productive ecosystem. In the canopy of the tropical rain forest, which may be more than 30 meters above the ground, we find small mammals, birds, snakes, frogs, and insects. All these animals show remarkable adaptations to this treetop environment.

23-19. Alpine tundra may be found on the tops of the highest mountains. Many of the plants that occur in the arctic tundra are also found in the alpine tundra.

Coniferous Rain Forests

In some areas in the temperate zone, much rain falls. Such an area exists on the western side of the Olympic Mountains in Washington State. The ecosystem produced by abundant rain and moderate temperatures supports magnificent specimens of temperate-zone coniferous trees. As in the tropical rain forests, the trees of coniferous rain forests often are covered with epiphytes. In the coniferous rain forests many mosses and liverworts hang from the branches of the trees (Figure 23-18). Only shade-tolerant plants such as ferns and mosses survive on the forest floor.

TUNDRA ECOSYSTEMS

In our trip across the mountains, above the timberline (Figure 23-19) we encounter the tundra zone, where no large trees can grow. We do find miniature willows and birches that never grow taller than five to ten centimeters. This area, with its lichens, mosses, and tiny flowering plants (Figure 23-20) is sometimes called the alpine tundra ecosystem.

23-20. A variety of plant life occurs on alpine tundra rocks and slopes. Most of these plants were photographed on Mt. Rainier. Lupine, red mountain heath, saxifrage, sedge, phlox, and grasses.

23-21. Tundra. Beyond the northernmost reaches of tree growth, the arctic tundra begins.

On some mountains, the alpine tundra is subjected to winds greater than 160 kilometers per hour and temperatures down to −57°C or lower. At this high altitude, visible light and ultraviolet radiation are intense. Water in the form of snow is abundant and, when thawing occurs, the surface soil usually is boggy. Summers provide only a month or two of favorable weather.

Some animals, especially birds such as the ptarmigans and piñon jays, live only in certain tundra areas. Pikas and some marmots are well adapted to the rugged environment above timberline. Larger mammals such as deer and elk move up or down from one area to the other with seasonal changes in the availability of food. Most of the plants here are small perennials that can reproduce rapidly.

Similar specialized organisms grow far away in the vast tundras of North America, above the Arctic Circle (Figure 23-21). The abiotic factors of these two widely separated ecosystems—the alpine and arctic tundras—are very similar. Temperatures are low, winds are strong, and the growing season is short.

Among the producers in the arctic tundra ecosystem are mosses, miniature birches and willows, and other tiny flowering plants. Few trees grow here, making the arctic tundra one of the most desolate of all places. A gray-green blanket of mosses and other small plants covers the thin layer of tundra soil. In the one or two summer months, marshes and lakes form as the snow melts. These make ideal breeding grounds for ducks, geese, and other water fowl.

As we might expect, tundras, in comparison with most forest ecosystems, can support many fewer species of plants and animals.

Primary consumers such as hares, lemmings, caribou, reindeer, and musk oxen move from one place to another through the ecosystem. They obtain a limited amount of food from vast expanses of vegetation.

Although humans have occupied almost every kind of ecosystem, very few can live in the tundra. Only the Eskimos of North America and the Lapps of northern Europe have managed to survive in this part of the biosphere.

GRASSLAND ECOSYSTEMS

Water is a necessity for all plants, but the requirements vary among species. For example, much water is required to support a forest. Where there is less than enough water to support a forest in the temperate zone or the tropics, a grassland ecosystem develops (Figure 23-22). Grasslands occur where yearly rainfall averages about 250 to 750 millimeters.

But more than the *amount* of rain is involved. Grasslands are regions of frequent, and often severe, drought. These lands might

23-22. A prairie grassland in Nebraska. Vast regions of the central United States, between the eastern woodlands and the Rocky Mountains, were grasslands during the time of the Plains Indians. Today in much of this area cultivated grasses, such as wheat, rye, oats, and barley, have replaced the native grasses.

SCIENCE AT WORK: A Front Lawn Ecosystem

Not all investigations in science take place in the laboratory with test tubes or trained rats. Ecologists are scientists who study the interactions of organisms in their environment. Such work need not be done in wilderness areas far from cities and towns. Sometimes, ecological studies await the investigator very close to home.

One ecologist studied a lawn and found a complex ecosystem there. Although the plants—grass and weeds— were most evident, a closer look showed many species of animals as well. With a vacuum cleaner, the ecologist collected pill bugs, aphids, leaf hoppers, ants, and beetles. Spiders, mites, snails, and worms also inhabited the lawn. Although birds were not full-time residents of the lawn, they played a role in the ecosystem. They consumed insects. And their wastes on the lawn supported a variety of decomposers.

The grass and weeds were the producers in this ecosystem. And, as you might imagine, these plants must store a great deal of energy to be able to support such a wide variety of animal life. To measure just how much, the ecologist mowed the grass, waited several weeks, and mowed again. The grass from the second time was dried and then it was weighed. This gave the ecologist a measure of how much plant material was produced over a certain length of time.

Through a series of calculations, the ecologist was able to determine how much energy the lawn plants were fixing. These plants were fixing as much energy as a typical corn field of the same size. Furthermore, the lawn plants yielded 7 percent more energy than was used in caring for it. These calculations included human labor, the energy in fertilizer, and the energy of the gasoline used in the lawn mower.

Thus, even your front lawn, or a nearby grassy lot, is a complex ecosystem. How do you think you fit into these ecosystems?

23-23. The "dust bowl" of the American midwest in the 1930s.

receive enough water for tree growth for several years. Then a dry period will come that kills the trees. Such conditions could not support a forest. The relationship of water to the type of plants supported is often vividly illustrated in a grassland. Only grass can grow in most of the area. Along stream beds, however, a narrow ribbon of trees clearly indicates that, given a little more water, trees could grow.

Grasslands have been the home of grazing animals for millions of years. The Great Plains of North America at one time were the home of the American buffalo. The grasslands of Africa are the home of many species of antelope. Kangaroos are common grazing animals in Australian grasslands. We now use many of these grasslands to pasture cattle and sheep that provide our meat, hides, and wool.

Compared with forest ecosystems, grasslands are ecosystems of stress. Forests provide shade and protection from winds. They release tremendous amounts of water into the atmosphere through transpiration. Transpiration also moderates the temperature in a forest. Humidity, always important for land animals and plants, is lower in the grassland than in the forest.

Grasses and other short plants in grasslands do not offer much protection to the other organisms living there. Winds sweep across a grassland. The organisms are exposed to the sun and to greater temperature variations than organisms in a forest.

In the United States, large areas that once were grassland have been converted to crop land. Sometimes these areas have enough rain in the "good years" for abundant crops, especially wheat and corn. When the rains fail, however, the crops fail. The soil then loses its protective cover of permanent grass, and it erodes through the action of wind and water. Grasslands that are used only as pasture often are overgrazed. The grass mat is destroyed and erosion begins. Many grasslands throughout the world have been reduced to near-desert conditions by human activities (Figure 23-23).

23-24. A waterhole in a tropical savanna of Africa. Grasses are the most abundant plants but widely spaced trees and shrubs are also found. Water is scarce and the animals congregate in the few places where it occurs. Impala, wildebeest, and zebra are shown here.

SAVANNA ECOSYSTEMS

Those of us living in North America are not familiar with this kind of ecosystem. Yet it is very important in much of South America, Africa, Australia, and parts of Asia, especially India.

The savanna might be described as a semiforested grassland (Figure 23-24). Often a savanna is very arid, with a few scattered waterholes in a setting of tall grasses and thorny, flat-topped trees. In parts of Africa the savanna is populated by zebras, giraffes, antelopes, and rhinoceroses. Lions, jackals, hyenas, and vultures also live there. The photosynthetic activity of plants (and, therefore, the food supply) is greatest during the rainy season. It becomes lower when dry, hot weather arrives.

DESERTS

Grasslands cannot support many trees. But conditions are favorable enough for a continuous cover of vegetation. Deserts are the next step. In a desert, only isolated plants grow with much bare ground between them.

In a true desert, there are no constantly flowing streams, although some deserts have rivers flowing through them. The Nile, for example, flows through the desert of Egypt, but it originates in the well-watered areas of the interior of Africa. A desert stream bed contains torrents of water after a heavy rain, but it is usually a dry gully. Deserts generally average less than 250 millimeters of rain per year. Rain may be frequent in one year, however, and there may be almost none for the next several years. This unreliability is a condition to which many desert plants and animals are suited (Figure 23-25).

Without a protective cover of vegetation, the desert environment is highly variable. In other ecosystems, plants limit the amount of runoff of rain water. In a desert, runoff is rapid. What little rain does fall is not available for very long. The humidity of a desert is always low. Winds sweep across the land. Frequently, deserts are hot by day and cold at night. It is not unusual for the temperature to change 30 degrees in 24 hours.

The great deserts are found on all continents except Europe and Antarctica. The Sonoran Desert is in the American southwest and north central Mexico. Near-desert conditions exist in the United States as far north as the central part of Washington State. The Sahara Desert stretches across north Africa and through the Arabian Peninsula. Deserts also occur, with interruptions, across central Asia almost to the Pacific coast. The center of Australia is a desert. In South America much of the west coast, from central Chile to Ecuador, is desert or nearly so.

A desert ecosystem does not support the abundance of organisms found in most forests or grasslands. There is far less food than in the other major terrestrial ecosystems since the producers are less numerous. Water is a limiting factor for plants as well as for animals. The plants and animals that do live in the desert show numerous adaptations for obtaining and conserving water.

Some desert plants have very shallow roots that extend over a wide area. When rains come, the plants quickly take up water and

23-25. The Mojave Desert of California. Most of the plants you see are growing in the lowland area between the foreground and the hills in the distance. *Try to suggest why.*

store it in specialized tissues. Cacti and euphorbias store large amounts of water in their fleshy stems. This water is then available to the plants in long periods of drought. Other desert plants have deep root systems that take up water far below the surface.

Plants that survive in deserts have fewer and smaller leaves than plants in other ecosystems. There is nothing on a cactus plant that you will immediately recognize as a leaf (Figure 23-26). Actually, the spines are modified leaves. This modification reduces water loss. The green stem carries out photosynthesis. Its thick covering also protects against water loss. When more typical leaves are present

23-26. (Lower right) The ocotillo of the Sonoran Desert has very small leaves. During periods of drought, its leaves (enlarged above) are shed. When rains come, new leaves are grown. Compare this adaptation with that of the saguaro cactus at the left. **(Lower left)** The stem of the saguaro absorbs water during rains. A close-up of a cactus is shown above. The green stem carries out photosynthesis. Modification of leaves into spines protect against water loss.

23-27. The desert in bloom. When the rains come the desert shrubs begin to flower. Seeds of many species of flowering plants quickly germinate and flower.

on desert plants, they usually are small and thick. They may be either leathery or easily shed in time of drought. Often stomates are sunk in little pits.

The structure, physiology, and behavior of desert animals are adaptations to the desert ecosystem. Many species live in burrows, where humidity is greater and temperature changes are less extreme than on the surface. Many desert animals are active only at night when the temperature is lower and the humidity is greater. Desert animals often need far less water than do similar species living in grasslands and forests. Many never drink at all. They rely on water from their food. They also rely on moisture that is produced during respiration.

A barren desert may become a luxuriant garden after a heavy rain (Figure 23-27). Plants grow, flower, and produce seeds during the brief periods when enough moisture is available. Animals become very active. The desert is transformed into a colorful living world. When the rains pass, however, and the water evaporates, the desert relapses into its usual condition.

In Brief

Many different kinds of ecosystems occur throughout the biosphere. Various producers and consumers are well adapted to survive in the environmental conditions of these diverse areas.

Light is available rather uniformly in the upper layers of ocean waters. Green plants in these layers produce food for a wide assortment of consumers. In the depths of the ocean, where the light does not reach, producers cannot grow. The consumers here rely on material that drifts down from the richer surface layers.

Environmental conditions on land are much more varied than those of the ocean. Adaptations in land organisms provide for conservation of water, which is not as readily available on land as it is in the oceans. In a land ecosystem, the relationships among such factors as the amount of water, variations in temperature, and altitude determine the kinds of life that will survive.

In some cases, the land and climate can support a tropical rain forest. Here, a thick growth of trees, shrubs, and smaller plants produces food for many kinds of animals. In other areas, sandy soil, drying winds, and very little rainfall support only a few, well adapted species. Other ecosystems have varying conditions to support different numbers and kinds of producers and consumers.

Guide Questions and Problems

1. In what ways does the ecosystem in the surface waters of the open ocean resemble land ecosystems?

2. In what ways are environmental conditions in the oceans more stable than those on land? What adaptations occur in land plants and animals that would not be necessary in aquatic plants and animals?

3. In what ways do land plants and animals conserve water?

4. Describe the ecosystems that occur in your community. You may have difficulty in finding one in your book that corresponds with one in your community. What are some of the reasons for this?

5. The photograph in Figure 23-28 shows a place in the American southwest where the rainfall is not sufficient to support a forest. Nevertheless, there are some trees—they show up as the darker patches of green. The view is from the west. Notice that the trees are clustered on the south side of the two valleys on the side of

23-28.

the mountain. That is, the trees are on the slopes that face north. Suggest some hypotheses that might explain this distribution of trees.

Ideas for Independent Study

1. Over the course of many generations, people have found ways to survive in each one of the great ecosystems on earth. You may find it interesting to study the pre-industrial life patterns developed in each of two or three different ecosystems—desert, temperate deciduous forest, and tundra, for example. Your investigation should include, if possible, nutrition, clothing, shelter, transportation, and recreation.

2. If you live in a large town or city, you are in an ecosystem quite different from any that we have described in this chapter. Today, most people in the United States inhabit such urban environments, and many organisms (people and other forms) live out their entire lives there. You might wish to undertake a study of this type of ecosystem. Find out what organisms occur here and in what numbers. What are the sources of nutrients? How is waste disposed of? What variations in physical conditions, such as light, temperature, and rainfall, occur in this ecosystem?

3. What methods have been used in designing an artificial ecosystem for long voyages into space? What are some of the problems that must be considered in building such a closed system? People in space will need sources of oxygen and food. At the same time, some means of getting rid of carbon dioxide and other body wastes must be devised. How has this been done? How is light energy supplied and waste energy (heat) removed? You might carry these ideas further, considering in detail all aspects of survival in such a closed ecosystem.

Related Readings

BOOKS

Engel, L. *The Sea.* New York: Time-Life, 1968.

Fanning, O. *Man and His Environment: Citizen Action.* New York: Harper & Row, 1975.

Farb, Peter. *The Forest.* New York: Time-Life, 1969.

Gildea, R. Y., Jr. *Arsenic and Old Lead: A Layman's Guide to Pollution and Conservation.* Jackson, Miss.: University Press of Missouri, 1975.

Leopold, A. Starker. *The Desert.* New York: Time-Life, 1967.

Lucas, J. and P. Critch. *Life in the Oceans.* New York: Dutton, 1974.

Peterson Field Guide Series. Edited by R. T. Peterson. Houghton Mifflin Co., Boston. Volumes on birds, mollusks, butterflies, mammals, reptiles, amphibians, insects.

ARTICLES

Grosvenor, G., F. Leydit, and F. Grehan. "America's Wilderness: How Much Can We Save." *National Geographic,* February 1974, page 151.

24 Interactions in Ecosystems

In nearly any natural ecosystem that is not disturbed by human activities, only slight changes occur with the passage of time. Year after year, about the same number of frogs live in a woodland pond, and about the same number of flowers bloom nearby. The number and kinds of insects and soil microorganisms also remain about the same.

The life of an individual organism is a striking contrast to this stable life. An individual exists as part of a **community**—a group of plants and animals living together in the same place. Often the individual goes through a complex growth pattern during which its activities may vary tremendously. It matures and produces young. Eventually the organism dies, but the community continues to exist. The stability of the natural community is the result of the interactions of individual organisms in the community with each other and

24-1. An animal in and out of its environment.

with the environment. The study of these interactions is called **ecology.**

The importance of the environment to our understanding of organisms is suggested by the obvious differences between the two drawings in Figure 24-1. Both show the same deer. But the drawing at the right conveys nothing of this animal's way of life. It does not show where the animal lives or what it eats. It does not indicate whether the deer's habitat is warm or cold, densely overgrown by plants or partly open and sunny. What would be the effect of the color of its coat when the deer is in a thicket of trees?

As part of a community, an individual is involved in many complex interactions. Ecologists study communities of interacting organisms in different ecosystems. The ecosystem may be a square meter of prairie, the edge of a pond, a tidal pool, an alpine meadow, or a few cubic meters of surface water of the ocean. It may even be a closed bottle containing some algae and protozoans. Even the simplest ecosystems have so many interactions that it is difficult to study them as a whole.

Figure 24-2 shows some of the important factors that influence an ecosystem. We can classify them as abiotic factors and biotic factors. We will consider these factors and then discuss more complex interactions of organisms in their environments.

ABIOTIC FACTORS OF AN ECOSYSTEM

Each ecosystem has a particular set of abiotic, or physical, factors. These influence the biotic, or biological, factors that exist in that ecosystem. Let's look at some of these abiotic factors.

Temperature

As gases, liquids, or solids in ecosystems absorb radiant energy, their temperatures rise. The heat energy then flows from warmer to cooler parts of the environment. For example, if the temperature of the soil or ocean is higher than the temperature of the surrounding atmosphere, heat energy will flow into the atmosphere.

The body temperature of most organisms changes with the temperature of the surroundings. A few insects, fish, and reptiles have systems that modify body temperatures to a certain extent. Only birds and mammals, however, can maintain body temperature at a constant level. When you are swimming in a cold lake, at 15°C for

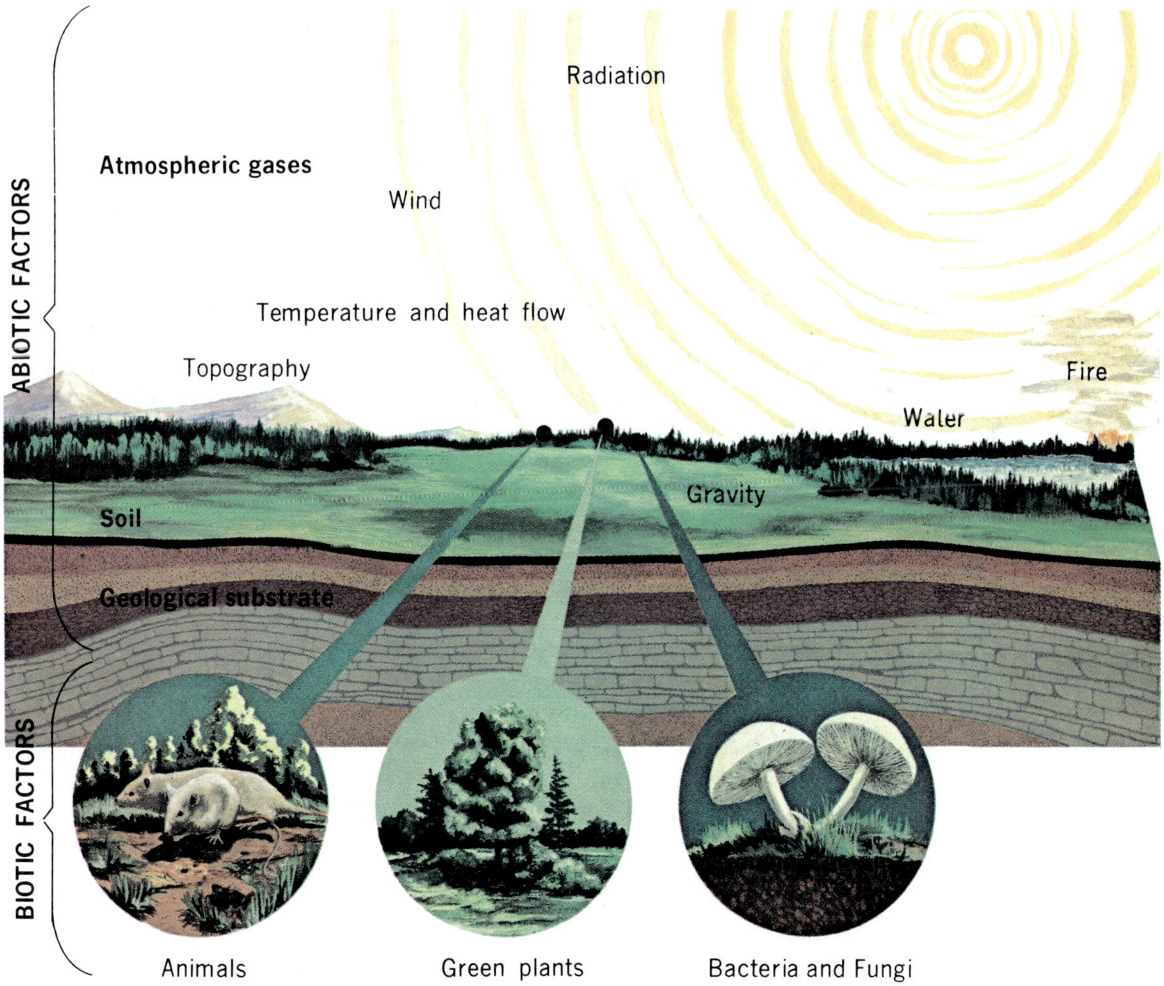

24-2. Interaction of plants and animals with the environment.

example, your body temperature will be 37°C. That of a fish in the same lake would be about 15°C. If you both swam into a colder pocket of water, your temperature would stay at 37°C, but the temperature of the fish would drop. Similarly, if you both swam into a warm, shallow inlet, your temperature would remain at 37°C but the fish would now be much warmer.

Changes in the water temperature, then, would have a much greater effect on the fish than on you. Your metabolism might increase slightly in the colder water and might decrease somewhat in the warm water. On the other hand, the body functions of the fish would occur at a much lower rate in cold water and at a much higher rate in the warm water. Changes in environmental temperature affect the metabolism of organisms that do not have internal mechanisms that control temperature.

Most species of organisms are limited in the range of temperatures in which they can survive. The arctic, temperate, and tropical zones of the earth each support a variety of animals and plants. Yet most of the species found in one zone could not survive in the others.

Water

The water needed for life on land comes ultimately from rain and snow (Figure 24-3). Moisture in clouds condenses to form rain or snow, which then falls to the earth. Some of this water enters the soil and is taken up by plants. Excess water runs off the land into bodies of water, such as ponds, rivers, and oceans. Evaporation returns part of the water to the atmosphere. When the concentration of water in the air becomes great enough, clouds form again. This cycle of precipitation and evaporation is so balanced that the amount of moisture available at any place on earth is roughly constant, year after year.

Water is one of the most important factors in an ecosystem. Water is essential for all living organisms, but it is not distributed evenly over the earth. The abundance of water, or the lack of it, has a striking effect on ecosystems. For example, for the state of Washington, the primary source of water is the Pacific Ocean. Water vapor from the ocean is carried in from the northwest by storm systems that originate over the Aleutian Islands and the Gulf of Alaska. The moisture-laden Pacific air moves inland over the Olympic Mountains along the coast. At this point, the air rises, expands, cools, and then condenses. As a result, the average annual precipitation in certain places on the west side of the mountains is about 3,500 millimeters. This abundant rainfall has helped to support a coniferous rain forest in this area.

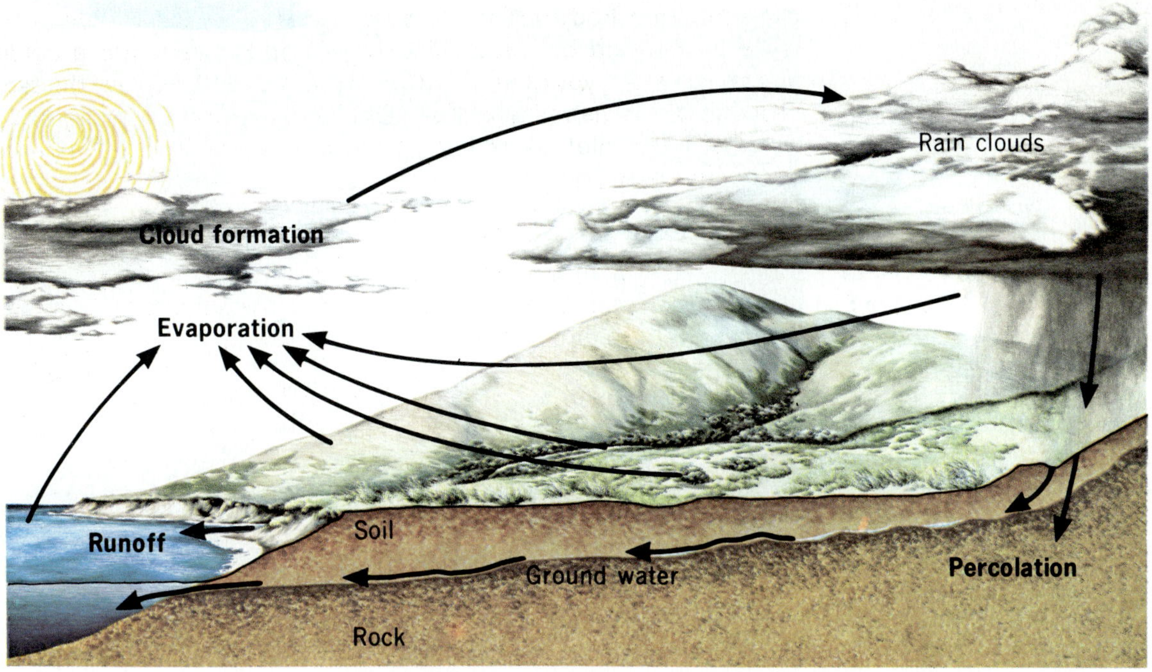

24-3. The water necessary for the life of plants and animals that live on land comes from rain and snow.

After moving inland over the Olympic Mountains, the air loses some of its moisture. It then descends into the trough formed by Puget Sound. As the air descends, it is warmed. Because warm air tends to hold its moisture, Seattle receives only 875 millimeters of rainfall a year. As the warm air continues eastward from the sound, it encounters the Cascade Mountains. Once again the air rises, cools rapidly, and its moisture condenses into rainfall. But the air contains less moisture than when it passed over the Olympics. The annual rainfall in the Cascades is 2,000 to 2,500 millimeters, much less than in the Olympics.

By the time the air mass crosses the Cascades, it has lost most of its water vapor. When the air descends the east side of the Cascades into the Columbia River Valley, it contains little or no moisture. In the Columbia River Valley, there may be less than 180 millimeters of rainfall a year. You might guess that the valley has the characteristics of a desert ecosystem (Figure 24-4). Here we see outcroppings of volcanic rocks, scattered sagebrush, and a few tufts of drought-resistant bunch grasses. Nowhere can we see any resemblance to the coniferous rain forests west of the Olympic Mountains.

The availability of water in an ecosystem is related to many factors. One of these is the rate of runoff—water not absorbed by the

24-4. A portion of the Columbia River Valley which lies east of the Cascades in the state of Washington.

soil. This in turn may be determined by the slope of a hill, the kind of soil, or the type of vegetation growing in the ecosystem. If the soil contains sand and large rocky particles, water will drain rapidly into gullies and dry stream beds. Little of it is held by the soil. Fine-grained soil that contains many decaying plants soaks up and holds water like a sponge. Sandy, rocky soils often occur in desert ecosystems. Soils rich in organic material usually are found in grasslands and forests.

Atmosphere and Wind

Because we cannot see the atmosphere, we may overlook it as an important part of an ecosystem. The atmosphere is, however, a storehouse of some very important substances. The carbon dioxide required for photosynthesis and the oxygen so vital in respiration are both found here. Some nitrogen from the atmosphere becomes part of the living community when it is changed to nitrogen compounds. This is accomplished either by lightning or by the action of nitrogen-fixing bacteria in the soil. These bacteria are the only living things that can use atmospheric nitrogen.

Atmosphere in motion is wind. Air currents created by heat flow change the distribution of materials in the atmosphere. By increasing the rate of evaporation and transpiration, the air currents increase the amount of water vapor in the atmosphere. This water vapor then condenses and falls to the earth as precipitation. Wind affects the form and distribution of organisms (Figures 23-15 and 23-16). Hurricanes and tornadoes, in particular, have drastic effects on ecosystems.

Fire

Of all the physical factors that may bring about major changes in an ecosystem, fire is the most common. A fire of a few minute's duration can change drastically an ecosystem that has been centuries in the making. Human activities and lightning are the most frequent causes of fire. Ecosystems in which fires are frequent, contain many fire-resistant trees and plants.

We usually think of fire as a destructive agent, as indeed it is. But the destruction also can play a part in the continued vitality and renewal of aging forests. The cones of lodgepole pine trees do not open and release seeds unless they are exposed to high temperatures. In a burned area, the cones open. The seedlings begin to reforest the burned area. Later, the lodgepoles are replaced by species that germinate in the shade of the lodgepoles—primarily, spruce and subalpine fir.

Gravity

Gravity is one of the most constant abiotic factors in an ecosystem. The roots of plants grow down and the stems grow up in response to gravity. Many of the structural adaptations of plants and animals are related to overcoming the pull of gravity. Striking examples are birds flying and tall trees lifting a water column more than 115 meters against the force of gravity (Figure 21-6).

The gravitational pull of the moon has an indirect effect on ecosystems, especially those in tidal regions along ocean shores. As the tide ebbs, organisms that were under water often are exposed to the harsh environment of the atmosphere.

Topography

We have already pointed out how the topography—the shape of the earth's surface—may affect an ecosystem. The slope of a mountain affects the rate of water runoff. The height of a mountain determines the amount of precipitation, solar radiation, wind, and water that organisms of the ecosystem receive. The furrows in a cornfield provide a microtopography where there may be more moisture, thus more weeds.

Geological Substrate and Soil

The soil and geological substrate—the underlying rocks—are closely related. Both provide a reservoir of water and minerals needed by the organisms of the ecosystem. Compounds of iron, calcium, magnesium, and other elements necessary for the growth of organisms are found here. Too much, or too little, of any of these may determine which organisms can survive in the ecosystem.

The geological substrate is eroded by wind and rain. Erosion breaks the rocks into particles of silt, sand, or clay. Water and air move into the spaces between the particles. Countless microorganisms, mainly bacteria and fungi, decompose dead organic material. This organic material mixes with the silt, sand, or clay to form part of the soil.

Although we have looked at these abiotic factors separately, all are active at the same time in setting the conditions for life in real ecosystems.

BIOTIC FACTORS OF AN ECOSYSTEM

Your studies of animals, green plants, fungi, and microorganisms have revealed the major roles that each plays in the balance of nature. Let's recall these major roles.

The green plants and a few other organisms are unique. Only they can convert light energy and simple molecules from the nonliving world to chemical energy and complex organic molecules. Because they synthesize their own food, green plants are producers. Their bodies—alive, dead, and decaying—are the primary source of energy for other organisms.

The animals, fungi, and most microorganisms are unable to use the molecules of the nonliving world. They require the complex molecules produced by the green plants. If these organisms live directly on green plants, we call them primary consumers. We can also call such animals herbivores. A grasshopper feeding on grass is a primary consumer eating a producer.

Not all animals feed directly on plants. The carnivores feed on other animals. The omnivores eat both plants and animals. A carnivore that feeds on a primary consumer is considered a secondary consumer. An example of a secondary consumer would be the bird that eats the grasshopper. An animal that ate the bird would be a tertiary consumer.

24-5. PRODUCERS AND CONSUMERS

We also can think of some animals, fungi, and microorganisms as decomposers, since they break down complex molecules in their food—other organisms—into simple inorganic substances. Green plants then use these simple substances in producing their food. All of these interrelations are shown in Figure 24-5. In the figure, bird **c** eats seeds, bird **b** eats insects, and bird **a** is a predator on other birds and small mammals. Which organism in the figure is a producer? Which is a herbivore? a carnivore? Is there a primary consumer present? a secondary consumer?

The interactions of organisms in building up and breaking down molecules are in balance. In photosynthesis, plants use carbon dioxide and release oxygen. In cellular respiration, both plants and animals consume oxygen and release carbon dioxide. Yet the concentrations of these gases remain in balance, and the atmosphere remains about the same, year after year.

In any ecosystem, the same atoms are used over again. Sugar contains carbon, hydrogen, and oxygen. After your body uses sugar, it releases carbon dioxide and water. Green plants use these products in making more sugar, and so on.

The same atoms flow through an ecosystem over and over again. Atoms are built into complex compounds that later are decomposed. These repeating patterns in the movement of materials in the ecosystems are called **cycles.** Let's look at some of these cycles more closely.

CYCLES IN AN ECOSYSTEM

Living organisms are composed chiefly of carbon, hydrogen, oxygen, and nitrogen. In the balance of nature, the flow of these four elements, as they cycle through living organisms, is especially important. Because the cycles of carbon, hydrogen, and oxygen are so closely related, we will consider the flow of these three elements as the carbon-hydrogen-oxygen cycle. It is sometimes called just the carbon cycle for short. The nitrogen cycle has special features that deserve separate attention.

The Carbon-Hydrogen-Oxygen Cycle

All organic compounds—the building materials of all living things—contain carbon. We have already seen that carbon is present in carbohydrates, fats, proteins, and nucleic acids, discussed in earlier chapters. But where does all this carbon come from? In the nonliving world there are only three common sources of this element. They are carbon dioxide in air (0.03 to 0.04 percent) and dissolved in water; rocks, such as limestone, containing carbonate; and coal and petroleum.

When coal and petroleum are burned, one of the products is carbon dioxide. When carbon dioxide dissolves in water, some of it reacts to form carbonic acid. This reaction yields carbonate and bicarbonate ions. Today the richest store of carbon in the waters of the earth is these ions. These ions, as well as dissolved carbon dioxide, are used by marine organisms in their life activities.

The cycle that involves carbon, hydrogen, and oxygen, as it occurs for land organisms, is shown in Figure 24-6. Respiration and decomposition release carbon dioxide and water. These compounds are used by plants in photosynthesis. During photosynthesis, organic compounds are produced, and oxygen is released to the atmosphere. These, in turn, are used by animals in respiration and decomposition. What changes would be necessary

24-6. CARBON–HYDROGEN–OXYGEN CYCLE ON LAND

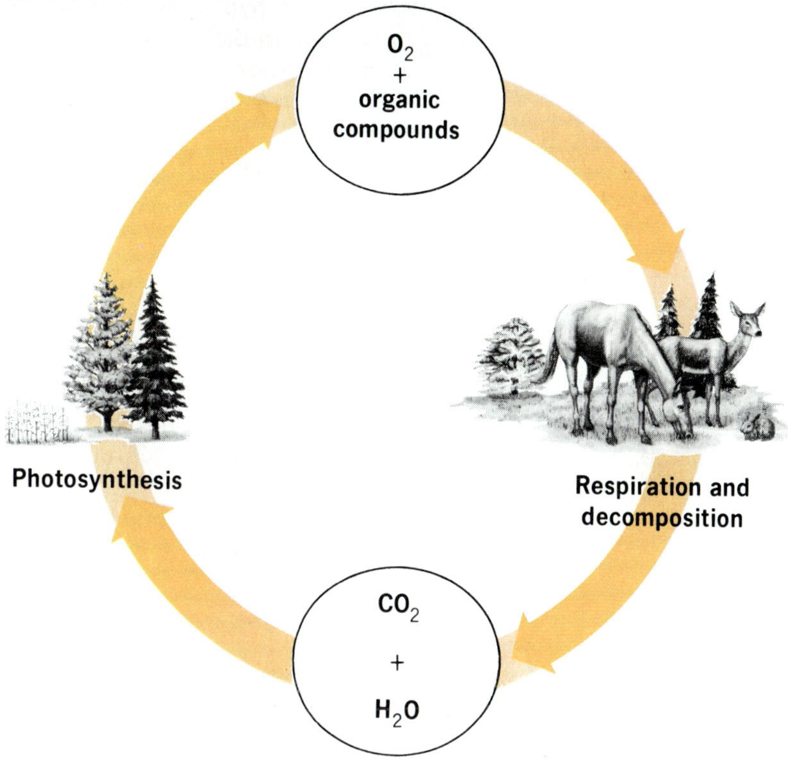

to make the diagram also apply to the ocean? Try to think of the actual organisms involved in these processes and what role each is playing. Also look at the other illustrations of various ecosystems and organisms in this chapter. Try to interpret what is happening in terms of the carbon-hydrogen-oxygen cycle.

The Nitrogen Cycle

Nitrogen is found in all proteins and nucleic acids. It is an essential element in living substances. Since four fifths of our atmosphere is pure nitrogen, one might suppose that organisms obtained nitrogen directly from the air. But this is not so. Animals must have nitrogen in the form of amino acids. Plants need soluble nitrogen salts, which are combined with carbohydrates in making proteins and nucleic acids. When amino acids are broken down in the bodies of animals, or when animals die, the products generally cannot be used by green plants. Microscopic decomposers convert these compounds into ones that the green plants can use for synthesizing proteins.

24-7. NITROGEN CYCLE ON LAND

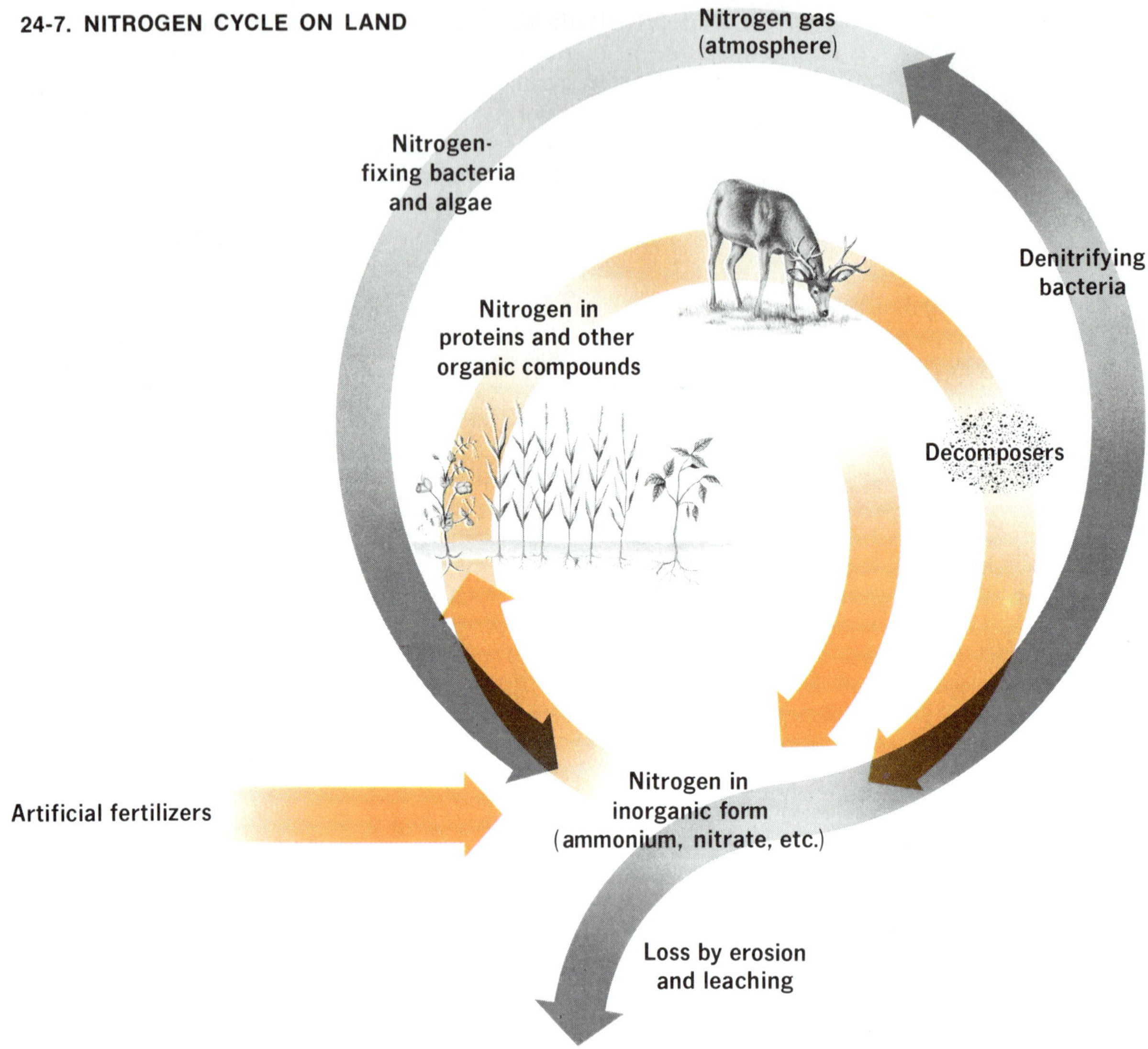

The nitrogen cycle involves five main steps (Figure 24-7). First, when a plant or an animal dies, when leaves fall from a tree, or when an animal excretes its waste products, nitrogen compounds pass into the soil or water. Next, bacteria in the soil or water begin to break down these organic nitrogen compounds. This process converts much of the organic nitrogen to ammonia. (In freshwater communities and in the oceans, other forms of life also produce ammonia as an end product of metabolism.) In the third step, ammonia gas is changed to a soluble salt. The salt ionizes to produce ammonium ions, as well as some other ions. Some green plants can take ammonium ions directly into their roots. The fourth step is carried out in the soil by nitrifying (NY·truh·fy·ing) bacteria. Some of these bacteria change ammonium ions in water to nitrites (NY·tryts), salts containing nitrite ions. Others oxidize nitrites into nitrates (NY·trayts). Nitrates are salts that contain nitrate ions. In the fifth step, the nitrates, dissolved in soil water, are taken up by plant roots.

As the bacteria change ammonia to nitrites and nitrites to nitrates, they obtain energy. The bacteria use the energy in their own activities. Nitrates are the waste products of the bacteria. But these nitrates can be used by green plants.

Another way in which nitrogen is made available for protein synthesis is by nitrogen fixing. That is, atmospheric nitrogen is converted into nitrogen compounds by nitrogen-fixing bacteria in soil. These compounds can be used by photosynthetic plants. Some kinds of nitrogen-fixing bacteria grow in the roots of plants called legumes (LEG·yoomz). Examples of legumes are garden beans, soybeans, peas, alfalfa, and clover. After these bacteria invade the roots of the host they form little growths, called nodules, on the roots (Figure 24-8). These bacteria actually improve the growth of the plants. They convert atmospheric nitrogen into nitrogen compounds. These compounds go into the proteins and nucleic acids manufactured by the legumes. Nitrogen-fixing bacteria obtain the energy to fix nitrogen from carbohydrates and oxygen in the plants. Thus, both the green plant and the bacteria benefit from this close association.

24-8. The nodules on the roots of clover contain nitrogen-fixing bacteria.

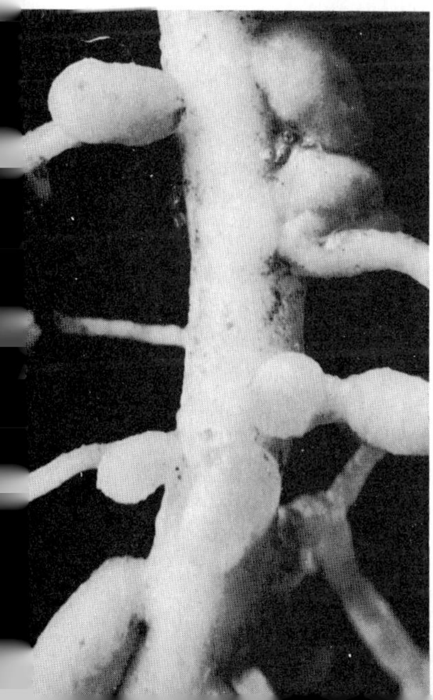

To a limited extent, humans supply green plants with nitrogen in commercial fertilizers. In the agricultural regions of North America, farmers apply ammonia or nitrates to their fields to increase the yield and protein content of crops. As a result, green plants have at least three possible sources of nitrogen. But where do you, I, and other animals get nitrogen that our bodies need to build proteins? We get it directly from plants—when we eat a salad, some green vegetable, peanuts, beans, or such. We get it indirectly from plants when we eat meat or cheese, or when we drink milk. The protein in meat, cheese, and milk is formed from amino acids. The amino acids contain nitrogen that was in alfalfa, corn, or other plants eaten by the animal.

There are some "leaks" in the nitrogen cycle. One such leak occurs when certain soil bacteria convert some of the nitrates into nitrous oxide, a gas that is lost to the atmosphere. Another leak involves loss of nitrogen-fixing bacteria if soil erosion occurs. Still another leak occurs when nitrogen-containing compounds are washed away before they can be absorbed by plant roots. In spite of these and other losses the cycle remains in balance. Why do you think this is so?

24-9. SUMMARY OF ENERGY RELATIONSHIPS IN LIFE

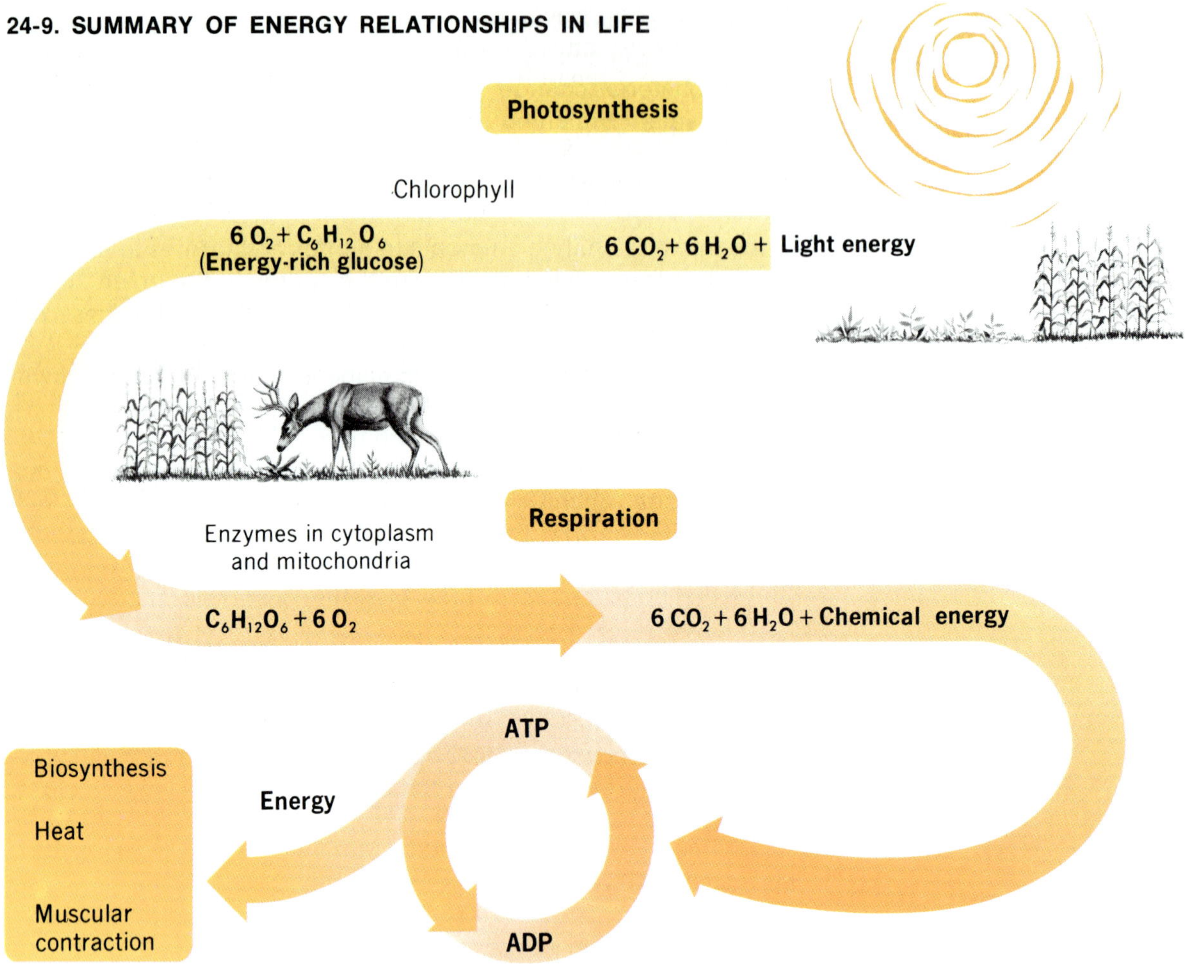

ENERGY FLOW IN AN ECOSYSTEM

Living things certainly are affected by their environment. But organisms also modify the environment. Our surroundings—atmosphere, soil, oceans, fresh waters—possess many of their characteristics because of the presence of earlier plants and animals. And life today indirectly affects future generations of living things.

The cycles we have already discussed provide striking examples of interaction between parts of an ecosystem. Carbon, hydrogen, oxygen, nitrogen, and other elements move from one organism to another and to and from the biotic and abiotic parts of the ecosystem. All of these interactions require a great deal of energy. What is the source of the energy? What form or forms does it have? Is it cycled in nature as are the elements we have talked about?

Perhaps the best way to answer these questions is to review some basic principles of energy relationships in an ecosystem. Figure 24-9 reminds us that the ultimate source of nearly all energy is the sun. Photosynthesis stores light energy from the sun in the form of chemical energy in glucose molecules. Follow the arrows in the diagram. You will see that an animal breaks down some of the organic compounds stored by the plants it eats and obtains energy that is used in its own activities. Since animals, fungi, and most bacteria cannot carry on photosynthesis, they rely on energy stored in green plants for their life activities. Much of the energy transferred by cellular respiration is used to build ATP molecules. The energy in ATP is used in many ways. Some is used in the formation of chemical compounds—biosynthesis. Some is used in muscle contraction, and some is released into the environment as heat. The diagram shows that energy is not cycled in the ecosystem as are carbon, hydrogen, and oxygen. Energy flows in one direction only. The rate at which the energy flows through the ecosystem will determine the number and kinds of organisms found there. Why do you think this is so?

A basic concept of science is that energy may be changed from one form into another, but it is never created or destroyed. The transformation of energy always involves a decrease in the amount of energy available to do work. Thus, in any energy conversion, some energy is always changed to heat (Figure 24-10). In this form, it is lost as far as further use by organisms is concerned. The useful conversion of one kind of energy into another is never 100 percent efficient. We might say there are many leaks in the system and "new" energy must be added continually. The source of the new energy is, of course, solar radiation.

Let us take the example of one square meter of an ecosystem (Figure 24-10). Suppose, in one day, it receives 3,000 calories of energy in the form of light. In a 100 percent efficient system, all of this energy would be absorbed and converted into chemical energy. Part of the total amount of light, however, never enters into the light reactions of photosynthesis. Some light energy is reflected from the surface of the plant, some is transmitted through the leaf, and some is dispersed in the form of heat.

Of the 3,000 calories of total light energy, perhaps only half, or 1,500 calories, is absorbed by producers. Of the 1,500 calories, only 15 to 75 are converted into chemical energy (food). Much of the chemical energy in food is used by the corn in its own life activities. But green plants absorb and store more chemical energy than they can use. It is this energy that becomes available to animals and other organisms that feed on plants.

Herbivores, such as the rabbit in Figure 24-10, are primary consumers. Of the energy available to the primary consumers in the

24-10. ENERGY FLOW IN ECOSYSTEM

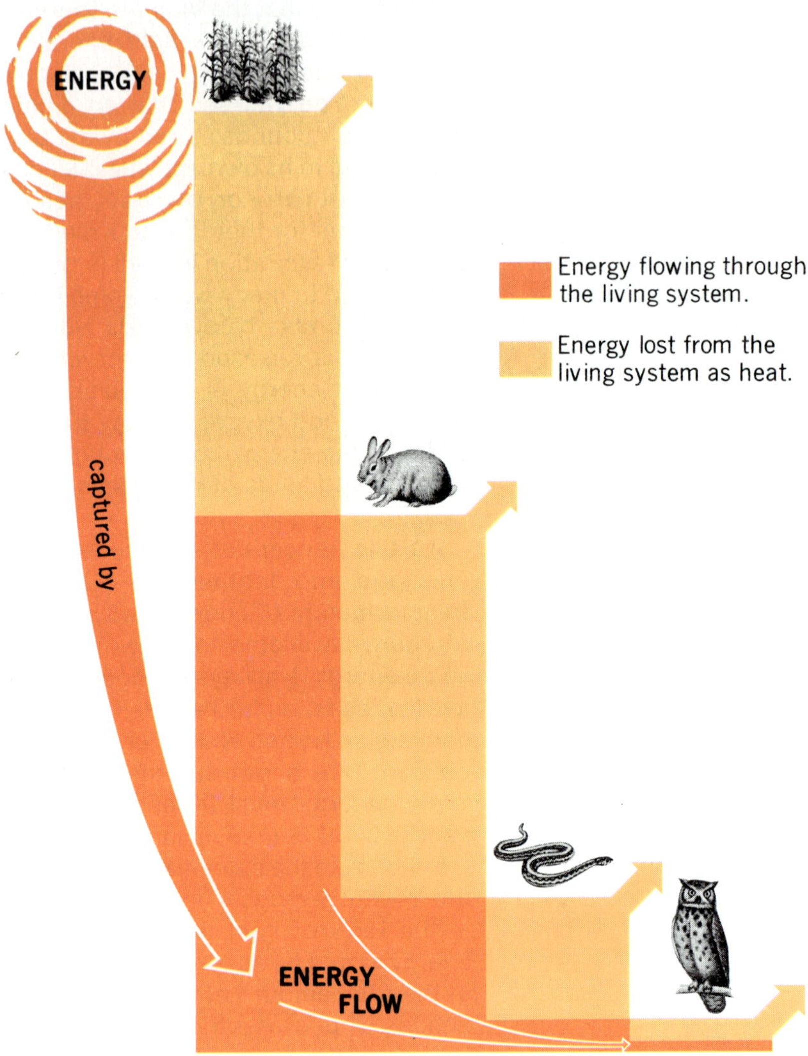

ecosystem, much is dispersed. Again, some of this energy is lost as heat. Some is converted into other forms of energy by the animals' activities.

As the energy flow passes through the secondary consumers of the ecosystem, it is once more reduced by loss of heat and the energy expended through the activities of animals. How many of the original 3,000 calories would be available to the snake? How much would be available to the tertiary consumer, the owl, should it eat the snake?

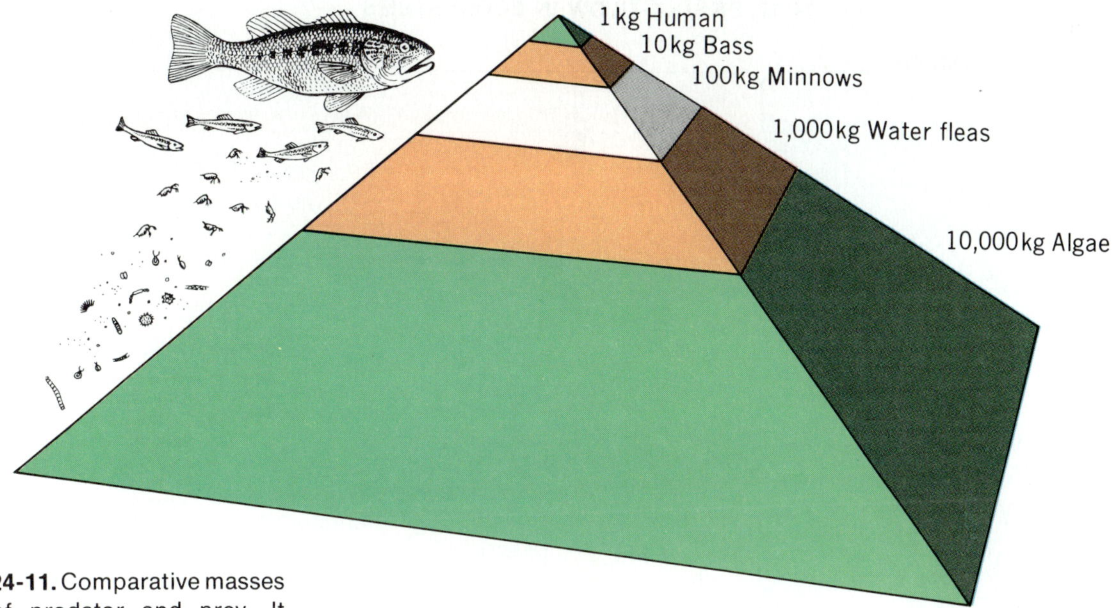

24-11. Comparative masses of predator and prey. It takes about 10 units of prey to support 1 unit of predator.

Food Chains

The sequence of events and changes in energy levels illustrated by Figure 24-10 are the events and changes that occur in a **food chain.** Another way of showing these events is illustrated by Figure 24-11. Here we see a food chain of a freshwater ecosystem. In this case, the producers are algae. The primary consumers are rotifers, *Paramecium,* water fleas, and other small invertebrates. Secondary consumers are small fish that eat the invertebrates. A tertiary consumer is a large fish that eats smaller fish. At the top is the fourth-level consumer, the person who caught the large fish and ate it. Note how many kilograms of producers at one end of the chain are required to support one kilogram of a person at the other end of the chain.

The fewer steps in the food chain, the greater is the availability of energy. The longer the food chain, the more energy is dispersed as heat or is used in the metabolism of the various consumers.

Productivity

An extremely important characteristic of an ecosystem is its **productivity.** Productivity is a measure of the amount of energy available to an ecosystem through the photosynthetic activities of the producers. We calculate the productivity of an ecosystem by first

24-12. PRODUCTIVITY IN ECOSYSTEMS

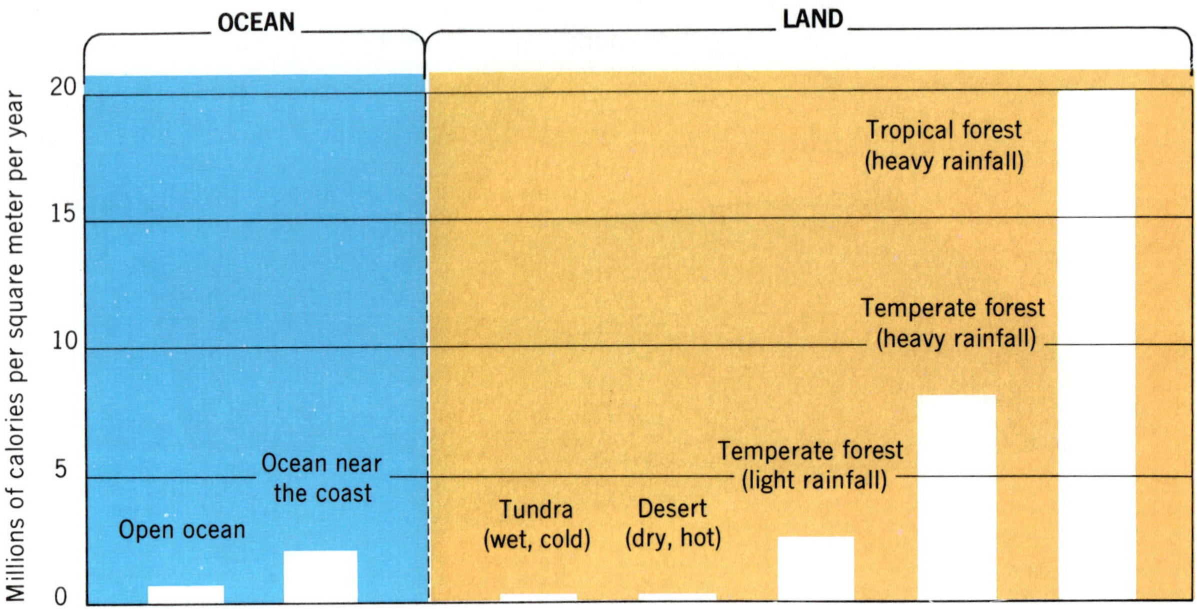

Total production of organic compounds by photosynthesis in different habitats

measuring the rate at which light energy is converted to chemical energy in photosynthesis. We then must determine how much of this chemical energy actually is available to consumers.

Ecosystems vary considerably in productivity. The productivity of an alpine environment per year is lower than that of a tropical rain forest. In the alpine ecosystem, photosynthesis is carried on only about two months of the year. In a tropical rain forest, photosynthesis occurs all year. A higher productivity can maintain a greater amount of living material in the ecosystem. Thus, the amount of living material, or **biomass,** is directly related to the amount of energy stored by the producers of the ecosystem.

Figure 24-12 shows the average production of organic compounds in a few ecosystems. The values shown in the graph are based on total production of organic compounds by photosynthesis. They are given in terms of energy (calories) stored in the compounds. What factors might account for the difference in productivity between open ocean and ocean near the coast? Compare the five land ecosystems. What factors might account for the differences in their productivity? If you wished to increase productivity in a desert, what might you do? Suggest a factor (or several factors) that might account for the difference between temperate forests and tropical forests with heavy rainfall.

Food Webs

The interactions among organisms in an ecosystem often are more complex than simple food chains. For example, a bat ate a mosquito that had bitten a coyote that had eaten a mouse that had eaten a leaf. These organisms make up a food chain. This sequence is incomplete. Other animals eat mice and mosquitoes. Coyotes and bats eat and are eaten by other organisms. Thus there are many different producers and primary consumers. One producer species might be eaten by different primary consumers. And secondary consumers may eat a variety of primary consumers. Thus, it is usually better to think of **food webs** (Figure 24-13).

24-13. A FOOD WEB

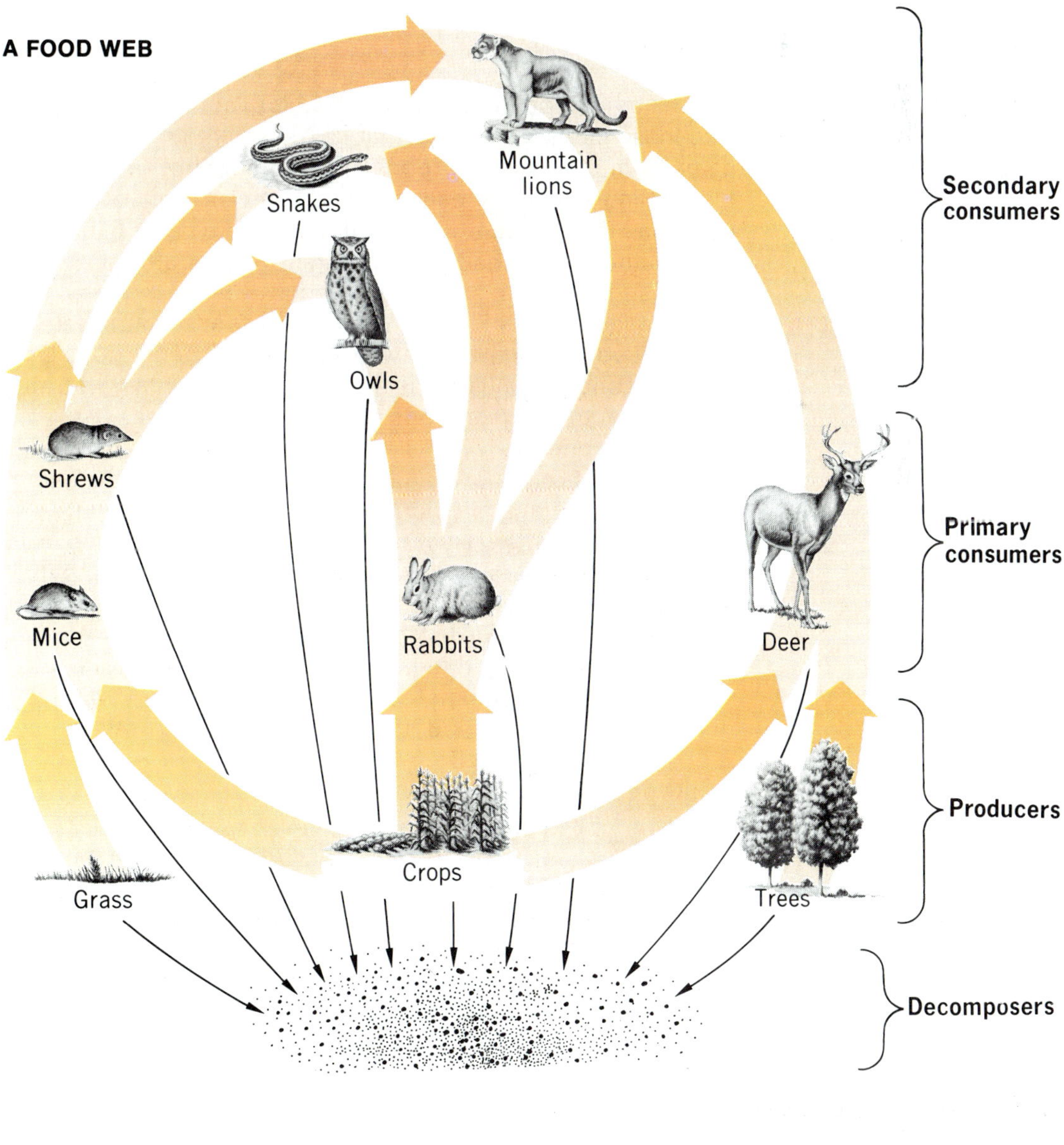

A food web shows the many relationships between producers and consumers—that is, who eats whom. You can learn more about food webs by completing **Inquiry 24-1.**

INQUIRY 24-1 Building a Food Web

MATERIALS

pencil and paper

Procedure

List the main food items that you had for your last lunch. Now separate these items into foods that came from animals and those that came from plants. Many foods are combinations of different foods. List the ingredients of each food separately. Then indicate whether each was from a plant or animal—and the type of plant or animal. For instance, if you had cake for lunch you should list: flour—plant, wheat; sugar—plant, sugar beets; eggs—animal, chicken. For every animal that you have listed, list several foods that it eats. For example, if you had milk with your cake, you listed a cow. Grass and corn also should be listed as the food the cow eats. **1.** Are any of the items you listed as foods for animals also foods that you could eat? Which ones?

Across the bottom of a clean piece of paper, write the names of all the plants you have mentioned. In another row, above this row, list all those animals (herbivores) that eat any of the plants in the first row. From each plant, draw a line to every animal that eats it.

Above the herbivore line, enter the names of all those animals (carnivores) from your list that eat other animals. From each herbivore, draw a line to the carnivores that eat that herbivore.

You have now completed a food web showing the interrelationship of the plants and animals that made up your lunch.

Food Webs in Ecosystems

The variety of pathways in a food web helps maintain the stability of the ecosystem. For example, owls prey on rabbits and mice. If a disease reduces the rabbit population, fewer plants are consumed. The large plant population produces more fruits and seeds, which, in turn, support a larger mouse population. The increased number of mice becomes the major food source for the owls. The rabbit population gradually increases, and these primary consumers again become a food source for the owls.

Only a few of the possible offspring of a plant or an animal survive to reproduce. Of all the seeds a plant forms, all but a few are eaten by animals or are killed by diseases or adverse weather conditions. Otherwise, an ecosystem could not support all the offspring that might be produced. For example, a pair of meadow mice, born early in the year, usually produces 13 litters before the winter—a total of about 78 young. These young mature very quickly. Several generations of mice soon would be living in the same meadow. The original pair of mice would have about 50,000 descendants in one year, if all their offspring survived and reproduced.

Just one mouse requires about ten kilograms of plant food per year and ruins far more than this. Thus, 50,000 mice would eat or destroy more than 1,500 metric tons of vegetation in a year. (One metric ton equals 1,000 kilograms.) Actually, a meadow mouse population remains fairly stable. About two offspring actually survive from each pair of parents. The others die of diseases or are caught by hawks, owls, foxes, snakes, and other secondary consumers.

Without fruits, seeds, young plants, and foliage, primary consumers could not exist. On the other hand, without primary consumers, many plants would die of overcrowding that resulted in a lack of nutrients and light. If the secondary consumers did not keep the populations of primary consumers under control, the plants would soon be gone. Thus, the various populations of organisms in an ecosystem are held in check, and the ecosystem remains stable.

Tolerance in Food Webs

When an environmental feature, such as freezing temperatures, limits the number of species that can live in an ecosystem, the food web is simpler. A tundra on a mountaintop or in the Arctic (Figure 23-21), for example, can support only a few species through the year. When the weather is warm on the tundra, however, some species of insects and small plants reproduce in astonishing numbers. Nowhere on earth do mosquitoes, blackflies, and bees reach greater abundance. Then the cold weather kills them or they become dormant in the winter.

On the other hand, the water of a hot spring is a living space for plants and animals that can tolerate high temperatures. Here bacteria and blue-green algae (Figure 24-14) form crusts, often with bright pastel colors, in waters as hot as 85°C. At high elevations this is not much below the boiling temperature of water. Few animals or plants, except the bacteria and blue-green algae, can live at temperatures above 50°C.

24-14. A thermal spring in Yellowstone National Park, so hot that steam is rising from the surface. The white areas are deposits of salts. The colors are growths of algae and bacteria that are adapted to the extreme environment of the pool.

From these examples, we see that ecosystems vary from place to place and time to time. Systems of checks and balances exist within ecosystems. These checks and balances are natural responses to changes produced by natural forces. These natural forces include the activities of living things.

SUCCESSION

Occasionally, living things spread into new regions. A mountain slope may be exposed after a landslide. A new beach may form at the edge of a pond or lake. A forest fire, flood, or glacier may change the land. Some kinds of organisms can occupy such an area quickly. Others gradually follow as conditions improve and eventually crowd out the first organisms. These changes in the kinds of species that occupy an area are known as **succession.**

To understand how succession can occur over time, imagine yourself walking inland from the beach toward the small dunes along the shore of Lake Michigan (Figure 24-15). The clean white sand, unmarked by any vegetation, is called the middle beach. In the summer, the sand here is hot and dry. It is constantly shifted by the wind, making life difficult for any organisms except a few burrowing insects and spiders.

Beyond the middle beach we may find the first evidence of producers. These are perennial grasses—hardy plants, indeed. They resist drying and require a minimum of mineral elements. They hold the sand that blows over them from the middle beach. These pioneer plants form the windbreak around which small dunes are formed. As the roots and leaves of the plants die, organic material is added to the sand. The wind gradually shifts the small dunes higher on the beach. The grasses increase in number, and their roots hold more sand. The dune slowly moves up the beach and finally stops.

After a dune stops, birds bring in seeds of sand cherry, cottonwood, and willow that grow in the moister spots of the dunes. With the addition of these plants, the productivity of the changing ecosystem increases. More species of insects, rodents, birds, and predators move in. The plants and animals of this stage add to the organic material accumulating in the sand.

The next change in the ecosystem is the pine forest. Before we reach the dense pine forest, however, we go through a zone of old cottonwoods and pine seedlings. The accumulation of leaves from broad-leaf plants and needles from the pines increases the organic material in the soil. This sets the stage for the development of an oak forest.

Again, the kinds of animals change and their numbers increase. Hoards of insects and tree-living animals inhabit the productive oak-forest ecosystem. Groundhogs, foxes, wood mice, and squirrels

24-15. Sand dunes along the shore of Lake Michigan. Over a long period of time succession can change this ecosystem into a much more productive one.

24-16. SUCCESSION IN A DUNE ENVIRONMENT

are abundant. On the forest floor, under rotting logs and dead leaves, we find earthworms, crickets, slugs, snails, and millipedes.

Continuing our walk through the woods away from the beach, we notice that the forest seems more shaded. The dense shade is provided by the foliage of beech and maple trees that form a dark canopy. The shade slows the evaporation of water from the forest floor. As we walk over the dead logs and fallen leaves, we have the impression of walking on a spongy surface.

The shade has another effect. It prevents the growth of most plants other than mosses, lichens, a few ferns, and the seedlings of beech and maple. This stage in succession remains essentially the same if nothing changes in the environment to upset the balance. Because it is a stable stage in succession, the beech-maple forest is considered to be the **climax** stage for this region.

From what we know of productivity, there must be an increasingly efficient use of energy in each stage of this succession. For example, on the barren beach, most of the radiant energy is lost as heat. The beech-maple forest probably is the most efficient ecosystem in using energy. The forest also makes maximum use of water and increases soil fertility.

In our walk from the beach to the forest, we have passed through the changing ecosystems that occurred where the beech-maple forest now grows. In only a few minutes, we walked through a succession of changes that happened over a period of hundreds of years in that area. Figure 24-16 shows the main stages in a succession that occurred at one place along the shores of Lake Michigan.

Succession is going on in ecosystems of all kinds, everywhere in the world. If you live near a deciduous forest, you may have seen the succession that occurs after a forest fire. This is more rapid than the succession on the beach. After only one season has passed, the burned area again may be green with blackberry bushes, fireweed, and other plants whose seeds are dropped by birds. Winds and birds

bring the seeds of quick-growing trees such as poplar, birch, and sumac. These soon dominate the lower vegetation of bushes and grasses. Gradually oak and beech or evergreens rise above these trees. They provide so dense a shade that the quick-growing trees can no longer survive. As the vegetation changes, so do the kinds and numbers of animals. In its final climax stages, the ecosystem as a whole is fairly stable. An equilibrium has been established in the interactions of the biotic and abiotic components of the ecosystem. The succession of plants and animals has reached a climax.

Succession can be observed even as a layer of life edging across a bare rock. Succession of this sort might have required decades or even centuries before the climax stage was reached. You can follow the stages of a more rapid and easily observed succession in the laboratory.

INQUIRY 24-2 Biological Succession

MATERIALS

Part A

400 ml of pond water or other culture
compound microscope
graduated cylinder, 100-ml
2 glass slides with 4 or 6 etched 1-mm squares
2 cover glasses
3 manometers made of the following materials:
- *3 erlenmeyer flasks, 500-ml*
- *3 rubber stoppers, 2-hole*
- *6 pieces of glass tubing, 7.5 cm long, to fit stoppers*

6 pieces of rubber tubing to fit glass tubing
3 lengths of small-bore, glass tubing, 60 cm long, bent for use in a manometer
3 clamps for rubber tubing
3 sheets of graph paper (mm ruling)
3 cardboard backings
cellophane tape
food dye (red, blue, or green) in water
petroleum jelly
ink marker
200 ml distilled water

Part B (Optional)

graduated cylinder, 100-ml
test tube
acetone
spectrophotometer
centrifuge

Procedure (Part A)

Oxygen-Carbon Dioxide Exchange. Swirl the culture. Place a drop of culture on the etched glass slide. Cover it carefully with a cover glass. Count the number of organisms per square millimeter of field with either low or high power, or both. Count organisms that touch the upper or right borders of the square, but not those touching the lower and left borders. Count any organisms that move across a square during the count. Determine the producers (green plants) and consumers. Record the number of producers and consumers per square millimeter.

With an ink marker, label the 500-ml flasks 1, 2, and 3. Set up your team's manometers as shown in Figure 24-17. To Flasks 1 and 2, add 200 ml of culture; to Flask 3 add only 200 ml of distilled water. Be sure that the rubber stopper and the inside lip of the flask are dry when stoppering the flask. All other connections should be airtight. A small amount of petroleum jelly may be used as a seal, if necessary. Close the clamps when the colored water in both arms of each manometer is at the same height. Record this height. **1.** What is the purpose of the third manometer?

Over the next three to five days, observe the movement of the liquid in the manometers. Record the level of the liquid at the same time each day. **2.** If the liquid moves away from Flasks 1 and 2 during the day, what is happening in the ecosystem of the two flasks? **3.** If it moves toward the ecosystem, what probably is happening? **4.** How will you know when the system reaches an equilibrium? **5.** How would more frequent readings affect your data?

24-17. ASSEMBLED MANOMETER

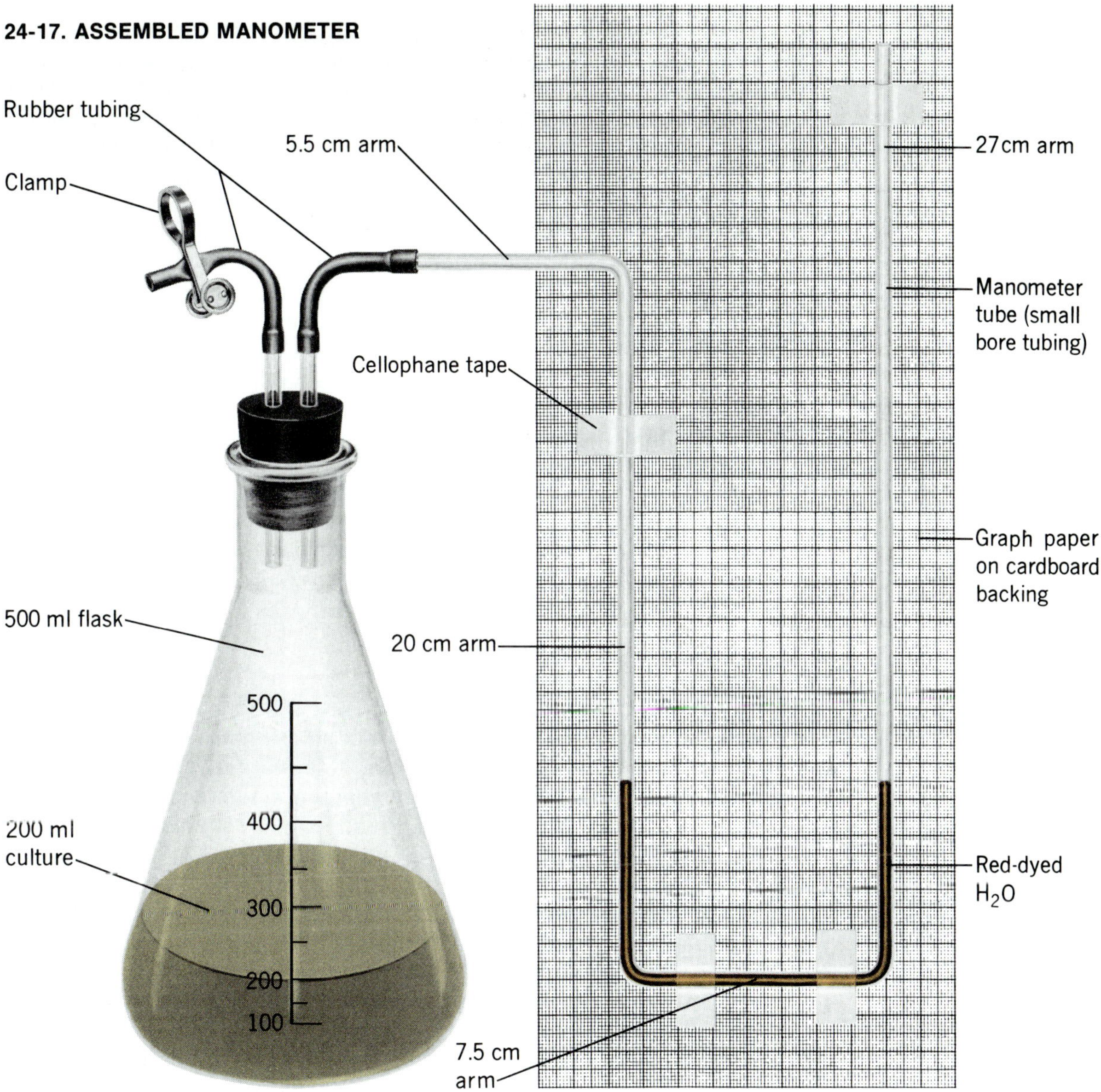

When the system has reached equilibrium, swirl the contents of Flasks 1 and 2. Remove and examine a sample from each. Place the samples under the microscope one at a time. Calculate the number of producers and consumers per square millimeter. Record changes from the initial populations. **6.** What is the relationship between manometer readings and the composition of the population?

Now that you understand the composition of the system in equilibrium, it is time to vary one factor in Flask 1. Decide on your variable. You may want to introduce a new species of producer or consumer, or increase the number of a species already present. You might vary the hours of light per day, the distance from a light source, or the temperature. If you change to a three-hole stopper, you can attach a flask containing yeast cells and 5 percent dextrose to the experimental system. This will supply additional carbon dioxide to the system's atmosphere. Or, you may choose some other variable. **7.** For what new purpose can you now use Flask 2?

Sample the populations over several days and note changes in the manometer. Record your data and graph the results, with "time in days" on the horizontal axis and "species count" or "consumer-producer ratio" on the vertical axis. Calculate the consumer-producer ratio (C-P) by dividing the total number of consumer organisms into the total number of producers. **8.** What would happen if the observations were carried on for an additional ten days?

Briefly summarize the stages of succession through which equilibrium was reached. Include the time required for this succession. **9.** How has this laboratory study been similar to a field study?

Procedure (Part B) (Optional)

Study of Pigments. You can use a spectrophotometer to determine how consumer-producer relationships change over time. Each day, centrifuge a 5-ml sample from the culture at high speed. Pour off the liquid portion. Add 3 ml of acetone to the residue in the bottom of the tube. Shake well to speed up pigment extraction. Centrifuge again or let the cell debris settle out.

Use the spectrophotometer to read and chart the yellow pigments at 430 nanometers and the chlorophyll green at 665 nanometers. These are the wavelengths at which the pigments absorb the most light. **10.** What is the ratio of the two pigments?

Graph your results. Place either the ratio or the percentage of transmission (or absorption) of each pigment on the vertical axis and the "time in days" on the horizontal axis. **11.** What is the relationship between the two pigments in a young colony? in a mature colony? **12.** How do the results of the spectrophotometer readings relate, if at all, to the manometer readings or to the population changes you noted?

In Brief

Interaction and coordination exist at all levels of organization. The living system is usually in balance with its physical and biological

SCIENCE AT WORK: An Island Is Born

On November 14, 1963, 12 miles off the southern coast of Iceland, volcanic activity on the ocean floor began to form a new island. By the next day, a narrow black ridge broke the surface of the ocean, and the island began to grow. By March of 1965, a mass of lava, ash, cinders, and pumice, 2.8 kilometers square, had formed. The people of Iceland named this newly formed island Surtsey.

This newest land on earth has been declared a sanctuary. The emergence of Surtsey from the ocean floor has provided scientists with a unique opportunity. They can study the natural succession of life on a new land. Many questions about succession have resisted answers in laboratories around the world. Some of the answers now may be found on this small piece of land that formed just a few years ago in the North Atlantic.

Two graduate students in biology from the University of Iceland in Reykjavik watch over Surtsey each summer. They live in a simple wooden house on the island. The house and the biologists' footprints are the only signs of human occupation on this new land.

Each day that the weather permits, the biologists walk around the entire island. They observe the seaside sandworts, the hardy succulents that have replaced Surtsey's first flowering plants, the sea rockets. They also check their insect traps. Tiny flies, called midges, are the only insects that reproduce on the island. However, many other insects, including butterflies, are carried to Surtsey by winds from surrounding islands.

The only birds now nesting on Surtsey are the fulmars. They have built nests high in the cliffs on the west side of the island. Each nest contains a single egg. Occasionally, some of the nests are robbed by black-backed gulls.

The biologists do not interfere with any of the natural events that occur on the island. To do so would be to destroy the succession they are studying. They must even resist the temptation to cover the exposed roots of a dying plant.

On Surtsey, scientists have a rare opportunity to study succession at its beginning – to see what comes first, what is destroyed, and what survives. They will be able to chart the entire history of animal and plant life on a fresh, new land.

environment. Populations exist in ecosystems made up of many kinds of organisms interacting with one another and the environment. Consumers live on the energy captured by producers. Decomposers return the materials of the dead to the atmosphere, fresh water, seas, and soil.

The cycles of nature are evident in the movements of carbon, hydrogen, oxygen, nitrogen, and other elements. These elements pass through the bodies of living things and reenter the environment. Energy flow in the ecosystem is one-way, with a loss of energy in the form of heat. Energy flow starts with producers and ends with the highest order of consumers.

Food webs are one way of looking at the interdependence of living organisms. Different species assume the roles of producers, consumers, and decomposers in different food webs. Life may be destroyed or badly disturbed in an ecosystem. It gradually returns through a succession of relationships, leading up to a climax stage. The invasion of a new volcanic island by living things, or the renewal of life in an area devastated by fire, offers an example of succession (Figure 24-18).

Every organism within the biosphere affects the life of every other, directly or remotely. Humans, for example, cannot live without the bacteria in the soil, the green plants on land and in the sea, and even the scavengers of dead material. And what we do to the environment affects all other living creatures.

Guide Questions and Problems

1. How are producers and consumers different?

2. In the course of evolution, a few species of plants have lost both chlorophyll and the ability to carry out photosynthesis. They live as parasites on the roots of other plants. Are such plants producers or consumers? Explain your answer.

3. Sunlight is the basic energy source for all ecosystems. What communities exist in perpetual darkness? How is energy fed into such communities? In what way is a city similar to an ecosystem in perpetual darkness?

4. In comparing the consumers in two ecosystems, an investigator found that one ecosystem had only primary and secondary consumers. The other also had third-, fourth-, and fifth-level consumers. Which of the two ecosystems would be more stable? Why?

5. An ecosystem in a valley on the leeward side of a mountain range in the western United States probably consists of sagebrush and

24-18. Fire raged over this hillside and destroyed all parts of the plants that were above ground. After 28 months, lush new growth surrounds the blackened branches. Depending on environmental conditions, a climax community may be restored quickly or slowly.

cactus plants. Why do we not find a rich growth of conifers or hardwoods in this place? Why are conifers and hardwoods more likely to be on the windward side of the mountain range?

6. In countries where the density of the human population is very high, the diet tends to be more vegetarian than in countries where the population is small compared with available food supplies. How can you explain this observation?

Ideas for Independent Study

1. Natural lakes are abundant in the northern United States and in Canada. They are rare in the southeast. Find out what factors may account for this.

2. Try to discover if there are areas near your home that were formerly used by humans but no longer are. These might be abandoned farms, vacant lots, abandoned railroad right-of-ways, or others. Describe what is happening in these areas.

3. Select several areas near your home, such as the edge of a pond, a rocky hillside, a city park, or some other area. Make a census of the main plant species in the area. (You do not need to know their proper names.) How many of the species are found in all of the habitats? If some species are restricted, try to find out what may be the limiting factors.

4. The photograph in Figure 24-19 was taken near Trail Ridge Road in Rocky Mountain National Park in Colorado. The altitude is about 3,100 meters. The small sign on the post reads "Old Quarry Road." You are walking on the remnant of a road used many years ago to haul rock for constructing Trail Ridge Road. Although the

24-19.

road is no longer used, not even by hikers, revegetation has been very slow. This is typical of plant growth in tundra regions. Try to suggest some hypotheses to explain why plant growth is slow.

Related Readings

BOOKS

Ehrlich, P. R., et al. *Ecoscience: Population, Resources and Environment.* San Francisco: Freeman, 1977.

Fridricksson, S. *Surtsey: Evolution of Life on a Volcanic Island.* New York: John Wiley & Son, 1975.

Kormondy, E. *Concepts of Ecology.* 2nd ed. Englewood Cliffs, N. J.: Prentice-Hall, 1976.

Worster, D. *Nature's Economy: The Roots of Ecology.* San Francisco: Sierra Club, 1977.

ARTICLES

Borman, F. H. and G. E. Likens. "The Nutrient Cycles of an Ecosystem." *Scientific American,* October 1970, page 92.

Horn, H. S. "Forest Succession." *Scientific American,* May 1975, page 90.

National Geographic. "Our Ecological Crisis." December 1970. Section includes three articles: "Threat to Man's Only Home," page 738; "The World and How We Abuse It," page 782; and "The Fragile Beauty All About Us," page 785.

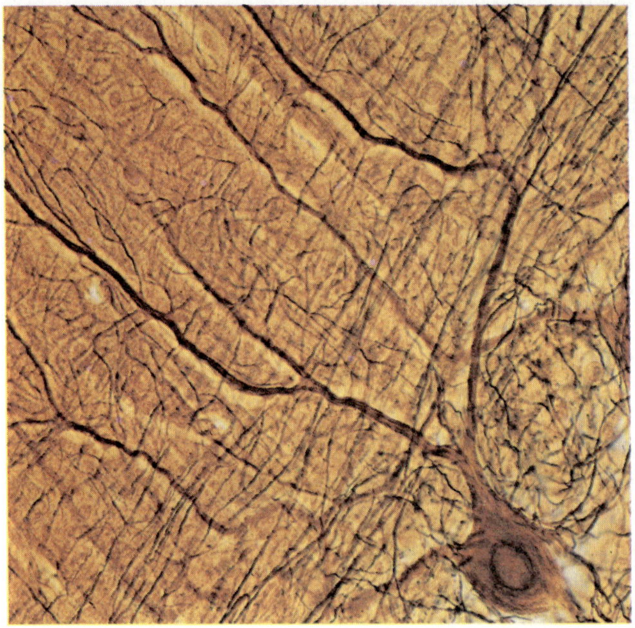

25 Challenges in Biology – Today and Tomorrow

The laboratory is the place where the work of science is done, where its spirit lives within those who work there, where its methods are transmitted from one generation to the next. One does not really learn science from books; one learns science by asking nature the right questions. And the laboratory is the place where one learns most readily what questions can be asked fruitfully, and how they must be put. It is where one learns why science insists on precise measurements, accurate observations, and conciseness and clarity in communication.

That statement, written by an eminent biologist, H. Bentley Glass, is as true today as it was when it was written 20 years ago. It is equally true no matter how we define the laboratory – as a room or a building, or as the natural environment where biologists investigate living things on land, in water, and in air. Because biology is the study of life, the laboratory of a biologist may be as small as a drop of water or as large as our complex biosphere.

Today the work of science goes on in backyards and basements, in business and industry, in schools and universities all over the world. In this chapter we will look at some of this work and at some of the questions being investigated. We also will consider some of the difficult issues biologists and concerned citizens are wrestling with.

The inquiries into life that biologists make are seldom random. They look for answers to those questions that are important in their time. Furthermore, scientists are citizens. They participate in public debates about issues of importance to us all—environmental quality, health and disease, as well as problems of population, food supplies, and land use. Sometimes scientists disagree. Although such disagreements may sometimes confuse other citizens, we must remember that facts seldom speak for themselves. Scientists and citizens must interpret scientific facts. They must reach decisions based on the best of what they know and what they can try to predict.

The cooperation of scientists and other citizens in making decisions that affect us all is extremely important in a democratic society. It also is very important that every citizen, no matter what his or her interests or work, understands the role of science in our society. Every citizen should participate fully in making public policy. In this chapter, we will consider some of the important areas of biological inquiry today. We also will look at some of the difficult problems we must deal with in our complex and rapidly changing world.

ENVIRONMENTAL QUALITY

In nature, there is a balance between living and nonliving things. Organisms constantly carry out processes that deplete the resources necessary for life. For example, animals remove oxygen from the atmosphere. Plants also remove oxygen, as well as carbon dioxide. Yet the atmosphere does not lose all its oxygen or carbon dioxide. Animals take in oxygen from the air and return carbon dioxide. Plants take in carbon dioxide and return more oxygen than they use.

All the materials necessary for life are parts of natural cycles. The driving force—the energy of the sun—comes from outside the earth. Apart from this, the earth is a self-contained unit. We are on our own spaceship. There is no chance of taking on additional resources and no realistic opportunity for finding another planet on which we could live. Obviously, our future depends on how wisely we use the earth's resources now.

If we look at the headlines in our daily newspapers, we find that the ecological balance between living and nonliving things is being

25-1. A sampling of headlines that reflect our concern with the environment.

upset. Many problems are associated with disruption of this natural balance (Figure 25-1). Why is this so? If the cycles of nature have been going on since the origin of life—for several billions of years—why worry about them now? Part of the answer lies in human population growth and human activity. There is evidence that suggests that we are placing too heavy a burden on the earth's natural cycles and using its resources too fast. What are some of the problems? And what can we do to solve them?

Renewable and Nonrenewable Resources

Human life can be maintained for only a few minutes in the absence of air, for a few days in the absence of water, and for a few weeks in the absence of food. Air, water, and food are said to be *renewable resources.* That is, natural cycles replace them when they are consumed. The interactions of green plants and animals maintain the balance of oxygen and carbon dioxide in the atmosphere. The cycle of evaporation and precipitation maintains a supply of water on the earth.

There are other sorts of resources that are said to be *nonrenewable.* These are resources that cannot be replaced. If consumed or destroyed, they are gone forever. The **fossil fuels**—coal, oil, and natural gas—are examples of nonrenewable resources. They were formed from decomposing organisms long ago. The conditions that were present when they were formed are not present on the earth today. Thus the fossil fuels that are being rapidly consumed today are not being replaced.

Before reading further, try making a list of all the resources that are important in your life. Do not restrict this list only to the things animals need to stay alive, such as food. Include also the resources that make your life more pleasant. Which of the items on your list are renewable resources and which are nonrenewable?

The Air We Breathe

The blanket of air that surrounds the earth extends outward for many kilometers. Human beings can live in this blanket of air between sea level and an altitude of about 6,000 meters. Because of gravity, the air becomes thinner (less dense) at increasing elevations above sea level. Above 6,000 meters, a person generally needs supplementary oxygen.

25-2. In some places, there are many green plants and the air is still pure **(left).** In other places, people, automobiles, and factories have been crowded into small spaces. Natural cycles are overloaded. They can no longer maintain the purity of the air **(right).**

25-3. Air pollution can do more than simply soil a building such as Westminster Abbey in London shown here. Note the color differences that result from the cleaning operation on the right. For example, the marble of the Lincoln Memorial in Washington, D.C. is deteriorating rapidly. Sulfuric acid in the air reacts with the marble and carbon dioxide is given off. One engineer said "the Monument is like a giant Alka-Seltzer—you can almost hear it fizz when it rains."

During most of human history, the air was clean (Figure 25-2, left). What happened to change the picture? During the Industrial Revolution of the late eighteenth century, the use of coal to power the new factories increased greatly (Figure 25-2, right). Smoke poured from the factory chimneys and from the nearby homes of the workers, often shutting out the sun. However, few people complained. With the polluted air came more jobs and more products.

As time went on, however, people realized that polluted air had serious effects on their health and on economics. Polluted air contains particles that can clog breathing passages and irritate the eyes. These particles also soil clothes, houses, and buildings (Figure 25-3).

Polluted air also contains harmful gases like carbon monoxide. Carbon monoxide is formed when fossil fuels are burned in the presence of limited quantities of oxygen. This happens in factories, furnaces, and automobiles. Carbon monoxide is a dangerous poison. One percent of carbon monoxide in the air would kill a person in five minutes.

In addition to carbon monoxide, polluted air contains the oxides of nitrogen and sulfur. Sulfur oxides can combine with moisture in the air to produce sulfuric acid. This strong acid can damage things as delicate as the lining of our lungs or as durable as stone (Figure 25-3).

SCIENCE AT WORK: Tragedy of the Commons

Sometimes advances in science are not the result of collecting data in the lab or the field. New ideas, new insights, and new ways of looking at facts can be revolutionary in themselves. Such was the case with the paper "Tragedy of the Commons" published by Garrett Hardin in 1969 (photo above). In this paper, Hardin used as an illustration the common pasture (the "commons") that many villages had long ago. Every homeowner in the village had a right to let livestock graze in the commons. The system worked so long as each person restricted the number of animals that grazed. But what happened when a few individuals decided that they could increase their wealth, at no additional feeding cost, by greatly enlarging their herds? Soon there were more animals than the commons could support. The natural cycle of balanced production by plants and consumption by animals was disrupted. The enlarged herds overgrazed the commons. What was the result? The Tragedy of the Commons.

Hardin used this example from history to examine some modern problems. Let's consider air pollution as an example. A manufacturer must produce a product at the lowest possible cost in order to increase profits. If a manufacturer installed expensive devices to control air pollution, production costs would increase. The manufacturer would then be unable to compete successfully with a less public-spirited manufacturer who made no effort to reduce smoke belching from factory chimneys.

Consider the consequences of a manufacturer's inaction. The air is polluted, and because of increases in illness, health care costs rise. But the increased costs are shared by *everyone*. The manufacturer's expense is only a fraction of the total cost – and a great deal less than the cost of air pollution control devices. So everybody loses a little – but the manufacturer gains a lot. Thus, the best interest of the individual (in this case the manufacturer's profit) is opposite to the best interest of the entire population.

Now we have a problem of conflicts of human rights. Why shouldn't a manufacturer say "The air is free, I can use it as I wish"? At the same time, you can ask yourself, "Do I have a right to breathe pure air?" Most of us probably think that it is important to have as much freedom as possible. Nevertheless, it is not possible for both you and the manufacturer to have your freedoms – the freedom to produce smoke and the right not to have to breathe smoke. Freedom for one is freedom denied the other.

25-4. The eastern part of the Los Angeles Basin on a clear day **(left)** and on a smoggy day **(right)**. The automobile is the major source of air pollution in Los Angeles and in many other cities.

Much of the time, in many of the cities of the United States, the air can be described as unsatisfactory for human consumption. The same verdict can be given for major cities throughout the world. The factory, however, is no longer the main offender. The new offender is the automobile's internal combustion engine. This new source of pollution first became a problem in the United States in the 1940s. At that time, there was a rapid increase in the standard of living for many Americans. More people could afford to buy automobiles. In addition, many people migrated from the country to cities and suburbs. Thus, more people with more cars became crowded into less space. The air over metropolitan areas soon became overburdened with gases produced by automobiles (Figure 25-4).

In 1970, Congress passed the Clean Air Act. The Act set target dates for meeting certain air quality standards and required each state to develop a master plan for pollution control and abatement. Numerous delays and problems plagued achievement of the national goals. Target dates were not met in most areas. Nevertheless by 1977, the nation's worst air pollutants were reduced by as much as one-third. The Council on Environmental Quality predicts that air in most urban areas will meet health standards by the late 1980s. Despite these encouraging steps, the problem is far from being solved. Some 4,000 deaths and four million sick days each year are caused by automobile pollution alone.

Air seems so vast in quantity and so readily available that it has always been "free." That is, no "breathing tax" has ever been levied.

25-5. (Left) This stream of clear water cascading down the Rocky Mountains is eventually stored in a reservoir for domestic use. **(Right)** Run-off from abandoned mining operations can introduce large quantities of iron compounds into the water causing a red color as in the stream shown here.

Nevertheless, we all are now paying indirectly (in higher taxes, higher costs of maintenance, additional medical expenses, and even in higher death rates) for the not-so-clean air that many of us breathe.

The Water We Drink

The United States and parts of Canada have a relatively high average rainfall—about 750 millimeters per year. Half of this rainwater evaporates. One-sixth sinks into the ground and flows to the sea in underground streams. One-sixth flows over the surface in rivers. The remaining sixth is absorbed by vegetation. Most of the water absorbed by plants is lost to the air through transpiration.

A person must drink, or have in his or her food, about two liters of water every day. Only a hundred years ago, nearly everyone in America drank water directly from wells or springs. For washing clothes and bathing, they used rainwater stored in cisterns.

Today, two liters of pure drinking water are still needed by each person. There are, however, twice as many people. Moreover, we use more water in our daily lives. Every time we take a shower, we use 115 liters of water. Ten or more liters are needed to flush a toilet. We require large quantities of water for air conditioning.

In the United States, domestic use of water accounts for less than 10 percent of the total water used. The rest is about equally divided between irrigation and industrial use. Transportation of wastes, including industrial wastes, also is a major use of water. We also need large reserves of water for fighting fires.

The world is running out of fresh water, just as it is running out of clean air. The reasons are much the same. Townspeople used to simply dump sewage into a nearby river or lake. Wastes emptied into the nearby river or lake were consumed by microorganisms. A few kilometers downstream, the water was pure again. So long as the towns were small, this system worked effectively. Because rivers and lakes were not overburdened with wastes, they could be used by town after town as a source of pure water as well as a means of removing wastes.

This natural cycle of purification finally broke down as population and technology increased. So many wastes were produced that the microorganisms could not recycle them before the water reached the next city. The downstream city then had to use elaborate methods to make the water useable. Industries also used vast quantities of water. Much of this was polluted when returned to the river. It contained wastes of slaughterhouses, chemical plants, oil refineries, steel mills, and so on.

Often, chemical wastes are extremely toxic. They either kill the microorganisms that purify water or inhibit their growth. It is not surprising, then, that many rivers become sewers—ill-smelling channels that can no longer purify themselves (Figure 25-5).

Chemical wastes that find their way into our waters also are harmful to other living things. DDT and other pesticides, which are nearly indestructible, accumulate in tissues of organisms around the world. Even penguins in Antarctica have small amounts of DDT in their tissues. Accumulation of DDT has caused the decline of many bird populations, especially fish-eating birds such as ospreys. Ospreys with significant amounts of DDT in their tissues lay eggs with shells that are so fragile they often break. In other species, the young often lack the strength to emerge from the egg.

In Japan, a dreadful disease was caused by mercury compounds released into Minamata Bay by a chemical company. The mercury became concentrated in the fish that people ate. In the United States, harmful asbestos fibers have been discharged into Lake Superior by iron-mining companies. Polychlorinated biphenyls (PCBs), which are very toxic chemicals produced in the manufacture of vinyl, have been found in our rivers. The Hudson and Ohio rivers are carrying dangerous loads of PCBs. Fishing there and in other waters, such as Lake Ontario, is now restricted.

Most of these problems can be overcome by close monitoring of our fresh waters. Potential problems can be spotted promptly and

25-6. Centuries of erosion by wind and water formed the Bryce Canyon in Utah.

the discharge of toxic materials can be stopped. Of course, such actions increase production costs. They might even make it impossible to produce certain items. Thus, we must ask ourselves how much clean water is worth and how much we are willing to pay. How do you think we, as citizens, should react to these problems?

Water quality problems have been recognized for decades and even centuries. It was not until 1977, however, that the United States established national standards for the quality of drinking water. The country's 40,000 community water systems are now under federal supervision. By July of 1977, 3,600 of the nation's 4,000 industrial polluters had taken effective steps to control their pollution output. And billions of dollars had been spent by cities and towns on sewage treatment. Some of the nation's most polluted rivers—including the Ohio, the Buffalo, and the Susquehanna—are showing marked improvements.

On the other hand, sewage treatment is largely ineffective in removing chemicals such as plastics from the water. And such chemicals do not break down quickly, if ever. Another major source of water pollution is runoff from urban and agricultural land. We have taken few steps to control such runoff, primarily because of cost. It would cost about $54 billion just to build storm sewers to carry urban runoff through treatment plants. As you can see, much still remains to be done to prevent and control water pollution.

The Land We Live On

Land is abused in many ways. One example is erosion. Erosion by wind and water is a natural process. It has been going on for thousands of years in Bryce Canyon National Park in Utah (Figure 25-6). The result of this erosion is spectacular.

Erosion of valuable cropland also can occur. Much of this loss of land is due to poor agricultural methods. Horses, sheep, and cattle

25-7. The pasture in the foreground has been overgrazed by sheep. The exposed soil may become eroded.

25-8. Strip mining is an easy and rather inexpensive way of getting coal—a much needed energy source. Yet such mining destroys the land. A 1977 federal law requires that some strip-mined land be reclaimed.

grazing in pastures may destroy the plant cover and expose the soil. The soil then may be eroded by wind and water (Figure 25-7). In addition, intensive farming removes nutrients from the soil faster than they can be replaced naturally.

Another way that land is abused is by "strip mining," a common method of mining coal (Figure 25-8). What is happening to the earth? What do you imagine will happen to this land during the next century?

But questions of land use go much deeper than the misuse of the land. They bear directly on the quality of life for everyone. Where and how we build factories and highways determine, for example, how clean or how dirty the air and water will become. Such decisions also affect our daily life styles. They determine how crowded we are, what access we have to wilderness areas, or how beautiful our surroundings are.

We also must consider the pressures that our population places on the land. People in this country and all around the world need more food. Yet we are rapidly removing prime agricultural land from production. We are using it for the growth of cities, towns, and suburbs (Figure 25-9).

25-9. Vast areas of prime agricultural land have been converted to cities and towns.

Local and state governments have begun to confront the fact that zoning and land management plans of the past are often inadequate to meet present dilemmas. Appropriate balances between the rights of private ownership and the desires of public interest must be determined. Such decisions are seldom popular and are often controversial. Is this still another example of the "tragedy of the commons"?

Recently, land use decision making has received governmental attention at all levels. In 1961, Hawaii initiated a program of zoning and planning for all state lands. Under the National Environmental Policy Act of 1970, environmental impact statements must be prepared before such projects as power plants, highways, bridges, and jet runways can be started. These statements must describe what effects the project might have on the environment. The act itself requires statements only for projects receiving federal funds. Many communities and states, however, now require similar procedures for many kinds of construction and development projects.

Regions and states also have established various land use planning groups. Examples of such groups include Vermont's State Environmental Board, Oregon's State Land Conservation and Development Commission, the Colorado Land Use Commission, and the California Coastal Zone Conservation Commission. These groups are a demonstration of the increasing importance of making wise decisions about land use.

Wildlife Around Us

If we examine more closely the intricate relationships of the natural world such as those discussed in Chapters 23 and 24, we find that "everything affects everything else." An environment that destroys animal and plant life is dangerous to human life as well.

Much of the loss of wildlife in this country comes about when natural habitats are destroyed. When we build dams, drain swamps, and excavate mountains for coal and highways, we destroy the places where wild organisms live and reproduce. Too often, our decisions about protection of wild things involve difficult tradeoffs.

Recently, studies of the effects of a major project on the wildlife in the area have been made. In 1977, construction of a big ditching project in the Cache River Basin in Arkansas was delayed so that environmental impact studies could be made. It happens that the project would destroy the wintering grounds of the largest concentration of mallard ducks on the continent. As of this writing, the decision about what to do on the Cache River has not been made. And the decision is a hard one. How important are ducks to the quality of life of all of us, not just the residents of Arkansas?

Species that are so reduced in their population numbers as to be on the verge of extinction are said to be endangered (Figure 25-10).

a

b

c

d

e

f

g

25-10. Endangered species: **(a)** tiger **(b)** white naped crane **(c)** maned wolf **(d)** California condor **(e)** orangutan **(f)** scimitar-horned oryx **(g)** Furbish lousewort.

Laws protecting endangered species and controlling water quality have done much to preserve some kinds of wildlife. At the same time, such laws have caused bitter arguments over what should come first—the protection of endangered species or human interests in modifying the environment. During 1977, for example, a massive dam project in Tennessee was halted by a federal judge because a little-known fish, the snail darter, was endangered. In Maine, some people were concerned about a rare plant called the Furbish lousewort. Its habitat would be destroyed if plans to build a dam on the St. John River were carried out. In North Dakota, a highway project was stopped because the draining of wetlands would have destroyed prime waterfowl habitat. Have there been similar problems in your state? How have they been resolved?

ENERGY

Perhaps you wonder why biologists should even worry about energy problems. The energy shortage that we all have felt in the last decade has been a result of rapidly decreasing supplies of fossil fuels—coal, oil, and natural gas. (These are called fossil fuels because they are the remains of plants and animals that lived millions of years ago.) We depend on these fuels for about 95 percent of all our energy. Many plans have been put into action to accomplish three objectives: (1) to make the fossil fuels we have last longer by using less; (2) to find new supplies of fossil fuels in this country to reduce our dependence on imported oil; and (3) to substitute other sources of energy for fossil fuels.

It might seem that these problems are best left to chemists, engineers, and politicians. But biologists will continue to have an important voice as we face our energy dilemmas and attempt to find solutions. Any energy decisions that we make will be linked to questions of human survival and of the quality of life for all living things. Energy decisions often involve consideration of human health and environmental impact.

Fossil Fuels

Today, most of the things we do and the products we use rely on considerable energy input. Virtually all that input comes from one source—fossil fuels.

Coal was formed 300 to 500 million years ago from the remains of plants—probably forests of giant mosses and ferns. Oil and natural gas were formed much more recently from both plant and animal material that sank to the bottom of a sea. The discovery of fossil fuel deposits and their usefulness has led to extensive drilling and mining operations around the world. It also has led to great dependence on fossil fuels as a primary source of energy in this country and elsewhere.

Some experts estimate that the United States has enough reserves of coal to last for about 300 years, if we continue to use coal at the present rate. But there are problems in relying on coal as a source of energy. Burning coal is a serious source of air pollution. Most high-grade coal has already been mined. The coal that remains produces smoke that is dirtier and contains more sulfur. And surface, or strip, mining can destroy valuable forest land, drainage areas, and wildlife habitats. Strip-mined lands can be "reclaimed." This means that, after all the coal is removed from an area, topsoil and vegetation are restored. Of course, reclaiming strip-mined land is costly and time consuming.

Because the supply of fossil fuels is fixed and nonrenewable, it will run out—sooner or later. What effects this will have on the lives of future citizens depends on a number of questions. Can another source of energy be used? Can we change our life styles so that we don't use as much energy?

Nuclear Energy

When the nucleus of a radioactive uranium atom breaks down, or decays, heat is released. In theory, this heat can be used to generate electricity. However, the decay of uranium takes place very slowly, and the rate of decay cannot be changed. But scientists have found a way of "splitting the atom." This process, known as fission, releases tremendous amounts of energy. Fission is the basis for nuclear power plants that heat water to generate electricity or to propel a submarine.

Another kind of nuclear energy is fusion. In some ways, it is the opposite of fission. When two small atoms join, energy is released. The hydrogen bomb is a fusion reaction; the atomic bomb is a fission reaction. Scientists have had far less success with fusion than with fission. Tremendously high temperatures are required for fusion. Also, this process produces a hot plasma (ionized gas) that is difficult to contain.

Nuclear energy has much to recommend it. Supplies of nuclear fuels are plentiful. Nuclear power plants can be built to operate safely. But there are some disadvantages in relying on nuclear energy. Nuclear power plants release excess heat into the air and water, which can upset natural ecosystems. Another problem is that the waste products from nuclear power plants are radioactive. Many of these substances remain radioactive for a very long time.

25-11. Ginna Nuclear Power Station in New York State.

At present, most nuclear wastes are held in liquid form. But even the best containers corrode and leak. The wastes might be solidified in ceramic blocks and stored in abandoned salt mines that are free of moisture. The technology to accomplish this, however, has not been developed.

Fission reactors (Figure 25-11) currently supply less than 1 percent of the nation's power. But many new nuclear power plants are either planned or are actually under construction. Nuclear power, for all its potential, has some serious problems. Nuclear fuels release radiations that can disrupt and destroy living tissue. Substantial measures, such as thick lead shielding, are required to protect human and other life from the fatal effects of radioactivity in the environment. Nuclear fuel is relatively easy to mine and transport. The possibilities of theft and accident, however, are real and potentially disastrous.

Geothermal Energy

The source of geothermal energy is essentially a huge nuclear "reactor"—the earth's core. Here, natural radioactive decay produces heat that sometimes comes to the surface as hot water, hot rock, or steam. In California, geothermal power plants use hot springs for generating electricity. Even in areas where the heat does not reach the surface, it is possible to drill deep into the earth to harness it. Although geothermal plants cost less to build and operate than conventional power plants, they are not without problems. Such plants create considerable air and thermal pollution. They also produce very salty waste water that can harm the environment and corrode equipment.

Solar Energy

The sun's energy is, for all practical purposes, an infinite source of energy. We now have two methods for using solar energy. One way is to capture sunlight as heat and use it directly for heating and cooling a home, school, or office building (Figure 25-12). Some scientists believe that, within five years, such units will be available commercially at prices that can compete with those of conventional furnaces and air conditioners.

The other method involves using the sun's energy to generate electricity that can be stored in batteries. The solar cells for this process are much too expensive to make the process economical now, but some scientists and designers feel that cost reductions will be possible soon. Solar-heated houses and schools now exist and many solar-powered devices are planned. This technology, however, has run into some difficult problems. Solar heating units, for example, cost more to install—although less to operate—than conventional furnaces. Building contractors and potential home buyers are unlikely to welcome higher costs for houses. A miniature solar power plant is already operating in Senegal, West Africa. But many questions concerning the economics and distribution of solar energy are still unanswered.

25-12. Advances in technology have made solar heating and cooling inexpensive enough for use in some homes like this one in Colorado.

Solid Wastes

Manure, food wastes, and industrial wastes can be turned into gas or oil. Any one of three processes—hydrogenation, pyrolysis (destructive distillation), or bioconversion—can be used to accomplish this end. The first two processes are similar to those that can be used to change coal into synthetic oil and gas. Bioconversion relies on digestion of waste materials by bacteria. Methane, which can be burned as a fuel, is a product of digestion.

The technology available for using solid waste in fuel production is far from perfect. Sorting and transporting solid waste materials can be complex and expensive, as can the treatments. Hydrogenation is expensive, handling of hydrogen is complex, and high pressure must be maintained. No pressure is needed for pyrolysis, but the end result of the process is a mixture of fuels that is hard to separate and market. Problems of sulfur emission, polluted water, and disposal of the organic sludge produced by bioconversion have yet to be solved.

Wind

People have used wind as a power source for generations. Windmills are a common sight in some European countries. Until recently, they were used in the United States, especially in the Midwest. There are some who suggest that we should return to windmills for some fraction of our electricity needs. Problems here include high costs, changes in weather, and the visual pollution of great numbers of unsightly generators.

Ocean Thermal Gradients

Water near the ocean's surface, especially in the tropics, is considerably warmer than water at greater depths. A device has been suggested that would draw water from different depths, using the difference in temperature to drive a turbine that would generate electricity. But there are many disadvantages in using ocean thermal gradients as a source of energy. The equipment probably would become corroded by seawater. In addition, many such power plants could change ocean temperatures. The outcome of such changes is as yet unknown. Even if these problems were solved, the distance that the electricity would have to be transmitted could mean great losses. An experimental unit of this type was installed off the coast of Africa in 1956. It was shut down soon after because of mechanical failures.

25-13. Hoover Dam, on the Colorado River along the border of Arizona and Nevada, provides flood control, power generation, and water conservation for irrigation and recreation.

Hydroelectric Power

The energy of falling water can be used to generate electricity (Figure 25-13). Unlike fossil fuels or nuclear fuels, the supply of moving water will not run out. Hydroelectric power is inexpensive and fairly clean, but natural ecosystems are destroyed when dams are built. Most of the useable sites for hydroelectric plants in the United States have already been developed. These plants currently supply about 4 percent of the energy used in the United States. However, there is great potential for hydroelectric development in South America, Africa, and Southeast Asia.

Tidal Power

The gravitational pull of the moon causes tides. A bay that is filled and emptied at high and low tides can be dammed. And, much like a traditional hydroelectric plant, the force of the water can turn turbines and generate electricity. Only in a few places could the power of the tides be harnessed. The cost of a tidal plant is greater than that of a regular hydroelectric plant.

Conservation

There is another energy "source" that is different from the ones already mentioned. That is the potential to save energy—to use the sources we have wisely so that supplies can last longer. Some of the steps each of us can take toward conservation have been well publicized. Examples include installing storm windows and insulation to decrease energy use for home heating and cooling. Public transportation and car pools are good ways to decrease energy use in transportation. Indeed, the potential for saving energy can make a significant difference. A homeowner in Los Angeles can, for example, save 12 percent on the amount of fuel used to heat a home by turning the thermostat down from 21°C (70°F) to 18°C (65°F) at night. And two car poolers cut their individual gasoline consumption in half every time they ride together.

POPULATION AND FOOD

The continuous growth of photosynthetic plants provides us with a renewable resource—food. But, unlike other renewable resources such as air and water, food has always been limited in amount. Let's look into this relationship between food supply and population growth.

25-14. THE GROWTH OF THE HUMAN POPULATION

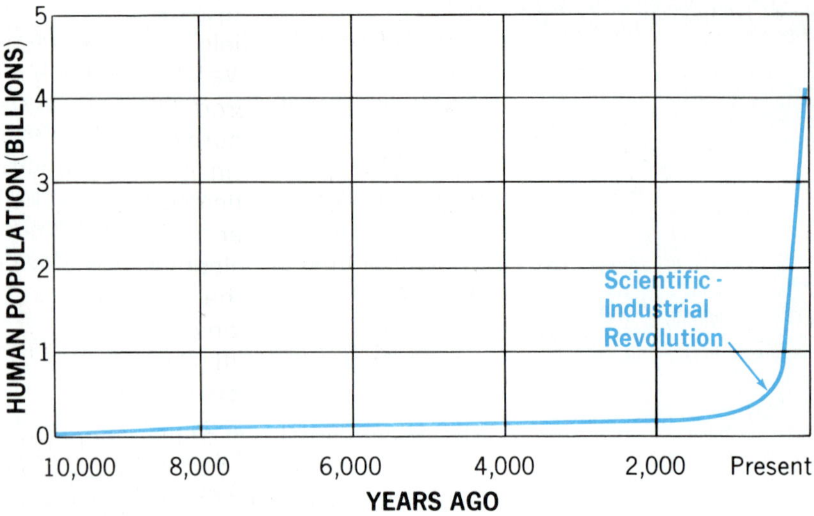

The human population grows when more people are born each year than die. For many thousands of years, the world's population grew very slowly (Figure 25-14). Between 12,000 B.C. and 1650 A.D., the population increased from about 10 million people to about 545 million. In those early years of human history, the population doubled about every 2,000 years. After 1850, the population doubled every 200 years. From 1850 to 1950, availability of food and better sanitation and medical care decreased the doubling time to less than 100 years. The present rate of doubling is about once every 35 years.

The Food We Eat

This very rapid growth rate is now a matter of concern to many. They believe the rate at which our numbers are increasing threatens to overwhelm the ability of the natural cycles to provide us with the resources necessary for life. Is there a real danger that we will destroy the ecosystem that supports us?

We can begin to answer this question by asking another question. How do people affect their ecosystems? One major way is through agriculture. Whenever a new field is prepared for cultivation, the natural ecosystem is changed. In some methods of cultivation there is only a single species where there were many species before. This single species might be potatoes, rice, wheat, corn, or soybeans. Many of these crops can be eaten by people directly. Others—corn, for example—are fed to livestock. The livestock use the food to make their own proteins. We in turn eat the livestock, using their proteins to make our own.

Weeds in a cornfield use water that otherwise would be available to the corn plants. Weeds also shade the young seedlings, impairing their growth and productivity. Insects, birds, rats, and mice consume some of the corn. How do people affect the normal interactions of organisms in this food web? We get rid of the unwanted ones. We use insecticides and herbicides to get rid of pests, weeds, and other consumers. Thus, we eliminate the food web and establish a fairly efficient food chain. Little energy is diverted into unwanted parts of the web.

In making alterations in our food webs, what do we do to the balance of nature? Cultivation changes the physical properties of the soil. Unless cultivated soil is carefully managed, erosion and other undesirable changes occur. Insecticides may eliminate part of an insect population, but the survivors are insects that are resistant to the chemical sprays. These resistant insects soon take the place of those that have been eradicated. They may be even more difficult to eliminate.

We kill foxes, wolves, and birds that prey on domestic animals, such as sheep. This increases the numbers of undesirable primary consumers—woodchucks, rabbits, rats, and mice—that normally are kept under control by the secondary consumers. In these and other ways, we place stress on our ecosystem as we try to increase the amount of energy available to us.

We have other ways of increasing food production besides modifying food webs. Plant and animal geneticists are responsible for tremendous improvements in the productivity of crops and in the livestock that uses the crops for food. Among these improvements are hybrid corn and disease-resistant strains of wheat, rice, and other cereal crops. We also use fertilizers that add more nitrogen, thus more proteins, to our plants. We also have improved the primary consumers—the pigs and cows—so they make better use of the producers. That is, a greater proportion of the available energy is stored in the tissues of the animal. Thus, the energy is available for human consumption. All of this is done to channel the maximum amount of energy into our foods.

Getting Enough to Eat

The recommended diet for a 14- to 16-year-old female of average size contains 2,400 Calories per day. But Calories of food energy are not the only important things in a diet. It also should include a reasonable balance of different kinds of foods (Figure 25-15). People need carbohydrates, fats, and proteins, as well as minerals such as iron and calcium, and vitamins, such as vitamins A, B_1, B_2, B_{12}, and C. The recommended daily consumption for a 15-year-old female of average size is 55 grams of protein and 1.3 grams each of calcium and phosphorus.

25-15. Food values per kilogram for cooked foods.

	Calories	Proteins (grams)	Fats (grams)	Carbohydrates (grams)	Calcium (grams)	Phosphorus (grams)
Beef (hamburger)	3883	235	321.2	0	0.097	1.685
Milk (whole)	726	37	41.8	53	0.125	0.992
Corn (sweet)	906	27	6.6	216	0.053	0.554
Rice (white)	1269	26	1.1	279	0.086	0.480
Potatoes (boiled, peeled)	887	22	1.1	205	0.117	0.598
Wheat (whole meal)	768	29	3.3	169	0.097	0.887

Protein is the most important nutrient needed for normal growth. Even after growth has stopped, protein is needed daily for repair and maintenance of body tissues. Humans can synthesize ten of the 20 or so amino acids they need. The remaining ten must be present in protein foods. Animal proteins are the most efficient sources of the amino acids we cannot synthesize. Plant proteins contain these amino acids too, but not always in proportions that we need. A vegetarian diet must include plant proteins that provide balances of amino acids similar to those that are found in animal proteins. For example, beans alone or rice alone are inadequate proteins. But when they are eaten together in the correct quantities, the protein is complete.

It has been estimated that at least 20 percent of the people living in developing countries are not getting the minimum *quantity* of food needed. At least 60 percent are not getting the minimum *quality* of food needed for good health. This means that about one third of the world's population is suffering from malnutrition, if not downright hunger or starvation.

Some Possible Solutions

A hectare of land yields 15 to 24 times more dietary Calories when the crops are consumed directly by humans than when the crops are fed to animals to produce meat. Figure 25-16 contains some data on the average yields of a hectare of land when used for different kinds of crops. Expressing food requirements in terms of yield per hectare gives us some very useful information. *Food is a renewable resource but the land needed to produce it is limited.* Some agricultural ex-

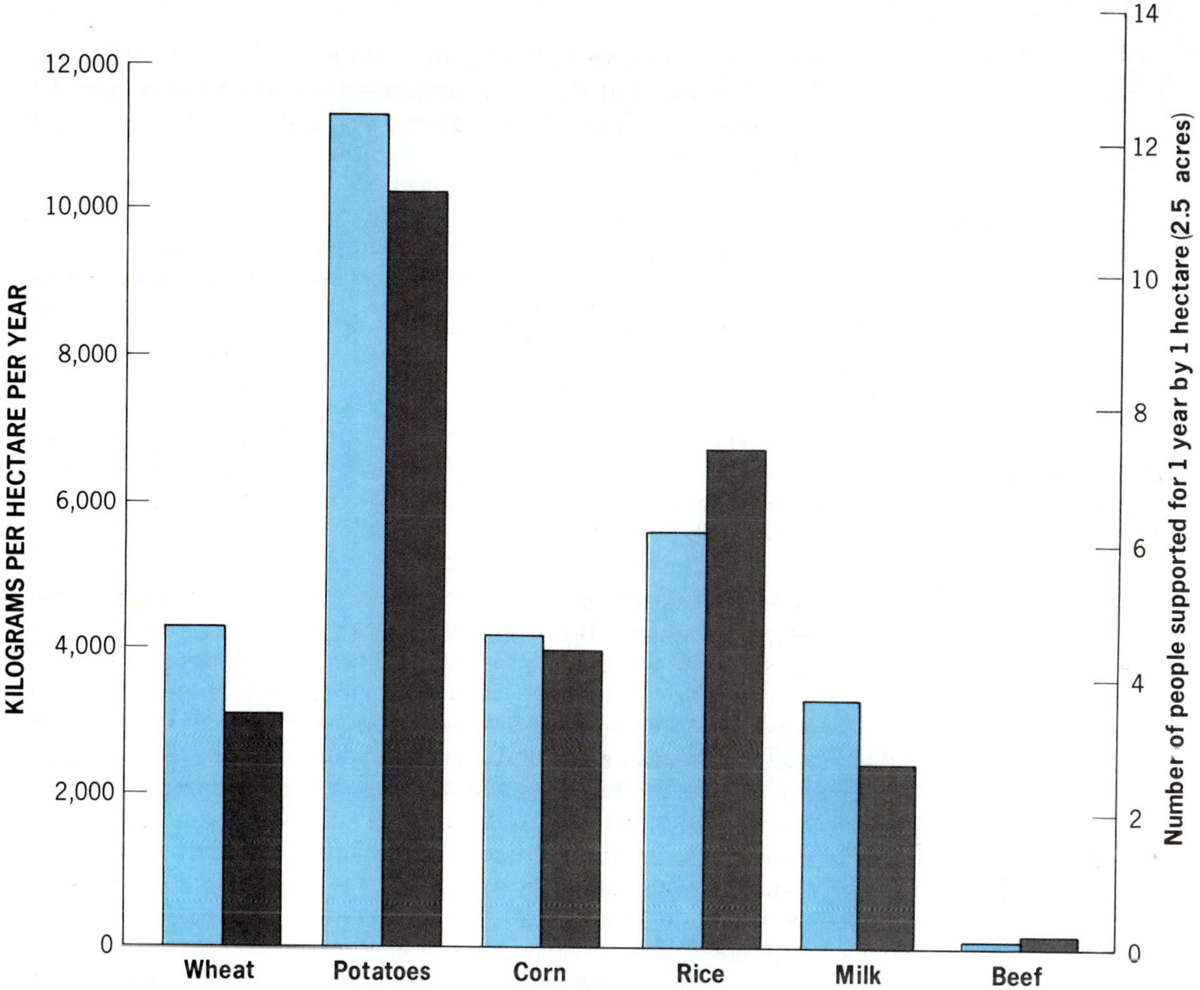

25-16. A hectare of land will yield different amounts, depending on what crops are grown. The blue bars in this graph show kilograms of food that might be produced on one hectare of farm land planted with different cereals, potatoes, or hay fed to cattle to produce milk. For comparison, production of beef on range land is also shown. The gray bars show the number of people that could be fed by the different foods produced on a single hectare. Use the scale on the right of the graph to read the gray bars. *Which crops are the most energy-efficient? the least?*

perts believe that nearly all the land most suitable for agriculture is already being farmed. In addition, the growth of cities and deserts is decreasing the amount of land available for agriculture.

Putting new lands into production requires massive amounts of energy in the form of fertilizer; human, animal, or mechanical labor; and irrigation. Today, few farmers around the world have the knowledge or money needed to put these forces to work.

Scientists are investigating other sources of food energy that are not used extensively at this time. For example, the Japanese and other people living in coastal regions eat the seaweed kelp found in great beds along the ocean shores. Some biologists think that, in the future, kelp farms will be cultivated and fish will be raised like cattle in such areas. Furthermore, researchers also have found that energy stored in unpalatable plants and animals can be made into edible forms. For example, a high-protein flour can be made from fish that we would not otherwise eat.

In the deep oceans, nutrients accumulate near the bottom, where they are of no use to photosynthetic organisms. Here, populations of plankton and fish are small and the potential for farming is not encouraging. But in certain places—along the coast of Peru, for example—the surface waters of the ocean contain large amounts of nutrients. Here, the ocean's productivity is quite high and farming may well be possible.

Much of the food that is produced is lost through waste. In American homes, for example, we often serve more food than we can eat. Waste also occurs in harvesting and storing food. In some parts of the world, 30 percent of the harvest may become spoiled. In many parts of the world, there is little or no refrigeration. Inadequate storage, rodents, insects, and poor harvesting techniques also contribute to the waste. Cutting waste would increase our food supply.

Many corn, wheat, rice, and other grain plants also are lost each year to plant diseases. One of these diseases, wheat rust, is caused by a fungus. The fungus has a very complex life cycle that involves two hosts, wheat and the common barberry plant. By eliminating the barberry plant, we have partially controlled wheat rust. In addition, rust-resistant varieties of wheat have been developed from mutations of wheat genes. Often, however, the development is followed by the rapid evolution of mutant forms of the fungus to which the wheat is susceptible. Thus far, plant breeders have stayed just one jump ahead of the wheat rust fungus.

What actually will happen to the balance between our population size and the food supply required to support it is just a guess. Even if ways of providing an adequate food supply for the growing world population are found, there are other problems. Our planet provides us with only a limited amount of space in which to live. Many of our cities are already crowded (Figure 25-17). Such crowding is associated with other problems—air and water pollution, unemployment, poverty, and the loss of human dignity. For these reasons, many people believe that we should attempt to slow the rate of population growth.

This opinion is not new. In many countries, nationwide programs have brought about positive results. The Japanese, for example, are trying to stabilize their population at about 100 million. Their

25-17. A street scene in Madurai, India.

birthrate has dropped by one-half during the last 20 years. India has one of the fastest growing populations in the world. Family planning centers have been established in cities and villages throughout that country. China has reduced the growth of its population by a variety of social measures. Most of these programs are under the control of doctors who serve local populations.

Such measures have slowed the rate of population growth in many parts of the world. For example, by 1977, the birthrate in the United States had fallen below the "replacement level" of two children per couple. But this does not mean that the population is no longer growing in the United States. Why? Primarily because of the age structure of the population. There are so many young adults of child-bearing age that even at an average of two children per family, the population will continue to grow well into the next century. Also, we cannot be sure that the low birthrates of the late 1970s will continue. Some experts point out that birthrates parallel economic conditions. When money is scarce, couples tend to decide against having children. But in "good times," birthrates can increase dramatically. Thus, even in our own country, the effects of population growth still need to be confronted. We are faced with hard questions of how many people the resources of our land can support.

And the problem goes beyond mere numbers. We must also deal with matters of how much of the earth's resources each of us uses. For example, it has been estimated that an American consumes as much as ten people in India consume. That American also causes 50 times as much pollution of the environment. How might we begin to solve problems of population growth and resource use?

HEALTH AND DISEASE

People would not wish to see the size of the human population limited by the raw forces of disease and starvation. Modern medical practice attempts to insure the physical and mental health of every individual. How successful have these attempts been? How successful might they be in the future?

Control of Infectious Diseases

By the time you read this, smallpox, a highly contagious, viral disease, probably will have been eradicated. In 1978, there were only a few people who had smallpox. They were in Mogadisho, the capital and major port of Somalia in eastern Africa. Everyone who came in contact with the diseased people was vaccinated. Doctors hoped these would be the last cases of smallpox.

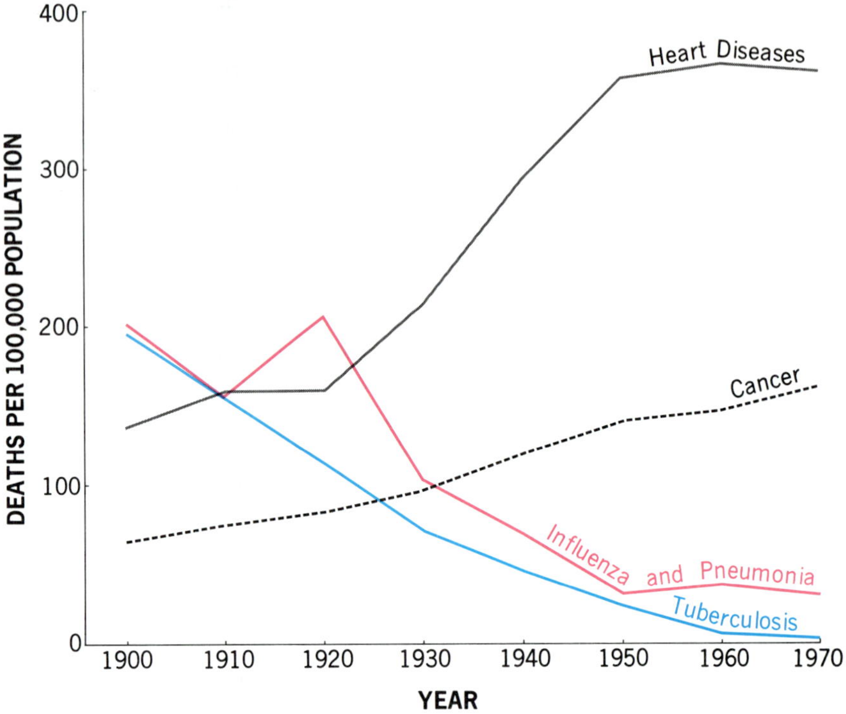

25-18. In 1900, infectious diseases (pneumonia, influenza, and tuberculosis) were the major causes of death. Today, heart diseases and cancer head the list. *What changes since 1900 might account for these trends?*

This apparently successful fight was the result of a ten-year campaign, begun in 1967, by the World Health Organization of the United Nations. At that time, the disease was present in 30 countries. Smallpox has since been eliminated in western and central Africa, Brazil, Indonesia, southern Africa, Pakistan, India, and Bangladesh.

The eradication of smallpox will demonstrate what can be achieved when scientists throughout the world join in such a program. Vaccines also are available for diptheria, whooping cough, tetanus, measles, poliomyelitis, and tuberculosis. A logical next step would be to begin similar campaigns to immunize everyone against these diseases as well.

Today's Health Problems

The major health problems now confronting industrialized nations are not infectious diseases, but rather degenerative diseases (Figure 25-18). In 1900, six of the ten leading causes of death in the United States were either infectious diseases or disabilities related to them. Today, however, influenza, pneumonia, and certain diseases of early infancy are the only infectious disorders among the ten major causes of death.

25-19. Changing pattern of births, deaths, and life expectations in the United States, 1900-1970.

	1900	*1910*	*1920*	*1930*	*1940*	*1950*	*1960*	*1970*
Live births per thousand of population	32.3	30.1	27.7	21.3	19.4	24.1	23.7	18.2
Deaths per thousand of population	17.1	14.7	13	11.3	10.8	9.6	9.5	9.4
Excess of births over deaths	15.2	15.4	14.7	10	8.6	14.5	14.2	8.8
Expectation of life at birth	47.3	50	54.1	59.7	62.9	68.2	69.7	70.8
Average increase in longevity since 1900	—	2.7	6.8	12.4	15.6	20.9	22.4	23.5

The diseases that now head the list are heart disease, cancer, cerebrovascular accidents (strokes), obstructive lung disease, diabetes, arteriosclerosis, and cirrhosis. These are often associated with increased life expectancy (Figure 25-19). Because people in the United States live longer, they develop diseases that never showed up when people died at a young age. Many of these diseases are associated with the environment of industrialized societies. Their causes may be physical, in the form of environmental pollutants, or emotional, in the form of stress and anxiety. Personal habits, such as not getting enough exercise, not eating the right foods, smoking, and drinking alcohol, also contribute to the increased incidence of these diseases. In addition, these factors interact with the individual's genetic makeup.

In the times when infectious disorders were the major health care problem, it was relatively simple to pinpoint a "cause." It was obvious that some bacteria, protozoans, or other microorganisms had invaded the body and upset normal functioning. When we consider today's major diseases, the causes are not so clear-cut. We know, for example, that smoking increases an individual's risk for developing lung cancer or emphysema. But other factors are involved as well. Some individuals may be more susceptible to cigarette-related lung cancer than others who smoke just as heavily.

Unfortunately, we know very little about the effective treatment or cure of today's major diseases. We can treat some forms of cancer effectively with drugs, surgery, or radiation, or combinations of these treatments, but other forms respond poorly to such treatments.

Prevention of Disease

Further research may shed new light on the causes and the cures of heart disease, cancer, and today's other major diseases. Meanwhile, our best treatment is to prevent these diseases. Given some understanding of the interaction of genotype and environment as it affects health, individuals can begin to make decisions about their own health.

In 1974, the Canadian government proposed the "Health Field Concept" as a way of looking at health problems. The four factors this approach considers are human biology, environment, life style, and the health care organization. The report states: "Until now most of society's efforts to improve health, and the bulk of direct health expenditures, have been focused on the Health Care Organization. Yet, when we identify the present main causes of sickness and death in Canada, we find that they are rooted in the other three elements of the concept: *Human Biology, Environment, and Life-Style.* It is apparent, therefore, that huge sums are being spent treating diseases that could have been prevented in the first place."

What the Canadian government was suggesting is that our health care system—doctors, nurses, and hospitals—is already doing about as much as we can expect. Numbers of patients are increasing and costs continue to soar. We must look for other ways of taking care of our health. We must aim to *prevent* diseases. How can that be accomplished? We have no hard and fast answers, but the following points deserve some consideration:

Nutrition. A balanced diet is important for good health. Information on good nutrition is available, but many people ignore or overlook opportunities for a healthy diet.

Improving our diets is not at all difficult. With a little attention each day, we all can improve our diets and feel better for it (Figure 25-20). The key lies in making wise choices. We can, for example, choose broiled fish instead of hamburger. Although both are good sources of protein, the fish has much less fat. We can choose margarine instead of butter. Butter is high in saturated fat, which increases the risk of a heart attack if consumed in quantity over many years. Margarine, especially the kind made with corn, safflower, or sunflower oil, is low in saturated fat. It is, therefore, a wiser choice than butter. We can select whole grain breads and cereals rather than those made from refined flour. Many vitamins and minerals are lost in the refining process. We can select fruits and vegetables as snacks, instead of high-Calorie, low-nutrition "junk" foods. And fruit and vegetable juices are better nutritional choices than soft drinks and carbonated beverages.

We would do well, according to the experts, to get more of our daily energy from carbohydrates rather than from fats. Why? Carbohydrates—whole grain breads, cereals, rice, and enriched spaghetti—are good sources of energy. And carbohydrates are less harmful than the fats found in red meat, butter, and milk. Carbohydrate foods also are a good source of fiber, which is important in proper digestion and elimination.

Rest. Most people need between six and nine hours of sleep a night. A regular pattern of sleeping is healthier than varying patterns such as a great deal of sleep one night and very little the next.

Experts agree that the average American should consume

25-20.

Smoking. Smoking is a major risk factor for heart disease, some forms of cancer, emphysema, strokes, and even increased danger of automobile accidents. It is estimated that each cigarette smoked removes about 12 minutes from an individual's life expectancy.

Drug Abuse. Drug abuse contributes to mental and physical illnesses. Drugs play a role in accidents, homicides, and suicides.

Alcohol Consumption. Drinking is a major cause of automobile accidents and of cirrhosis of the liver, our ninth leading cause of death.

Tension and Stress. Each of us is subjected to a certain amount of stress in our work, school, and home. Overeating, drinking, or drug abuse in response to stress compounds the problem. It is much better for us to learn to relax through a hobby or exercise.

Exercise. Inactive life styles contribute to high blood pressure, obesity, and heart disease. Regular hard physical exercise strengthens the heart, lungs, and arteries.

The Power of Advertising. Many of the products we are encouraged to buy do little more than ruin our health. Sugary cereals, alcoholic beverages, and cigarettes are examples. Before you buy, ask yourself if the product is truly something you need and want.

"An Injury is No Accident." Accidents are the leading cause of death among males ages 15 to 24. Most of us assume that everyone's chances of an accidental injury are the same. Such is not the case. Risks for automobile accidents are highest for young male drivers who drink and drive in the late evening or early morning hours. The chances of an accident occurring are higher on roads without median dividers and for individuals who choose not to use seat belts. Similar risk factors contribute to injuries that occur in the home or on the job.We can prevent accidents by consciously controlling our behavior and by being aware of hazards in our environment.

Crime. Homicide is the second leading cause of death among males ages 15 to 24. The causes are complex and poorly understood. We do, however, know some of the factors that seem to contribute to high crime rates. Poverty and crowded living conditions are at least partly to blame. Unemployment, low educational levels, and limited access to improving one's personal and economic status are also involved. We need to examine the problem of crime and exercise effective citizenship in seeking solutions.

We can make decisions about our personal life styles and our health. But at first glance, it might seem that individuals have very little control over the effect of the environment on their health. However, effective citizen action can make a difference. Let's look at one example. A few years ago, it became evident that a hazard resides in the large-scale use of fluorocarbons. Fluorocarbons are our most common type of refrigerant and, until the mid-1970s, were the propellant in millions of spray cans.

The fluorocarbon gas released into the air was thought to be quite harmless. Yet this gas reacts with ozone in the upper levels of our atmosphere. The reaction weakens the protective layer of ozone that prevents much ultraviolet radiation from reaching the earth. Too much ultraviolet radiation can cause skin cancer. It can even kill many small forms of life.

As a result of pressures from scientists and from consumers, there may be no such spray cans on the shelf of any store in the United States by the time you read this. Citizens *can* control industrial pollution and can prevent the introduction of untested chemicals or the widespread use of harmful radiations.

NEW FRONTIERS OF SCIENTIFIC RESEARCH

What lines of inquiry should biologists follow for the future? We can be sure that environmental and energy concerns, investigations into problems of population and food, and new efforts to improve human health and prevent disease will continue to be important. Discussions among scientists and citizens about how to solve our complex problems also will continue.

New areas of potentially important research are hard to predict. A few years ago, for example, studies of insects that prey on other insects were little more than a scientific curiosity. Now, such research is beginning to yield some real answers for controlling insect pests without polluting the environment. By the time you read this chapter, the newspapers and magazines probably will be full of new problems, new discoveries, and new questions. On the following pages, we will look at only a few of the new challenges in biology for today and tomorrow. We also will look at some of the people who are attempting to meet these challenges.

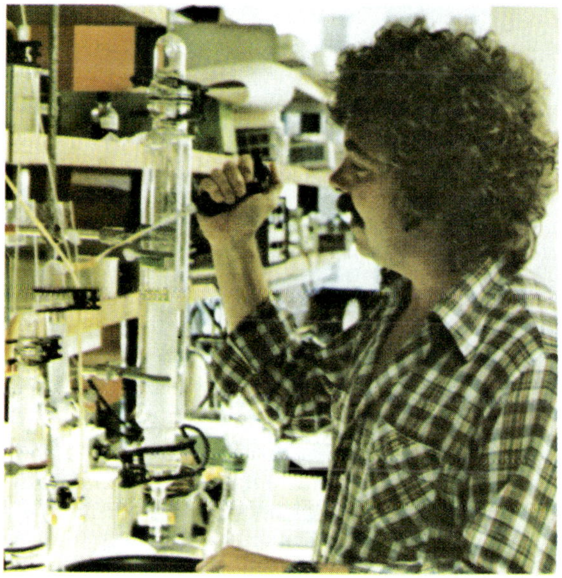

25-21. In the mid-1970s, scientists developed techniques for attaching genes to the DNA of bacteria. DNA transformed in this way is called recombinant DNA. Researchers are now working on methods to activate the attached genes to effect a change in the compounds produced by the bacteria. The work of a team of scientists in California resulted in a bacterial cell that actually produced a hormone found in the human brain. Other scientists are applying the techniques of recombinant DNA to studies of simple and economical ways to make organic drugs. Shown here is Herbert Boyer, a leader in the research on recombinant DNA. Recombinant DNA techniques will undoubtedly continue as an important area of research and applied science.

25-22. The models of gene action, DNA replication, and messenger RNA production described early in this book are accepted, at least in principle, by most scientists. But the details are far from complete. In fact, researchers have found themselves in for a few surprises. Shown here is Louise T. Chow of the Cold Spring Harbor Laboratory in New York. She and others working in the field have discovered some unexpected pairings of messenger RNA with segments of the master DNA molecule. It appears that some messenger RNA molecules may share "leader" sequences on the DNA molecule. The exact nature of such unexpected findings undoubtedly will be the subject of much research and interpretation as the studies of gene action continue.

25-23. Some scientists are more active in government and politics than in the laboratory. And their contributions are just as important. More than 60,000 chemical products are available. These, typically, are used in homes and industries. The safety of these products is often suspect, and the task of checking their effects on health and safety is monumental.

Shown here is a chemist for the Consumer Product Safety Commission testing consumer products for the presence of hazardous chemicals. In 1977, the Commission researched reports that the chemical benzene causes cancer. Since benzene is a chemical that is found in many common household products and since millions of people are exposed to benzene in these products each year, the Commission set out to investigate the possibility of banning products containing benzene. *Can you find out if the ban was actually instituted?*

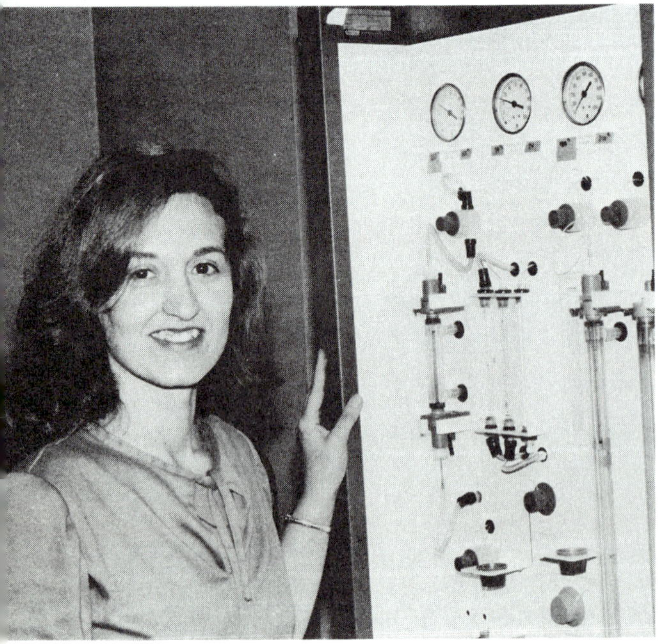

25-24. Investigations into what causes abnormal human behavior are increasing. One fascinating area is the role of certain brain chemicals in diseases of the nervous system (such as Parkinson's Disease) and in acute mental illnesses (such as schizophrenia). So far, we know very little about the chemistry of the brain. But new facts and hypotheses are emerging every day.

Shown here is Roberta Palmour at the University of California at Berkeley. She and Frank Ervin of UCLA worked together on research that examined the proteins in the blood of schizophrenic patients who received a series of blood-filtering treatments. They found that the concentration of a certain protein dropped after the treatments. It rose again in some of the patients who experienced a relapse. When injected into laboratory animals, the protein caused abnormal brain patterns.

These results, however, are not conclusive. The protein may be an effect of the disease and not its cause. Research into the causes of nervous disorders and mental illnesses continues.

25-25. The collecting, describing, and classifying of living things on planet Earth is by no means a finished task. Scientists continue to probe the highest mountains and the deepest seas for previously unknown life forms. In the winters of 1976-77 and 1977-78, divers working below the permanent sea ice of Antarctica recovered two animals never before seen by humans. The animals are a type of foraminifera (Chapter 12). Both wear an armor of sand grains attached to their surfaces by organic cement.

The research team was led by Jere H. Lipps, professor of geology at the University of California, Davis. The team established their camp on the semi-permanent sea ice in McMurdo Sound. Dr. Lipps sliced frozen seal meat for use as bait in traps set below the Ross Ice Shelf. Numerous organisms were caught, such as a perfectly preserved sponge thought to be several thousand years old.

25-26. Ecology was a respected science long before the "environmental movement" of the 1960s. Ecologists study the interactions of organisms and environments in every imaginable location and for a great variety of reasons.

Shown here is Mary Jean Currier, a graduate research assistant in wildlife biology at Colorado State University. She earned her masters degree conducting the field work on a study of mountain lions. Currier began by estimating the size of the mountain lion population in Colorado. She then tried to determine the age of each lion she captured (the one shown above is very young). She searched for the lions with a mountain guide, tracking dogs, and a tranquilizer gun. Currier's investigation showed that the mountain lion is not endangered in Colorado, nor is it placing a stress on the environment by overpopulating the region. The study of lions is continuing.

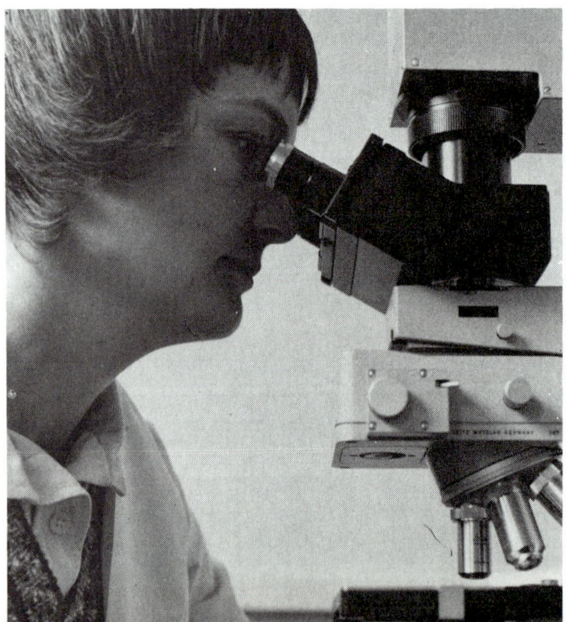

25-27. Our exploration of space has provided many opportunities for new research in biology. For example, the United States and the Soviet Union joined together, in the summer of 1975, in the Apollo-Soyuz mission.

The two spacecrafts were connected (docked) while they were in orbit around the earth. The 34 experiments conducted by this cooperative research program were selected from 161 proposals submitted by scientists in nine different countries. One of these experiments was designed to detect changes in the immune systems of the astronauts during their mission in space. Shown here is the principal investigator for the experiment, B. Sue Criswell of the Baylor College of Medicine in Houston, Texas.

25-28. Very often scientific inquiry results in new technologies. We see these on TV and in newspapers every day. Sewage is a "waste" that is disposed at great cost, mainly because it contains disease-causing organisms. But sewage also represents a vast resource of nutrients that might be useful as fertilizer and even as feeds for cattle and sheep. Sandia Labs in Albuquerque has developed a way to eliminate pathogens in sewage by using *gamma*-rays from "waste" radionuclides. Dr. Stan Smith is an agricultural scientist at New Mexico State University. His experiments show that nutrients in radiation-treated sewage can be useful both as fertilizer and as feed supplements. Currently, Dr. Smith is studying the effectiveness and safety of the new procedure.

25-29. Computers are becoming increasingly complex and are being used in many areas. They have taken over much of the drudgery of such things as keeping track of how many television sets are stored in a warehouse. They also sort and organize the vast amounts of data obtained during space flights. Computers can handle, in seconds, calculations that would require people weeks or months to complete. And the memory banks of computers can store billions of pieces of information that can be recalled in an instant. The development of computer science is useful to governments, scientists, businesses, and industries. It affects each of us in many ways every day.

One of the greatest benefits of computer technology may come through ways of helping people learn. This student is using a system known as PLATO. This system is used by the University of Illinois, Champaign-Urbana. The machine teaches 150 subjects, ranging from Swahili to rocketry.

In Brief: Looking Back on this Course

Perhaps your experience in this course has given you some idea of how biologists work and the kinds of questions they try to answer. As you reconsider your own experiences as a student of biology, consider the following points:

1. *The raw material of science is observation.* The observations you made in the field and in the laboratory should have served as a basis for questioning, further observation, and experimentation. Perhaps you can now see why we say that knowledge is *constructed* rather than *discovered.*

2. *Scientific knowledge is tentative and incomplete.* Science is not merely a collection of hard, fast, and proven facts. The scientific knowledge of today is the *best* explanation for what we know *now.* Actually, there usually are several possible explanations for experimental results. Explanations that apply in one case may not apply in another. And the search for answers invariably leads to new questions.

3. *Biology is inquiry into the natural world.* The questions of biology most often relate to living things and their interactions with each other and their environment. Humans form an important part of this study.

4. *Science is one way of seeking answers to questions.* In different areas of human experience, we use different means of obtaining answers. The techniques used in a subject like music are not the same as those applied by the working biologist. Scientists apply certain patterns of thought that seem to be useful in finding answers to certain kinds of questions. Not all questions are scientific, and not all scientific questions can be answered with the methods we have now.

5. *Scientific inquiry is a search for order.* We seek the overlying principles that will help us to explain a body of facts. We look for unifying concepts that can serve both as explanations for what we already know and as signposts toward exploring the unknown.

6. *Science is an active, human enterprise.* Scientific activities are seldom the clear-cut, logical exercises that some people imagine. The most promising hypothesis is not supported by the data. The most predictable experiment yields the most unexpected findings. People make mistakes. Organisms respond in variable ways. The scientist does not stumble over new and earth-shattering discoveries. He or she gradually pieces together observations that lead to new knowledge.

7. *Science is a creative act.* Scientific research is as creative as the finest painting or the most evocative poem. Of course, science has methods and procedures different from those applied to art or writing. But science can be as imaginative and challenging as any other form of human expression.

The writers of this book hope that at least some of the ideas now have more meaning for you. And we hope that you will continue your interest and your inquiries in the biological sciences.

Guide Questions and Problems

1. List ways in which your own environment could be made better. How can you personally help to bring about these improvements?
2. One of the dangerous pollutants of water is oil. Yet oil is an important energy resource. Why should oil have such contrasting roles in our ecosystem?
3. For a week, keep a record of everything you eat and drink. Determine your Calorie intake, and the amounts of protein, fat, and carbohydrate in your diet. How nutritionally sound are your eating habits? What can you do to improve your daily nutrition?
4. What are some of the limiting factors that prevent the world from supporting a human population of ever-increasing size? Is increasing food production a solution to the problem of an increasing population? Why or why not?
5. If you were the president of a country where population growth exceeded the productive capacity of your country, how would you attempt to solve the problem? What difficulties might you encounter?
6. What is meant by "recycling"? How do recycling campaigns in your community relate to this idea? Are there limits on the effectiveness of recycling? If so, what are they?
7. What happens to the ecosystem when national parks and wildlife areas are established? What happens when new roads, dams, or nuclear power plants are built? What human needs are met (or fail to be met) as a result of such actions?
8. Many communities have established standards for air quality. They have set up agencies to see that these standards are met. Is this true in your community? What are the dangers, the safeguards, and the unsolved problems of the air supply in your community?

9. Collect clippings from newspapers and magazines on new developments in science and technology. What areas seem most interesting and important to you? What can you find out about the people who are doing the work?

10. Review the section "In Brief: Looking Back on this Course" at the end of this chapter. Write your own short paper describing what you believe science to be. Use examples from your experience in this course to explain your ideas.

Ideas for Independent Study

We have not listed any special ideas for independent study at the end of this chapter, because the whole chapter is built up of such ideas. We hope that you are already well on your way in your search for knowledge about, and solutions to, the major problems besetting our species. It's hard to imagine a more interesting or useful topic of study.

Related Readings

BOOKS

Asimov, I. *Science Past—Science Future.* Garden City, N.Y.: Doubleday, 1975.

Baxter, W. F. *People or Penguins: The Case for Optimal Pollution.* New York: Columbia University Press, 1974.

Biological Sciences Curriculum Study. *Energy and Society: Investigations in Decision Making.* Northbrook, Ill.: Hubbard, 1977.

Biological Sciences Curriculum Study. *Environmental Resource Papers.* Menlo Park, Calif.: Addison-Wesley, 1975.

Brown, J. E. *The Sea's Harvest: The Story of Aquaculture.* New York: Dodd, Mead, 1975.

Brown, L. R., P. L. McGrath, and B. Stokes. *Worldwatch Paper 5 Twenty-Two Dimensions of the Population Problem.* Washington, D.C.: Worldwatch Institute, 1776 Massachusetts Ave., N.W., 1976.

Fowler, K. N. *Hunger: The World Food Crisis.* Washington, D.C.: National Science Teachers Association, 1724 Connecticut Ave., N.W., 1977.

Fowler, K. N. *Population Growth: The Human Dilemma.* Washington, D.C.: NSTA, 1977.

Lappe, F. M. *Diet for a Small Planet.* Westminster, Maryland: Ballantine, 1975.

Marks, J. and D. Hendin. *The Genetic Connection.* Caldwell, N.J.: Morrow & Co., 1978.

Marshall, J. *Going, Going, Gone? The Waste of Our Energy Resources.* New York: Coward, McCann, and Geoghegan, 1976.

Newton, D. E. *Science and Society.* Boston: Holbrook Press, 1974.

Page, L. W. and T. Page. *Apollo-Soyuz Pamphlet No. 7: Biology in Zero-G.* Washington, D.C.: NASA, 1977.

Riccuiti, E. R. *To the Brink of Extinction.* New York: Harper & Row, 1974.

Vicker, R. *This Hungry World.* New York: Scribner's, 1975.

Vicker, R. *Human Population, A Scientific American Book.* San Francisco: W. H. Freeman & Co., 1974.

ARTICLES

Canby, T. Y. "Can the World Feed Its People?" *National Geographic,* July 1975, page 2.

Croce, C. M. and Koprowski, H. "The Genetics of Human Cancer." *Scientific American,* February 1978, page 117.

Henderson, D. A. "The Eradication of Smallpox." *Scientific American,* October 1976, page 25.

Horowitz, N. H. "The Search for Life on Mars." *Scientific American,* November 1977, page 52.

Langer, W. L. "Checks on Population Growth: 1750–1850." *Scientific American,* February 1972, page 92.

McEwen, B. S. "Interactions between Hormones and Nerve Tissue." *Scientific American,* July 1976, page 48.

Nathanson, J. A. and P. Greengard. " 'Second Messengers' in the Brain." *Scientific American,* August 1977, page 108.

Raff, M. C. "Cell-Surface Immunology." *Scientific American,* May 1976, page 30.

(continued from page iv)

Children's Research Hospital, Memphis, Tenn.; 95, Jan Marc Orenstein and Emma Shelton; 96, Werner A. Loewenstein; 98, Keith Porter; 101, George E. Palade; 103, Drs. J. M. Oliver and J. A. Krawiec, University of Connecticut Health Center; 104, Birgit Satir; 105(both), Dr. John Heuser; 108(both), Leukemia Society of America; 111(b), Phillip G. Coleman, APSA; 112(top), Phillip G. Coleman, APSA; 113, HBJ Photos; 115(r), American Cancer Society; 119, Margery Shaw, M.D.; 122, Margery Shaw, M.D.

Part two: Page 127(b), Nathan Cohen; 135, Burpee Seeds; 136(both), Burpee Seeds; 152, U.S. Department of Agriculture; 153, The Internationl Rice Research Institute; 155(tl), art adapted from material supplied by Victor A. McKusick; (tr), A. Iino; (b), Roy D. Schmickel, Golder N. Wilson, David A. Jackson, University of Michigan Medical School; 167(both), Margery Shaw, M.D.; 170, Ram S. Verma; 179, Margery Shaw, M.D.; 182, Courtesy of A. M. Winchester; 184, T. F. Anderson, E. L. Wollman, and F. Jacob; 185(t), Prof. Giuseppe Penso; (b), H. Wunderli, J. V. D. Broek, E. Kellenberger; 190, Margery Shaw, M.D.; 191, Margery Shaw, M.D.; 192, Margery Shaw, M.D.; 197(l), Collection, The Museum of Modern Art, New York, A. Conger Goodyear Fund; 201(r), Courtesy of Thomas F. Anderson, The Institute for Cancer Research, Philadelphia, Pa.; 208, Prof. M. H. F. Wilkins, Biophysics Dept., King's College, London, England; 216(both), Courtesy March of Dimes and Denver Children's Hospital, Dr. Favara; 223(l), Dr. Gobind Khorana; 229, L. B. Shettles; 235, Courtesy of Minicourse Development Project, Purdue University; 237, John A. Moore; 238, John A. Moore; 240, Carnegie Institution of Washington, Dept. of Embryology, Davis Division; 241(both), Chester F. Reather, FBPA, Carnegie Institution of Washington; 243(all), Roberts Rugh; 247(l), The Electron Microscopy Group of Biozentrum, Basel, Switzerland; 249, John A. Moore; 254(both), Leukemia Society of America, Inc.; 256, John A. Moore; 261, Carl Lillequist; 262 (l), Vermont Travel Division; 265(both), American Museum of Natural History; 271(r), Marg and Bill Staley; 274(r), American Museum of Natural History, HBJ Photos; 276(both), NASA; 280(l), Edmund Appel; (r), George Harrison/Grant Heilman; 281(both), Gordon Uno; 299(both), J. A. Bishop; 312(tl), American Shorthorn Association; (tr), American Brahman Breeders Association; (b), Santa Gertrudis Breeders International; 316(r), American Airlines; 317, Transvaal Museum, Pretoria; 319(all), American Museum of Natural History; 320(tl,tr), American Museum of Natural History; (c,b), British Museum; (r), Irven De Vore, Anthro-Photo; 322(tl,tr), American Museum of Natural History; 322(b), Musee du Louvre; 323-24, Dr. Elwyn Simons, Baher El Koshab, and David Pilbeam, Duke University; 325, Rick Search, Chicago Zoological Park; 327(tl, tr), San Diego Zoo; 331, James A. Tuck; 332(both), James A. Tuck; 337, Annan Photo Features—Lyn.

Part three: Page 344, Hicks, "The Peaceable Kingdom," Photography by Sandak, Inc.; 347(tr), Don Pitts, Appaloosa News; (bl), Palomino Horse Breeders of America; 349, Utah Zoological Society; 359, Garland Johnson; 363(tl,rc,bl,br), HBJ Photos; (bc), Elso S. Barghoorn, Harvard University; 364(b), HBJ Photos; 365(br), Eric Gravé; 366(tl,bl), HBJ Photos; 367(tl), Macmillan Science Co., Inc.; (tr,bl), Donald Baird); (br), HBJ Photos; 368(bl,br), John A. Moore; 369(t,bl), John A. Moore; (br), Donald Baird; 370(tl), Donald Baird; (tr), Grant Heilman; (bl,br), John A. Moore; 371(tr,br), Werner W. Schulz; (tl,tc,cr,bl), John A. Moore; 372(tr), Castle and Cooke, Inc. (bc), Breck's; (br), Farmland Industries, Inc., Kansas City; 373(lc,tr,bl), John A. Moore; (c), American Society for Microbiology/Dr. T. L. Maunder; (br), John A. Moore; 375(t), Werner W. Schulz; (bl), Bob Gilpin, Courtesy of Leland Shanor; (br) Constantine Alexopoulos; 377(br), Miami Seaquarium; 378(tl), HBJ Photos; (tr), Manfred Kage/Peter Arnold; (br), Bergman Associates; 379 (br), Bergman Associates; 380(tl, tr), D. P. Wilson; (bl), Donald Baird; 381(tl), David Fritts; (tc), Jeff Stanton; (cl, cr), Miami Seaquarium; (bl), Parks, Canada; 382(cr, all b), Edward S. Ross; 383 (l 3rd from t), John G. Shedd Aquarium; (bl,br), Miami Seaquarium; (tc,tr), John A. Moore; 384(t), Donald Baird; (bl), John A. Moore; 385(tl), John A. Moore; (bl), Miami Seaquarium; (r), HBJ Photos, Courtesy of American Museum of Natural History; 386(tl), Donald Baird; (tc), Marc Bekoff; (bl), American Museum of Natural History; 387(tr,br), HBJ Photos, Courtesy of the American Museum of Natural History; (lc), John A. Moore; (bl), San Diego Zoo; 388(tl), David G. Houston/Bruce Coleman Inc; (cr), New York Zoological Society; (br), Richard Stabile/DPI; 389(all), San Diego Zoo; 390(tl, tc), San Diego Zoo; (bl,bc), Judd Cooney; 391(bl), Carl Lillequist; (br), San Diego Zoo; 392(tl,tr,bl), David Fritts; 393(t), Donald Baird; 397(t), New York State Museum and Science Service; (b), American Museum of Natural History; 398-99, American Museum of Natural History; 400(l), American Society for Microbiology, George Zeikus; (r), Warren N. Marble; 410(l), Miami Seaquarium; 413, John A. Moore; 416(l,r), Rothamstead Experimental Station; 418, HBJ Photos; 438(r), Brown Brothers; 445(r), Jim Brock; 460(l), Belle Bonfils Memorial Blood Center; (r), American Museum of Natural History; 464, George I. Schwartz; 476, Dr. Rolland C. Reynolds; 477(l), Charles F. Pfizer Inc.; 493(l), Michael D. Sullivan; (r), Dr. Brian Bracegirdle; 498(l), Georgenne Tomlinson, (r), Elaine Brock; 517, John A. Moore; 524(l), HBJ Photos; 526, Prof. Luciano Zamboni; 528(tl), HBJ Photos; 548(r), David Fritts; 550, L. K. Shumway; 557(both), J. C. Allen and Son; 560(l), Harold V. Greene; 572(tl), Dr. H. K. Phinney; (bl), HBJ Photos; 573(lc,tr,br), Eric Gravé; 580(l), Florida Game and Freshwater Fish Commission; (r), Miami Seaquarium; 587, Gordon Uno; 588(t), Dr. H. K. Phinney; 589(bc,br), John A. Moore; 591, John A. Moore; 593(both), John A. Moore; 595(b), Hugh Spencer; 598(both), John A. Moore.

Part four: Page 601(t), David Fritts; 602(both), David Fritts; 611(both), Thomas Eisner; 619(r), Edward S. Ross; 621(r), James O. Sneddon, University of Washington; 616, Dr. David Premack; 628(t), David Fritts; (bl), Paul Linger; (r), Jen and Des Bartlett/Bruce Coleman Inc.; 630(t), Edward S. Ross; 636, John A. Moore; 638(l), Elaine Brock; (r), Burton McNeely; 640(t), John A. Moore; 644, NASA; 654(l), Vermont Agency of Development and Community Affairs; 655(all), Werner W. Schulz; 656(tl,tr,bl), John A. Moore; 657(tl), John A. Moore; 658(all), John A. Moore; 659(br), John W. Marr; (others), John A. Moore; 661, Nebraska Game and Parks Commission; 662(l), Werner W. Schulz; 663, USDA, Soil Conservation Service; 664, David Fritts; 665, John A. Moore; 666(tl,tr), John A. Moore; (bl), Saguaro National Monument; (br), Willis Peterson; 667, John A. Moore; 668, John A. Moore; 671(l), David Fritts; (r), Peter Behrendt; 676, John A. Moore; 683, Hugh Spencer; 692, Yellowstone National Park; 693, Wilson N. Stewart; 699(both), Surtsey Research Society; 701(all), John A. Moore; 703(both), Manfred P. Kage/Peter Arnold; 706(l), John A. Moore; (r), Yoram Lehman/Peter Arnold; 708(l), Kay Shuper; 709(both), John A. Moore; 710(l), John A. Moore; (r), U.S. Bureau of Land Management; 712(tl), John A. Moore; (b), USDA, Soil Conservation Service; 713(both), U.S. Bureau of Land Management; 715(tr), International Crane Foundation; (br), Dr. L. W. Macior; (others), Paul Linger; 718, Rochester Gas and Electric Corp., 721, U.S. Bureau of Reclamation; 733(t), M. C. Betlach, (b), Louise T. Chow; 734(t), Consumer Product Safety Commission; (b), Andrei Laszlo; 735(all), Jere H. Lipps; 736(tl), Mary Jean Currier; (tr), Edward G. Currier; (b), Baylor College of Medicine; 737(t,c), New Mexico State University-Sandia Labs; (b), University of Illinois at Urbana-Champaign.

Index

Boldface numbers indicate illustrations

D

B 0
C 1
D 2
E 3
F 4
G 5
H 6
I 7
J 8